Biomathematical and Biomechanical Modeling of the Circulatory and Ventilatory Systems

Volume 7

For further volumes:
http://www.springer.com/series/10155

Marc Thiriet

Diseases of the Cardiac Pump

Springer

Marc Thiriet
Project-team INRIA-UPMC-CNRS REO
Laboratoire Jacques-Louis Lions, CNRS UMR 7598
Université Pierre et Marie Curie
Place Jussieu 4
75252, Paris Cedex 05
France

ISSN 2193-1682 ISSN 2193-1690 (electronic)
Biomathematical and Biomechanical Modeling of the Circulatory and Ventilatory Systems
ISBN 978-3-319-12663-0 ISBN 978-3-319-12664-7 (eBook)
DOI 10.1007/978-3-319-12664-7

Library of Congress Control Number: 2014959283

Springer Cham Heidelberg New York Dordrecht London

Printed on acid-free paper

Springer is part of Springer Science+Business Media (www.springer.com)

Preface

Me vero primum dulces ante omnia Musae, Quarum sacra fero ingenti perculsus amore, Accipiant, caelique vias et sidera monstrent; Defectus Solis varios, Lunaeque labores: Unde tremor terris, qua vi maria alta tumescant Obicibus ruptis, rursusque in seipsa residant: Quid tantum Oceano properent se tingere soleshiberni: vel quae tardis mora noctibus obstet. [But most beloved, ye Muses, at whose fane, Led by pure zeal, I consecrate my strain, Me first accept! And to my search unfold,Heaven and her host in beauteous order rolled, The eclipse that dims the golden orb of day, And changeful labour of the lunar ray; Whence rocks the earth, by what vast force the main Now bursts its barriers, now subsides again; Why wintry suns in ocean swiftly fade, Or what delays night's slow-descending shade.]
(P. Vergilius Maro [c.-70–c.-19], Georgics, Book II)

The first volumes including Vols. 1–5, review main events and their associated signal transduction processes in cells and biological tissues, that is, at the nano-, micro-, and mesoscopic scales in normal conditions. Because of the amount of accumulated data on the physiology and pathophysiology of the circulatory and ventilatory apparatus, the initially scheduled one-volume book (referred to as Vol. 6 *Circulatory and Ventilatory Conduits in Normal and Pathological Conditions* in previous volumes of the series *Biomathematical and Biomechanical Modeling of the Circulatory and Ventilatory Systems*) has been split. Volume 6 mainly targets the organ scale, that is, the macroscopic scale, in normal conditions (i.e., anatomy and physiology of blood circulation and the body's ventilation), although it reviews histological and biochemical data for a better understanding of macroscopic scale processes. Volumes 7 and 8 focus on pathological conditions that affect the heart (Vol. 7), vasculature, and the respiratory tract (Vol. 8) in a modeling perspective rather than a clinical point of view. They mainly deal with diseases of the cardiovascular and ventilatory apparatus investigated using biomechanical modeling. Some diseases have a mechanical origin; most are associated with a mechanical disorder. Local flow disturbances can actually trigger pathophysiological processes or, conversely, result from diseases of the pump or conduit walls or their environment. The ability to model these phenomena is essential to the development of a personalized medicine and manufacturing of medical devices, which incorporates a stage of numerical tests in addition to experimental procedures.

Programs in computational medicine rely on a software-coupling platform aimed at incorporating structural (anatomical) and functional data at various length and time scales to simulate functioning of the physiological apparatus in both normal and pathological conditions, thereby promoting personalized, predictive, and, if possible, preventive healthcare. Modeling supports computer-aided diagnosis and treatment planning, as well as prediction of therapeutic outcomes and prognosis.

Modeling is associated with numerical simulations of physical and chemical processes. Problem formulation and mathematical analysis lead to the construction of solution algorithms. Approximations result from inherent errors of mathematical models of natural processes that arise from partial understanding of these phenomena and their random nature, as well as measurement uncertainties. Moreover, model equations cannot be commonly solved analytically. Problem equations are converted to a set of algebraic equations (numerical approximations). Numerical procedures introduce truncation errors associated with numerical approximations and round-off errors, as a finite number of digits are used to represent numbers.

On the other hand, biomedical informatics aims at integrating the collection of biochemical data, physiological signals, medical images, biological rhythms, epidemiological factors, clinical history events, and simulation results into usable information to foster the creation of new diagnostic and therapeutic methods and improve both life quality and healthcare cost.

Bioinformatics supports the development of knowledge and data management platforms to identify predisposition to diseases and biomarkers for early diagnosis and therapy adaptation in a framework of individualized medicine. Tasks incorporate data mining and information retrieval, indexing, and representation.

> At nunc per maria ac terras sublimaque caeli multa modis multis varia ratione moveri cernimus ante oculos..." [But by seas, lands, and heights of heaven many things move in many ways for various reasons before our eyes]
> (T. Lucretius Carus [c.-99–c.-55], De Rerum Natura [On the Nature of Things], Book I)

Physiological flow behavior in pathological conditions is described by a set of conservation equations. The Navier–Stokes equations (Vols. 1, Chap. 1, Cells and Tissues; and 9, Chap. 1, Hemodynamics), derive from the continuum mechanics theory. This equation set predicts transient flow behavior of a fluid of given properties in a given domain of the highly deformable cardiac pump using proper boundary conditions and values of flow-governing dimensionless parameters.

Mathematical modeling also targets maladaptive tissue growth that causes mural pathologies, especially in the heart (hypertension-initiated cardiac hypertrophy) or results from medical device implantation. Tissue remodeling can be investigated at a macroscopic scale and all micro- and mesoscopic scale phenomena lumped in parameters that are incorporated in a system of nonlinear, coupled, parametric, partial differential equations. This type of equation set is used in continuous-type models that rely on mixture theory. However, this solving procedure necessitates an efficient identification stage to estimate involved parameters.

Like Vol. 6, the present book mainly refers to the macroscopic scale, although nano- and microscale pathophysiological mechanisms are described. It includes nine chapters. Chapters 1 and 2 describe the context, factors, and processes involved in pathogenesis of cardiac diseases. Chapter 3 is devoted to adverse cardiac remodeling, that is, fibrosis and hypertrophy, Chap. 4 to cardiomyopathies, Chap. 5 to rhythm and conduction alterations, Chap. 6 to cardiac valve diseases, Chap. 7 to heart failure, Chap. 8 to coronary artery disease and myocardial infarction, and Chap. 9 to interventional medicine and surgery of cardiac diseases.

Acknowledgments

The author acknowledges the patience of his wife Anne, daughter Maud, sons Alrik and Damien, and their respective French (Julien, Jean, and Louis), American (Raphaëlle, Matthieu, and Alexandre), and Polish (Joanna and Frédéric) families.

Contents

Chapter 1
Pathogenesis of Cardiac Diseases

> Ο βιοσ βραχυσ, η δε τεχνη μακρη, οδε καιροσ οξυσ, η δε πειρα σφαλερη, η δε κρισισ χαλεπη.
> *[Life is short, art long; opportunity fleeting; experience perilous, and judgment difficult.]*
> (Hippocrates of Cos [c. – 460–c. – 370], Aphorisms, Sect. I)

In the present handbook, the nouns *ailment*, *disease*, *illness*, *malady*, *pathology*, *sickness*, and *syndrome*, are often used as synonyms to avoid multiple repetitions of the same word. However, these terms can have distinct meaning.

A *syndrome* is a collection of signs (observed by individuals external to the patient), symptoms (reported by the patient), and phenomena or features that frequently appear together.[1] For example, a metabolic syndrome is not a disease; it underlies a set of diseases and dysfunctions, such as obesity, type-2 diabetes, insulin resistance, dyslipidemia, chronic inflammation, and hypertension, with possible history of stroke and coronary heart disease. A syndrome is, at least originally, less understood than a disease, but can be later explained, or, at least, various pathophysiological aspects involved in the context of a syndrome can be handled (e.g., signaling pathway dysregulation and immunity abnormalities). In other words, the noun usage often continues to be utilized even after an etiology has been successful or when numerous causes that generate the same combination of symptoms and signs have been detected. In fact, some syndromes have a single cause; others have multiple possible causes; the cause of a third category of syndromes remains unknown. Many syndromes are eponymous, i.e., named after the namegiver physicians that first reported the association.

A *disease* is a disorder in a physiological apparatus or organ that affects the bodily function. It corresponds to any disturbances of the structure–function relation of a tissue, organ, or physiological apparatus manifested by a characteristic set of symptoms and signs and whose etiology and prognosis can be usually identified and predicted. A disease is a condition characterized usually by at least two criteria among three basic elements: (1) recognized etiologic agent (cause); (2) identifiable group of signs and symptoms; and (3) consistent anatomopathologic alterations.

[1] συνδρομος: running together; συνδρομη: tumultuous concourse.

M. Thiriet, *Diseases of the Cardiac Pump*, Biomathematical and Biomechanical Modeling of the Circulatory and Ventilatory Systems 7, DOI 10.1007/978-3-319-12664-7_1

Some diseases can cause a particular syndrome. The term disease is more concrete than illness, which can also include mental aspects.

An *illness* is an abnormal process in which aspects of the social, physical, emotional, and/or intellectual function of a person are impaired compared with previous behavior. It is a malady of either body or mind, the symptoms of which may be physically unobservable. Illness encompasses the subjective characteristics of the disturbance as a whole rather than only objective symptoms.

A *sickness* commonly results from difficulty of adjusting to environmental conditions (e.g., altitude, sea, car, decompression, and radiation sickness).

Cardiovascular disease refers to any dysfunctioning or lesion that affects the cardiovascular apparatus, the cardiac pump, as well as the irrigation and drainage circuits, mainly those of the heart, brain, and kidneys.

1.1 Gene Transcription in Cardiovascular Diseases

The transcriptional control of the cardiovascular function involves programs of gene activation and suppression. Transcription factors, chromatin remodelers, and histone modifiers, that is, a set of coactivators and corepressors cooperates to influence chromatin state and hence controls accessibility of nucleosomal DNA, as well as directly regulates gene expression in the heart.

During cardiogenesis, chamber specification is controlled by temporal and spatial expression of cardiac transcription factors such as heart and neural crest derivatives expressed factor HAND, GATA-binding protein GATA4, myocyte enhancer factor MEF2, cardiac-specific NK2 homeobox-encoded factor NKx2-5, and T-box factor TBx5 [1]. Transcription factors GATA4, MEF2c, MEF2d, and myogenic differentiation factor MyoD, are involved in transcriptional programs launched by signals that control pathological gene expression in the heart [1].

1.1.1 Chromatin Remodelers

Chromatin architecture and its modification is involved in gene regulation as well as DNA repair, recombination, and replication. In the cell nucleus, DNA is packaged by histones to form structured nucleosome units. Chromatin remodeling is governed by two categories of enzymes: histone modifiers and ATP-driven nucleosome repositioners.

Transcriptional controllers encompass:

- *Catalyzers of chemical modifications* (i.e., acetylation and methylation) of histone proteins within the chromatin by histone acetyltransferases (HATs), deacetylases (HDAC), and methyltransferases (HMT);
- *Conditioners of the chromatin structure and topology* such as chromatin remodelers and helicases of the switch/sucrose nonfermentable(Swi/SNF) ATPase

Table 1.1 Chromatin modifications and histone deacetylase (HDAC)-dependent pathological gene expression in the heart. (Source: [1])

Type	Disease	Target genes
HDAC1	Cardiac hypertrophy	Ncx1 BECN1 (autophagy related beclin-1)
HDAC2	Cardiac hypertrophy, fibrosis	Anp, Bnp Klf, Yy1
	Hsp70, ATG5, BECN1	Inpp5f, Gsk3β
HDAC3	Cardiac hypertrophy, fibrosis	Tnfsf1
HDAC4	Heart failure, ischemia	SUV39H1 HIF1A
HDAC5	Cardiac hypertrophy Ischemia	NKX2.5, p300 Ncx1
HDAC6	Cardiac hypertrophy (induced by angiotensin-2; pro)	
HDAC9	Cardiac hypertrophy	FOXP3

complex implicated in cardiogenesis and heart diseases, in particular, Swi/SNF-related, matrix-associated, actin-dependent regulator of chromatin SMARCa2 and SMARCa4;[2] and
- *Noncoding RNAs* (e.g., Braveheart) that interact with chromatin.

HDACs and HMTs are associated with certain cardiac pathologies (Tables 1.1 and 1.2). Mutations in genes encoding HMTs may contribute to congenital heart disease.

Class-1 nuclear histone deacetylases (HDAC1–HDAC3 and HDAC8), class-2 HDACs, which shuttle between the cytosol and nucleus and can further be subdivided into subclass-2a (HDAC4, HDAC5, HDAC7, and HDAC9) and -2b (HDAC6 and HDAC10), and class-4 HDACs (HDAC11), which localize to the nucleus and cytosol and depend on zinc. Class-2a HDACs repress cardiac hypertrophy. On the other hand, class-3 HDACs, or sirtuins (SIRT1–SIRT7), rely on nicotinamide adenine dinucleotide (NAD^+). They reside in both nucleus and cytoplasm.

The class-2b cytoplasmic histone deacetylase HDAC6, which lodges in striated (cardiac and skeletal) myocytes and deacetylates cytoskeletal proteins, is involved in adverse remodeling induced by chronically signaling angiotensin-2 [2]. HDAC6^{null} mice develop cardiac hypertrophy and fibrosis, but without cardiac function alteration, in response to angiotensin-2 administration during up to 8 weeks, whereas

[2] Also known as Brahma homolog Brm and Brahma-related gene product BrG1, respectively. SMARCa4 supports binding to target gene promoters of cardiac transcription factors GATA4, NKx2-5, and TBx5 [1]. In addition, SMARCa4 can assist linkage of the DNA-binding protein poly(ADP-ribose) polymerase PARP1 with chromatin, especially in adverse cardiac hypertrophy.

Table 1.2 Chromatin modifications associated with histone methylation and demethylation in cardiac diseases. The MYH6 and MYH7 genes encode α- and β-cardiac myosin heavy chain, respectively. (Source: [1])

Type	Function	Pathology (role)	Target gene
ASxL2	$H3K_{27}me^3$	Ischemic and idiopathic DCM	MYH6/7
KMT1c	$H3K_9me^3$	Ischemia (pro)	Sirt1
KMT3c	$H3K_{36}me$	ND	ND
KMT3d	$H3K_4me$	Arrhythmia (pro)	ND
KMT4	$H3K_{79}me$	DCM (anti)	DMD (dystrophin)
KMT6	$H3K_{27}me^3$	ND	Anp, Bnp, MYH6/7
KDM3a	ND	ND	ND
KDM4a	$H3K_9$ demethylase	Cardiac hypertrophy, HCM, ICM (pro)	Anp Fhl1
KDM4b	$H3K_9$ demethylase	DCM, ICM (pro)	Anp, Bnp
KDM6a	$H3K_{27}$ demethylase	Hypoxia (pro)	Nos3
KDM6b	$H3K_{27}$ demethylase	Hypoxia (pro)	Nos3
PaxIP1	Cofactor of KMT2c/d ($H3K_4me^3$)	Ventricular arrhythmia (anti)	KCNIP2
HDM	$H3K_4me^3$	Heart failure	DUX4
	$H3K_9me^3$		
	$H3K_{36}me^3$		

ASxL additional sex combs-like protein, *HDM* histone demethylase, *KDM* histone Lys (K)-specific demethylase, *kKMT* histone Lys (K) ^{N}methyltransferase, *PaxIP* paired box (Pax)-interacting with transcription-activation domain protein, *anti* impedes disease, *pro* favors disease, *DCM* dilated cardiomyopathy, *HCM* hypertrophic cardiomyopathy, *ICM* ischemic cardiomyopathy, *DUX* double homeobox gene, *FHL* four and a half LIM domain-containing protein, *ND* not described

wild-type mice develop systolic dysfunction. Inhibition of HDAC6 preserves systolic function.

Histone acetylation and methylation that result from developmental programs of the heart dictate cardiac gene expression. Histone Lys ^{N}methyltransferase KMT6[3] and paired box (Pax) transactivation domain-interacting protein PaxIP1 (or PTIP) are two HMTs involved in chromatin condensation. The former participates in cardiac myosin heavy chain (MHC) gene regulation, as it interacts with a long noncoding RNA in chromatin [1].

[3] Also known as Enhancer of zeste homolog EZH2, a polycomb group HMT. Polycomb group members are connected to several regulatory complexes that promote chromatin modification and maintain gene silencing. Methylation of $H3K_{27}$ by KMT6 is mandatory in cardiomyocyte lineage specification during mouse embryogenesis.

Trimethylation $H3K_{36}me^3$ is typically associated with transcriptional elongation and DNA methylation that cooperate to regulate the expression of double homeobox Dux4 gene in the failing human heart [1].

Histone-3 trimethylation at Lys4 (e.g., $H3K_4me^3$) and some types of acetylation ($H3K_9ac$ and $H4K_{14}ac$) are generally correlated with permissive gene transcription, as they weaken the connection of histone proteins and DNA.

In particular, $H3K_4me^3$ enables the expression of the KCNIP2 gene that encodes K_V channel-interacting protein KChIP2 involved in repolarization [1]. Expression of the KCNIP2 gene associated with reduced $H3K_4me^3$ at its promoter falls in heart failure. cAMP-responsive element-binding protein (CREB)-binding protein (CBP) and P300 are two HATs detected on activated genes in the left and right ventricles of the heart. Binding of cardiac transcription factors, such as SRF and GATA4, may be mediated at least partly by P300 acetyltransferase [1].

The histone modifiers, additional sex comb-like protein ASxL2 and histone Lys ^{N}methyltransferase KMT6, interact at MHC promoters [1].

On the other hand, methylation of $H3K_9$ and $H3K_{27}me^3$ and deacetylation are correlated with gene transcription suppression, as they cause chromatin condensation (i.e., transcriptionally repressive chromatin conformation) [1]. In mouse models, ischemic reperfusion provokes $H3K_9me^3$ at the Sirt1 promoter, hence suppressing the Sirt1 gene transcription.

However, whereas deacetylation of histone tails is linked to transcriptional suppression, deacetylation of nonhistone proteins often activates gene expression.

The Polycomb repressive complexes PRC1 and PRC2 cause gene silencing via intrinsic HMT activity. The PRC2 holoenzyme is a complex of HMTs, which leads to $H3K_{27}me^3$ modification.

The regulation of gene expression involves the coordinated actions of the transcriptional machinery and regulators (i.e., activators or repressors at gene promoters). Gene transcription can result from a loss of inhibitory methylation and a gain of stimulatory acetylation in a given Lys residue (e.g., K_{27}) of histone H3.

1.1.2 Gene Transcription and Mechanical Factors

At a given station, the velocity and possibly the direction of blood flow change (e.g., in limb arteries, but not in cerebral arteries) during the cardiac cycle, in addition to spatial variations along the vascular circuit. The endothelium is sensitive to hemodynamic shear stresses applied at its luminal surface (Vol. 5, Chap. 9. Vascular Endothelium).

Blood flow via hemodynamic stress exerted on the wall-wetted surface as well as within the wall can promote expression of genes that encode antiatherogenic, antithrombotic, and anti-inflammatory factors [3].

1.1.2.1 Sheared Endothelium

The vascular endothelium is a thin monolayer of cells that line the luminal surface of all blood and lymph vessels. This barrier controls the exchange of water, electrolytes, macromolecules, and cells between the vascular lumen and surrounding tissue. It regulates leukocyte adhesion and transendothelial migration, as well as platelet aggregation via the expression of adhesion molecules. It supervises smooth myocyte function via vasoactive substances (e.g., nitric oxide, prostacyclin, and endothelin-1).

Inside the vascular wall, cyclic strain and stretch result form the normal hemodynamic stress, blood pressure. The latter indeed cyclically generates axial and circumferential tension applied to both intimal endothelial and medial smooth muscle cells.

Shear is the mechanical force created by blood particles decelerated (slowed) by the immobile solid wall or by adjacent slower blood particle (in the local normal direction to wall surface) on its wall-facing surface and accelerated (dragged, i.e., gaining momentum) by adjacent quicker blood particle. At the wetted endothelial surface, the moving blood particle exerts a tangential force and cyclically shears the luminal border of endotheliocytes.

Shear stress sensors of endotheliocytes include ion channels, plasmalemmal receptors (GPCRs and RTKs), adhesion molecules connected to the cytoskeleton and matrix fibers, and caveolae, as well as the glycocalyx and primary cilia. Mechanoreceptors convert mechanical cues into chemical signals that control cell function and fate via stored molecules and regulation of gene expression.

Cultures of endotheliocytes in the absence of mechanical stress do not mimic the physiological condition. Once endotheliocytes are exposed to shear stress (the common situation in vivo), expression of some genes increases with respect to that at rest (unphysiological condition), such as those that encode NOS3, thrombomodulin, cytochrome-P450, heme oxygenase-1, diaphorase-4, PECAM1, regulators of G-protein, and KLF2 and KLF4 [4]. On the other hand, the expression of genes encoding inflammatory and thrombogenic agents (e.g., VCAM1, E-selectin, TGFβ, and BMP4) is downregulated.

Transcription factors of the Krüppel-like factor family regulate responses to shear stress and engender anti-inflammatory and anticoagulant transcripts. The Krüppel-like factor KLF2 elicits expression of atheroprotective genes, such as those encoding NOS3, thrombomodulin, MAPK9, and von Willebrand factor, and impedes transcription of proatherogenic genes, such as those encoding NFκ B, BMP4, interleukin-8 (CXCL8), and Serpin-E1) [4]. However, endothelial signature genes, such as those encoding VEGFR1 and VEGFR2, TIE1 and TIE2, PDGF, VEGF, and cadherin-5 are not significantly influenced by KLF2. KLF2 and KLF4 are regulated via the MAP2K5–ERK5 pathway.

1.1.2.2 Hemodynamic Stress and Vascular Remodeling

Hemodynamic stress field on and in the vascular mural domain influences not only the vascular tone, that is, the local size of the vascular lumen and, hence, the flow resistance, but also outward and inward vascular remodeling (Vol. 5, Chap. 11. Tissue Development, Repair, and Remodeling). Atherosusceptibility refers to an adaptive chronic low-level inflammatory state that ensures the endothelial function with an increased likelihood of atherogenesis. Atherosclerotic plaques do not distribute evenly over the arterial bed, but localize to predilection sites with low-magnitude velocity at bifurcations, side branches, and some regions of curved arteries, where a reduced blood flow velocity supports uptake and activation of molecules and inflammatory cells. In addition, some shear stress patterns provoke specific plaque compositions in a high-cholesterol environment, a low and oscillatory shear stress favoring plaque stabilization with respect to a low steady regime [4].

1.1.2.3 Mechanical Stress and Stem Cells

The adult heart is not a fully terminally differentiated organ, as a cardiomyocyte turnover exists with a rate estimated at about 1 %/year in young adults and 0.5 %/year in the elderly.

Mechanical stresses influence vascular injury and repair. In particular, blood flow impacts maturation of circulating or resident stem cells into functional vascular cells. A heterogeneous population of stem and progenitor cells may exist in different layers of the vessel wall.

In healthy walls, stem cells are quiescent and localize to parietal stem cell niches. SCFR+, Sca1+, Lin− or Sox17+, Sox10+, and S100β+ multipotent stem cells can differentiate into many cell types. HCAM+, NT5E+, Thy1+, CD34−, and PTPRc− mesenchymal stem cells constitute a second stem cell population. The third population is formed by CD34+, NG2+, and PDGFRβ+ pericytes [5].

Vascular stem cells can differentiate into endotheliocytes or migrate across the vessel wall and subsequently differentiate into intimal smooth myocytes [5].

Endothelial progenitor cells reside within the endothelium as well as in the intimal subendothelial layer, where they can form neovessels, although they are normally dormant [5].

Pericyte progenitor cells lodge around adventitial vasa vasorum. They support angiogenesis [5].

In the blood vessel wall, *multipotent vascular stem cells* can differentiate into neurons and mesenchymal stem cell-like cells that subsequently differentiate into smooth myocytes [5]. At least some lesional smooth myocytes with a proliferative and synthetic phenotype derive from these stem cells rather than smooth myocytes that dedifferentiated and migrated from the media.

Resident stem cells may be released into the blood circulation. They can differentiate into vascular cell lineage to repair damaged vessels or form new microvessels in damaged tissues. In addition, bone marrow stem cells can generate circulating endothelial progenitor cells. Smooth myocyte progenitors also flow in blood and can contribute to the pathogenesis of vascular diseases [5].

Vessel-resident (intimal, medial, and adventitial) and bloodborne vascular progenitor cells differentiate according to their location within the vascular wall into smooth myocytes when stretch is the main actor and endotheliocytes when shear is the major player, despite its lower magnitude, as the shear sensor density may be higher than that of stretch transducers [5].

1.1.3 Gene Transcription in Cardiomyocytes and Glucocorticoids

The stress response relies on the hypothalamic–pituitary–adrenal axis. This activated axis triggers the release of glucocorticoids from the adrenal gland. Glucocorticoids have potent anti-inflammatory and immunosuppressive effects. These stress hormones connect to the glucocorticoid receptor (GR), a member of the nuclear receptor superfamily of ligand-dependent transcription factors (NR3c1).

Expression of genes involved in GR-deficient hearts include [6]: (1) the maintenance of cardiac contractility (DMD [dystrophin] and RYR2 [cardiac ryanodine receptor]); (2) repression of adverse cardiac hypertrophy (Klf15 [Krüppel-like factor-15 and TRIM63 [Tripartite motif-containing protein-63; also known as Muscle-specific RING finger ubiquitin–protein ligase (MURF1)]); (3) promotion of cardiomyocyte survival (Ptgds [prostaglandin-D_2 synthase]); and (4) inhibition of inflammation decreases (LCN2 [lipocalin-2], and Zfp36 [Zinc finger protein-36 or tristetraprolin]). The dysregulated glucocorticoid-responsive target genes (lowered expression) encode many transcription factors, plasmalemmal receptors, and effectors of G protein-coupled receptor (GPCR) and mitogen-activated protein kinase (MAPK) signaling [6]. A glucocorticoid signaling deficiency to cardiomyocyte thus causes maladaptive cardiac hypertrophy, heart failure, and death [6].

In addition to GR, cardiomyocytes express the mineralocorticoid receptor (MR; NR3c2) that binds aldosterone and glucocorticoids with similar high affinity. Glucocorticoids typically circulate at much higher levels than aldosterone, and cardiac-specific MR may be primarily occupied by glucocorticoids [6]. However, the contribution of MR is negligible.

Patients who bear a single nucleotide polymorphism (A3669G) in the GR gene are characterized by an increased expression of the dominant-negative receptor variant GRβ and concomitant glucocorticoid resistance. They have an augmented risk of coronary artery disease, adverse cardiac hypertrophy, systolic dysfunction, and heart failure [6]. In addition, patients with Addison's disease identified by a decreased production of glucocorticoids have heart failure.

1.2 Myocardial Regeneration Capacity

Regeneration of the cardiac wall requires the recreation of the myocardial structure with its stroma, nodal tissue, and vasculature and restoration of the contractile function of quasi-synchronized cardiomyocytes due to distinct electrophysiological features according to regional and intramural localization. The cardiac regeneration may be achieved by stimulating the proliferation of cardiac progenitors as well as cardiomyocytes.

Whereas the mammalian fetal or neonatal heart can repair and generate a new contractile myocardium after damage, after a later myocardial loss, fibrosis predominates. The capacity of the heart to repair after damage remain thus restricted.

1.2.1 Cardiomyocytes Before and After Birth

The heart is the first organ built during embryo- and fetogenesis. During the fetal life, differentiated cardiomyocytes enter the cell cycle. The primitive heart formed in the early stages of cardiac specification mostly consists of quiescent cells. As development progresses, cardiomyocytes at the outer curvature resume proliferation to form the future chambers. On the other hand, cardiomyocytes at the inner curvature do not enter the cell cycle and will form the conduction system.

Embryonic and neonatal cardiomyocytes are capable of synthesizing, assembling, and disassembling their contractile proteins via the assembly of contractile structures for chromosome segregation and cytokinesis when they proliferate [7].

Proliferation and differentiation are not mutually exclusive in the developing as well as adult heart in some animal species that can regenerate their cardiac pump. For example, the damaged zebrafish heart reactivates a developmental program involving the expansion of GATA4+ cardiomyocytes to achieve regeneration [7].

However, the cardiomyocyte proliferation is limited in mammals. In humans, the cardiomyocyte stops dividing early after birth, possibly due to intracellular activation of the cyclin-dependent kinase inhibitors CKI1b and CKI2c, when breathing brings a sudden and marked increase in oxygenation (the arterial oxygen partial pressure $p_{A_{O_2}}$ ranges from 2.4 to 3.7 kPa [18–28 mmHg] which, during the fetal life, rises to 13.3 kPa [100 mmHg]) [7]. At birth, oxygen free radicals and reactive oxygen species (ROS) may stop cardiomyocytes proliferation and favor their terminal differentiation.

The neonatal cardiomyocytes lack a developed T-tubular network (microscopic invaginated segments of the sarcolemma); the myofibrils are located near the sarcolemma. In growing cardiomyocytes, T-tubules develop within a few days after birth.

Caveolin-3 localized to the nanoscopic invaginations of the sarcolemma, caveolae, transiently associates with primitive T-tubules. The location of various K^+, Na_V1, and $Ca_V1.2$ channels relies on caveolae.

Bridging integrator BIn1 (or ^{M}amphiphysin-2) is a major agent of T-tubule genesis and transfer of Ca^{2+} handling proteins in T-tubules. This membrane scaffold initiates T-tubule genesis in striated myocytes, as it induces membrane curvature. In addition, it determines the delivery of microtubule-transported membrane proteins. In particular, it forms BIn1–$Ca_V1.2$ clusters and tethers microtubules to membrane, thereby supporting $Ca_V1.2$ delivery to the T-tubular membrane [8]. It especially anchors microtubules and $Ca_V1.2$ at *junctional membrane complexes*. These specialized nanodomains connect ion channel clusters between transverse tubule and sarcoplasmic reticulum that are involved in excitation–contraction coupling that links membrane depolarization to mechanical contraction. A small entry of extracellular Ca^{2+} through $Ca_V1.2a$ causes a large Ca^{2+} release from its major intracellular store through RyR2 channels. This Ca^{2+}-induced Ca^{2+} release (CICR) inactivates $Ca_V1.2a$ for repolarization. Altered splicing and mutations of Bin1 are involved in myopathies.

In cardiomyocytes, junctophilin-2 encoded by the JPH2 gene is a major structural protein in T-tubules that forms junctional membrane complexes. Junctophilin-2 spans the junctional distance (~12 nm cleft) and tethers the plasma membrane to that of the sarcoplasmic reticulum to facilitate Ca^{2+} handling [9]. Its N-terminus is attached to the T-tubular sarcolemma and its C-terminal transmembrane domain is embedded in the sarcoplasmic reticulum. Junctophilin-2 is necessary for the development of postnatal T-tubules and stabilization of junctional membrane complexes in mature cardiomyocytes. Junctophilin-2 also interacts with caveolin-3 [9]. Its deficiency by mutations or expression downregulation is linked to hypertrophic cardiomyopathy, arrhythmias, and heart failure.

1.2.2 Oxygen Concentration and Cardiomyocyte Proliferation

The epicardium, which can be considered as a *cardiac stem cell niche*, is subjected to a relative hypoxia that yet enables the proliferation of cardiac progenitor cells. Epicardial progenitor cells are activated during embryogenesis and generate cardiomyocytes, endotheliocytes, and vascular smooth myocytes, as well as in adulthood after injury, once they are stimulated by thymosin-β4 [7]. The latter activates epicardial progenitors by upregulating antioxidative enzymes, thereby attenuating oxidative stress in these cells. In addition, a low oxygen partial pressure promotes the proliferation of human cardiomyocyte progenitors [7].

1.2.3 Mechanical Load and Cardiomyocyte Proliferation

In addition, cardiomyocytes are subjected to an increased mechanical load after birth. Cells of the inner wall surface, which sense shear in addition to stretch, are the first to exit cell cycle [7]. Cardiomyocytes are equipped with stress and strain sensors in Z disks and titin filaments that can modify the transcriptional activity. Stretch can provoke the production of growth factors and cytokines, such as IL6 and IGF1, which serve as paracrine controllers of cardiomyocyte proliferation [7].[4] Mechanical strain increases the intracellular level of ROS that promote cardiomyocyte differentiation and block their division.

[4] The heart of fishes, amphibians, and reptiles bears low pressure and high cardiac regenerative capacity.

1.2.4 Growth Factors and Cardiomyocyte Proliferation

Neighboring fibroblasts, endotheliocytes, smooth myocytes, adipocytes, and inflammatory cells influence cardiomyocyte proliferation, as they release paracrine messengers, in particular growth factors and cytokines.

In neonatal cardiomyocytes, fibroblast growth factor FGF2 promotes DNA synthesis via activated protein kinase-C [7]. In the adult myocardium, the FGF2 activity is potently inhibited by FGF16, which is preferentially expressed in the postnatal heart.

In neonatal cardiomyocyte cultures, platelet-derived growth factor supports cell proliferation, as it represses the G1-phase inhibitor CKI1b, activates PKB, and inhibits GSK3β [7].

Inflammatory meukocytes secrete the TNFSF12 cytokine that helps cardiomyocyte proliferation upon binding to TNFRSF12a receptor expressed by neonatal cardiomyocytes and downregulated in adulthood, except in pathological conditions [7].

Interleukin-6, which is abundantly secreted in the heart by both epicardiocytes and hypoxic adipose stromal cells, augments the proliferation rate of cultured slow-dividing neonatal cardiomyocytes [7].

Neuregulin-1, a member of the EGF family, is an inducer of adult cardiomyocyte proliferation via HER4 receptor [7]. Neuregulin-1 is first expressed by the endocardial endothelium of neonatal hearts, where it participates in trabeculation and cushion formation, and later by endotheliocytes. Periostin is another secreted molecule that may increase proliferation of both neonatal and adult cardiomyocytes via integrins and the PI3K pathway [7].

1.2.5 Extracellular Matrix and Cardiomyocyte Proliferation

The composition of the extracellular matrix contributes to the control of the rate of cardiomyocyte proliferation before and after birth. The cardiomyocyte division in the developing heart correlates with the regulated synthesis of matrix proteic constituents and integrins. In the fetal myocardium, the extracellular matrix that is enriched in fibronectin supports cardiomyocyte proliferation [7].

On the other hand, an elevated concentration of collagen-1 impedes cardiomyocyte proliferation after birth [7]. Focal adhesion kinase (FAK) that assists fibronectin binding to integrins facilitates cardiomyocyte proliferation during cardiogenesis heart development. FAK-related nonkinase (FRNK), the noncatalytic FAK C-terminus, is transiently expressed in the postnatal heart with peak levels occurring just prior to the withdrawal of cardiomyocytes from the cell cycle [7]. Therefore, FRNK may dampen FAK-dependent cardiomyocyte cell cycle progression and facilitate cell cycle exit in the postnatal myocardium.

1.2.6 MicroRNAs and Cardiomyocyte Proliferation

Cardiogenesis as well as cardiac homeostasis in adulthood is controlled by several different microRNAs [7]. These genome-encoded small RNAs mediate posttranscriptional silencing, thereby preventing unwanted gene expression.

The expression of microRNA-195, a member of the MIR15 family, is induced when cardiomyocytes exit the cell cycle. MiR29a and miR133 also suppress cardiomyocyte proliferation.

On the other hand, miR199a-3p and miR590-3p are able to cause cell cycle reentry and cytokinesis of cultured adult cardiomyocytes [7]. In vivo, these miRs specifically promoted cardiomyocyte (but not fibroblast) proliferation in both neonatal and adult hearts after myocardial infarction.

MicroRNA-499 promotes neonatal cardiomyocyte proliferation via Sox6 and cyclin-D1 [7]. The microRNA-17–92 cluster is also involved in cardiomyocyte proliferation, as its overexpression engenders cardiomyocyte proliferation in embryonic, postnatal, and adult hearts.

1.2.7 Genetic Manipulations and Cardiomyocyte Proliferation

The cardiomyocyte proliferation can be stimulated by different types of extracellular molecules. In the mouse adult heart, a slow-rate cardiomyocyte renewal is sustained by the division of preformed cardiomyocytes [7].

The exogenous administration of selected growth factors is capable of inducing neonatal and, in some instances, also adult cardiomyocyte proliferation [7]. The cardiac regeneration in the adult heart may also be achieved by reactivating the cardiogenesis program.

In addition, certain diffusible factors can regulate the proliferation and cardiac commitment of endogenous or implanted stem cells. The type of cardiomyocytes (nodal, atrial, or ventricular) can be identified by electrophysiological properties (e.g., shape of action potential waveform and duration and amplitude of generated action potentials). The characterization of action potentials enables automated categorization of cardiomyocyte populations from embryoid bodies (i.e., aggregates of pluripotent stem cells formed from embryonic or induced pluripotent stem cells by reprogramming somatic cells) for optimal implantation [10]. The Notch pathway operates between adjacent cells expressing Notch ligands (Jag1 and Jag2 and DLL1 and DLL3–DLL4) and receptors (Notch-1–Notch-4; Vol. 3, Chap. 10. Morphogen Receptors). In the developing heart, Notch modulates cardiomyocyte survival, cardiac stem cell differentiation, and angiogenesis and regulates trabeculation, cardiomyocyte proliferation, and valve formation [11]. In the postnatal heart, the Notch axis controls cardiac precursor expansion and differentiation, thereby sustaining cardiomyocyte proliferation.

However, in adult mice hearts, reactivation of the Notch pathway does not assist cardiac regeneration after myocardial infarction [7]. Therefore, at least in adult rodent hearts, cardiac regeneration does not necessarily mimic that in embryonic hearts.

On the other hand, in the adult heart of mice overexpressing the Notch ligand Jagged-1 on cardiomyocytes, activated Notch limits adverse cardiac remodeling [11]. Notch actually impedes cardiomyocyte apoptosis as well as cardiac hypertrophy in adult mouse hearts subjected to pressure overload and fibrosis mediated by transforming growth factor-β and its effector connective tissue growth factor [11]. Hence, the Notch pathway acts on cardiac mesenchymal stromal cells (cardiac fibroblasts and precursor cells). It reduces the proliferation of myofibroblasts. It also stimulates the expansion of SCA1+ and NKx2-5+ cardiac precursor cells [11].

Overexpression of cyclin-A2 that enables the cell cycle progression through the G2–M checkpoint (Vol. 2, Chap. 2. Cell Growth and Proliferation) can engender cardiomyocyte proliferation [7]. Cyclin-D1 causes a sustained DNA synthesis, but abnormal patterns of multinucleation. Overexpressed cyclin-D2 supports regenerative proliferation.

In transgenic mice, overexpression of CDK2 kinase also transiently raises cardiomyocyte proliferation [7]. Less-differentiated and mononucleated cardiomyocytes abound.

The transcription factor and regulator of cardiogenesis Meis homeodomain-containing protein Meis1 acts as a modulator of cell proliferation after birth.

The genetic Meis1 deletion extends cardiomyocyte proliferation after birth and reactivates cell division in adult hearts [7].

The postnatal cardiomyocyte proliferation is controlled by many signaling cascades, such as the Hippo pathway, and modulated by microRNAs.

The Hippo kinase cassette intervenes in myogenesis, organ size control, and regeneration of the heart [12]. Kinases and scaffold proteins of the Hippo axis suppress cell proliferation. The Hippo pathway regulates the activity of TEA domain-containing transcription factors (TEAD1–TEAD4) mainly via phosphorylation of the transcriptional coactivators YAP1 and WWTR1 that activate TEAD factors (Vol. 4, Chap. 10. Other Major Types of Signaling Mediators).

The cardiomyocyte proliferation involves the activation of IGF1 receptor, subsequent PKB activation, and inactivation of GSK3β, which stabilizes β-catenin, a promoter cardiac development [7].

The main effectors of the Hippo pathway are the transcriptional coactivators YAP1 and YAP2 that derepress the genes activated by the Wnt–β Ctnn pathway. Genetic deletion of the upstream inactivating effectors (STK4, WW45 [Sav1], and LaTS1/2) causes cardiac hyperplasia [7]. On the other hand, transgenic mice overexpressing STK4, LaTS1, or LaTS2 die postnatally with dilated cardiomyopathy without compensatory ventricular myocyte hypertrophy.

1.2.8 Growth Factors and Cardiac Progenitor Proliferation in the Adult Heart

Various types of adult progenitor cells have been injected into damaged hearts or into coronary circulation, such as mesenchymal stromal cells, hematopoietic stem cells, endothelial progenitor cells, skeletal myoblasts, and cardiac progenitor cells to ensure cardiac protection, metabolism, contractility, proper remodeling, and neovascularization via stem cell paracrine signaling using growth factors, cytokines, and chemokines (FGF1–FGF2, HGF, IGF1, PDGF, VEGF, Nrg1, angiopoietin-1, IL6, and CXCL12), as well as differentiation and repopulation.

1.2.9 Heat Shock Proteins and Cardiac Repair

The member HSPa12b of the HSP70 family is predominately expressed in endotheliocytes of the heart, brain, lung, kidney, and adipose tissue, among others, but not in liver sinusoids [13]. In the heart and brain, the vascular endothelium of vessels of all sizes produces HSPa12b, whereas in lungs and adipose tissue, it is mainly synthesized in capillaries. It is required for angiogenesis, specifically in adhesion, migration, and tube formation.

The heat shock protein HSPa12b possesses 22 client proteins such as aryl hydrocarbon receptor nuclear translocator (ARNT). Among them, A-kinase-anchoring protein AKAP12 and podocalyxin-like protein PodxL are implicated in cell adhesion.

HSPa12b protects endotheliocytes against lipopolysaccharide-induced dysfunction and inflammation, as well as cerebral ischemia–reperfusion injury via the PI3K–PKB pathway possibly stimulated by angiopoietin-1 [13].

Transgenic mice expressing human HSPa12b specifically in endotheliocytes (without change in expression of other HSPs [i.e., HSPb5, HSP25, HSP60, HSP72, and HSP90]) have an improved cardiac function and remodeling assessed by left ventricular enlargement, wall thinning, and fibrosis, with a lower rate of cardiomyocyte apoptosis and higher capillary and arteriolar densities, from 1 to 4 weeks after myocardial infarction compared with wild-type mice [14]. The HSPa12b chaperone yields a protective and reparative response to ischemic insult in the heart.

Under ischemia, levels of proteins involved in cardiomyocyte survival and angiogenesis, such as NOS3 (without significant change of NOS2), Ang1, VEGF, and BCL2, are further elevated [14]. Angiopoietin-1 has an anti-inflammatory and protective actions on endotheliocytes via the PI3K–PKB axis, a decreased expression of cell adhesion molecules ICAM1 and VCAM1 and hence immunocyte recruitment, and an increased expression of antiapoptotic proteins such as BCL2. The VEGF factor augments NOS3 expression, thereby enhancing nitric oxide synthesis. The latter messenger acts against oxidative stress, inflammation, hypertrophy, and fibrosis, thereby reducing adverse cardiac remodeling. Protected cardiomyocytes, in turn, secrete VEGF and paracrine factors that protect endotheliocytes and support angiogenesis.

1.3 Molecular Chaperones in Cardiac Diseases

The continuous activity of cardiomyocytes requires the maintenance of *proteostasis*, that is, a stringent control of synthesis, folding, and turnover of proteins (Vol. 1, Chap. 5. Protein Synthesis), especially in sarcomeres, endoplasmic reticulum (ER), and mitochondria.

Table 1.3 Molecular chaperones in pathophysiology. (**Part 1**) General chaperones and cochaperones. (Source: [15])

Chaperone	Targets	Pathophysiological role
HSP90	Protein folding PKB–NOS3 axis Raf–ERK axis	Linkage to immature $K_V11.1$ ↓ ischemia–reperfusion injury
HSP70	Protein folding Ubiquitination of misfolded proteins Intracellular signaling	Production induced by ischemia Linkage to immature $K_V11.1$
HSP40	Mitochondrial genesis	Avoids dilated cardiomyopathy
CHIP (STUB1)	Unfolded proteins moved to proteasome	Limit infraction extension

CHIP C-terminus of HSC70-interacting protein (which inhibits the ATPase activity of the chaperones HSC70 and HSP70; this ubiquitin–protein ligase is also called STUB1 [STIP1 homology and U-box-containing protein])

Molecular chaperones are devoted to regulating protein folding and maintaining the correct balance between synthesis and degradation. Molecular chaperones, heat shock proteins (HSP), prevent accumulation of damaged proteins either by supporting their refolding or by targeting them to proteasomal or autophagosomal degradation. In addition, chaperones participate in intracellular signaling, as they control conformational changes required for activation and deactivation of signaling mediators and their assembly in signalosomes.

In normal conditions, cells react to chemical and adverse mechanical stresses using molecular chaperones of the salvage machinery. However, high and persistent mechanical overload, oxidative stress, shifts in pH and temperature, as well as genetic mutations, can provoke accumulation of misfolded proteins. Insoluble aggregates of misfolded proteins that then form are toxic for the cell. Accumulation and aggregation of misfolded protein can be involved in cardiac diseases, such as dilated cardiomyopathy, arrhythmias, and heart failure [15].

Stimulation of chaperone synthesis has a cardioprotective role in ischemia–reperfusion injury, heart failure, and arrhythmias [15]. On the other hand, mutations in genes encoding chaperones generate different forms of cardiomyopathies.

The chaperone machinery consists of numerous proteins that cooperate. Chaperones constitute a large set of proteins that can be subdivided into three groups (Tables 1.3, 1.4, and 1.5) [15]:

1. The general chaperones HSP90 and HSP70;
2. The small chaperones (sHSP); and
3. The ER chaperones

Table 1.4 Molecular chaperones in pathophysiology. (**Part 2**) Small chaperones. (Source: [15])

Chaperone	Targets	Pathophysiological role
HSPb1 (HSP25)	Protein folding, aggregation	Protection against atrial fibrillation
HSPb2 (HSP27)	Protein folding, aggregation	Reduction of ischemia–reperfusion damage
HSPb5	Desmin and titin folding	Reduction of oxidative stress Attenuation of apoptosis and necrosis Opposition to cardiac hypertrophy
HSPb6 (HSP20)	PKB/BAD/Casp3 MAP3K5–JNK/P38	Cardioprotection DCM (gene mutation)
HSPb7	Cytoskeletal protein folding (?)	Protection against atrial fibrillation DCM (gene SNP)
HSPb8	Autophagy	Reduction of infarct size
HSP22	BMPR2–MAP3K7–PI3K–PKB STAT3	Opposition to fibrosis
Melusin	FAK; ERK1/2, PKB–GSK3β	Preservation of contractility

DCM dilated cardiomyopathy, *SNP* single-nucleotide polymorphism

Table 1.5 Molecular chaperones in pathophysiology. (**Part 3**) Endoplasmic reticulum chaperones and stress response proteins. (Source: [15])

Chaperone	Targets	Pathophysiological role
HSPa5	Protein folding	Resistance to ischemia
PDI	Protein folding	Reduction of infarct size
DDIT3 (CHOP)	Apoptosis	Heart failure Cardiac hypertrophy, fibrosis
XBP1	ER chaperone synthesis	Dominant-negative XBP1 augments apoptosis
ATF6	ER chaperone synthesis	Production of HSPa5 and HSP90β1 Left ventricle protection upon pressure overload

ATF6 activating transcription factor-6, which improves endoplasmic reticulum chaperone synthesis, *DDIT3* DNA damage-inducible transcript-derived protein-3; also known as C/EBP-homologous protein (CHoP), a proapoptotic transcription factor, *PDI* protein disulphide isomerase, *XBP1* X-box-binding protein-1, a transcription factor enhancing endoplasmic reticulum chaperone synthesis

1.3.1 General Chaperones

The function of HSP90 and HSP70 is tightly controlled by cochaperones that regulate the specificity and dynamics of a chaperone reaction, as they recognize specific target proteins and control ATP binding and hydrolysis. Cochaperones also participate in the decision toward degradation of misfolded proteins.

For example, HSP70 binds to its substrate assisted by its cochaperone HSP40. The additional cochaperone HSP70/HSP90-organizing protein homolog (HOP)[5] promotes the transfer of the substrate to HSP90 that enables conformational changes of the substrate. Other additional cochaperones, such as the HSP90 cochaperone P23, which is the cytosolic prostaglandin-E_2 synthase PGeS3, and prolyl isomerase, regulate HSP90 ATPase activity and protein folding [15].

1.3.2 Small Chaperones

Small chaperones (sHsp or HSPb) constitute a family of 11 members that can homo- and hetero-oligomerize. This polymerization can be regulated by transient phosphorylation in response to stress [15].

Small chaperones are devoid of ATPase activity. Nonetheless, they can prevent misfolded protein aggregation and regulate intracellular signaling, sometimes, in cooperation with HSP70 and HSP90.

1.3.3 Chaperones of the Endoplasmic Reticulum

The ER of cardiomyocytes consists of a vesicular compartment mainly in the perinuclear region involved in protein synthesis, and another compartment organized around myofibrils and T-tubules, which controls excitation–contraction coupling via calcium cycling. The two networks are interconnected.

Hypoxia, blood pressure overload, and drug-induced insults activate the ER chaperone machinery that maintains the proper balance between protein folding and degradation in the ER.

The ER chaperone machinery that assists protein folding and catalyzes disulphide bond formation includes:

- Calnexin and calreticulin that control Ca^{2+} homeostasis;
- Chaperones HSPa5 and HSP90β1;
- Protein disulphide isomerase; and
- Thiol oxidoreductase-like protein ERP57.

These molecular chaperones act in concert with three ER membrane complexes, PERK, IRE1α, and ATF6, which activate the ER stress response [15]. In unstressed conditions, PERK (or eIF2α K3 kinase), IRN1 (or inositol-requiring Ser/Thr protein kinase and endoribonuclease IRe1 or IRe1α), and ATF6 bind to HSPa5 in the ER lumen, which inhibits their function. Upon ER stress, HSPa5 connects to misfolded proteins in the ER lumen, dissociating from PERK, IRN1, and ATF6 that become

[5] Also known as suppression of tumorigenicity ST13.

activated. The PERK (pancreatic eIF2α ER kinase) represses protein synthesis, as it phosphorylates the eukaryotic initiation factor-2α. Endoplasmic reticulum-to-nucleus signaling protein ERN1 generates a spliced XBP1 mRNA encoding a potent transcription factor that activates the expression of ER stress response genes. X-box-binding protein XBP1 upregulates genes involved in ubiquitin-dependent protein proteasomal degradation. Activating transcription factor ATF6 translocates to the Golgi body upon activation and undergoes cleavage by specific peptidases, yielding a soluble cytosolic transcription factor that enters the nucleus and induces further transcription of ER stress response genes.

Prolonged or extreme ER stress induces the expression of the transcription factor CHOP that triggers synthesis of proapoptotic proteins and ER-mediated cell death.

1.3.4 Protein Degradation and Autophagy

C-terminus of HSP70-interacting protein (CHIP) is a ubiquitin ligase highly expressed in the heart that, in cooperation with HSP70 and BCL2-associated athanogene (BAG) family molecular chaperone regulator BAG1, controls degradation of sarcomeric proteins by the proteasome [15].

In the absence of CHIP, ischemic damage leads to an increased infarct size. When misfolded proteins accumulate, proteic aggregates form that can activate autophagy by lysosomal hydrolases. Autophagy is observed in heart failure and ischemic heart disease [15].

The cochaperones BAG3 and HSPb8 operate in autophagic clearance in the heart (*chaperone-assisted selective autophagy*) [15]. Mutations of the Bag3 gene cause dilated cardiomyopathy.

1.3.5 Molecular Chaperones and Signal Transduction

Chaperones and scaffold proteins promote the formation of proteic complexes and stabilize the interaction of signaling components. The HSP90 substrates include the kinases Src, Raf, PKB, PI3K, and PDK1, as well as NOS3 [15]. Interaction of HSP90 with kinases involves the cochaperone protein Cdc37 (also called HSP90 chaperone protein kinase-targeting subunit).

Most signaling mediators are degraded upon inhibition of HSP90. The latter operates in VEGF-dependent NOS3 activation in endotheliocytes that leads to the disruption of the caveolin–NOS3 complex and promotes the NOS3–HSP90 association. The NOS3–HSP90 dimer recruits PKB that phosphorylates NOS3 (Ser1177) [15].

In addition to HSP90, sHSPs and cochaperones (e.g., HSPb6, HSPb8, melusin, and BAG3) regulate signaling pathways that control cardiomyocyte survival and calcium cycling. The small HSPb6 chaperone promotes cardiomyocyte survival via PKB and inhibition of the MAP3K5–JNK/P38MAPK modules. XS in response to β-adrenergic stimuli, HspB6 that binds to PP1 phosphatase, hence increasing phospholamban phosphorylation, is phosphorylated by PKA (Ser16), thereby activating SERCA2a pump [15].

Morgana and Melusin cooperate with HSP90 to regulate signal transduction in pathophysiological processes. Morgana is ubiquitously expressed, whereas Melusin is a striated muscle-specific chaperone [16]. Morgana controls genomic stability by regulating the centrosome cycle via the RoCK2 kinase. Melusin, or integrin-β_1-binding protein Itgβ1BP2 organizes ERK signaling in cardiomyocytes and regulates cardiac hypertrophy in response to pressure overload.

Melusin, in association with HSP90, promotes the organization of the Melusin signalosome that regulates ERK1, ERK2, and PKB signalings in response to different stresses. The Melusin signalosome comprises the scaffold and RhoGAP IQGAP1 that binds to FAK as well as the kinases of MAPK modules cRaf, MAP2K1, MAP2K2, ERK1, and ERK2.

1.3.6 Molecular Chaperones and Gene Mutations

Chaperones can tether to polypeptide variants encoded by mutated genes and support their refolding, hence restoring their function. Moreover, HSP90 can prevent transposon-induced genetic variations via piRNAs involved in transposon silencing [15].

The Hspb5 gene is mutated (R120G) in certain types of cardiomyopathies. The R157H mutation that affects the interaction of HSPb5 with titin can be detected in dilated cardiomyopathy. Conversely, overexpression of HSPb5 prevents formation of proteic aggregates of desmin variants that cause cardiomyopathies. A heterozygous Hspb6 point mutation (P20L) can also be identified in some cases of dilated cardiomyopathy.

1.3.7 Cardioprotective Chaperones

The mitochondrial chaperone HSP40 contributes to the prevention of dilated cardiomyopathy. Overexpression of HSPb1 ensures protection against pacing-induced cardiomyocyte damage, as it preserves sarcomeric constituents from proteolysis by calpain [15]. HSPb6–HSPb8 also protect against tachycardia-induced atriomyocyte remodeling.

1.4 Control of Cardiomyocyte Fate and Function

1.4.1 Nitric Oxide Synthases—Mechanochemotransduction

In cardiomyocytes, stretch and hence contraction trigger the production of reactive oxygen and nitrogen species that target Ca^{2+}-signaling proteins (*mechanochemotransduction*). This process involves nitric oxide synthases (NOS1 more than NOS3), NADPH oxidase NOx2, and Ca^{2+}-calmodulin-dependent kinase CamK2 [17].

The NOS1 subtype is situated closer to the Ca^{2+} release sites, that is, to ryanodine receptors. The NOS1 isoform, but not NOS3, operates in afterload-induced Ca^{2+} sparks via elevated ryanodine receptor sensitivity, thereby enhancing contractility to counter mechanical load [17].

On the other hand, spontaneous Ca^{2+} sparks during diastole can be arrhythmogenic. Both CamK2 and NOx2 also contribute to afterload-induced Ca^{2+} sparks. However, mechanotransduction launched by NOS1 and CamK2 does not depend on NOx2, as inhibition of NOS1 and CamK2, but not NOx2, in cardiomyocytes eliminates Ca^{2+} sparks, at least in a mouse model of familial hypertrophic cardiomyopathy with enhanced mechanotransduction [17].

1.4.2 StIM1–Orai1-Mediated Store-Operated Calcium Entry

Cycle of Ca^{2+} influx and outflux from the cytosol relies on Ca^{2+} store refilling. The latter process is carried out by Ca^{2+} reuptake from the cytosol and activation of Ca^{2+} influx using the slow store-operated Ca^{2+} entry (SOCE) mechanism. The latter is delayed by more than 30 s after store depletion.

In cardiomyocytes, SOCE coexists with Ca^{2+}-induced Ca^{2+} release (CICR) from the ER mediated by the couple formed by plasmalemmal voltage-gated Ca^{2+} channels and ryanodine receptors on the apposed ER membrane.

In response to binding of angiotensin-2 and endothelin-1 to their respective GPCRs, the newly produced inositol trisphosphate (IP_3) binds to its receptor (IP_3R) on the ER membrane, causing Ca^{2+} release from its main store.

1.4.2.1 Involved Molecules

In cardiomyocyte, store-operated Ca^{2+} entry mediated by stromal interaction molecule StIM1 in conjunction with plasmalemmal Ca^{2+} release-activated Ca^{2+} (CRAC) channels Orai1 and canonical transient receptor potential protein (TRPC) contributes to Ca^{2+} handling and hence excitation–contraction coupling.

The StIM1–Orai1 complex mediates highly Ca^{2+}-selective and nonvoltage-gated current. On the other, the StIM1–TRPC complex conveys Ca^{2+}-selective and - nonselective store-operated currents, as TRPC is a relatively nonselective cation channel conducting inward currents carried by both monovalent and divalent cations (e.g., Na^+ and Ca^{2+}) [18]. Among TRPCs, TRPC1, TRPC3, TRPC4, and TRPC6 are implicated in SOCE-mediated hypertrophic signaling in cardiomyocytes.

The Orai1–StIM1 complex localizes in particular to the transverse tubule of skeletal myocytes [19]. Its rate of activation is regulated by the RyR1 receptor.

Stromal Interaction Molecule The STIM family comprises two proteins (StIM1–StIM2). The most widely studied StIM1, a ubiquitous type-1 membrane protein, predominantly localizes to the ER membrane where it acts as an Ca^{2+} sensor that detects changes in ER Ca^{2+} concentration. In response to Ca^{2+} depletion in the ER, Ca^{2+} dissociates from StIM1 that then oligomerizes and translocates to subplasmalemmal punctae.

StIM1 At rest, StIM1 dimerizes. In the absence of store depletion, StIM1 clusters in cytosolic regions rather than at portions near the plasma membrane. Upon ER Ca^{2+} depletion, StIM1 oligomerizes and moves to regions close to the plasma membrane (ER–PM junctions) and links to Orai1 (mainly) or TRPC channels. Oligomerization of StIM1 depends on phosphorylation.

StIM1 is also a microtubule plus-end-binding protein serving in microtubule extension. However, in contrast to all other cytosolic plus-end-tracking proteins, StIM1 is attached to the ER membrane. It binds to microtubule-associated protein of the RP/EB family MAPRE1 at sites where polymerizing microtubule ends contact the ER [20]. ER tubule elongates with the MAPRE1+ end of a growing microtubule. Therefore, microtubule growth-dependent concentration of StIM1 in the ER membrane enables ER remodeling.

Stanniocalcin-1 is a secreted glycoproteic auto- or paracrine regulator of calcium and phosphate homeostasis. It prevents Ca^{2+} uptake. Stanniocalcin Stc2 is an ER stress protein as a component of the unfolded protein response, the expression of which is induced by oxidative stress and hypoxia. It colocalizes with StIM1 and behaves as a SOCE inhibitor [21].

In addition, StIM1 oligomerization and SOCE are precluded by the ER-resident protein disulfide isomerase PDIa3 [22].

Other modulators of StIM1 and Orai1 include EF-hand calcium-binding domain-containing protein EFCaB4b, or CRAC regulator CRACR2a. The cytosolic Ca^{2+} sensor CRACR2a stabilizes CRAC channels, as it promotes StIM–Orai binding at low Ca^{2+} concentrations. The ternary Orai1–StIM1–CRACR2a complex dissociates at elevated Ca^{2+} levels [23].

Another ER integral membrane protein Surf4 (Surfeit locus protein-4 or simply Surfeit-4) binds to StIM1 and modulates StIM1-dependent SOCE [24]. This modulator impedes StIM1 clustering upon store depletion.

The ubiquitous ER membrane protein, SOCE-associated regulatory factor SARAF [25, 26]: (1) connects to the StIM1–Orai1 activation region (SOAR) of StIM1 domain (which activates Orai1); (2) facilitates slow Ca^{2+}-dependent inactivation of Orai1 via the C-terminal inhibitory domain (CTID) of StIM1 (which hinders spontaneous StIM1 clustering and full Orai1 activation in the absence of store depletion and determines access of SARAF to the SOAR motif to regulate both FCDI and SCDI); and (3) protects cells from calcium overload.

At rest, SARAF binds to the SOAR sequence of the StIM1 protein. Activation of Orai1 requires dissociation of SARAF from the SOAR domain that then can tether to Orai1.

StIM2 In mice, global deletion of the STIM1 gene is lethal within 2 days after birth, whereas lack of StIM2 enables survival until 4–5 weeks [18]. The StIM2 subtype has a different N-terminus with respect to StIM1. Once it is linked to Orai1, StIM2 causes both store-dependent and -independent Ca^{2+} entry.

StIM1L In addition, the high-molecular weight splice variant StIM1L also lodges in cardiomyocytes [18]. It forms a permanent cluster with actin and colocalizes with Orai1 [27]. The unique behavior of the CRAC flux in skeletal myocytes may result from the coupling of Orai1 channels with StIM1L [19].

Orai The ORAI family includes three members (Orai1–Orai3). The StIM1 protein also contributes to store-independent Ca^{2+} entry via the arachidonic acid-regulated Ca^{2+} channel that is a heteromer consisting of both Orai1 and Orai3 channels, whereas the classical SOCE flux relies on StIM1 connected to an Orai1 homohexamer [18].

At rest, Orai1 homodimerizes or homotetramerizes. Upon activation, it hexamerizes. Orai1 is a highly Ca^{2+}-selective, inward-rectifying channel prominently regulated (inactivated) by Ca^{2+} ion. The half time of the 2 modes of inactivation, fast (FCDI) and slow (SCDI) Ca^{2+}-dependent inactivation, ranges from 10 to 100 ms and 2 to 3 ms, respectively [26]. The former depends on StIM1 and calmodulin binding to the Orai1 channel [25].

1.4.2.2 SOCE Goal, Regulation, and Function

In response to IP_3R- and/or RyR-mediated Ca^{2+} release from sarcoplasmic reticulum stores, SOCE enables Ca^{2+} influx from the extracellular space and a sustained increase in cytosolic Ca^{2+} concentration to regulate gene transcription (Vol. 4, Chap. 11. Signaling Pathways). SOCE is also aimed at rapidly refilling the depleted store. Although StIM1 and Orai1 alone can reconstitute SOCE activity, other proteins are involved (e.g., SERCA).

Once SOCE is activated, it is subjected to various regulatory processes that determine the duration and magnitude of the Ca^{2+} influx. In particular, the StIM1–Orai1 complex is regulated by the SGK1 and AMPK kinases that may reduce unnecessary SOCE when energy reserves are low [18]. On the other hand, in pulmonary arterial smooth myocytes, PDGF activates the PKB–TOR pathway and, subsequently, enhances the SOCE process and cell proliferation [28].

The SOCE process is specifically involved in the IP_3-mediated α-adrenergic signaling, but does not transmit the β-adrenergic message [18].

1.4.2.3 Pathophysiological Role of SOCE

SOCE and Metabolic Perturbations Hyperglycemia is associated with a decrease in total ER Ca^{2+} content and StIM1 level at least in endotheliocytes [18]. In smooth myocytes, SOCE is inhibited by hyperglycemia. In platelets from type-2 diabetic patients, SOCE is reduced because of attenuated association between StIM1 and Orai1, TRPC1, and TRPC6 channels.

The SOCE process may be regulated by $^{O}Glc^{NAc}$ attachment (O-GlcNAcylation). A sustained posttranslational protein modification by ^{O}linked ^{N}acetylglucosamine contributes to the adverse effects of hyperglycemia [18]. Glucosamine, an amine sugar that selectively fuels the hexosamine synthesis pathway, mimickes the effects of hyperglycemia [18]. Inhibition of glucose entry into the hexosamine synthesis pathway suppresses the effects of hyperglycemia on SOCE. Angiotensin-2-induced increase in cytosolic Ca^{2+} level is attenuated in cardiomyocytes by increasing the $^{O}Glc^{NAc}$ level [18]. O-GlcNAcylation targets sites that are frequently subjected to phosphorylation. O-GlcNAcylation of StIM1 actually prevents its phosphorylation and lowers StIM1-associated subplasmalemmal puncta formation and subsequently SOCE-mediated Ca^{2+} flux [18].

SOCE and Arrhythmias On the other hand, Orai1 and Orai3 may be linked to arrhythmias in both atrial and ventricular myocytes [18]. The TRPC channel is produced in pacemaker cells of the sinoatrial node. Moreover, elevated Orai-mediated SOCE can initiate atrial and ventricular arrhythmias.

SOCE and Cardiac Hypertrophy StIM1 is necessary and sufficient in the occurrence of cardiomyocyte hypertrophy due to the pressure overload in the adult heart. Calcium-mediated cardiomyocyte hypertrophy such as that primed by angiotensin-2 depends on SOCE, but neither on $Ca_V1.2a$ nor Na^+–Ca^{2+} exchanger [18]. Both TRPC3 and TRPC6 contribute to the development of angiotensin-2-induced hypertrophy. A TRPC4-dependent sustained increase in cytosolic Ca^{2+} activates the PP3–NFAT pathway implicated in cardiac hypertrophy. Endothelin-1 activates SOCE, upregulates TRPC1 expression, and provokes NFAT nuclear translocation and cardiomyocyte hypertrophy. In addition, the concentration of the large StIM1 splice variant StIM1L rises in response to hypertrophic stimuli.

SOCE and Ischemia–Reperfusion Injury

Cardiomyocyte During ischemia–reperfusion injury, calcium overload was assumed to result from the coupling of Na^+–H^+ exchanger with Na^+–Ca^{2+} exchanger operating in its reversal mode and, to a lesser extent, $Ca_V1.2a$ channel [18]. In fact, SOCE inhibition protects the heart against Ca^{2+} overload induced by acute ischemia–reperfusion injury.

Hypoxia stimulates StIM1 accumulation at subplasmalemmal punctae possibly due to the action of ROS. However, the accompanying acidosis uncouples the StIM1–Orai1 complex, thereby preventing Ca^{2+} overload [18].

Upon reperfusion, the acidosis is quickly resolved and SOCE may be again activated. In addition, the Ca^{2+} concentration in the sarcoplasmic reticulum decays rapidly during reperfusion.

In cardiomyocytes subjected to ischemia–reperfusion injury, overexpressed TRPC3 increases apoptosis and calpain activation with respect to control cardiomyocytes [18].

Endotheliocyte Pharmacological inhibition of SOCE abolishes Ca^{2+} overload in cardiac microvascular endotheliocytes subjected to ischemia–reperfusion injury [18].

1.4.3 Sphingosine 1-Phosphate

Circulating sphingosine 1-phosphate signals via RhoA GTPase that promotes cardiomyocyte survival and PLCε to prime the phosphorylation (activation) of protein kinase-D1, thereby ensuring cardioprotection against ischemia–reperfusion injury [29]. The PKD1 kinase phosphorylates (inhibits) the cofilin phosphatase slingshot homolog SSH1, thereby impeding cofilin-2 translocation to mitochondria in response to oxidative stress or ischemia–reperfusion injury. Cofilin-2 links to the proapoptotic protein BAX. Therefore, S1P preserves mitochondrial integrity upon oxidative stress and supports cardiomyocyte survival.

1.4.4 Forkhead Box Transcription Factors

The member of the forkhead box family of transcription factors FoxO1 is involved in oxidative stress response, immunity, pluripotency in embryonic stem cells, and regulation of cell metabolism, proliferation, and death (Vol. 4, Chap. 10. Other Major Types of Signaling Mediators).

The FoxO1 factor promotes gluconeogenesis, as it stimulates synthesis of hepatic glucose 6-phosphatase and phosphoenolpyruvate carboxykinase [30]. In skeletal muscle, FoxO1 raises expression of pyruvate dehydrogenase kinase, thereby decreasing glucose oxidation. In the heart, FoxO1 regulates glucose and fatty acid metabolism, as it targets the genes that encode pyruvate dehydrogenase kinase PDK4 and nitiric oxide synthase NOS2 [30].

In the heart, cellular insulin-stimulated glucose uptake through the GluT1 (basal glucose uptake) and mainly GluT4 (stimulated glucose uptake) and glycolysis leads to the formation of pyruvate that undergoes oxidation in the mitochondria to yield ATP. In cardiomyocytes, insulin resistance is linked to PDK4 overexpression, hence with lowered glucose and heightened fatty acid oxidation due to nuclear FoxO1 action [30].

The FoxO1 factor also contributes to the regulation of the supply and oxidation of fatty acids. Dietary fatty acids are delivered to the heart from lipoprotein lipase-mediated lipolysis of triglyceride-rich lipoproteins as well as by albumin-bound fatty acids. The uptake of fatty acids by cardiomyocytes across the plasma membrane is mediated by three protein-fatty acid translocases: plasmalemmal and mitochondrial glutamate oxaloacetate transaminase GOT2, fatty acid transport protein (FATP), and the ScaRb3 receptor. AcylCoA synthase enhances fatty acid uptake. The FoxO1 agent is implicated in lipid supply and storage, as its nuclear localization is correlated with an elevated plasmalemmal density of ScaRb3.

The FoxO1 factor is involved in the regulation of the oxidative stress response. It increases expression of antioxidant genes such as those that encode superoxide dismutase and the stress sensor GADD45, thereby promoting ROS scavenging and preventing DNA damage [30]. It collaborates with SIRT1, hence activating expression of antioxidant genes such as those that encode ^{Mn}SOD and catalase.

In embryonic stem cells, FoxO1 can maintain their pluripotency, a site acts on promoters of the OCT4 and SOX2 genes.

1.4.5 Regulation by β-Adrenoceptors

β-adrenoceptor regulation of cardiac contraction and gene expression is mediated by protein kinase-A. The PKA activity is also controlled by several other messengers acting via Gs-coupled receptors. The velocity and specificity of β-AR signaling are determined by the localization of the PKA holoenzyme and associated A-kinase anchoring proteins as well as the spatiotemporal pattern of cAMP signaling.

The second messenger cAMP is small diffusible molecule. The cAMP–PKA couple is organized into signaling complexes organized by A-kinase anchoring proteins that tether inactive PKA holoenzymes and other signaling molecules and distributed in discrete compartments in cardiomyocytes [31]. From these cellular foci, specific cAMP signals propagate to determined subcellular loci. In cardiomyocytes, AKAP6 anchors PKA to multiple subcellular compartments, including the nucleus and nuclear membrane [32]. In the nucleus, AKAP6 also connects to PDE4d3 that controls the cAMP level.

On the sarcoplasmic reticulum in cardiomyocytes, PKA regulates the calcium cycling for excitation–contraction coupling that is dysfunctional in heart diseases. The direct activation of adenylate cyclases or application of a cAMP analog provokes a much slower and smaller increase in PKA activity than stimulated adrenergic receptors [31]. The signaling primed by adrenoceptors has a preferential access to the sarcoplasmic reticulum, the major calcium store. Calcium ions are released through ryanodine receptors and taken up through sarco(endo)plasmic reticulum Ca^{2+} ATPase. The triggered cAMP–PKA signaling leads to phosphorylation of these ion carriers and/or associated regulatory proteins, thereby raising Ca^{2+} flux through these channels to adapt cardiac contraction.

In the absence of cAMP, the PKA holoenzyme is a heterotetramer consisting of two catalytic subunits bound and inhibited by a dimer of regulatory subunits. Activation of adenylate cyclases increases intracellular cAMP concentration. This second messenger dissociates (activates) the catalytic subunits.

Then, PKA phosphorylates many substrates in various subcellular compartments, such as the ion channels involved in excitation–contraction coupling, sarcolemmal $Ca_V1.2a$ and sarcoplasmic reticulum RyR2, the SERCA inhibitor phospholamban (Pln), and the sarcomeric components cardiac myosin-binding protein-C (cMyBPc) and troponin-I (cTnnI; Sect. 4.1.1).

In the nucleus of cardiomyocytes, PKA targets the transcription factors of the CREB family and class-2 histone deacetylases HDAC4 and HDAC5.

The cAMP concentration depends on its production by adenylate cyclases and degradation by phosphodiesterases (PDE1–PDE4 and PDE8) in cardiomyocytes. Members of the cAMP-specific PDE4 family (PDE4a–PDE4b and PDE4d encoded by three genes) are major inhibitors of the cAMP messenger. At the nuclear envelope, PDE4d complexes with AKAP6. It is involved in the regulation of cardiomyocyte hypertrophy. The loss of PDE4d can cause dilated cardiomyopathy, at least in mice. AKAP-anchored PDE4 controls the extent of PKA released upon βAR stimulation and transferred into the nucleus from a cytosolic pool [32].

In the cytoplasm, phosphatases PP1 and PP2 contribute to the termination of PKA action, whereas only PP1 acts in the nucleus [32].

In adult rat ventriculomyocytes, β-adrenoceptor stimulation raises quickly the cytoplasmic and nuclear cAMP concentration, whereas PKA activity elevation is prompt in the cytoplasm, but slower in the nucleus [32]. Similar slow kinetics of nuclear PKA activation is observed upon adenylate cyclase activation or phosphodiesterase inhibition. Pulse stimulation causes maximal PKA activation and phosphorylation of myosin-binding protein-C in the cytoplasm, but has a slight impact on PKA activation and phosphorylation of cAMP response element-binding protein in the nucleus.

The nuclear PKA activity is temporally dissociated from that of nuclear cAMP elevation and cytoplasmic PKA activation. Once PKA holoenzyme is activated in the cytoplasm, its catalytic subunit translocates to the nucleus [32].

In cardiomyocytes, the upstream inhibitors PDE3 and PDE4 and downstream inhibitors PP1, PP2, and PP3 can participate in the cytoplasmic and nuclear PKA response to βAR stimulation.

The PDE3 and PDE4 enzymes represent the major PDE activity in cardiomyocyte nuclei. They primarily localize at the nuclear envelope. The PDE4d subtype controls the nuclear as well as cytosolic PKA activity, whereas PDE3 has no effect [32].

The PP1 and PP2 phosphatases contribute differently to PKA target sites in the cytoplasm and nucleus. They counteract the action of the βAR–cAMP–PKA axis, especially the phosphorylation of mediators of the electromechanical coupling, but PP3 has no effect [32]. The PP3 phosphatase may play a minor role when PP1 and PP2 are active. However, PP3 intervenes in pathogenesis, especially cardiac adverse hypertrophy resulting from pressure overload. In particular, PP3 dampens β AR effect on Pln in spontaneously hypertensive rats. Moreover, it is detected in the nucleus of cardiomyocytes in the diseased myocardium, but not in the case of healthy hearts [32].

In addition, cAMP can also activate $_{\text{cAMP}}$GEFs (RapGEF3–RapGEF4) at the plasma membrane and nuclear and perinuclear compartment area in cardiomyocytes [32]. These effectors increase nuclear Ca^{2+} concentration as well as CamK2-

dependent nuclear export of HDAC5 and subsequent derepression of the MEF2 transcription factor. Unlike CamK2- and PKD-mediated phosphorylation, PKA phosphorylates HDAC5 in the nucleus, thereby causing its nuclear retention and the repression of the hypertrophic gene program.

A sustained βAR activation, as that occurring during intense and prolonged physical exercise or after myocardial injury, increases nuclear PKA activity and CREB phosphorylation [32]. Transcription factors of the CREB family can contribute to βAR-dependent maladaptive remodeling.

1.5 Interactions Between Cardiac Cell Populations

The heart is composed of different cell types that interact to maintain the cardiac and bodily homeostasis, in addition to the dynamical partnership created by the nervous, renal, and cardiovascular system.

The largest cardiac volume is occupied by cardiomyocytes. Cardiofibroblasts have the highest density. Other cardiac cell types include nodal cells, epicardiocytes, endocardial and vascular endotheliocytes, coronary smooth myocytes, leukocytes, myofibroblasts, adipocytes, mesenchymal stem cells, and sympathetic and parasympathetic neurons (Vol. 5, Chap. 6. Heart Wall).

Cardiac cell types need a well-handled communication to optimize signaling resulting from numerous messengers. The intercellular communication encompasses direct between-cell contact, cell–matrix interaction, and long-range signals, using chemical (short peptides, proteins, lipids, nucleotides, and RNAs) and electrochemical (ions) cues. In addition to secreted molecules, cells release vesicles that contain proteins, RNAs, and sometimes DNA.

Interactions between different cardiac cell types enable proper myocardial contractility, sufficient perfusion, adequate myocardial stiffness, and controlled functioning of immunity. However, heterotypic intercellular interactions are also involved in disease progression.

Interactome refers to gene and protein interactions within a pathway as well as between distinct axes (genic and proteic interactomes), as well as cell communications inside a population of a given cell type (isotypic interactions, i.e., connections between cells of the same bulk phenotype) and between different populations of cell types (heterotypic signaling, i.e., data transmission between cells of different phenotypes) in a spatiotemporal regulated manner and in a given physiological or pathological context (cellular interactome; Tables 1.6 and 1.7).

A subpopulation of a given cell type such as fibroblast with its given functional characteristics can influence the activity of other isotypic subpopulations, in addition to other cell types.

Acute ischemic injury and maladaptive cardiomyocyte hypertrophy are the most common acute and chronic insults to the heart. After cardiac injury, the spatial arrangement of cells is disrupted and the cellular composition is altered, as different cell populations are temporally recruited to the heart. These cell types interfere using many signaling pathways.

Table 1.6 Examples of messengers involved in interactions between cardiac cells

Source	Target	Messengers
CMC	CFB EC	ANP, BNP, TGFβ, CTGF, IL1β/6, miR133a Nrg1, CSF3, TNFSF1
CFB	CMC	CNP, ET1, CTGF, FGF2, PDGF, TGFβ, TNFSF1, IL6/10/17/33, LIF, CT1
EC	CMC SMC	ET1 MiR126/143/145
SMC	CFB	CTGF
Leukocytes	CMC CFB	TNFSF1, IL17 TNFSF1
Stem cells	CMC EC	ATP SHh

CFB cardiofibroblast, *CMC* cardiomyocyte, *EC* endotheliocyte, *SMC* smooth myocyte, *A(B/C)NP* A(B/C)-type natriuretic peptide, *CSF* colony-stimulating factor, *CT* cardiotrophin, *CTGF* connective tissue growth factor, *ET* endothelin, *FGF* fibroblast growth factor, *IL* interleukin, *LIF* leukemia-inhibitory factor, *miR* microRNA, *Nrg* neuregulin, *PDGF* platelet-derived growth factor, *SHh* sonic Hedgehog, *TGF* transforming growth factor, *TNFSF* tumor-necrosis factor superfamily member

1.5.1 Examples of Signaling Messengers

After cardiac injury, multiple signaling pathways, such as the Notch, Wnt, TGFβ, and BMP axes, are activated.

The Notch-primed signaling exerts a negative feedback on the Wnt–β Ctnn pathway; the intracellular domain of Notch receptor can bind to Wnt signaling effectors such as Disheveled, thereby inhibiting Wnt signaling.

Calcium ions are major second messengers in both the cardiac pump and vasculature. Localized changes in Ca^{2+} ions encompass waves, oscillations, sparks, sparklets, puffs, transients, and flashes, which can correspond to different triggering events.

1.5.1.1 Hedgehog

During myocardial ischemia, the Hedgehog pathway promotes neovascularization, thereby increasing cardiomyocyte survival. The canonical Hedgehog pathway activates gene transcription via transcription factors of the Gli family upon derepression of the GPCR Smoothened.

In neonatal rat ventriculomyocytes, sonic Hedgehog activates Gli1 via Smoothened and Gi proteins [34]. The Gli1 transcription factor then targets a subset of 37 cardiomyocyte-specific genes, such as those encoding mediators of the PKA and nucleotide pathways.

Table 1.7 Short- and long-distance bidirectional communication between different cell types in the heart. Cardiac myocytes (85 % of the cardiac mass) and fibroblasts (~ 2/3 the total number of cardiac cells), vascular endothelial and smooth muscle cells, and epicardiocytes constitute the main cell populations in the healthy adult heart. Extra- and intracellular signals that regulate cardiomyocyte proliferation during the postnatal life include growth factors and microRNAs. In diseases, these heterocellular connections reorganize and initiate adaptive and maladaptive responses. Exosomes secreted from stem and progenitor cells yield beneficial effects in the injured myocardium. (Source: [33])

Type of communication	Mediators	Role
Healthy heart		
Myocyte–endotheliocyte	Neuregulin ROS	Angiogenesis Cardiomyocyte survival, cardiomyocyte hypertrophy
Fibroblast–immunocyte	Cytokines	Feedback control of fibroblast and leukocyte activation
Diseased heart		
Myocyte–endotheliocyte	Neuregulin ROS	Angiogenesis Cardiomyocyte survival, cardiomyocyte hypertrophy
Myocyte–macrophage (M1)	Cytokines DAMPs	
Myocyte–fibroblast	MicroRNAs	Tissue repair
	lncRNAs	Tissue stiffness Contractile dysfunction
Fibroblast–stem cell	Cytokines	Suppression of myofibroblast differentiation
Fibroblast–adipocyte	Adipokines Activin-A	Myofibroblast differentiation Fibrosis
Fibroblast–macrophage	Cytokines	
Endotheliocyte–macrophage (M2)		Neovascularization Myocardial infarction healing

DAMP damage-associated molecular pattern molecules, *lncRNA* long noncoding RNA, *ROS* reactive oxygen species

1.5.1.2 Wnt

The Wnt–βCtnn axis with the transmembrane-frizzled receptors and related antagonists is used during cardiogenesis and cardiac injury response. The Wnt messengers constitute a family of 19 lipophilic proteins that bind to ten types of transmembrane receptors of the frizzled family and 2 types of coreceptors lipoprotein receptor-related proteins LRP5 and LRP6 (Vol. 3, Chap. 10. Morphogen Receptors). They trigger β-catenin-dependent canonical and noncanonical pathways.

The Wnt signaling is hindered by two categories of antagonists. Secreted frizzled-related protein (sFRP1–sFRP5) either directly binds to extracellular Wnt or competitively to Fz receptor, thereby preventing Wnt binding to Fz receptor in

Table 1.8 Expression of Wnt messengers and antagonists in the normal heart and after ischemia (Source: [35]). The secreted frizzled (Fz)-related protein (SFRP) family comprises five glycoproteins in humans (sFRP1–sFRP5) that act as extracellular signaling ligands that can bind Wnt proteins and Fz receptors. The interaction between sFRP and Wnt proteins prevents the latter from binding Fz receptors. Signaling by Wnt mesengers can be downregulated by the formation of an inhibitory complex with the frizzled receptors. The Dickkopf (Wnt signaling inhibitor) family of secreted protein encompasses four members (Dkk1–Dkk4) that antagonize Wnt–βCtnn signaling, as they interact with the Wnt coreceptors LRP5 and LRP6. They are also high-affinity ligands for the transmembrane proteic modulators of Wnt signaling Kremen-1 and -2

Normal heart	Ischemia
Involved molecules	
Wnt2/5a–5b/7b/9a/11	Wnt1/2/4/7/10b/11
sFRP1/2	sFRP1/2/4
Dkk3	Dkk1/Dkk2
Cells responsive to canonical Wnt signaling	
Valve mesenchymal cells	Epicardial-derived cells of subepicardial space
Subsets of smooth myocytes	Fibroblasts, smooth myocytes and endotheliocytes in the injury region
Endotheliocytes	

both cases. Dickoppf proteins (Dkk1–Dkk3) competitively preclude Wnt binding to LRP5 and LRP6 coreceptors.

The Wnt family members Wnt2, Wnt5a, Wnt5b, Wnt7b, Wnt9a, and Wnt11 as well as their antagonists sFRP1, sFRP2, and Dkk3 are expressed in the normal heart [35]. Mesenchymal cells in cardiac valves and subsets of endothelial and smooth muscle cells scattered throughout the myocardium respond to Wnt (Table 1.8). The Wnt morphogens regulate interactions between fibroblasts and other cardiac myocyte and nonmyocyte cell types.

Endotheliocytes possess most types of frizzled receptors (Fz1–Fz2, Fz4–Fz7, and Fz9–Fz10). The Wnt1, Wnt3a, and Wnt5a proteins regulate endotheliocyte proliferation and migration. The sFRP molecules exert both pro- and antiangiogenic effects. Subtype sFRP1 promotes migration and tube formation, but reduces endotheliocyte proliferation [35]. In hearts of patients with dilated cardiomyopathy and coronary artery disease, sFRP3 and sFRP4 levels are elevated [35].

1.5.1.3 Signal Transducer and Activator of Transduction

Signal transducers and activators of transcription constitute a family of transcription factors (STAT1–STAT4, STAT5a–STAT5b, and STAT6) that are all expressed in the myocardium. They homo- and heterodimerize.

STAT Family The STAT1 agent is activated in ischemia–reperfusion injury. It upregulates the expression of caspase-1, TNFSF6, and TNFRSF6a, thereby promoting apoptosis in cardiomyocytes [36].

The STAT2 factor heterodimerize with the STAT1 agent. The STAT3 factor targets genes encoding proteins and microRNAs. It intervenes especially in cardiac physiology and pathophysiology. The STAT4 factor modulates inflammation in autoimmune and infectious myocarditis [36].

Both STAT5 and STAT6 participate in effects of the renin–angiotensin–aldosterone axis. They are rapidly phosphorylated during angiotensin-2 stimulation in the nonischemic region and bind to the STAT consensus sequence in the angiotensinogen promoter, thereby causing a sustained autocrine activation of angiotensin-2-mediated signaling.

STAT3 The STAT3 factor targets genes encoding proteins and microRNAs. It is involved in mitochondrial energy production and in cell survival, proliferation and differentiation, and metabolism in cardiomyocytes, fibroblasts, endotheliocytes, progenitor cells, and various types of inflammatory cells, as well as oxidative stress and protective and stress-induced adverse cardiac remodeling [36].

The STAT3 factor participates in intercellular communication between cardiomyocytes, endotheliocytes, fibroblasts, and cardiac progenitor cells, as well as infiltrated immunocytes (Table 1.9).

STAT3 Signaling The STAT3 factor is activated by multiple receptors. Its activity depends on dimerization and posttranslational modifications (acetylation [Lys49, Lys87, and Lys685] and phosphorylation [Ser727]). It is phosphorylated by PKCδ, PKCε, ERK1, ERK2, P38MAPK, CDK5, DAPK3, and TOR kinases, thereby potentiating the STAT3 transcriptional activity via recruitment of transcriptional cofactors, such as the P300 and CBP HATs.

The related signaling cascades are associated with positive and negative feedback loops. It participates in the production of the vasodilators, nitric oxide and prostacyclin. It mediates proangiogenic signaling via VEGF and erythropoietin.

Suppressor of cytokine signaling forms a negative feedback of STAT signaling. In particular, SOCS3 transcription is upregulated by STAT3 [36]. Both SOCS1 and SOCS3 interact with various JaK proteins or the GP130 receptor, thereby hindering STAT phosphorylation. Both PTPn6 and PTPn11 also target cytokine receptors and dephosphorylate JaK proteins, thereby impeding STAT3 phosphorylation.

Proteic inhibitors of activated STAT (PIAS1 and PIAS3) preclude DNA binding by dimerized phosphorylated STAT1 and STAT3, thereby diminishing transcriptional activation of STAT target genes. Protein Tyr phosphatase receptor PTPRt dephosphorylates STAT3, thereby, controlling target gene expression and cellular localization.

STAT3 can inhibit gene transcription. Acetylation (Lys685) involved in STAT3 interaction with DNA methyltransferase-1 causes methylation (silencing) of various gene promoters [36]. Deacetylation by histone deacetylases (e.g., HDAC1 and HDAC4) leads to STAT3 degradation.

Table 1.9 Intercellular communication relying on the STAT3 factor. The STAT3 factor is activated by numerous liganded receptors (Source: [36])

Cell type	Targeted cellular function
Cardiomyocyte	Sarcomere organization (miR199a)
	Cardiomyocyte architecture (PRMT1–ADMA)
	Secretome
	Survival, growth
	Oxidative stress (ROS)
Fibroblast	Secretome
	Extracellular matrix
	Survival, proliferation, differentiation
Endotheliocyte	Adhesion
	Survival, proliferation, differentiation (apoptosis and reduced proliferation and migration in PPCM due to $_{N}$Prl and miR146a)
	Oxidative stress (ROS)
Progenitor cell	Survival, proliferation, differentiation
Immunocyte	Secretome
	Differentiation
Mediators	
Receptors	GP130 (IL6R), IL10R, CCR2, CSF3R, EpoR, Nrg1–HER, VEGFR, leptin R, AT_1
Cytokines	EPo, VEGF, Opn, Tsp1
MicroRNAs	MiR21/146a/199a

ADMA asymmetric dimethylarginine, *EPo* erythropoietin, *miR* microRNA, *Opn* osteopontin, $_{N}$*Prl* antiangiogenic and proapoptotic N-terminal 16-kDa prolactin fragment, *PPCM* peripartum and postpartum cardiomyopathies, *PRMT* arginine ^{N}methyltransferase, *ROS* reactive oxygen species, *Tsp1* thrombospondin-1, *VEGF* vascular endothelial growth factor

MicroRNA-199a that alters the sarcomere organization is transcriptionally suppressed by STAT3 [36]. MicroRNA-199a targets the ubiquitin conjugases UbE2i and UbE2g1, thereby disturbing the protein turnover by the ubiquitin–proteasome axis and subsequently reducing expression of the genes that encode the sarcomeric α- and β-myosin heavy chain.

STAT3 in Cardiomyocyte–Endotheliocyte Communication STAT3 is involved in the communication between cardiomyocytes and the myocardial vasculature. In cardiomyocytes, STAT3 provokes VEGFa expression in particular upon stimulation by IL6 and erythropoietin, especially in ischemia [36].

In endotheliocytes, STAT3 contributes to the expression of cytokines (e.g., IL6), chemokines (e.g., CCL2 and CX_3CL1), and cell adhesion molecules (ICAM1 and VCAM1). It promotes the differentiation of cardiac progenitor cells into endotheliocytes.

Juxta- and paracrine communication between cardiomyocytes and between cardiomyocytes and endotheliocytes also relies on neuregulin-1 and colony-stimulating factor CSF3. The CSF3–STAT3 signaling improves cardiomyocyte survival and vascularization after myocardial infarction [36].

In addition, STAT3 antagonizes *arginine* ^{N}*methyltransferase* PRMT1 that increases the synthesis of *asymmetric dimethylarginine* (ADMA), an inhibitor of NO synthesis and promoter of ROS production in endotheliocytes [36]. The PRMT1 enzyme is activated in response to disturbed protein turnover and accumulation of damaged proteins, thereby severely damaging the cardiomyocyte structure and function and simultaneously impairing the endothelium in a paracrine manner, and hence ultimately priming initiation and progression of heart failure.

The endothelial differentiation capacity of resident cardiac SCA1+ progenitor cells that express both the erythropoietin and chemokine CCR2 receptors requires STAT3 [36]. Cardiomyocyte-derived erythropoietin, production and secretion of which depends on STAT3, influences the expression of CCL2. In addition, the secretion of cytokines and chemokines from cardiac cells, such as IL11, CT1, and LIF, engenders endothelial differentiation of cardiac progenitor cells via the GP130–STAT3 axis. The mobilization of bone marrow endothelial progenitor cells into the blood circulation relies on an IL10–STAT3 pathway [36].

Interference between cardiomyocytes and bone marrow-derived mesenchymal stem cells affects the myogenic conversion [36]. The reduction of miR124 in bone marrow-derived mesenchymal stem cells generated by cardiomyocytes increases expression of STAT3 and STAT3-associated myocytic markers, such as atrial natriuretic peptide, troponin-T, and α-myosin heavy chain, as well as the cardiac potassium channel currents.

STAT3 in Cardiomyocyte–Fibroblast Communication Fibroblasts and cardiomyocytes synthesize various matrix constituents as well as matrix-regulatory enzymes (matrix metallopeptidases and their inhibitors TIMPs). In fibroblasts, STAT3 promotes cell survival and proliferation, as well as the synthesis of matrix components (e.g., collagen). STAT3 is a major player in communication between cardiomyocytes and fibroblasts.

In cardiofibroblasts, STAT3 regulates not only the synthesis of genes encoding matrix constituents but also that of cytokines, as well as connective tissue growth factor, thrombospondin-1, tissue inhibitor of metallopeptidases TIMP1, osteopontin, tenascin-C, and plasminogen activator inhibitor-1 (or serpin-E1) [36].

Cardiac stress factors, such as ischemia, angiotensin-2, and pressure overload, induce the expression of IL6 family cytokines (i.e., IL6, IL11, leukemia inhibitory factor [LIF], oncostatin-M [OSm], and cardiotrophin-1 [CT1], among others) in both cardiomyocytes and fibroblasts. Communication between cardiomyocytes and fibroblasts relies on the IL6Rα–GP130 receptor. Other stress factors such as β-adrenoceptor agonists provoke production of IL6 cytokines only in cardiofibroblasts, but not in cardiomyocytes [36].

1.5.1.4 Tumor-Necrosis Factor TNFSF1

The proinflammatory cytokine tumor-necrosis factor-α (TNFSF1) connects to its cognate TNFR1 (TNFRSF1a) and TNFR2 (TNFRSF1b) receptors on cardiomyocytes.

In mice, acute stimulation with a low TNFSF1 dose (10 μg/kg) raises cytosolic calcium transient amplitude and cardiac contractility [37]. In mice, TNFSF1 overexpression causes dilated cardiomyopathy. On the other hand, exposure of adult cardiomyocytes to TNFSF1 during 12 h protects against hypoxic injury.

The Ras interaction/interference protein RIN1, afadin, and Ras association domain-containing protein family member RASSF1a modulate TNFSF1 signaling in cardiomyocytes. RASSF1a does not interact with TNFRSF1b. This adaptor facilitates the recruitment of TRADD that links TNFRSF1a and RIPK1, and TRAF2 to form the TNFRSF1a complex. It is required for signal transmission from the TNFRSF1a complex to effectors, such as cytoplasmic phospholipase-A2 and protein kinase-A, but not calcium–calmodulin-dependent kinase CamK2. It enables activation of $Ca_V1.2a$ and RyR2 channels [37]. In addition, RASSF1a also complexes with plasma membrane calcium ATPase PMCA4, a calcium extrusion pump located in caveolae, as is TNFRSF1a.

However, TNFSF1 is a proapoptotic cytokine. The TNFSF1–TNFRSF1a couple alters mitochondrial function and Ca^{2+} handling and provokes cell death. In acute ischemia, TNFSF1 activates caspase-8 that alters mitochondrial functioning, increasing production of ROS, and causes Ca^{2+} leak from the sarcoplasmic reticulum [38].

In murine adult ventriculomyocytes exposed to TNFSF1, caspase-8 is activated, thereby increasing the mitochondrial production of ROS and priming poly^{ADP}ribose polymerase PARP1 activation and hence production of adenosine diphosphoribose (^{ADP}ribose) and subsequent poly(ADP-ribosyl)ation of proteins. In particular, oxidative stress stimulates Ca^{2+}-permeable TRPM2 channel that can be activated by the intracellular second messenger ^{ADP}ribose, in addition to nicotinamide adenine dinucleotide (NAD^+) and cyclic adenosine diphosphate–ribose (cADPR) [38]. The TRPM2 channel conveys a nonspecific cationic current that mediates TNFSF1-induced ventriculomyocyte death. Its inhibition thus protects against apoptosis.

1.5.1.5 MicroRNAs

The effects of microRNAs depend on the cell type and context, that is, the expression of target transcripts.

Once synthesized, microRNA precursors are processed initially in the nucleus by the Drosha microprocessor complex (primiR cleavage) and, after being exported to the cytoplasm, by Dicer (premiR cleavage) and its cofactors (e.g., Argonaute, protein kinase interferon-inducible double-stranded RNA-dependent activator [PKRA], and TAR RNA-binding protein [TARBP] [39]; Vol. 1, Chap. 5. Protein Synthesis).

MicroRNAs and Chemical Stressors Adverse environmental conditions and chemical stressors such as excessive amounts of ROS trigger an adaptive response aimed at reducing cell damage. Most mRNAs are repressed, the translation machinery (e.g., translation preinitiation complex with the small ribosomal 40S subunit, RNA-binding proteins that regulate mRNA translation and stability, agents involved in mRNA splicing, transport, and metabolism) is stored in stress granules, and cellular functions focus on damage repair [39].

Table 1.10 MicroRNAs involved in the regulation of auto- and paracrine factors involved in crosstalk between cardiomyocytes and cardiofibroblasts. (**Part 1**) After angiotensin-2 (ATn2) stimulation. (Source: [40])

MicroRNA type	Effect
Cardiomyocyte	
MiR1	Suppression upon ATn2 stimulation
MiR21/22	Upregulation after ATn2 stimulation
MiR30	Downregulation after ATn2 stimulation
MiR132/212	Upregulation after ATn2 stimulation
Cardiofibroblast	
MiR21/29b	Induction after ATn2 stimulation
MiR101a/b	Suppression after ATn2 stimulation
MiR132/146b/212	Induction after ATn2 stimulation

Specific microRNA species are involved in the stress response and hence in stress signaling in diseases. In particular, loss of miR208a does not affect cardiovascular function, but abrogates stress-responsive cardiac remodeling [39].

MicroRNAs and Age-Related Diseases MicroRNAs participate in aging and age-related diseases, as dysregulation of microRNA function can permit activation of aberrant pathways that are repressed by normal microRNA activity. Dicer dysfunction can be observed in cerebromicrovascular endotheliocytes of aged rats, reduced Dicer activity, and/or concentration lessening stress tolerance [39].

MicroRNA Messengers MicroRNAs can operate as messengers in intercellular communication [40]. They are detected in body fluids (e.g., blood, urine, and breast milk). In the circulation, secreted miRNAs are protected from degradation by ribonuclease once they are tethered to RNA-binding proteins such as Argonaute-2, linked to lipoproteins, such as HDLs and LDLs, or packaged into vesicles, such as exosomes, microvesicles, or apoptotic bodies (Sect. 3.2.3.6).

Once they are released into the extracellular space, microRNAs can transmit signals between cells and affect gene expression of target mRNAs in the recipient cells. MicroRNAs usually cause posttranscriptional gene silencing. A given microRNA can target several transcript types and a single transcript can be a substrate for multiple microRNAs. In particular, microRNA moderators of crosstalk in the myocardium target transcripts, the products of which are messengers transmitted from fibroblasts to cardiomyocytes and conversely belong to intracellular signaling pathways launched by paracrine factors (Tables 1.10, 1.11, 1.12, 1.13 and 1.14).

In the heart, microRNAs are secreted by multiple cell types. These communicators then serve as paracrine mediators. MicroRNAs can be involved in crosstalk between cardiac cells either by affecting the secretion of growth factors and cytokines or by exerting a direct signaling.

MicroRNAs in Myocyte–Fibroblast Communication Certain microRNAs are deregulated in hypertrophic or fibrotic hearts. In particular, microRNAs participate in the myocyte–fibroblast communication. The myocyte-specific miR133a affects fibroblast function [40]. In the cardiomyocyte, miR133a regulates the synthesis and secretion of the profibrotic CTGF growth factor. Once it is secreted, CTGF provokes

Table 1.11 MicroRNAs involved in the regulation of auto- and paracrine factors involved in crosstalk between cardiomyocytes and cardiofibroblasts. (**Part 2**) TGFβ signaling. (Source: [40])

MicroRNA type	Effect
Cardiomyocyte	
MiR29b	Reduction of TGFβ production, prevention of cell growth
MiR208a	Prohypertrophic effect via autocrine TGFβ release
Cardiofibroblast	
MiR21	Promotion of FGF2 secretion upon induction by TGFβ
	Activation of TGFβ via Tβ R3
MiR24	Repression of furin and subsequently TGFβ
MiR29	TGFβ-mediated repression, supporting fibrosis
MiR101a	Repressor of TGFβ via Fos
MiR133a	Overexpression directly decreases TGFβ
MiR590	Overexpression directly decreases TGFβ

Table 1.12 MicroRNAs involved in the regulation of auto- and paracrine factors involved in crosstalk between cardiomyocytes and cardiofibroblasts. (**Part 3**) Other growth factors. (Source: [40])

Messenger type	MicroRNA type	Cell type	Effect
CTGF	MiR30c	CMC, CFB	Decreased CTGF level
	MiR133a	CMC	Decreased CTGF level
FGF2	MiR21	CFB	Elevated release by derepressing MAPK
IGF1	MiR1	CMC	Regulation of IGF1 expression
	MiR29b	CFB	Overexpression reduces IGF1 level
	MiR30c	CFB	Overexpression raises IGF1 level
PDGF	MiR29b	CMFB	

CFB cardiofibroblast, *CMC* cardiomyocyte, *CMFB* cardiomyofibroblast, *CTGF* connective tissue growth factor, *FGF* fibroblast growth factor, *IGF* insulin-like growth factor, *MAPK* mitogen-activated protein kinase, *PDGF* platelet-derived growth factor

cardiac fibrosis. Overexpression of miR133 reduces apoptosis and fibrosis in mice after transverse aortic constriction. In addition, miR133a also regulates TGFβ expression and secretion. The CTGF production and secretion is also controlled by miR30, which is highly expressed in cardiac fibroblasts.

MicroRNAs in Monocytes and Macrophages Circulating microRNAs serve as prognostic and diagnostic markers in numerous heart diseases, especially coronary artery disease and myocardial infarction [40].

Although the miR155 plasma concentration in patients with coronary heart disease may not be markedly elevated, miR155 expression is upregulated in CD14+ monocytes [41]. In addition, miR155 levels rise in both plasma and atherosclerotic

Table 1.13 MicroRNAs involved in the regulation of auto- and paracrine factors involved in crosstalk between cardiomyocytes and cardiofibroblasts. (**Part 4**) Cytokines. (Source: [40])

Messenger type	MicroRNA type	Effect
TNFSF1	MiR146a	Overexpression lowers I/R-induced TNFSF1 increase
IL1β	MiR146a	Cardioprotection via IRAK1 suppression
IL6 family	MiR17/21/199	Increased expression in hearts of LIF-treated mice
	MiR29b	Overexpression reduced LIF level in CFB
IL17	MiR101	Induced by IL17a Promotes CFB migration and proliferation
IL-10	MiR27a	Overexpression decreasesd IL10 level in CMC

CFB cardiofibroblast, *CMC* cardiomyocyte, *IL* interleukin, *IRAK* IL1 receptor-associated kinase, *LIF* leukemia-inhibitory factor, *TNFSF* tumor-necrosis factor superfamily member

Table 1.14 MicroRNAs involved in the regulation of auto- and paracrine factors involved in crosstalk between cardiomyocytes and cardiofibroblasts. (**Part 5**) Other regulator types. (Source: [40])

Messenger type	MicroRNA type	Effect
ANP	MiR26b	Overexpression suppresses ANP transcript
	MiR34 family	Decreased ANP level (pressure overload)
	MiR425	Repression of ANP production
ET1	MiR1	Inhibition of ET1 gene expression
	MiR23a	Elevated level in hypertrophy
	MiR132/212	Increased upon ET1 infusion

ANP atrial natriuretic peptide, *ET* endothelin

macrophages of ApoE$^{-/-}$ mice. Several miRs, such as miR33, miR125b, miR146, and miR155, operate in immunity and lipid metabolism. MicroRNA-146 and -155 contribute to the regulation of inflammatory signaling in macrophages.

In atherosclerosis, the unlimited uptake of oxidized low-density lipoproteins through scavenger receptors in macrophages is followed by the storage of cholesterol esters in lipid droplets due to the restricted efflux of free cholesterol. The latter can also accumulate in the membrane of the ER, thereby triggering ER stress and activating inflammation. In addition, cholesterol crystals can form and stimulate the NLRP3 inflammasome and IL1β production [42]. Oxidized low-density lipoproteins may bind to TLR4 and, in cooperation with Ifnγ, provoke miR155 expression (from the genomic region B-cell integration cluster [BIC]) in macrophages.

In macrophages, miR155 suppresses inhibitors of inflammatory cytokines (e.g., the SOCS1 factor, SHIP1 phosphatase, and BCL6, an inhibitor of the CCL2 chamokine) [42]. On the other hand, miR155 can reduce the secretion of cytokines, such as IL6 and TNFSF1, in oxLDL-stimulated macrophages, as it inhibits MAPK modules. Its sister strand miR155* has opposite effects in dendritic cells [42]. Moreover, miR155 also affects the expression of several other miRs.

MicroRNA-155 suppresses the production of the transcriptional repressor HMG box-containing protein HBP1 that impedes the function of macrophage migration inhibitory factor (MIF) and NADPH oxidase organizer NOxO2 [41]. Whereas MIF increases oxLDL uptake by macrophages, the derepression of NOxO2 promotes ROS production.

Several transcriptional factors (e.g., AP1, MyB, and NFκ B) upregulate the miR155 expression in immunocytes. Yin Yang YY1 is a ubiquitous transcription factor that interacts with HATs and HDAcs to activate or suppress gene transcription. It binds to the promoter of the primiR155 gene and acts as a transcriptional repressor [41]. The YY1–HDAC2–HDAC4 complex hence favors HBP1 synthesis. In addition, PKB1 also precludes miR155 expression; its transcript is a substrate of miR342-5p [42].

Control of Natriuretic Peptides by MicroRNAs Atrial and brain natriuretic peptides act not only remotely, but also locally at synthesis sites. Cardiac wall stretch and paracrine agents, such as ET1, increase natriuretic peptide production in and exocytosis from cardiomyocytes [40]. Cardiac myocytes and fibroblasts express the natriuretic receptors.

In fibroblasts, ANP impedes collagen synthesis and, in cooperation with BNP, cell proliferation [40]. Furthermore, ANP and BNP exert an antifibrotic effect. C-type natriuretic peptide secreted by fibroblasts attenuates cardiac fibrosis and prevents ET1-induced myocyte hypertrophy.

MiR425 represses directly ANP production [40]. Furthermore, the inhibition of members of the MIR34 family may protect the heart from adverse remodeling by regulating ANP among other hypertrophy-associated factors. When it is overexpressed, miR26b opposes prohypertrophic proteins and elevates the ANP level.

Control of Endothelin-1 by MicroRNAs The paracrine regulator ET1 is mainly formed in endotheliocytes. ET1, as well as its two receptors, is produced in cardiac myocytes and nonmyocytes. It is induced by TGFβ. It may act in the ATn2–TGFβ pathway to promote fibroblast activation and cardiac fibrosis, as well as hypertrophy [40].

Endothelin-1 may initiate cardiac hypertrophy via a miR23a-dependent pathway and ATn2-mediated ET1 receptor activation, as well as increased miR132 and miR212 levels [40]. On the other hand, miR1 represses ET1 expression.

Control of Angiotensin-2 by MicroRNAs The paracrine messenger angiotensin-2 is a vasoactive agent and a strong inducer of cardiac hypertrophy and fibrosis, which operates in crosstalk between cardiac nonmyocytes and myocytes via its cognate receptors (mainly AT_1) expressed by all cardiac cell types. It provokes the release of auto- and paracrine factors (TGFβ, ET1, and IL6) from cardiac myocytes and fibroblasts.

Several miRNAs are involved in angiotensin-2-mediated cardiac hypertrophy and fibrosis. Upon stimulation of cultured cardiofibroblasts by ATn2, several miRNAs related to fibrosis are deregulated (miR21, miR29b, miR101a–miR101b, miR132, miR132* [sister strand of miR132], miR212, and miR146), thereby affecting matrix synthesis and turnover, as well as fibroblast proliferation.

Angiotensin-2-treated cardiomyocytes exhibit reduced levels of antihypertrophic microRNAs (e.g., miR30 and miR123) [40]. On the other hand, the expression of hypertrophy-promoting microRNAs (e.g., miR22, miR132, and miR212) is upregulated.

Control of Transforming Growth Factor-β by MicroRNAs The cytokine TGFβ regulates the response to pressure overload and myocardial injury, such as cardiac hypertrophy and fibrosis. Among three isoforms (TGFβ1–TGFβ3) encoded by distinct genes, TGFβ1 is the predominant and ubiquitous subtype produced by many cell types, such as cardiac myocytes and fibroblasts. It binds to TβR1 and TβR2 receptors on both cardiac myocytes and fibroblasts.

TGFβ affects the phenotype and function of cardiofibroblasts, supporting their transition to myofibroblast and enhancing the production of matrix proteins and tissue inhibitors of metallopeptidases. It causes expression of profibrotic miR21, especially in fibroblasts during cardiac remodeling [40]. MicroRNA-21 promotes interstitial fibrosis at least partly by repressing Sprouty homolog-1, thereby derepressing MAPK signaling that enhances FGF2 secretion from cardiofibroblasts. It also stimulates endothelial–mesenchymal transition. In addition, miR21 targers TβR3 receptor that augments the TGFβ signal.

TGFβ reduces the level of microRNA-29, an antifibrotic agent, as it attenuates matrix formation and myofibroblast differentiation [40]. On the other hand, miR24 and miR101a repress the TGFβ expression via the transcription factor Fos and the peptidase furin, respectively. MicroRNA-133a and miR-590 may directly repress TGFβ.

Cardiomyocytic TGFβ signaling that involves SMADs can preserve cardiomyocytes from hypertrophy. Mechanical stretch enhances the expression of prohypertrophic miR208a in cultured cardiomyocytes possibly via autocrine TGFβ [40]. On the other hand, TGFβ1 alters the expression of prohypertrophic miR27b and prevents cardiomyocyte growth.

Control of Connective Tissue Growth Factor by MicroRNAs Connective tissue growth factor, a TGFβ target, is involved in cardiac fibrosis and hypertrophy. It is predominantly produced in cardiofibroblasts. However, in cardiac remodeling, CTGF is also released from cardiomyocytes. It raises fibroblast proliferation and migration as well as matrix formation.

The CTGF expression and secretion are repressed by miR30c expressed in both cardiac myocytes and fibroblasts and miR133a synthesized in cardiomyocytes [40].

Control of Fibroblast Growth Factor-2 by MicroRNAs The FGF2 agent is another mediator of cardiac hypertrophy and fibrosis. It is synthesized by cardiac myocytes and fibroblasts. Once it is secreted by fibroblasts, FGF2 acts in a paracrine manner and promotes cardiomyocyte hypertrophy via the MAPK module. As an autocrine factor, FGF2 engenders fibroblast proliferation and secretion of further prohypertrophic factors.

Secretion of FGF2 is influenced by miR21 that prevents MAPK signaling, thereby lowering FGF2 release from cardiofibroblasts [40].

Control of Insulin-Like Growth Factor-1 by MicroRNAs Insulin-like growth factor-1 is produced and released by cardiofibroblasts, but not cardiomyocytes [40]. It promotes collagen synthesis in fibroblasts.

The Igf1 transcript is targeted by miR1 [40]. In hypertrophy, miR1 is repressed. In addition, IGF1 expression is further affected by miR29b and miR30c [40].

Control of Platelet-Derived Growth Factor by MicroRNAs The PDGF expression is elevated in cardiac fibrosis and hypertrophy. Interstitial cells such as fibroblasts are the main PDGF source. The PDGF signaling enables fibroblast proliferation and migration, as well as matrix deposition.

The PDGF agent causes fibroblast proliferation and promotes cardiac hypertrophy via the secretion of growth factors and cytokines. As for IGF1, miR29 regulates the PDGF expression [40].

Control of Cytokines by MicroRNAs Cytokines are messengers in inflammation as well as cardiac remodeling and wound healing. Short-term activation of proinflammatory cytokines ensures cardioprotection, but long-term exposure induces a maladaptive response.

Tumor-Necrosis Factor-α Tumor-necrosis factor-α (TNFSF1) is synthesized by and secreted from infiltrating immunocytes as well as cardiac myocytes and fibroblasts after certain cellular stresses. Interaction of these cardiac cell types enables differential cytokine expression.

In cultured cardiomyocytes, TNFSF1 induces cell hypertrophyic and apoptosis and impairs cardiac contractility [40]. Furthermore, TNFSF1 promotes fibroblast proliferation, secretion of metallopeptidases, collagen synthesis, and proinflammatory cytokine release.

MicroRNA-146a may mediate inflammatory signaling, but prevents an adverse response, as it precludes TNFSF1 expression [40].

Interleukins Interleukins contribute to cardiac remodeling and may influence crosstalk between myocytes and nonmyocytes to ensure reparative inflammation and then cardioprotection. Several miRNAs are involved in inflammation in the failing heart.

IL1β IL1β is a proinflammatory cytokine that has an overlapping action with that of TNFSF1. It is expressed by cardiomyocytes in response to cardiac injury and stress. However, cardiofibroblast is the main source. It assists cell migration, impairs fibroblast proliferation, and matrix remodeling via reduced collagen expression and MMP secretion [40]. In cardiomyocytes, IL1β provokes hypertrophy.

MicroRNAs affect IL1β signaling. MicroRNA-146a is cardioprotective, as it targets IRAK1 kinase that becomes associated with the IL1 receptor upon stimulation [40].

IL6 Family Members Cytokines of the IL6 family can alter cardiac cell survival and growth. They are secreted by cardiac myocytes and fibroblasts in response to cardiac injury. The IL6 family members operate synergistically.

Interaction between cardiac cells supports IL6 release. Leukemia inhibitor factor (LIF) and cardiotrophin-1 (CT1) produced by fibroblasts mediate prohypertrophic effects of angiotensin-2 [40]. In addition, LIF and CT1 promote fibroblast proliferation. Moreover, LIF impairs myofibroblast transition and collagen accumulation.

Long-term LIF delivery in vivo slightly induces the expression of miR17, miR21, and miR199 [40]. On the other hand, miR29b may directly regulate leukemia inhibitor factor.

IL10 The anti-inflammatory cytokine IL10 prevents the production of prohypertrophic cytokines. Its production rises in response to TNFSF1. It limits pressure overload-induced cardiac remodeling [40]. Cardiomyocytes are an IL10 source, but nonmyocytes constitute the major source. Fibroblasts release IL10 that then protects cardiomyocytes by antagonizing apoptosis mediated by TNFSF1. IL10 reduces MMP9 activity and fibrosis.

The profibrotic miR27a targets the Il10 transcript, thereby abrogating the antifibrotic effect of IL10 [40].

IL17 Family Members Members of the IL17 family are mainly linked to the T-cell-dependent immunity. They influence cardiac fibrosis, hypertrophy, and dilated cardiomyopathy.

Cardiofibroblasts secrete IL17 and express the corresponding receptor. Paracrine communication engenders cardiomyocyte apoptosis, thereby impairing contractility as well as hypertrophy [40]. As an autocrine regulator, IL17a provokes synthesis of collagen and MMPs. Furthermore, IL17a induces miR101, thereby supporting fibroblast migration and proliferation [40].

IL33 Interleukin-33 may serve as a paracrine messenger between cardiac myocytes and fibroblasts during pressure overload. It seems to be primarily synthesized by cardiac fibroblasts. The cytokines TNFSF1 and IL1β promote IL33 production [40]. Interleukin-33 binds to the IL1R-like receptor-1 (ILR1L1) that is expressed by cardiomyocytes. It may antagonize prohypertrophic stimuli.

1.5.1.6 Material Exchanges via Vesicles

Cellular communication relies on [43]:

1. Intercellular contacts via gap junctions, cell–cell adhesions, and contacts between plasmalemmal proteins of apposed cells that facilitate signal propagation;
2. Cell–matrix interactions, especially via integrins, which not only sense and transmit mechanical stimuli from and to the extracellular matrix, but also secreted messengers that enable auto- and paracrine signaling;
3. Remote (endocrine) signals that circulate in the blood stream (e.g., hormones);
4. Electrochemical signals (i.e., ions), such as nodal and nerve impulses;
5. And vesicles that carry various types of molecules.

The vesicular vehicles for local and remote transmission are usually classified according to their intracellular origin. They comprise [43]:

1. *Apoptotic bodies* (size 1–5 μm);
2. *Aicrovesicles* (size 100–1000 nm); and
3. *Exosomes* (size 40–100 nm).

These vesicles have a lipid bilayer and contain a cell-specific cargo of proteins, lipids, and genetic material.

Vesicles are released upon different stress signals, such as thrombin, intracellular calcium, extracellular ATP, lipopolysaccharide, and hypoxia, as well as upon DNA damage [43].

Vesicular cargo uptake results from binding with proper plasmalemmal receptors and endocytosis, as well as fusion with the recipient plasma membrane and subsequent delivery.

In the myocardium, endothelial vesicles target smooth myocytes, fibroblasts, and myocytes, as well as immunocytes. Nonendothelial vesicles also facilitate a proper response to environmental changes and maintain myocardial homeostasis. Stem cell-derived vesicles may stimulate repair.

Cardiomyocytes secrete several hormones, growth factors, cytokines, and chemokines (e.g., ANP, BNP, TGFβ, and TNFSF1), as well as microvesicles and exosomes called *cardiosomes*.

Apoptotic Bodies Apoptosis, a clearing of damaged cells, is characterized by cell contraction, nuclear condensation, fragmentation, and release of apoptotic bodies (Vol. 2, Chap. 4. Cell Survival and Death).

Apoptotic bodies are released from the plasma membrane as blebs when cells undergo apoptosis, and contain several intracellular fragments, histones, DNA segments, and cellular organelles.

Their secretion depends on RoCK1 kinase and myosin ATPase [43]. The former is necessary and sufficient for formation of membrane blebs and reallocation of fragmented DNA into blebs and apoptotic bodies.

Microvesicles Microvesicles, also called membrane particles, microparticles, and ectosomes, are generated by outward budding or blebbing of the plasma membrane.

Microvesicles are created and shed during cellular differentiation and senescence, increased intracellular calcium, and exposure to high mechanical stress and proinflammatory and prothrombotic stimuli [43].

Activated Ca^{2+}-sensitive peptidases, such as calpain and gelsolin, enable vesiculation, as they detach membrane proteins from the intracellular cytoskeleton [43].

Microvesicles can contain plasmalemmal receptors, cell adhesion structures, growth factors, cytokines, chemokines, and RNA species (e.g., mRNA and miRs). The nature of cell activation and cellular environment determines microvesicle composition and behavior in intercellular communication [43].

Exosomes Exosomes originate from the inward budding of the membrane of multivesicular bodies (components of the late endosomal compartment) that form intraluminal vesicles using the endosomal sorting complex required for transport (ESCRT) for content sorting into the endosomal membrane and its inward budding (Vol. 1, Chap. 9. Intracellular Transport).

Different sorting mechanisms determine whether these intraluminal vesicles are destined for degradation when the multivesicular body membrane fuses with that of lysosomes or are released into the extracellular medium. These intraluminal vesicles are exported out of the cell after fusion of the multivesicular body membrane with the plasma membrane.

The formation and functioning of these MVB intraluminal vesicles rely on clathrin, flottilin, tetraspanins, Rab GTPases, ALIX, VPS23,[6] and HSP70, among others [43]. Tetraspanins (Tspan28–Tspan30) play an important role in fusion and cell penetration. Small Rab GTPases regulate exosome docking and membrane fusion.

Exosomes also contain annexins, metabolic enzymes, ribosomal proteins, signaling mediators, adhesion molecules, ATPases, cytoskeletal components, ubiquitin, growth factors, cytokines, mRNA, and microRNAs. Messenger RNAs carried by exosomes can be translated into proteins in the target cell. MicroRNAs transferred to recipient cells silence transcript expression.

In addition to molecules shared between exosomes from multiple cell types, exosomes carry specific proteins from their parental cell type. Exosome production, at least in some cell types, partly depends on P53 transcription factor and lipid mediators such as diacylglycerol [43].

Constitutive and inducible releases of exosomes from cells depend on the synthesizing cell type [43]. The constitutive secretion is mediated by specific Rab GTPases, hetrotrimeric G protein, and protein kinase-D. The inducible secretion results from increased intracellular Ca^{2+} level, DNA damage, and stimulations by thrombin, extracellular ATP, hypoxia, and lipopolysaccharide.

TNFSF1+ exosomes from hypoxic cardiomyocytes may trigger cell death in other cardiomyoctes. Circulating TNFR1+ exosome-like vesicles may modulate inflammation [43].

Communication via Endotheliocyte-Derived Vesicles Endotheliocytes interact with smooth myocytes during cardio- and angio- as well as atherogenesis. The former secrete miR126 that targets smooth myocytes.

Argonaute-2-bound miR126 is transferred to smooth myocytes, in which it represses FOXo3, Bcl2, and Irs1 transcripts [43]. It is also transmitted freely and supports neointima formation. A part of miR126 lodges in vesicles, but does not affect smooth myocytes.

Endothelial apoptotic bodies can also contain miR126 that may be internalized by neighboring endotheliocytes, thereby changing the CXCL12 and RGS16 levels and recruiting progenitor cells to stabilize and reduce atherosclerotic lesions [43].

[6] Also known as endosomal sorting complex required for transport ESCRT1 complex subunit tumor susceptibility gene protein TSG101.

Endotheliocyte-derived vesicles are created upon Krüppel-like factor KLF2 activation that increases miR143 and miR145 levels [43]. These microRNAs are exported via vesicles to smooth myocytes in which they change gene expression into an atheroprotective phenotype.

Differences in endothelial vesicle content result from distinct exposures of cellular stresses, such as hypoxia, TNFSF1 stimulation, and high glucose and mannose concentrations [43]. Although the content varies, the size distribution of secreted vesicles does not differ significantly with the environmental conditions.

Communication via Immunocyte-Derived Vesicles Exosomes are released from immunocytes (B and T lymphocytes, mastocytes, and dendritic cells [43]. Immunocyte-derived exosomes assist antigen presentation, immunoregulation, and signal transduction.

Immunocyte-derived exosomes intervene in the tolerance and survival of cardiac allografts. Exosomes released by immature dendritic cells and regulatory T cells can suppress T-cell activation [43].

1.5.2 Mitochondrial Signals

Mitochondria (approximately one-third of the cardiomyocyte volume) operate in energy production, metabolic regulation, ROS signaling, and calcium homeostasis, as well as apoptosis.

1.5.2.1 Mitochondrial Structure and Functions

Mitochondrion[7] is a double membrane-bound organelle (size 0.5–1.0 μm). The outer mitochondrial membrane (OMM) has a smaller surface area than the inner mitochondrial membrane (IMM) that is enlarged by cristae.

Cardiomyocytes contain spatially distinct pools of mitochondria with different shapes, sizes, and cristae arrangement:

- *Subsarcolemmal mitochondria* (ssM; 0.4–3.0 μm) below the plasma membrane, which have variable shape (oval, spherical, polygonal, and U-shaped) and mainly broad flat (lamelliform) cristae;
- Abundant elongated *interfibrillar mitochondria* (ifM; 1.5–2.0 μm) organized in rows between sarcomeres (usually one mitochondrion per sarcomere), which occupy the entire space limited by Z lines and junctional compartments of the sarcoplasmic reticulum (which is facing the T-tubule) and possess curved, mainly tubular (or lamelliform) cristae (with a lower content of ceramide than that of ssM cristae); and
- *Perinuclear mitochondria* (pnM) at the nucleus poles, mostly spherical (size 0.8–1.4 μm) with large curved cristae, and hence reduced between-crista space.

[7] μιτ ος: thread of the warp; χ ονδριαω: swell with clots of milk; χ ονδρος: granule.

Mitochondria of the perinuclear region may generate ATP that drives metabolism close to the nucleus, interfibrillar mitochondria may supply ATP to sarcomeres for contraction, and subsarcolemmal mitochondria may primarily deliver ATP for active transport of ions and metabolites across the sarcolemma. These mitochondrial subpopulations may differ by the functioning mode of oxidative phosphorylation and tricarboxylic acid cycle, possibly related to variations in posttranslational modifications [44].

The dynamics of the mitochondrion (fusion and fragmentation) govern its function, such as the regulation of cell survival. Optic atrophy protein OpA1, a manganese-inhibited, large dynamin-like GTPase and mitochondrion-shaping (fusion) protein of the inner mitochondrial membrane, and mitofusins Mfn1 and Mfn2 promote mitochondrial fusion. On the other hand, dynamin-related protein DRP1 interacts with outer mitochondrial membrane fission-related protein Fis1 and mitochondrial fission factor (MFF), thereby participating in mitochondrial fission. The OpA1 concentration decays in heart failure [45].

ROS modulator ROMo1 is a redox-regulated protein required for mitochondrial fusion and normal crista morphology [46]. It is involved in OpA1 oligomerization.

The mitochondrial populations respond to pathological conditions in different fashion [44].[8]

Nitric oxide participates in the regulation of mitochondrial genesis and dynamics in many cell types. Nitric oxide synthase NOS3 supports formation of complexes of the mitochondrial electron transport chain as well as that of the outer mitochondrial membrane porin voltage-dependent anion channel VDAC1, and manganese superoxide dismutase via transcriptional regulators cAMP response element-binding protein and peroxisome proliferator-activated receptor-γ coactivator PGC1α [47].

Mitochondrial DNA Mitochondria contain their own DNA (mtDNA) that represents less than 1 % of total cellular genetic material. It is organized as a double-stranded DNA. The heavy (H) and the light (L) strand are enriched in guanine and cytosine, respectively. The H and L strand encodes 28 and 9 genes, respectively (13 genes encoding polypeptides, 22 transfer RNA, and 2 small and large subunits of ribosomal RNA).

Mitochondrial DNA is packed into aggregates, the nucleoids. The most common component of nucleoids is the mitochondrial transcription factor-A (mtTFa) encoded by the TFAM gene that compacts and packs mtDNA similarly to histones.

Mitochondrial dysfunction is related to decreased ATP production, increased ROS generation, calcium dysregulation, and mitochondrial DNA damage.

Mitochondria have a DNA repair capacity. Moreover, mitophagy and the ubiquitin–proteasome system remove inadequate mitochondria. Nevertheless, mitochondrial DNA is often damaged during oxidative stress, such as that occurring in diabetes, hypertension, and atherosclerosis.

[8] Neurons also contain structurally and functionally different pools of mitochondria situated in distinct neuronal regions (dendritic, somatic, axonal, and presynaptic compartments), where energy demands and calcium signaling dynamics can differ.

Mitochondrial ^{N}Formyl Peptides Mitochondria employ ^{N}formyl methionyl tRNA as an initiator of mitochondrial protein synthesis. Mitochondria synthesize only 13 proteins that participate in electron transport chain. They all contain an ^{N}formyl group at the beginning of the amino acid sequence.

^{N}Formyl peptides derived from mitochondria have a powerful relaxant effect in resistance arteries [48].

Mitochondrial Energy Production Adenosine triphosphate must be continually generated at a high rate to sustain the myocardial function (contraction–relaxation cycles and ionic fluxes), in addition to basal metabolism. In the normal adult heart, more than 95 % of ATP is produced by mitochondrial oxidative phosphorylation, the remainder being engendered by glycolysis and the tricarboxylic acid cycle. The heart has a high rate of ATP hydrolysis (~30 μmol/mn/g at rest) [49].

The inner mitochondrial membrane contains the phospholipid cardiolipin. Cardiolipin anchors cytochrome-C to the inner membrane and supports oxidative phosphorylation. It also possesses the oxidative phosphorylation machinery (ADP being phosphorylated to ATP using the energy of hydrogen oxidation) and electron transfer chain, an enzymatic series of proteic complexes with electron donors and acceptors. The electron transport chain is coupled to oxidative phosphorylation by a proton gradient across the inner mitochondrial membrane for ATP production.

During oxidative phosphorylation, electrons are transferred from electron donors to electron acceptors in redox reactions. Oxygen serves as the terminal electron acceptor. The transfer of electrons generates a proton gradient that results from an active proton pumping into the intermembrane space.

The composition of small molecules in the intermembrane space is similar to that of the cytosol due to OMM porins, but with a different proteic content. The latter comprises in particular cytochrome-C, a component of the electron transport chain.

The mitochondrial matrix contains pyruvate generated from glycolysis and fatty acid oxidation that is transformed into acetylCoA, the primary substrate of the tricarboxylic acid cycle. This pathway produces $FADH_2$ and NADH that feed the *electron transfer chain* (ETC; ^{ETC}complex-I–^{ETC}complex-IV, i.e., NADH dehydrogenase [or NADH:ubiquinone oxidoreductase], succinate dehydrogenase, cytochrome-BC1, and cytochrome-C oxidase; the respirasome consisting of ^{ETC}complex-I, -III, and -IV).

The classical respirasome contains the mitochondrial ETC complexes linked by carriers coenzyme-Q (CoQ) and cytochrome-C (CytC). However, free respiratory complexes can exist and ETC complexes, except ^{ETC}complex-II, can associate in respiratory supercomplexes that include *mitochondrial respiratory supercomplex* mtRSC1 (^{ETC}cI–^{ETC}cIII–^{ETC}cIV), mtRSC2 (^{ETC}cI–^{ETC}cIII), and mtRSC3 (^{ETC}cIII–^{ETC}cIV) [50]. These supercomplexes define CoQ and CytC pools.

^{ETC}Complex-III connects preferentially to CoQ-containing supercomplexes mtRSC1 and mtRSC2 with preferential electron flux from ^{ETC}complex-I to ^{ETC}complex-III. Two functional CoQ populations exist: (1) CoQ involved in electron transfer from NADH (CoQ^{NADH}) in ^{ETC}complex-I-containing supercomplexes and

(2) free CoQ in the inner mitochondrial membrane linked to ^{ETC}complex-II and other enzymes that use FAD ($\mathrm{CoQ}^{\mathrm{FAD}}$).

^{ETC}Complex-IV-containing supercomplexes also possess cytochrome-C. Cytochrome-C oxidase subunit-7a polypeptide 2-like (CcOx7a2L)[9] is observed in ^{ETC}complex-IV+ supercomplexes mtRSC1 and mtRSC3, but not in free ^{ETC}complex-III and -IV [50].

Three ^{ETC}complex-IV populations can thus be defined: (1) a first pool that receives electrons exclusively from NADH (^{ETC}complex-$\mathrm{IV}^{\mathrm{NADH}}$); (2) a second from FAD-dependent enzymes (^{ETC}complex-$\mathrm{IV}^{\mathrm{FAD}}$); and (3) a third from both NADH and FAD. Incorporation of ^{ETC}complex-IV mediated by CcOx7a2L into respiratory supercomplexes minimizes competitive inhibition of respiration between pyruvate and succinate [50].

The assembly of these supercomplexes optimizes the use of available substrates (i.e., mitochondrial respiration via alternate routes to ^{ETC}complex-IV). The substrates of the mitochondrial electron transport chain, glutamate and malate, generate intramitochondrial reduced form of nicotinamide adenine dinucleotide (NADH) to feed electrons to ^{ETC}complex-I, thereby eliciting respiration via ^{ETC}complex-I, -III, and -IV [50]. The other substrate, succinate, feeds electrons to ^{ETC}complex-II via flavin adenine dinucleotide (FAD), priming mitochondrial respiration via ^{ETC}complex-II, -III, and -IV [50].

Adenosine triphosphate synthase in the IMM uses the reentry of protons in the matrix as a source of energy for ADP phosphorylation and ATP production. In brown adipocytes, mitochondrial respiration is uncoupled from ATP synthesis by uncoupling protein-1, and electron transfer serves as heat generation.

Fatty Acid β-Oxidation In the adult normal heart, 50–70 % of its ATP is derived from fatty acid β-oxidation (Vol. 6, Chap. 3. Cardiovascular Physiology). Fatty acids utilized in cardiac fatty acid β-oxidation primarily originate from plasma free fatty acids (FFA; concentration 0.2–0.6 mmol/l) bound to albumin. As the majority of circulating FFAs are incorporated in triacylglycerol (TAG or TG) in lipoproteins, they are released from exogenous TAG contained in chylomicrons (mainly) and very-low-density lipoproteins (via VLDL–apoE receptors) upon action of the primary tissue lipase, adipose triacylglycerol lipase (ATGL), and lipoprotein lipase (LPL) on the capillary endothelial surface [49].[10]

The majority of fatty acids undergoing β-oxidation are mono- (e.g., oleate, the most abundant fatty acid in blood) and polyunsaturated fatty acids.

---

[9] Also known as supercomplex assembly factor-1 (SCAF1).

[10] Inactive monomeric proenzyme from the ER of the cardiomyocyte is activated between the ER and the Golgi body and then secreted as an active homodimer. It binds to cardiomyocyte surface heparin sulfate proteoglycans (HSPG). The HSPG–LPL complex is transferred to endotheliocytes [49]. Fasting augments LPL activity that partly results from transport from cardiomyocytes to endotheliocytes, upon stimulation of AMP-activated protein kinase (AMPK). In adipose tissue, LPL secretion decreases, angiopoetin-like protein-4 supporting conversion of active dimerized LPL to the inactive monomer.

The ratio of NADH:FAD electrons feeding the mitochondrial electron transport chains is higher when the substrate is glucose than using fatty acids [50]. Nevertheless, the adjusted assembly of mitochondrial respiratory supercomplexes ensures an efficient oxidation of available substrates. Separate electron routes enable the shift to the FAD pathway upon switching to fatty acid degradation.

The activated sympathetic nervous system can also rapidly increase the plasma FFA concentrations, mainly from β-adrenoceptor-mediated stimulation of hormone-sensitive lipase in the adipose tissue.

Fatty acids are taken up by cardiomyocytes by diffusion and through the scavenger receptor ScaRb3 (or fatty acid translocase [FAT]; chiefly; resides also in the mitochondrial membrane) and SLC27a1 and SLC27a6 transporters (or fatty acid transport proteins FATP1 and FATP6]). The plasmalemmal isoform of fatty acid-binding protein (pmFABP) concentrates fatty acids priorto uptake [49].

Inside the cytosol of cardiomyocytes, fatty acids bound to fatty acid-binding proteins (FABP) are esterified to fatty acylCoA by fatty acylCoA synthase (FACS). The fatty acylCoA can then be esterified to complex lipids or the acyl group transferred to carnitine. Carnitine palmitoyltransferase CPT1, which is inhibited by malonylCoA (synthesized from acetylCoA by acetylCoA carboxylases [ACCα and (predominantly) ACCβ] and degraded by cytosolic, peroxisomal, and mitochondrial malonylCoA decarboxylases [MCD]), converts long-chain acylCoA to long-chain acylcarnitine [49]. The latter is then transferred to mitochondria by carnitine–acylcarnitine translocase (CACT), where it is converted back to long-chain fatty acylCoA by CPT2 located on the matrix side of the inner mitochondrial membrane.

The majority of fatty acylCoA enters the fatty acid β-oxidation, hence producing acetylCoA, NADH, and $FADH_2$. In certain conditions, mitochondrial thioseterase (MTE) can cleave long-chain acylCoA to fatty acid anions (FA^-), which can leave the mitochondrial matrix via uncoupling proteins [49].

Long-chain acylCoA can also be converted into complex lipids such as TAG, diacylglycerol (DAG), and ceramides that can be implicated in the development of insulin resistance, myofibrillar disorganization, and heart failure.

The myocardium has labile TAG stores (~ 3 mg/g myocardium [49]). Cytosolic long-chain acylCoA can be converted to TAG by glycerolphosphate acyltransferase [49]. Fatty acids deriving the intramyocardial TAG pool may represent 36 % of the energy expenditure in hearts perfused only with glucose and 11 % when palmitate is added to the perfusate. Intramyocardial TAG synthesis rises in diabetes and fasting, and TAG degradation catalyzed by hormone-sensitive lipase activated by cAMP upon adrenergic stimulation.

Fatty acid β-oxidation is controlled by:

- The myocardial work, and hence cardiac energetic demand;
- Control of mitochondrial function (i.e., not only of fatty acid β-oxidation but also of tricarboxylic acid cycle and electron transport chain activity);
- Transcriptional control of enzymes of the fatty acid metabolism and mitochondrial genesis;

- Fatty acid source, concentration, and supply to the heart;[11]
- Fatty acid uptake, esterification to coenzyme-A, and mitochondrial transfer;
- 3Competing energy substrates (glucose, lactate, ketones, and amino acids);
- Hormonal milieu; and
- Oxygen supply to the heart.

A β-oxidation cycle produces acetylCoA that enters the TCAC, flavin adenine dinucleotide ($FADH_2$), and nicotinamide adenine dinucleotide (NADH), and a 2-carbon shorter chain fatty acid.

Fatty acid β-oxidation involves four enzymes [49]:

- AcylCoA dehydrogenase,
- EnoylCoA hydratase,
- HydroxyacylCoA dehydrogenase, and
- 3-ketoacylCoA thiolase (3KAT).

In the heart, different isoforms are related to fatty acid chain length. Each of these enzymes experiences a negative feedback by the products of the reaction, including $FADH_2$ and NADH.

β-oxidation of mono- or polyunsaturated fatty acids is facilitated by auxillary enzymes, (2,4)-dienoylCoA reductase and enoylCoA isomerase [49].

The amount of fatty acid β-oxidation enzyme is transcriptionally controlled. The transcription factors peroxisome proliferator-activated receptors (or nuclear receptors NR1c), heterodimerize with retinoid X receptor upon fatty acid binding. The PPAR–RXR heterodimer then translocates into the nucleus where it binds to specific response elements of genes that encode regulators of fatty acid storage (e.g., diacylglycerol acyl transferase [DGAT]) as well as enzymes of fatty acid oxidation (e.g., medium-chain acylCoA dehydrogenase [MCAD]) and glucose metabolism (e.g., pyruvate dehydrogenase kinase PDK4).[12] They are assisted by the transcriptional coactivators PGC1α and PGC1β.

On the other hand, 30–50 % comes from pyruvate oxidation. Pyruvate derives in approximately equal amounts from glycolysis and lactate oxidation [49]. Pyruvate formed from glycolysis is converted to lactate, decarboxylated to acetylCoA by pyruvate dehydrogenase in the mitochondrial matrix, or carboxylated to oxaloacetate or malate.

---

[11] The arterial fatty acid concentration is the primary determinant of the rate of myocardial fatty acid uptake and oxidation [49]. Chronic or acute increases in circulating FFA concentrations affect the rates of cardiac fatty acid uptake and β-oxidation. Circulating FFA levels are chronically elevated in obesity and diabetes. The sympathetic nervous system activity rises during and after a myocardial ischemic insult or in chronic heart failure.

[12] The nuclear receptor NR1c1 (PPARα) is a major transcriptional regulator of fatty acid metabolism. Its target genes encode ScaRb3, SLC27a1, FABP, FACS, glycerol 3-phosphate acyltransferase, diacylglycerol acyltransferase, MCD, CPT1, very-long-chain, long-chain, and medium-chain acylCoA dehydrogenase, 3KAT, mitochondrial thiesterase MTE1, uncoupling proteins UCP2 and UCP3, and PDK4.

The glucose–fatty acid (Randle) cycle describes the reciprocal relation between fatty acid and glucose metabolisms. Elevated generation of acetylCoA derived from fatty acid β-oxidation diminishes glucose oxidation, as it activates pyruvate dehydrogenase kinase (PDK), hence phosphorylating (inhibiting) pyruvate dehydrogenase. High rates of fatty acid β-oxidation also inhibit phosphofructokinases PFK1 and PFK2 via citrate [49].

Conversely, increased amounts of acetylCoA derived from glucose oxidation prevent fatty acid β-oxidation. 3-KetoacylCoA thiolase is indeed inhibited by acetylCoA [49]. Moreover, cytosolic acetylCoA is a substrate for acetylCoA carboxylase (ACC) that can generate malonylCoA, an inhibitor of CPT1 [49].

In obesity and diabetes, fatty acid uptake and β-oxidation increase [49]. A high cardiac fatty acid β-oxidation rate may contribute to the development of cardiomyopathies. Ischemia also alters fatty acid β-oxidation. Heart failure is linked to a repression of the transcription of numerous metabolic enzymes [49].

Mitochondria and Calcium Homeostasis Mitochondria contribute to the maintenance of calcium homeostasis. At rest, mitochondrial matrix Ca^{2+} concentration equals about 100 nmol/l [53].

Calcium uptake into the mitochondrial matrix influences energy production, in addition to the control of cellular signaling and apoptosis. Large mitochondrial influx can actually alter intracellular Ca^{2+} signals and initiate cell death. However, mitochondrial Ca^{2+} uptake varies greatly among cell types.

The spatiotemporal Ca^{2+} signaling relies on the transfer through carriers (channels, exchangers, and pumps) from two main Ca^{2+} sources:

1. The extracellular medium (level $\sim$ 1 mmol/l [54]) and
2. Intracellular stores (level $>$ 100 μmol/l [54]) that encompass the ER, mitochondria, golgi body, endosomes, and lysosomes.

The sarcoplasmic reticulum, the primary intracellular Ca^{2+} storage organelle, releases Ca^{2+} during every heartbeat, the cytosolic Ca^{2+} concentration transiently rising globally from about 100 to 500 nmol/l (exposure time to high local $[Ca^{2+}]_i \sim 10$ ms) and in the subdomain near intermyofibrillar mitochondria close to Ca^{2+} release units between the transverse tubule and junctional sarcoplasmic reticulum membranes (transfer distance 50–100 nm) up to about 20 μmol/l during the release phase [55].

Calcium ions are then removed from the cytosol through the sarco(endo)plasmic reticulum Ca^{2+} ATPase (SERCA), mitochondrial Ca^{2+} uniporter (mtCU), and the sarcolemmal Na^+–Ca^{2+} exchanger (NCX) in about 500 ms. Most of the cytosolic Ca^{2+} ions are resequestered into the sarcoplasmic reticulum and exported to the extracellular milieu, the mitochondrial uptake accounting for approximately 1 % of cytosolic Ca^{2+} extrusion [55].

Mitochondria can accumulate large amounts of Ca^{2+} ions via an energy-linked process. The Ca^{2+} uptake is an alternative to ATP synthesis, a similar energy quantity being required to phosphorylate one ADP molecule and to transfer two Ca^{2+} ions into the mitochondrion [56]. Two dehydrogenases of the tricarboxylic acid cycle and pyruvic acid dehydrogenase phosphate phosphatase are sensitive to Ca^{2+} ion.

Calcium transfer into the mitochondrial matrix is supported by cytosolic Mg^{2+} in normal concentrations. Phosphate anion is transferred with Ca^{2+} across the mitochondrial envelope and can precipitate it in the matrix as an insoluble salt [56]. Calcium in mitochondria can prime H^+ export.

The bulk Ca^{2+} cytosolic concentration does not reflect its concentration in the immediate vicinity of mitochondria, where it can actually reach higher levels. Near-mitochondrion cytosolic calcium nanopools generated by Ca^{2+} release from vicinal ER compensate the poor affinity of the mitochondrial Ca^{2+} carrier (uniporter) involved in Ca^{2+} uptake. Calcium carrier activity depends on extramitochondrial Ca^{2+} concentration. It has a low open probability at rest that rises upon cell stimulation to ensure rapid mitochondrial Ca^{2+} uptake and activates oxidative metabolism.

In fact, mitochondrial Ca^{2+} concentration results from activity of several mitochondrial Ca^{2+} carriers. The outer mitochondrial membrane is permeable to small solutes and ions. It is enriched in voltage-dependent anion channels. Their density controls mitochondrial Ca^{2+} level. The voltage gating and ion selectivity and permeability of three VDAC isoforms are influenced by several interactors and metabolites.

Calcium transfer across the inner mitochondrial membrane is strongly regulated by the highly selective mitochondrial Ca^{2+} uniporter. It forms an active channel as an oligomer. The MCUB gene encodes the less selective carrier subunit mtCUb. In the heart, the MCU/MCUb ratio is much lower (3:1) than in the skeletal muscle (40:1) [57].

The mtCU carrier interacts with regulators such as mitochondrial Ca^{2+} uptake protein MiCU1 [54]. Mitochondrial Ca^{2+} uniporter is a proteic complex composed of the channel-forming subunits mtCU and mtCUb, as well as interactors MiCU1, MiCU2, mitochondrial calcium uniporter regulator mtCUR1,[13] its isoform coiled-coil domain-containing protein CCDC90b, essential mtCU regulator (EMRe), and Ca^{2+}-binding mitochondrial phosphate carrier SLC25A23 [57, 58]. The ion-conducting pore mtCU, which in fact has a high open probability and Ca^{2+} affinity [53], is hence controlled by its regulatory partners that are responsible for the resulting low Ca^{2+} affinity.

The MICU family comprises three members (MiCU1–MiCU3) that modulate mtCU open probability. Both MiCU1 and MiCU2 are widely expressed, but MiCU3 resides mostly in the brain [57]. The MiCU1 subtype that undergoes large conformational changes upon Ca^{2+} binding is a mtCU activator and MiCU2 an inhibitor, preventing Ca^{2+} overload in the mitochondrial matrix. The MiCU1 paralog MiCU2 interacts with MiCU1. The MiCU1–MiCU2 dimer may exert an opposite effect according to external Ca^{2+} concentration. It may keep mtCU close and hence prevent mitochondrial Ca^{2+} uptake when the extramitochondrial Ca^{2+} level is low (< 3 μmol/l). It may promote mtCU activity when the cytosolic Ca^{2+} level rises.

[13] Also known as coiled-coil domain-containing protein CCDC90a. This membrane protein interacts with mtCU, but not with its MiCU1 partner.

Calcium-dependent MiCU2 inhibition and MiCU1 activation enable a rapid mitochondrial Ca^{2+} influx. In addition, the MiCU1–MiCU2 heterodimer interacts also with the EMRe interactor that can mediate MtCU–MiCU interactions [53].

On the other hand, uncoupling proteins UCP2 and UCP3 regulate Ca^{2+} loading of the ER rather than that of mitochondria.

Calcium influx via mtCU is counteracted by efflux through low-affinity, mitochondrial Na^+–K^+–Ca^{2+} (mtNCX)[14] and H^+–Ca^{2+} (mtHCX)[15] exchangers.

In normal conditions, Ca^{2+} flux across the inner mitochondrial membrane through mtCU is smaller than that through other cytosolic Ca^{2+} extrusion carriers (single mtCU conductance ~ 105 mmol/l) [55]. Ca^{2+} transfer through mtCU is modulated by cytosolic Ca^{2+} concentration. However, mitochondrion is not a significant buffer of cytosolic Ca^{2+} concentration in physiological conditions.

Numerous molecules generate and decypher variations of Ca^{2+} concentration and localization within the cytosol. Calcium carriers cooperate with Ca^{2+} sensors and buffers to control the spatiotemporal pattern of Ca^{2+} signals that operate either directly (via Ca^{2+}-binding sites) or indirectly (via Ca^{2+}-dependent kinases, phosphatases, and scaffold proteins).

Mitochondria possess calcium-signaling, synapse-like apposition nanodomains with the ER and plasma membrane. These sites ensure fast Ca^{2+} transfer. Mitochondrial Ca^{2+} ion buffers cytosolic Ca^{2+} concentration, thereby regulating Ca^{2+} carrier function.

Some mitochondria lodge nearby the sarcoplasmic reticulum endowed with ryanodine and IP_3-sensitive Ca^{2+} channels. They operate as fast sensors of Ca^{2+} signals and localized buffers. They enable a high rate of mitochondrial Ca^{2+} uptake through the IMM despite the mtCU low affinity. In cardiomyocytes, mitochondrial Ca^{2+} uptake contributes to the control of the excitation–contraction coupling.

In particular, mitochondria are linked to the ER via *mitochondrion-associated endoplasmic reticulum membrane* (MAERM). This apposition is tethered by proteic filaments. Mitofusin Mfn2 is involved in MAERM tethering via both homo- and heterotypic interactions with Mfn1 [54]. These synapses prevent mitochondrial Ca^{2+} overload.

These contact sites (gap < 200 nm) are nanodomains of high Ca^{2+} concentration, which are maintained by proteins and chaperones [54]. The ER chaperones and Ca^{2+} buffers calreticulin and calnexin control the stability or sorting of signaling proteins in these apposition sites. Calnexin activity is regulated by phosphofurin acidic cluster sorting protein PACS2 [54]. On the ER side, TRIM19 complexes

[14] Also known as NCKX6, SLC24a6, and SLC8b1. It electrogenically countertransports four Na^+ for one Ca^{2+} and one K^+ ion. Calcium extrusion by mtNCX depends on mitochondrial membrane potential; it is facilitated by a negative membrane potential. It is predominantly active in excitable cells [59].

[15] The electrogenic H^+–Ca^{2+} exchanger operates at nanomolar cytosolic Ca^{2+} concentration [55]. At elevated mitochondrial matrix Ca^{2+} concentrations and low cytosolic pH, it extrudes Ca^{2+} ions. It is preferentially active in nonexcitable cells [59].

with IP_3R, PKB, and PP2 and hence modulates IP_3R activity. Heat shock protein HSPa9 mediates the interaction of VDAC1 with IP_3R and facilitates mitochondrial Ca^{2+} uptake. Members of the family of long-chain fatty acid–CoA ligases (FACL) involved in lipid metabolism abound in these sites.

This specialized membrane is implicated in the transfer of ions such as calcium as well as proteins and metabolites, in addition to synthesis of cholesterol and ceramide used for maintenance and phospholipid production and nonvesicular transport (e.g., long-chain fatty acylCoA synthase). It thus participates in dynamical changes in mitochondrial morphology (fusion and fission).

Voltage-dependent anion channels in the OMM of MAERMs and calcium uniporter in the IMM enable rapid mitochondrial uptake of calcium. The mitochondrial calcium concentration is further regulated by calcium efflux via the Na^+–Ca^{2+} exchanger in the IMM and permeability transition pore in the OMM, as well as by IP_3R and SERCA on the ER side of MAERMs [48].

Mitochondrial Ca^{2+} ion also stimulates aerobic metabolism and ATP production. Three matrix dehydrogenases of the tricarboxylic acid cycle are activated by Ca^{2+}: pyruvate dehydrogenase (PDH), which is regulated by a Ca^{2+}-dependent phosphatase, as well as α-ketoglutarate (α KGDH) and isocitrate (ICDH) dehydrogenases, which are controlled by Ca^{2+} binding [54]. Moreover, aspartate–glutamate and malate–citrate antiporters of the IMM possess a Ca^{2+}-binding domain exposed in the intermembrane space, which allow to raise the metabolite transport.

In normal feeding conditions, mitochondrial Ca^{2+} signals preclude autophagy [54]. During starvation, cytosolic Ca^{2+} signals support prosurvival autophagy. Calcium–calmodulin-dependent kinase Cam2Kβ (or Cam2K2) activates the AMPK–TOR axis and autophagy.

On the other hand, mitochondrial Ca^{2+} accumulation can trigger cell apoptosis or necrosis by a transient or sustained opening of the high-conductance permeability transition pore (mtPTP). Calcium ions can exit the mitochondrial matrix through the mitochondrial permeability transition pore. The latter consists of proapoptotic BCL2 family members BAX and BAK on the OMM and ATP synthase on the IMM under the control of ADP, peptidylprolyl isomerase-D (or cyclophilin-D), and cyclosporine-A that prevent pore forming and opening. More precisely, the C subunit of the F_1F_O ATP synthase forms the adenine nucleotide (ADP and ATP)- and voltage-sensitive mtPTP pore at the IMM [60].

The intrinsic pathways of apoptosis are initiated by the release of apoptosome components such as cytochrome-C from the mitochondria. Calcium ion cooperates with various apoptotic signals to liberate proapoptotic mitochondrial components.

In ischemia–reperfusion events, mitochondrial Ca^{2+} overload, in conjunction with ROS, causes a prolonged mtPTP opening and mitochondrial swelling.

Mitochondrial Connexin-43 Connexin-43 participates not only in the propagation of the electrochemical wave and metabolic coupling between neighboring cardiomyocytes but also outside gap junctions in paracrine signaling, as it forms hemichannels at the plasma membrane, as well as in the modulation of gene expression and cell growth, as Cx43, or its C-terminus can translocate to the nucleus [51].

Moreover, Cx43 lodges at the inner mitochondrial membrane of subsarcolemmal mitochondria of cardiomyocytes (but not in interfibrillar mitochondria, or at very low levels with respect to subsarcolemmal mitochondria). Its localization to mitochondria depends on its interaction with heat shock protein HSP90 and translocase of the outer membrane (TOM) complex via its component TOMM20. In normal conditions, calcium capacity is lower in subsarcolemmal mitochondria than interfibrillar mitochondria.

MtCx43 modulates mitochondrial K^+ uptake, ^{ETC}complex-I function, and radical oxygen species generation [51]. It ensures cardioprotection, especially in ischemia–reperfusion events.

In fact, Cx43 of subsarcolemmal mitochondria mediates the cardioprotective effect of fibroblast growth factor FGF2 [52]. It attenuates calcium-induced opening of mitochondrial permeability transition pore upon phosphorylation (Ser262 and Ser368) by protein kinase-Cϵ.

Subsarcolemmal mitochondria are more responsive than interfibrillar mitochondria to the FGF2 factor. The protective effect of FGF2 is associated with increased levels of PKCϵ and translocase of outer mitochondrial membrane homolog TOMM20 in both mitochondrial populations, and GSK3β in interfibrillar mitochondria [52].

Mitochondria and Cell Apoptosis Mitochondria are involved in the intrinsic apoptosis pathway caused by oxidative stress, DNA damage, ER stress, high (toxic) calcium concentration, and hypoxia (Vol. 2, Chap. 4. Cell Survival and Death). Activation of proapoptotic BCL2-antagonist killer (BAK) or BCL2-associated X protein (BAX) leads to OMM permeabilization and leakage of cytochrome-C into the cytoplasm. Cytochrome-C then binds to apoptosis peptidase-activating factor APAF1 and forms the apoptosome.

Mitochondrial ROS Production A certain amount of electrons escapes the electron transfer chain, particularly at ^{ETC}complex-I and -III, reacts with molecular oxygen, and generates superoxide anion. In normal conditions, ROS formation is involved in physiological processes (Vol. 4, Chap. 10. Other Major Types of Signaling Mediators). The small resulting ROS quantity contributes to normal cell signaling.

Nitroxyl triggers cardiac preconditioning to ischemia–reperfusion injury (Sect. 8.8) via oxidative and/or nitrosative stress-related mechanisms. Several messengers, such as acetylcholine, bradykinin, and opioids, launch a preconditioning-like protection during the first few minutes of myocardial reperfusion following ischemia via a mitochondrial K_{ATP}–ROS-dependent process [61]. Moreover, ischemic postconditioning also ensures cardioprotection via redox signaling.

However, excessively produced amounts of mitochondrial ROS due to a dysfunctional respiratory chain (e.g., unrestrained NADH levels and low ATP production), and/or other mitochondrial sources (e.g., monoamine oxidase and Src homology (SHC)-2 domain-containing transforming protein SHC1 [P66SHC])[16] overwhelm

[16] In particular, SHC1 (P66SHC) catalyzes the formation of H_2O_2 and causes reperfusion damage [62]. The Shc1 gene encodes a cytoplasmic adaptor protein with several splice variants (P46SHC,

the antioxidant system within the matrix (e.g., ROS scavengers manganese superoxide dismutase and glutathione peroxidase). The resulting oxidative stress damages mitochondrial DNA (mtDNA) and provokes autophagy, apoptosis, and necrosis. The imbalance between ROS generation and removal, that is, oxidative stress intervenes in cardiac diseases, especially in myocardial injury caused by reperfusion after ischemia. Mitochondria are the most relevant site for ROS formation in cardiomyocytes.

1.5.2.2 Mitochondria in Immunity and Inflammation—Mitochondrial Alarmins

Alarmins, or damage-associated molecular patterns, originated from the cell include constituents of the plasma membrane, nucleus, cytosol, ER, and mitochondria.

Inflammation is an initial response of the innate immunity, which subsequently stimulates the adaptive immunity to trigger a robust and selective defense.

Immunological response is triggered by exogenous stimuli (e.g., bacteria, viruses, and fungi), whereas immunological tolerance discriminates between healthy self elements and exo- and endogenous alarmins. Immunity can be activated by mitochondrial damage-associated molecular patterns (mtDAMP).

Mitochondrial alarmins, that is, ^{N}formyl peptides and mtDNA, released during cell injury and/or death, are immunological activators recognized by specific pattern recognition receptors (PPR) of the innate immunity, in particular those expressed in the cardiovascular apparatus, that is, formyl peptide (FPR) and Toll-like receptor (TLR), respectively. When mitochondrial components are liberated into blood circulation upon plasma membrane rupture and improper degradation within the cell, they are sensed by PRRs and can prime a systemic inflammatory response.

Mitochondrial DNA Unlike nuclear DNA, mitochondrial DNA contains inflammatogenic unmethylated cytosine–phosphodiester–guanine (CpG) dinucleotides. Mitochondrial DNA that escapes degradation is a ligand for the TLR9 receptor at the ER of immune as well as endothelial and vascular smooth muscle cells. Upon stimulation by mtDNA, TLR9 translocates to endosomal membranes. Activation of TLR9 provokes the synthesis of proinflammatory cytokines and activation of the MPAK module. In particular, released mtDNA by pressure overload launches a TLR9-mediated inflammatory response in cardiomyocytes and can engender myocarditis and dilated cardiomyopathy [48].

P52SHC, and P66SHC) involved in intracellular signaling from activated protein Tyr kinases to small Ras GTPase. SHC1 (P66SHC) is not involved in Ras-mediated signaling, but in ROS metabolism regulation. A fraction of P66SHC localizes within the mitochondrial intermembrane space, where it oxidizes reduced cytochrome-C. Molecular oxygen is then partly reduced to hydrogen peroxide by P66SHC–cytochrome-C action. Hydrogen peroxide contributes to the opening of the mitochondrial permeability transition pore that triggers apoptosis.

Once it is freed from injured cells, mtDNA remains associated with mtTFa and promotes endosomal signaling via the PI3K–PKB and ERK axes in plasmacytoid dendritic cells. Immune response mediated by TLR9 is amplified by mtTFa.

Mitochondrial ^{N}Formyl Peptides ^{N}Formyl peptides, such as ^{N}formyl methionine, produced by the degradation of cells and bacteria are potent chemoattractants that have a high-affinity binding for the formyl peptide receptor involved in chemotaxis.

Formyl peptide receptors (FPR1–FPR3) constitute a subcategory of GPCRs. They are expressed at high levels on neutrophils and monocytes, as well as dendritic cells and natural killer cells that kill tumoral and virally infected cells. They are also detected in endothelia, epithelia, and smooth muscles, as well as lung, liver, spleen, and skeletal muscles [48].

Once they are activated by ^{N}formyl Met–Leu–Phe (fMLP), they stimulate NADPH oxidase, hence generating ROS, as well as cell chemotaxis and release of proteolytic enzymes.

Activated FPR triggers rearrangement of the cytoskeleton, thereby facilitating cell migration as well as chemokine synthesis. Binding of ^{N}formyl peptides to their specific plasmalemmal receptors initiates various cascades of chemical reactions:

1. G protein stimulates phospholipase-C and subsequently the IP_3–Ca^{2+} and DAG–PKC pathways.
2. FPR also activates small Ras GTPase and subsequently the Raf–MAPK axis (i.e., P38MAPK, JNK, and ERK1 and ERK2 kinases).
3. Liganded FPR can also excite cADP–ribose hydrolase at the plasma membrane of many immune leukocytes (CD4+ and CD8+ T and B lymphocytes and natural killer cells).

NAD^{+} that enters the cytosol is converted into cADPR, a second messenger that interacts with ryanodine receptors, thereby further raising cytosolic Ca^{2+} concentration.

Mitochondrial Phospholipid Cardiolipin Mitochondrial dysfunction can lead to cardiolipin release. Cardiolipin is then rapidly oxidized and converted into an alarmin. In atherosclerotic lesions, oxidized cardiolipin in apoptotic cells is implicated in adverse immune response [48].

Mitochondrial ATP Once released, ATP binds to the plasmalemmal inotropic P2X nonselective cation channels and G-protein-coupled P2Y receptors. Intravascular release of ATP contributes to tissue damage via inflammation associated with recruitment of circulating neutrophils.

In human microvascular endotheliocytes, a high concentration of extracellular ATP causes the release of IL6 as well as CCL2 and CXCL8, and increases ICAM1 expression, thereby favoring vascular inflammation [48].

Mitochondrial Cytochrome-C Cellular stress releases cytochrome-C from mitochondria into the cytosol, where it can act as an intracellular alarmin and activate caspase family members.

MicroRNAs Once they are secreted from tumoral cells in exosomes, miR21 and miR29a activate TLR7 and TLR8 in immunocytes, thereby activating NFκ B and supporting inflammatory cytokines production [48].

MicroRNAs are ligands for TLRs. Pre-microRNAs and microRNAs can be detected in human mitochondria, at least in some cell types. They are overexpressed in carotid arteries after angioplasty.

1.5.3 Epicardium

The epicardium is an epitheliocyte monolayer that surrounds the heart. The proepicardium arises from extracardiac precursor cell aggregate at the venous pole of the developing heart. It a source of signals and cells for the myocardium and coronary vasculature. The embryonic epicardium generates the majority of interstitial and perivascular cardiofibroblasts and coronary smooth myocytes via mobilization and differentiation of epicardium-derived cells controlled by auto- and paracrine signals.

In particular, epicardial FGF induces Sonic Hedgehog signaling that regulates the formation of coronary arteries and veins [63].

Notch signaling regulates smooth myocyte differentiation of epicardium-derived cells once they have reached a perivascular position.

Whereas the canonical Wnt signaling is used in myocardial progenitors, it is not involved in epicardial epithelial–mesenchymal transition and smooth myocyte differentiation.

The Hedgehog–Smoothened pathway is implicated in vasculo- and angiogenesis in embryos and adults via cardiac and vascular endotheliocytes. Both SHh and Ptch1 are synthesized in the epicardium, but the Hh pathway is not relevant in the epicardium. Moreover, the epicardially activated Hh pathway is deleterious for cardiogenesis [63].

Fibroblast growth factor receptors FGFR1 and FGFR2 are involved in proliferation, migration, and differentiation of numerous cell types. In mice, epicardial deletion of the CTNNB1, SMO, Fgfr1 and Fgfr2 gene expression does not affect epicardial cell mobilization and differentiation [63].

On the other hand, platelet-derived growth factor receptor PDGFRα is involved in epicardiogenesis. The loss of PDGFRA reduces epithelial–mesenchymal transition and prevents the differentiation of epicardium-derived cells into mature fibroblasts [63].

After ischemic injury, the epicardium expands, gives rise to epicardial-derived cells that undergo an epithelial–mesenchymal transition and engender cardiofibroblasts and myofibroblasts [35]. The Wnt–β Ctnn pathway contributes to the regulation of the fate of epicardial-derived cells after ischemic injury.

1.5.4 Cardiac Cell–Endotheliocyte Interactions

The cardiac endothelium includes the endocardium and coronary endothelium. They share common features in signal transduction launched by neurotransmitters, hormones, and mechanical stresses, but differ in their embryological origin as well as developmental, morphological, and functional properties [64]. They also have distinct properties.

Endocardial (eEC) and coronary microvascular endotheliocytes (cmivEC), which have different embryological origin, interact directly with neighboring cardiomyocytes, hence being able to modulate myocardial performance and growth. On the other hand, endotheliocytes of coronary arteries, large arterioles and venules, and veins interact with smooth myocytes.

1.5.4.1 Endocardium

The luminal surface of the mature heart comprises furrows, cylinder- and sheet-like trabeculae, and papillary muscles. The cavitary surface of the cardiac wall is lined by the endocardium, a sheet of connected endotheliocytes with a central nuclear bulge covered by a dense electrically charged glycocalyx. The endocardium is a paracrine regulator of myocardial performance, particularly in the right ventricle [65].

Endocardial endotheliocytes are larger than their vascular homologues. Their luminal surface possesses numerous microvilli [65]. Gap junctions contain connexin Cx43, Cx40, and Cx37 between endocardial endotheliocytes abound, hence form a syncytium. On the other hand, the concentrations of intercellular adhesion molecule ICAM1 and antigens of the class-1 and -2 major histocompatibility complex are lower than those in myocardial capillary endotheliocytes.

Numerous plasmalemmal ion channels (e.g., inwardly rectifying K^+, Ca^{2+}-activated K^+, background Cl^- and cation, volume-activated Cl^-, and stretch-activated cation channels) as well as pumps (e.g., Na^+–K^+ ATPase) are asymmetrically distributed in the luminal and abluminal cell surface [65]. For example, the predominant α1 type of Na^+–K^+ ATPase is confined to the luminal membrane of the endocardial endotheliocyte. The transendocardial electrical resistance is two to five times higher than that in other endothelia.

The intercellular cleft between endocardial endotheliocytes is three to five times deeper than that between myocardial capillary endotheliocytes (most clefts are shallow having few tight junctions with many interruptions) and is often highly tortuous. Most clefts contain one or two tight junctions and zonula adherens interacting with a circumferential actin filament band and several connecting proteins [65].

Moreover, the endocardial endothelium that exhibits an extensive intercellular overlap provides a *blood–heart barrier* (BHB) that ensures an active transendothelial physicochemical gradient of various ions [65]. Tight junctions localize to the luminal side of the intercellular clefts. The glycocalyx is more developed at the luminal side than below the tight junctions.

Table 1.15 Gene expression in cultured endocardial and aortic endotheliocytes. The voltage-gated sodium channel $Na_V2.1$ (also known as Na_x, NaG and SCL11 in rats, and $Na_V2.3$ in mice) encoded by the SCN7A gene is a sodium sensor in the subfornical organ. (Source: [66])

Endocardial endotheliocyte	Aortic endotheliocyte
GATAGT2	Connexin-26, VCAM1
TGFβ2	Decorin
PTHR, IP_3R	V_{1A}
ScaRe1, ApoE	STAR
Nav2.1	
Lysozyme, creatine kinase-B, fatty acid translocase, ATP1β1, GuCy1β3, MME, CyP7b	
Sortilin-1, β A4-crystallin	
Preprocomplement-C3	
Cyclin-D2	

ATP1β1 Na^+–K^+-dependent ATPase β1 subunit, *GATAGT* GATA-binding protein, *MME* membrane metalloendopeptidase, *PTHR* parathyroid hormone receptor, *STAR* steroidogenic acute regulatory protein, V_{1A} vasopressin receptor

In the subendocardial space, endocardial endotheliocytes can interact with the subendocardial terminal *Purkinje fiber* network and with the extensive subendocardial *nervous plexus*.

Cardiac endotheliocytes differ from aortic endotheliocytes in their transcriptional activity [66]. Sets of genes are preferentially expressed in cultured endocardial endotheliocytes than in cultured aortic endotheliocytes. However, behavior of cultured endotheliocytes can differ from that of cells within the body. Moreover, behavioral difference may be specific to the cellular phenotype. Nonetheless, the endothelial genetic diversity observed in culture reflects partly the physiological variability.

The analysis shows that 299 genes are preferentially expressed in EECs from rat right and left ventricles ($\geq$ twofold increase, 100 genes with a $\geq$ tenfold relative overexpression) and 201 genes are preferentially transcribed in aortic endotheliocytes (Table 1.15) [66].

1.5.4.2 Role of Endocardial and Vascular Endotheliocytes

Endotheliocytes of the endocardium and myocardial vessels participate in the regulation and maintenance of cardiac function, both types being close to cardiomyocytes. Cardiac development and growth rely on endothelial–myocytic interaction. Mediators governing this interaction, such as angiopoietin, neuregulin, and vascular endothelial growth factor, maintain phenotype and survival of cardiomyocytes.

Cardiac and vascular endotheliocytes synthesize, activate, and release auto- and paracrine agents, such as vasoconstrictors and -dilators (e.g., angiotensin-2, endothelin-1, nitric oxide, and prostaglandin-I_2), pro- and anticoagulant (pro- and

Table 1.16 Paracrine messengers in redox signaling between endotheliocytes (*EC*) and cardiomyocytes (*CMC*; Source: [64]). Neuregulin-1 (*Nrg1*), a member of epidermal growth factor (*EGF*) family and ligand for receptor Tyr kinases of the HER family, is produced by endotheliocytes. Hydrogen peroxide causes Nrg1 release from ECs as well as activates the HER4–PKB signaling

Agent	Source	Signaling in source cells	Signaling in target cells	Effects
Nrg1	EC	Stimulated release	HER2/4, PKB	Myocyte hypertrophy and survival
		$O_2^{\bullet -}$	cGMP–PKG	Altered relaxation and stiffness
NO	EC	NOS3	RGS4 degradation PP3–NFAT	Myocyte hypertrophy
VEGF	CMC	NOx4–HIF1–VEGF NOx4–ROS–GATA4	Angiogenic signals	Angiogenesis
H_2O_2	CMC	NOx4	NOS3 Angiogenic signals	Increased perfusion

antithrombotic) factors, and growth and antigrowth factors that contribute to angiogenesis and tissue remodeling, as well as to inflammation and immunity. These factors influence cardiac metabolism, growth, contractility, and rhythmicity of the adult heart.

1.5.4.3 Communication Between Endotheliocytes and Cardiomyocytes

Cardiomyocytes (30 % of cardiac cells) are intimately arranged within the coronary microvasculature (capillary density 3000–4000/mm^2) for adequate blood supply. Endotheliocytes communicate with cardiomyocytes over a short distance (distance between cardiomyocytes and endotheliocytes < 2–$3\ \mu$m).

Endotheliocytes support cardiomyocyte differentiation and cardiogenesis. Conversely, signaling agents secreted by cardiomyocytes, which influence endotheliocytes, are required for the proper cardiac development.

Both the endocardial and myocardial vascular endothelia can directly modulate the contractile state of adjoining cardiomyocytes using paracrine factors, such as nitric oxide, endothelin-1, prostanoids, natriuretic peptides, and cytokines, among other agents [64]. Nitric oxide assists myocardial relaxation via the cGMP–PKG-mediated phosphorylation of troponin-I and reduction in myofilament Ca^{2+} sensitivity [64]. In addition, the NO–cGMP–PKG signaling reduces cardiomyocyte stiffness, as it phosphorylates titin. Several paracrine messengers participate in redox signaling between endotheliocytes and cardiomyocytes (Table 1.16).

In the postnatal and adult heart, cardiac endotheliocytes can affect cardiac function in various ways. The contribution of endocardial endotheliocytes must be distinguished from that of vascular endotheliocytes, and within the latter category, between myocardial and epicardial sources as well as between vascular compartments (arteries, microvessels, and veins).

The vascular endothelium in the coronary conductive and resistive arterial compartments controls coronary blood supply to the myocardium. In adult rats, cardiac endotheliocyte-to-myocyte volume ratio equals 0.04–0.05 [65]. The intercapillary distance equals 20.2 μm in the ventricular wall and 15.6 μm in papillary muscle of normal rat heart.

The closest distance from endocardial endotheliocytes to adjacent cardiomyocytes depends on animal species and cardiac site (from < 1 μm in small mammals to 10–30 μm in the ventricle and > 50 μm in atria of larger mammals such as humans) [65].

1.5.4.4 Angiogenesis

Cardiac myocytes and fibroblasts interact with endotheliocytes to regulate angiogenesis. Cardiofibroblasts produce both angiogenic (e.g., VEGF) and antiangiogenic (e.g., CTGF) molecules [35]. Cardiomyocytes release multiple paracrine messengers such as VEGFa to regulate the coronary vasculature [64].

After cardiac injury, fibroblasts also produce matrix metallopeptidases that support endotheliocyte migration and vascular sprouting. The Wnt proteins induce several MMP types. Cardiofibroblasts also synthesize tissue inhibitor of metalloproteinases that exerts both pro- and antiangiogenic effects and operates via the Wnt–β Ctnn signaling. Macrophages limit aberrant or excessive angiogenesis via VEGFR1 and a noncanonical Wnt pathway [35].

The Wnt1 morphogen supports angiogenic ability of endothelial progenitor cells [35]. The Wnt antagonists Dkk1 and Dkk2 promote mobilization of endothelial progenitors from the bone marrow.

1.5.4.5 Microvasculature–Myocardium Communication

Neuregulins enable communication between the cardiac microvasculature and cardiomyocytes [67]. Neuregulins constitute a family of similarly functioning growth factors encoded by four NRG genes. All Nrg types possess a receptor-binding EGF-like domain, the C-terminus of which differs in Nrgα and more potent Nrgβ isoforms, each existing in diverse variants [67]. The NRG1 gene encodes various isoforms (Nrg1-1–Nrg1-6 according to distinct N-termini). The NRG2–NRG4 genes give rise to a much lower number of isoforms (two N-terminal sequence variants [e.g., Nrg2-1A–Nrg2-1B]).

Neuregulins are synthesized in various cell types (Table 1.17). Only NRG1-1 isoform is expressed in the adult heart (endocardium and coronary microvasculature) [67].

Neuregulins are cleaved either by ADAM17, and act as para- and juxtacrine messengers, or BACE1 peptidase [67]. However, Nrgβ3 is directly secreted into the extracellular space. The ADAM17 sheddase liberates the ectodomain. Once neuregulin is cleaved by the BACE1 β-secretase, its intracellular domain translocates to the nucleus to target antiapoptotic genes (*reverse Nrg signaling*).

Table 1.17 Neuregulin isoforms. (Source: [67])

Gene	Isoform	Receptors	Localization
Nrg1	Unspecified	HER3/4	Pulmonary epitheliocytes, fibroblasts, Golgi-2 neurons, cholinergic cells of basal forebrain
Nrg1	Nrg1-1		Neurons, astrocytes, cardiac microvascular endotheliocytes
	Nrg1-2		Neurons, astrocytes
	Nrg1-3		Neurons, astrocytes, motoneurons
	Nrg1-4		Hippocampus
	Nrg1-5		Prefrontal cortex
	Nrg1-6		
Nrg2		HER3/4	Neurons, granule cells, Purkinje cells

Neuregulins bind to receptor Tyr kinases of the HER family (HER1–HER4), either triggering receptor dimerization and phosphorylation (*canonical* Nrg–HER *forward signaling*) or receptor cleavage and internalization (*noncanonical* Nrg–HER *forward signaling*). Receptors HER1 and HER2 do not bind Nrg1–Nrg4 isoforms; HER3 can bind both Nrg1 and Nrg2, but needs to heterodimerize for signal transmission; liganded HER4 can autonomously respond, though it also heterodimerizes [67]. Different Nrg subtypes may activate distinct signaling cascades, such as the ERK1, ERK2, PI3K–PKB, and JaK–STAT pathways.

Both HER2 and HER4 are expressed in the pre- and postnatal heart. The HER3 receptor is detected in the invading mesenchyme and endocardial cardiac cushions of the developing heart (but not in the adult heart) [67]. In the fetal heart, Nrg1 and Nrg2 synthesized by cardiac microvascular endotheliocytes bind to HER4 on cardiomyocytes that dimerizes with HER2, thereby launching a signaling cascade that elicits proliferation and differentiation.

In the adult heart, Nrg1 has cardioprotective and -regenerative functions (e.g., protection against apoptosis as well as proliferation of postnatal cardiomyocytes; Table 1.18). It may also intervene in cardiac hypertrophy likely via feedforward and feedback signaling [67].[17] Endothelin-1 and elevated mechanical stress increase Nrg production, whereas angiotensin-2 lowers it [67]. Integrins expressed by cardiomyocytes act as sensors of mechanical stress and likely noncanonical Nrg1 receptors.

In congenital heart disease patients, HER4 density is correlated with defects of the left ventricular outflow tract. In heart failure, patients have normal Nrg levels, but extremely decreased HER2 and HER4 levels. After injury, endothelium-derived Nrg1 promotes angiogenesis; it also acts on cardiofibroblasts to attenuate the scar size.

[17] Other growth factors have cardioprotective and -regenerative effects. Whereas FGF1 and IGF1 promote cardiac hypertrophy, periostin and Nrg1 have an opposite effect [67]. Nrg1, IGF1, FGF1, FGF2, VEGF, TGFβ1, urocortin, and cardiotrophin reduce apoptosis.

Table 1.18 Effects of neuregulin-1 (*Nrg1*) isoforms in the adult heart. (Source: [67]). Recombinant human neuregulin-1 (*rhNRG1*) is also called neucardin. Anthracyclines are drugs used in chemotherapy

Type	Effects	In vitro	In vivo
Nrg1	Improvement of myocardial function in diabetic cardiomyopathy		+
	Protection against cardiomyocyte apoptosis in diabetic cardiomyopathy	+	
	Prevention of ATn2-induced diastolic dysfunction	NA	+
	Reduction in ATn2-induced cardiac hypertrophy	+	+
	Reduction in ATn2-induced myocardial fibrosis	+	+
Nrg1α	Prevention of anthracycline-induced myofilament injury	–	
Nrg1α2	Negative inotropy	+	
Nrg1β	Prevention of anthracycline-induced myofilament injury	+	
	Protection against anthracycline-induced cardiotoxicity		+
	Cardiomyocyte proliferation	+	+
	Improved systolic function	NA	+
	Reduced hypertension	NA	+
Nrg1-1β	Reduction in anthracycline-induced alterations of ECC	+	
	Angiogenesis	+	
Nrg1-2β	Cardiomyocyte survival, proliferation, and hypertrophy	+	
	Protection of cardiomyocytes from anthracycline-induced apoptosis	+	
Nrg1β1	Angiogenesis	+	
	Improved Ca^{2+} handling	+	
Nrg1β2a	Protection against anthracycline-induced cardiotoxicity		+
Nrg1β3	Angiogenesis	+	+

ATn2 angiotensin-2, *ECC* excitation–contraction coupling, *MI* myocardial infarction +(–): positive (negative) effect, *NA* not available

1.5.4.6 Redox Signaling Between Endothelial and Other Cardiac Cells

The redox signaling (i.e., signaling via oxidation–reduction modification of mediators) within and between endotheliocytes and cardiomyocytes is a component of communication between these cell types. It influences myocyte contractility and hypertrophy, and angiogenesis, as well as cardiac remodeling and fibrosis [64].

In general, redox crosstalk between cardiomyocytes and cardiac endotheliocytes results from [64]: (1) diffusion of ROS and nitric oxide; (2) action of ROS produced in cardiomyocytes or endotheliocytes on the extracellular matrix; and (3) ROS-dependent alteration of the paracrine release from endotheliocytes of various cytokines and growth factors involved in cell communication.

ROS involved in redox signaling include ROS, such as superoxide anion ($O_2^{\bullet -}$), hydroxyl radical ($OH^{\bullet}$), and hydrogen peroxide (H_2O_2), and reactive nitrogen species, such as nitric oxide ($NO^{\bullet}$) and peroxynitrite ($ONOO^-$).

The ROS effects depend on the generated molecule type, its concentration, subcellular localization, and the endogenous antioxidant status. When the production of small ROS amounts in response to stimuli remains tightly regulated and spatially confined, ROS reversibly modulates the activity of molecular targets such as ion pumps and channels, protein phosphatases and kinases, and other types of signaling effectors.

On the other hand, when overall ROS production overwhelms cellular antioxidant defense, the cell experiences oxidative stress and, hence, cellular damage, dysfunction, energetic deficit, and death due to irreversible modifications of membrane lipids, proteins, and nucleic acids. ROS generated by NOx enzymes and their interactions with NO are implicated in redox signaling during the development of heart failure [64].

In the heart, ROS modulate the activity of components of cardiac remodeling. In cardiac cells, the ROS sources include mitochondria, xanthine oxidase, uncoupled NO synthases, and NADPH oxidases (NOx1–NOx5 and DuOx1–DuOx2).

Nitric Oxide Radical Nitric oxide is synthesized by constitutive nitric oxide synthases, NOS3 mostly in endotheliocytes and, to a lesser extent, in caveolae of cardiomyocytes, and NOS1 predominantly in the sarcoplasmic reticulum and possibly mitochondria, as well as inducible NOS2.

Nitric oxide influences cellular functions via the sGC–cGMP pathway as well as ^{S}nitrosylation of cysteine residues of effector proteins.

Excess $O_2^{\bullet -}$ interacts with NO extremely rapidly to form peroxynitrite. In addition, once they are uncoupled (due to depletion of the BH4 cofactor or NOS oxidation), NOSs generate superoxide anion instead of nitric oxide.

Elevated $O_2^{\bullet -}$ and $ONOO^-$ levels amplify NOS uncoupling by oxidizing BH4 cofactor, thereby causing maladaptive cardiac remodeling induced by pressure overload as well as increasing myocardial damage during ischemia–reperfusion injury.

NADPH Oxidases In cardiomyocytes and endotheliocytes, NOx2 and NOx4 are the predominant isoforms [64]. They differ according to the structure, activation mode, function, intracellular localization, and generated ROS types.

The NOx2 subtype that synthesizes $O_2^{\bullet -}$ is activated by GPCR agonists (e.g., angiotensin-2, endothelin-1, and α-adrenoceptor agonists), thrombin, growth factors and hormones (e.g., insulin), cytokines (e.g., TNFSF1), metabolic factors (e.g., glucose), and mechanical forces.

Constitutively active NOx4 mainly manufactures H_2O_2 that has a higher stability and diffusibility than those of superoxide anion. It increases in cardiomyocytes and other cardiac cell types in response to pressure overload [64].

Table 1.19 Molecular signals released by fibroblasts and cardiomyocytes and then taken up by the target cell (Source: [40])

Communication direction	Messengers
Reciprocal CMC–FB communication	ATn2, ET1, TGFβ, TNFSF1, IL6F
Predominantly fibroblast to myocyte	CTGF, FGF2, and IL1β IL10/33 (inhibitory)
Cardiofibroblast to myocyte	PDGF, IGF1, and IL17
Cardiomyocyte to fibroblast	ANP, BNP (inhibitory)

AN(B)NP atrial (brain) natriuretic peptide, *ATn2* angiotensin-2, *CTGF* connective tissue growth factor, *ET1* endothelin-1, *FGF* fibroblast growth factor, *IGF* insulin-like growth factor, *IL* interleukin, *IL6F* IL6 family member, *PDGF* platelet-derived growth factor, *TGF* transforming growth factor, *TNFSF* tumor-necrosis factor superfamily member

1.5.4.7 Communication Between Fibroblasts and Cardiomyocytes

Adverse cardiac hypertrophy and fibrosis that result from mechanical, metabolic, and genetic stress can lead to heart failure depending on crosstalk between cardiac cell types.

Cardiac myocytes and fibroblasts secrete hormones, growth factors, and cytokines, as well as microRNAs and long noncoding RNAs that influence cardiomyocyte and fibroblast activation in an auto- and paracrine manner.

Cardiomyocytes change with increased cell size, modified sarcomeric assembly, altered gene expression with possible reinduction of the fetal gene program, abnormal Ca^{2+} handling, and accelerated cell death.

Cardiofibroblasts proliferate and differentiate to a myofibroblast phenotype characterized by contractile smooth muscle markers and an elevated production of extracellular matrix components that alters the cardiac microstructure and forms a barrier between cardiomyocytes, thereby impairing electrical coupling and elevating nutrient transfer distances, hence engendering hypoxia.

Fibroblasts thus influence cardiomyocyte metabolism, performance, and size. Conversely, cardiomyocytes affect fibroblast phenotype and function. Crosstalk between cardiac cells is related to messenger exchange via gap junction (juxtacrine regulation) and secretion of soluble molecules (auto- and paracrine regulation).

Paracrine signals that mediate crosstalk between cardiomyocytes and fibroblasts are released by both cardiac cell types (angiotensin-2, TGFβ, endothelin-1, IL6 family members, and TNFSF1) and exclusively (PDGF, IGF1, and IL17) or predominantly (CTGF, FGF2, and IL1β) by fibroblasts, the two latter categories also having an autocrine function [40]. These messengers support cardiac adverse hypertrophy and fibrosis. Other molecules impede hypertrophy (IL10 and IL33) and fibrosis (IL10, IL33, ANP, and BNP; Table 1.19).

1.5.5 Adipocyte and Structural Remodeling

The thickness of epicardial adipose tissue, a major source of cytokines, is related to the incidence and severity of atrial fibrillation, the most frequent cardiac arrhythmia [68].[18]

The epicardial adipose tissue secretome has a paracrine effect on the neighboring atrial myocardium. The epicardial adipose tissue has a greater capacity to metabolize and secrete free fatty acids for the myocardial energetic metabolism, but the glucose use is lower than in other visceral fatty depots (Table 1.20) [68]. In addition to its role in energetic and lipid metabolism, adipose tissue produces numerous inflammatory mediators (i.e., inflammatory cytokines [e.g., TNFSF1 and IL6] and chemokines [e.g., CCL2]) and adipocytokines.

Activin-A, a member of the TGFβ superfamily abundantly produced by epicardial adipocytes during heart failure or diabetes causes fibrosis in the atrial myocardium. Adipocytes may also infiltrate the atrial myocardium. Progenitor cells that abound in the epicardial adipose tissue may be a source of matrix-producing myofibroblasts. Activin-A has also antihypertrophic and -apoptotic effects on the myocardium experiencing ischemia–reperfusion events and pressure overload injury [68].

In ischemic cardiopathy, the epicardial adipose tissue contains more ROS than the subcutaneous adipose tissue, as the activity of the antioxidant enzyme catalase that protects cells against hydrogen peroxide is reduced [68].

1.5.6 Interactions Between Cardiac Cells and Resident and Recruited Leukocytes and Platelets

Cardiomyocytes contribute to the inflammatory response triggered by myocardial damage. Injured cardiomyocytes release alarmins, or damage-associated molecular pattern molecules, such as high-mobility group box HMGB1, DNA fragments, heat shock proteins, and matricellular proteins, which instruct surrounding healthy cadiomyocytes to produce inflammatory mediators [69]. These mediators comprise cytokines (mainly interleukins IL1β and IL6 and TNFSF1) and chemokines (e.g., CCL2) which, in turn, activate signaling cascades in surviving cardiomyocytes and trigger leukocyte activation and recruitment.

---

[18] The term *pericardial adipose tissue* can in some studies define epicardial adipose tissue (i.e., inside the pericardial sac) or in other works the sum of both epi- and paracardial adipose tissues. In fact, the pericardial brown adipose tissue comprises: (1) the paracardial adipose tissue located outside the visceral pericardium and (2) the epicardial adipose tissue situated between the visceral pericardium and epicardium. Intramyocardial fatty deposits are dedicated to triglyceride storage. Some white adipose tissue can be detected in the atrioventricular and interventricular grooves of the adult heart.

Table 1.20 Activity of the human epicardial adipose tissue (Source: [68])

Function factors	Effects
Metabolic activity	
Lipolysis (high) Glycolysis (low)	Source of free fatty acids
Heat production (UCP1)	Protection against hypothermia
Angiogenesis	
Angiogenin, endostastin, VEGF, thrombospondin-2, angiopoietin	Cell adhesion, cell proliferation, cell migration
Tissue development and remodeling	
TGFβ1–TGFβ3, activin-A, follistatin	Fibrosis, myocyte calcium signaling
MMP1/2/3/8/9/13	Extracellular matrix remodeling
Adipocytokine signaling	
Adiponectin, leptin	Increased insulin sensitivity, anti-inflammatory activity
Resistin, visfatin, omentin	Inflammation, atherosclerosis
Inflammatory cytokines and chemokines	
IL1β/6, TNFSF1, IL6/7 soluble receptor, serpin-E1	Atherosclerosis
CCL2	Coronary artery disease
Miscellaneous	
Adrenomedullin	Vasodilation, anti-inflammatory property
FABP4	Negative inotropy
Phospholipase-A2	Atherosclerosis

FABP fatty acid-binding protein, *MMP* matrix metallopeptidase, *TNF* tumor-necrosis factor, *VEGF* vascular endothelial growth factor. Serpin-E1 is also called plasminogen activator inhibitor PAI1

Fibroblasts traditionally recognized as quiescent cells that produce the extracellular matrix interfere with immunocytes. They are stimulated by cells of the innate and adaptive immunity and conversely modulate immunocyte behavior, as they interfere with the cytokine environment [70]. Cardiofibroblasts also orchestrate infiltration of inflammatory leukocytes in the heart.

As in the arterial wall, the healthy myocardium contains a high number of macrophages. The cardiomacrophage population is the fourth cardiac resident cell set after fibroblasts, cardiomyocytes, and endotheliocytes. Macrophages interact with cardiomyocytes, endotheliocytes, and fibroblasts.

The Wnt morphogens may modulate inflammatory responses. Stimulation of TLRs in macrophages induces Wnt5a production which, in turn, upregulates the expression of proinflammatory cytokines, such as interleukins IL1β and IL6 and chemokine CXCL8 (or IL8) [35]. The Wnt5a–Fz–CamK pathway is involved in macrophage activation that is inhibited by sFRP1 and sFRP5 molecules.

1.5.6.1 Cardiomyocyte–Leukocyte Interactions

Coordinated interactions between cardiomyocytes and immunocytes launch three successive events during myocardial injury and subsequent inflammation [69]:

1. Production of inflammatory mediators (mainly cytokines and chemokines) by stressed and damaged myocardial cells;
2. Transmigration of inflammatory cells to the damaged tissue leading to secondary cytokine amplification, wound healing, and tissue remodeling; and
3. Release of anti-inflammatory signals (e.g., GDF15) to restrain leukocyte invasion and terminate inflammation and wound healing.

Many signaling pathways are sequentially activated in cardiomyocytes and leukocytes to ensure proper spatiotemporal orchestration of postinjury myocardial remodeling.

Stressed and damaged cardiomyocytes further contribute to the establishment of a proinflammatory environment in the myocardium, as they produce different types of cytokines and chemokines. Hypoxia, abnormal mechanical stresses, and infection stimulate cardiomyocytes that then mobilize inflammatory mediators and signaling pathways that are silent or minimally active in the healthy adult myocardium.

Abnormal mechanical stimuli are sensed by mechanosensors, such as integrins, sarcolemmal and cytoskeletal proteins. Foreign and intra- or extracellular alarmins, also known as danger signals and damage-associated molecular pattern molecules (DAMP), are recognized by pattern recognition receptors such as TLRs localized within the signaling cell or neighboring cardiomyocytes.

Activated receptor triggers signal transduction cascades that involve the MAPK modules, the JaK–STAT axis, and PP3-dependent pathways. These cascades stimulate nuclear transcription factors, mainly NFκ B and AP1, which are required for the transcription of most cytokine and chemokine genes (Tnfsf1, IL1B, and Il6, as well as CCL2) [69]. Cytokines and chemokines can upregulate synthesis of adhesion molecules on endotheliocytes (ICAM1 and VCAM1), thereby assisting immune leukocyte extravasation. Chemokines also support leukocyte directional migration by binding to leukocyte plasmalemmal receptors (e.g., CXCR4). Phagocytes remove apoptotic, necrotic, and infected cardiomyocytes, and clear cellular debris.

1.5.6.2 Fibroblast–Leukocyte Interactions

Fibroblasts constitute a heterogeneous population of stromal cells characterized by a spindle shape and an oval nucleus. Among cardiac cells, fibroblasts constitute the main source of inflammatory signals in ischemic hearts, whereas endotheliocytes primarily release inflammatogenic molecules in response to pressure overload [69].

Fibroblasts Diversity Fibroblast subtypes differ according to the proliferation capability, collagen and matrix metallopeptidase production, contractility, and immunomodulatory function. These various fibroblast phenotypes are related to the tissue type.

In addition, the diversity of fibroblasts within a given tissue can be explained by distinct origins. Fibroblasts mainly originate from [70]: (1) mesenchymal stromal cells; (2) fibrocytes that are defined as circulating monocyte-derived cells; (3) epitheliocytes after epithelial–mesenchymal transition (as in the kidney and liver); (4) endotheliocytes after endothelial–mesenchymal transition (as in the lung and heart); and (5) circulating cells.

Fibroblast Function Fibroblasts are aimed at synthesizing and remodeling the extracellular matrix. The inflammation resolution requires the elimination of the major part of immunocytes that were recruited and proliferated during the acute phase of inflammation by emigration and death. A persistent activation of fibroblasts and myofibroblasts causes chronic inflammation.

Fibrosis is a consequence of scarring and complication of inflammation. Fibrosis is characterized by excess deposition of collageneous and noncollagenous matrix due to the accumulation, proliferation, and sustained activation of fibroblasts and myofibroblasts. Fibrosis disrupts tissue structure and function.

Cardiofibroblasts (60–70 % cardiac cells) communicate with cardiomyocytes (30–40 % cardiac cells), thereby assisting electrical coupling between cardiomyocytes and contraction coordination, as well as allowing mechanical force distribution throughout the myocardium and contributing to angiogenesis.

Fibroblast Activation Cells of the innate and adaptive immunity activate fibroblasts, as they release growth factors, cytokines, and enzymes that target fibroblasts and myofibroblasts (Table 1.21). Fibroblasts secrete and respond to growth factors, cytokines, and chemokines.

Fibroblasts possess various pattern recognition receptors such as TLRs. Once these receptors are liganded, fibroblasts are activated and can differentiate into collagen-producing myofibroblasts.

The TGFβ factor is a profibrotic agent. It has several isoforms (TGFβ1–TGFβ3) that are synthesized as precursors bound to latent TGFβ-binding proteins (LTBP1 and LTBP3–LTBP4). The latter are cleaved extracellularly, thereby releasing active TGFβ. The TGFβ protein induces the expression of profibrotic genes, such as those that encode collagen-1 and connective tissue growth factor (CTGF) via the SMAD3 factor. It also promotes collagen synthesis via the MAPK modules (P38MAPK, JNK, and ERK). The concentrations of TGFβ1 and TGFβ3 rise in cardiac fibrosis.

Fibroblast–Immunocyte Crosstalk Fibroblasts assist the recruitment of immunocytes and regulate their behavior, retention, and survival in damaged tissues. Crosstalk between fibroblasts and leukocytes depends on the interaction between TNFRSF5 on fibroblasts and TNFSF5 on immunocytes [70]. This interaction upregulates ICAM1 and VCAM1 production in fibroblasts.

Fibroblasts support activation of the endothelium and enable the homing of circulating leukocytes in response to tissue injury. They produce constitutive and cytokine-induced chemokines (e.g., CCL2 to CCL5, CXCL2, and CXCL10) and express chemokine receptors [70].

Table 1.21 Activation of fibroblasts and myofibroblasts by cells of the innate and adaptive immunity (Source: [70])

Cell type	Messengers
B lymphocyte	TGFβ, IL6, TNFSF1/13b
Eosinophil	TGFβ, MBP Lysosomal hydrolases, peroxidases
Macrophage	Arginase, NOS2 MMP1/7–9/12, TIMP IL4/10/13, TGFβ, TNFSF1, IL1
Mastocyte	TGFβ, IL4 Tryptase, chymase
Neutrophil	ROS MMP, cathepsins
Platelet	Coagulation factors (FVII, FIX, FX) PDGF, TGFβ
T_{H1}	Ifnα/γ
T_{H17}	IL17/22, CXCL8
T_{H2}	IL1/4/5/13, TNFSF1
T_{Reg}	TGFβ, IL10

Ifn interferon, *IL* interleukin, *MBP* major basic protein, *MMP* matrix metalloproteinase, *NOS* nitric oxide synthase, *PDGF* platelet-derived growth factor, *ROS* reactive oxygen species, *TGF* transforming growth factor, *TIMP* tissue inhibitor of matrix metalloproteinase, *TNF* tumor-necrosis factor

The CCL2 chemokine upregulates the expression of collagen and TGFβ in fibroblasts (auto- and juxtacrine stimulation) [70]. The CCL2–CCR2 axis enhances vascular cell adhesion molecule VCAM1 expression in human fibroblasts that promotes monocyte adhesion to fibroblasts (reciprocal enhancement of monocyte–fibroblast adhesion and chemokine production).

Activated fibroblasts increase the CXCR4 level on T cells via TGFβ and express its ligand CXCL12, thereby supporting the local retention of infiltrated leukocytes [70].

1.5.6.3 Platelets

Platelets also express TLRs that explain their contribution as immunocytes during inflammation and infection. Platelets promote systemic and cardiac inflammatory responses as well as ventricular remodeling [70]. Activated platelets release several growth factors that support healing, such as chemotactic PDGF and TGFβ that stimulate matrix deposition. Coagulation factor-X has a profibrotic effect; a deficiency of clotting factor VII can cause cardiac fibrosis.

1.5.6.4 Eosinophils

Eosinophils secrete the profibrotic agents TGFβ and major basic protein MBP1, as well as lysosomal hydrolases and eosinophilic peroxidase that are implicated in tissue remodeling. Eosinophilic peroxidase, TGFβ, and MBP provoke epithelial–mesenchymal transition, further contributing to myofibroblast generation [70].

1.5.6.5 Neutrophils

Neutrophils are the first cells attracted to the injured site followed by monocytes and finally lymphocytes and mastocytes. They initiate an acute inflammatory response to engulf dead cells and tissular debris to facilitate tissue repair.

Neutrophils release large amounts of ROS (respiratory burst) via NADPH oxidase. They also secrete proinflammatory cytokines and enzymes, such as MMPs, elastase, and cathepsins.

1.5.6.6 Mastocytes

Mastocytes synthesize peptidases (e.g., tryptase and chymase), growth factors (e.g., TGFβ), cytokines, and vasoactive agents. Many released molecule types activate fibroblasts.

1.5.6.7 B Lymphocytes

B lymphocytes release the profibrotic cytokine IL6 [70]. In addition, TNFSF13b (or B-cell-activating factor) is inducer of collagen, TIMP1, MMP9, α SMA expression in human fibroblasts, as well as of that of proinflammatory and profibrotic cytokines TGFβ and IL6 and CCL2 chemokine [70].

1.5.6.8 Helper T Lymphocytes

T_{H2} cells secrete interleukins IL4, IL5, and IL13 involved in wound healing and fibrosis. In addition, IL4 and IL13 cooperate to elicit the phenotypic transition of human fibroblasts to myofibroblasts using JNK kinase [70]. Moreover, IL13 hampers MMP synthesis by fibroblast and hence matrix degradation, causing an excessive collagen deposition. It influences T_{H17}-mediated inflammation and T_{H2}-driven fibrosis, as well as the profibrotic activity of myofibroblasts via IL13Rα1 and IL13Rα2 receptors [70].

T_{H17} lymphocytes produce IL17a and IL22 that have both profibrotic and antifibrotic activities in the heart [70]. In addition, these cells attract neutrophils. Furthermore, IL17 promotes MMP1 expression in cardiofibroblasts via NFκ B, AP1, and C/EBPβ and engenders cardiac fibrosis via the PKCβ–ERK1/2–NFκ B pathway [70]. On the other hand, IL17a decreases the expression of CTGF and collagen-1 in fibroblasts of healthy patients.

1.5.6.9 Regulatory T Lymphocytes

As does T_{H17} cells, regulatory T lymphocytes can suppress or promote fibrosis. Interleukin-10 precludes collagen synthesis by cardiofibroblasts via the reduction of STAT3 activity and inhibition of the NFκ B pathway [70].

1.5.6.10 Macrophages

Macrophages are large extravascular immunocytes with diverse phenotypes with distinct regulatory and effector functions. They are actually important modulators and effectors of immunity. They are the main source of several MMP types (MMP1, MMP7–MMP9, and MMP12) as well as TIMP suppressors. They produce the profibrotic TGFβ agent.

Cardiac macrophages have a cardioprotective function especially in pathological conditions. They carry out numerous tasks in wound healing, regeneration, and tissue remodeling. Resident macrophages may have distinct functions than those of monocyte-derived macrophages.

As a component of the innate immunity, classical, phagocytic, and antigen-presenting macrophages provide a first line of defense in injury and infection. Classical cytotoxic macrophages deliver oxidative bursts and degrade using secreting peptidases.

Macrophages secrete cytokines involved in wound healing and tissue remodeling and orchestrate the immune response.

Many stimuli combine to determine the phenotype of macrophages. The development of acquired immunity relies on reciprocal interactions between macrophages and activated T and B lymphocytes. Cytokines and other stimuli are categorized into M1 and M2 according to their roles in the development, maturation, and activation of macrophages. The concept of classical[19] and alternative[20] activation termed M1 and M2, respectively (see below), mimics T_H-cell nomenclature, but it is not entirely appropriate. However, signaling, genetic, and functional signatures acquired during maturation and activation are usually matched to the M1/M2 model

---

[19] Macrophage classical activation refers to, in an infection context, the antigen-dependent, nonspecifically enhanced microbicidal activity of macrophages toward bacteria upon secondary exposure to pathogens. This type of activation relies on interferon-γ.

[20] Macrophage alternative activation refers to the observation that the mannose receptor is selectively excited by the T_{H2}-type cytokines IL4 and IL13 in murine macrophages. They trigger a strong endocytic clearance of mannosylated ligands, increase expression of major histocompatibility complex (MHC) class-2 molecules, and decrease proinflammatory cytokine secretion.

of macrophage polarization. Macrophages activated with T_{H1} and T_{H2} cytokines differ in the metabolic pathway: M1 macrophages produce nitric oxide redical, whereas M2 macrophages synthesize trophic polyamines [75].

M1 macrophages that originate from macrophages stimulated by TLR ligands and interferon-γ influence tissue remodeling and contribute to inflammation resolution [33].

Cardiac wall-resident, M2 macrophages that originate from macrophages stimulated by IL4 and IL13 are cardioprotectors [33]. They inhibit CD4+ T cells and release anti-inflammatory cytokines. M2 macrophages release arginase-1 that controls production of proline used in collagen synthesis by activated myofibroblasts.

Macrophages are endowed with various plasmalemmal receptors for lineage-determining growth factors and helper T-cell cytokines, as well as B-cell, host, and microbial products. Once they are mature and activated by a combination of these stimuli, they acquire a specialized functional phenotype.

Cardiac inflammation and fibrosis are exacerbated by TLR4+, Casp1+, IL1β+ M2 macrophages and attenuated by T-cell (or transmembrane) immunoglobulin and mucin domain-containing molecule TIM3+ M2 macrophages [70].[21]

Sources of Macrophages Different sources of macrophages exist according to the cellular function and phenotype. In normal conditions, the majority of macrophages, except intestinal macrophages, derive from local progenitors that arise from the embryonic yolk sac [73].

Macrophages produced by local progenitors reside in their destination tissues prior to birth (e.g., microgliocytes in the brain, pulmonary macrophages, Langerhans cells in the skin, and Kupffer cells in the liver). However, intestinal macrophages derive from circulating monocytes.

During inflammation, most macrophages derive from inflammatory monocytes ($Ly6C^{high}$ in mice; CD14+, CD16− in humans) that are recruited to the site of inflammation from blood. These monocytes originate from hematopoietic stem cells and progenitors in the bone marrow.

In the hematopoietic niche, regulatory and supplying cells that include mesenchymal stem cells, endotheliocytes, macrophages, neurons, and osteoblasts, control the

[21] At least, two distinct populations of M2 macrophages (TLR4+ and TIM3+) regulate inflammation and fibrosis in the heart [71]. They crossregulate each others expression on CD11b+ (α_M-integrin+) cells. TLR4 and TIM3 are expressed on GR1+ (lymphocyte antigen-6 complex locus G [Ly6G]), EMR1+, IL4R+, and M2 macrophages from the heart during acute myocarditis. M2 macrophages that express significantly more TIM-3 and less IL1β reduce inflammation. Proteins of TIM family are expressed by multiple immunocyte types. Type-1 transmembrane (or T-cell) immunoglobulin and mucin domain-containing proteins (TIM1–TIM4) in mice, the genome of which contains eight predicted Tim genes, among which only four encode functional proteins, but TIM2 is absent in the human TIM family, as the human genome only contains three Tim genes (Tim1 and Tim3–Tim4) [72]. The TIM1 molecule has stimulatory and costimulatory effects on T cells. On the other hand, TIM3 inhibits T_{H1}-cell response, but activates T-cell responses. It is constitutively expressed by dendritic and microglial cells and, once it is liganded, it increases production of costimulatory receptors and cytokines.

Table 1.22 Selected M1 and M2 functions in macrophages, M1 and M2 macrophage polarization and effects being here limited to Ifnγ and IL4 (Source: [75])

	M1	M2
Activators	Ifnγ	IL4/IL13
Markers	Fcγ R1, SOCS1, CXCL10	MRc1, TGm2, Fcϵ R1, CCL22
Phagocytosis	↑ Phagocytosis of *Candida albicans*	↓ Phagocytosis of particles
	↓ Fc-mediated phagocytosis	↑ Inflammatory cytokine production
	↓ Complement-mediated phagocytosis	
Autophagy	Autophagy in tuberculosis	↓ Autophagy in tuberculosis
Fusion	Increases fusion	Induces fusion
	Fusion induction in alveolar macrophages	⊖⟶ Ifnγ-induced fusion
Nitric oxide	Mycobacteria killing via NO	Favors arginase-1 wrt. NOS2 (Arg1+ macrophages suppress T_{H2}), inflammation and fibrosis

↑ increase, ↓ decrease, ⊖⟶ inhibition Fcγ(ϵ)R Fc receptor of IgG(E), *Ifn* interferon, *IL* interleukin, *MR* mannose receptor, *NOS* nitric oxide synthase, *SOCS* suppressor of cytokine signaling, *TGm* transglutaminase

blood cell production in the bone marrow by delivering messengers (e.g., SCF, CSF3, CXCL12, and angiopoietin-1) to hematopoietic stem cells. In addition, hematopoietic stem cells possess receptors (e.g., Toll-like and interferon receptors) to sense circulating danger signals [73].

The spleen may also contribute to the blood monocyte pool. When the splenic reservoir of monocytes empties, the spleen produces new monocytes, as it can host extramedullary hematopoiesis [73]. β3-adrenoreceptors on cells in the bone marrow niche stimulate hematopoietic cells and liberate hematopoietic progenitor cells that then migrate to the spleen. Monocytes are released from the spleen independently of CCR2, but caused by angiotensin-2.

The source of macrophages in the healthy heart remains to be determined, but most likely derive from local progenitors. In addition, patrolling monocytes travel in small coronary arterioles [73].

M1 and M2 Macrophages In mice, cardiac macrophages are classified into two main categories (Table 1.22) [74]:

1. $Ly6C^{high}$, CD204− (macrophage scavenger receptor MSR− or ScaRa1−), CD206− (mannose receptor MRc1−), classically activated, M1 macrophages; and
2. $Ly6C^{low}$, CD206+, CD204+, less inflammatory, alternatively activated M2 macrophages.

Imbalances between these two types of macrophages induce adverse cardiac remodeling.

Macrophage activation is primed by combinations of stimuli that commonly comprise interferon-γ and/or lipopolysaccharide, tumor-necrosis factor-α (TNFSF1), immune complexes, colony-stimulating factor CSF2, and Myd88 (*M1 stimuli*), as well as CSF1 and interleukin-4 (*M2 stimuli*), among many other possible combinations. M2 subtypes are also defined, **M2a** being associated with CSF1 and IL4 stimuli, **M2b** with Fc receptors and immune complexes, and **M2c** with glucocorticoids.

Stimuli governing macrophage activation are organized according to their role in immunity. Four levels can be identified [75]: (1) *maturation* (e.g., CSF1 and CSF2) as well as *survival* and *recruitment* (e.g., chemokines and adhesion molecules) factors, in addition to vitamin-D3, retinoic acid, and PPARγ ligands; (2) *interaction with lymphoid and myeloid cytokines*, among others; (3) *interaction with pathogens* directly (via TLRs, NODs, NLRs, RLRs, and nucleic acid sensors) and or via humoral recognition receptors, such as complement, lectins, ficolins, and B-cell-derived immunoglobulins (IgA/E/G); and (4) *resolution* with systemic (e.g., glucocorticoids) and local (e.g., ATP, resolvins, maresins, matrix proteoglycans, and other mediators with general anti-inflammatory properties) factors.

However, the M1/M2 classification has some limitations. In the healthy heart, macrophages that interfere with cardiomyocytes and endotheliocytes are weakly inflammatory. They guard against infection and contribute to angiogenesis regulation and matrix turnover. They sparsely express the surface marker Ly6C and, at high levels, proteins associated with M2 phenotype, as well as some inflammatory agents (e.g., interleukin-1β) [73].

Cardiac Resident Macrophages Cardiac resident M1 macrophages are aimed at removing debris of damaged cardiomyocytes and launching proinflammatory effects after heart injury, neovascularization and debris clearance being mediated via VEGFa and TGFβ, respectively [74]. The M1 macrophage activates CD4+ T cells via IL12. Conversely, it is activated by CD4+ T cells. Interaction between activated CD4+ T cells and M1 macrophages induces a proinflammatory phenotype in the macrophage. The M1-macrophage polarity is associated with proinflammatory cytokines (IL1β, IL6, and TNFSF1).

The tasks of cardiac resident M2 macrophages are limiting excessive inflammation, priming adaptive response, and ensuring tissue maintenance [74]. Its differentiation relies on STAT3 factor and its activation on T_{H2}-type cytokines. Interleukin-13, rather than IL4, induces the alternative activation of resident and newly recruited monocyte-derived macrophages [74]. The M2 macrophage activity is inhibited by CD4+ T cells. Conversely, it activates CD4+ T cells, as it secretes IL12, which stimulates interferon-γ-producing T cells. Interaction between M2 macrophages and activated CD4+ T cells enables M2 macrophage differentiation and proliferation during cardiac stresses. The M2 macrophage exerts anti-inflammatory effect against excessive fibrosis. Once it is activated, M2 macrophage impedes the procardiac injury and profibrotic functions (by IL1β, IL18, and Ifnγ) of CD4+ T cells and granulocytes.

Serum- and glucocorticoid-inducible kinase SGK1 supports proliferation and activation of cardiac M2 macrophages via STAT3 and IL13 (not IL4) stimulation, thereby mediating at least partly cardiac fibrosis and hypertrophy caused by angiotensin-2.

Monocyte-Derived Macrophages Two distinct monocyte-derived cardiomacrophage types contribute to healing or injury. In postnatal heart, the myocardium remodeling is assisted by numerous macrophages, as in acute heart injury such as myocardial infarction. In the early stage of cardiac injury, $Ly6C^{high}$ monocytes are recruited from the bone marrow and spleen to injured myocardium using the CCR2 chemokine receptor [74]. They differentiate into monocyte-derived cardiac M1 macrophages. The latter produce proinflammatory cytokines (IL1β, IL6, IL12, and TNFSF1) and chemokines (CCL2 and CCL5). It also exhibits phagocytic and proteolytic activities.

$Ly6C^{high}$ monocytes can differentiate into both M1 and M2 macrophages according to whether the mineralocorticoid receptor or the ScaRa scavenger receptor operates, respectively. Macrophage class-A scavenger receptor (ScaRa), a modulator of inflammation, actually shifts the cardiomacrophage polarity toward the M2 phenotype. On the other hand, $Ly6C^{low}$ monocytes recruited during inflammation only evolve to M2 macrophages.

The mineralocorticoid receptor (subclass-3 aldosterone nuclear receptor NR3c2), a transcription factor, in cardiac macrophages regulates macrophage polarity [74]. It indeed assists M1-macrophage polarity, thereby elevating the M1/M2 activated macrophage ratio and supporting production of M1-type proinflammatory cytokines and chemokines.

In the later stage of cardiac injury, $Ly6C^{low}$ monocytes are recruited to the myocardium using the CX_3CR1 chemokine receptor [74]. They differentiate into monocyte-derived cardiac M2 macrophages. The latter express the TGFβ and IL12 cytokines. The scavenger receptor ScaRa on M2 macrophages enables the maintenance of the M2 phenotype. This macrophage type contributes to anti-inflammatory response, angiogenesis, and myofibroblast activation during the healing process.

Refined Classification Markers of tissue-resident macrophages include CD45 (PTPRc), EGF-like module-containing, mucin-like, hormone receptor-like sequence protein EMR1 (in humans; F4/80 in mice), major histocompatibility complex (MHC) class-II molecule, macrophage surface antigens (Mac1–Mac3), that is, $\alpha_M\beta_2$-integrin (CD11b), galactoside-binding, soluble lectin LGalS3, and lysosomal-associated membrane protein LAMP2, respectively, CD11c (α_X-integrin), CD68 (ScaRd1), and CD115 (CSF1R).[22]

The classification of cardiac macrophages can be refined using class-II major histocompatibility complex molecules (MHC2), CD11c, and CCR2 (Table 1.23).

[22] Macrophages and fibroblasts frequently appear together in fibrosis following inflammation. Markers are then needed to identify monocytes or macrophages. S100 calcium-binding protein S100a4, or fibroblast-specific protein FSP1, is a specific marker of fibroblasts as well as in epithelia undergoing epithelial–mesenchymal transition to form fibroblasts and in metastatic tumor cells [76]. EMR1+ fibroblasts in fibrotic tissue also express vimentin and HSP47, or occasionally α-smooth muscle actin, but these proteins are not specific for fibroblasts. About 12 % of fibroblasts arise from the blood stream after being released by the bone marrow4. FSP1+ fibroblasts from the bone marrow and blood are Mac1+ and $Gr1^{low/-}$.

Table 1.23 Refined classification of cardiac macrophages at least in mice (Source: [74]). Cardiac macrophages abound in the healthy heart as well as after myocardial infarction. Resident macrophages promote myocardial health. Proinflammatory macrophages of infarcted zones can derive from circulating monocytes; they are implicated in tissue remodeling and resolution of inflammation during postmyocardial infarction healing. A simplified categorization defines proinflammatory M1 macrophages working in the early healing phase and anti-inflammatory and profibrotic M2 macrophages involved in the late phase. The proinflammatory M1 macrophage can become an anti-inflammatory M2 macrophage when phagocytosis of apoptotic cells occurs

Type	Response to myocardial infarction and angiotensin-2 administration
	Role
Ly6C^{high}	
CCR2−	↑
CCR2+	↑, inflammation
Ly6C^{low}	
CCR2−, CD11c^{low}	
MHC2^{high}	↓, immunosurveillance (antigen processing and presentation)
MHC2^{low}	↑ ↑, phagocytosis
CCR2+, CD11c^{high}	
MHC2^{high}	↑, immunosurveillance, inflammation, IL1β production

↑ increase, ↓ decrease, *CCR2* chemokine receptor, *CD11c* α_X-integrin, *MHC2* class-II major histocompatibility complex molecule

1.5.7 Interactions Involving Cardiac Progenitors

Cardiofibroblasts intervene in the adaptation and/or repair during and after cardiac aging, stress, and injury that involve the transdifferentiation of fibroblasts into myofibroblasts in the infarct border zone or stressed myocardium.

Myofibroblasts produce contractile proteins (e.g., α-smooth muscle actin) and secrete profibrotic and anti-inflammatory agents. In addition, myofibroblasts influence the function of transplanted stem cells aimed at minimizing adverse myocardial scarring. In addition to myofibroblast–stem cell coupling, matrix rheology is an important factor [77].

During cardiogenesis, Wnt3, Wnt3a, and Wnt8a can stimulate cardiomyogenic differentiation of embryonic stem cells. A decreased β-catenin activity in cardiomyocytes improves cardiac repair after myocardial infarction [35]. Cardiac stem cells, or cardiac progenitor cells, can counteract myocardial scarring and fibrosis, thereby limiting evolution of myocardial scarring to heart failure and cardiac arrhythmias [77].

Cardiac progenitor cells from the myocardium and epicardium release matrix metallopeptidases (MMP2, MMP9, and MMP14) that degrade collagen fibers, the main constituent of myocardial scars. Simultaneously, the amount of tissue inhibitor of matrix metallopeptidase TIMP4 decays. Therefore, cardiac progenitors can invade the scarred area and form new cardiomyocytes and vasculature.

Transplanted noncardiac (exogenous) stem cell derived from the bone marrow (endothelial progenitor cells and hematopoietic and mesenchymal stem cells) can differentiate into endotheliocytes, fibroblasts, pericytes, myofibroblasts, smooth myocytes, and possibly cardiomyocytes. In addition, bone marrow-derived mesenchymal stem cells secrete numerous cytokines and chemokines. However, bone marrow-derived stem cells have only a modest inhibitory effect on cardiac fibrosis [77].

Nevertheless, several factors may have antifibrotic effects, like FGF2, HGF, IGF1, adrenomedullin, and the TGFβ-neutralizing proteoglycan biglycan [77]. The IGF1 factor reduces the production of proapoptotic miR34a. Bone marrow-derived cells transplanted near the infarcted myocardium attenuate the expression of the profibrotic miR21. The HGF factor suppresses profibrotic signaling from miR155.

1.6 Short Peptides and Cardiac Development and Function

Short peptides (< 100 AAs) are important regulators of bodily development and physiology.[23] Small peptides include neuropeptides, peptide hormones (e.g., insulin), secreted messengers (e.g., FGF growth factor), and intracellular signaling regulators such as those associated with ion carriers.

The majority of these small peptides are encoded as large preproteins that undergo posttranslational cleavage and modification. Others are encoded by small open reading frames (smORF) that are short DNA sequences.

1.6.1 Apela

The Apela gene (apelin receptor early endogenous ligand; also known as Toddler and Elabela) encodes a 32-amino acid hormone in human embryonic stem cells, the earliest ligand for the G-protein-coupled apelin receptor during cardiogenesis [78].[24]

This regulator is ubiquitous in naive ectodermal cells of the embryo. It supports cell movement during gastrulation as a nondirectional signal to assist the internalization and motion of ventrolateral mesendodermal cells [78]. In zebrafish embryos, loss of apela causes the development of a rudimentary heart or the absence of heart.

The Apela gene encodes a short secreted peptide apela that acts as a local activator of internalization and signaling of the apelin receptor to promote cell motility required for cardiogenesis [79].

[23] Small transmembrane proteins produced by viral genomes (often < 50 amino acids) intervene in virus replication and virulence.

[24] The apelin receptor participates in the regulation of cardiovascular development and physiology and control of fluid homeostasis, as well as acts as a coreceptor for HIV infection. Nodal activates the expression of the apelin receptor.

1.6.2 Small Open Reading Frames and Short Peptides

Thousands of small open reading frames are able to encode small peptides of less than 100 amino acids. Some short peptides (< 30 amino acids) regulate calcium transport, hence influencing regular myocardial contraction being implicated in cardiac pathologies [80].

1.6.2.1 Short Peptides and Calcium Handling

Whereas RyR1 and Ryr2 liberate Ca^{2+} ions from the sarcoplasmic reticulum of skeletal and cardiac myocytes, respectively, thereby engendering myocyte contraction, SERCA1a and SERCA2a pumps translocate calcium ions from the cytosol into the lumen of the sarcoplasmic reticulum of skeletal and cardiac myocytes, respectively, thereby initiating muscle relaxation. The SERCA pump is a single polypeptide corresponding to the α subunit of the dimeric Na^+–K^+ and H^+–K^+ ATPases. Two Ca^{2+} ions are transported for each hydrolyzed ATP molecule.

Two single-pass membrane SERCA inhibitors, phospholamban (Pln), a 52-residue integral membrane protein, which lowers the apparent calcium affinity of the ATPase, and sarcolipin (Sln), a 31-amino acid integral membrane proteolipid with a variable expression level, which uncouples ATP hydrolysis from accumulated Ca^{2+} ions, slightly reducing Ca^{2+} affinity of SERCA, bind the Ca^{2+} ATPases SERCA1a and SERCA2a. The Sln–SERCA1a complex closely resembles the Pln–SERCA1a complex.

SERCA inhibition is relieved by phosphorylation of phospholamban by protein kinase-A (Ser16) and/or Ca^{2+}–calmodulin-dependent protein kinase (Thr17) with different physiological effects. Sarcolipin can be phosphorylated (Thr5) by the STK16 kinase [81].

1.6.3 Pseudo-Noncoding RNAs

Conventional protein-encoding genes endowed with a single, long, conserved encoding sequence account only for a fraction of RNA transcribed from the genome generating a protein (> 100 amino acids). Noncoding RNAs (ncRNA), that is, all RNAs except mRNAs, have structural roles, such as ribosomal RNAs (rRNA), transfer RNAs (tRNA), small nucleolar RNAs (snoRNA), among others. They also include small nuclear RNAs (snRNAs), microRNAs (miR), and long noncoding RNAs (lncRNAs). Noncoding RNAs are characterized by their RNA sequences and structure. Eukaryotic ncRNAs can be categorized into small (sncRNA; 20–30 nucleotides; e.g., miRs), intermediate (incRNA; 30–200 nucleotides; e.g., snRNAs), and long noncoding RNAs (lncRNA; > 200 nucleotides) [83]. Noncoding RNAs undergo chemical modifications. Both their structure and chemical modifications determine their function. Noncoding RNAs can assemble with proteins and operate as ribonucleoproteins (RNP). Many ncRNAs are transcribed by RNA polymerase-2, but some transcripts are processed into miRs to exert posttranscriptional gene silencing. Other transcripts are retained in the nucleus to form specific foci and/or mediate epigenetic regulation.

However, certain types of small functional peptides are directly translated from long intergenic noncoding RNAs (lincRNA). Long intergenic noncoding RNAs operate in [84]:

1. Chromatin modification (e.g., chromosome X inactivation);
2. Transcription regulation with enhancer-like effect;
3. MicroRNA genesis as microRNA precursors;
4. Control of posttranscriptional gene expression by sequestering microRNA repressors; and
5. Production of small functional polypeptides (10–100 AAs).

Multiple open reading frames generate protein-encoding transcripts, among which some are lincRNAs. In humans, many widespread transcripts that lack the classical coding sequence of eukaryotic protein-encoding genes, originated from micropeptide-encoding genes such as lincRNAs, contain short open reading frames (ORFs or smORFs; both in the 5′ and 3′ UTR) and encode small polypeptides [85]. In cardiomyocytes, the SERCA2a pump replenishes the calcium store by resequestering Ca^{2+} ions into the sarcoplasmic reticulum from the sarcoplasm, enabling sarcomere relaxation and ventricular filling with blood (positive lusitropic effect). The SERCA activity is modulated by several soluble substances (e.g., histidine-rich calcium-binding protein [HRC], calreticulin [CalR], and S100a) and governed by two small transmembrane proteins, phospholamban and sarcolipin [86]. Both sarcolipin and phospholamban bind to the SERCA2a ATPase, thereby repressing its activity. Phospholamban inhibits SERCA, but its phosphorylation relieves its inhibition. Sarcolipin precludes SERCA activity by uncoupling ATP hydrolysis by SERCA from calcium transport. Phospholamban is encoded by a small open reading frame within a single exon of a large spliced transcript that engenders a 52-amino acid protein [86]. Sarcolipin is also encoded by a single exon of a spliced transcript that generates a 31-amino acid protein.

The lincRNA sarcolamban expressed in myocytes contains 2 functional smORFs that encode small 28 and 29 amino acids peptides related to sarcolipin and phospholamban [80]. Similarly to sarcolipin and phospholamban, the small peptides sarcolamban-A and -B are involved in the regulation of Ca^{2+} uptake in the sarcolemmal compartment of the ER. Nevertheless, sarcolambans cannot compensate for the loss of phospholamban and sarcolipin. On the other hand, overexpression of either sarcolamban, phospholamban, or sarcolipin induces cardiac arrhythmia and dampens calcium transient amplitude and decay rate. Sarcolamban may thus cooperate with phospholamban and sarcolipin.

1.7 Pregnancy and Cardiac Function

During pregnancy, the heart can be subjected to diseases such as peripartum cardiomyopathy (Sect. 4.17) in addition to preexisting cardiac pathologies, such as valvular failure and cardiomyopathy.

Developmental insults (i.e., maternal undernutrition) or stressors (i.e., exposure to stress-related hormones such as cortisol) during pregnancy raise the risk of developing metabolic (i.e., obesity, poor glucose tolerance, diabetes, and dislipidemia), neurological, reproductive, and cardiovascular (e.g., coronary artery disease) disorders in subsequent years of life [87].

Maternal undernutrition during pregnancy leads to endothelial dysfunction with impaired endothelium-dependent relaxation and increased sensitivity to vasoconstrictors in the offspring. In humans, low birth weight is associated with impaired endothelium-dependent relaxation in infants, children, and young adults [87]. In sheep offsprings, maternal nutrient restriction provokes impaired endothelium-dependent and -independent relaxation of femoral arteries. In rat, maternal undernutrition alters not only vasodilation of offspring mesenteric arteries and aortas, but also vasoconstriction in offspring femoral and carotid arteries [87].

In sheep, vasodilation of fetal coronary arteries is altered by maternal nutrient restriction during the last two-thirds of pregnancy. Maternal undernutrition during mid- to late-gestation impairs relaxation in fetal coronary arteries in response to the endothelium-dependent vasodilator bradykinin [87]. Bradykinin launches vasodilation via multiple endothelium-dependent pathways. In coronary arteries of fully developed animals, bradykinin causes the release of nitric oxide (NO) and endothelium-derived hyperpolarizing factor (EDHF). The latter is linked to the release of endothelium-derived epoxyeicosatrienoic acid (e.g., [14,15]EET) that activates the large-conductance, Ca^{2+}-activated channel ($BK_{V,Ca}$) in coronary vascular smooth myocytes. Maternal nutrient restriction does not affect the response to (14,15)EET. In addition, potassium flux through $BK_{V,Ca}$ channels is similar in coronary smooth myocytes from fetuses of control and undernourished ewes. The bradykinin-induced relaxation is resistant to inhibitors of NOS3 and cyclooxygenase in fetal coronary arteries from control animals, the response depending on the endothelium-derived hyperpolarization of vascular smooth myocytes rather than NO and prostacyclin. On the other hand, bradykinin-induced relaxation of fetal coronary arteries from nutrient-restricted animals is abolished completely by inhibition of NOS3, and hence depends only on nitric oxide [87].

1.7.1 Pregnancy Preparation—Menstrual Cycle and Hormonal Control

The menstrual cycle can be divided into three phases controlled by the endocrine system. The *ovarian cycle* consists of the follicular phase, ovulation, and luteal phase. The *uterine cycle* comprises menstruation and proliferative and luteal secretory phase.

1.7.1.1 Ovarian Follicle

The *ovarian follicle* is the basic female reproduction unit. This cellular aggregate contains a single immature *oocyte*. The latter is surrounded by avascular *granulosa*. A cavity, the antrum, is hollowed inside this cellular agglomerate during the growth of the ovarian follicle. Granulosa cells are enclosed by the follicular basement membrane and the *theca* interna and externa.

The granulosa cell density increases in response to heightened levels of circulating gonadotropins. Granulosa cells produce peptides involved in ovarian hormone synthesis regulation.

At the midpoint of the menstrual cycle, ovulation liberates the dominant mature oocyte. After ovulation, the oocyte can only survive for about 24 h without fertilization. After ovulation, vessels of the endocrine theca interna invade the granulosa. Granulosa cells form the *corpus luteum*. They take up lipids and become endocrine cells. In the absence of fertilization, the corpus luteum is invaded by a connective tissue and transforms into *corpus albicans*.

1.7.1.2 Hormonal Control

Hypothalamic–Pituitary–Gonadal Axis The hypothalamic–pituitary–gonadal axis is a pathway that links by direct action and feedback three endocrine glands: the hypothalamus, pituitary gland, and gonad (ovary or testis).[25]

- The hypothalamus produces gonadotropin-releasing hormone (GnRH);
- The anterior pituitary gland (adenohypophysis) follicle-stimulating (FSH) and luteinizing (LH; or lutrophin) hormones;[26] and
- Gonads estrogens and androgens.

Follicle-stimulating and luteinizing hormones are called gonadotropins because they stimulate the gonad. They are synthesized and secreted by gonadotroph cells of the anterior pituitary gland. Most gonadotrophs secrete only LH or FSH, but some secrete both hormones.

These glycoproteins are composed of α and β subunits. The common FSH and LH α subunit is identical to that of chorionic gonadotropin and thyrotropin. The specific receptor-binding β subunit characterizes hormone function.

Gonadotropin-releasing hormone travels through the hypophyseal portal circuit and binds to receptors on the secretory cells of the adenohypophysis. Gonadotropin-releasing hormone then increases the Fsh and Lh gene transcription.

---

[25] γ ουα: generation; γ ουη: offspring.

[26] Five types of endocrine cells secrete hormones: corticotropes (corticotropin or adrenocorticotropic hormone [ACTH]), gonadotropes (LH and FSH); prolactin (Prl)-releasing cells; somatotropes (somatotropin or growth hormone [GH]); and thyrotropes (thyrotropin [TSH]).

Both FSH and LH activate endocrine ovarian cells that then synthesize estrogens and inhibins, thereby regulating the ovarian and menstrual cycles. The regulatory loop between the hypothalamus, adenohypophysis, and gonads leads to the pulsatile secretion of GnRH, LH, and, to a lesser extent, FSH. The pulse frequency is variable.

Estrogens form a negative feedback, as they preclude the GnRH production in the hypothalamus. In addition, inhibins inhibit activins that stimulate GnRH-producing cells. On the other hand, the positive feedback between estrogen and LH and FSH prepares the ovarian follicle for ovulation and the uterus for blastocyte implantation. Therefore, estrogens exert:

1. A negative feedback during the follicular phase, when their levels are still low; and
2. A positive feedback at high concentrations near the end of the follicular phase.

Progesterone prevents GnRH, FSH, and LH secretion via its cognate receptor in the hypothalamic neurons and secreting cells of the pituitary gland. It thus stops the estrogen–LH positive feedback loop. Progesterone impedes FSH secretion upon 5α-reduction [88].

Hence, the cycle of hormone secretion from the hypothalamic–pituitary–gonadal axis comprises the following stages:

1. GnRH release from the hypothalamus;
2. FSH and LH liberation from the anterior pituitary;
3. Estrogen and progesterone (in small amount) secretion from ovarian follicle;
4. LH and FSH (to a lesser extent) surge at midcycle, causing ovulation;
5. Progesterone and estrogen (in small amount) secretion from corpus luteum;
6. Inhibited production of FSH and LH;
7. Decline in progesterone and estrogen release; and
8. Reliberation of FSH and LH.

FSH stimulates the maturation of germ cells and initiates follicular growth, specifically targeting granulosa cells. Follicle-stimulating hormone provokes the synthesis of luteinizing hormone receptor in granulosa cells. When circulating LH binds to its cognate receptors, cell proliferation stops.

The LH surge launched by GnRH triggers ovulation and initiates the conversion of the residual follicle into a corpus luteum. Luteinizing hormone supports thecal cells that synthesize androgens and hormonal precursors of estradiol. Both FSH and LH act synergistically. Their concentrations reach a peak at ovulation.

When pregnancy occurs, the LH level decays and the luteal function is maintained by human chorionic gonadotropin secreted from the placenta.

Estrogens Estrogens, especially estradiol, are secreted by cells of the vascularized theca interna. As do all steroid hormones, estrogens diffuse across the cell membrane.

Once inside the cell, they bind to and activate estrogen receptors that are transcriptional nuclear factors (estrogen receptors ERα–ERβ [NR3a1–NR3a2]; Vol. 3, Chap. 6. Receptors).

The estrogen-related receptors ERRα to ERRγ are also nuclear receptors (NR3b1–NR3b3) closely related to the estrogen receptors. They share target genes, coregulatory proteins, ligands, and sites of action with the estrogen receptors [89]. They influence the estrogenic response. The ERRα subtype is an effector of the transcriptional coactivator PGC1α involved in oxidative phosphorylation and mitochondrial genesis [90]. The transcription of ERRα-regulated genes is downregulated in *insulin resistance*. These transcriptional coactivators are targeted by various signaling pathways. The ERRα isoform is an effector in signaling cascade launched by growth factor receptor protein Tyr kinases (e.g., HER2 and IGF1R) [91].

In addition, estrogens activate a G-protein-coupled estrogen receptor (GPER or GPR30; Vol. 3, Chap. 7. G-Protein-Coupled Receptors).

The estrogen concentration increases progressively during the preovulatory part of the menstrual cycle to reach a peak before ovulation and then falls abruptly. A second estradiol level elevation of smaller amplitude forms a plateau during the postovulatory part of the menstrual cycle.

Estradiol acts as a growth hormone in the reproductive tract. It supports walls of the fallopian tubes, uterus, and vagina, as well as the cervical glands and the myometrial growth.

Among other effects, estrogens (estrone [E_1], estradiol [E_2]), and estriol [E_3], according to the number of hydroxyl groups):

- Trigger via a positive feedback the luteinizing hormone surge by the hypothalamopituitary axis and, subsequently, trigger ovulation (estrogen surge);
- Improve the coronary arterial blood flow;
- Accelerate metabolism and increase lipidic storage;
- Increase hepatic production of some proteins;
- Stimulate endo- and myometrial growth (thus antagonizing the myometrial-suppressing activity of progesterone) and thicken the vaginal wall;
- Stimulate ductal and alveolar growth in mammary glands; and,
- In late gestation, induce expression of myometrial oxytocin receptors, thereby preparing the uterus for parturition.

Progesterone Progesterone, another ovarian steroid hormone (progestational steroidal ketone), is produced by granulosa cells of the corpus luteum. It prepares the endometrium to receive and nourish an embryo.

The progesterone concentration is relatively low and stable during the preovulatory phase of the menstrual cycle. During the postovulatory (luteal) phase, the progesterone concentration progressively rises, reaches a peak, and progressively decays.

In the absence of embryo implantation, the corpus luteum involutes causing sharp drops in progesterone and estrogen levels.

Progesterone operates partly via the intracellular ligand-activated nuclear receptor NR3c3. In fact, three subtypes (PRa–PRc) arise from the the PGR gene. The progesterone receptor isoforms PRa and PRb are generated using separate promoters and translational start sites. A special transcription activation function TAF3 exists in PRb, but not in PRa. Both isoforms (PRa–PRb) are generally expressed at similar levels. When PRa remains transcriptionally inactive, it acts as a progesterone-dependent inhibitor of estrogen receptor [92]. The PRc isoform is truncated at the N-terminus. It acts as a transcriptional potentiator, as it enhances the activity of the larger PR proteins [92].

In humans, ovarian granulosa and thecal cells synthesize three PR isoforms. The PR receptor is produced by steroidogenic, granulosa, and luteal cells, as well as stromal fibroblasts of the corpus luteum in the human ovary [92].

In the endometrium, progesterone downregulates its own receptors [92]. On the other hand, estradiol upregulates PR expression. Both PRa and PRb isoforms are expressed in comparable amounts in the glandular epithelium during the proliferative phase. The PRb production persists during the midsecretory phase, but the PRa synthesis decreases from the proliferative to the secretory phase of the menstrual cycle. In the endometrial stroma, PRa shows a predominant expression throughout the menstrual cycle [92]. Progesterone regulates the expression of the calcitonin, histidine decarboxylase, amphiregulin, and lactoferrin genes, thereby supporting blastocyte implantation as well as decidualization of uterine stromal cells. Whereas PRa is expressed in the decidualizing stroma with a gradual decline, PRb expression is downregulated during decidualization.

In humans, a progesterone receptor of different size lodges on the spermatozoon membrane [92]. This testicular protein can serve as a marker of the fertilizing potential of the spermatozoa. In humans, progesterone operates via a plasmalemmal receptor (mPR) and Ca^{2+}, IP_3, and PLCγ, in addition to its transcriptional activity [92]. Progesterone rapidly primes activation of the Src–Ras–Raf–ERK pathway via the mPR receptor from the inner side of the plasma membrane [92]. This nongenomic axis is mediated by the membrane progesterone receptor that belongs to the PAQR (progestin and adiponectin receptors) category [93]. It activates inhibitory G protein subunit (Gα_i), suggesting that they are GPCRs. However, PAQRs more closely resemble proteins of the alkaline ceramidase family and they may possess enzymatic activity. The GPCR and alkaline ceramidase modes of operation are not necessarily mutually exclusive [93].

Membrane-bound progesterone receptor localizes to human aortic endothelial cells, hepatocytes, brain cells, granulosa cells, and spermatozoa, as well as the mammary gland and ovary [92]. The nongenomic progesterone actions regulate relaxation of intestinal and uterine smooth muscle [92].

Among other effects, progesterone supports ovulation, mammary gland development, and establishment and maintenance of pregnancy. It:

- Inhibits secretion of the pituitary gonadotropins luteinizing (LH) and follicle stimulating (FSH) hormones;
- Induces protein synthesis and hypertrophy in cardiomyocytes;
- Modifies intracellular calcium concentration, provoking a rapid influx of extracellular calcium in T cells, but attenuating the cytosolic calcium concentration in myometrial cells and vascular smooth myocytes;
- Provokes ovulation;

Table 1.24 Uterine secretions (Source: [94]). Predecidual cells secrete numerous substances that can have a nutritious, metabolic, or immunosuppressive function

Hormones and growth factors	Prolactin, relaxin, renin FGF1/2
Fertilization regulators	PAEP
Neuropeptides	β-endorphin, Leu-enkephalin
Growth factor partners	IGFBP1/2, pappalysin-1 (PAPPa), IGF2BP3
Structural components	Mucins, heparan sulfate proteoglycan, collagen-4, entactin, fibronectin, laminin, integrins
Chaperones	HSPb1
Immunity components	Lactoferrin, uteroglobin
Miscellaneous	Plasminogen activator, plasminogen activator inhibitor, diamine oxidase, progesterone-dependent carbonic anhydrase

IGFBP insulin-like growth factor-binding protein, *IGF2BP3* insulin-like growth factor 2 mRNA-binding protein-3, *PAPP* pregnancy-associated plasma protein, *PAEP* progestagen-associated endometrial protein

- Remodels the endometrium prepared by estrogens for embryo implantation, stimulating its secretory and vascular activity, and then maintains the endometrium thickening and decreases contractility of the uterine smooth muscle to prevent spontaneous motions of the uterus;
- Strengthens the cervical mucus plug to prevent sperm penetration and infection;
- Decreases the maternal immunity to enable tolerance of the embryo.
- Stimulates the growth of breast, but impedes milk production during pregnancy; and
- Maintains the placental function.

1.7.2 Uterus

Implantation of the fertilized egg is the very early stage of pregnancy. The *blastocyst* adheres to the uterine wall.

Pregnancy depends on synchronized and successive events managed by communication based on nervous, local and remote (endocrine) hormonal, and immune signals between mother and embryo and then fetus.

Uterine secretions nourish the preimplantation embryo, promote growth, and prepare it for implantation. Proteins, glycoproteins, and peptides released by the endometrial glands during pregnancy that regulate the timing and occurrence of the appropriate sequence of events in the fertilization and promote a uterine environment suitable for pregnancy are given in Table 1.24.

Pregnancy-associated protein comprises various electrophoretic proteic fraction (β-lipoprotein, α2-macroglobulin, posttransferrin species, and albumin) in early pregnancy. Progesterone is required for the development of decidual tissues.

Table 1.25 Substances secreted by the blastocyst (Source: [94])

Immunoregulators	CSF, TNFSF6, IL1α/6/8, LIF, Ifnγ, PGe_2, PAF, HSPe1
Peptidases	MMP2–MMP3, MMP8–MMP11, MMP13 Serine peptidases
Growth factors and hormones	CG, FGF, IGF2, TGFβ, inhibins, E_2
Miscellaneous	Plasminogen activator and its inhibitors

CG chorionic gonadotropin, *CSF* colony-stimulating factor, E_2 estradiol, *EHRF* embryo-derived histamine-releasing factor, *HSP* heat shock protein, *Ifn* interferon, *IL* interleukin, *LIF* leukemia inhibitory factor, *MMP* matrix metallopeptidase, *PAF* platelet-activating factor, *TNFSF* tumor-necrosis factor superfamily member

Progesterone-regulated proteins produced by the endometrial epithelium include progesterone-associated endometrial protein (PAEP),[27] insulin-like growth factor-binding protein IGFBP2, crystalloglobulin, uteroglobin, integrins, and type-1 mucins, as well as prostaglandins PGd_2 and $PGf_{2\alpha}$ secreted by stromal cells such as endometrial specialized fibroblasts [95].

Lipocalins are small extracellular proteins, many of which bind small hydrophobic molecules, such as retinol and steroids. Among members of the lipocalin family, progestagen-associated endometrial protein is a glycoprotein mainly synthesized in secretory decidualized endometrial glands after progesterone exposure [96]. Many distinctly glycosylated forms exist. Glycosylation dictates the PAEP activity. Some of the alternatively spliced mRNAs lack the sequences encoding glycosylation sites and/or the lipocalin signature motif. Insulin-like growth factor-binding protein IGFBP1,[28] is also synthesized by the decidualized secretory endometrium. Progesterone stimulates its secretion [97].

1.7.3 Blastocyte

Embryomaternal communication that relies on various secreted molecules enables embryo implantation and maintenance. The attachment between the uterine luminal epithelium and the blastocyst trophectoderm is followed by stromal decidualization and luminal epithelial apoptosis at the site of blastocyst implantation.

Numerous substances are liberated by the blastocyst (Table 1.25). Immunoregulators prevent rejection. Serine peptidases and metallopeptidases allow invasion of the trophoblast into the endometrium. Chorionic gonadotropin serves as an autocrine growth factor.

[27] Also known as glycodelin, placental protein PP14, and pregnancy-associated endometrial α2-globulin (PAEG).

[28] Also known as placental protein PP12.

Embryo-derived histamine-releasing factor (EHRF),[29] is a secretagogue for colony-stimulating factor CSF2 and tumor-necrosis factor TNFSF1 by uterine mastocytes during the preimplantation period [98]. Histamine-releasing factor functions as a growth-regulating guanine nucleotide-exchange factor for the RHEB GTPase implicated in the TSC1/2–TOR pathway, a guanine nucleotide-dissociation inhibitor for the elongation factors EF1a, a small GTPase, and EF1bβ, a guanine nucleotide-exchange factor, and an antiapoptotic protein, as it inserts into the mitochondrial membrane and prevents BAX dimerization [99].

Histamine is implicated in ovulation, blastocyst implantation, placental blood flow regulation, lactation, uterine contractile activity, and pregnancy maintenance. The uterine histamine interacts with the embryonic H_2 receptor.

1.7.4 Placenta

The placenta,[30] the origin of which is mostly fetal, invades maternal uterine wall early in pregnancy and release hormones and other factors to reprogram maternal organ function, in particular changing cardiac metabolism and hemodynamics. Maternal metabolism is also modified to support requirements of embryo- and fetogenesis.

The placenta is a fetomaternal organ with two compartments:

1. The fetal placenta, the *chorion*,[31] the innermost membrane surrounding the embryo, which develops from the blastocyst that forms the embryo and then fetus; and
2. The maternal placenta, the *decidua*,[32] a thick layer of modified mucous membrane that lines the uterus during pregnancy, which develops from the maternal uterus and is shed after birth.

The amnion[33] is the innermost membrane of the extraembryonic membrane that envelops the embryo and contains the amniotic fluid.

---

[29] Also known as translationally controlled tumor protein (TCTP) and fortilin. It is also implicated in late-phase allergic reactions and chronic allergic inflammation. It can stimulate histamine release and interleukins IL4 and IL13 production from IgE-sensitized basophils and mastocytes. It is secreted by macrophages among other cell types. Mastocytes and basophils are major effector cells for IgE-dependent allergic inflammations. These cells secrete preformed proinflammatory mediators (e.g., histamine, nucleotides, peptidases, and proteoglycans) as well as de novo synthesized lipids (e.g., leukotrienes and prostaglandins) and polypeptides (e.g., cytokines and chemokines). HRF dimers and oligomers interact with IgE, crosslink IgE to the high affinity receptor for IgE (Fcϵ R1), and can launch asthma [100].

[30] πλακας: floor; πλακι ον: small slab; πλακις: couch, sofa, divan; πλακαυς: flat cake. Latin *placenta*: cake.

[31] κ ορι ον: membrane that encloses the foetus.

[32] Latin *decido* (*deciduus/a/um*): fall (fallen); cut in slice, detach.

[33] αμνας: ewe lamb.

Table 1.26 Placental hormones and neuromediators (Source: [94]))

Peptidic hormones	CRH, GnRH, GHRH, TRH, TSH, CG, CSH1/2, GH2, inhibins, PTHLH, somatostatin
Neuropeptides	Neuropeptide-Y, neurotensin, substance-P
Steroid hormones	Estrone, estradiol, estriol, progesterone

CG chorionic gonadotropin, *CRH* corticotropin-releasing hormone, *CSH* chorionic somatomammotropin hormone, *GH* growth hormone, GHRH growth hormone-releasing hormone, *GnRH* gonadotropin-releasing hormone, *PTHLH* parathyroid hormone-like hormone, *TRH* thyrotropin-releasing hormone [thyroliberin], *TSH* thyroid-stimulating hormone [thyrotropin]

In addition to its role in transferring molecules between mother and fetus, the placenta is a major endocrine organ (Table 1.26). It synthesizes numerous types of hormones and cytokines that influence ovarian, uterine, and mammary, as well as fetal functions.

1.7.4.1 Placental Steroid Hormones

The placenta produces both progesterone, the pregnancy hormone involved in establishment and maintenance of pregnancy, and estrogens during pregnancy. The concentration of these hormones rises during pregnancy.

The major estrogen produced by the placenta is estriol from 16-hydroxydehydroepiandrosterone sulfate ([16OH]DHEAS), an androgen steroid. The placenta produces pregnenolone and progesterone from circulating cholesterol. Pregnenolone is converted in the fetal adrenal gland into dehydroepiandrosterone (DHEA), subsequently sulfonated to dehydroepiandrosterone sulfate (DHEAS). The latter is converted to [16OH]DHEAS in the fetal liver and adrenal glands.

Progesterone enriches the uterus with a vascular network to support the growing embryo and then fetus.

1.7.4.2 Placental Peptidic Hormones

Chorionic Gonadotropin Placenta proteic hormones include chorionic gonadotropins produced by the syncytiotrophoblast that generates the placenta. The human chorionic gonadotropin (hCG), the signal for maternal recognition of pregnancy, is produced by fetal trophoblast cells. It binds to the luteinizing hormone receptor on cells of the corpus luteum, thereby precluding luteal regression.

This glycoprotein is composed of 237 amino acids is a heterodimer composed of an α subunit similar to luteinizing (LH), follicle-stimulating (FSH), and thyroid-stimulating (TSH) hormone, and β subunit.

It interacts with the LHCG receptor of the blastocyst and promotes the maintenance of the corpus luteum during the beginning of pregnancy. The corpus luteum secretes progesterone during the first trimester.

Growth Hormone and its Related Hormones The cluster of genes located on chromosome 17q22-24 encodes the human growth hormone (hGH). It contains five highly related genes. From 5′ to 3′ end, they encode:

1. GH1 (somatotropin), or (h)GHn ([human] growth hormone normal) expressed by somatotropes in the anterior pituitary;
2. CSHL1 (chorionic somatomammotropin hormone-like-1 hormone), also known as (h)CSL ([human] chorionic somatomammotropin-like hormone);
3. CSH1 (chorionic somatomammotropin hormone-1), also known as (h)PL ([human] placental lactogen) and (h)CSa ([human] chorionic somatomammotropin-A);
4. GH2 (placental growth hormone-2), or (h)GHv ([human] growth hormone variant), also called chorionic somatotropin; and
5. CSH2 (chorionic somatomammotropin hormone-2), also known as (h)CS2 ([human] chorionic somatomammotropin-2) and (h)CSb ([human] chorionic somatomammotropin-B).

These five genes are expressed according to a mutually exclusive tissue distribution, GH1 in pituitary somatotropes and the four remaining genes in placental villous syncytiotrophoblasts.

The genes of the GH cluster have a developmentally coordinated pattern of expression, but differ in their tissue distribution-dependent expression, levels of expression, and patterns of alternative splice-site selection [101].

The predominant splicing pattern shared by Gh1, Gh2, Csh1, and Csh2 transcripts involves the ligation of five common exons. Some pituitary Gh1 transcripts use an alternate splice-acceptor site in exon 3. The Gh2 transcripts can undergo an alternative splicing in which intron 4 is retained in the processed transcripts; they produce GH2v2 isoform. The Cshl1 pseudogene splicing happens between exons 2 and 3.

The parallel rise in circulating concentrations of CSH1 and GH2 during gestation may reflect placental growth or increase in gene expression. The elevation in GH2, CSH1, CSH2, and CSHL1 levels between 8 and 20 weeks is followed by a plateau until term [101].

Placental lactogen (or chorionic somatomammotropin; 191 amino acids) secreted by the syncytiotrophoblast is related to prolactin and growth hormone. It modifies the metabolic state of the mother during pregnancy to facilitate the energy supply of the fetus. It may participate in the mammary gland development prior to parturition.

Its metabolic effects encompass:

- Reduction of maternal insulin sensitivity;
- Attenuation of maternal glucose utilization; and
- Elevation of lipolysis and release of free fatty acids.

It has similar actions to those of growth hormones, but much weaker.

Relaxins Relaxin acts synergistically with progesterone to maintain pregnancy. It relaxes pelvic ligaments at the end of gestation. It is produced by either the placenta, the corpus luteum, or both, according to mammalian species.

The relaxin-like peptide family belongs in the insulin superfamily. It comprises seven members (relaxins Rln1–Rln3 [H1–H3] and insulin-like peptides InsL3–InsL6). They are produced not only by the corpus luteum and breast, but also, during pregnancy, by the placenta (chorion and decidua).

In humans, relaxins Rln1 to Rln3 are composed of two chains (α–β), as in insulin. Relaxin-3 binds to and activate Rln1, but not Rln2, in vitro [103]. In women, the RLN2 gene is expressed in the corpus luteum, endometrium, placenta, and breast, whereas the RLN1 transcript is detected in the placenta only [573]. The highest Rln2 level is measured in pregnancy and during the second phase of the menstrual cycle.

The relaxin/insulin-like family peptide GPCRs include four subtypes ($RxFP_1$–$RxFP_4$). $RxFP_1$ can be coexpressed with $RxFP_2$ at least in some cell types. $RxFP_1$ is significantly more produced in the decidua than in the amnion; it is expressed at relatively low levels in the chorion [102]. The major splice $RxFP_1$ variant and $RxFP_2$ are undetectable in the placenta and fetal membranes.

1.7.5 Global Metabolic Changes During Pregnancy

During early gestation, maternal metabolism focuses on anabolism in preparation for the upcoming demands, as the metabolic need of the fetus is maximal in the third final trimester.

Late in gestation, maternal metabolism becomes catabolic, bringing nutrients to the rapidly growing fetus [104]. The delivery of nutrients to the placenta dictates the rate of fetal growth. Glucose is the preferred substrate of the fetus.

The liver processes glycerol and, to a lesser extent, amino acids to synthesize glucose for the fetus and consumes lipids, thereby generating ketones used by the brain, muscle, and fetus. Adipose tissue releases fatty acids for consumption by both the liver and muscle. The fetus incorporates amino acids, lipids, and glucose for its growth as well as glucose for its energy need. The maternal basal metabolic rate rises up at least to 60 % during the second half of pregnancy, reaching about 250 kcal/day near term [104].

Nearly half of the energetic cost of pregnancy is taken from maternal adipose tissue storage mainly built during the anabolic first half of pregnancy [104].

Glucose handling changes during pregnancy. Maternal insulin resistance develops usually early in pregnancy to limit maternal glucose consumption and allow its transfer to the fetus. A least 80 % decrease in insulin sensitivity can be observed in the late stage of pregnancy [104].

Insulin resistance in the maternal liver increases gluconeogenesis by about 30 %. Simultaneously, insulin resistance in muscle, the largest sink for glucose in the body, limits glucose consumption by the mother. Insulin resistance in adipose tissue increases lipolysis, providing fatty acids as an alternative fuel for maternal consumption and glycerol as a preferred gluconeogenic substrate for the liver during pregnancy, thereby preserving amino acids for fetal growth.

The α chain of the hormone dimer human chorionic gonadotropin is first launched, followed by human chorionic somatomammotropin hormone (CSH), placental growth hormone-2, estrogen, progesterone, prolactin, and cortisol, all causing insulin resistance [104]. The placenta also liberates numerous cytokines (e.g., TNFSF1), among which many contribute to insulin resistance during diabetes. In addition, elevated plasma level of fatty acids can also contribute to insulin resistance.

Molecular changes in the skeletal muscle are similar to those observed in diabetes: binding by insulin is not altered, but phosphorylation of the insulin receptor decays and inhibitory phosphorylation of IRS1 rises [104].

Insulin concentration rises during pregnancy from the second trimester and threefold in late pregnancy. Pregnancy influences the pancreatic β-cell secretion flux and density (10–20 % increase due to cell hypertrophy and hyperplasia) [104].

Pregnancy generates an average transport between mother and fetus of 1 kg of proteins [104]. Amino acids are spared from gluconeogenesis in the liver, consumption of branched chain amino acids decreases, and protein synthesis is elevated by 25 % in the third trimester.

Serum lipid levels increase during pregnancy. Triglycerides are elevated two to four times, total cholesterol by 25–50 %, and LDL level by 50 %. The lipidic balance shifts from lipogenesis in early gestation to lipolysis in late gestation with higher turnover of glycerol.

Fatty acids become the chief source of maternal fuel during late pregnancy. Free fatty acids liberated by lipolysis are processed in the liver. They are oxidized or packaged as triglycerides in VLDLs. The high flux of fatty acids in the liver renders pregnant women prone to ketosis. Placental lipoprotein lipase can provide lipids from circulating lipoproteins for fetal use.

The growing fetus also appropriates other nutrients. Fetal red blood capsules require iron ($\sim$ 500 mg during late gestation), necessitating at least a fivefold increase in daily maternal absorption of iron. Hepcidin, which inhibits gut absorption of iron, disappears completely in late pregnancy. The fetal skeleton becomes mineralized in the last few weeks of gestation, requiring up to 300 mg/d of calcium near the term. Maternal intestinal absorption of calcium increases.

1.7.6 Cardiac Metabolism Before and During Pregnancy

Uninterrupted cycles of cardiac contraction and relaxation require a high nutrient input. As the ATP reserve is minimal and ATP turnover completes in approximately every 10 s, the heart depends on a continuous energy supply. The heart uses different types of energy substrates (carbohydrates, lipids, ketone bodies, and amino acids) according to environmental conditions and substrate availability.

In normal conditions, the main myocardial fuel is mainly constituted by fatty acids ($\sim$ 70 %) supplemented by glucose. Fatty acids and glucose arise via the triglyceride and glycogen pools, respectively. Substrate catabolism converges on the generation of acetylCoA in mitochondria via lipid β-oxidation and glycolysis.

The PI3K–PKB2 pathway serves in glucose uptake, as it triggers the translocation of glucose transporter GluT4 to the plasma membrane.

Various fatty acids bind to PPARα (nuclear receptor NR1c1) and stimulate transcription of genes involved in fatty acid import (e.g., ScaRb3), transport into the mitochondria (e.g., carnitine palmitoyltransferase CPT1b),[34] and β-oxidation (e.g., medium-chain acylCoA dehydrogenase [MCAD]).[35] Mitochondrial oxidative phosphorylation is coupled to cardiac contractile function. Other important transcriptional regulators of metabolism are PPARγ (nuclear receptor NR1c3) coactivator PGC1α and PGC1β encoded by the PPARGC1A and PPARGC1B genes. In the heart,

---

[34] Carnitine ^{O}palmitoyltransferase (CPT) is a mitochondrial transferase enzyme involved in the metabolism of palmitoylcarnitine into palmitoylCoA. Four CPT isoforms exist in humans (CPT1a–CPT1c and CPT2). Carnitine palmitoyltransferases connect carnitine to long-chain fatty acids that can then cross the inner membrane of mitochondria. Inside mitochondria, carnitine is removed and fatty acids can be processed to produce energy. Carnitine palmitoyltransferase-1A is the hepatic subtype. More than 20 mutations in the CPT1A gene severely reduce or eliminate the activity of this enzyme. Fatty acids cannot enter mitochondria and be converted into energy. Fatty acids accumulate in cells and damage the heart, brain, and liver. Moreover, reduced energy production leads to hypoglycemia and hypoketosis. Three main types of CPT2 deficiency exist: a lethal neonatal, severe infantile hepatocardiomuscular, and myopathic form. The CPT1b subtype supports transfer of long-chain fatty acylCoAs from the cytoplasm into mitochondria.

[35] Fatty acids are an important source of energy for the body, especially during periods of fasting. Fatty acids are transported into cells and then mitochondria, the energy-producing centers in cells, to be catabolized. Fatty acid oxidation disorders constitute a category of inherited metabolic anomalies that lead to an accumulation of fatty acids and a decrease in cell energy metabolism. Each fatty acid oxidation disorder is associated with a specific enzyme defect in the fatty acid metabolism. Fatty acid oxidation disorders (FAOD) include:

- Carnitine–acylcarnitine translocase deficiency (CACTD);
- Carnitine palmitoyl transferase type-2 deficiency (CPT2D);
- Carnitine palmitoyl transferase type-1A deficiency (CPT1AD);
- Carnitine uptake defect (CUD);
- Glutaric aciduria type-2 (GA2) or multiple acylCoA dehydrogenase deficiency (MADD);
- IsobutyrylCoA dehydrogenase deficiency (IBCD);
- Medium-chain acylCoA dehydrogenase deficiency (MCADD);
- Long-chain 3-hydroxyacylCoA dehydrogenase deficiency (LCHADD);
- Short-chain acylCoA dehydrogenase deficiency (SCADD);
- Medium- and short-chain ^{L}3-hydroxyacylCoA dehydrogenase deficiency (M/SCHAD);
- Trifunctional protein deficiency (TFPD);
- Very-long-chain acylCoA dehydrogenase deficiency (VLCADD).

Medium-chain acyl CoA dehydrogenase (MCAD) is one of four mitochondrial acylCoA dehydrogenases that carry out the initial dehydrogenation step in the β-oxidation cycle. Its deficiency impairs oxidation of dietary and endogenous fatty acids of medium chain length (6–12 carbons). It is an autosomal recessive disorder of β-oxidation of fatty acids. Associated hypoglycemia results occur from an inability to match gluconeogenic requirements during fasting despite activation of alternate proteolysis. It causes cerebral edema and fatty liver, heart, and kidneys. It provokes sudden death in infants, most often after periods of fasting or vomiting. Mitochondrial fatty acid β-oxidation can be studied by incubating stable isotope-labeled fatty acid probes with human fibroblasts in the presence of ^{L}carnitine [106]. The ratios of octanoylcarnitine to decanoyl- or decenoylcarnitine appear specific of MCAD deficiency.

PGC1α overexpression increases mitochondrial genesis, β-oxidation, and expression of genes involved in oxidative phosphorylation. The PGC1α subtype binds to and coactivates PPARs, estrogen-related receptors (nuclear receptor NR1b1–NR1b3), and numerous other transcription factors, thereby coordinating multiple pathways such as fatty acid consumption (largely via PPARs), mitochondrial genesis (via nuclear respiratory factors NRF1 and NRF2 [GABP] and NR3b1), and vascular density (via NR3b1) [104].

During strenuous exercise, the heart predominantly utilizes fatty acids and lactate, the blood concentrations of which are elevated. During exercise, the high lipolysis rate in adipose tissue heightens circulating levels of triglycerides and free fatty acids. Exercise stimulates AMPK that promotes GluT4 translocation to the plasma membrane and hence glucose uptake, as well as glycolysis. Activated AMPK also increases fatty acid import and oxidation, as it favors the translocation of the fatty acid transporter ScaRb3.

During prolonged fasting, the heart and brain consume ketone bodies.

During pregnancy, cardiomyocytes increase and decrease utilization of fatty acids and glucose, respectively. Cardiac metabolism is modified to accommodate both fetal needs and augmented cardiac work.

The selected fuel used by the heart changes during pregnancy. Glucose utilization declines early in pregnancy and is continuous (drops 75 % in late pregnancy). Decayed glucose oxidation is not related to inhibited pyruvate dehydrogenase, the activity of which remains normal. During pregnancy, GluT4 production diminishes. In addition, nitric oxide, the production of which heightens, promotes fatty acid uptake, instead of glucose uptake. Mediator isoform contribution may vary during pregnancy. Attenuated PKB2 levels may depress glucose uptake, whereas elevated PKB1 level favors growth [104].

However, between-species difference can be observed. In pregnant dogs, glucose oxidation in late pregnancy was less markedly reduced than that in rats. Anyway, the generation of ATP in late pregnancy results almost exclusively from lipid utilization in the heart exhibiting a relative insulin resistance.

Pregnancy affects the maternal hormonal milieu, increasing levels of estrogen, progesterone, prolactin, and placental hormones. Many of these messengers cause insulin resistance. In particular, estrogens augment cardiac fatty acid utilization, but do not modify glucose metabolism.

1.7.7 Cardioprotective and Hypertrophic Pathways During Pregnancy

The plasma calcium concentration varies during pregnancy. It is markedly higher in the first and second trimester and lower in the third trimester with respect to nonpregnant subjects. Calcium-activated phosphatase PP3 is involved in cardiac hypertrophy. It dephosphorylates nuclear factor of activated T cells (NFAT) that then translocates to the nucleus. During pregnancy, its concentration and activity rise in the early stage and diminish in the late phase [109]. Progesterone, but not estradiol, increases the PP3 concentration as well as NFAT activity. Activation of PP3 in early pregnancy by progesterone is required for adaptive cardiac hypertrophy. The PP3 stimulation leads to ERK1, ERK2, and PKB activation. Progesterone induces cellular hypertrophy via activated ERK1 and ERK2 kinases. On the other hand, reduced PP3 level and activity in late pregnancy may result from estradiol [109]. Therefore, the regulation of PP3 activity in pregnancy differs from that in response to pressure overload. Pregnancy is a physiological state with augmented oxidative stress related to high metabolic turnover and elevated oxygen requirements. Oxidative stress can be, at least partly, caused by activation of the SHC adaptor and impaired FoxO3a signaling. During pregnancy, the heart undergoes a reversible adaptive enlargement in response to mechanical stress and chemical stimuli. The PI3K–PKB pathway activated by estrogens and increased mechanical stimuli participates in cardioprotection during pregnancy. Activation of the STAT3 pathway is aimed at protecting the heart. The STAT3 factor promotes myocardial angiogenesis and can mediate cardiomyocyte hypertrophy. In addition, STAT3 protects the heart against oxidative stress, as it prevents overproduction of ROS by upregulating manganese superoxide dismutase [108]. Both STAT3 and PGC1α are needed in peripartum to protect the maternal heart from oxidative stress and circulating antiangiogenic factors via overexpression of antioxidative enzymes (e.g., ^{Mn}SOD) and proangiogenic factors (e.g., VEGF) [107].

Mice lacking the STAT3 gene specifically in cardiomyocytes develop peri- and postpartum cardiomyopathy after delivery. Two-third of STAT3-deficient mice died after delivering their second litters [108].

The STAT3 pathway is activated by prolactin produced by the pituitary gland. The prolactin full-length 23-kDa form can be cleaved by cathepsin-D into a potent antiangiogenic, proapoptotic, and proinflammatory 16-kDa fragment (Prl16). Prolactin-16 is a powerful destroyer of endothelium, thus decreasing cardiac capillary density. It not only destroys the cardiac microvasculature but also reduces the cardiac function and cardiomyocyte metabolism, even in the absence of pregnancy. In addition, this fragment also prevents NOS3 activation. The decrease in PKB activity in the postpartum maternal heart may result from the rapid decline in circulating estrogens and the cardiac unloading after delivery. Postpartum PKB activation is detrimental for the heart as it lowers antioxidative defense. In combination with low STAT3 levels, a high PKB level favors cardiac inflammation and fibrosis in the postpartum heart [107].

1.7.8 Hemodynamic Changes During Pregnancy

1.7.8.1 Functional Cardiac Changes

Cardiac output increases by 20–50 % from gestational week 5, as the cardiac frequency rises by 15–30 % from gestational week 5 and the stroke volume by 15–25 % from gestational week 8 [104]. Elevated preload and contractility and reduced afterload combine to magnify cardiac output. Cardiac work (commonly calculated by the product of cardiac output and afterload) is amplified, but only by 20–30 %, because mean arterial pressure drops during pregnancy.

Cardiac mean oxygen consumption matched by an increased coronary blood flow rises by about 15 % [104]. Cardiac nitric oxide synthase NOS3 activity is intensified during pregnancy, hence supporting coronary arteriole dilation.

Both blood volume and red blood capsule mass increase, thereby amplifying the preload, whereas afterload diminishes during pregnancy [104].

1.7.8.2 Structural Cardiac Changes

Pregnancy produces a normal volume overload in the maternal circulation due to the demands of the growing fetus. In the last pregnancy trimester, all four chambers and valves are enlarged. Atrial enlargement begins in early pregnancy and reaches a maximum around gestational week 30. Left ventricular end-diastolic dimension heightens at gestational week 24 by about 10 % [104]. The end-diastolic posterior wall and septal thickness increase in the same proportion (eccentric hypertrophy).

The left ventricular mass elevates up to 50 % of the antepregnancy value with a peak in the course of the third trimester. The morphological changes reverse to antepregnancy values at postpartum month 6. Pregnancy-induced changes in leaflet size of the mitral valve are associated with structural modifications of the collagenous matrix [105]. Small-angle light scattering assesses changes in internal fiber architecture (degree of fiber alignment and direction). Collagen concentration rises (16 %). Collagen fibers are less aligned with an 11.5° rotation of fiber orientation toward the radial axis. The fibrosa thickens (53 %).

Peripartum cardiac remodeling is regulated by signaling pathways related to MAPKs, protein kinase-B, and nitric oxide synthase NOS3 [110]. In adult female pregnant rats, the left ventricular mass, mass/volume ratio (0.7 ± 0.02–1.28 ± 0.02 g/ml), and ejection fraction (64 ± 3–74 ± 2 %) increase. Whereas the left ventricular mass and mass/volume ratio return to their prepregnancy values in postpartum, the ejection fraction is low (53 ± 3%). Cardiac hypertrophy during pregnancy relies on downregulation of several antihypertrophic kinases (e.g., P38MAPK, JNK, and PKB) without change in ERK, a progrowth kinase [110].

1.7.8.3 Vascular Flow Variables

Blood flow increases in the uteroplacental circulation. Vascular resistance decreases down to 30 % of the antepregnancy value at gestational week 8 due to the release of vasodilators, such as nitric oxide and prostaglandins [104].

Expression of endothelial and myocardial nitric oxide synthase NOS3 is elevated in the myocardium of pregnant rats and returns to control levels at 4 days after birth. Nitric oxide is a major mediator of the vascular resistance fall. Activation of NOS3 by PKB is counteracted by ERK in some vascular beds [110].

Systolic blood pressure usually remains unchanged, whereas diastolic blood pressure initially decays, reaching its lowest value in the second trimester, and increasing back to baseline during the last trimester. Consequently, the mean arterial pressure initially lowers and augments at the end of pregnancy.

1.7.8.4 Delivery and Puerperium

During labour and delivery, hemodynamic fluctuations can be profound. Stroke volume increases and, consequently, the cardiac output. Blood loss and resulting anemia during delivery can further compromise the hemodynamic state.

The hemodynamic changes during the postpartum are mainly due to uterine contractions and relief of caval vein compression. After pregnancy, a rapid reversal of cardiac hypertrophy is possibly related to reduced preload. The left ventricular mass returns to normal by 12–24 weeks postpartum.

The cardiovascular changes usually return to baseline within 3–4 weeks following delivery, but sometimes are completely resolved only at week 12 after delivery.

MAPKs (P38MAPK, JNK, and ERK) support cardiac hypertrophy via the PI3K–PKB pathway. In particular, ERK1 and ERK2 activated principally by mechanical stretch and mitogenic stimuli regulate smooth muscle contraction, in addition to promoting cardiac hypertrophy.

The expression of antihypertrophic kinases (P38MAPK, JNK, and PKB) decreases during pregnancy and normalizes during postpartum, except JNK, the level of which is higher than normal levels [110]. Activation of P38MAPK and JNK is correlated with a lower left ventricular mass/volume ratio. The NOS3 concentration is also higher than baseline levels in postpartum; NOS3 expression is upregulated at 14 days postpartum. Therefore, the rapid reversal of hypertrophy in postpartum is accompanied with a delayed functional recovery.

1.7.9 Pregnancy-Associated Cardiac Diseases

Eclampsia[36] is an acute life-threatening hypertensive complication of pregnancy. It is associated with seizures and possibly coma (hypertensive encephalopathy) in the absence of preexisting brain disorders. Women with preexisting hypertension, diabetes, nephropathy, and thrombophilic diseases have a higher risk of developing preeclampsia and eclampsia.

Placental hypoperfusion leads to hypersensitivity of the maternal vasculature to vasoconstrictors that causes vasospasm and irrigation drop of maternal organs. Adrenomedullin, a potent vasodilator, is produced in reduced quantities by the placenta. Moreover, activation of the coagulation cascade induces microthrombus

[36] ϵκλαμψς: brightness, here in the sense of sudden occurrence.

formation, hence aggravating the hypoperfusion. Preeclampsia is associated with marked elevations of circulating soluble VEGFR1, a decoy receptor and VEGF inhibitor, secreted from the late gestational placenta [104]. The heart is more vulnerable during pregnancy. Cardiac ischemia and reperfusion injury damage the myocardium at a much greater extent. Late pregnancy and the postpartum period are also characterized by a higher blood coagulability. Peri- and postpartum cardiomyopathy (PPCM) is another rare life-threatening cardiac complication of pregnancy with left ventricular systolic dysfunction occurring in the last month of pregnancy or the first 5 months postpartum. Prolactin (Prl) is a ubiquitous and pleiotropic hormone. This pituitary hormone is a 23-kDa protein. It experiences posttranslational modifications (glycosylation and phosphorylation and proteolytic cleavage). Its synthesis and secretion are regulated by inhibitors and activators. Prolactin-releasing factors (PRF) include thyrotropin-releasing hormone, vasoactive intestinal peptide, and serotonin; prolactin release-inhibiting factors (PIF) comprise dopamine. Prolactin is produced in increasing amounts during pregnancy and suckling. It acts primarily on the mammary gland, where it triggers and maintains lactation in the postpartal period.

This late-gestational nursing hormone can be aberrantly cleaved in several tissues (myocardium, retina, and mammary gland) by peptidases (e.g., cardiac cathepsin-D) upon oxidative stress into the Prl16 fragment (Sect. 1.7.7). Plasma levels of cathepsin-D and Prl16 are elevated in PPCM patients. In addition, Prl16 causes vascular secretion of miR146a that decreases metabolic activity, or even triggers apoptosis in adjoining cardiomyocytes. Circulating levels of miR146a are markedly elevated in PPCM women [104].

Chapter 2
Context of Cardiac Diseases

In addition to aging, lifestyle, and environmental conditions, numerous structural and functional changes can alter the cardiovascular function. Health management is aimed at preventing or, at least, attenuating evolution rate by targeting risk factors, that is, supporting healthy eating and exercise and avoiding smoking tobacco, as well as screening populations at risk.

Research is thus devoted to: (1) discover the genetic causes of congenital heart diseases; (2) find new functional indices and exploration techniques for early diagnosis; (3) prevent evolutive complications; and (4) improve therapy using new efficient drugs with limited side effects and optimized medical devices and surgical procedures, of diseases of the heart and vasculature as well as the lung and blood, including sleep disorders.

2.1 Frequency of Cardiac Diseases

Cardiovascular diseases are the leading causes of death in rich countries that provoke about 12 million deaths per year worldwide. They account for more than half of all deaths in Europe of people over 65 years old. They are responsible for 33 % of all deaths in France. In France, about 4 people in every 10,000 die prematurely from heart disease, about 13 in every 100,000 from ischemic cardiopathies, and about 7 in every 100,000 from stroke. The rate can be increased by a factor 6 or more in other European countries.[1]

The largest contribution comes from coronary heart disease that can be, at least in some countries, responsible for approximately 1 in every 5 deaths in men and 1 in every 6 deaths in women.

Cardiac pathologies remain the number one cause of death from congenital malformations in infancy (45–50 % of postnatal deaths due to congenital anomalies).

[1] Sources: Cité de la Science in Paris and European Health Status publication of the European Commission.

M. Thiriet, *Diseases of the Cardiac Pump*, Biomathematical and Biomechanical Modeling of the Circulatory and Ventilatory Systems 7, DOI 10.1007/978-3-319-12664-7_2

The majority of congenital heart defects arise from abnormal development of the valvuloseptal compartment.

Heart failure results from leakage or narrowing of heart valves (i.e., valvular insufficiency and stenosis), cardiac contractility insufficiency, disturbed electrical activation of the myocardium by the nodal tissue, among other causes.

2.2 Modeling of Cardiac Diseases

Two major arterial pathologies encompass aneurysms and stenoses due to atherosclerosis, which have been targeted by physical and mathematical modeling.

The major objective of validated computational models is to describe evolving quantities of interest, display fields of mechanical variables, demonstrate the role of governing parameters, and to predict short- and long-term behavior.

Animal models are aimed at understanding physiological and pathological processes, exhibiting factors involved in complex biological processes at various length scales, via histological and immunohistochemical analyses and genetic approaches, and evaluating drug effects, hence, improving therapeutic strategies. For example, apolipoprotein-E $(ApoE)^-$ and low-density lipoprotein receptor $(LDLR)^-$ mice can develop atherosclerosis both with and without high-fat feeding, hence their selection as a representative model of human situation. However, lesions that develop over a period of weeks do not mimic accurately those that evolve over decades in humans, in particular fibrous plaques as well as lesions with calcification, ulceration, hemorrhage, and/or thrombosis. Whatever the animal species, some mediators can lack or intervene at a different concentration. Therefore, the process in a given species can vary in comparison with that in another species. For example, cholesteryl ester transferase exists at a much greater level in humans than in mice. Conversely, HDL concentration is higher in mice than humans. In any case, all models are a simplification of the reality.

2.3 Risk Factors

Genetic and environmental risk factors contribute to the variability in disease susceptibility for cardiovascular diseases. The occurrence of cardiovascular diseases also depends on the subjected pathophysiological ground and habits.

The two most important risk factors are smoking addiction and abnormal ratio of blood lipids (apolipoprotein-B/apolipoprotein-A1 ratio). In fact, major cardiovascular risk factors include diabetes, hypertension, obesity, and the familial context (family members with heart diseases), in addition to hypercholesterolemia and smoking. Minor cardiovascular risk factors are age, high-stress job, sedentary habits, lack of daily consumption of fruits and vegetables, and high resting cardiac frequency ($\sim$1.25 Hz [75 beats/min]).

Table 2.1 Age score. (Framingham point score; Source: National Institutes of Health—National Heart, Lung, and Blood Institute)

Age	Point	
	Men	Women
20–34	−9	−7
35–39	−4	−3
40–44	0	0
45–49	3	3
50–54	6	6
55–59	8	8
60–64	10	10
65–69	11	12
70–74	12	14
75–79	13	16

The Framingham risk scores for men and women (Tables 2.1–2.5) are aimed at estimating the 10-year cardiovascular risk of an individual, more precisely, development of coronary heart diseases within the next decade in particular from plasma concentrations of lipids (total cholesterol and HDL–cholesterol [mg/dl]), value of the systolic blood pressure, and smoking habit. In fact, the LDL score is a better index than HDL and triglyceride scores [111].

2.3.1 *Lifestyle*

Lifestyle plays a major role in the prevention of complications of numerous chronic diseases. Although epidemiology does not give proof for the existence of a cause-and-effect relation, but provides statistical links between an investigated disease and explored parameters.

2.3.1.1 Inactivity

Inactivity can diminish life expectancy because it influences aging and predisposes to aging-related diseases. In particular, a relation was found between leukocyte telomere length and physical activity.

Table 2.2 Smoking habit score. (Framingham point score; Source: National Institutes of Health—National Heart, Lung, and Blood Institute)

	Points									
	20–39	40–49	50–59	60–69	70–79	20–39	40–49	50–59	60–69	70–79
NS	0	0	0	0	0	0	0	0	0	0
S	8	5	3	1	1	9	7	4	2	1

NS nonsmoker, *S* smoker

Table 2.3 Systolic blood pressure (BP) score (Framingham point score; Source: National Institutes of Health—National Heart, Lung, and Blood Institute)

Systolic BP (mmHg)	Men		Women	
	Untreated	Treated	Untreated	Treated
< 120	0	0	0	0
120–129	0	1	1	3
130–139	1	2	2	4
140–159	1	2	3	5
≥ 160	2	3	4	6

Moreover, combined to overnutrition, sedentary life causes obesity. Normal values of the *body mass index*, the ratio between the weight (kg) and the square height (m), range between 18.5 and 25. A value between 25 and 30 means moderate overweight; greater than 30, obesity. Obesity associated with adipose tissue dysfunction, characterized by infiltration of inflammatory cells and aberrant production of adipokines, yields an increased risk for metabolic and cardiovascular disorders, especially insulin resistance.

Sedentary lifestyle is associated with impaired endothelial functions exhibited by elevated plasma levels of endothelial microparticles [112]. Months of bed rest increase markers of endothelial dysfunction, such as circulating endotheliocytes and endothelial microparticles, which are released from the membrane of activated, injured, or apoptotic endothelial cells (Vol. 8—Chap. 5. Adverse Wall Remodeling), as well as impaired endothelial vasodilatory capacity (reduced cutaneous acetylcholine-induced vasodilation) [112].

2.3.1.2 Diet

Diet recommendations aimed at optimizing lipid and lipoprotein profiles, blood pressure values, glycemia, and body weight, currently rely on the consumption of plant-based foods, vegetables and fruits, and beverages that contain essential nutrients, such as vitamins (e.g., vitamin-B, -C, and -E), potassium, and magnesium,

Table 2.4 Total cholesterol score. (Framingham point score; Source: National Institutes of Health—National Heart, Lung, and Blood Institute)

	Points									
	Men					Women				
Age	20–39	40–49	50–59	60–69	70–79	20–39	40–49	50–59	60–69	70–79
<160	0	0	0	0	0	0	0	0	0	0
160–199	4	3	2	1	0	4	3	2	1	1
200–239	7	5	3	1	0	8	6	4	2	1
240–279	9	6	4	2	1	11	8	5	3	2
≥280	11	8	5	3	1	13	10	7	4	2

Table 2.5 HDL score. (Framingham point score; Source: National Institutes of Health—National Heart, Lung, and Blood Institute)

HDL (mg/dl)	Points	
	Men	Women
≥60	− 1	− 1
50–59	0	0
40–49	1	1
< 40	2	2

and healthy phytochemicals, such as flavanols, a subset of flavonoids.[2] In addition, vegetables and fruits have a low lipid and high fiber content, as well as an adequate sodium/potassium ratio.

Long-chain polyunsaturated ω3-fatty acids such as docosahexaenoic acid (DHA) that abound in oily fish (e.g., anchovy, herring, mackerel, and salmon) protect the immune, nervous, and cardiovascular systems. In vascular smooth myocytes, DHA (but not its ethyl ester derivative) directly, rapidly, potently, and reversibly activates large-conductance, voltage- and Ca^{2+}-activated K^+ channels (BK or $K_{Ca}1.1$)[3] that acts as a vasodilator, thereby lowering blood pressure [113]. The BK channel is activated by intracellular Ca^{2+} and depolarization. It keeps the membrane hyperpolarized (negative feedback on cellular excitability). This proteic complex operates as a high-affinity receptor for DHA without needing Ca^{2+} ions. Lipid DHA is released from the plasma membrane by G-protein-activated, Ca^{2+}-dependent phospholipase-A2. The concentration required to activate BK channel is about 20 times lower than that needed to stimulate GPR120 involved in anti-inflammatory action of DHA [113].

Ester-conjugated ω3-fatty acids refer to esterification with ethanol (i.e., ethyl esters) or with glycerol as triglycerides. Various ω3 and ω6-fatty acids as well as their ethyl and glycerol ester derivatives have distinct effects on BK channels and blood pressure when acutely applied. In particular, ethyl ester of DHA fails to reduce blood pressure. Oral or parenteral administration of these products thus has different clinical impact. In addition, the physiological response of healthy individuals and patients to various types of fatty acids may differ appreciably.

Flavonoids Flavonoids share a common basic chemical structure. Important flavonoid subsets encompass:

[2] Flavonoids (Latin *flavus*: yellow [golden or reddish yellow]) constitute a set of plant ketone-containing metabolites that comprise: (1) flavones derived from 2-phenylchromen 2-phenyl (1,4)-benzopyrone; (2) isoflavonoids derived from 3-phenylchromen 3-phenyl (1,4)-benzopyrone; and (3) neoflavonoids derived from 4-phenylcoumarine 4-phenyl (1,2)-benzopyrone. Flavanoids, a.k.a. flavan-3-ols and catechins, are nonketone polyhydroxy polyphenol compounds that share with flavonoids a 2-phenyl (3,4)-dihydro 2H-chromen 3-ol skeleton. Flavonoids (flavonols and flavanols) may have antiallergic, anti-inflammatory, antioxidant, antimicrobial, and anticancer effects.

[3] These channels are composed of the pore-forming Slo1 and an auxiliary subunit. DHA activates vascular Slo1–$_{BK}\beta1$ and neuronal Slo1–$_{BK}\beta4$ channels [114]. Stimulation by DHA has a much smaller effect on Slo1–$_{BK}\beta2$ and Slo1–$_{BK}\gamma1$ channels [114].

- *flavonols* (e.g., kaempferol, myrestin, and quercetin [in onions]);
- *flavanols* (e.g., catechin and epicatechin [in cocoa, red wine, green tea, and apples]);
- *flavones* (e.g., apigenin and luteolin [in peppers]);
- *isoflavones*;
- *flavanones* (e.g., eriodictyol, hesperetin, and naringenin [in oranges]); and
- *anthocyanidines* (e.g., cyanidin, delphinidin, malvedin, and pelargonidin [in blueberries]), which are defined by the chemical residues attached to the basic flavonoid structure.

Flavanols are monomers and procyanidins are oligomers of flavanols.

Major flavanol sources include green tea (up to 300 mg/infusion), red wine, cocoa (up to 920–1220 mg/100 g), fruits, such as grapes, pears, berries, and especially apples (up to 120 mg/200 g). However, the flavanol profile (e.g., $^{-}$catechin and $^{+}$catechin as well as $^{-}$epicatechin and $^{+}$epicatechin) can vary considerably according to the food type.

Most of the flavanols in foods exist as oligomers (procyanidins). Only monomers and dimers are absorbed and rapidly metabolized. In plasma, $^{-}$epicatechin (5–250 nmol) represents a minor part of total plasma flavanols (<3 μmol; short half-life) [115]. Epicatechin metabolites include methylated, glucuronidated, and sulfated adducts. Monomeric flavanols are further processed in the liver.

Endothelium-dependent (nitric oxide-mediated) vasodilation is related to the intake of flavanols in healthy subjects with cardiovascular risk factors (smoking, diabetes mellitus, hypertension, and hypercholesterolemia) [116]. In addition to the recovery of endothelial function, flavanol-rich diets improve insulin sensitivity, decrease blood pressure, and reduce platelet aggregation [115].

Regular, moderate consumption of some red wine reduces the risk of coronary heart disease. Red wines are sources of low levels of resveratrol and larger quantities of procyanidins. Procyanidins constitute a subclass of flavonoids. These vasoactive polyphenols lower blood pressure [117]. However, consumption of red wine alone cannot explain the French paradox (low rate of cardiovascular mortality in France despite high saturated fat consumption).

Chocolates and apples contain the largest procyanidin content (164.7 and 147.1 mg, respectively) with respect to red wine and cranberry juice (22.0 and 31.9 mg, respectively) [118]. However, the procyanidin content varied greatly between apple species with the highest amounts in Red Delicious (average 207.7 mg/serving) and Granny Smith apple (average 183.3 mg/serving) and the lowest amounts in Golden Delicious (average 92.5 mg/serving) and McIntosh apple (average 105.0 mg/serving). Flavonoid-rich cocoas contain monomeric flavanols (epicatechin and catechin) and oligomeric procyanidins formed from monomeric units. Both monomers and oligomers support cardiovascular health.

Cocoa and, hence, dark chocolate contain various active compounds, such as flavanols, theobromine, and caffeine, among others. The cocoa content varies greatly between chocolates; the flavanol profile and content differ strongly between cocoas. An inverse relation between flavanol-rich chocolate consumption and cardiovascular disease risk (myocardial infarction and stroke) was observed in a middle-aged (35–65 years) German population of both sexes, without complications of cardiovascular diseases at the time of enrollment [119]. In the group with the highest chocolate consumption (7.5 g/d), both systolic and diastolic blood pressure are 1 mmHg lower on average than that in the low-chocolate consumption group. The group with the highest chocolate intake is also the group with the lowest vegetable diet.

Sulfides Excess salt intake over many years can lead to hypertension. On the other hand, garlic consumption is correlated with the reduction of risk factors of cardiovascular diseases. Garlic is rich in organosulfur compounds.[4] Garlic-derived organic polysulfides are converted by red blood capsules and vascular cells into hydrogen sulfide [120]. The gasomediator hydrogen sulfide decreases blood pressure, protects against ischemic reperfusion damage, and induces O_2-dependent vasodilation.

Cheese and Fungus Metabolites Molded cheese, especially blue cheeses such as Roquefort, may favor cardiovascular health due to the presence of metabolites produced by *Penicillium roqueforti* of the Trichocomaceae set and other fungi, such as andrastin-A to andrastin-D and roquefortine, which can prevent cholesterol synthesis and bacterial growth [121].

Mediterranean Diet and Nitrofatty Acids The Mediterranean diet is characterized by a high consumption of unsaturated fatty acids, especially from olive oil and fish rich in oleic and linoleic acids, with vegetables rich in nitrite and nitrate, resulting in endogenous formation of reactive nitrofatty acids.[5] In particular, linoleic acid and nitrite are major constituents of the Mediterranean diet that elevates concentrations of thiol-reactive electrophilic nitrofatty acids. The acidic and low-oxygen environment in the stomach enables an efficient nitration of such unsaturated fatty acids by nitrite. Nitro-oleic acid (or 9- and 10-nitrooctadeca 9-enoic acid [NO_2OA]) and other free and esterified fatty acid nitroalkenes are formed at elevated levels when unsaturated fatty acids are ingested together with a source of nitrite [123].

---

[4] Allicin, the main organosulfur compound, is produced from amino acid alliin by alliinase and rapidly decomposes mainly into diallyl sulfide, disulfide, and trisulfide, and ajoene. After consumption, these compounds are rapidly metabolized.

[5] The free radical nitric oxide can be added to unsaturated and hydroperoxy fatty acids. Nitrated fatty acids result from the reaction of nitrogen dioxide with unsaturated fatty acids. The $NO_2^\bullet$ radical can form, according to the local oxygen concentration, nitronitrite and nitroallyl derivatives as well as lipid hydroperoxide. Nitrated derivatives of unsaturated fatty acids are formed under oxidative and nitrative stresses. In biological tissues, nitrofatty acids are produced by nonenzymatic reactions not only with $NO^\bullet$ and $NO_2^\bullet$, but also with peroxynitrite ($ONOO^-$). Nitrofatty acids lodge in membrane phospholipids. Nitrated derivatives of arachidonic, eicosapentaenoic, linoleic, linolenic, oleic, and palmitoleic acids and their nitrohydroxy derivatives (nitrohydroxy oleate, linoleate and linolenate) can be also detected in human plasma and urine. In plasma, nitrofatty acids are free, bound reversibly to thiol-containing proteins, and esterified (as cholesterol esters). In plasma, nitrofatty acids are stabilized by incorporation into lipoproteins. Nitrofatty acids modulate macrophage activation, prevent leukocyte and platelet activation, and promote vascular relaxation [122].

The endogenous lipid electrophile nitro-oleic acid and other fatty acid nitroalkenes signal via pleiotropic mechanisms, such as activation of peroxisome proliferator-activated receptor (PPAR)γ (nuclear receptor NR1c3), the stress sensor and NFE2L2 sequestrator Kelch-like erythroid cell-derived protein with CNC homology (EHC)-associated protein KEAP1,[6] and nuclear factor erythroid-derived-like factor NFE2L2-regulated antioxidant response genes, and inhibition of proinflammatory gene expression regulated by NFκB [123]. Nitrofatty acids also inhibit lipopolysaccharide-induced cytokine expression and induce HO1 expression via activation of NFE2L2-regulated gene expression. Therefore, electrophilic lipid derivatives can control gene transcription that overall engenders an anti-inflammatory response.

Nitroalkenes react with nucleophiles, such as cysteine and histidine in various target proteins. In particular, they target transient receptor potential (TRP) channels in the central and peripheral nervous system [124]. Nitroalkene fatty acid derivatives can activate TRP channels on capsaicin-sensitive afferent nerve terminals, leading to increased smooth myocyte contractility via release of neuropeptides (neurokinins) and activation of $Ca_V1.2b$ channel. Nitro-oleic acid activates TRPA1 and TRPV1 channels in sensory neurons involved in neurogenic inflammation and pain induced by noxious chemicals or thermal stimuli, thereby provoking calcium influx, membrane depolarization, and firing. In addition, high concentrations of nitro-oleic acid suppress firing in dorsal root ganglion neurons, hence contributing to anti-inflammatory effects.

Nitrofatty acids that can be detected in healthy human urine are produced at heightened levels during metabolic stress and inflammatory conditions (from nmol to μmol concentrations) [124].

Nitro-oleic acid is a member of the category of electrophilic nitroalkenyl fatty acids formed by reactions between unsaturated fatty acids, nitric oxide (NO)- and nitrite (NO_2^-)-derived nitrogen dioxide (NO_2), these reactions being favored by the prooxidative condition of inflammation. In addition, when pH is low enough (< 6) to protonate NO_2^- to nitrous acid (HNO_2), this condition also yields the nitrating species NO_2. Therefore, nitrogen dioxide is both a product of oxidative inflammatory reactions involving nitric oxide and nitrite and acidic conditions in the presence of nitric oxide or nitrite. Similarly, nitroalkenes are produced by addition of the radical nitrogen dioxide ($NO_2^{\bullet}$) to one or more of the olefinic carbons of unsaturated fatty acids [123].

[6] In unstressed conditions, NFE2L2 is permanently ubiquitinated by the Cul3–$(KEAP1)_2$ protein–ubiquitin ligase complex and rapidly degraded in proteasomes. The KEAP1 homodimer binds to a single NFE2L2 molecule. Upon exposure to electrophilic and oxidative stresses, KEAP1 is modified and its ubiquitin ligase activity declines, thereby stabilizing NFE2L2 to induce cytoprotective gene transcription.

Linkage of electron-withdrawing nitro group to alkenyl groups confers a potent and reversible electrophilic reactivity to fatty acids. Fatty acid nitroalkenes can modify proteins covalently, thereby associating metabolic and redox signaling with the posttranslational regulation of target proteins.

Lipidic electrophiles mediate antihypertensive signaling, as they connect to soluble epoxide hydrolase (sEH; at Cys521 proximal to its catalytic site), thereby precluding its activity, especially hydrolysis of its vasoactive substrates epoxyeicosatrienoic acids (EET) [123]. EETs are metabolites of arachidonic acid processed by cytochrome-P450 epoxygenase. They are hydrolyzed into corresponding dihydroxyepoxyeicosatrienoic acids (DHET) by sEH. Upon sEH inhibition, EETs accumulate and provoke vasodilation, hence lowering blood pressure. The EET/DHET ratio is elevated in plasma.

Vasodialation is unaffected by inhibition of soluble guanylate cyclase. Nitro-fatty acids not only relax vascular wall, but also attenuate platelet activation and inflammation via a cGMP-independent mechanism [123].

Ketone Bodies Ketone body and anaplerotic metabolisms have been introduced in Vol. 6, Chap. 3. Cardiovascular Physiology.

Ketone Bodies in Cardiovascular Diseases Defective ketone body synthesis and catabolism are pathogenic factors. Conversely, some diseases disturb ketone body metabolism. Hepatic ketogenesis is suppressed at later stages of hyperinsulinemic obesity [125]. Ketone bodies participate in the regulation of mitochondrial metabolism, energetics, and ROS production.

Impaired cardiac energetics cause cardiomyopathy. Conversely, cardiomyopathies are associated with changes in energetic metabolism. In dilated and hypertrophic cardiomyopathies, the contribution of ketone bodies to cardiac energetics is augmented [125]. Decayed myocardial ketone body oxidation causes pathological outcomes.

In addition, hepatic ketogenesis is reinforced and circulating ketone body concentration rises during the development of heart failure. Ketone body metabolism may contribute to myocardial adaptation to ischemia–reperfusion injury, at least in rodents [125].

Ketogenic Diet Ketogenic diet raises blood levels of cholesterol and free fatty acids [125]. Anaplerotic 5-carbon ketone bodies and their precursor odd-chain fatty acids are proposed for the treatment of long-chain fatty acid oxidation (LCFAO) disorders. Ingestion of odd-chain fatty acids promotes hepatic C5-ketogenesis; β-oxidation of odd-chain fatty acids yields propionylCoA, an anaplerotic substrate that, in the liver, can also be packaged into C5 ketone bodies [125]. In extrahepatic cells, C5 ketone bodies are oxidized by CoA transferase SCOT, hence regenerating propionylCoA.

Dietary Phosphate A deranged calcium–phosphate metabolism and elevated phosphate concentrations are correlated with reduced life expectancy and cardiovascular events because of endothelial dysfunction, vascular calcification, and myocardial hypertrophy [126]. Phosphate intake (by ingestion) and phosphatemia are indices of cardiovascular risk, not only in chronic kidney disease, but also in individuals with intact renal function [127].

Sources of Dietary Phosphate Sources of dietary phosphate include: (1) phosphate additives, i.e., inorganic phosphate salts (i.e., sodium phosphate) that serve as preservatives, flavor enhancers, color stabilizers, sweeteners, antioxidants, and emulsifiers, and (2) natural phosphate (phosphoproteins, phospholipids, phosphate esters, and phytates from milk products, fish, meat, and vegetables, respectively), i.e., organic phosphates in unprocessed foods.

Intestinal absorption is regulated by calcitriol. Natural phosphates are not only slowly and incompletely hydrolyzed but also slowly and incompletely (30–60 %) absorbed in the digestive tract in the presence of adequate vitamin-D levels [126]. On the other hand, inorganic phosphate is absorbed in the gut in larger proportions (80–100 %).

Calcium–Phosphate Metabolism and its Regulation Average western diets provide a daily phosphate intake of 1000–1700 mg according to food composition. The major determinant of plasma phosphate concentration is renal phosphate excretion; phosphaturia balances oral intake and intestinal absorption. Calcium and phosphate are regulated by parathyroid hormone (PTH), active vitamin-D, or calcitriol ([1, 25]$(OH)_2D_3$),[7] and fibroblast growth factor FGF23.

The major fraction of total body calcium localizes in bones, only a small part of calcium circulates in plasma as free (ionized) calcium (50 %) and binds to proteins (mostly albumin) as well as citrate, sulfate, and phosphate.

A reduced plasma Ca^{2+} concentration rapidly provokes secretion of PTH that inhibits renal calcium excretion and stimulates renal hydroxylation of 25OH-vitamin-D to calcitriol. The latter triggers intestinal calcium absorption (Table 2.6). In addition, calcitriol and PTH cause bone resorption by osteoclasts. Conversely, hypercalcemia inhibits PTH secretion via parathyroid calcium-sensing receptors.

FGF23 The master regulator of renal phosphate handling is fibroblast growth factor FGF23 liberated by osteocytes upon high calcitriol and phosphate levels, rather than PTH. Factor FGF23 is mainly synthesized and secreted by osteoblasts stimulated by calcitriol, PTH, and hyperphosphatemia.

Target cells of FGF23 comprise renal and parathyroid gland cells as well as cardiomyocytes [126]. In the kidney and parathyroid glands, FGF23 interacts with β-glucoronidase Klotho coreceptor. Membrane-bound Klotho selectively localizes

---

[7] Humans regularly exposed to sunlight synthesize adequate concentrations of vitamin-D. Solar ultraviolet-B waves convert 7-dehydrocholesterol to previtamin-D_3 (or precholecalciferol), which spontaneously isomerizes to vitamin-D_3 (cholecalciferol). Otherwise, few food components contain sufficient vitamin-D, such as fish liver, fatty fish, and egg yolks. Vitamin-D receptor lodges in myocardial and vascular cells. Calcitriol represses the renin–aldosterone axis, causes insulin secretion and sensitivity, prevents proliferation of vascular smooth myocytes and hypertrophy of cardiomyocytes, and can exert immunoregulatory functions.

Table 2.6 Calcium–phosphate metabolism. (Source: [127])

Organ	Effect	Regulators
Gut	Ca^{2+} absorption	Calcitriol (⊕)
	Phosphate absorption	Calcitriol (⊕)
		FGF23 (⊖)
Kidney	Ca^{2+} excretion	PTH (⊖)
		Hypercalcemia (⊕)
	Phosphate excretion	PTH (⊕)
		FGF23 (⊕)
		Hyperphosphatemia (⊕)
	Calcitriol synthesis	PTH (⊕)
		Hypercalcemia (⊖)
		Hyperphosphatemia (⊖)
Bone	Ca^{2+} release	PTH (⊕)
		Calcitriol (⊕)
	Phosphate release	PTH (⊕)
		Calcitriol (⊕)
PTG	PTH secretion	Hypercalcemia (⊖)
		Hyperphosphatemia (⊕)
		Calcitriol (⊖)

⊕ stimulation, ⊖ inhibition, *PTG* parathyroid gland, *PTH* parathyroid hormone

to the kidney, parathyroid gland, and choroid plexus. The FGF23–Klotho dimer binds to the fibroblast growth factor receptor, a receptor Tyr kinase, thereby causing its autophosphorylation and triggering signaling via 3 major pathways: PI3K–PKB, PLCγ–PKC, and Ras–MAPK. Factor FGF23 regulates phosphate balance via expression of genes involved in PTH, vitamin-D, and phosphate metabolism. In cardiomyocytes, FGF2 uses heparan sulfate proteoglycans as coreceptors, FGFR, and primarily the Ras–MAPK pathway, whereas FGF23 signals primarily via the PLCγ–PP3 pathway.

Factor FGF23 impedes calcitriol synthesis and promotes its degradation. It indeed represses renal 1α-hydroxylase, hence the renal synthesis of calcitriol [126]. Proximal tubule cells of the nephron produce fibroblast growth factor receptors FGFR1, FGFR3, and FGFR4, FGFR1 being the predominant receptor for the FGF23 hypophosphatemic action [128]. In the proximal tubule, FGF23 reduces production and activity of 2 sodium–phosphate cotransporters NaPi2a and NaPi2c (SLC34a1 and SLC34a3), hence augmenting phosphaturia. In addition, FGF23 suppresses PTH synthesis in the parathyroid glands.

Endothelial Dysfunction Even transient hyperphosphatemia, such as that during the postprandial period, can cause endothelial dysfunction and imbalance between nitric oxide and reactive oxygen species. Increased influx of phosphate via Na^+–Pi cotransporters leads to inhibitory phosphorylation of NOS3 synthase.

Vascular Calcification Intake of calcium salts (acetate or carbonate) can lead to calcium tissue deposition, in particular vascular calcification, even in the absence of hypercalcemia. In the case of overload, calcium–phosphate crystals may dispose in the vessel wall and myocardium.

High phosphatemia initiate *osteogenic transdifferentiation*, i.e., the evolution of vascular smooth myocytes toward a osteochondrogenic phenotype, vascular smooth myocytes differentiating into osteoblast-like cells [127]. Extracellular phosphate is actively taken up by Na^+–Pi cotransporters into the cell. Calcium ion stimulates Na^+–Pi cotransporter SLC34a1 on vascular smooth myocytes, thereby permitting intracellular accumulation of phosphate ions.

After its uptake into vascular smooth myocytes, phosphate primes synthesis of proteins involved in matrix mineralization and bone formation and downregulates vSMC-specific transcription factors. Vascular smooth myocytes subsequently evolve from a contractile into osteochondrogenic phenotype, and release membrane-enclosed matrix vesicle-like structures (size 100–400 nm) from the plasma membrane [127]. Furthermore, phosphate can induce apoptosis in vascular smooth myocytes, which subsequently release apoptotic bodies. Moreover, Ca^{2+} supports liberation of matrix vesicle-like particles of living vSMCs and apoptotic bodies from apoptotic vSMCs that act as nuclei for extracellular calcium–phosphate precipitation [127]. Calcium also reduces the expression of calcification inhibitors by vascular smooth myocytes.

Myocardial Hypertrophy Elevated FGF23 levels cause maladaptive left ventricular hypertrophy via vascular stiffening and subsequent elevated cardiac afterload, in addition to calcifications. Maladaptive cardiac hypertrophy results from the activation of the PP3–NFAT pathway [127]. Moreover, elevated phosphatemia triggers secretion of FGF23 to raise phosphaturia, but in parallel activates cardiomyocytes, thereby further exaggerating maladaptive cardiac hypertrophy. In addition, intake of calcium supplements for bone protection against osteoporosis can cause myocardial infarction [127].

Nutrition and Regulator Hormones The *arcuate nucleus* of the *hypothalamus* controls the metabolic rate, hunger, and satiety. It contains two cell types: (1) hunger NPY/AgRP[8] cell and (2) satiety POMC cell.[9] *Ghrelin* and *leptin* are two major hormonal controllers of hypothalamic hunger and satiety cells. They are antagonistic and complementary, responding to acute and chronic changes in energy balance. Their effects are mediated by hypothalamic neuropeptides, such as *neuropeptide-Y* and *agouti-related peptide*. Endocrine and nervous (vagal afferent) signals contribute to actions of ghrelin and leptin.

[8] NPY: neuropeptide-Y; AgRP: agouti-related protein homolog.

[9] POMC: proopiomelanocortin.

The hunger hormone, ghrelin, a member of the motilin-related category of regulatory peptide[10] and a ligand of the G-protein–coupled growth hormone secretagogue (GHS) receptor,[11] is a 28-amino acid hunger-stimulating hormone. In addition to stimulating GH secretion and gastric motility, ghrelin wakes up appetite and induces a positive energy balance with body's weight gain. This orexigenic and adipogenic peptide is produced mainly by endocrine P/D1 cells of oxyntic glands[12] in the fundus of the stomach and ϵ cells of the pancreas. It circulates (plasma concentration 117 ± 37 fmol/ml) and is conveyed to the hypothalamus where it activates hunger cells and inhibits satiety cells. Although about 90 % of ghrelin is made in the stomach, duodenum, and jejunum, it is also synthesized in the pancreas, pituitary gland, kidney, and various regions of the brain such as hypothalamus.

The satiety hormone, leptin, is produced by adipocytes. Once it is liberated into blood, it travels to the hypothalamus where it activates satiety cells and inhibits hunger cells. High leptin concentrations in combination with high insulin levels prevent ghrelin production. Low ghrelin concentrations render hunger cells hypersensitive to ghrelin. On the other hand, high leptin concentrations cause insensitivity to leptin of satiety cells. *Obestatin*, a preproghrelin-derived peptide, is a hormone that stops the hunger sensation. This anorectic peptide is produced by ghrelin-producing cells of the gastrointestinal tract. It antagonizes growth hormone secretion and food intake induced by ghrelin [131].

Ghrelin improves memory and concentration. Moreover, ghrelin favors restorative sleep and dreams that preclude leptin production.

2.3.2 *Inflammatory Intestine, Gut Microbiome, and Cardiovascular Disease*

The intestinal barrier determines the nutrient uptake. The intestinal microbiota within the alimentary canal influences the intestinal barrier and hence the nutritional and metabolic status of the organism.

2.3.2.1 Intestinal Microbiota

The gut microbiome (1000–1500 bacterial species) interacts with the intestinal mucosa and may affect the function of other organs, such as the heart, lung, and

[10] Ghrelin and motilin are coproduced in the same cells in the duodenum and jejunum and stored in all secretory granules in these cells before being cosecreted [129]. In addition, preproghrelin not only gives rise to orexigenic ghrelin, but also obestatin, a ghrelin antagonist [130].

[11] Ghrelin is a potent GHS, once it is processed by ghrelin ^{O}acyl transferase.

[12] Four types of endocrine cells, D, enterochromaffin (EC), enterochromaffin-like (ECL), and P/D1 cells, have been identified in the oxyntic mucosa. Major products stored in granules of D, EC, ECL, and P/D1 cells are somatostatin, serotonin, histamine and uroguanylin, and ghrelin, respectively,

lymphatic circuit [132]. Diet, bacterial composition of the environment, and host genetics, influence the composition of the microbiome in a given subject.

The commensal biota provides numerous nutrients and small molecules. In addition, certain bacterial species such as Firmicutes contribute to a higher uptake of molecules such as short-chain fatty acids. Moreover, these commensals contribute to immunity, as it directly influences the cytokine production of epitheliocytes and innate immunocytes.

Many acute and chronic disorders that affect the heart, such as obesity and metabolic syndrome, are linked to inadequate postnatal microbiome acquisition or environmental microorganism exposure during early childhood [132]. In obese patients, the gut contains different bacterial species, especially Firmicutes. Transplantion of the microbiome of obese mice in lean mice provokes a weight gain in the absence of diet change (but the transplantation of the microbiome of lean mice into obese mice does not engender a weight loss).

2.3.2.2 Oral Drugs and Alimentary Tract

The gut determines not only nutrient uptake but also absorption of drugs after oral administration. Conversely, drugs influence the intestinal function. Macrolide antibiotics inhibit the cytochrome-P450 isozyme CyP3a that is constitutively expressed in small intestinal villi and contributes to prehepatic metabolism of drugs [132]. Statins are metabolized by CyP3a4 and CyP3a5, hence influencing their pharmacokinetics; conversely, they increase CyP3a expression [132].

2.3.2.3 Altered Intestinal Barrier and Bacteria

An altered intestinal barrier with elevated permeability for bacterial products (lipopolysaccharide, bacterial DNA with CpG motifs, and peptidoglycans [e.g., muramyl dipeptide]) can contribute to atherosclerosis and chronic heart failure. An acute prominent inflammatory response to bacteria facilitates innate immune defense, but can increase the risk of atherosclerosis. Microbial components (toxins and DNA) as well as factors secreted by intestinal epithelial and dendritic cells can intervene in the pathogenesis.

Conversely, impaired cardiac function in chronic heart failure impacts intestinal microcirculation and can cause a barrier defect of the intestinal mucosa, thereby favoring bacterial invasion. Toll-like receptor-4, the receptor for lipopolysaccharide of Gram− bacteria, is expressed on cardiomyocytes and foam cells, among other cell types. Once they are recognized by Toll-like receptors, microbial products signal to neutrophils, activating NFκB and transcription of proinflammatory genes. Single nucleotide polymorphism in the Tlr4 gene replacing Asp299 with Gly (D299G) and Thr399 with Ile (T399I) in TLR4 causes lipopolysaccharide hyporesponsiveness. These two genetic variants are linked to various infectious and noninfectious diseases.[13] Carriers of 1 or 2 alleles with TLR4 polymorphisms (Asp299Gly and Thr399Ile) are more susceptible to bacterial infections. In particular, patients with the Asp299Gly TLR4 allele have lower levels of certain proinflammatory cytokines (e.g., IL6), acute-phase reactants, and adhesion molecules. On the other hand, the Asp299Gly TLR4 polymorphism that attenuates TLR4 signaling and inflammatory response to Gram− bacteria decreases risk of atherosclerosis [134].

In addition, bacterial lipopolysaccharides can interact with low-density lipoproteins, hence influencing lipoprotein metabolism and contributing to the development of atherosclerosis [132]. Furthermore, lipopolysaccharides damage endotheliocytes and support the production and release of superoxide anions and the oxidation of low-density lipoproteins.

2.3.2.4 Inflammatory Bowel and Risk for Cardiovascular Diseases

The intestine is associated with metabolic diseases. The autoimmune disorder of the small intestine *celiac disease* and the regional immunity-related chronic enteritis *Crohn syndrome* (caused by a combination of environmental, immune, and bacterial factors in genetically susceptible individuals) disturb the absorption of nutrients and drugs in the small intestine. In addition, celiac disease is characterized by a decreased expression of some cytochrome-P450 isozymes such as CyP3a. Patients with inflammatory bowel diseases have a higher risk for coronary artery disease, despite a lower exposure to classical risk factors [132].

The metabolism by the gut flora of phosphatidylcholine that generates three metabolites, choline, trimethylamine ^{N}oxide, and betaine, predicts risk for cardiovascular disease [132].

Patients with chronic heart failure, a state of chronic inflammation, have elevated levels of soluble CD14 (shed from the plasma membrane), a component of the bacterial lipopolysaccharide receptor. Chronic heart failure favors bacterial migration through a congestive intestinal mucosa. The altered mucosal perfusion indeed raises intestinal mucosal permeability, hence facilitating the penetration of bacteria and their product. Lipopolysaccharides trigger catecholamine release by granulocytes and phagocytes [132].

Moreover, CHF patients have morphological and functional alterations of the intestine. The large bowel wall thickens. The mucosal permeability for the sugar alcohol mannitol, nondigestible lactulose, and artificial sucralose increases in both the small and large intestine [132]. The passive carrier-mediated transport for ^{D}xylose lowers. In addition, the bacterial density in the sigmoidal mucosal biofilm and the extent of their adherence heighten with respect to healthy subjects.

The small and large intestine is affected by hypoxia caused by chronic heart failure. Hypoxia raises sympathetic activity and production of inflammatory cytokines,

[13] TLR4 polymorphism is not associated with the risk of developing asthma, but genetically determined hyporesponsiveness to lipopolysaccharide can increase atopy severity [133].

leukotrienes, and prostaglandins, thereby engendering intestinal dysfunction. Increased sympathetic tone and resulting vasoconstriction redistribute blood flow away from the splanchnic circulation. Moreover, the venous stasis further increases mucosal hypoxia [132].

2.3.3 *Psychological Context*

Depression, acute stress, phobia, and anxiety increase the probability of fatal coronary heart disease. This psychological context favors inflammation induced by secretion of cytokines by immunocytes. Myocardial infarction indeed has a strong inflammatory component. Conversely, laughter, happiness, and self-esteem lower the expression of some proinflammatory cytokines.

2.3.4 *Smoking*

Smokers have a two to threefold higher risk of developing arterial disease than nonsmokers. Smokers have a higher risk to develop heart failure than ex-smokers and nonsmokers. Ex-smokers have a 30 % lower mortality than smokers 2 years after smoking cessation.

Nicotine, the addictive ingredient of tobacco, stimulates the sympathetic nervous system, thus increasing blood flow rate, and causes endothelial dysfunction. Cardiac frequency determines myocardial O_2 consumption. An increased resting cardiac frequency reduces cardiac performance and ischemic threshold.

Nicotine also exaggerates postinjury intimal hyperplasia and contributes to the formation of intracranial [135] and abdominal aortic aneurysm [136]. Smoking is a major risk factor in abdominal aortic aneurysm, because of oxidation of serpin-A1 (or α1-antitrypsin), which is carried by high-density lipoproteins at low levels in dyslipidemia, among other factors.

Smoke exposure also provokes damage of nuclear and mitochondrial DNA. Mitochondrial DNA damage can precede atherogenesis and, then, be exacerbated by impaired antioxidant activity [137]. Moreover, cardiac adenine nucleotide translocase activity needed for ATP synthesis can decrease.

Smoking can contribute to endothelial dysfunction with a disturbed balance between nitric oxide and oxygen free radicals. Like hypertension, smoking alters the expression of endothelins. Like hypertension, hypercholesterolemia, and oxidative stress, smoking promotes NFκB activation. As with other cardiovascular risk factors (diabetes, dyslipidemia, hypertension, and renal disease), smoking is associated with a decreased number and function of bone marrow-derived endothelial progenitor cells, which participate in endothelial regeneration and angiogenesis. Smoking (1) increases levels of proinflammatory compounds (e.g., tumor-necrosis factor-α and endothelial intercellular adhesion molecule-1), and reactive oxygen species, (2) decreases concentrations of anti-inflammatory, antioxidant HDL–cholesterol (HDL^{Cs}), which promotes reverse cholesterol transport, and adiponectin, and (3) activates leukocytes and platelets.

Tobacco smoke elicits production of intra- and extracellular superoxide ($O_2^{\bullet -}$) as well as peroxynitrite ($ONOO^-$) and causes oxidation (inactivation) of nitric oxide NOS3 synthase [138]. Smokers have a drop of endothelium-mediated vasodilation.

Proliferation of vascular smooth myocytes results from hypoxia, arterial injury, and stimulation by angiotensin-2, which lead to activation of early growth response EGR1 factor [139]. Factor EGR1 stimulates transcription of proinflammatory genes, such as those that encode the cytokines TNFSF1 and IL2, chemokine CCL2, and adhesion molecule ICAM1, as well as expression of growth factors, such as FGF2, PDGF, and TGFβ, and tissue factor [140].

Proatherosclerotic and prorestenotic nicotine increases vascular smooth myocyte proliferation via nonneuronal nicotinic acetylcholine receptors,[14] ERK1 and ERK2, and phosphorylated ELk1 factor, which in turn upregulates the EGR1 expression [139].

2.3.5 *Resting Tachycardia*

An increased, sustained resting heart frequency reduces cardiac performance. Moreover, resting tachycardia is an accelerator of atherosclerosis development [141].

When the cardiac frequency is reduced, the sensitivity of the baroreflex increases (Table 2.7). The baroreflex is aimed at short-term controlling sympathetic activity, cardiac frequency and flow rate, as well as blood pressure via the vasomotor tone.

Activation of endothelial β2-adrenergic receptor raises endothelial cytosolic concentration of free calcium ion and NOS3-dependent NO production and release. β1-adrenoceptor does not reside on endothelial cells [143].

Nitric oxide exerts a stronger effect on the regulation by the baroreflex of cardiac frequency than on the renal sympathetic nerve activity aimed at regulating blood volume to normalize arterial blood pressure.

2.3.6 *Hypertension*

Arterial hypertension is a major cardiovascular risk factor. Arterial blood pressure is a complex genetic trait with heritability estimates of 30–50 %. Arterial hypertension affects at least 25 % of adults in industrialized societies.

[14] Among nicotinic acetylcholine receptors, which are pentameric ligand-gated ion channels, in vascular smooth myocytes, the homomeric α7-nAChR possesses high Ca^{2+} permeability and rapid onset of desensitization.

Table 2.7 Baroreflex. The processing nervous path from arterial baroreceptors in the carotid body and aortic arch to the medulla, spinal cord, and blood vessel walls (sympathetic nerve activity; Source: [142]). When arterial blood pressure increases, baroreceptors are activated and stimulate the nucleus of the solitary tract (NTS) that activates the caudal ventrolateral medulla (CVLM), an inhibitory medullary site, which, in turn, inhibits the sympathoexcitatory site rostral ventrolateral medulla (RVLM), thus inhibiting the sympathetic branch of the autonomic nervous system via the intermediolateral nucleus of the spinal cord (ILNSC), thereby decreasing arterial blood pressure. Conversely, a low arterial blood pressure increases the sympathetic tone via disinhibition, i.e., activation of the rostral ventrolateral medulla. The NTS also sends excitatory fibers to the nucleus ambiguus (vagal component) that regulates the parasympathetic nervous system, assisting decreasing sympathetic activity during hypertension. Baroreceptor activation thus inhibits the sympathetic nervous system and stimulates the parasympathetic nervous system. The baroreflex maximizes arterial blood pressure reduction, as it couples sympathetic inhibition and parasympathetic activation in response to hypertension. Conversely, sympathetic activation coupled with parasympathetic inhibition enables the baroreflex to elevate arterial blood pressure in response to hypotension. Nitric oxide reduces sympathetic nerve activity, hence modulating the sympathetic effect on arterial blood pressure

	Afferents and central processors
1	Arterial baroreceptors
2	Excitatory (glutamatergic) afferents neurons
3	Excitatory (glutamatergic) neurons of NTS
4	Inhibitory (gabaergic) neurons of CVLM
5	Excitatory (glutamatergic) neurons of RVLM
	Efferents
6	Sympathetic preganglionic (cholinergic) neurons of ILNSC
7	Sympathetic postganglionic (noradrenergic) neurons
8	Adrenergic receptors of vascular endothelial and smooth muscle cells
	Adrenergic receptors of cardiac cells
	Adrenergic receptors of renal cells

Arterial blood pressure ($p_a = \mathtt{R}q$) is controlled by peripheral vascular resistance ($\mathtt{R}$) and blood flow rate (q).[15] Vascular resistance depends on the vasomotor tone under local and remote control. Blood flow rate depends on cardiac performance and blood volume, itself related to salt content.

Constitutive and environmental factors that influence blood pressure include sympathetic tone, dietary salt intake, alcohol consumption, age, body mass index, and physical activity. Arterial hypertension depends in particular on the renal sodium handling, steroid hormone metabolism, and mineralocorticoid receptor activity.

Treatment of hypertension involves numerous types of drugs. Blockers of β-adrenergic and angiotensin AT_1 receptors abolish the action of adrenaline and

[15] The pressure–flow relation is an Ohm-like law. In electricity, the Ohm law states that the current (i) through a conductor between two points is proportional to the potential difference (u) between these two points. The constant of proportionality is the resistance ($u = \mathtt{R}i$).

angiotensin-2. β-blockers inhibit β-adrenergic receptors, thereby decreasing cardiac frequency and contractility. β-adrenoceptors initiate signaling cascades upon phosphorylation by AMPK, CamK2, PKA, PKB, and TOR kinases. These enzymes target in particular ion channels such as $K_V7.1$ and transporters that mediate ion fluxes at high cardiac frequency [144]. G-protein-coupled receptors that regulate myocardial contractility are also substrates. Inhibitors of angiotensin-converting enzyme prevent formation of angiotensin-2. Diuretics hinder increase in plasma volume. Calcium antagonists treat vasoconstriction.

High salt intake raises blood volume and vascular smooth myocyte contraction, thereby increasing heart load. Natriuresis with increased glomerular filtration and inhibited sodium reabsorption restores normal osmotic pressure.

The midbrain produces natriuretic signals that operate via: (1) the adrenal gland, which produces cardioactive steroids, and (2) the heart, which synthesizes atrial natriuretic peptide.

The vascular smooth myocyte tone determines the total peripheral vascular resistance and arterial blood pressure. Most forms of hypertension result from vasoconstriction. Hormone levels increase in response to a high-salt diet. Hormone binding to GPCRs located in caveolae of smooth myocytes with Na^+–K^+ ATPase triggers Ca^{2+}-dependent and -independent pathways.

Calcium-dependent mechanism functions via Gq/11, PLCβ, and IP_3, thereby inducing Ca^{2+} release from intracellular stores and myosin light chain phosphorylation by activation of myosin light chain kinase.

Calcium-independent process impedes phosphorylated myosin light chain degradation via subunits of the G12/13 subclass, RhoGEF12 guanine nucleotide-exchange factor,[16] Rho GTPase, and RoCK kinase that inhibits myosin light chain phosphatase. Phosphorylated myosin light chain allows myosin to interact with actin and generate smooth myocyte contraction.

The dual regulation of myosin light chain phosphorylation using both signaling cascades produces vasoconstriction. The Gq/11 pathway is responsible for the maintenance of basal blood pressure and intervenes in the development of salt-induced hypertension [145]. The G12/13 subunit-primed pathway only yields salt-induced hypertension. In response to sodium, hormones may bind to Na^+–K^+ ATPase and launch Ca^{2+} influx, as well as inhibit myosin light chain dephosphorylation via Rho GTPase to prevent smooth muscle relaxation.

2.3.7 *Diabetes*

Diabetes mellitus,[17] or simply diabetes, is a chronic disease associated with hyperglycemia. Three types of diabetes mellitus exist. *Type-1 diabetes* (a.k.a. insulin-dependent and childhood-onset [juvenile] diabetes) in which the pancreas fails to

[16] Protein RhoGEF12 is activated by the interaction of $G\alpha_{12}$ with $G\alpha_{13}$ subunit.

[17] διαβητησ: compass (Latin: *diabetes*).

produce the proper amount of insulin, an hormone that assists glucose uptake by cells to yield energy. This autoimmune disease is characterized by loss of insulin-producing β cells of islets of Langerhans in the pancreas. *Type-2 diabetes* (a.k.a. insulin-independent, obesity-related, and adult-onset diabetes) results from insulin resistance, as cells do not respond to the insulin. Sometimes, it combines resistance to insulin action, inadequate insulin secretion, and inappropriate glucagon secretion. *Gestational diabetes* occurs when pregnant women develop hyperglycemia. Classical symptoms include polyuria, polydipsia, and polyphagia. Long-term complications include microangiopathy, diabetic neuropathy, chronic renal failure, diabetic retinopathy, and cardiovascular disease such as aggravated atherosclerosis.

Diabetes mellitus is also characterized by endothelial dysfunction. In gestational diabetes, the adenosine–^{L}arginine–nitric oxide pathway is activated [146]. Second messengers involved in adenosine signaling include PKC, ERK1, and ERK2 that activate ^{L}arginine ingress through SLC7a1,[18] but preclude adenosine import through the SLC29a1 carrier.[19] Subsequent extracellular accumulation of adenosine activates the A_{2A} receptor, increases transcription of the Nos3 and SLC7A1 genes, hence NO synthesis. Cultured human umbilical vein endotheliocytes from gestational diabetic pregnancies or subjected to hyperglycemia produce higher NO levels. However, the NO-dependent downregulation of the SLC29A1 gene transcription lowers uptake of vasodilatory adenosine [146]. Nitric oxide supports formation of the complex made of DNA-damage-inducible transcript DDIT3, or CCAAT/enhancer-binding protein (C/EBP) homologous protein-10 (CHOP or CHP10),[20] and C/EBPα factor [147].[21] In the nucleus, like the transcription factor Specific protein SP1 in hyperglycemia, the DDIT3–C/EBPα heterodimer represses Slc29a1 gene transcription (as well as glucose transporter GluT4) [147].

2.3.8 *Metabolic Syndrome*

Metabolic syndrome is the set of risk factors that includes hyperlipidemia, hypertension, chronic inflammation, obesity, and type-2 diabetes; Vol. 8—Chap. 2. Metabolic Syndrome). These perturbations that cause atherosclerosis result from failure to sense and respond to metabolic cues properly.

[18] A.k.a. cationic amino acid transporter CAT1 and system y^+ basic amino acid transporter.

[19] A.k.a. equilibrative (nitrobenzylmercaptopurine riboside [NBMPR]-sensitive) nucleoside transporter ENT1.

[20] A.k.a. growth arrest and DNA-damage-inducible protein GADD153.

[21] Nitric oxide modulates expression and activity of transcription factors, such as CHOP–C/EBPα, NFκB, GATA, Activator protein-1, and interferon regulatory factor IRF1 in endotheliocytes [147]. Nitric oxide represses Vcam1 gene transcription via NFκB in human aortic and saphenous vein smooth myocytes.

Elevated circulating lipids, especially low-density lipoproteins, is a major risk factor. Transfer of LDLs into the vessel wall and subsequent oxidation by intracellular lipoxygenases or action of reactive oxygen species initiate atherogenesis.

The heart and blood vessels are surrounded by epicardial[22] and perivascular adipose tissues, respectively, which secrete numerous adipokines, or adipocytokines. In addition to storage of energy, adipose tissue actually acts as a metabolic sensor and endocrine organ that participates in the regulation of glucose and lipid metabolism as well as insulin sensitivity, among other roles.

Adipokines function as endo- and paracrine messengers that support the *adipocardiovascular axis*, as they mediate interferences (crosstalk) between adipose tissue depots, heart, and vasculature. Some adipokines are proinflammatory, others protect the cardiovascular apparatus.

Adipokines secreted by adipocytes include: (1) hormones, such as adiponectin, angiotensin-2, apelin, chemerin, hepcidine, leptin, omentin, resistin, retinol-binding protein RBP4, vaspin, and visfatin; (2) chemokines such as CCL2; (3) cytokines, either proinflammatory, such as interleukin-6 and tumor-necrosis factor-α, in addition to IL1β, IL4, IL8, and IL18, or anti-inflammatory such as interleukin-10; (4) serpins; such as serpin-E1, or plasminogen activator inhibitor PAI1, and serpin-F1, or pigment epithelium-derived factor (PEDF); (5) fatty acid-binding proteins (FABPs) such as adipocyte FABP (aFABP, or FABP4); (6) lipocalin-2; and (7) adhesion glycoproteic molecules such as thrombospondin-1.

Many adipokines, such as TNFSF1, resistin, aFABP, and lipocalin-2, are proinflammatory, thereby causing endothelial and cardiac dysfunction.

On the other hand, adiponectin has beneficial effects, especially during caloric restriction and improvement of left ventricular function as well as antiapoptotic activity and reduction of infarct size. Adiponectin activates AMPK and PKB kinases as well as ceramidase, thus producing antiapoptotic sphingosine 1-phosphate and subsequently inhibiting caspase-8 (Table 2.8) [148]. Adiponectin also stimulates lipoprotein lipase via the RhoA–RoCK axis and actin remodeling as well as VEGF production.

Obesity is associated with the production of proinflammatory molecules and recruitment of immunocytes to metabolic organs, particularly adipose tissue, liver, pancreas, and hypothalamus [149, 150]. Resulting derangements of local metabolism lead to insulin resistance, type-2 diabetes, nonalcoholic fatty liver disease, and dyslipidemia.

Obesity increases JNK activity in adipose tissue and liver. Obesity-induced stress response in adipose tissue depends partly on JNK1 in adipocytes. In fact,

[22] Two cardiac adipose tissue depots include [148]: (1) *epicardial adipose tissue* between the myocardium and visceral pericardium, which localizes to both ventricles in the atrioventricular and interventricular grooves extending to the apex and along the coronary arteries, and (2) *pericardial adipose tissue* outside the inner layer of the pericardium, which covers 80 % of the heart surface. The volume of epicardial fat increases in patients with coronary artery disease and unstable angina with respect to healthy individuals.

Table 2.8 Cardioprotection ensured by adiponectin. Adiponectin is synthesized mainly by adipocytes, but also by cardiomyocytes, triggers multiple actions via adiponectin receptors AdipoR1 and AdipoR2, which both lodge on cardiomyocytes, and cadherin-13: (1) an antiapoptotic effect by activating ceramidase, PKB, and AMPK, ceramidase producing antiapoptotic sphingosine 1-phosphate (S1P); (2) an antioxidative and -nitrative stress effect via decreased synthesis of inducible NOS2 and NOx2 subunit of NADPH oxidase; (3) an anti-inflammatory effect by activating sphingosine kinase-1 (SphK1) and cyclooxygenase-2 (COx2); (4) lipid uptake stimulation via AMPK-mediated upregulation of ScaRb3 scavenger receptor; and (5) glucose uptake promotion via PKB-dependent translocation to the plasma membrane of glucose GluT4 transporter; (6) angiogenesis via VEGF. (Source: [148])

Process	Pathway and effect
Angiogenesis	VEGF
Apoptosis	AMPK, ceramidase–S1P Inhibition of caspase-8
Glucose uptake	PKB GluT4 transfer to the plasma membrane
Inflammation	SphK1–COx2 Inhibition of TNFSF1
Lipid uptake	AMPK–ScaRb3
Oxidative and nitrative stresses	Inhibition of NOS2 Inhibition of NOx2

adipocyte hypertrophy and hyperplasia enable JNK1 and JNK2 activation in resident macrophages. In adipose tissue and muscle (but not liver), JNK1 intervenes during the development of insulin resistance. In hepatocytes, JNK1 reduces hepatic steatosis and insulin resistance [149]. Tissue-resident macrophages belong to 2 main populations. Classically activated macrophages (**M1**) induced by interferon-γ or endotoxin promote interleukin-12-mediated helper T_{H1} lymphocyte-based immunity. Activated macrophages by IL4 or IL13 (**M2a**), immune complexes (**M2b**), and anti-inflammatory cytokines IL10 or TGFβ (**M2c**) support T_{H2}-based immunity implicated in wound healing, tissue repair, and resolution of inflammation. Obesity favors macrophage polarization to the proinflammatory M1 phenotype [150]. Kinase JNK is required for the differentiation of proinflammatory macrophages. In myeloid cells, JNK1 and JNK2 are involved in obesity-induced recruitment of macrophages and inflammation in adipose tissue, without affecting other myeloid cell populations (eosinophils and neutrophils) [150].

Metabolic syndrome can be caused by mutation in the Lrp6 gene that encodes LDLR-related protein-6, a coreceptor of the Wnt pathway, thus impairing Wnt signaling [151]. The cellular metabolic sensor PAS domain-containing protein Ser/Thr kinase (PASK, or STK37) integrates multiple cues to monitor cellular energetic status. Nutrient-responsive PASK is activated by glucose. Activated PASK contributes to insulin secretion in pancreatic β cells, increases the synthesis and storage of triglycerides in hepatocytes, and decreases ATP generation both from carbohydrate and fatty acid oxidation in skeletal muscle [152].

2.3.9 *Altered Sleep*

Sleep is constituted of two main stages, rapid eye movement (REM) and nonrapid eye movement (NREM) sleep. More precisely, sleep is characterized by three (formerly four) stages of increasingly deep dreamless sleep and a stage of dreaming and REMs.

Over the course of the night, sleep cycles are repeated between NREM and REM stages. Sleep timing is controlled by the circadian clock (Vol. 2, Chap. 5. Circadian Clock).

Sleep is involved in various bodily functions, such as energetic metabolism, signaling from endocrine glands, memory processing, and waste clearance from the brain, as well as immunity and wound healing.

Sleep, as for wakefulness, is influenced by various neurotransmitters in the brain as well as foods and medicines that change the balance of these signals [153]. Caffeinated drinks and some drugs can cause insomnia. Heavy smokers have reduced amounts of REM sleep. Alcohol supports light sleep, but deprive REM and restorative stages of sleep. Many antidepressants suppress REM sleep.

2.3.10 *Sleep Stages*

Sleep usually proceeds in four to five REM–NREM cycles per night. The order is normally NREM1–NREM2–NREM3–NREM2–REM. The proportion of REM sleep rises in the last sleep part. The first REM sleep period usually occurs about 70–90 min after the sleep onset [153]. A complete sleep cycle takes 90–110 min on an average.

Stage-1 sleep (lightest sleep) at the beginning of the night represents 5 % of the total sleep time (TST) and stage-2 sleep (fairly light) 55–60 % TST. Deep sleep, or slow-wave sleep, in stages 3 and 4 accounts for about 20 % TST.

The latencies between the sleep onset and the initial time of stage 4 and REM sleep can be calculated. Sleep stage shifts are incidents of sleep stage changes. Arousals (interruptions of sleep lasting 3–15 s) occur spontaneously or as a result of sleep-disordered breathing or other sleep disorders.

NREM stage 1 (somnolence or drowsy sleep) is a phase between sleep and wakefulness during which the skeletal muscles (with possible sudden contractions, *hypnic myoclonia*) are active and eyes roll slowly, opening and closing moderately. In the brain, fluctuations of the electrical activity measured by electroencephalography (EEG) corresponds to the transition from α waves (frequency 8–13 Hz) to θ waves (frequency 4–37 Hz).

NREM stage 2 is characterized by a slowing cerebral electrical activity with occasional bursts of rapid waves, *sleep spindles* (frequency 11–16 Hz), and K complexes. Eye movements stop. The muscular activity and perception of environmental stimuli lowers.

NREM stage 3 (slow-wave sleep [SWS]) can incorporate stage 4. In the brain, δ activity, that is, high-amplitude waves (frequency < 3.5 Hz), begins to appear, interspersed with smaller and faster waves.

NREM stage 4 (deep sleep) consists quasi-exclusively of δ activity.

REM stage (paradoxical sleep) occurs approximately every 90 min. It is triggered by acetylcholine secretion and inhibited by serotonin. Arousal is less easily obtained than at any other sleep stage. The EEG pattern is characterized by high-frequency waves.

2.3.10.1 Insufficient Sleep

Repeated nights of insufficient sleep (i.e., short sleep duration [≤ 6 h]) influence balance between energy expenditure and source. Insufficient sleep increases total daily energy expenditure by about 5 % as well as weight, despite changes in levels of ghrelin, leptin, and peptide-YY, which signal excess energy stores [154].[23] Insufficient sleep and disturbed circadian rhythm are associated with impaired vigilance and cognition as well as obesity, diabetes, cardiovascular disease, and, at the nano- and microscale, perturbations of gene transcription regulation and cellular metabolism as well as altered inflammatory, immune, and stress responses [155]. In addition, insufficient sleep delays the circadian melatonin phase and primes an earlier circadian wake phase [154]. Conversely, recovery sleep decreases intake of nutrients, especially lipids and carbohydrates, hence causing a weight loss.

Insufficient sleep up- or downregulates 711 genes [155]. Genes, the expression of which is upregulated during acute total sleep loss, encode proteins involved in synaptic remodeling and chaperoning such as heat shock proteins. On the other hand, genes, the expression of which is downregulated during sustained wakefulness, encode proteins implicated in molecular synthesis and energy production. Insufficient sleep reduces the number of genes with a circadian expression profile from 1855 to 1481 as well as their amplitude response, but increases the number of genes that respond to total sleep deprivation from 122 to 856 [155].

Genes affected by insufficient sleep are associated with [155]: (1) the circadian rhythm (PER1–PER3, CRY2, CLOCK, NR1D1–NR1D2, RORA, DEC1, and CSNK1E); (2) sleep homeostasis (CAMK2D, KCNV2, IL6, and STAT3); (3) oxidative stress (PRDX2 and PRDX5); and (4) metabolism (ABCA1, SLC2A3, SLC2A5, and GHRL).

During routine periods of sufficient sleep, about 9 % of genes expressed in blood cells have a circadian expression profile. Genes with a circadian rhythm-dependent expression include the core clock genes (PER1–PER2, NR1D1–NR1D2, and ARNTL) as well as genes involved in metabolism (e.g., ABCA1, SLC2A3–SLC2A14, and PYGL) and regulation of gene expression (e.g., ELAV1, ELP2,

[23] The 36-amino acid peptide Tyr–Tyr (PYY) is secreted by the neuroendocrine cells of the mucosa of the gastrointestinal tract, especially the ileum and colon, in response to a meal.

HNRNPU, SNRPF, and TCEA3). Gene transcripts characterized by a peak concentration during the night ($\sim 58\,\%$) encompass most canonical clock genes as well as those encoding agents of nucleic acid binding, gene transcription and translation, RNA metabolism and binding, ribosome genesis, and cellular metabolism (peak at 01:30 AM) [155]. Gene transcripts characterized by a peak level during the day ($\sim 32\,\%$) comprise antiphase clock genes (ARNTL and PROK2) as well as those that encode factors of cytokine receptor activity, NFκ B signaling, and response to cellular stress, inflammation, and immunity [155]. Blood contains markers for diagnosis or prognosis. Blood cells produce neurotransmitter receptors and transporters as well as opsins.

2.3.10.2 Sleep Fragmentation

The sympathetic overactivity that accompanies the majority of sleep disorders is a risk factor for hypertension. In sleep-disordered breathing, respiratory events end with a rise in vascular and cardiac sympathetic activity. The respective roles of hypoxemia and arousal in sympathetic overactivity during sleep are difficult to separate.

Sleep fragmentation results from arousals systematically associated with sympathetic surges. The central nervous system reactivity is objectivized by electroencephalographic activation and concomitant changes in cardiovascular parameters under autonomic control, such as RR intervals and blood pressure. Repeated sympathetic arousals responsible for sleep fragmentation assessed by the *autonomic arousal index* (AAI) derived from the pulse transit time affect blood pressure that is continuously measured using 24-h ambulatory monitoring and cardiac frequency variability indices [156]. The *pulse transit time* (PTT; i.e., the time taken by the pressure wave to propagate through a given length of the arterial circuit; typically 200–250 ms) is used to detect sleep arousals.[24] Sympathetic activation increases the vascular tone, hence shortening the PTT and elevating blood pressure. The *cardiac frequency variability* (CFV) derived from RR intervals serves to assess sympathetic (low-frequency power [LF]) and parasympathetic (high-frequency power [HF]) activity. Dimensionless CFV indices such as the LF/HF ratio are utilized as markers of sympathovagal balance. In healthy elderly subjects, sleep fragmentation raises diurnal and nocturnal systolic blood pressure, independently of confounders, such as sleep-disordered breathing, body mass index, sex, diabetes, hypercholesterolemia, and sleep duration and quality [156]. This augmentation is engendered by a nocturnal and diurnal sympathetic overactivity.

Increase in upper airway resistance that do not cause complete pharyngeal airway collapse have a clinical importance similar to that of apneas for producing sleep

[24] The onset of the R-wave yields the initial time at the aortic valve. The peripheral measurement station is usually the fingertip, where the pressure wave is detected by a photoplethysmographic probe. The virtual base point of the photoplethysmographic signal corresponds to the intersection point between the tangent to the pulse wave at the point with the maximal slope during the systolic rise and the horizontal line crossing the minimal pressure point.

fragmentation. This increase leads to *obstructive hypopneas* with a marked reduction in air flow and *upper airway resistance episodes* (UARE) with high inspiratory resistance without a significant fall in air flow or arterial oxygen saturation (S_{aO_2}). Ostructive hypopneas and UAREs are grouped together under the term of *obstructive nonapneic respiratory events* (ONARE), which cannot be easily differentiated from centrally induced episodes [157].[25]

Pulse transit time that can detect surges in blood pressure due to microarousals is used as a noninvasive surrogate marker of inspiratory effort with respect to esophageal pressure measured invasively using an esophageal catheter that alters the pharyngeal airway dynamics as well as the sleep quality. This parameter has a high sensitivity and specificity at differentiating obstructive from central upper airway apneas and hypopneas [157].

2.3.10.3 Sleep Indices and Grades

Numerous sleep indices are used in clinical practice (Tables 2.9 and 2.10). The *sleep efficiency* is the ratio between the TST and the total recording time (TRT). *Sleep latency* is related to the first 30 s of sleep and *sleep onset* to the first 90 s of uninterrupted sleep. The *sleep pressure score* (SPS) relates the apnea–hypopnea index and respiratory (RAI), spontaneous (SAI), and total (TAI) arousal indices:

$$\mathrm{SPS} = \frac{\mathrm{RAI}}{\mathrm{TAI}} \times \frac{\mathrm{SAI}}{\mathrm{TAI}}. \tag{2.1}$$

Index grades can be defined. The respiratory arousal index (RAI) and apnea–hypopnea index (AHI), among others, are considered

- normal (≤ 5 events/h),
- mild (5–15 events/h),
- moderate (15–30 events/h), and
- severe (> 30 events/h).

A *sleepiness scale* lower or equal to 10 means an absence of noticeable daytime somnolence; ranging from 10 to 15 indicates daytime somnolence; and greater than 16 excessive daytime somnolence.

2.3.10.4 Molecular Context

GABA The neurotransmitter γ-aminobutyric acid (GABA) promotes sleep. The mitochondrial enzyme GABA transaminase (GABAT) expressed in glial cells catabolizes GABA to succinic semialdehyde, which in turn is transformed into succinic

[25] Hypopneas are defined as a clear and discernible reduction in flow ($> 50\,\%$) that ended with an arousal and fall in S_{aO_2} of at least 3 %. UAREs refer to as flow limitation without flow reduction occurring concurrently with an increasing esophageal pressure (or surrogate PTT), ending with an arousal, and followed by a return of the esophageal pressure to its resting level.

Table 2.9 Sleep indices (**Part 1**). Arousal is an interruption of sleep of duration greater than 3 s and an observable episode of α rhythm. Apnea is defined by the cessation of air flow at the nose and mouth lasting at least 10 s; hypopnea is a 50 % decrease in the amplitude of oronasal flow associated with either an arousal or a 3 % oxygen desaturation. Apnea–hypopnea index scores an apnea when the amplitude of a breath is reduced by at least 80 % during at least 10 s and a hypopnea when the breath amplitude is reduced by 50 % during at least 10 s. Conventional indices of sleep fragmentation include EEG microarousals, apnea–hypopnea index and oxygen saturation dip rate. Autonomic indices comprise cardiac frequency and blood pressure rises. Respiratory-effort related arousals (RERA) are arousals from sleep due to an increasing respiratory effort caused by air flow limitations without significant O_2 desaturation for at least 10 s. Periodic leg motions are bursts of limb muscle contraction of 0.5–5.0 s duration and amplitude of at least 25 % of bursts recorded during calibration occurring with a periodicity of 5–120 s between each movement. Average number is the total number (N) of events divided by the actual TST ([mn]) multiplied by 60 ($N \times 60$/TST)

Index	Alias	Definition
Autonomic arousal index	AAI	Average number of arousals triggered by the sympathetic system per hour of sleep (vegetative activation associated with a transient EEG pattern different from a conventional arousal)
Apnea–hypopnea index	AHI	Average number of episodes of apneas and hypopneas per hour of total sleep time
Breathing-related arousal index	BRAI	Number of arousals due to apneas, hypopneas, snorings, and RERAs multiplied by the number of hours of sleep
Central apnea index	CAI	Average number of episodes of apneas without breathing (duration $\geq$ 10 s) per hour of sleep effort
Central hypopnea index	CHI	Average number of episodes of hypopneas without breathing effort (duration $\geq$ 10 s) per hour of sleep
Desaturation index	DI	Average number of arterial oxygen desaturations (drop of 3 % below average saturation per hour of sleep)

BAI behavioural arousal index, *MAI* microarousal index, *VAI* vegetative arousal index

acid by succinic semialdehyde dehydrogenase. Repression of GABAT expression hence increases the GABA amount and boosts total daily sleep [158]. Another sleep-promoting molecule downstream of GABAT activity is γ-hydroxybutyric acid (GHB).

Leukotrienes Obstructive sleep apnea (OSA) is characterized by early atherosclerosis and an increased prevalence of cardiovascular events. The extent of vascular remodeling is related to the amount of nocturnal oxygen desaturation.

Obstructive sleep apnea is also associated with an activation of the leukotriene pathway in a context of chronic intermittent hypoxia and obesity. The production of leukotrienes that are lipid mediators derived from arachidonic acid liberated from cell membrane phospholipids by cytosolic phospholipase-A2 is correlated with OSA severity [159]. Urinary excretion of leukotriene-E_4 and leukotriene concentration in exhaled breath condensate from OSA patients rise. In OSA patients, the percentage of time spent with an oxygen saturation (S_{aO_2}) lower than 90 % and mean and minimal

Table 2.10 Sleep indices (**Part 2**)

Index	Alias	Definition
Obstructive apnea index	OAI	Average number of complete air flow cessation (duration ≥10 s) per hour of sleep
Obstructive hypopnea index	OHI	Average number of partial airway obstruction (duration ≥10 s) per hour of sleep
Periodic leg movement arousal index	PLMAI	Number of periodic muscle motion-related arousals multiplied by the number of hours of sleep
Respiratory arousal index	RAI	Number of apneas, hypopneas, and snoring-related arousals multiplied by the number of hours of sleep
Respiratory disturbance index	RDI	Average number of apneas, hypopneas, and respiratory effort-related arousals per hour of total sleep time
Spontaneous arousal index	SAI	Number of arousals in the absence of respiratory events, limb movements, and snoring multiplied by the number of hours of sleep
Total arousal index	TAI	Number of all arousals multiplied by the number of hours of sleep
Upper airway resistance episodes index	UAREI	Average number of arousals caused by upper airway obstruction per hour of sleep

nocturnal S_{aO_2} as well as apnea–hypopnea index are correlated with increased neutrophil leukotriene-B4 production. The desaturation–reoxygenation cycle in OSA patients engenders oxidative and nitrosative stresses.

In inflammatory cells, once arachidonate 5-lipoxygenase (ALOx5) that interacts with nuclear membrane-bound ALOx5-activating protein (ALOx5AP) is activated, arachidonic acid is converted to leukotriene-A_4. In intermittent hypoxia, the HIF1α and NFκB factors activate transcription of ALOx5AP upon binding to hypoxia-response and NFκ B-binding motif in the ALOX5AP promoter [159].

In neutrophils, monocytes, and macrophages, LTa_4 is converted by LTa_4 hydrolase (LTa_4H) into LTb_4, a gene transcription modulator and potent chemoattractant. Leukotriene-B_4 also stimulates release from neutrophils of lysozyme, myeloperoxidase, and matrix metallopeptidases (Table 2.11). Leukotriene-A_4 is also secreted to mediate auto- and paracrine effects via the LTa_4R1 (BLT_1) and LTa_4R2 (BLT_2) receptors. Leukotriene-A_4 can be conjugated with glutathione by LTc_4 synthase (LTc_4S) to produce leukotriene-C_4. The latter can be converted to leukotriene-D_4 by γ-glutamyl transpeptidase (γ GT). Leukotriene-D_4 can then be metabolized by dipeptidase into leukotriene-E_4.

Cysteinyl leukotrienes (CysLTs; i.e. LTc_4–LTe_4) can be synthesized in inflammatory cells, such as macrophages, mastocytes, and platelets, and vascular cells due to transcellular metabolism of LTa_4. These agents targets the $CysLT_1$ and $CysLT_2$ receptors on leukocytes and vascular cells.

Table 2.11 Effects of leukotrienes in obstructive sleep hypopnea–apnea syndrome on inflammatory leukocytes and vascular cells, hence related to vascular inflammation and remodeling and atherosclerosis. (Source: [159]; ↑: increase)

Target cell	Effects
Granulocyte	Chemotaxis, delayed apoptosis ↑ MMP9 production
Monocyte, macrophage	↑ production of IL1β/6, CCL2, MMP9
T lymphocyte	Chemotaxis ↑ production of Ifnγ, IL17
Endotheliocyte	↑ vascular permeability ↑ production of CXCL2, P-selectin
Smooth myocyte	Chemotaxis, proliferation Coronary vasoconstriction ↑ production of MMP2, serpin-B2

2.3.10.5 Obstructive Sleep Hypopnea–Apnea Syndrome

Obstructive sleep apnea syndrome is defined by repetitive cycles of hypopnea and/or apnea during sleep that are combined with diurnal sleepiness. Inspiratory airflow that is partly (hypopnea) and completely (apnea) occluded during sleep causes oxygen desaturation and sleep disruption. In Europe and North America, approximately 20 and 7 % of adults have mild and moderate to severe obstructive sleep apnea, respectively. About 30 % of the adult population in western countries may be affected by asymptomatic OSA and approximately 2–4 % by symptomatic obstructive sleep hypopnea–apnea syndrome (OSHAS) [160].

Resulting hypoxemia and hypercapnia excite chemoreceptors that, after signal processing by the central nervous system, increase sympathetic vasoconstrictory activity. Furthermore, strenuous inspiratory effort combined with nocturnal hypoxemia elicits a set of neural and humoral responses with vascular, inflammatory, and metabolic consequences. Epidemiology has shown that obstructive sleep hypopnea–apnea syndrome is a risk factor for defective organ perfusion (angina, myocardial infarction, and stroke; Vol. 8, Chap. 1. Diseases of Large Blood Vessels—Context).

OSHAS is characterized by repetitive partial or complete pharyngeal closure during sleep due to transient collapse of the soft pharyngeal walls. Pharyngeal obstruction leads to arterial oxygen desaturation, persistent diurnal sympathetic activity, which increases during sleep, hypertension, and cardiac brady- or tachyarrhythmias.

The diagnosis of sleep hypopnea and apnea uses nocturnal polysomnography, which records cardiac frequency, respiratory rhythm, electroencephalogram, eye motions, muscle activity, and oxygen saturation.

Continuous positive airway pressure (CPAP) is the usual therapy that reduces blood pressure, sympathetic activity, and inflammation in patients with OSHAS. Positive airway pressure generated by a pump maintains an open pharynx. Duration of CPAP greater than 4 h per night for several weeks can improve endothelial function and decrease oxidative stress and abnormal lipid peroxidation, as well as blood levels of proinflammatory interleukin-6 and anti-inflammatory interferon-γ [160]. In addition, treatment improves the sympathovagal balance, as it reduces catecholamine levels and sympathetic nerve activity and increases baroreflex sensitivity, an index of cardiac vagal responsiveness. The CPAP treatment also enhances levels of circulating nitric oxide, nitrate, and nitrite, as well as endothelial progenitor cell density, hence ameliorating endothelial function [160]. However, approximately 30 % of patients with OSHAS do not tolerate CPAP in the long term. Moreover, some patients do not use CPAP for a sufficient duration (~4 h per night). Alternative treatments are needed for nonsleepy patients with obstructive sleep apneas reluctant for CPAP therapy.

Sleep-disordered breathing not only can elicit cardiac and vascular damages, increasing the occurrence probability of impaired organ perfusion, but also resistance to therapies. Obesity, hypertension, and aging are usual features of patients with sleep hypopneas and apneas, although they can occur in patients who are not obese. Obstructive sleep hypopneas and apneas can be associated with other risk factors of cardiovascular diseases, such as insulin resistance.

Adults that had apnea of prematurity, hence intermittent hypoxemia, have an increased risk of sleep-disordered breathing and hypertension. Intermittent hypoxia in preterm infants may enhance hypoxic sensitivity and cause an exaggerated response to hypoxia by the carotid body and adrenal chromaffin cells, causing irregular breathing with apneas and hypertension [161]. Associated oxidative stress results from a decreased expression of genes encoding antioxidant enzymes such as the Sod2 gene that encodes superoxide dismutase-2, due to DNA hypermethylation.

2.3.11 Air Pollution

Exposure to *indoor air pollution* from solid fuels can cause many diseases, in particular pneumonia in children and chronic respiratory diseases in adults (Vol. 8, Chap. 12. Respiratory Obstructive Disorders). *Outdoor air pollution* is also a threat to health.

Human beings are increasingly exposed to chemicals and their harmful effects as well as electromagnetic fields and ionizing radiations, especially workers facing *occupational hazards*, in addition to large-scale and global *environmental hazards*. Air pollution affects particularly densely populated metropolitan areas and megalopolis.

Practical lifestyle recommendations to reduce exposure to airborne pollutants comprise:

1. to favor public transport;
2. to avoid walking and cycling in streets with high traffic intensity;
3. to run and exercise in parks, not near pollution sources;
4. to limit infant exposure to outdoor pollution (i.e., during rush hours and in proximity to industrial sources); and
5. to take in a diet rich in fruits and vegetables containing antioxidants.

Pollutants are categorized into:

1. *primary pollutants* produced directly from vehicle traffic, industrial emissions, and domestic heating (e.g., nitrogen oxides [NO_x] and sulfur dioxide [SO_2]), and
2. *secondary pollutants* that originate from primary pollutants as a consequence of chemical reactions in the atmosphere (e.g., ozone).

2.3.11.1 Indoor Air Quality

In addition to dirty air conditioning vents and filters, dust dislodged during renovations, and aftermath of water damage to interiors, molds (mainly from *Alternaria*, *Aspergillus*, *Cladosporium*, and *Penicillium*) in indoor environments are implicated in allergies.

The sick-building syndrome (irritation of eyes and respiratory tract) does not have identified specific etiologic factors but can be defined by a context (poor ventilation, office and cleaning supply chemicals, damp indoor environment due to water damage, and microbial contamination).

On the other hand, building-related illness has known causal factors (e.g., legionellosis and asbestosis).

2.3.11.2 Particulate Matter

Particulate matter (PM) in the inhaled air, primary or secondary in origin, comprises:

1. fine ($PM_{2.5}$; $0.1 <$ aerodynamic diameter ≤ 2.5 μm) and
2. ultrafine ($PM_{0.1}$; aerodynamic diameter ≤ 0.1 μm) particles,

which penetrate not only pulmonary alveoli, but also the blood circulation.

Although the average chemical composition of ultrafine particle samples collected in rural and urban sites is similar, the average concentration can be higher, at least in some regions, at the rural (520 ng/m^3) than in the urban site (490 ng/m^3) [162].

The most abundant components are ammonium, nitrate, sulfate, and Ca^{2+} ion. Particle-bound water-soluble ions include chloride, sulfate, nitrate, methane sulfonate, oxalate, malonate, succinate, glutarate, ammonium, sodium, potassium, magnesium, and calcium.

Metals usually form cations through electron loss, reacting with oxygen in the air to form oxides over various time scales. The most important metals are calcium (Ca), iron (Fe), potassium (K), sodium (Na), and zinc (Zn) at both sites. Heavy metals comprise copper (Cu), lead (Pb), nickel (Ni), and vanadium (V).

Organic anions oxalate and methane sulfonate contributed similarly at both urban and rural sites.

The average contribution of ultrafine mass to the fine particle ($PM_{2.5}$) mass was about 7 % at the urban and 8.5 % at the rural site. At both sites the contribution of ultrafine to fine was especially high for boron (B), selenium (Se), silver (Ag), and nickel (10–20 %) and, at the rural site, cobalt (Co; 20 %), Ca^{2+} (16 %), and molybdenum (Mo; 11 %).

Air pollution developed in London during the great smog between 5 and 9 December 1952 was due to a combination of cold weather, anticyclone around a central region of high atmospheric pressure, and absence of wind and rain.[26] The resulting intensive domestic heating with low-quality coal formed a thick layer of smog over the city and caused a marked mortality peak (about 12×10^3deaths).

The World Health Organization (WHO) manages a database on urban outdoor air pollution monitored during the period from 2003 to 2010 in about 1100 cities in 91 countries [163]. It was estimated that urban outdoor air pollution and smoking caused 1.34×10^6 and 4.8×10^6 premature deaths worldwide, respectively. Respiratory and cardiovascular diseases can be triggered by acute and chronic inhalation of gaseous and corpuscular airborne substances.

2.3.11.3 Impact on Health

Air pollution, particularly ground-level ozone, is associated with heart diseases. Breathing air containing elevated levels of fine particles and ozone can constrict arteries. Exposure to particulate matter $PM_{2.5}$ supports atherosclerosis. Exposure to ultrafine particles $PM_{0.18}$ inhibits the anti-inflammatory capacity of high-density lipoprotein, thereby augmenting oxidative stress [164].

Particulate matter exposure potentiates angiotensin-2-induced hypertension, cardiac maladaptive hypertrophy, and collagen deposition via the RhoA–Rho kinase axis [165]. Ozone interacts with cholesterol of the plasma membrane of respiratory epithelia as well as of blood cells to produce metabolites such as atheronals that affect artery wall rheology. In addition, leukocytes involved in inflammation of arterial walls can produce ozone and afterward atheronals-A and -B.

Dysregulation of the autonomic nervous system, inflammation, atherosclerosis, and thrombosis interfere. Animals exposed to PM have an overproduction of reactive oxygen species, which are implicated in the initiation and progression of atherosclerosis [166]. A higher mortality rate is observed in patients who suffered from an acute coronary syndrome and were exposed to higher levels of pollution in England and Wales, with a 20 % increased risk of deaths for any 10 $\mu g/mm^3$ $PM_{2.5}$ elevation [167].

Rodents exposed to a particulate matter dose comparable with that to which people are exposed in metropolitan areas develop platelet dysfunction and hemostatic changes, ultimately provoking intravascular thrombus formation [166]. The hypercoagulable state following exposure to PM results from a shortening of coagulation time and higher plasma concentrations of the prothrombotic amino acid homocysteine, particularly in smokers. In the lung, activated inflammatory cells (endotheliocytes, macrophages, and circulating neutrophils) raise local levels of cytokines such as interleukin-6, which may then act as a stimulus for subsequent systemic inflammation that can cause hypercoagulability and enhanced thrombogenesis. In healthy

[26] Air pollution is lower on rainy rather than on sunny days.

people and patients with coronary artery disease, the finest PM contained in diesel exhaust impedes vasodilatory response to proper stimulants and supports hypofibrinolysis [166]. Markers of atherosclerosis (carotid wall thickness and calcium content in coronary arterial walls) are correlated with $PM_{2.5}$ concentration.

In addition, PM exposure decreases cardiac frequency variability, a risk factor for arrhythmias and sudden death [166]. Augmented sympathetic control can result from activation of pulmonary nervous reflex arcs and direct effects of pollutants on cardiac ion channels.

Long-term exposure to fine PM and night-time traffic noise are both independently associated with subclinical atherosclerosis and may both contribute to the association of traffic proximity with atherosclerosis [168]. Higher concentrations of particulate air pollution (PM10 and PM2.5) and greater traffic noise at night are each independently associated with a higher risk of developing thoracic aortic calcifications.

However, air pollution and ambient noise share common sources and their effects on cardiovascular outcomes can be entangled [169]. Noise, even at moderate intensities, may activate stress responses with detrimental cardiovascular effects, both during daytime activity and during sleep [169].

Long-term particulate air pollution and traffic noise alters health. In addition, short-term increases in pollution ranging from hours to days affects the cardiovascular apparatus at low levels of exposure of ambient air pollution, particularly for the risk of acute events, such as myocardial infarction and heart failure [169].

Air pollution includes a variety of gaseous species as well as suspended PM of various sizes. Many isolated pollutants have adverse effects on the cardiovascular apparatus at high doses as well as at lower levels of exposure as mixtures linked to ambient air pollution. Relatively low levels of fine particulate exposure from either air pollution or passive cigarette smoke are sufficient to induce adverse biological effects [170]. However, the exposure–response relation between cardiovascular disease mortality and PM2.5 is relatively steep and nearly linear at low levels of exposure, consistent with low air pollution and very light active smoking (1–2 cigarettes/day), and flattens out at higher exposures. Exposure to both ambient air pollution and active or passive cigarette smoke do not raise the cardiovascular risk at high concentrations (but not attenuating it) due to the nonlinear relation with a plateau above a concentration threshold [170].

2.3.11.4 Aloft Toxins—Mycotoxins

Chemical pollutants comprise organic and inorganic compounds and microbes of tropospheric winds. Environmental agents such as windborne pathogens and toxins (e.g., aloft bacterial and fungal toxins) are seasonal and nonseasonal etiologic factors. They travel via regional winds and large-scale atmospheric circulation. They enter through the mucosa of the upper respiratory tract.

Fungi are major plant and insect pathogens. Growth of fungi on animal hosts produces mycoses (by primary [e.g., *Coccidioides immitis* and *Histoplasma capsulatum*] and opportunistic pathogens [e.g., *Aspergillus fumigatus* and *Candida albicans*]. Dietary, respiratory, dermal, and other exposure types to toxic fungal metabolites generate mycotoxicoses. Whereas microfungi make mycotoxins, macroscopic fungi and mushrooms produce mushroom poisons.

Many cereal species harbor fungal spores. Inhalation of spore-associated mycotoxins[27] that are secondary metabolites of various chemical pathways cause diverse sets of symptoms in humans. The liver, kidneys, and brain are prominent targets [171].

Mycotoxins produced by microfungi that are toxic to vertebrates in low concentrations include [172]:

- *aflatoxins* produced by *Aspergillus* species, often associated with crops of cotton, peanuts, spices, pistachios, and maize (they are difuranocoumarin derivatives synthesized by a polyketide pathway by many *Aspergillus* species);[28]
- *citrinins* manufactured by *Penicillium* and *Aspergillus* species, which is associated with many human foods (barley, corn, oat, rye, rice, wheat) and are nephrotoxins;
- *fumonisins* produced by *Fusarium* species, which can affect the nervous system;[29]
- *trichothecenes* (e.g., diacetoxyscirpenol and deoxynivalenol) mainly synthezised by *Fusarium* species, which constitute a family of more than 60 sesquiterpenoid metabolites, are potent inhibitors of eukaryotic protein synthesis and can provoke chronic toxic effects, in addition to their cytotoxic and immunosuppressive activity;
- *zearalenone*, another metabolite from *Fusarium* species with potent estrogenic activity, hence being a nonsteroidal mycoestrogen and not a mycotoxin (misnomer);
- other *Fusarium* toxins, which include beauvercin, butenolide, equisetin, enniatins, and fusarins;
- *ochratoxins* (ochratoxin-A–ochratoxin-C) made by *Aspergillus* and *Penicillium* species, which is a potent nephrotoxin;
- *patulin* from *Aspergillus*, *Penicillium* (e.g., *Penicillium expansum* associated with a range of moldy fruits and vegetable), and *Paecilomyces* species, an antimicrobial (antibacterial, antiviral, and antiprotozoal) agent, but also a toxic to both plants and animals; and
- *ergot alkaloids* are indole alkaloids from the sclerotia of *Claviceps* species, which are common pathogens of various grass species.[30]

---

[27] μκης: fungus, τοξικον: poison on an arrow (τοξεμα). A given fungus can produce many types of mycotoxins. A given mycotoxin can be produced by several species of fungi. These toxic secondary metabolites colonize crops.

[28] Cytochrome-P450 enzymes convert aflatoxins to the reactive (8,9)-epoxide form that can bind to DNA and proteins, hence their carcinogenicity.

[29] *Fumonisins* interfere with sphingolipid metabolism in animals.

[30] Several ergot alkaloids induce smooth muscle contractions. Ergotism comprises gangrenous and convulsive forms. In animals, ergotism causes gangrene, abortion, suppression of lactation, hypersensitivity, convulsions, and ataxia [172].

They are consistent with an agricultural source region from which they are lofted into the atmosphere and there is a short incubation time between exposure and onset of fever. Kawasaki disease, the most common cause of acquired acute coronary artery vasculitis in children, results from infection by *Candida* species as the dominant fungal species that operate via windborne toxins rather than organisms requiring replication. They are collected in aerosols transported by tropospheric winds from cereal (corn, rice, and spring wheat) croplands in the densely and intensively cultivated region of northeastern China to Japan and even to the West Coast of the United States [171]. The incubation period is very short (< 24 h) after exposure.

2.4 Genetic Background

Human genomics is aimed at predicting people at risk for specific diseases, especially at determining the genetic background of cardiovascular diseases, and at identifying individuals in a target population that can benefit from specific therapies. A positive family history (heritability) is one of the strongest cardiovascular risk factors.

Transcriptional variations and posttranscriptional changes that often modify transcript stability, transcription rate, RNA export from the nucleus, and alternative splicing, as well as any processes that affect expression levels, lead to alterations in protein structure and state, hence in biological processes.

Candidate genes carrying mutations or variants affecting their function are those encoding proteins involved in the pathogenesis of cardiovascular diseases. Risk-associated alleles, which are carried by most individuals, are inherited factors that act independently of traditional risk factors and confer a modest hazard. Each risk allele indeed mildly increases (< 30 % per allele in the case of myocardial infarction [173]) the probability of occurrence of a given pathology. However, individuals who are homozygous for a risk allele have a much stronger risk (~ 60 % risk increase in the case of myocardial infarction [173]) with respect to subjects who do not carry this allele. Molecular mechanisms associated with these chromosomal variants may explain genetic susceptibility for a given disease.

Genome wide affected sib-pair linkage analyses assume that siblings who are affected by a phenotype are more likely to share the chromosomal region on which the responsible gene is located. Yet, the sequencing of the 3×10^9 base pairs of the human genome and the cataloguing of *single-nucleotide polymorphisms* (10^6 SNPs), i.e., of DNA sequence variations by changes in a single nucleotide, at these bases support *genome wide association* (GWA) studies of complex traits in unrelated subjects.[31]

A small part of the human genome encodes proteins. Noncoding regions of DNA variants can also be associated with diseases. In addition, most commonly occurring variations in DNA are only weakly related to occurrence of cardiovascular diseases,

[31] SNPs constitute a large fraction of variations of the human genome. Two alleles of each SNP have different frequencies in the population. Additional variations comprise insertions, deletions, and structural variations, as well as epigenetic modifications.

whereas rare genetic variants can have strong effects such as those involved in familial hypercholesterolemia [111].

2.4.1 Genetic Background and Metabolism

Certain gene mutations cause severe congenital metabolic disorders. Genetic variations influence metabolism (i.e., production, uptake, release, and clearance of chemicals from given organs) and disease. However, metabolism is governed by genetic factors of variable impact as well as nongenetic effects.

Metabolic intermediates are assigned to metabolic groups (amino acids, peptides, carbohydrates, lipids, nucleotides, and cofactors and vitamins) or to energetic and xenobiotic metabolism (e.g., gene products associated with drug targets, drug-metabolizing enzymes, and transporters).

Genetic variants have effects ranging from loss-of-function alleles in metabolic disorders to common polymorphisms with moderate consequences in multifactorial diseases, taking into account variations within the normal range of healthy individuals [174].

Metabolic intermediates can identify connectivity between genes and diseases [174]. The genetic variants can have an effect in a specific tissue when the corresponding protein is expressed and active in this tissue. For example, the bradykinin–kininogen–kinin axis is related to the cardiovascular apparatus and, hence, its disorders.

Most loci map preferentially in or near genes that encode enzymes, metabolite transporters, and regulators of metabolism. Genetic influences on human blood metabolites that can serve as markers are first analyzed. Metabolite and transcript concentrations have greater statistical power than more complex traits (e.g., high- and low-density lipoprotein–cholesterol levels) [174]. Metabolite-associated single nucleotide polymorphisms affect metabolite concentrations via variation in the transcript level of corresponding causal genes. The allele associated with an altered metabolite level is linked to abnormal gene expression in at least one biological tissue. Combination of genetics with metabolomics enable identification of 145 genetic loci linked to 400 blood metabolites [174].

2.4.2 Genetic Background and Cardiovascular Diseases

Mutations in the SLC2A10 gene, which encodes glucose transporter GLUT10, associated with upregulation of the TGFβ pathway in the arterial wall, cause aortic aneurysms and arterial tortuosity [175].[32]

[32] Arterial tortuosity syndrome is an autosomal recessive disorder characterized by tortuosity, elongation, stenosis, and aneurysm in major arteries that results from disruption of medial elastic fibers in the arterial wall.

Increased arterial stiffness raises the cardiovascular morbidity because of an elevation of afterload and altered coronary perfusion. Thirty-five gene expressions[33] are correlated with the pulse wave velocity chosen as an index of arterial stiffness in human aortic specimens [176]. Two distinct groups of genes, associated either with cell signaling or with interactions between the cytoskeleton, plasma membrane, and extracellular matrix[34] are involved.

Most common forms of hypertension begin by repeated mild or intermittent elevations in arterial pressure associated with vasoconstriction. As hypertension progresses with age, blood vessels remodel. Hypertension can occur without any known cause. Essential hypertension, the most common cardiovascular disease, is a major risk factor. Multiple approaches are used to determine the genetic background of essential hypertension. Certain specific mutations indeed lead to hypertension (others to hypotension), via defective electrolyte transport in the nephron.

Many genomic regions are implicated in the regulation of blood pressure. Clusters of blood pressure-related traits are mainly located on chromosomes-1 and -3 [177]. Blood pressure quantitative trait loci can colocalize with those for renin activity and sodium excretion. A cluster of metabolism-related traits, including indices of obesity, are also mapped to chromosome 1.

The LIPA gene (lysosomal acid lipase-A) is identified as a candidate gene for coronary artery disease [178]. Osteoglycin (Ogn) is a protein that regulates left ventricular mass in humans and endonuclease-G (Endog) regulates mitochondrial genesis and cardiac hypertrophy.

2.4.3 Genetic Background and Neurotransmission

Noradrenaline is involved in sympathetic neurotransmission in: (1) the brain, where it supresses central sympathetic activity; as well as (2) postganglionic sympathetic neurons to the heart, vasculature, and kidney, thereby increasing the cardiac frequency and contractility, vascular tone, activity of the renin–angiotensin axis, and renal sodium reabsorption. Therefore, noradrenaline has sympatholytic action in the central nervous system that tends to lower blood pressure, whereas its stimulatory effect in the peripheral nervous system raises blood pressure.

Approximately 80–90 % of noradrenaline released is taken up again through the Na^+- and Cl^--dependent SLC6a2 transporter, or noradrenaline transporter (NET1),

[33] Strong correlations are found with the catalytic subunit of the myosin light chain phosphatase, A-kinase anchoring protein AKAP9, regulatory subunit $^{P85\alpha}PI3K_{r1}$ of phosphoinositide 3-kinase, protein kinase-Cβ1, and synaptojanin-1.

[34] Concentrations of integrin-α_{2B}, α_6, β_3, and β_5 differ between stiff and distensible aortas. Certain proteoglycans, decorin, osteomodulin, aggrecan-1, and chondroitin sulfate proteoglycan-5 (neuroglycan-C), and related proteins, dermatopontin (a decorin-binding proteoglycan), have different amounts between stiff and distensible aortas.

a neuronal reuptake neurotransmitter transporter of the monoamine transporter superfamily. Uptake is driven by an inwardly directed Na^+ gradient maintained by Na^+–K^+ ATPase. Noradrenaline is then repackaged into vesicles through the vesicular monoamine transporter VMAT2 or degraded by monoaminooxidase. A smaller proportion of released noradrenaline spills out of the synaptic cleft and is catabolized by catechol ^{O}methyltransferase.

Alternatively spliced variants of the SLC6A2 transcript (hence SLC6a2 gene product) exist. In addition, the SLC6A2 gene is subjected to epigenetic modifications. Altered functioning of the SLC6a2 transporter is implicated in numerous cardiovascular diseases [179]. Postural tachycardia syndrome (PoTS) is defined by elevated cardiac frequency equal to 0.5 Hz upon standing without orthostatic hypotension, orthostatic symptoms persisting more than 3 months, and a usual augmented plasma noradrenaline concentration. It primarily affects women. It relies on mutations of the SLC6A2 gene (Ala457Pro)[35] with lowered plasmalemmal density. SLC6a2 mutants oligomerize with normal SLC6a2, thereby decreasing SLC6a2 amount at the cell surface and hence noradrenaline uptake. Patients and family members heterozygous for dominant negative A457P mutation have PoTS symptoms; PoTS patients without the A457P mutation also have decreased SLC6a2 expression in leukocytes and forearm veins [179]. Most polymorphisms encode for SLC6a2 variants (R121Q, N292T, A369P, and Y548H)[36] with decayed noradrenaline affinity. On the other hand, the $SLC6A2^{F528C}$ variant is associated with increased SLC6a2 function (elevated noradrenaline uptake 30 %), which is insensitive to PKC-mediated downregulation. [179].

Chronically altered SLC6a2 function is associated with heart diseases. Noradrenaline uptake from the synaptic cleft through SLC6a2 is an energy-dependent process. Limited oxygen supply can then impair noradrenaline ingress. Conversely, a defective noradrenaline import and excessive synaptic noradrenaline concentration such as that observed in hypertrophic cardiomyopathy can promote heart disease and predispose to cardiac arrhythmias. Excessive adrenergic activity in the heart can be detected in myocardial infarction, unstable ischemic heart disease, stress-induced cardiomyopathy, and congestive heart failure that can benefit from β-adrenoceptor blockers [179].

---

[35] The Ala457Pro exchange, or A457P variation, is defined by a Pro457 mutant and Ala457 wild-type forms.

[36] The A369P polymorphism (relatively high allele frequency) leads to intracellularly retention and hence lack of noradrenaline transport. Like A457P polymorphism, A369P and N292T polymorphisms exert a dominant negative effect on wild-type SLC6a2 transporter [179]. Other gene variants (T182C and A3081T) alter SLC6A2 promoter activity; they are associated with pressor response during exercise, but not with cardiac frequency and plasma catecholamine changes.

2.4.4 Genetic Background in Lipidic Metabolism

Gene expression in adipose tissues is strongly correlated with obesity [180]. Medical genetics aims at identifying specific genes associated with a severe risk of diseases and associations of gene variants, each providing a moderate risk.

Heterozygous familial hypercholesterolemia is due to monogenic disorder associated with one defective allele coding for the LDLR. The homozygous disease is defined by both defective alleles, resulting in nonfunctioning LDL receptors.

Loci near the ANGPTL3[37] and MLXIPL genes[38] are significantly associated with triglyceride concentrations [181]. Endothelial dysfunction is associated with hypoalphalipoproteinemia, a genetic disorder with low plasma levels of HDL^{Cs} and ApoA1 and high coronary heart disease risk. The GALNT2 gene[39] is substantially related to the HDL^{Cs} concentration.

The chromosomal region 1p13.3 is related to a risk of myocardial infarction.[40] It is associated with LDL^{Cs}; in European populations, the minor allele is linked to lower levels of LDL^{Cs}, hence a smaller risk of coronary artery disease. This chromosomal region harbors 4 genes that encode proline/serine-rich coiled coil protein PSRC1, cadherin, EGF LAG 7-pass G-type receptor CELSR2, myosin-binding protein H-like molecule (MyBPhL), and sortilin-1 (Sort1). The latter is a transmembrane receptor that binds to various ligand types. It is involved in the endocytosis and intracellular degradation of lipoprotein lipase, a rate-limiting enzyme of triglyceride hydrolysis, as well as internalization of apolipoprotein-A5–containing chylomicrons.

The hepatic levels of Psrc1, Celsr2, and SORT1 transcripts (mRNAs) correlate with LDL^{Cs} plasma concentration. The coronary artery disease (CAD) risk allele is associated with lower levels of CELSR2 and SORT1 and a higher LDL^{Cs} level.

In addition to the chromosomal 1p13 locus near the Celsr2, Psrc1, and SORT1 genes, the 19p13 locus near the CILP2 and PBX4 genes and 8q24 locus near the TRIB1 gene are significantly associated with the LDL^{Cs} concentration.

2.4.5 Congenital Cardiac Malformations

Congenital cardiac malformations (CCM), a.k.a. congenital heart diseases or defects (CHD),[41] caused by dysregulation of heart development before birth, is the leading

---

[37] The ANGPTL3 gene encodes angiopoietin-like protein-3 that influences triglyceride metabolism.

[38] The MLXIPL gene encodes carbohydrate-response element-binding protein. This transcription factor connects hepatic carbohydrate flux to fatty-acid synthesis.

[39] The GALNT2 gene encodes ^{N}acetylgalactosaminyltransferase-2 involved in O-linked glycosylation.

[40] The chromosome 1 is the largest human chromosome, with more than 249×10^6 nucleotide base pairs [182].

[41] This acronym also refers to as coronary heart disease. It is thus avoided.

Table 2.12 Heart development during embryogenesis. The first and second heart fields contribute to the left and right ventricle, respectively, as well as for the latter to the outflow tract, sinus venosus, and left and right atria. During heart maturation, the cardiac cushions give rise to the atrioventricular valves, left and right ventricular myocardium to the septum, the growth of primary and the secondary septa to the atrial septum, and outflow tract septation separates the common outflow tract into the aorta and pulmonary artery. (Source: [318])

Stage	Time (day)	Events
Early heart development stages		
Cardiac crescent	15	Cardiac differentiation, migration to midline
Linear heart tube	20	Heart tube formation, first heart beats, anteroposterior and dorsoventral patterning
Looping heart	28	Early chamber formation, looping to right
Chamber formation	32	Chamber formation, trabeculation, cushion formation, outflow tract septation, early conduction system formation
Heart maturation		
Ventricular septation	50–90	Ventricular septum
Atrial septation	60–270	Atrial septum
Outflow tract septation	60–270	Aorta and pulmonary artery

cause of infant morbidity in high-income countries (19–75 per 10^3 live births according to types of malformations, excluding cardiomyopathies, conduction tissue abnormalities, and laterality defects [318]). Congenital cardiac malformations range from simple to severe defects, such as complete absence of one or more chambers or valves (7–8 per 10^3 births).

Congenital cardiac malformations result from mutations in genes that encode regulators of heart development and environmental agents. Prenatal exposure to angiotensin-converting enzyme inhibitors increases the risk of congenital malformations.

The earliest stage of heart formation (cardiac crescent) is characterized by 2 pools of cardiac precursors (Table 2.12):

1. the *first heart field* that contributes to the left ventricle, and
2. the *second heart field* devoted to the right ventricle and later to the outflow tract, sinus venosus, and left and right atria.

Secreted molecules, such as fibroblast growth factor, bone morphogenetic proteins, and Wnt proteins, among others, interacting transcription factors,[42] histones and components of chromatin-remodeling complex such as BAF60C of the Swi/Snf-like complex BAF control the fate of cardiac progenitors [318]. Histone methyltransferase SMYD1 (or BOP) regulates cardiac chamber growth and differentiation. Notch-1

[42] Transcription factors Tbx5 and Nkx2-5 interact and synergistically activate their downstream targets. Defective interactions between GATA4 and NKX2-5 and between GATA4 and TBX5 might underlie congenital heart diseases caused by GATA4 mutations.

is used for epithelial-to-mesenchymal transition and valve formation. Morphogen Wnt11 signals downstream from the transcription factor Paired-like homeobox gene product Pitx2 to regulate outflow tract morphogenesis via transforming growth factor-β2.

2.4.5.1 Classification

Congenital heart diseases can be classified into 3 main categories:

1. *cyanotic heart diseases* with mixing of oxygenated and deoxygenated blood;
2. *left-sided obstruction defects*; and
3. *septation defects*.

Cyanotic heart defects include transposition of the great arteries, tetralogy of Fallot, tricuspid atresia, pulmonary atresia, Ebstein's anomaly of the tricuspid valve, double outlet right ventricle, persistent truncus arteriosus, and total anomalous pulmonary venous connection.

Left-sided obstructive lesions comprise hypoplastic left heart syndrome, mitral stenosis, aortic stenosis, aortic coarctation, and interrupted aortic arch. Septation defects can target atrial or ventricular septation or atrioventricular septum formation. Other types of congenital malformations consist of bicuspid aortic valve and patent ductus arteriosus.

2.4.5.2 Genetic Background in Structural Heart Defects

Most congenital heart diseases result from genetic changes, focal mutations, deletion, or addition of DNA segments. A repository of human genetic data for congenital cardiac malformations describes both the phenotype and genetic lesions [183]. Large chromosomal anomalies that cause structural heart defects encompass trisomies 13, 18, and essentially 21. Atrial septal defect can be associated with a mutation of the chromosomal segment 14q12 [184].[43] Microdeletion of the long arm of chromosome 22 (22q11) is responsible for the velocardiofacial syndrome (or DiGeorge syndrome) characterized by conotruncal malformations (tetralogy of Fallot, interrupted aortic arch, ventricular septal defect, and persistent truncus arteriosus) associated with abnormal facies, thymic aplasia, cleft palate, and hypoparathyroidism. Mutations of the TBX1 gene are involved in this syndrome.

Other small chromosomal anomalies comprise microdeletion of the long arm of chromosome 1 (1q21) and of the short arm of chromosome 8 (8p23), among others.

The first identified single-gene mutation that generates atrial and ventricular septation defects and conduction tissue alterations was in the T-box transcription factor TBX5 gene. Deficiency in TBx5 in ventriculomyocytes causes the *Holt–Oram syndrome* [185]. Alterations of the atrioventricular node conduction are associated with

[43] The shorter chromosome arm is called *p arm* (from French *petit*: small) and longer arm *q arm* (in alphabetical order for increasing size).

impaired relaxation, i.e., diastolic dysfunction. The TBx5 factor can indeed activate the Atp2a2 promoter of the gene that encodes sarco(endo)plasmic reticulum Ca^{2+} ATPase SERCA2a, thereby promoting relaxation.[44] Whereas mutations in the MYH6 gene can engender atrial septal defects, mutations in the genes that encode the cardiac transcription factors regulating expression of the cardiac myosin heavy chain-α isoform, such as the Gata4, TBX5, NKX2-5 genes, provoke both atrial and ventricular septal defects. In addition, mutations of the NKX2-5 and TBX5 genes provoke anomalies of the action potential conduction. In particular, mutations in the NKX2-5 gene induce atrial septation defects and an atrioventricular block.

Mutations in the TFAP2B[45] and MED13L[46] genes are sources of heart malformations [318]. Mutations in the NOTCH2 gene are detected in Alagille syndrome. Mutations in a set of genes that encode proteins of the Ras–MAPK pathway (PTPN11, KRAS, RAF1, and SOS1) cause Noonan syndrome. Calcification of the aortic valve is one of the leading causes of heart disease in adults. Mutations in the signaling and transcriptional regulator Notch-1 cause a spectrum of aortic valve anomalies, from early developmental defect in the aortic valve to later valve calcification [186].

Mutations of genes encoding components of the mitogen-activated protein kinase module cause cardiofaciocutaneous syndrome with cardiac defects among other developmental abnormalities [187].

2.4.5.3 Numerical Simulations of Surgical Repair

Reconstructive procedures for congenital cardiac malformations create new vessel circuits and connections. For instance, the Fontan operation treats complex congenital cardiac malformations, the serial systemic and pulmonary circulations being driven by the single available (anatomical or functional) ventricle.

The Fontan procedure was originally aimed at curing tricuspid atresia by an atriopulmonary connection. As a better design of the vascular reconstruction can avoid progressive atrial dilatation, the operation was thus modified to the lateral tunnel technique, the superior vena cava was sutured directly to the right pulmonary artery [188]. This anastomosis (bidirectional Glenn) carried out at an early age is an intermediate step before the final Fontan circulation. A recent modification of the technique replaces the intraatrial routing of the venous blood by the insertion of an extracardiac conduit between the inferior vena cava and the right pulmonary artery.

The surgically created circuit must not only provide adequate blood supply, but also avoid areas of high stress generated by the flowing blood that can damage the

[44] The ventricle relaxation has 2 components: (1) an active component during which cardiomyocytes relax after each contraction with regulated Ca^{2+} removal from the cytosol, and (2) a subsequent passive component during which the ventricle distends due to blood influx from the atria.

[45] The TFAP2B gene encodes transcription factor activating enhancer-binding protein-2β.

[46] The MED13L gene encodes Mediator complex subunit-13-like protein, or thyroid hormone receptor-associated protein THRAP2.

wall wetted surface. Imaged-based numerical simulations can be used to manage surgical repairs of congenital cardiac malformations to limit possible postoperative sequelae. Anatomical and functional data are acquired for each patient prior to surgical reconstruction. Inflow and outflow boundary conditions match available measurements.

Computational blood flow models are also aimed at optimizing the surgical procedure according to hemodynamics [189, 190]. Proper blood flow distribution into the lungs can be provided by fitting a hemodynamic design of cavopulmonary connection, as demonstrated by a 3-D finite element model coupled to a lumped-parameter model of the pulmonary circulation [191]. In particular, numerical tests suggest that the common practice of imposing a right/left pulmonary artery flow split of 55/45 should be avoided [192]. Moreover, alternative surgical designs can be tested [193].

In any case, hemodynamic input data uncertainties on simulation results must be explored as it was done for 5 cavopulmonary connections [192]. A sensitivity analysis investigates the impact of input data.

2.5 Hypoxia and Inflammation

Tissular hypoxia results from an imbalance between the oxygen supply necessary to satisfy metabolic requirements and blood perfusion. It happens during physiological (e.g., fetal development and exercise) and pathophysiological (e.g., ischemia and tumorigenesis) processes. Hypoxia-inducible factor-1α is a transcription factor that ensures cardioprotection. It keeps the mitochondrial polarization during anoxia and increases compensatory glycolytic capacity (possibly, but not necessarily, with larger preischemic glycogen reserve, glycolytic flux, and ATP preservation during ischemia) [194].

Hypoxia and inflammation are 2 interdependent processes, as hypoxia primes inflammation and, conversely, inflammatory diseases are frequently characterized by tissular hypoxia and/or stabilization of hypoxia-dependent transcription factors such as hypoxia-inducible factor. For example, ischemia–reperfusion injury is characterized by an inflammatory response (*hypoxia-induced inflammation*). On the other hand, pulmonary inflammation such as in acute lung injury is associated with metabolic alterations leading to HIF1α stabilization (*inflammatory disease characterized by hypoxia or HIF activation*).

Hypoxia is a prominent feature of chronically inflamed bodily tissues due to an increased metabolic activity that is unmatched by an impaired local perfusion, hence a decreased oxygen supply at the inflamed site. Hypoxia affects the regulation of neutrophil and macrophage survival, macrophage and T-lymphocyte differentiation, and dendritic cell function.

The cellular adaptation to hypoxia is regulated by the HIF family of transcription factors, especially hypoxia-inducible factor-1 (HIF1α–HIF2α). Hypoxia-inducible factor activates transcription of multiple genes that encode proteins involved in angiogenesis, erythropoiesis, pH regulation, and glucose metabolism. The hypoxia-induced reprogramming leads to an increased dependence on glycolysis over oxidative phosphorylation; this metabolic shift is called the *Warburg effect*.[47] Activators of HIF are aimed at yielding tissular protection.

Hypoxia can dampen tissular inflammation. Hydroxylases link hypoxia to inflammation. Oxygen-sensing hydroxylases control the cell adaption to hypoxia, especially transcriptional adaptation to hypoxia via hypoxia-inducible factor and nuclear factor-κ B, both regulating inflammation. These cellular oxygen sensors indeed intervene in the posttranslational regulation of hypoxic and inflammatory pathways. They control the stability of the α subunit of the HIF factor.

Hydroxylases comprise 4 enzymes: *prolyl hydroxylases* PHD1 to PHD3 and a single asparaginyl (asparagine) hydroxylase called *factor-inhibiting HIF* (FIH). In normoxia, PHDs control the elimination of HIF, as they target HIF for proteasomal degradation upon proline hydroxylation, whereas FIH renders HIF transcriptionally less active, as it finely tunes HIF activity by regulating interactions with the transcriptional coactivators CREB-binding protein (CBP) and P300 [196, 197]. They require oxygen as a cofactor for hydroxylation; when oxygen level lowers, they are inactive and their inhibition of HIF is hence relieved, thereby launching the HIF-regulated gene program that includes production of anti-inflammatory molecules, such as adenosine and netrin-1 [196].

In hypoxia, the HIFα–HIFβ heterodimer in the nucleus binds to hypoxia response promoter element (HRE) on genes that encode nuclear factor-κB and Toll-like receptors. The former is a transcriptional activator of the Hif gene (mutual activation). Once it is activated, NFκB initiates the transcription of inflammatory genes as well as the Hif gene, hence genes involved in tissue protection and homeostasis.

Hydroxylases, which are inhibited by hypoxia, modulate inflammation via posttranslational modifications in the IL1β pathway [197]. Interleukin-1β is a major proinflammatory cytokine associated with multiple inflammatory diseases. Interleukin-1β engenders NFκB action in vivo at the level of or downstream from the tumor-necrosis factor receptor (TNFR)-associated factor TRAF6 complex via hydroxylases, more precidely via the combined activity of PHD1 and FIH, in addition to its signaling via the IKK complex [197]. The ubiquitin conjugase for TRAF6 ubiquitin ligase is composed of ubiquitin conjugase-E2 variant-1 (UbE2V1)[48] and ubiquitin conjugase-E2N (UbE2N) that are substrates of PHD1 hydroxylase. On the other hand, OTU domain-containing ubiquitin aldehyde-binding deubiquitinase-1

[47] Among 6 long noncoding RNAs (lncRNA) and large intergenic noncoding RNAs (lincRNA) according to whether they are encoded intra- or intergenically, respectively, that possess hypoxia response elements (HRE) in their gene promoters, only lincRNAp21 is boosted by hypoxia [195]. It enables hypoxia-induced changes in metabolism, favoring hypoxia-enhanced glycolysis. Moreover, it protects HIF1α against its degradation, as it interacts with both HIF1α and VHL ubiquitin ligase and competitively disrupts the VHL–HIF1α interaction, thereby assisting HIF1α stability in response to hypoxia. Conversely, HIF1α supports the hypoxia-induced increase in lincRNAp21 level, which is a HIF1α transcriptional target (positive feedback).

[48] A.k.a. TRAF6-regulated IKK activator-1β.

(OTUB1) that antagonizes the UbE2V1–UbE2N complex is hydroxylated by the FIH enzyme. Downstream from the TRAF6 complex, Iκ Bβ is also hydroxylated by the FIH enzyme. Moreover, several proteins in the IL1β pathway are hydroxylated by PDH and/or FIH hydroxylases. Therefore, these hydroxylases can regulate IL1β signaling and subsequent inflammatory gene expression. Hydroxylase inhibitors reduce inflammation in vivo, as they suppress IL1β-induced NFκ B activity.

2.6 Mitochondrion and Inflammation

Mitochondria can trigger inflammation, as released mitochondrial DNA is an alarmin. On the other hand, Toll-like receptors affect the mitochondrial activity. Altered mitochondrial Ca^{2+} handling intervenes in the pathogenesis of metabolic disorders and cancers. The calcium signal links the endoplasmic reticulum and mitochondria. Calcium signaling is translated within mitochondria into increased ATP production, release of apoptotic cofactors, and energetic collapse linked to necrosis [198].

Mitochondrial Ca^{2+} uptake indeed controls both the activity of mitochondrial matrix dehydrogenases of the tricarboxylic acid cycle, thus increasing electron feeding to the oxidative phosphorylation chain and ATP production, as well as the sensitivity to apoptotic and necrotic cues. An augmented Ca^{2+} ion level in the mitochondrial matrix favor the opening of the permeability transition pore (PTP), hence inducing mitochondrial morphological and functional alterations that lead to apoptosis and necrosis. Moreover, heightened feeding of electrons to the cellular respiratory chain by stimulated Ca^{2+}-dependent matrix dehydrogenases can increase ROS production [198]. Reactive oxygen species contribute to the stabilization of the HIF1 factor. Toll-like receptors constitute a family of receptors that includes 10 paralogs in humans. Upon binding of pathogen-associated molecular patterns (PAMP) of bacterial, viral, or fungal origin, a signaling cascade is activated that triggers the transcription of genes encoding inflammatory mediators (e.g., TNFSF1 and IL6). In addition to microbial PAMPs, TLRs can also sense alarmin (or damage-associated molecular pattern [DAMP]) released from infected or stressed cells. These TLR ligands comprise nuclear structural components (e.g., HMGb1), heat shock proteins (HSP60 and HSP70), and mitochondrial components (e.g., mtDNA). Upon tissue damage, mitochondrial DNA enriched in unmethylated cytidine^{P}–guanosine (CpG) oligodeoxynucleotide (CpGODN) motifs is secreted.

Toll-like receptors are not necessarily associated with pro-inflammatory effects, as small PAMP doses can attenuate inflammatory response to subsequent larger PAMP doses and to injury, possibly via transcription of genes encoding inhibitors of the TLR–NFκB pathway [198]. Among TLR ligands capable of initiating an anti-inflammatory response, unmethylated CpGODN is a potent TLR9 ligand. It attenuates the acute inflammatory cardiac dysfunction primed by lipopolysaccharide as well as ischemia–reperfusion injury, as it inhibits the NFκB pathway in ventriculomyocytes [198].

In addition to the canonical TLR–NFκB axis, an alternative anti-inflammatory TLR9-launched cascade is activated in parenchymal cells of inflamed tissues [199]. The TLR9 receptor temporarily reduces energy substrates and activates AMP-activated protein kinase (AMPK) to induce cellular protection from stress and obviate canonical inflammatory signaling.

In immunocytes (e.g., macrophage), the chaperone-like protein Unc93 homolog Unc93b1 assists TLR9 transfer from the endoplasmic reticulum to the endosomal–lysosomal compartment, where CpGODN-bound TLR9 is cleaved (N-terminus shedding) [199]. The subsequent immune-prone receptor form then initiates the canonical MyD88-dependent pro-inflammatory pathway.

In cardiomyocytes and neurons that have a poor regenerative capacity, Unc93b1 is expressed at low levels, and TLR9 is mainly retained in the endoplasmic reticulum [199]. In this organelle in cardiomyocytes (but not in cardiofibroblasts), CpGODN-bound TLR9 then inhibits the sarco(endo)plasmic reticulum Ca^{2+} ATPase SERCA2, thus reducing Ca^{2+} transfer from the cytosol to mitochondria and aerobic metabolism, thereby increasing cell resistance in the inflammation site [199].

In addition, inositol trisphosphate receptor (IP_3R), a Ca^{2+} release channel of the endoplasmic reticulum that enables Ca^{2+} efflux from mitochondria and Ca^{2+}-dependent stimulation of aerobic metabolism, hence the maintenance of ATP level and attenuation of AMPK signaling due to an elevated ATP/AMP ratio, also hinders autophagy [198].[49]

Therefore, signaling from TLR9 protects cardiomyocytes (and neurons among other cell types) from stress-induced cell death, as it impedes Ca^{2+} influx in mitochondria and modulates energy metabolism, thereby favoring cell survival.

2.7 Oxidative and Nitrative Stresses and Cardiac Diseases

Oxidative stress results from an imbalance between the production of reactive oxygen species and scavengers of these toxic intermediates. Oxidative stress is a major contributor to the development of heart failure. Oxidative and nitrative stresses are shared by many cardiac disorders such as ischemia–reperfusion, among other pathologies.

Reactive oxygen species activate signaling via MAPK modules (i.e., ERKs or JNKs). Whereas low ROS concentrations generally stimulate cell proliferation, high ROS concentrations cause cell apoptosis and death. Low ROS concentrations engender phosphorylation (activation) of ERKs [200]. High concentrations of ROS provoke phosphorylation (activation) of JNKs. MAP3K11 phosphorylates JNKs and determines the relative phosphorylation of ERKs and JNKs upon ROS exposure, promoting JNK activation and lowering ERK activation.

[49] Oncogenes reduce the Ca^{2+} level in the endoplasmic reticulum. Cancer-related microRNAs lower the expression of the mitochondrial Ca^{2+} uniporter, thus reducing sensitivity to apoptosis. On the other hand, tumor suppressors have the opposite effect [198].

On the other hand, ATP-binding cassette transporter ABCg2+ cells are implicated in cardioprotection against oxidative stress [201]. Parkinson disease-susceptibility gene product Park7 is an antioxidant agent. It is highly expressed in the normal heart, but is markedly reduced in heart failure [202]. In cooperation with P53, it yields proper capillary density. It also enables adequate mitochondrial genesis.

2.7.1 Altered Posttranslational Modifications of Sarcomeric Constituents

Oxidative stress disturbs the pump function because of changes in production and/or function of regulators of cytosolic Ca^{2+} concentration. Cardiodepressant action of reactive oxygen species depends on posttranslational modifications of sarcomeric protein and ROS-activated enzymes that regulate cardiac contractility. Oxidation of actin (Cys374) enables maximum actomyosin ATPase activity and actin filament sliding velocity [203].

Cycles of sarcomere contraction and relaxation are regulated by between-protein interactions that lead to Ca^{2+} influx and conversion of chemical energy associated with ATP into mechanical energy followed by Ca^{2+} efflux. In the healthy heart, these between-protein interactions are controlled by phosphorylation. In cardiac disorders, alterations in posttranslational modifications of sarcomeric constituents cause contractile dysfunction. Myosin heavy chain is a sarcomeric redox sensor. Redox modifications of myosin heavy chain (Cys697 and Cys707) decrease myosin ATPase activity [203].

Redox modifications of actin and tropomyosin provoke impaired actin–myosin cross-bridge formation and altered actin filament activation by Ca^{2+} ions [203]. Glutathionylation of actin (Cys374) may decrease tropomyosin–actin binding. Oxidation of tropomyosin (Cys190) by ROS leads to dimerization and alters tropomyosin flexibility and tropomyosin–actin interaction in ischemic pig hearts. Titin scaffold that recruits signaling mediators is sensitive to reactive oxygen species. Oxidative stress decreases titin extensibility and increases its passive tension [203]. Desmin, an intermediate filament protein, forms a network around sarcomeric Z discs, which links neighboring myofibrils and connects myofilaments to other cellular structures (nucleus, cytoskeleton, and mitochondria). Oxidized or nitrated desmin accumulates in insoluble aggregates that disrupt the sarcomeric lattice [203].

2.7.2 ROS-Dependent Regulation of Sarcomeric Protein Phosphorylation

Several sarcomeric constituents, such as cardiac troponin-I[50] and -T,[51] cardiac isoform of myosin-binding protein-C (cMyBPc or MyBPc3), and titin, are phosphorylated by ROS-sensitive enzymes (Table 2.13). Therefore, a disturbed intracellular redox state and a shift in the kinase–phosphatase balance can alter cardiac contractility.

Myosin-binding protein-C bridges myosin and the M-band part of titin. Its phosphorylation (Ser273, Ser282, and Ser302) by PKA decreases myosin–actin interactions and increases force generation [203]. In addition, PKCδ, * PKCϵ, and PKD target MyBPc3 (Ser302) to regulate sarcomere contractility.

Phosphorylation of titin (Ser469) by PKA or PKG decreases its passive tension [203]. Protein kinase-G also phosphorylates titin at other sites, thereby influencing interactions with its partners, but not its mechanical properties.

Sarcomeric proteins are generally dephosphorylated by PP1 and PP2 protein phosphatases [203]. The latter colocalizes with cTnnT and cTnnI as well as with ROS-sensitive enzymes at Z discs (PKCϵ, PKCζ, PAK1, and P38MAPK).

Oxidative stress typically increases protein phosphorylation, as protein phosphatases are inhibited (e.g., PP3) and kinases stimulated (especially PKC and PKD [203]. However, oxidative stress increases PP1 and/or PP2 activity via ROS-activated kinases (PKCζ, PAK1, and P38MAPK). ROS-activated kinases typically phosphorylate multiple constituents of the sarcomere.

Oxidative modifications of sarcomeric proteins lead to a decrease in force generation, whereas sarcomeric protein phosphorylation by ROS-activated enzymes decreases myofilament Ca^{2+} sensitivity.

2.7.2.1 MAP3K5

Stress-activated ROS-regulated MAP3K5, which abounds in cardiomyocytes, acts as a redox sensor that controls cell death. Oxidation of thioredoxin-1 dissociates the MAP3K5–TRx1 and MAP3K5–14-3-3 complexes, hence relieving MAP3K5 inhibition. Then, MAP3K5 undergoes autophosphorylation and oligomerization (activation).

---

[50] Troponin-I is phosphorylated (Ser23/Ser24) by PKA upon β-adrenergic receptor stimulation, enabling positive lusitropy. Protein kinase-G, various PKC isoforms, and PKC-activated enzymes, such as ^{P90}RSK and PKD also target this phosphorylation site. Protein kinase-C phosphorylates other sites (Ser43/Ser45 and Thr144).

[51] Troponin-T phosphorylation (Thr206) by PKC or cRaf decreases maximum force and myofilament Ca^{2+} sensitivity. Both PKC and MAP3K5 phosphorylate cTnnT at other sites. Protein kinase-A and -G phosphorylate cTnnI only when anchored to cTnnT.

Table 2.13 ROS-mediated modifications of cardiac sarcomeric proteins. (Source: [203])

Target	Modification	Effect
α-Actn	Oxidation	Longitudinal force transmission ↓
Actin	Oxidation	Myosin ATPase activity ↓
		Actin filament sliding velocity ↓
		TMy-actin binding ↓
		Actin depolymerization ↑
Desmin	Oxidation	Myofibrillar disarray, aggregate formation
		Proteasome degradation
MHC	Oxidation	Myosin inhibition
		Maximum force ↓
MyBPc3	Oxidation	Contractility ↓
	Phosphorylation	Actin–myosin interaction ↓
	[PKA, PKC, PKD]	Force generation ↓
Titin	Oxidation	Extensibility ↓
		Passive tension ↑
	Phosphorylation	Passive tension ↓
	[PKA, PKG]	
	[PKCα]	
TMy	Oxidation	Contractility ↓
		Flexibility ↓
		Binding to actin ↓
cTnnI	Phosphorylation	Calcium dissociation from TnnC ↑
	[PKA, PKC, PKG, PKD,	Calcium sensitivity ↓
	^{P90}RSK]	Relaxation rate ↑
	[PKC]	Maximal force ↓
	[PKC, STK4]	Calcium sensitivity ↓
	[STK4]	Altered conformation
	[PAK3]	Calcium sensitivity ↑
cTnnT	Oxidation	Contractility ↓
	Phosphorylation	Maximal force ↓
	[PKC, Raf]	Calcium sensitivity ↓
	[PKC, MAP3K5]	Phosphorylation potentiation

↑ increase, ↓ decrease, *α Actn* α-actinin, *MHC* myosin heavy chain, *MyBPc* myosin-binding protein-C, *TMy* tropomyosin, Tnn troponin

Pressure overload or myocardial infarction activate MAP3K5 via hydrogen peroxide (H_2O_2) or liganded G-protein–coupled receptors that cause ROS accumulation [203]. Activated cardiodepressant MAP3K5 contributes to ventricular remodeling, as it stimulates the JNK and/or NFκB pathways. In the sarcomere, MAP3K5 phosphorylates cTnnT (Thr197 and Ser201).

2.7.2.2 STK4

ROS-activated protein Ser/Thr kinase STK4 activates P38MAPK and JNK as well as caspase-dependent mechanisms that trigger apoptosis. It phosphorylates cTnnI (Thr32, Thr52, Thr130, and Thr144) and cTnnT, thereby influencing binding affinity of cTnnI for cTnnT and cTnnC [203].

2.7.2.3 PKA

Protein kinase-A regulates cardiac contractility, myocardial metabolism, and gene expression. Heterotetrameric PKA holoenzyme remains inactive due to 2 ubiquitous, cAMP-binding regulatory subunits that target PKA to different subcellular compartments using PKA-anchoring proteins. In general, PKA activation results from the β AR–cAMP pathway that dissociates inhibitory subunits.

Oxidation (Cys199) of ROS-dependent PKA, which assists dephosphorylation (Thr197), decreases PKA catalytic activity [203]. Its substrates comprise cTnnI (Ser23/Ser24). In fact, low-intensity oxidant stress amplifies PKA response, as it inactivates phosphatases that counteract PKA; high-magnitude oxidative stress inactivates PKA [203].

In cardiomyocytes subjected to oxidative stress, particularly during short-duration ischemia and reperfusion, expression of A-kinase-interacting protein AKIP1[52] is upregulated [204]. It protects against ischemic injury via enhanced mitochondrial integrity. Scaffold AKIP1 interacts with mitochondrial apoptosis-inducing factor (AIF),[53] which acts both in mitochondria and the nucleus, under normal conditions as well as oxidant stress, and increases PKA activity. Other PKA substrates, Drp1 and ChChD3, participate in the maintenance of mitochondrial integrity. Another PKA substrate, ATP synthase-α, limits AIF translocation into the nucleus during oxidant stress.

[52] In humans, AKIP1 has 3 splice variants, full-length protein (AKIP1a), a subtype encoded by a transcript lacking exon 3 (AKIP1b), and another by a transcript lacking exon-3 and -5 (AKIP1c). Isoform AKIP1b recruits the histone deacetylase sirtuin-1 upon neddylation that represses gene transcription. On the other hand, AKIP1a recruits NFκ B in a PKA-dependent manner and enhances gene transcription. In the nucleus, AKIP1 enhances the PKA–NFκ B–SIRT1 axis. Another role takes place at mitochondria.

[53] In mitochondria, the apoptosis-inducing factor may stabilize ^{ETC}complex-I. This death effector is released from mitochondria and translocates to the nucleus. In the nucleus, it causes chromatin condensation and DNA fragmentation. Protein AKIP1 interacts with AIF and sequesters AIF to mitochondria, thereby preventing cell death.

2.7.2.4 PKC

Protein kinase-C are activated by growth factors via the PLC–DAG pathway. Conventional PKCα, PKCβ1, PKCβ2, and PKCγ are diacylglycerol- and calcium-sensitive isoforms. In cardiomyocytes, novel PKCs that include PKCδ and PKCϵ isozymes bind to lipids such as diacylglycerol. Atypical PKCζ and PKCι connect to PI(1,4,5)P_3 or ceramide, but not diacylglycerol.

Oxidation mediated by reactive oxygen species activates PKC by relieving autoinhibition [203]. In addition, ROS-dependent phosphorylation by Src (Tyr311) specifically activates PKCδ (but not other PKC isoforms). Whereas allosterically activated PKCδ phosphorylates cTnnI at a single Ser cluster (Ser23/Ser24), ROS-activated PKCδ phosphorylates cTnnI also at a Thr residue (Ser23/Ser24 and Thr144), in addition to cMyBPc substrate. Phosphorylated PKCδ by Src reduces maximum tension and cross-bridge kinetics [203].

2.7.2.5 PKD

Diacylglycerol activates PKD, as it drives PKD phosphorylation (Ser744 and Ser748) by novel PKCs; PKD then autophosphorylates (Ser916). Its substrates include cTnnI (Ser23/Ser24) and cMyBPc (Ser302). Protein kinase-D accelerates cross-bridge kinetics via phosphorylated cMyBPc [203].

Oxidative stress stimulates PKD via successive actions of protein Tyr kinases Abl (Tyr463) and Src (Tyr95) followed by that of ROS-activated PKCδ [203].

2.7.2.6 PKG

The stimulated NO–cGMP axis causes a conformational change that relieves autoinhibition, thereby activating homodimeric PKG enzyme. Isoforms PKG1α and PKG1β are alternatively spliced variants, whereas PKG2 is encoded by a different gene. In cardiomyocytes, PKG1 is the major subtype. Isoform PKG1α binds cGMP with a tenfold higher affinity than PKG1β variant.

Whereas PKG1β is not ROS-sensitive, PKG1α experiences a ROS-regulated dimerization [203]. Dimer PKG1α accumulates during oxidative stress. Agents cGMP and ROS increase the kinetics of PKG1α reaction with and affinity for its substrates such as cTnnI, respectively.

2.7.3 Nitrative Stress

Calcium-dependent nitric oxide synthases synthesize nitric oxide in cardiomyocytes, endocardial and coronary endotheliocytes, and cardiac neurons.

2.7.3.1 Nitric Oxide

Nitric oxide participates in the regulation of coronary vasodilation, inhibition of platelet activation and neutrophil adhesion as well as cardiac oxygen consumption, and modulation of cardiac contractility. Nitric oxide stimulates soluble guanylate cyclase and cGMP-dependent protein kinase-G to decrease cytosolic Ca^{2+} concentration.

In addition, NO contributes to the termination of lipid radical reactions caused by oxidative stress, thereby protecting against ischemia [205]. Nitric oxide interacts with components of the mitochondrial electron transport chain, hence modulating the generation of reactive oxygen species by mitochondria.

Yet, at high concentration, or in the presence of reactive oxygen species, NO can exert cytotoxic effects via the formation of peroxynitrite ($ONOO^-$), thereby causing myocardial and vascular dysfunction associated with oxidative and nitrative stresses [205].

2.7.3.2 Peroxynitrite

Peroxynitrite primes apoptosis of cardiomyocytes and vascular endothelial and smooth muscle cells, attenuates cardiomyocyte contractility, and irreversibly inhibits the mitochondrial electron transport chain (Table 2.14) [205]. The myofibrillar isoform of creatine kinase, an energetic controller of cardiomyocyte contractility, undergoes a peroxynitrite-induced nitration (inactivation). Peroxynitrite also inactivates myocardial aconitase, an enzyme of the citric acid cycle that resides in mitochondria and cytosol. Furthermore, myocardial α-actinin, which is used in the maintenance of the Z line and integrity of sarcomeres, and sarcoplasmic reticulum Ca^{2+} ATPase are subjected to peroxynitrite-induced nitration. Peroxynitrite activates ERK, which is linked to heart hypertrophy, and inhibits NFκB, activation of which is triggered by inflammatory stimuli. Moreover, it increases the density of adhesion molecules on the endothelial wetted surface.

2.7.4 Sympathovagal Imbalance

Reactive oxygen and nitrogen species, such as superoxide ($O_2^{\bullet -}$), free radical nitric oxide ($NO^{\bullet}$), and peroxynitrite ($ONOO^-$), contribute to cardiac sympathovagal imbalance in the brainstem and peripheral cardiac neurons as well as cardiomyocytes, which all experience oxidative stress [206]. Nitric oxide depresses the sympathetic activity. Sympathetic hyperactivity and parasympathetic hypoactivity observed in many cardiovascular diseases can trigger arrhythmias. Noradrenaline causes endothelium-dependent relaxation via stimulation of endothelial α1- and α2-adrenoceptors, thereby counteracting its vasoconstrictory effect. Enzyme NOS2 is expressed throughout the central and peripheral nervous system. Nitric oxide is a neurotransmitter in both the central and peripheral autonomic nervous systems that can modulate levels of sympathetic nerve activity and subsequently blood pressure.

Table 2.14 Selected effects of peroxynitrite in the cardiovascular apparatus. (Source: [205])

Inhibition of BH4-dependent, NAD-dependent, antioxidant enzymes Depletion of antioxidants (e.g., glutathione, cysteine) Uncoupling of NO synthase (production of O_2^- rather than NO) Increased lipid peroxidation
Impairment of mitochondrial respiration and energy metabolism (inactivation of cytochromes and NADH–CoQ1 reductase) Promotion of Ca^{2+}-dependent mPTP opening Inhibition of myocardial aconitase
Inactivation of myofibrillar creatine kinase (impairment of contractility) Inactivation of α-actinin in cardiomyocytes
Inactivation of SERCA in cardiac and vascular smooth myocytes (Ca^{2+} dysregulation)
Alteration of α1-, β1-, and β2-adrenoceptors, vasopressin V_{1A}, and angiotensin receptors
Activation of ERK kinases Activation or inhibition of NFκB Nitration of PKCϵ (promoting translocation and activation)
Activation or inactivation of K_{ATP} channel Inactivation of K_V and K_{Ca} channels in coronary arterioles
Alteration of vascular relaxation Inactivation of prostacyclin synthase
Upregulation of adhesion molecules in endotheliocytes Endothelial glycocalyx disruption Enhanced neutrophils adhesion
Pro- or antiaggregatory effects on platelets depending on context
Activation of myocardial matrix metallopeptidases (^{S}glutoxidation of prometallopeptidases)
Apoptosis of cardiomyocytes, vascular endothelial and smooth muscle cells

ATn2 angiotensin-2, *BH_4* tetrahydrobiopterin, *CoQ* coenzyme Q [ubiquinone, NADH–CoQ reductase is the electron transport chain complex-I], *mPTP* mitochondrial permeability transition pore, *NFκB* nuclear factor-κB, *NO* nitric oxide, *PARP* poly^{ADP}ribose polymerase, *SERCA* sarco(endo)plasmic reticulum Ca^{++} ATPase

It has a sympathoinhibitory effect on the paraventricular nucleus, nucleus of the solitary tract (NTS), rostral ventrolateral medulla, carotid body, and renal nerves (Table 2.7). It can increase the sympathoinhibition exerted by γ-aminobutyric acid and decrease sympathoexcitation yield by angiotensin-2 and glutamate [142].[54]

Nitric oxide is not only an inhibitor of sympathetic nerve activity, but also a modulator of vasoconstriction exerted by angiotensin-2 and vasopressin in regional blood

[54] Glutamate released from the presynaptic neuron activates $GluN_1$ (^{NMDA}Glu) receptor and provokes a calcium influx that stimulates NOS2 linked to DLg4 anchoring protein attached to $GluN_1$ via calmodulin. Nitric oxide produced in the postsynaptic neuron can then diffuse to the presynaptic neuron or neighboring cells (neurons or astrocytes) to activate sGC that synthesizes cGMP (in both pre- and postsynaptic neurons) [142]. Messenger cGMP can then influence activity of ion channels, phosphodiesterases, and protein kinases.

circulation. Neuronally produced NO plays a more important role than endothelially synthesized NO in some regional vascular beds. Neuronal NO inhibition of noradrenaline release is stronger in small arteries. Neuronal NO can regulate regional blood flows at the supraspinal, spinal, and ganglionic levels. In rats, nitric oxide suppresses sympathetic vasoconstriction in the mesentery and kidney at the spinal level and in hindquarters at supraspinal and ganglionic levels [207].

2.8 Myocarditis

Myocarditis is an an acute or chronic inflammation of the myocardium associated with the immune response to a viral (mainly enteroviral coxsackievirus-B3, adenovirus, parvovirus-B, and human herpes virus), bacterial, or protozoan infection, or hypersensitivity, autoimmunity, or cardiotoxicity.

2.8.1 Viral Myocarditis

Myocarditis is mainly caused by viral infections or autoimmune reactions. Two cardiac-enriched miRNAs, miR208b and miR499-5p, are specifically elevated in patients with acute viral myocarditis [237]. They reflect disease severity, as they are correlated with troponin-T levels and ejection fraction.

2.8.2 Cardiac Cell Interactions In Myocarditis

Cytotoxic T lymphocytes and natural killer cells destroy infected cardiomyocytes, whereas innate immunity cells, such as neutrophils, macrophages, and mastocytes, contribute to cardiac damage via ROS overproduction. Uncontrolled inflammation causes cardiac wall stiffening and impaired contractilility, and can lead to dilated cardiomyopathy.

Early after infection, cardiomyocytes expose on their plasma membrane the viral antigens complexed with class-I major histocompatibility complex molecules. The MHC–AG complex is recognized by T-cell receptors that trigger leukocyte-activation signaling. The subsequent massive release of soluble factors by cardiomyocytes and immunocytes contribute to attract leukocytes to the infected myocardium.

Numerous signaling pathways operate in cardiomyocytes and leukocytes in myocarditis (Table 2.15). In viral myocarditis, immunocytes are actively recruited to the myocardium by MHC–antigen complexes exposed on the membrane of infected cardiomyocytes. Once they are recruited, cytotoxic T lymphocytes and NK cells eliminate infected cardiomyocytes, whereas macrophages and mastocytes exacerbate cardiac injury via uncontrolled ROS production or indirectly via stimulation of cytokine secretion by cardiomyocytes [69].

Table 2.15 Signaling pathways associated with cardiomyocyte–leukocyte communication in myocarditis (Source: [69])

Signaling in cardiomyocytes	Mediator	Effect	Signaling in leukocytes
IL6	CCL2, TNFSF1, IL10	Immunity activation	
Virus	CCL2	Macrophage and neutrophil recruitment	CXCR2
Ifnγ	CXCL10	Macrophage, T- and NK-cell recruitment	CXCR3
	Osteopontin	T-cell expansion and recruitment	
CAR–JNK–P38MAPK–ERK1/2 and CAR–βCtnn		Macrophage and T- and NK-cell recruitment	

CAR coxsackie- and adenovirus receptor, *Ctnn* catenin, *ERK* extracellular signal-regulated protein kinase, *Ifn* interferon, *IL* interleukin, *JNK* Jun N-terminal kinase, *MAPK* mitogen-activated protein kinase, *TNF* tumor-necrosis factor

Several proinflammatory cytokines (TNFSF1, IL1β, IL6, and Ifnγ) as well as anti-inflammatory agents (e.g., IL10) are secreted in the myocardium early after viral infection [69].

Interkeukin-6 that has a proinflammatory function in the acute phase of viral myocarditis operates as an inhibitor of TNFSF1 that reduces the production of the CCL2 chemoattractant and the immunomodulatory IL10 cytokine, thereby limiting immunocyte recruitment and anti-inflammatory signaling. Infected cardiomyocytes partly orchestrate the IL6-dependent regulation of the immune responses to viral invasion.

Cardiomyocytes overexpress the CXCL10 chemokine, thereby priming an early immune response, recruiting macrophages, T lymphocytes, and NK cells, and limiting viral replication [69]. On the other hand, overexpression by cardiomyocytes of osteopontin that promotes cytotoxic T lymphocyte activation and expansion leads to chronic myocarditis [69].

On the cardiomyocyte membrane, the coxsackievirus and adenovirus receptor, a member of the immunoglobulin superfamily, enables the entry of viruses. It also triggers the production of several inflammatory mediators (TNFSF1, IL1β, IL6, IL12, Ifnγ, and CCL2), thereby allowing infiltration of macrophages, T lymphocytes, and NK cells. It also permits disruption of cardiomyocyte adherence junctions and activation of β-catenin hypertrophic signaling [69].

Chapter 3
Adverse Cardiac Remodeling

Adverse cardiac remodeling results usually from abnormal cardiac load or injury, that is, upon and/or after

- pressure overload (e.g., due to aortic stenosis [Chap. 6] and hypertension [Vol. 8, Chap. 4. Hypertension]);
- myocarditis (Sect. 2.8);
- cardiomyopathies (Chap. 4);
- valvular regurgitation and subsequent volume overload (Chap. 6); and
- myocardial infarction (Chap. 8).

Although the etiology and time course of events differ, these diseases share several chemical and mechanical processes [208]. Maladaptive cardiac remodeling is influenced by hemodynamic load and neurohumoral stimulation. Adverse cardiac remodeling can be defined as an altered genomic expression that provokes molecular, cellular, and interstitial changes (Table 3.1; [208, 209]).

The cardiomyocyte is the major cell involved in the pathological remodeling. Other implicated components include fibroblasts as well as the interstitium with collagen fibers and coronary vasculature. Relevant processes encompass ischemia and subsequent cell apoptosis and necrosis. Clinical manifestations comprise variations of size, shape, and function.

Processes occurring in ventricular remodeling include [208]:

1. Cardiomyocyte lengthening, hypertrophy, and loss
2. Inflammation
3. Excessive accumulation of collagen in the cardiac interstitium
4. Ventricular wall thinning and dilation and reshaping of the left ventricle that becomes more spherical
5. After myocardial infarction, reabsorption of necrotic tissue, scar formation, and continued infarct expansion rather than extension

The neurohormonal context implicates especially the sympathetic nervous system and the renin–angiotensin axis as well as oxidative stress that describes the imbalance between production and/or activity of oxygen-derived free radicals and those

M. Thiriet, *Diseases of the Cardiac Pump*, Biomathematical and Biomechanical
Modeling of the Circulatory and Ventilatory Systems 7, DOI 10.1007/978-3-319-12664-7_3

Table 3.1 Cellular changes during maladaptive cardiac remodeling. (Source: [209])

Wall constituent	Changes
Myocytes	Hypertrophy, lengthening
	Reexpression of fetal gene program
	Altered excitation–contraction coupling
	Impaired energy metabolism
	Defective myofibrillar content and function
	Apoptosis, necrosis
Vasculature	Endothelial dysfunction
	Intimal thickening
	Smooth myocyte hyperplasia
	Rarefication of capillaries
Interstitium	Activation of matrix metallopeptidases
	Increased collagen synthesis, fibrosis
	Collagen isoform shift

of antioxidant factors. Various substances are involved, such as noradrenaline, aldosterone, renin, angiotensin-2, potent vasoconstrictor endothelin, atrial natriuretic peptide, nitric oxide, and cytokines (tumor necrosis factor TNFSF1 and interleukins).

The myocardium consists of myocytes tethered and supported by a connective tissue composed largely of fibrillar collagen. When cardiomyocytes stretch, the local activity of noradrenaline, angiotensin-2, and endothelin rises, thereby provoking adverse myocyte hypertrophy. In addition, fibroblasts proliferate. They synthesize and degrade the extracellular matrix. Moreover, elevated activation of aldosterone and cytokines can also stimulate collagen synthesis, thus engendering fibrosis. On the other hand, collagenase present as an inactive proenzyme in the cardiac wall can be activated after myocardial injury [208].

Cardiac dilation can happen without hypertrophy. Myocardial stress then increases and can generate further dilation of the heart.

Adverse cardiac remodeling causes a progressive worsening of cardiac function and then heart failure. Hence, therapy is aimed at slowing and even reversing maladaptive remodeling, thus enhancing cardiac function (end-diastolic and end-systolic volume and ejection fraction). The antineuroendocrine therapy with angiotensin-converting enzyme inhibitors and β-adrenergic blockers improve ejection fraction and ventricular volumes [209].

Cardiac-impaired structure with inflammation, oxidative stress, and remodeling (fibrosis) and ventricular dysfunction is also observed during the development of cardiomyopathies.

3.1 Cardiac Fibrosis

Fibrosis affects the architecture and structure of the heart. It arises during cardiac remodeling. Fibrosis is the feature of numerous types of cardiomyopathies (e.g., hypertrophic and inflammatory cardiomyopathies). Myocardial fibrosis can engender diastolic (abnormal left ventricular filling and reduced compliance and elevated diastolic pressure) and/or systolic dysfunction and ultimately heart failure. Inflammation and fibrosis are correlated with diastolic dysfunction in heart failure with a preserved ejection fraction.

Fibrosis is mainly engendered by collagen-secreting myofibroblasts, which originate not only from cardiofibroblasts but also from other cell sources, such as endothelial and circulating cells.

Fibroblasts are pleiomorphic and pleiotropic cells of connective tissues responsive to numerous profibrotic factors. Conversely, cardiac fibrosis triggers endothelial–mesenchymal transition that contributes to about 30 % of activated fibroblasts.

Cardiofibroblasts ensure the maintenance of the extracellular matrix, as they modulate the production of matrix constituents (e.g., collagen-1 and collagen-3 to collagen-6) and degraders (e.g., matrix metalloproteinases). They also secrete growth factors and cytokines that exert auto- and paracrine effects on cell fate (e.g., proliferation and apoptosis). Under pathological conditions, in response to proinflammatory cytokines (e.g., TNFSF1, IL1, IL6, and TGFβ), cardiofibroblasts have a higher proliferation, migration, and differentiation rate and secrete more collagens, MMPS, and cytokines. These agents contribute to the development of perivascular and interstitial fibrosis. Cardiofibroblasts contribute to cardiac fibrosis via many synergistic factors (e.g., ATn2, ET1, TGFβ, and CTGF).

Collagen production and collagen fiber formation involve multiple stages. Following assembly into a triple helix structure, procollagen is transported in specialized CoP2+ vesicles and secreted into the extracellular space. Procollagen is then processed by peptidases that cleave the N- and C-propeptide-forming collagen that self-assembles into fibrils.[1]

For example, once secreted, procollagen-1 is converted to mature collagen-1 by adamlysin with thrombospondin motifs ADAMTS2 that excises the N-terminal propeptide to generate ^{pC}collagen-1 and by bone morphogenic protein BMP1 that excises the C-terminal propeptide to produce ^{pN}collagen-1 [211]. The collagen on primary murine dermal fibroblast surfaces comprises four forms: procollagen-1, ^{pC} and ^{pN}collagen-1, and collagen 1. In some cases, propeptides are retained following incorporation into fibrils.

Procollagen lysine (1,2)-oxoglutarate 5-dioxygenase PLOD1 (or lysyl hydroxylase LH1) converts lysine into hydroxylysine and lysyl oxidase (LOx) catalyses

[1] The α1 and α2 subunits of procollagen-1 are synthesized from N- and C-propeptides enzymatically released by specific peptidases. The absence of processing of propeptides of procollagen-1 following incorporation into fibrils (diameter 20–40 nm) as well as premature processing of N-propeptide in procollagen-1 precludes fibril fusion and limits expansion.

cross-linking of collagen fibrils and formation of collagen fibers via oxidative deamination of lysine and hydroxylysine.

The matricellular proteins,[2] secreted protein acidic and rich in cysteine (SPARC)[3] and secreted phosphoprotein 1 (SPP1),[4] are major contributors to posttranslational modification, stabilization, and deposition of collagen in the heart.

Collagen-binding SPARC is produced by cardiofibroblasts. It is required for aggregation of collagen fibrils (diameter 60–70 nm), an intermediate step in collagen fiber assembly [212]. It mediates the tethering of procollagen-1 to cells as well as its subsequent processing and incorporation into the extracellular matrix. In the absence of SPARC, the amount of cell surface-associated collagen and the proportion of total collagen-1 without propeptides on SPARC-null fibroblasts increase [213].

Secreted acidic phosphoglycoprotein SPP1 functions in tissue repair, matrix organization, and collagen fibrillogenesis. It intervenes in the development of collagen-1-induced cardiac fibrosis and dysfunction in human dilated cardiomyopathy [214]. Collagen-1 upregulates SPP1 expression, but not fibronectin, laminin, or vitronectin [215]. It is detected in atherosclerotic plaques. It connects to collagen-1 to -3 and -5 synthesized by vascular smooth myocytes as well as collagen-4 in basement membranes [216]. It is targeted by widespread intra- and extracellular Ca^{2+}-dependent tissue transglutaminase that catalyzes the formation of high-molecular-mass complexes of its proteic substrates, as it cross-links glutamine and lysine residues [216]. As is SPARC, SPP1 is linked to fibrosis. In $SPP1^{-/-}$ mice, angiotensin-2-induced myocardial fibrosis is hindered. On the other hand, the cytokine IL1β potentiates angiotensin-2-induced SPP1 expression by fibroblasts. Macrophages and neutrophils also produce SPP1.

3.1.1 Differentiation of Cardiofibroblasts into Myofibroblasts

Fibroblasts originated from diverse organs are heterogeneous according to size, shape, proliferative capacity, growth factor production rate, and collagen secretion rate.

Fibroblasts represent the most numerous cardiac cell type. Myofibroblasts appear in pathological conditions. Fibroblast differentiation is caused by the cytokine

[2] That is, linked to the extracellular matrix, but not structural matrix components.

[3] A.k.a. basement membrane protein BM40 and osteonectin. It is involved in the regulation of cell proliferation and migration as well as growth factor and matrix metallopeptidase activity [211]. It binds to collagen-1 to -5. It influences basal lamina assembly via collagen-4. It is cleaved by MMPs. It has a counteradhesive activity, as it interferes with cell–matrix interactions and affects integrin-linked kinase action. It also has antiproliferative activity in cell cultures [212]. SPARC binds nascent monomeric collagen-1 and/or procollagen-1 and delays its incorporation into nascent fibrils until propeptide processing and/or engagement of proteoglycans (e.g., fibromodulin, lumican, and decorin) occur in a precise spatial and temporal sequence in the pericellular space [211].

[4] A.k.a. osteopontin.

transforming growth factor-β1 (TGFβ-1) produced by inflammatory cells and mechanical stress.

Differentiation of cardiac fibroblasts into myofibroblasts supports collagen deposition and connective tissue construction during myocardial remodeling. It is also involved in augmented production of angiotensin by angiotensin convertase ACE1.

Differentiation of cardiofibroblasts consists of two stages [217]:

1. Differentiation to protomyofibroblasts with ^{G}actin polymerization into ^{F}actin stress fibers
2. Further differentiation into myofibroblasts in which stress fibers are decorated with α-smooth muscle actin (SMA)

Polymerization of stress fibers prevents sequestration of myocardin-related transcription factors in the cytosol by globular actin. In addition, TGFβ1 and the RoCK kinase increase the ^{F}actin/^{G}actin ratio, thereby favoring MRTF translocation to the nucleus. Inside the nucleus, MRTFs support activation of α-SMA synthesis [217]. Suppression of α-SMA diminishes arrhythmias in cardiomyocytes cocultured with cardiac myofibroblasts.

Adult rat cardiac fibroblasts cultured on a stiff plastic substratum spontaneously differentiate to proliferating myofibroblasts characterized by stress fibers connected to α-SMA [217].

Fibroblasts in collagen matrices do not exhibit any contraction, whereas quiescent and proliferating myofibroblasts contract. Developed cell tension can provoke TGFβ1 production. Transforming growth factor-β1 promotes differentiation into α-SMA+ quiescent myofibroblasts via the Tβ R1 receptor [217].

Fibroblasts produce small amounts of collagen, but high levels of interleukin-10 [217]. Quiescent myofibroblasts synthesize collagen, tissue inhibitor of metalloproteinases TIMP1, and the CCL2 chemokine at high concentrations. The transcriptome shows that different gene networks are activation in myofibroblasts (e.g., paxilin and PAK) and quiescent myofibroblasts (cell cycle regulators) [217]. In particular, MRTFa required for activation of myofibroblast differentiation is distinctly regulated in fibroblasts and quiescent and proliferating myofibroblasts. The expression of MRTFa is low in quiescent myofibroblasts [217].

Dedifferentiation of proliferating myofibroblasts associated with stress fiber depolymerization, but not of quiescent myofibroblasts, results from Tβ R1 receptor inhibition and mechanical strain reduction [217]. Loss of strain in the culture medium also reduces α-SMA expression in noncardiacmyofibroblasts and collagen-1 production.

3.1.2 *Profibrotic Signaling*

Cardiac fibrosis is characterized by fibroblast proliferation, migration, and differentiation and extracellular matrix turnover.

3.1.2.1 Selective Atrial Fibrosis by TGFβ

Selective atrial fibrosis relies on TGFβ. The latter is an important activator of chemokines, thereby attracting monocytes into the cardiac wall. When the TGFβ production increases to a similar level in both atrio- and ventriculomyocytes, only the atria respond to TGFβ by interstitial fibrosis, although levels of signaling mediators (Tβ Rs, SMADs, and AP1) are similar in atria and ventricles [218]. Atria are indeed characterized by an enhanced receptor binding and receptor kinase activity and hence phosphorylation of SMAD2 and SMAD3 as well as reduced synthesis of inhibitory SMAD7 with respect to ventricles [218].

Improved TGFβ1–Tβ R1/2 connection is favored by increased expression of connective tissue growth factor, Tβ R3 (endoglin), and atrial natriuretic peptide [219]. The enhanced activation of Tβ R together with lowered SMAD7 expression increase receptor-regulated SMADs and their phosphorylation.

Receptor-regulated rSMADs subsequently complex with SMAD4, move into the nucleus, and bind to target genes. Many genes are differentially expressed in atria and ventricles, among which those that encode mediators of fibrosis (e.g., collagen and other matrix proteins, [218]). Other AP1-regulated genes (e.g., *COL1*, *FN1*, *PAI1*, and *LOX*) are upregulated in the atria, but not in the ventricle.

3.1.2.2 Platelet-Derived Growth Factor-D

Platelet-derived growth factor-D and its receptor PDGFRβ, which is primarily expressed by fibroblasts, are implicated in the development of cardiac fibrosis [220]. The profibrogenic action of PDGFd is mediated by activation of the TGFβ1 pathway. Moreover, TGFβ1 exerts a positive feedback on PDGFd synthesis.

The PDGFd subtype augments [220]:

- Cardiofibroblast proliferation and myofibroblast differentiation
- Synthesis and secretion of collagen-1, matrix metallopeptidases (MMP1–MMP2 and MMP9) and their inhibitors (TIMP1–TIMP2)
- TGFβ1

3.1.2.3 Vascular Endothelial Growth Factor-C (VEGFc)

The vascular endothelial growth factor subtype VEGFc is involved in lymphangiogenesis (Vol. 5, Chap. 10. Vasculature Growth). In addition, the VEGFc–VEGFR3 axis participates in fibrogenesis during cardiac repair. The VEGFR3 receptor is expressed not only in lymph ducts but also in myofibroblasts.

The VEGFc isoform activates the TGFβ1 pathway and primes ERK phosphorylation, promotes myofibroblast proliferation and migration as well as production of collagen-1 and -3, matrix metallopeptidases MMP2 and MMP9, and tissue inhibitor of metallopeptidases TIMP1 and TIMP2 [221].

3.1.2.4 Tissue Inhibitors of Metallopeptidases

Tissue inhibitors of metallopeptidases and hence the TIMP–MMP balance participate in adverse myocardial remodeling (hypertrophy and fibrosis) as well as termination of the immune response [222]. Cardiac pressure overload provokes severe cardiac fibrosis and hypertrophy in mice lacking TIMP2 and TIMP3, but not in TIMP4 [223].

Among tissue inhibitors of matrix metallopeptidases, antihypertrophic TIMP2 and antiinflammatory, antifibrotic TIMP3 differ in their action in adverse cardiac remodeling engendered by the fibrogenic and hypertrophic messenger angiotensin-2 [223]. In response to angiotensin-2 infusion during 2 weeks, TIMP2$^{-/-}$, TIMP3$^{-/-}$, and wild-type (WT) mice still have a preserved left ventricular ejection fraction. However, both TIMP2$^{-/-}$ and TIMP3$^{-/-}$ mice are characterized by left ventricular diastolic dysfunction. Furthermore, TIMP2$^{-/-}$ mice exhibit an impaired relaxation and a larger degree of myocardial hypertrophy without fibrosis, whereas TIMP3$^{-/-}$ mice have an increased left ventricular stiffness and inflammation associated with an excessive fibrosis without hypertrophy.

Adult wild-type mouse cardiomyocytes undergo hypertrophy upon angiotensin-2 exposure only when cocultured with cardiofibroblasts, except in TIMP3$^{-/-}$ cardiomyocytes [223]. Inflammation and fibrosis characterize cultures of quiescent and cyclically stretched cardiofibroblasts of TIMP3$^{-/-}$ hearts. Fibrosis results from collagen deposition by the matrix proteins SPARC and SPP1 (Sect. 3.1). On the other hand, attenuated collagen deposition in TIMP2$^{-/-}$ hearts exposed to angiotensin-2 is associated with reduced levels of the cross-linking enzymes lysyl oxidase and procollagen-lysine (1,2)-oxoglutarate 5-dioxygenase PLOD1 [223].

Expression of TIMP1 increases upon angiotensin-2 infusion. Elevated TIMP1 levels are frequently linked to fibrosis. However, its augmentation in TIMP2$^{-/-}$ angiotensin-2-subjected hearts does not cause marked fibrosis despite an elevated collagen mRNA level with respect to excessive fibrosis in TIMP3$^{-/-}$ hearts without increased synthesis of collagen. In cultured quiescent mouse cardiofibroblasts, angiotensin-2 usually induces collagen production. However, TIMP2 raises collagen production by cardiofibroblasts and TIMP3 provokes cardiofibroblast apoptosis [223].

TIMP2 is the main inhibitor of MMP2 that degrades inflammatory and profibrotic cytokines, thereby favoring inflammation and fibrosis. Both LOx and PLOD1 elicit collagen cross-linking and fibre formation; their level lowers in TIMP2$^{-/-}$ angiotensin-2-subjected hearts [223].

The TIMP3 subtype is the single TIMP species that can inhibit ADAM17 sheddase, hence reducing activity of proinflammatory cytokines (e.g., TNFSF1 and IL6), as well as HER ligands (TGFα, amphiregulin, and HBEGF) via P38MAPK, ERK1 and ERK2 pathways. In addition, it attenuates action of profibrotic TGFβ and of collagen cross-linking stimulators SPARC and SPP1, thereby preventing the posttranslational processing and deposition of collagen fibers and hence the perivascular and interstitial myocardial fibrosis.

3.1.2.5 Eicosanoids and Their Inhibitors

Eicosanoids are potent modulators of immunity. They derive from arachidonic acid processed by cyclooxygenases, lipoxygenases, and cytochrome-P450 epoxygenases. The latter generates cardioprotective epoxyeicosatrienoic acids (EET) that act as auto- and paracrine effectors. However, EETs are further metabolized by soluble epoxide hydrolase into corresponding dihydroxyeicosatrienoic acids (DHET).

Myocardial infarction causes cardiomyocyte necrosis and adverse cardiac remodeling with fibrosis. Several soluble epoxide hydrolase inhibitors of postinfarction cardiac remodeling improve the cardiac function, as they impede structural (cardiac fibrosis and adverse hypertrophy) and electrical remodeling [224]. The proliferative capacity of different populations of cardiofibroblasts as well as recruitment of fibroblasts originated from circulating bone marrow-derived cells assisted by chemokines (e.g., CCL2) drop.

3.1.3 Fibrosis and Between-Cell Crosstalk

Intercellular communication between several cardiac and recruited cell types is involved in fibrosis, in particular crosstalk between cardiac myocytes and fibroblasts.

Fibroblasts are a regulating and secreting cell type that is not only implicated in maintaining the extracellular matrix of myocytes but also in their fate. In particuler, the cytokines TNFSF1 and IL6 secreted from fibroblasts can cause cardiomyocyte hypertrophy and may modulate their electrophysiological behavior and contractility.

Moreover, cellular communication can change the cardiac cell phenotype, especially during fibrosis. An endothelial–mesenchymal transition generates fibroblasts using Snail transcription factor in endotheliocytes.

Furthermore, endotheliocytes can release connective tissue growth factor (CTGF) that provokes fibrosis, as it engenders the change from quiescent fibroblasts to secreting myofibroblasts.

3.2 Maladaptive Myocardial Hypertrophy

Maladaptive, or adverse, myocardial remodeling occurs in response to sustained hypertension, myocardial infarction, and exposure to infectious or cardiotoxic agents. Hypertensive heart disease corresponds to left ventricular hypertrophy caused by prolonged arterial hypertension. Although the left ventricle is a major target, the right ventricle is involved in structural congenital heart defects with pulmonary valve stenosis, pulmonary arterial hypertension, in addition to cardiomyopathy and ischemic heart disease.

Inflammation (Vol. 5, Chap. 11. Tissue Growth, Repair, and Remodeling) and fibrosis that occur during myocardial remodeling contribute to cardiac dysfunction. Fibrosis corresponds to an excessive formation of fibrous connective tissue as a reactive process. Aggravation results from a defective neurohumoral control (sympathetic nervous regulation and renin–angiotensin–aldosterone signaling; Vol. 8, Chap. 4. Hypertension).

Hypertensive myocardium is characterized by structural changes of cardiomyocyte and matrix, such as cardiomyocyte hypertrophy and excessive apoptosis, accumulation of interstitial and perivascular collagen fibers, and disruption of endomysial and perimysial collagen network.

This pathological remodeling facilitates the development of heart failure. Progression to cardiac failure is exhibited by increase in relative lung and right ventricular weights, cardiac function disorders, and overexpression of type-A and -B natriuretic peptides.

3.2.1 Cardiac Cell Interactions in Maladaptive Cardiac Hypertrophy

Cardiac hypertrophy is related to two distinct biological process, compensatory (physiological) and deleterious (pathophysiological) cardiac hypertrophy. The molecular mechanisms underlying different forms of hypertrophy depend on the type of stress and subsequent signaling. The PI3K–PKB–TOR cascade is preferentially activated in exercise- and IGF1-dependent adaptative hypertrophy (Vol. 5, Chap. 5. Cardiomyocytes). On the other hand, activation of the PP3 and MAPK pathways is related to adverse cardiac hypertrophy in response to pressure overload.

In adverse cardiac hypertrophy, an increased myocardial mass, fibroblast activation, deposition of an extracellular matrix, and neovascularization are slower than after acute injury. Maladaptive cardiac hypertrophy is characterized by an elevation of the cardiomyocyte size, fibrosis, and expression of a fetal gene program in affected myocytes.

Myofibroblasts contribute to the cardiac response to high blood pressure. M2 macrophages are also involved in cardiac hypertrophy.

Cardiac hypertrophy is associated with an augmented blood vessel density. In heart hypertrophy due to pressure overload, the HIF1–VEGF signaling enables the maintenance of the capillary density.

MicroRNA-155 produced in cardiomacrophages and T cells induces cardiac hypertrophy and inflammation via SOCS1 factor [74].

The GSK3β kinase regulates the canonical Wnt signaling. Overexpression of GSK3β attenuates cardiac hypertrophy and its inhibition augments hypertrophy in response to hypertrophic stimuli [35].

Nitric oxide has an antihypertrophic role in cardiomyocytes, as it impedes the PP3–NFAT pathway upon an elevated cGMP–PKG1 activity and/or reduces effect

Table 3.2 Signaling pathways associated with cardiomyocyte–leukocyte communication in hypertensive heart disease (**Part 1**; Source: [69])

Signaling axes in cardiomyocytes	Mediator effect
mtDNA–TLR9–MyD88–NFκB	IL1β/6 Macrophage recruitment
TLR4–TOR–NFκB	TNFSF1, IL1β/6 Macrophage recruitment
ATn2–HSP70–TLR4	CCL2 Macrophage recruitment
LPS–TLR4–PI3Kγ	HMGB1
βAR + IL1βR–cAMP–PKA	IL6 Macrophage activation
TLR4–P38MAPK/ERK—NFκB	TNFSF1 Mastocyte recruitment

AR adrenergic receptor, *ATn* angiotensin, *cAMP* cyclic adenosine monophosphate, *ERK* extracellular signal-regulated protein kinase, *HMGB* high-mobility group box protein [danger signal], *HSP* heat shock protein [DAMP], *IL* interleukin, *LPS* lipopolysaccharide, *MAPK* mitogen-activated protein kinase, *mtDNA* mitochondrial DNA [DAMP], *MyD88* myeloid differentiation primary response gene product-88, *NFκB* nuclear factor κ light chain enhancer of activated B cells, *PI3K* phosphatidylinositol 3-kinase, *PKA* protein kinase-A, *TLR* Toll-like receptor, *TOR* target of rapamycin, *TNF* tumor-necrosis factor

of cytoskeletal LIM protein [64]. In addition, subsequently to angiogenic stimulation, NO promotes the degradation of regulator of G-protein signaling RGS4 that induces cardiomyocyte hypertrophy via the Gβγ–PI3Kγ–PKB–TORC1 pathway.

Multiple signaling pathways operate in cardiomyocytes and leukocytes in hypertensive heart disease (Tables 3.2 and 3.3).

Hypertension affects cardiomyocyte mitochondria. During pressure overload, injured mitochondria that escape mitophagy release their DNA content. Mitochondrial DNA serves as alarmin (or danger-associated molecular pattern protein [DAMP]) recognized intracellularly by TLR9 on endosomes. The latter signals via MyD88 and primes the production of leukocyte recruiting factors such as IL6 [69].

Stressed cardiomyocytes also abundantly release heat shock protein HSP70 in the extracellular medium in response to angiotensin-2. The HSP70 signal binds to TLR4 on the membrane of surrounding cardiomyocytes, thereby initiating proinflammatory pathways [69].

The overexpressed TOR kinase hampers TLR4 signaling and secretion of TNFSF1, IL1β, and IL6 in cardiomyocytes [69]. Conversely, PI3Kγ promotes TLR4 signaling. This kinase is also an effector of β-adrenergic receptor that is constitutively engaged in pressure-overloaded hearts due to high levels of circulating catecholamines. In addition, the crosstalk between the β AR and IL1β cascades via the cAMP–PKA pathway provokes IL6 secretion [69].

Table 3.3 Signaling pathways associated with cardiomyocyte–leukocyte communication in hypertensive heart disease (**Part 2**; Source: [69])

Signaling axes in cardiomyocytes	Mediator effect
TNFSF1–NOx	IL1β/6 Macrophage activation
TNFSF11–TNFRSF11a–TRAF2/6–PLC–PKC–NFκB	TNFSF1, IL1α/β Macrophage activation
IL10–JaK–STAT3–NFκB	TNFSF1 Inhibition of monocyte recruitment
EMMPRIn–Itg via PI3K–PKB–IKK–NFκB and MAP2K7–JNK–AP1	IL18 Macrophage activation
IL18–MyD88–IRAK4–TRAF6–JNK–SP1	EMMPRIn MMP2 secretion by macrophage

AP Activator protein [transcription factor], *EMMPRIn* extracellular matrix metallopeptidase inducer [basigin], *IKK* IκB kinase, *IL* interleukin, *IRAK* IL1 receptor-associated kinase, *Itg* integrin, *JaK* Janus kinase, *JNK* Jun N-terminal kinase, *MyD88* myeloid differentiation primary response gene product-88, *NFκB* nuclear factor κ light chain enhancer of activated B cells, *NOx* NAD(P)H oxidase, *PI3K* phosphatidylinositol 3-kinase, *PKB[C]* protein kinase-B[C], *PLC* phospholipase-C, *SP1* specificity protein [transcription factor], *STAT* signal transducer and activator of transduction, *TNF* tumor-necrosis factor, *TRAF* tumor-necrosis factor receptor-associated factor

In cardiomyocytes stimulated by lipopolysaccharides targeting TLR4 also engaged by pressure overload-related DAMPs, TNFSF1 expression upregulation involves P38MAPK and ERK kinases, and subsequently activated NFκB. This pathway is supported by phospholipase-Cγ and histone deacetylase and impeded by NO-induced Ca^{2+} release, the JNK1–Fos axis, MAPK phosphatase MKP1, and GSK [69].

Cardiomyocyte-derived TNFSF1 initiates an autocrine signaling involving the NADPH oxidases Nox2 and Nox4 that ensure a sustained secretion of inflammatory mediators. The TNFSF1 agent can control matrix remodeling indirectly by attracting leukocytes and directly by promoting myocyte apoptosis and activation of matrix metallopeptidases [69].

Another member of the TNF superfamily, TNFSF11 is released by cardiomyocytes in response to angiotensin-2 stimulation and recognized by the cardiomyocyte TNFRSF11a receptor. The latter is coupled to the TRAF2/6–PLC–PKC cascade that can lead to the nuclear translocation of NFκB [69].

In pressure overloaded, cardiomyocytes can signal to leukocytes by upregulating basigin (Bsg)[5] due to high levels of catecholamines and reactive oxygen species [69].

[5] A.k.a. CD147, collagenase stimulatory factor, and extracellular matrix metallopeptidase inducer (EMMPrIn).

Basigin connects to integrins on surrounding cardiomyocytes, thereby triggering the Rac1-dependent PI3K–PKB–IKK–NFκB and MAP2K7–JNK–AP1 axes and subsequently synthesizing proinflammatory cytokines (mainly IL18). Conversely, IL18 stimulates the BSG gene transcription in cardiomyocytes via the MyD88–IRAK4–TRAF6–JNK–SP1 pathway (positive feedback), thereby favoring MMP expression in cardiomyocytes and monocytes [69].

Termination of pressure overload-dependent inflammation relies on IL10 that tethers to IL10R and signals via P38MAPK and the JaK–STAT3 pathway to antagonize NFκB activation and TNFSF1 generation [69].

3.2.2 Metabolic Remodeling

The dysfunctional heart is characterized early by disturbed metabolism and impaired protein synthesis. Metabolic changes prime and perpetuate cardiac structural remodeling. In addition, nutrients (e.g., Ca^{2+}, Mg^{2+}, Zn^{2+}, Se^{2+}, and vitamin-D) can contribute to myocardial remodeling.

Angiotensin-2 responsible for cardiomyocyte maladaptive hypertrophy, interstitial fibrosis, and apoptosis promotes inflammation. Angiotensin-2 represses fatty acid oxidation. This effect is mediated by enhanced synthesis of tumor-necrosis factor-α.

3.2.2.1 Cardiac Energetics

The energy provider adenosine triphosphate (ATP) is required for both cell viability and myocardial pump activity. Cardiomyocytes are characterized by large rates of ATP use and synthesis, mainly by fatty acid oxidation in mitochondria.

At rest, fatty acid oxidation covers more than 70 % of energy need [225]. The remainder comes from oxidation of carbohydrates, principally glucose. In the healthy heart, selection among lipids and carbohydrates is governed by plasma substrate availability and hormonal regulation (insulin and catecholamines).

When heart activity rises, additional sources of ATP synthesis, such as glycogenolysis, glycolysis, and phosphotransferase reactions catalyzed by creatine kinase and adenylate kinase, are involved. Production of ATP by phosphotransferase reactions owing to creatine kinase is about ten times faster than ATP synthesis in mitochondria ($\sim$0.7 mmol/s), which is approximately 20 times quicker than ATP generation via glycolysis [226]. Phosphocreatine is the primary energy reserve element in cardiomyocytes. Attenuated phosphocreatine concentration and elevated AMP level activate AMP-activated protein kinase that increases glucose transport and stimulates phosphofructokinase for glycolysis.

Mitochondria-derived ATP is mainly used for cardiac contraction, whereas ATP produced by glycolysis intervenes in activity of kinases and ion pumps and channels (e.g., Ca^{2+}–calmodulin-dependent kinase-2, ATP-sensitive K^+ channels, and Na^+–K^+ and Ca^{2+} ATPase [SERCA2]) [225].

The failing heart has limited *energy reserve*, i.e., capacity to generate ATP in excess of its normal rate of utilization. The failing heart also has impaired capacity to convert chemical energy into mechanical work. The ratio of phosphocreatine to ATP—an index of energy reserve—wanes.

Although glucose oxidation is more efficient with an ATP production-to-oxygen consumption ratio (ATP/O ~ 3.1) greater than that of fatty acid (ATP/O ~ 2.8), fatty acid oxidation provides much more ATP ($\sim$129) than glucose ($\sim$ 36) [225].

The *metabolic flexibility* (i.e., alternative selection of carbohydrates or lipids according to their circulating concentrations) is altered. Substrate preference shifts from fatty acids to glucose in maladaptive hypertrophy and failing heart, possibly due to dampened expression of fatty acid-handling genes.

In maladaptive hypertrophy and failing heart, three major disturbances in energy handling occur: (1) creatine concentration and creatine kinase activity (Vol. 1, Chap. 4. Cell Structure and Function) lower; (2) glucose uptake by insulin-independent glucose transporter GluT1 and insulin-dependent, dominant GluT4 and glucose metabolism are inadequate, although glucose becomes the main energy provider; and (3) fatty acid oxidation falls [226].

The complex formed by the transcription factor estrogen-related receptor ERRα (or nuclear receptor NR3b1) and peroxisome proliferator-activated receptor PPARγ (or NR1c3) coactivator PGC1α targets a set of genes that encode proteins involved in: (1) energy production, such as those involved in fatty acid and glucose uptake, tricarboxylic acid cycle, and electron transport chain; (2) transfer, such as mitochondrial creatine kinase and adenine nucleotide transporter; and (3) use [226]. In maladaptive hypertrophy and failing heart, cyclin-dependent kinases CDK7 and CDK9 repress PGC1α.

In summary, in advanced stages of ventricular maladaptive remodeling and heart failure, fatty acid oxidation is reduced and more glucose is oxidized. Multiple factors that control both fatty acid uptake and oxidation are inactivated and/or downregulated, such as peroxisome proliferator-activated receptors PPARα (NR1c1), PPARβ (NR1c2), and PPARγ (NR1c3), as well as PPARγ coactivator PGC1α [227].

3.2.2.2 Unsuitable Function of the Electron Transport Chain

Most ATP is synthesized by substrate oxidation in mitochondria that have a high density in cardiomyocytes. In the failing heart, a reduced mitochondrial respiration rate is caused by improper organization of electron transport chain complexes rather than reduced content and activity of complexes of mitochondrial respiratory chains [225].

Mitochondrial uncoupling is detected by a decline in the ratio of mitochondrial ATP production to oxygen consumption. Moreover, failing heart is characterized by decayed expression of mitochondrial genesis factors (PGC1α, nuclear respiratory factors NRF1 and NRF2 (GABP), and mitochondrial transcription factor-A [225].[6]

Mitochondrial electron transport is a major enzymatic source of oxygen radical generation in cardiomyocytes.[7] Altered mitochondria induce oxidative stress in the diseased heart. An elevated level of mitochondrial reactive oxygen species during hypoxia can activate hypoxia-inducible factor HIF1α, thus having beneficial effect [228].

3.2.3 *Gene Transcription and Translation*

Ventricular hypertrophy is characterized by changes in gene expression that include reactivation of the Anp, Bnp, and βMhc genes of the fetal gene program as well as the suppression of expression of genes that encode proteins implicated in adult heart function [1]. In particular, the αMhc and Serca2a genes expressed at high levels in healthy hearts have a repressed expression during cardiac hypertrophy.

3.2.3.1 Chromatin Remodeling in Cardiac Hypertrophy

Histone acetyltransferase (HAT) and deacetylases (HDAC) determine the expression of cardiac genes (Sect. 1.1.2). Suppression of gene expression also depends on switch/sucrose nonfermentable (Swi/SNF)-related, matrix-associated, actin-dependent regulator of chromatin SMARCa2–SMARCa4-associated HDAC corepressor complexes. The interaction of SMARCa2 with the proximal promoters of the Anp and Bnp genes enhances gene expression, but the recruitment of SMARCa4 to α Mhc promoter causes transcriptional suppression in models of cardiac hypertrophy.

Histone Acetyltransferases

Myocardial stress in the adult heart increases P300 histone acetyltransferase activity as well as histone $H3K_9$ and $H4K_{14}$ acetylations that provoke activation of genes linked to cardiac hypertrophy [1]. Inhibition of cyclin-dependent kinase CDK9, a mediator of P300–GATA4 complex, attenuates phenylephrine-induced hypertrophy in rat cardiomyocytes. The natural small molecule curcumin is a selective inhibitor of P300 that impedes cardiac hypertrophy-induced acetylation as well as the binding of P300–GATA4 complex at gene promoters.

---

[6] Mitochondrial transcription factor-A (TFaM) is a nucleus-encoded protein that regulates mtDNA transcription and replication.

[7] Reactive oxygen species are also produced by xanthine oxidase and/or NADPH oxidase in endothelial cells and NADPH oxidase in activated leukocytes [228]. Activated NADPH oxidase can also be involved in myocardial oxidative stress.

Histone Deacetylases

Histone deacetylases remove acetyl groups from lysine residues in various types of nuclear and cytosolic proteins, especially transcriptional regulators and nucleosomal histones.

Both class-1 (HDAC1–HDAC3 and HDAC8) and class-2 (HDAC4–HDAC7 and HDAC9–HDAC10) are involved in cardiac hypertrophy. In addition, HDACs interact with transcription factors such as myocyte enhancer factor MEF2 and coregulators of expression of genes implicated in cardiac hypertrophy and heart failure.

Different HDAC classes may mediate pro- and antihypertrophic signaling in the stressed myocardium. Class-1 HDAC inhibition can suppress cardiac hypertrophy upon angiotensin-2 infusion and transverse aortic constriction. On the other hand, class-2a HDACs (HDAC4, HDAC5, HDAC7, and HDAC9) repress expression of genes regulated by MEF2 that are involved in cardiac hypertrophy.

In cardiomyocytes, class-1 zinc-dependent histone deacetylases (HDAC1–HDAC3) repress cardiac hypertrophy, as they preclude expression of the gene that encodes dual-specificity phosphatase DuSP5, a nuclear phosphatase that hampers prohypertrophic signaling by dephosphorylation of nuclear ERK1 and ERK2 kinases [229]. Moreover, inhibition of DUSP5 by class-1 HDACs requires activity of the ERK kinase (*self-reinforcing mechanism*). The HDAC3 subtype regulates ERK1 and ERK2 in cardiomyocytes, whereas HDAC1 and/or HDAC2 controls JNK in these cells.

Phosphatases of the DUSP category[8] form the largest group of MAPK phosphatases. The DuSP1 subtype suppresses ERK, JNK, and P38MAPK signaling, thereby impeding cardiac hypertrophy in response to pressure overload [229]. Cardiac-specific overexpression of cytoplasmic DuSP6 promotes cardiac fibrosis and apoptosis in response to pressure overload.

On the other hand, hypertrophic stimuli concomitantly stimulate nuclear ERK1 and ERK2 phosphorylation and repress expression of DUSP5 via class-1 HDACs. Activated cardiac ERK1 and ERK2 cause a concentric cardiac hypertrophy and can be correlated with adverse hypertrophy due to pressure overload in humans [229]. Unlike MAP2K1 overexpression that engenders compensatory hypertrophy, expression of constitutively nuclear ERK2 favors pathological cardiac remodeling in response to pressure overload. Hence, nuclear ERK signaling leads to pathological hypertrophy, whereas phosphorylation of cytosolic substrates by ERK1 and ERK2 ensures cardioprotection.

[8] Dual-specificity phosphatases constitute three groups: (1) nuclear DuSPs; (2) cytosolic ERK-specific DuSPs; and (3) DuSPs that selectively inactivate stress-activated MAPKs (i.e., JNK and P38MAPK).

Both HDAC1 and HDAC2 stimulate cardiac hypertrophy, as they support autophagy [229]. Association of HDAC2 with the Ying Yang YY1 transcription factor promotes expression of BNP in response to a hypertrophic stimulus. In addition, HDAC1 stimulates sodium–calcium exchanger NCX1 expression during cardiac hypertrophy.

3.2.3.2 Reactivation of the Embryonic Program

Maladaptive cardiac hypertrophy is associated with reactivation of a fetal program of cardiac gene expression. The cardiac hypertrophy-associated increase in RNA synthesis is characterized by a reexpression of the NPPA and NPPB genes as well as the fetal myocardial β-myosin heavy chain encoded by the MYH7 gene in cardiomyocyte subsets [230].

β-Myosin heavy chain yields a marker of normal aging as well as pathological cardiac hypertrophy. βMHC+ cells in normal and old as well as hypertrophic hearts are predominantly located in subendocardial clusters within and surrounding fibrosis sites. β-Myosin heavy chain, thereby, is a marker of fibrosis rather than hypertrophy.

3.2.3.3 Elongation Phase of Transcription

The PKB axis promotes protein synthesis and angiogenesis, as it regulates, together with HIF1α, the proangiogenic VEGFa factor produced by cardiomyocytes. In adaptative hypertrophy, angiogenesis matches cardiomyocyte hypertrophy. When angiogenesis is impaired, the heart evolves more rapidly into failure.

The positive (productive) transcription elongation factor-b (pTEFb),[9] or cyclin T–cyclin-dependent kinase CDK9,[10] and its modulators such as the self-inhibitory

[9] Transcription is divided into preinitiation, initiation, promoter clearance, elongation, and termination. Elongation factors are recruited to RNA polymerase. Among elongation factors, pTEFb phosphorylates the second residue (Ser2) of the heptapeptidic CTD repeats of the largest subunit of the bound RNA polymerase-2, thereby assisting transition from transcriptional initiation to elongation as well as coordinating transcription elongation and RNA maturation, as it helps recruiting 5′-capping enzyme and the RNA splicing machinery. Shortly after initiation, RNA polymerase-2 is inhibited by the negative elongation factors (DRB sensitivity-inducing factor [DSIF] and nasal embryonic luteinizing hormone-releasing hormone [LHRH] factor [NELF]. pTEFb launches the transition to productive elongation, as it phosphorylates (inactivates) these two negative factors [231].

[10] The pTEFb factor is a heterodimer that consists of cyclin-dependent kinase CDK9 and a regulatory cyclin subunit of the T family (T1, T2, or K1). The CDK9 kinase is regulated by nuclear receptor corepressor (NCoR) and associated HDAC3. Once it is acetylated, its ability to phosphorylate the CTD of RNA polymerase-2 diminishes [231]. Both NCoR and NCoR2 complex with histone deacetylase HDAC3, G-protein pathway suppressor GPS2, transducin-β-like protein TBL1, and TBL1-related protein TBLR1. The HDAC3 enzyme is required, at least in the case of the thyroid hormone receptor, for transcriptional repression mediated by NCoR and NCoR2.

RNA kinase and transcriptional repressor *hexamethylene bisacetamide-inducible molecule* HexIM1[11] are implicated in cardiac hypertrophy [232]. A reduced expression of HexIM1 is accompanied with an activation of pTEFb. In the catalytically inactive state, PTEFb is complexed with HexIM1 and 7SK small nuclear RNA, among others. In response to various physiological and pathological stimuli such as hypertrophic signals, HexIM1 dissociates from cyclin. Subsequently, PTEFb binds to bromodomain-containing protein BrD4 and, upon phosphorylation of RNA polymerase-2, increases elongation activity and transcription of nascent mRNAs. The affinity of BrD4 to acetylated histones allocates the active pTEFb to specific regions of the genome epigenetically marked as transcriptionally active.

In addition, HexIM1 can directly bind to and modulate transcription factors, such as the estrogen receptor-α (NR3a1), CCAAT/enhancer-binding protein-α, and nuclear factor-κB.

Reactivation of HexIM1 in the adult heart upregulates the protein expression level, in particular that of HIF1α (in addition to MyC, GATA4, and PPARα, among others) and subsequently increasing that of the Vegf gene as well as of genes associated with fatty acid utilization, but decreasing that of genes related to glucose metabolism, without affecting that of stress genes, such as NPPA and NPPB genes [233]. Therefore, HexIM1 induces a cardiac hypertrophy that resembles that linked to exercise rather than that resulting from pressure overload.

3.2.3.4 Transcription Factor NFE2L2

The transcription factor nuclear factor erythroid-derived NFE2-related factor NFE2L2 is protective upon transient activation in response to stress (Vol. 8, Chap. 3. Endothelial Dysfunction). It coordinates the redox balance that conditions microvascular resistance and cardiomyocyte proteins homeostasis [234].

On the other hand, it is detrimental upon prolonged stimulation, engendering proteotoxic cardiac remodeling via reductive stress and toxic protein aggregates as well as maladaptive hypertrophy of cardiomyocytes [235]. Overcompensation of NFE2L2 during cardiac remodeling indeed favors reductive stress and protein ubiquitination and subsequent aggregation, especially that of chaperones and cytoskeletal components, and hence adverse cardiac hypertrophy [235].

A sustained activation of NFE2L2 develops in two phases, initially in response to ROS generation (∼3 months), and later, as a consequence of Kelch-like ECH-associated protein KEAP1 dysfunction (6 months) [235].

[11] A.k.a. cardiac lineage protein CLP1 and Ménage á quatre protein MAQ1. The expression of HEXIM1 is high during embryogenesis and the early postnatal period; it gradually lowers afterward. The pTEFb factor exists in two forms, the active CcnT–CDK9 heterodimer and inactive pTEFb–7SKsnRNA–HexIM1 complex. The balance between different pTEFb complexes is controlled by several phosphatases, such as PP3 and magnesium-dependent PPM1a, that dephosphorylate (deactivate) CDK9.

Table 3.4 In coronary atherosclerosis, microRNAs have been investigated in serum, plasma, whole blood, peripheral blood mononuclear leukocytes (lymphocytes and monocytes), or agranulocytes, and platelets by different teams of investigators. (Source: [237])

MicroRNAs	Source
MiR18b, miR22, miR92b, miR126, miR122, miR129-5p, miR-320a, miR423-5p, miR499, miR622, miR654-3p, miR1254	Plasma, serum
MiR19b, miR107, miR125b, miR139, miR142-3p, miR142-5p, miR497	Flowing agranulocytes

A NFE2L2 deficiency normalizes the excessive content of glutathione and prevents premature death. In NFE2L2$^{+/-}$ mice (i.e., a single functional copy of the Nfe2l2 gene), the expression of glutathione metabolic and antioxidative enzymes (e.g., NAD(P)H dehydrogenase quinone NQO1, catalase, glucose-6-phosphate dehydrogenase [G6PD],[12] and glutathione reductase [GSR]) lowers. Attenuated production of NFE2L2-dependent antioxidant enzymes dampens the reductive stress and reequilibrates the redox milieu [234]. It also restores the production of other redox-sensitive enzymes (e.g., HO1 and $^{(Cu,Zn)}$SOD and ^{Mn}SOD), which falls during reductive stress.

3.2.3.5 Hypoxia-Inducible Factor

Heart failure can arise when cardiac angiogenesis does not conform with cardiac hypertrophy [236]. Pressure overload initially promotes cardiac angiogenesis via hypoxia-inducible factor-1. Downregulation of HIF1α causes maladaptive hypertrophy during chronic pressure overload. In maladaptive cardiac hypertrophy, angiogenesis does not reach a sufficient level for the conservation of the cardiac function. Factor P53, the expression of which is upregulated, binds to HIF1α and favors its degradation.

3.2.3.6 Dysregulation in MicroRNA Activity

MicroRNAs are small noncoding RNAs that mediate posttranscriptional gene silencing. The microRNA content in the circulation may reflect the activation state of flowing cells. MicroRNAs circulate in all compartments of blood (plasma, platelets, red blood capsules, and leukocytes; Table 3.4). Circulating microRNAs are stable and resistant to degradation by RNAse. MicroRNAs can then be used as markers of cardiovascular diseases.

[12] Reduced glutathione (G^{SH}) is a tripeptide with a free sulfhydryl group that is required to combat oxidative stress. Oxidized glutathione (G^{SS}G) is reduced by NADPH generated by glucose 6-phosphate dehydrogenase in the pentose phosphate pathway.

Table 3.5 MicroRNAs in microparticles and their link to cardiovascular diseases. (Source: [239])

microRNA	Platelet microparticle	Monocyte microparticle	Endotheliocyte microparticle	Disease
Let7d		x	x	CAD
MiR17		x	x	CAD
MiR19	x	x	x	CH, CMP
MiR20a		x	x	CAD
MiR21	x	x	x	CAD
MiR27a		x	x	CAD
MiR92a1		x	x	CAD
MiR126		x	x	CAD, DAP
MiR130		x	x	CAD
MiR133	x	x	x	MI, CH
MiR143		x	x	CAD
MiR146a		x	x	Myocarditis
MiR146b		x		Myocarditis
				CAD
MiR155		x	x	Myocarditis
				CAD
MiR199			x	CAD
MiR221		x	x	CAD
MiR223	x	x		MI
				Myocarditis
MiR423		x	x	CF

CAD coronary artery disease, *CF* cardiac failure, *CH* cardiac hypertrophy, CMP cardiomyopathy, *DAP* diabetic angiopathy, *MI* myocardial infarction

Circulating microRNAs can reside in vesicles, such as exosomes (50–90 nm), microparticles (size range [100 nm–1.1 μm]), and apoptotic bodies (0.5–2 μm), or bound to lipoproteins (HDLs and LDLs) and proteic complexes (e.g., Argonaute-2 and nucleophosmin-1) [237, 238].

Several microRNAs with high levels in microparticles are involved in cardiovascular diseases [239]. Proinflammatory microparticles (plasma concentration ∼ 5000/μl) are transport carriers for numerous microRNAs, protecting them from degradation [239]. These microRNAs are packaged in producing cells, such as platelets, leukocytes, and endotheliocytes (Table 3.5). They contain cytosol and surface markers of their cells of origin. Once released into the circulation, microparticles bind and fuse with their target cells via receptor–ligand connections. These vectors mediate vascular inflammation and coagulation, due to their specific lipid composition (e.g., phosphatidylserine) as well as to transfer of inflammatory components and microRNAs synthesized in producing cells.

MicroRNA profiles of microparticles differ significantly according to producing cell type and whether the producing cell is stimulated and not, as well as between patients with stable and unstable coronary artery disease [239]. For example, miR21 is one of the highest upregulated miRs in stimulated vs. unstimulated monocytic microparticles. Its level augments under hypoxia in vascular smooth myocytes. Moreover, the miR content of microparticles significantly differs from that of their synthesizing cells.

Type-2 Diabetes

Type-2 diabetes mellitus characterized by chronic elevation of blood glucose levels is one of the major risk factors for cardiovascular disease. Plasma concentrations of miR9, miR15a, miR20b, miR21, miR24, miR28-3p, miR29a, miR29b, miR30d, miR34a, miR124, miR126, miR144, miR146a, miR150, miR191, miR192, miR197, miR223, miR320, miR375, miR486, and miR503 are significantly associated with diabetes mellitus, some being involved in insulin regulation [237].

Cardiac Maladaptive Hypertrophy

During adverse cardiac hypertrophy, several specific microRNAs, especially stress-inducible species, are dysregulated [240]. The expression of miR23a, miR27a, miR24.2, and miR195 is upregulated. In addition, miR21, which impedes apoptosis, is also overexpressed during cardiac hypertrophy. On the other hand, miR133, which represses serum response factor, is underexpressed during cardiac hypertrophy. Overexpression of these microRNAs in cardiomyocytes causes cardiac hypertrophy at least in vitro.

MicroRNA-23a, the production of which is regulated by nuclear factor of activated T cells NFAT3, is a prohypertrophic agent [241]. It targets the ubiquitin–protein ligase tripartite motif-containing protein TRIM63,[13] an antihypertrophic protein. MicroRNAs miR23a, miR24, and miR27a belong to the same cluster, but the upregulation of miR24 and miR27a expression occurs later than that of miR23a. In addition, miR23a can stimulate synthesis of early growth response EGR1 and pituitary tumor-transforming gene product PTTG1 (or securin), both involved in cardiac hypertrophy [241].

In addition, circulating levels of miR296-5p, let7e, and ^{hcmv}miR-UL112, a human cytomegalovirus-encoded microRNA that targets interferon regulatory factor IRF1, are altered in hypertension [237, 238].

[13] A.k.a. muscle-specific ring finger protein MuRF1.

Members of the microRNA-34 category (miR34a–miR34c) are upregulated in the heart in response to stress. Therapeutic inhibition of miR34s upregulates [242]: (1) vascular endothelial growth factor that supports coronary angiogenesis; (2) vinculin that preserves cardiac contractility and electrochemical transmission at intercalated discs; (3) protein ^{O}fucosyltranferase POFuT1 that acts in O-linked glycosylation of proteins, hence favoring cell survival; (4) Notch-1 that mediates cardiac regeneration after myocardial infarction; (5) B-cell lymphoma protein BCL6 that protects cardiomyocytes against inflammation; and (6) semaphorin-4B that prevents IL6 production by basophils. Therapeutic inhibition of miR34s attenuates pathological cardiac remodeling [242].

The microRNAs miR29b and miR30c are involved in the communication between cardiac fibroblasts and myocytes in myocardial hypertrophy [40]. MicroRNAs act as moderators in the paracrine myocyte–fibroblast crosstalk by mostly regulating the secretion of growth factors and cytokines.

MicroRNA-29b affects matrix protein deposition by fibroblasts as well as causes cardiomyocyte hypertrophy, as it regulates the secretion of messengers, such as IGF1, LIF, and pentraxin-3 [40]. However, a medium derived of miR29b-transfected cardiofibroblasts engenders cardiomyocyte atrophy [40]. On the other hand, miR30c only slightly influences the secretion of matrix constituents, its action being opposite to that of miR29b. A medium derived of miR30c-transfected cardiofibroblasts provokes cardiomyocyte hypertrophy, hence counteracting miR29b effect [40].

As does miR29b, miR133a that is highly expressed in cardiomyocytes targets collagen-1A1 transcripts (COL1A1). Therefore, microRNAs of both cardiac fibroblasts and myocytes synergistically affect COL1A1 mRNA expression, thereby regulating cardiac fibrosis [40].

3.2.4 Signaling Remodeling

Chronic maladaptive hypertrophy of cardiomyocytes can result from valvular dysfunction, persistent hypertension, and myocardial infarction. In particular, left ventricular maladaptive hypertrophy is a response to increased afterload due to systemic arterial hypertension, aortic stenosis, as well as myocardial infarction. Although it tends originally to restore pump function, it leads to heart failure.

The epicardium contains multipotent stem cells and releases paracrine factors for cardiac regeneration and repair. Injury of the adult heart reactivates an epicardial, developmental gene program. Transcription factors of the CCAAT/enhancer-binding protein (C/EBP) category supports neutrophil infiltration after injury and restoration of the cardiac function, especially cardiac remodeling following ischemia [243]. They assist the epicardial production and secretion of retinoic acid, cytokines, and chemokines. They operate with homeodomain-containing transcription factors that are contributors to developmental programs, Hox, Meis, and Grainyhead-like proteins.

In addition to chemical factors and pathways (PP3, CamK, MAPK, PKC, and PKD), mechanical stress triggers cardiac hypertrophy, partly, via activation of auto- and paracrine messengers (e.g., angiotensin and endothelin) that stimulate $G\alpha_{q/11}$-coupled receptors, including the α1-adrenergic receptors.

Pathological cardiac remodeling is characterized by an increase in cell size, protein synthesis, and sarcomere assembly, as well as reactivation of fetal genes.

3.2.4.1 Calcium Influx

Calcium not only mediates excitation–contraction coupling in cardiomyocytes but also supports the gene transcription that underlies cardiac hypertrophy.

Inositol Trisphosphate Receptors

In addition to ryanodine receptors (RyR), cardiomyocytes also produce inositol trisphosphate receptors (IP_3R), but to a lesser extent (approximately 50:1), IP_3R2 being the main isoform in cardiomyocytes. Nonetheless, Ca^{2+} release through IP_3Rs contributes to the inotropic, arrhythmogenic, and hypertrophic effect of Gq-coupled receptor agonists such as endothelin-1.

In cardiac hypertrophy, increased density of IP_3Rs in the junctional sarcoplasmic reticulum augments Ca^{2+} transients [244]. As IP_3Rs localize close to ryanodine receptors in the junctional sarcoplasmic reticulum, elevated Ca^{2+} release sensitizes RyRs and increases diastolic and systolic cytosolic Ca^{2+} concentration (Ca^{2+} sparks and puffs). These Ca^{2+} transients can trigger ventricular arrhythmias in cardiac hypertrophy.

However, IP_3R2 production does not rise in patients with heart failure caused by ischemic dilated cardiomyopathy [244]. On the other hand, atrial fibrillation is also associated with elevated IP_3R expression in atriomyocytes in humans.

Calcium Influx Through TRP Channels

The transient receptor potential canonical channels TRPC3 and TRPC6 are produced at low levels in normal conditions. Their expression rises in adverse cardiac hypertrophy. They support hypertrophy signaling triggered by angiotensin-2 or endothelin-1 [245]. A dual inhibition is necessary, as the deletion of the Trpc3 or Trpc6 gene alone does not protect against hypertrophy induced by pressure overload, but combined deletion is protective.

Cardiomyocyte activation by angiotensin-2 or endothelin-1 or by mechanical stress due to sustained hypertension stimulates the Ca^{2+}–calmodulin-dependent phosphatase PP3 that dephosphorylates the transcriptional regulator NFAT, thereby provoking nuclear translocation of NFAT and expression of maladaptive hypertrophic genes. The source of triggering Ca^{2+} ions is the TRPC3 and TRPC6 channels [245].

3.2.4.2 PP3–NFAT Axis

Calcium–calmodulin-dependent protein Ser/Thr phosphatase PP3 dephosphorylates nuclear factor of activated T-cells that can then enter the nucleus and interact with cardiac GATA4 transcription factor involved in cardiac hypertrophy under pathological conditions [265].

The calcineurin (PP3)–nuclear factor of activated T cells (NFAT) signaling operates not only in the nervous system (memory formation or apoptosis) and immunity but also in cardiovascular development (cardiac remodeling as well as cardiomyocyte differentiation and endotheliocyte activation during angiogenesis following inflammation) [246]. The NFAT factor intervenes in physiological and pathological conditions. Many NFAT target genes are involved in cell growth and proliferation as well as in inflammation.[14] NFAT signaling depends on the cell type as well as signaling magnitude.

The NFAT family of transcription factors encompasses five members. Among them, four are regulated by the Ca^{2+}-dependent protein Ser/Thr phosphatase PP3 (NFAT1–NFAT4). The osmotic stress-dependent NFAT5 subtype[15] does not depend on calcium ion.

In the cytosol, NFAT complexes with the long intergenic noncoding RNA for repressor of NFAT, IQGAP scaffold protein, and importin-β for nucleocytoplasmic transport [246]. Other NFAT partners include BMP2, VEGF, MAGI1, ADAMTS1, thrombin, and histamine.

The NFAT factors are substrates of dual-specificity Tyr phosphorylation-regulated kinase DYRK1a (NFAT-priming kinase), glycogen synthase kinase GSK3β and casein kinase CK1 as well as components of the MAPK module (MAP3K7, MAP3K14, P38MAPK, JNK, and ERK4). Phosphorylation (Ser112) of $DSCR1_S$ by MAPK leads to the subsequent phosphorylation (Ser108).

Upon dephosphorylation, NFAT undergoes a nuclear translocation. Diverse NFAT family members have redundant functions. The NFAT factors mainly cooperate with other transcription factors, such as AP1, GATA, and MEF2 family members. Otherwise, NFAT has two modes of DNA interaction: (1) binding on NFκB sites (e.g., Vcam1 and Il8) and forming homodimers and (2) binding on a NFAT recognition site as a monomer.

The PP3 phosphatase has many endogenous inhibitors, *calcipressins* (e.g., calcineurin-binding protein CaBin1, A-kinase anchoring protein AKAP5, and Down syndrome critical region DSCR1 [246]. The DSCR1 regulator prevents PP3 binding to NFAT as well as PP3 activity. Two major DSCR1 isoforms exist: long ($DSCR1_L$)

---

[14] In endotheliocytes, NFAT1 regulates expression of genes implicated in inflammation (CCL2, CXCL8, and CX_3CL1 chemokines), coagulation (tissue factor), leukocyte transmigration (E-selectin, VCAM1, and ICAM1), and transcription (EGR3 and NR1b1) [246].

[15] A.k.a. tonicity-responsive enhancer-binding protein (TonEBP) and osmotic-response element-binding protein (OREBP).

and short ($DSCR1_S$) variants expressed in the nervous system and in endotheliocytes, respectively, the latter being mainly regulated by the PP3–NFAT pathway and the former by the Notch and HES1-dependent pathways. In endotheliocytes, VEGF-inducible DSCR1 is an effector of a negative feedback loop that hinders NFAT-mediated cell activation and proliferation.

In endotheliocytes, NFAT targets the gene that encodes the autoinhibitory molecule Down syndrome critical region DSCR1 [246]. Expression of $DSCR1_S$ is primed by NFAT1 to NFAT3 as well as GATA2 and GATA3 factors. On the other hand, $DSCR1_L$ that lacks an NFAT consensus region is regulated by Notch and glucocorticoids.

Vascular endothelial cell growth factor and thrombin provoke DSCR1 expression. On the other hand, DSCR1 precludes VEGF-mediated angiogenesis (negative feedback) and monocyte adhesion. The constitutive DSCR1 expression may set endotheliocytes in a quiescent state after VEGF stimulation. In fact, $DSCR1_S$ impedes angiogenesis, whereas $DSCR1_L$ assists NFAT transcriptional activity and angiogenesis.

In smooth myocytes, angiotensin-2 favors DSCR1 activity, thereby promoting cell migration [246]. In macrophages, DSCR1 supports their migration toward oxidized low-density lipoprotein and their uptake.

A stable $DSCR1_S$ expression attenuates septic inflammatory shock as well as tumor growth and metastasis to lungs that have the highest endotheliocyte density among bodily organs.

The effect of DSCR1 on the PP3–NFAT cascade is related to posttranscriptional modification, such as phosphorylation and ubiquitination. Proteins DSCR1 and DYRK1a synergistically promote NFAT nuclear exclusion [246]. The DYRK1a kinase phosphorylates $DSCR1_L$ (Thr241) and $DSCR1_S$ (Thr192), thereby enhancing DSCR1–PP3 binding [246]. Phosphorylation of DSCR1 by ERK4 dissociates it from PP3. Phosphorylation of DSCR1 by MAP3K14 increases the stability [246]. Phosphorylation of $DSCR1_L$ (Ser149 and Ser191) converts it from a PP3 inhibitor to a facilitator, thereby enhancing NFAT1 nuclear occupancy. On the other hand, $DSCR1_L$ ubiquitination leads to its proteasomal degradation.

Protein DSCR1 is a inhibitor of the PP3–NFAT signaling, especially that involved in cardiac hypertrophy. In Dscr1 gene knockout mice, the constitutively active PP3-dependent cardiac hypertrophy is exacerbated, but adverse cardiac hypertrophy caused by pressure overload and chronic adrenergic stimulation is blunted [246].

3.2.4.3 Oxidative and Reductive Stress

The heart has a high metabolic rate and is hence particularly exposed to oxidative stress. Oxidative stress provokes adverse cardiac hypertrophy as well as P53-dependent decrease in capillary density, excessive DNA oxidation, and increased cardiomyocyte apoptosis, leading to heart failure.

Reactive Oxygen Species

Hydrogen peroxide (H_2O_2) is synthesized consequently to electron leakage during oxidative phosphorylation as well as by activated oxidases such as NADPH (reduced form of nicotinamide adenine dinucleotide phosphate) oxidases.

Its intracellular concentration rises in response to the activation of many signaling pathways, such as those triggered by stimulated growth factor receptors (EGFR and PDGFR), activated protein Ser/Thr kinases, and G-protein–coupled receptors [247]. Hence, numerous ligands (e.g., bradykinin, EGF, PDGF, TNFSF1, and thrombin) influence H_2O_2 signaling.

Hydrogen peroxide can directly modify proteins by oxidation of cysteine residues situated in a specific proteic environment, hence inactivating protein Tyr and Ser/Thr phosphatases (e.g., PTPn1 and PTen) [247]. Elevated H_2O_2 amounts exert different effects according to their concentration and intracellular location. At low concentrations, PTen inactivation and subsequent PKB activation cause cell proliferation signaling, in particular in cancers. At high concentrations, SHC that regulates mitochondrial swelling initiates cell apoptosis.

Superoxide ($O_2^{\bullet -}$) is produced by NADPH oxidase in endothelial and smooth muscle cells. Vascular NADPH oxidases are regulated by various compounds, such as growth factors (e.g., PDGF) and cytokines (e.g., TNFSF1 [16] and interleukin-1), as well as thrombin and oxidized LDL particles.

Superoxide is metabolized to hydrogen peroxide. It can also form hydroxyl ion (OH^-). Hydrogen peroxide modifies amino acids. Intracellular H_2O_2 can then regulate enzyme activity and signaling pathways. For example, it inactivates PTen phosphatase, leading to an increase in cellular phosphatidylinositol (3,4,5)-trisphosphate and subsequently activating protein kinase-B. Reactive oxygen species can protect cardiomyocytes because they activate transcription factors implicated in synthesis of antioxidant enzymes, such as manganese superoxide dismutase (^{Mn}SOD) and the nitric oxide synthase NOS3, thereby enabling ischemic preconditioning. Antioxidants encompass an enzyme group (e.g., superoxide dismutase, catalase, and glutathione peroxidase) and nonenzymatic molecules (e.g., vitamin-E and - C).

Reactive oxygen species also activate the MAPK module, especially ROS-sensitive MAP3K5, which stimulates both P38MAPK and Jun N-terminal kinase. Thioredoxin-1 binds to MAP3K5 and precludes its activation. Oxidized thioredoxin-1 dissociates from MAP3K5, which then undergoes autophosphorylation. Noradrenaline, angiotensin-2, and endothelin rapidly and transiently activate MAP3K5 via ROS agents. Kinase MAP3K5 is also involved in ligand-bound G-protein–coupled receptor activation of NFκB factor.

Vascular NADPH oxidase is stimulated by angiotensin-2 via the AT_1 receptor, in association with EGFR, PI3K, and Rac GTPase [249]. Moreover, reactive species can activate NFκB transcription factor.

[16] This cytokine recruits subunits to plasmalemmal NADPH oxidase [248]. It also activates NOS3 in plasmalemmal nanodomains. This dual activation locally produces superoxide and nitric oxide.

Nitric oxide quickly reacts with superoxide (O_2^-) characterized by the presence of an unpaired electron that can rapidly react to form peroxynitrite ($ONOO^-$). Also, nitric oxide hampers NADPH oxidase activity.

Reactive oxygen species act as second messengers that activate multiple targets, such as epidermal growth factor receptor; PKB, Src, and P38MAPK kinases; and Ras GTPase [250]. However, excessive exposure to H_2O_2 leads to oxidative stress.

Oxidative stresses associated with excessive production of reactive oxygen and nitrogen species lead to cardiovascular diseases. Reactive oxygen and nitrogen species are implicated in hypertension, intimal hyperplasia, and atherosclerosis [251].

Increased ROS generation that exceeds the capacity of antioxidants causes oxidation of membrane phospholipids and proteins, as well as mitochondrial DNA damage. Augmented ROS synthesis can be associated with decreased NOS3 activity as well as increased NOS2 function in macrophages.[17]

ROS Regulator PI5P and Its Controllers PIN1 and PIP4K

Correct management of reactive oxygen species prevents oxidative stress. Oxidative stress primes activation of proline-directed stress kinases (P38MAPK and JNK). Phosphorylation by these kinases leads to the recruitment of the peptidyl prolyl isomerase interacting with NIMA PIN1.

This isomerase recognizes substrates phosphorylated on serine or threonine residues situated next to proline residues. It binds to phosphorylated substrates using its WW domain and can induce a conformational change via its prolyl isomerase activity. It thus regulates the localization as well as catalytic activity, stability, phosphorylation state, and interaction with other partners of its subtrates.

Phosphatidylinositol 5-phosphate (PI5P) is a redox state- and PIN1-regulated second messenger that influences gene expression to determine the sensitivity of cells to ROS and cell response to oxidative stress. Concentration of PI5P rises in response to some physical and chemical stresses, especially hydrogen peroxide, thereby contributing to enhanced cell viability during exposure to oxidative stress. Increased PI5P stimulates the expression of genes involved in defense against oxidative stress and limits the accumulation of reactive oxygen species. In particular, PI5P heightens the transcriptional activity of the stress-dependent P53 factor.

Isomerase PIN1 controls the amount produced in response to hydrogen peroxide of the phosphoinositide second messenger PI5P [252]. Hydrogen peroxide provokes the PI5P synthesis at least partly by inducing the interaction between PIN1 with phosphatidylinositol 5-phosphate 4-kinase (PIP4K) that remove PI5P.

On the other hand, the PI5P amount can be increased by the stress-activated lipid kinase phosphatidylinositol 3-phosphate 5-kinase PI(3)P5K3[18] and myotubularins.

---

[17] Nitric oxide competes with oxygen for binding to cytochrome-C oxidase. Enhanced NO production causes elevated ROS synthesis. Angiotensin-2 increases mitochondrial ROS generation.

[18] A.k.a. FYVE finger-containing phosphoinositide kinase PIKFYVE.

At least in vitro, PIN1 precludes PI5P 4-kinase (PIP4K) activity [252]. The latter phosphorylates PI5P to produce $PI(4,5)P_2$, thereby lowering the PI5P amount. In fact, PIN1 rotamase inhibits phosphorylated PIP4K kinase. Activated P38MAPK phosphorylates (inactivates) PIP4Kβ (Ser326). The PIN1 isomerase interacts with either the high-activity PIP4Kα or the low-activity PIP4Kβ isoform [252]. Subtype PIP4Kβ can interact with and target PIP4Kα to the nucleus. Overexpression of PIP4Kα reduces the action of H_2O_2 on the Nfe2l2 gene implicated in the response to oxidative stress [252]. Interaction between PIN1 and PIP4Kβ may occur in the nucleus because both enzymes colocalize in nuclear structures.

In summary, oxidative stress, especially H_2O_2 exposure, promotes interaction between PIP4K and PIN1 enzymes [252]. However, the PIN1–PIP4K interaction is not the major mechanism for PI5P abundance in response to hydrogen peroxide. Accumulation of PI5P may result from PI(3)P5K3 and myotubularins.

Calcium Overload and Mitochondrial Permeability

Myocardial ischemia primes formation of reactive oxygen species and inflammatory cytokines (e.g., TNFSF1 and IL6). Oxidative stress is associated with elevated intracellular Ca^{2+} concentration, as ROS cause: (1) Ca^{2+} influx by increased membrane lipid peroxidation and opening of voltage-sensitive Ca^{2+} channels or Na^+–Ca^{2+} exchangers; (2) Ca^{2+} release from intracellular stores; and (3) Ca^{2+} uptake decay by SERCAs [253]. In addition, TNFSF1: (1) decreases Ca^{2+} sensitivity of myofibrils and (2) impedes SERCA2a production and activity.

Calcium overload induced by extensive ROS generation can enhance permeability of the mitochondrial membrane and cause mitochondrial depolarization. Mitochondrial leak and peptidase activation leads to energy depletion and contractile protein loss. Mitochondria also mediate cardiomyocyte apoptosis and necrosis. Enzyme MAP3K5 activated by ROS favors cardiomyocyte apoptosis. Reactive oxygen species and other cellular proteins released from necrotic cells after cardiac injury initiate inflammation.

Endothelial Dysfunction

Excess superoxide or superoxide-derived substances directly or indirectly promote lipid peroxidation and low-density lipoprotein oxidation. Superoxide inactivates nitric oxide, hence favoring: (1) vasoconstriction and smooth myocyte proliferation and migration, and (2) blood cell adhesion. These reactive species can act on inflammatory signaling.

Endothelial dysfunction is an early diabetes complication characterized by NOS3 uncoupling (i.e., a decrease in nitric oxide and a concomitant increase in superoxide). The SHC adaptor, which regulates ROS production during hyperglycemia, is involved in the cell response to oxidative stress associated with diabetic vascular diseases. Expression of the antioxidant heme oxygenase HO1 and NOS3 is upregulated in the absence of SHC, then avoiding ROS-dependent endothelial dysfunction induced by hyperglycemia [254].

Parkinson Disease Protein-7

Parkinson disease protein Park7 protects against oxidative stress. In the absence of oxidative stress, Park7 binds to and represses the translation of mRNAs encoding factors contending with oxidative stress, such as P53 factor, superoxide dismutase, and proteins involved in glutathione synthesis [202]. Oxidized Park7 dissociate from these transcripts, allowing their translation.

The Park7 protein is highly expressed in the normal heart. Its deficiency provokes adverse cardiac hypertrophy as well as compromised function of cytoplasmic and mitochondrial antioxidant systems, elevated ROS production, elevated cardiomyocyte apoptosis rate despite an exaggerated ERK1/2 activation via Src and Ras pathways, and a susceptibility to the development of heart failure [202].

3.2.4.4 Caveolae

Caveolae are plasmalemmal invaginations enriched in cholesterol, glycosphingolipids, and lipid-anchored proteins. They are the most abundant membrane nanodomains devoted to endocytosis (18–40 % of the plasma membrane) [255]. They constitute platforms for preassembled complexes of receptors, signaling components, and their targets at the plasma membrane, including their parts in T-tubules in cardiomyocytes.

Type-1 caveolae form endocytic vesicles, or *cavicles*, that carry cargos on microtubules to specialized endosomes, caveosomes. Cavicles can also travel to recycling endosomes. Type-2 caveolae, the most abundant type, are engaged in continuous cycles of fission and fusion at the cell surface. They can move rapidly in the cell cortex. They can serve in internalization and recycling of molecules as well as *potocytosis*, that is, delivery of small molecules (e.g., vitamins and ions) that are transported across the plasma membrane and deposited directly into the cytosol [255]. Type-3 caveolae form long, thin caveolin-1+ tubules that project from the plasma membrane deep into the cell interior.

Caveolae are composed of caveolin and cavin that are required for caveola genesis and functioning (Vol. 1, Chap. 7. Plasma Membrane). Cavins operate as caveolin adaptors that form the *caveolin–cavin complex* to regulate the caveola function.

Caveolin-1 and -2 are expressed in most cell types, including endotheliocytes, smooth myocytes, fibroblasts, and adipocytes. Caveolin-3 is expressed exclusively in smooth, skeletal, and cardiac myocytes.

Cavins are other structural components of caveolae. Four isoforms exist (cavin-1–cavin-4).[19] Cavin-1 and -2 are required for caveolar invagination and cavin-3 for caveolar budding to form caveolar vesicles [257]. Cavin-1 is required for caveola formation in various cell types (e.g., epitheliocytes and smooth and striated myocytes). Cavin-2, but not cavin-3, is involved in caveola formation in the endothelium of the lung, but not in that of the heart. Cavin-4 is dispensable for caveolar formation, but modifies the morphology of formed caveolae in cardiomyocytes [257].

G-protein-coupled receptors accumulate in caveolae of cardiomyocytes and signal from caveolae. Many GPCRs homo- or hetero-oligomerize, thereby influencing ligand binding, receptor activation, desensitization, transfer, and receptor signaling.

α1a- and α1b-Adrenoceptors colocalize with cavin-3 and -4 at the plasma membrane and partly within the cytoplasm in cardiomyocytes. α1-Adrenergic receptors engender synthesis of atrial natriuretic peptide and support myofibrillar organization in cardiomyocytes via cavin-4 [257]. The myocyte-specific caveola component Cavin-4 regulates α1a- and α1b-adrenoceptors. It facilitates ERK1 and ERK2 recruitment to caveolae in response to α1 AR stimulation that provokes concentric cardiac hypertrophy.

Caveolae possess an ERK-activation module. Other scaffold proteins involved in ERK signaling encompass kinase suppressor of Ras (KSR), MAP2K-interacting protein MAP2K1IP1,[20] MAPK organizer MOrg1, and β-arrestin

3.2.4.5 Nitric Oxide Synthase in Endothelial and Myocardial Cells

Nitric oxide synthase NOS3 contributes not only to myocardial and coronary adaptations to exercise but also to left ventricular and vascular maladaptive remodeling, especially in sustained hypertension. Endothelial cell dysfunction is involved in the establishment and progression of hypertensive heart disease.

---

[19] Cavin-1 is also called polymerase-1 and transcript release factor (PTRF), cavin-2 serum deprivation response protein (SDRP or SDR) and phosphatidylserine-binding protein, cavin-3 SDR-related gene product that binds to C kinase (SRBC), and cavin-4 muscle-related (or restricted) coiled-coil protein (MURC). PKCα controls type-2 caveola internalization [255]. Cavin phosphorylation by PKCα enables caveola internalization. Cavins are members of the STICK (substrates that interact with C-kinase) superfamily of PKC-binding proteins. All STICK proteins contain a phosphatidylserine (PS)-binding sites and reside at the interface between membranes and the cytoskeleton. They include myristoylated alanine-rich protein kinase-C substrate (MARCKS), annexin-1 and -2, desmoykin, vinculin, talin, α- and β-adducin, growth-associated protein GAP43 and A-kinase anchoring protein AKAP5. In particular, cavin-3 binds to PKCδ. It remains associated with caveolin when caveolae bud to form cavicles that travel on microtubules to different regions of the cell [255]. Cavin-3 dictates the balance between ERK and PKB signaling [256]. Cavin-3 increases ERK signaling at the expense of PKB. It facilitates signal transduction to ERK, as it anchors caveolae to the plasma membrane via myosin-1C.

[20] A.k.a. MAP2K-binding partner MP1, ragulator-3, and late endosomal/lysosomal adaptor, MAPK and TOR activator LAMTOR3.

Table 3.6 Coupled and uncoupled nitric oxide synthase NOS3

Status	Products	Pathway and effects
Coupled NOS3	NO	NO–sGC–cGMP–PKG
		MLCP phosphorylation (vasodilation)
		Calcium entry (SOCC, MSCC, $Ca_V 1$), transcription factor phosphorylation, gene transcription
		Opening of mK_{ATP}
		(ROS production at low levels [signaling])
		RyR⊕
		Troponin
Uncoupled NOS3	$O_2^{\bullet -}$	RyR⊖
	$ONOO^-$	PLb⊖
		Gene transcription⊖

⊕⟶ stimulation, ⊖⟶ inhibition, *cGMP* cyclic guanosine monophosphate, $O_2^{\bullet -}$ superoxide, *ONOO*$^-$ peroxynitrite, *MSCC* mechanosensitive Ca^{2+} channel, mK_{ATP} mitochondrial K_{ATP} channel, *PKG* protein kinase-G, *PLb* phospholamban [SERCA inhibitor], *RyR* ryanodine receptor, *SERCA* sarco(endo)plasmic reticulum Ca^{2+} ATPase, *sGC* soluble guanylate cyclase, *SOCC* store-operated Ca^{2+} channel

Structural and functional changes of the coronary microvasculature are associated with cardiac maladaptive hypertrophy and myocardial and perivascular fibrosis in hypertension [258]. Cardiac fibrosis is related to endothelial–mesenchymal transformation, activation of matrix metallopeptidases, and loss of endothelial cells [258].

In normal conditions, NOS3 protect against oxidative cytotoxicity, abnormal growth, and fibrosis. Nitric oxide synthase-3 ameliorates cardiac function and adverse remodeling via the PI3K–PKB–Rho–RoCK–ROS pathway in heart failure [259]. After myocardial infarction, NOS3 limits ameliorate cardiac dysfunction, pulmonary congestion, and interstitial fibrosis.

In an oxidative environment created by heart dysfunction, electron transfer from the NOS3 reductase domain to oxygenase domain can be altered and causes *NOS uncoupling*, i.e., decreases synthesis of NO and increases that of superoxide. In hypertensive cardiac hypertrophy, NOS3 uncoupling thus primes oxidative stress, thereby aggravating left ventricular dysfunction.

Cardiac maladaptive remodeling is limited by nitric oxide produced by coupled NOS3, but aggravated by superoxide formed by uncoupled NOS3 (Table 3.6) [260].

The detrimental influence of NOS3 in hypertensive hearts can be, at least partly, compensated by the beneficial influence of NOS3 in bone marrow-derived endothelial progenitor cells [260]. As cardiomyocyte-restricted restoration of NOS3 activity in mild cardiac maladaptive hypertrophy also has beneficial effect, NOS3 uncoupling may occur preferentially in coronary endotheliocytes.

Enzyme NOS3 is synthesized in coronary endotheliocytes and cardiomyocytes, as well as endothelial progenitor cells (or circulating angiogenic cells). In the heart, angiogenesis depends on nitric oxide. Angiogenesis can involve recruitment of endothelial progenitor cells from the bone marrow (Vol. 5, Chap. 10. Vasculature Growth).

Hypertension increases the number of endothelial progenitor cells in the bone marrow as well as in blood. Synthase NOS3 supports generation and migration of bone marrow-derived endothelial progenitor cells, hence myocardial capillarization, and ameliorates hypertension-induced left ventricular hypertrophy and fibrosis [258]. Conversely, in the absence of NOS3, left ventricular hypertrophy, capillary rarefaction, fibrosis, and pulmonary congestion are aggravated.

The imbalance between nitric oxide and endothelin causes cardiac defective remodeling [261]. Therefore, treatment of cardiac adverse hypertrophy is aimed at increasing the beneficial action of NOS3 and decreasing the activity of endothelin-converting enzyme or density of endothelin receptors. The ATP-sensitive potassium channel consists of an inward rectifier K^+ channel subunit ($K_{IR}6$) and regulatory sulfonylurea receptor subunit (SUR). The subunit types depend on the cell type. Cardioprotective mitochondrial K_{ATP} channels are involved in antihypertrophic signaling. The plasmalemmal SUR2b–$K_{IR}6.1$ channel is the main endothelial subtype. It contributes to the maintenance of the resting membrane potential. As K^+ efflux elicits Ca^{2+} influx, it regulates cytosolic Ca^{2+} concentration that stimulates NOS via Ca^{2+}–calmodulin, thereby favoring the release of nitric oxide. Activation (opening) of the plasmalemmal SUR2b–$K_{IR}6.1$ channel thus protects endothelium and restores NO–ET1 balance, hence endothelial function [261]. Stimulated sarcolemmal SUR2a–$K_{IR}6.2$ channel in cardiomyocytes shortens the action potential duration.

3.2.4.6 Heterotrimeric GTPases (G Proteins)

Proteins of the Gq/11 subclass operate in maladaptive cardiac hypertrophy in response to increased mechanical load. In particular, Gq subunit promotes an association between Gβγ and components of the MAPK module in hypertrophic cardiomyocytes that leads to ERK2 phosphorylation. In fact, the stimulated Raf1–MAP2K module activates ERK1 and ERK2 that translocate to the nucleus to phosphorylate nuclear substrates, whereas scaffold proteins, such as β-arrestins, retain active ERK1 and ERK2 in the cytoplasm. Signaling primed by the Gq subunit releases the Gβγ subunit from activated Gq and promotes Gβγ binding to Raf and ERK2, therefore acting as a scaffold that triggers ERK2 autophosphorylation (Thr188) and subsequent nuclear localization [262]. Nuclear accumulation of ERK1 and ERK2 can result from increased nuclear translocation or nuclear retention. In addition, ERK2 phosphorylation at the same site (Thr188) is also stimulated by receptor protein Tyr kinases such as HER2 receptor.

Angiotensin-2 causes cardiac hypertrophy by stimulating AT_1 receptors in the kidney, whereas activated AT_{1A} receptors in the heart induces cardiac fibrosis, but not hypertrophy. Angiotensin-2 generates a biphasic increase in cytosolic calcium concentration with an initial peak and a sustained elevation due to formation of inositol trisphosphate and cyclic ^{ADP}ribose (cADPR), respectively. Angiotensin-2 binds to Gq-coupled AT_1 receptor to sequentially activate Src kinase, phosphatidylinositol 3-kinase, protein kinase-B, phospholipase-Cγ1, and inositol trisphosphate. The latter provokes Ca^{2+} release from intracellular stores (sarcoplasmic reticulum) that activates ADPR cyclase, which forms cyclic ^{ADP}ribose. ^{ADP}Ribosyl cyclase produces cADPR from β NAD^+. Cyclic ^{ADP}ribose is a Ca^{2+}-mobilizing second messenger that causes a sustained Ca^{2+} influx via Ca^{2+}-induced Ca^{2+} release. The inhibitor of ^{ADP}ribosyl cyclase (2,2′)-dihydroxyazobenzene that precludes the sustained Ca^{2+} signal reduces angiotensin-2-induced cardiac hypertrophy [263]. Low-level AT_1 activity in cardiomyocytes induces adaptive cardiac hypertrophy, whereas increased stimulation induces maladaptive cardiac hypertrophy [264].

Chronic stimulation of Gq-coupled α1a-adrenoceptors also causes maladaptive hypertrophy of cardiomyocytes in pressure overload. Subunit Gq targets PLCβ, thereby generating IP_3 and activating diacylglycerol-dependent PKC isoforms involved in cardiac hypertrophy and failure [265].

In cardiomyocytes, Gq-coupled receptors can also transactivate some plasmalemmal protein Tyr kinases that are growth factor receptors. In cardiomyocytes, AT_1, ET_A, $P2Y_2$, and α1-adrenergic receptors can prime phosphorylation of epidermal growth factor receptor [265]. However, EGF, a strong activator of EGFR, is a very weak hypertrophic agent. Nevertheless, signaling from transactivated EGFR may differ from that of EGF-bound receptor. Once they are activated, these receptors recruit GRB2, PI3K regulatory subunit, and PLCγ, and can stimulate the MAPK module as well as PKB and $P70^{RSK}$ (S6K) kinases. Among the three major MAPK sets (ERK, JNK, and P38MAPK), ERKs are responsible for cardiac growth, whereas proapoptotic JNKs and P38MAPKs impede growth.

Ubiquitous members of the G12 subclass ($G\alpha_{12}$ and $G\alpha_{13}$) can be activated by angiotensin-2, endothelin-1, and α1-adrenoceptor ligands. Subunits of the G12/13 subclass participate in pressure overload-induced cardiomyocyte hypertrophy as well as cardiac fibrosis without the development of hypertrophy subsequent to pressure overload.

3.2.4.7 Small Monomeric GTPases

Cardiac hypertrophy is also accompanied by augmented activity of RNA polymerase-1, -2, and -3 that regulate synthesis of rRNA, mRNA, and tRNA, respectively. In addition, P70 ribomal S6 kinase and eukaryotic translation initiation factor-4E phosphorylation is elevated. Ras GTPases and RasGAP (Vol. 4, Chap. 9. Guanosine Triphosphatases and Their Regulators) operate in cardiac maladaptive hypertrophy due to pressure overload by enhancing the expression of promoters. Similarly to α1-adrenoceptor stimulation, Ras GTPase favors phosphorylation of RNA polymerase-2 and upregulates the expression of cyclin-dependent kinase-7 [266].

Concentration of small Rad GTPase[21] significantly decays in cardiac maladaptive hypertrophy induced by pressure overload and failing hearts [267]. GTPase Rad impedes cardiac maladaptive hypertrophy via CamK2δ. In cardiomyocytes, Rad inhibits CamK2 phosphorylation and autonomous activity.

In addition, Rem GTPase hinders the excitation–contraction coupling in striated myocytes by reducing the number of functional $Ca_V1.2$ channels [268].

Mutually exclusive binding of calmodulin and 14-3-3 protein to Gem and Kir GTPases modulates $Ca_V1.2$ transport to the plasma membrane. Nuclear transport controls the location of RGK GTPases in cells and thus their activity. Calmodulin binding hampers Kir and Gem translocation into the nucleus, as calmodulin interferes with importin-α5 binding that allows nuclear import [269].[22] On the other hand, protein 14-3-3 binding sequesters RGK GTPases in the cytoplasm. Nuclear transport into the nucleus of Kir and Gem GTPases is also regulated by phosphorylation (that inhibits importin binding) and dephosphorylation.

Cytoskeletal proteins can influence the transcriptional alterations associated with adverse cardiac remodeling. *Myozenin*-2 (or calsarcin-1), another Z-disc protein, precludes the PP3–NFAT pathway in cardiomyocytes [270]. The cytoskeletal molecule, four and a half LIM domain-containing protein, FHL2 prevents ERK2-mediated pro-hypertrophic signaling and serum response factor (SRF)-mediated transcription. The titin-associating complex also hampers SRF-mediated transcription, as it regulates the localization of the ubiquitin ligase MuRF2 that impedes SRF activity.

Muscle-specific *Actin-binding Rho-activating protein* (ABRA)[23] localized to the Z-disc and M-line activates SRF-dependent transcription. It supports the nuclear translocation of the SRF coactivators myocardin-related transcription factors MRTFa and MRTFb [270]. The Abra promoter contains a binding site for myocyte enhancer factor MEF2, a stress-responsive transcriptional activator. Protein ABRA thus functions as a cytoskeletal intermediary between MEF2 and SRF factors. Overexpression of ABRA in the heart sensitizes the heart to pressure overload and calcineurin signaling.

---

[21] Small Ras GTPase associated with diabetes Rad with related GTPases (Gem, Kir, Rem, and Rem2) form the RGK class of the RAS hyperfamily of small GTPases. This hyperfamily comprises many superfamilies, families, and subfamilies, such as RAS, RHO (Rho, Rac, and Cdc42), ARF, RAB, RAN, and RGK sets. Small RGK GTPases are inhibitors of voltage-dependent calcium channels. In the cardiovascular apparatus, small GTPases participate in the regulation of functioning of endothelial and smooth muscle cells as well as cardiomyocytes. Small Rad GTPase can inhibit insulin-stimulated glucose uptake in myocytes and adipocytes. Moreover, Rad and Gem modulate cytoskeleton remodeling via the Rho–RoCK pathway. In addition, Rad overexpression prevents SMC migration and suppresses neointimal formation after balloon injury [267].

[22] Small Rem GTPase also binds to importinα5, whereas Rad GTPase links to importins-α3, -α5, and -β [269].

[23] a.k.a. Striated muscle activator of Rho signaling (StARS).

3.2.4.8 Soluble Epoxide Hydrolase

Soluble epoxide hydrolase catalyzes the conversion of epoxyeicosatrienoic acids (EETs) into dihydroxyeicosatrienoic acids (DHETs).[24] Soluble epoxide hydrolase also participates in angiotensin-2-induced cardiac maladaptive hypertrophy due to hypertension. Soluble epoxide hydrolase is upregulated by vasoconstrictor angiotensin-2 to mediate its effect [271]. Angiotensin-2 increases expression of soluble epoxide hydrolase via mitogen-activated protein kinase and activator protein AP1 in vascular endotheliocytes. In addition, epoxyeicosatrienoic acids increase the opening for both sarcolemmal and mitochondrial cardioprotective ATP-sensitive potassium channels, whereas angiotensin-2 prevents their activity.

3.2.4.9 Nicotinamide Phosphoribosyltransferase

Maladaptive cardiac hypertrophy is associated with reduced levels of nicotinamide adenine dinucleotide (NAD). Intracellular nicotinamide phosphoribosyltransferase (iNAmPT) is a rate-limiting enzyme in NAD synthesis, thereby participating in the regulation of activity of several NAD-using enzymes. Production of NAmPT is upregulated in response to hypoxia; its expression is correlated with atherosclerosis. It is secreted and released into the blood circulation. Extracellular eNAmPT acts as a proinflammatory cytokine locally and remotely. Level of circulating NAmPT rises in obesity, type-2 diabetes, and chronic inflammatory diseases. Cardiomyocytes secrete NAmPT under stresses, thereby supporting development of adverse ventricular remodeling (hypertrophy and fibrosis) [297]. The NAmPT action is mediated by the PP3–NFAT pathway and activation of JNK1, P38MAPK, and ERK kinases.

3.2.4.10 Calpains

Cardiac remodeling is associated with increased levels of circulating catecholamines and angiotensin-2 subsequent to stimulation of the sympathetic nervous system and of the renin–angiotensin axis, respectively. These mediators produce oxidative stress and cytosolic Ca^{2+} overload and activate peptidases, such as matrix metallopeptidases, cathepsins, caspases, and calpains, which function cooperatively and degrade both intra- and extracellular substrates.

[24] Three enzymes, cyclooxygenase, lipoxygenase, and cytochrome-P450 epoxygenase, produce eicosanoids from arachidonic acid derived from membrane phospholipids (Vols. 4, Chap. 2. Signaling Lipids and 5, Chap. 7. Vessel Wall). Cytochrome-P450 epoxygenase produces four epoxyeicosatrienoic acids: (5,6)-, (8,9)-, (11,12)-, and (14,15)-EETs. The latter modulate several signaling cascades by increasing intracellular Ca^{2+} concentration and stimulating Ca^{2+}-activated K^+ channel. These molecules also operate as endothelium-derived hyperpolarizing factors to generate vasodilation.

Table 3.7 Calpain substrates in cardiac remodeling and failure. (Source: [272])

Target	Calpain effect
ABCa1	Degradation
Actin	Plasma membrane repair
Ankyrin, fodrin	Sarcolemma rupture
Ca_V1.2	Atrial fibrillation, disturbance of excitation–contraction coupling
Cadherin-5	Loss of adherence junctions, hyperpermeability
Ezrin, talin	Cytoskeletal remodeling, extravasation and migration of leukocytes (cardiac inflammation)
Fibronectin	Endotheliocyte migration and growth, endothelium repair
IκBα	NFκB activation, cardiac hypertrophy and inflammation
PKB-associated HSP90	PKB inactivation, limitation of cardiac hypertrophy
PP3	NFAT activation, cardiac hypertrophy
TGFβ-associated LAP	Collagen synthesis, cardiac fibrosis
Troponin-T	Sarcomere structure degradation

ABC ATP-binding cassette transporter, *HSP* heat shock protein, *IκB* inhibitor of NFκB, *LAP* latency-associated peptide, *NFκB* nuclear factor-κB, *NFAT* nuclear factor of activated T cells, *TGF* transforming growth factor

Hypertension characterized by increased blood pressure and total peripheral resistance causes cardiac remodeling and alteration of the calpain–calpastatin balance [272].

Calpains may contribute to myocardial hypertrophy and inflammation mainly via activated NFκB, as they degrade the inhibitor IκBα (Table 3.7). The signaling cascade downstream from G-protein-coupled receptors bound to their cognate ligands, such as angiotensin-2 (AT_1), endothelin, and noradrenaline, include successively PI(4,5)P_2 of the plasma membrane, PLC, IP_3, IP_3R of the endoplasmic reticulum, and Ca^{2+} release from its store into the cytosol, which may be amplified by chromogranin-B, thereby allowing calpain activation [272]. Factor NFκB promotes the transcription of numerous genes that encode apoptosis inhibitors (IAP1, BCL2, BCLxL) and mediators of cardiomyocyte hypertrophy. Calpains can also activate PP3 phosphatase responsible for the translocation of nuclear factor of activated T cells to the nucleus, where it induces the expression of prohypertrophic genes.

On the other hand, activated calpain limits the expression of prohypertrophic genes by degrading PKB-associated heat shock protein HSP90 that stabilizes PKB phosphorylated (activated) by phosphoinositide-dependent kinase PDK1 [272]. In the absence of calpain, phosphorylation of glycogen synthase kinase GSK3β by PKB represses its antihypertrophic effects.

In summary, calpain targets many antagonist signaling axes that either promote or preclude maladaptive hypertrophy. However, the balance between these pathways favors cardiac hypertrophy.

When the heart continuously experiences stress, synthesis of stress proteins, such as HSP10, HSP60, and HSP70 increases, being partly secreted. Binding of HSPs to Toll-like receptors at the surface of immune and cardiac cells causes the expression of cell adhesion molecules, chemokines, and chemokine receptors [272]. Subsequently, inflammatory leukocytes, such as monocytes, neutrophils, and lymphocytes, are recruited and activated.

Calpains also intervene in fibrosis partly by activating transforming growth factor-β. Extracellular calpain activates the latent form of TGFβ, as it cleaves latency-associated peptide (LAP). Intracellular calpain can also activate the latent TGFβ in the Golgi body of pulmonary arteriole smooth myocytes [272]. Stimulated fibroblasts are transformed into active myofibroblasts that liberate proinflammatory cytokines and collagens. Accumulation of collagen-1 and -3 is not balanced by degradation. The profibrotic action of calpains may be explained by an increase in collagen deposition rather than a decrease in collagen degradation [272]. Binding of angiotensin-2 to AT_1 receptor can transactivate EGFR, which activates calpains via increased intracellular calcium level and mitogen-activated protein kinase, thereby favoring matrix protein synthesis under TGFβ control. Factor TGFβ promotes via SMADs production of collagen-1, as its gene promoter contains a SMAD-binding site, as well as that of connective tissue growth factor CTGFβ implicated in collagen synthesis.

Pressure overload-promoted aggregates of ubiquitinated proteins in the heart can be substrates of autophagy; although ubiquitin–proteasome and autophagy–lysosome axes have distinct substrates, they cooperate [273].

3.2.4.11 Kinases

Calcium–Calmodulin-Dependent Protein Kinase CamK2

Calcium–calmodulin-dependent protein kinase CamK2δ phosphorylates class-2 histone deacetylases, causing HDAC nuclear export and thus relieving repression exerted by HDACs on transcription by myocyte enhancer factor MEF2, leading to altered gene expression, maladaptive hypertrophy, and heart failure [265].

Glycogen Synthase Kinase GSK3

During pressure overload, phosphorylation (inactivation) of glycogen synthase kinase GSK3α (Ser21) and GSK3β (Ser9) mediates heart maladaptive hypertrophy and failure, respectively [274]. Glycogen synthase kinase-3β is an inhibitor of hypertrophic signaling, as it phosphorylates NFAT and prevents its entry into the nucleus, hence precluding expression of hypertrophic genes. Under pressure overload, inhibition of phosphorylation of GSK3α (Ser21), but not that of phosphorylation in GSK3β (Ser9), as GSK3α is preferentially located in the nucleus, allows phosphorylation (inhibition) of G1 cyclins, and suppresses E2F and markers of cell proliferation, such as phosphorylated histone-3.

PIM Kinase

Protein Ser/Thr kinase PIM1 exerts potent cardioprotective effects in the myocardium downstream from protein kinase-B, as it weakens myocardial hypertrophy in response to pressure overload [275]. Overexpression of PIM1 in cardiomyocytes impedes ET1-induced hypertrophy, increases ANP expression, and reduces βMHC synthesis. Kinase Pim1 also induces SERCA2a pump and phospholamban expression. Other Pim1 salutary effects include decreased apoptosis, fibrosis, and necrosis.

Protein Kinase-C

Whereas the PI3K–PKB pathway induces adaptive cardiac hypertrophy, protein kinase-Cβ2 promotes maladaptive cardiac hypertrophy in the absence of active phosphoinositide 3-kinase [276]. The PI3K–PKCβ2 axis actually improves cardiac function, as PI3K reduces PKCβ2 synthesis. The PI3K kinase mainly controls adaptive development of cardiomyocytes and can modulate PKCβ2 activity to rescue cardiac function upon PKCβ2 overexpression.

Protein Kinase-G

Two major types of PKG exist: (1) a soluble, widespread PKG1, which lodges in smooth muscle and endothelial cells, leukocytes, in particular neutrophils and monocytes, and neurons, and (2) a membrane-bound PKG2 of more restricted location. Two isoforms of PKG1 are produced from alternative splicing of the Pkg1 transcript (PKG1α–PKG1β). They differ in their N-terminus, which encodes a dimerization and autoinhibitory domain. Subtype PKG1 targets cAMP response-element-binding protein that mediates the effect of the NO–cGMP–PKG axis on the Fos promoter.

In cardiomyocytes, PKG1 does not affect cardiac maladaptive hypertrophy caused by mild pressure overload in the absence of guanylate cyclase stimulators or phosphodiesterase inhibitors [277]. Antihypertrophic action of PKG1 relies on stimulation of its relevant targets (e.g., PP3 and Gq cascades); PDE5 inhibitor augments PKG1 activity in moderate and severe hypertension [278].

Raf–MAP2K Pathway

Small Ras GTPase is activated by many prohypertrophic stimuli. Overexpression of Ras causes adverse cardiac hypertrophy and heart failure. Activated Ras recruits the Raf kinase to the plasma membrane where it is activated and then triggers the MAPK cascade. In particular, activated bRaf–MAPK module causes cardiomyocyte hypertrophy.[25] The prohypertrophic Raf–MAP2K1/2 pathway is inhibited by reticulocalbin-1 (RCn1) that then reduces cardiomyocyte hypertrophy as well as expression of the hypertrophic factors ANP and BNP [279]. The antihypertrophic RCn1 reduces bRaf- and cRaf-induced MAP2K1/2 phosphorylation.

Reticulocalbin-1 belongs to the CREC protein family of low-affinity Ca^{2+}-binding proteins (RCn1–RCN3, stromal cell-derived factor SDF4, and calumenin) [280].[26]

3.2.4.12 Cyclin-Dependent Kinase Inhibitors

The cell cycle regulator cyclin-dependent kinase inhibitor CKI1b, the activity of which is impeded by antiapoptotic casein kinase-2 (CK2α′), hampers maladaptive enlargement of cardiomyocytes exposed to hypertrophic stimuli such as angiotensin-2 [281]. Regulator CKI1b is both an inhibitor and a substrate of CK2α′ enzyme. Upon stimulation by angiotensin-2, CK2α′ phosphorylates CKI1b for degradation, thereby favoring cardiomyocyte hypertrophy. Conversely, unphosphorylated CKI1b binds to and inhibits CK2α′ kinase. Despite a decline in the CKI1b amount in hypertension (as well as in acute and end-stage heart failure), residual CKI1b content can be sufficient to exert an inhibition on hypertrophy.

3.2.4.13 Adenosine

Extracellular adenosine reduces cardiac hypertrophy and failure in mice with chronic hypertension. Adenosine regulates the microtubule cytoskeleton, preferentially targeting stabilized microtubules that contain detyrosinated α-tubulin [282].[27] The reorganization of the cytoskeleton with accumulation of sarcomeric proteins, microtubules, and desmin promotes cardiac hypertrophy. Adenosine reduces density of stable microtubules.

---

[25] bRaf can bind and activate cRAF (transphosphorylation). bRaf phosphorylates MAP2K1 and MAP2K2 more efficiently than aRaf and cRaf.

[26] These proteins localize to the cytosol and in various compartments of the secretory pathway. They are also secreted to the extracellular space or lodge on the cell surface. Calumenin inhibits several proteins of the endoplasmic reticulum membrane, the vitamin-K1 (2,3)-epoxide (vitamin-K derivative) reductase (VKOR that yields the reduced form of vitamin-K called vitamin-K hydroquinone), γ-carboxylase, ryanodine receptor, and SERCA pump.

[27] Tyrosine is reversibly ligated to the C-terminus of tubulin by tubulin Tyr ligase and removed by tubulin carboxypeptidase. Microtubules grow and shrink from their functional plus-ends, whereas their minus-ends localize to the cell center at the microtubule organizing center. Tyrosinated ($tubulin^{Tyr}$) or detyrosinated ($tubulin^{Glu}$) α-tubulin represent dynamical and stable microtubules, respectively [283].

3.2.4.14 Erbin

On the other hand, the widespread molecule ErbB2 (HER2)-interacting protein (ErbB2IP, or *erbin*) has a cardioprotective role. It operates in compensated hypertrophy and prevents adverse cardiac hypertrophy. Its expression decays in heart failure. Erbin inhibits Ras-primed ERK activation by binding to suppressor of clear SOC2 homolog (SHOC2), a promoter of Ras–ERK signaling, as it serves as a scaffold for both the Ras GTPase and Raf kinase [284]. In the absence of erbin, Raf is phosphorylated and binds to SHOC2, thereby supporting ERK phosphorylation. Erbin inhibits various signaling pathways, not only the Ras–ERK axis but also the NFκB and TGFβ cascades. These 3 pathways contribute to cardiac hypertrophy. In addition, erbin complexes with HER2 and β2-adrenergic receptor in cardiomyocytes.

3.2.4.15 Morphogens—Dickkopf-3

Dickkopf-3 (Dkk3)[28] is both a cytosolic and secreted glycoprotein that antagonizes the Wnt–β Ctnn axis, as it interferes with Wnt coreceptors, LDLR-related proteins and Kremen-1 and -2 (i.e., Dickkopf receptors). Its expression is downregulated in patients with end-stage heart failure and mice with pressure-overloaded cardiac hypertrophy [285].

Dickkopf-3 impedes the activation of MAP3K5, thereby suppressing the activation of JNK and P38MAPK kinases [286]. The MAP3K5 enzymatic activity is suppressed by the reduced thioredoxin form within the MAP3K5 signalosome[29] that also contains the adaptors and Ub ligases TRAF2 and TRAF6 and the deubiquitinase ubiquitin-specific peptidase USP9x. On the other hand, ROS oxidize thioredoxin and induce autophosphorylation and oligomerization (activation) of MAP3K5 in maladaptive cardiac hypertrophy. Dickkopf-3 acts as a cardioprotective regulator of adverse cardiac hypertrophy.

3.2.4.16 Growth Factors

In myocardial infarction, the expression of the 3 TGFβ isoforms is upregulated. They participate in inflammation at an early phase and afterward cardiac remodeling.

[28] In German, *der Dickkopf* means obstinacy, stubbornness, mulishness (the expression "einen Dickkopf haben" signifies to be obstinate, stubborn, or mulish) as well as pighead and bullhead. Members of the Dickkopf family induce head formation in the Xenopus embryo. The DKK family includes Dkk1 to Dkk4 and Dkk3-related protein DkkL1 (Soggy). Dickkopf-3 has 2 isoforms, a 55-kDa cytosolic and 50-kDa extracellular subtype. In humans, the soluble form (45–65 kDa) has 2 splice variants of 45–55 and 40 kDa following deglycosylation.

[29] The ASK1 signalosome is activated in response to reactive oxygen species, endoplasmic reticulum stress, calcium overload, and inflammatory signals mediated by TNFSF1 and lipopolysaccharide [285].

Subtype TGFβ2 is the predominant isoform in cardiac fibrosis. Isoform TGFβ1 is involved in angiotensin-2-mediated pressure overload-induced cardiac hypertrophy.

In cardiomyocytes, VEGF stimulates cell growth via VEGFR2 and the PKB1 pathway. On the other hand, VEGF causes regression of cardiomyocyte hypertrophy. In cultured rat neonatal cardiomyocytes that undergo maladaptive hypertrophy, copper sulfate (5 μmol for 24 h) increases twofold the ratio of VEGFR1 to VEGFR2 [287]. Receptor VEGFR1 mediates Cu-induced regression of cell hypertrophy via cGMP-dependent protein kinase PKG1. Copper does not upregulate VEGFR1, but reduces VEGFR2 expression.

3.2.4.17 Cytokines

Certain cytokines, such as members of the interleukin-6 family, are implicated in cardioprotection as well as pathophysiology (cardiac hypertrophy, inflammation, and heart failure) via activation of the JaK–STAT pathway (Vol. 4, Chap. 4. Cytoplasmic Protein Tyrosine Kinases) and signaling termination by suppressors of cytokine signaling (SOCS). The latter constitute a family of eight members (cytokine-inducible SH2 domain-containing protein [CIS] and SOCS1–SOCS7).

Signaling starts with activation of the common receptor glycoproteic GP130 subunit [288]. All members of the IL6 family that homo- or heterodimerize GP130 activate JaK1, JaK2, and, to a lesser extent, TyK2 kinase. Stimulated cytokine receptors initiate the activation of Janus kinases. Activated JaKs phosphorylate the receptor cytoplasmic domains, thereby creating docking sites for signal transducers and activators of transcription. Once bound to the receptor, STAT proteins are phosphorylated by JaKs, dimerize, and translocate into the nucleus.

The GP130–JaK–STAT pathway in cardiomyocytes ensures cardioprotection, as it promotes the activity of antiapoptotic genes and ROS scavengers and upregulates the angiogenic factors. In fact, the JaK–STAT as well as Ras–ERK and PI3K–PKB pathways support hypertrophy and survival downstream from the common signal-transducing receptor GP130 subunit activated by IL6-type cytokines in response to pressure overload and ischemic injury. However, an excessive activation of this pathway induces oxidative stress and contributes to the cardiac dysfunction. Therefore, a well-balanced JaK–STAT signaling must be maintained.

In response to cardiac stresses, cardiomyocytes produce cytokines of the IL6 family (IL6, cardiotrophin-1, and leukemia inhibitory factor). Cytokines of the IL6 family stimulate SOCS1 and SOCS3 via the GP130–JaK–STAT axis. Both SOCS1 and SOCS3 suppress the GP130–JaK–STAT signaling (negative feedback). The SOCS proteins interact with elongin-B and elongin-C, Cullins, and the RING finger domain-only protein Rbx2 to form a ubiquitin ligase complex that mediates the degradation of bound partners of SOCS proteins. Moreover, both SOCS1 and SOCS3 can directly inhibit JaKs using their kinase inhibitory region.

The SOCS proteins can be stimulated by messengers independently of the JaK–STAT signaling. Inducers of SOCSs in the myocardium include, in addition to IL6 family cytokines, other cytokine types, such as interferon-γ and tumor-necrosis factor-α, as well as angiotensin-2 [288]. Myocardial SOCS3 expression is upregulated at the onset of the development of maladaptive hypertrophy by angiotensin-2 and by pressure overload. In the heart, SOCS1 intervenes in inflammation, having anti-inflammatory effects in autoimmune myocarditis, but the cardiac-specific SOCS1 overexpression increases susceptibility to viral infection of the heart [288]. Cardiotrophin-1 (CT1), a member of the IL6 family, is produced by cardiac myocytes and fibroblasts subjected to mechanical stress overload and/or exposed to excessive amount of angiotensin-2 [289]. Secreted CT1 interacts with heterodimeric receptor formed by GP130 and leukemia inhibitory factor receptor (LIFR) to initiate exaggerated cardiomyocyte growth, reduce calsequestrin expression, and impede formation of longitudinal bundles of cardiomyocytes.

Oncostatin-M (OsM) is a pleitropic cytokine of the IL6 family that signals via plasma membrane receptors that contain protein GP130 on various cell types. Type-1 and -2 oncostatin-M receptors are GP130–LIFR and GP130–OsMR complexes, respectively. Some cell types stimulated by OsM are irresponsive to LIFR, but respond to the OsMR receptor. Specific OsM activity is mediated by the type-2 receptor, whereas common functions of LIF and OsM depend on the type-1 receptor.

Expression of interleukin-1 receptor family[30] member IL1RL1 can be primed by mechanical stimuli in cardiomyocytes. Plasma concentration of soluble (secreted) $IL1RL1^S$ form rises in patients with chronic heart failure [290]. It can serve as a predictor of mortality or needed transplantation independently of B-type and proatrial natriuretic peptides.

Aberrant JaK–STAT signaling promotes evolution from maladaptive hypertrophy to heart failure [291]. When SOCS1 expression is upregulated in the decompensated phase, but not at the onset, of maladaptive hypertrophy induced by chronic pressure overload, it causes a transition to heart failure associated with sustained inhibition of the GP130 cytokine receptor. Whereas SOCS3 is mainly stimulated by IL6-type cytokines, SOCS1 is highly susceptible to Ifnγ stimulation.

3.2.4.18 Serotonin

Serotonin supports cardiomyocyte hypertrophy via the production of reactive oxygen species by monoamine oxidase-A and stimulation of $5HT_{2A}$ receptors. The hypertrophic $5HT_{2A}$–Ca^{2+}–PP3–NFAT pathway is independent of the hypertrophic MAOa–ERK axis, which is activated at a different serotonin concentration [292].

3.2.4.19 Urocortin

Urocortin, a corticotropin-releasing hormone-related peptide, is involved in cardiovascular and inflammatory responses, as it acts on central and peripheral CRH

[30] The interleukin-1 receptor family includes type-1 and -2 interleukin-1 receptor (IL1R1–IL1R2), and Il1R-like-1 and -2 receptors (IL1RL1–IL1RL2), as well as IL1R accessory protein (IL1RAcP) and IL1R-related protein (IL1RRP).

receptors, especially CRH_2 receptors in the heart (Vol. 1, Chap. 2. Cells of the Blood Circulation).

In cardiomyocytes, urocortin-2 has cAMP- and Ca^{2+}-dependent positive inotropic and lusitropic effects. In addition, urocortin-2 activates nitric oxide that raises cytosolic concentrations of NO, cAMP, and cGMP mediators. In isolated rabbit ventriculomyocytes, it increases phosphorylation of: (1) PKB (Ser473 and Thr308; activation) via corticotropin-releasing factor CRF_2 receptors and PI3K; (2) NOS3 (Ser1177) via the PI3K–PKB and cAMP–PKA axes; and (3) ERK1 and ERK2 (Thr202 and Tyr204) via MAP2K1 and MAP2K2 [293].

3.2.4.20 Galectin

Galectin-3 secreted by activated macrophages is a marker of cardiac maladaptive hypertrophy due to hypertension that evolves toward heart failure (Table 3.8).

Galectin-3 enhances macrophage and mastocyte infiltration, increases cardiac fibrosis, and promotes cardiac remodeling [294]. It augments TGFβ expression and SMAD3 phosphorylation. Its effects are partly or completely impeded by ^{N}acetyl-seryl–aspartyl–lysyl–proline.

3.2.4.21 Mucin-16

Mucin-16[31] is a glycoprotein that serves as a marker for heart failure and certain types of cancers. In fact, concentrations in B-type natriuretic peptide and Muc16 are significantly higher in subjects with heart failure than in the absence of heart failure as well as in severe rather than in mild heart failure [295]. Other tumor marker levels do not markedly vary. Concentrations in Muc16 and BNP that are correlated with heart ejection function decay after clinical improvement. Combined measurement of Muc16 and BNP levels thus evaluates short-term therapy efficiency.

3.2.4.22 Cystatin-3

Cystatin-C or cystatin-3[32] is a protein encoded by gene CST3 that can be detected in all tissues and body fluids (e.g., saliva, tear, milk, and semen). It is a potent inhibitor of lysosomal peptidases as well as extracellular cysteine peptidases

[31] A.k.a. carbohydrate antigen-125 and cancer antigen-125 (CA125).

[32] Cystatin-C is also called γ-trace, postγ-globulin, and neuroendocrine basic polypeptide. It is mainly used to assess the kidney function and as a marker of amyloid deposition-associated brain disorders. Cystatin-C belongs to the type-2 cystatin family that includes stefins, or type-1 cystatins, and kininogens. Cystatin-C dimerizes.

Table 3.8 Biochemical indices of myocardial maldaptive remodeling. (Source: [289]). Markers detected in blood circulation or in the heart by imaging are used in screening, diagnosis, prognosis, and therapeutic monitoring. Annexin-A5 is associated with excessive cardiomyocyte apoptosis. Collagen-derived peptides such as carboxy-terminal propeptide of procollagen-1 that is released into blood after cleavage of procollagen-1 into collagen-1 by peptidase is correlated with myocardial fibrosis. High blood concentration of matrix metallopeptidase-1 that reflects increased collagen degradation can serve as an index of progression to heart failure. In plasma, soluble (secreted) interleukin-1 receptor (IL1R)-like-1 receptor ($IL1RL1^S$) is a marker for neurohormonal activation in patients with heart failure. Cystatin-C, or cystatin-3, is a potent inhibitor of lysosomal peptidases

Molecules related to cardiomyocyte injury	
Hormones	Natriuretic peptides, urocortin
Growth factors	Insulin-like growth factor-1,
Cytokines	TNFSF1, TNFSF6, TNFRSF6a, interleukin-6, cardiotrophin-1, oncostatin-M, leukemia-inhibitory factor
Miscellaneous	Annexin-A5, mucin-16, neuregulin, myosin light chain-1, fatty acid-binding protein
Molecules related to extracellular matrix alteration	
Growth factors	TGFβ, connective tissue growth factor
Constituents	C-terminal propeptide of procollagen-1, osteopontin, thrombospondin, cystatin-C (Cst3)
Enzymes	MMP2, MMP9, TIMP4
Molecules related to inflammation and oxidative stress	
Cytokines	IL1α, soluble TNFRSF1a
Chemokines	CCL2
Enzymes	Myeloperoxidase, superoxide dismutase
Miscellaneous	Nitrotyrosin, 8-OH-2deoxyguanosine
Molecules related to neurohormonal activation	
Messengers	Adrenomedullin, neuropeptide-Y, urodilatin
Mediators	Soluble EGFR, soluble IL1RL1
Molecules secreted by leukocytes	
Miscellaneous	Galectin-3

3.2.4.23 Vasodilator-Stimulated Phosphoprotein

Vasodilator-stimulated phosphoprotein (VASP) is an actin-binding protein and a substrate of cyclic nucleotide-dependent kinases (PKA and PKG). In normal hearts, VASP abounds in intercalated discs. Its production is upregulated in hypertrophied hearts [296]. In cardiomyocytes, stimulated Gs-coupled β-adrenoceptor activates adenylate cyclase and cAMP-dependent pathway, thereby rapidly priming the reversible phosphorylation of VASP (Ser157 and Ser239).[33] On the other hand, activation of particulate guanylate cyclase by atrial natriuretic peptide as well as that of soluble guanylate cyclase by nitric oxide (mostly endotheliocyte-derived NO) cause VASP phosphorylation only at a single residue (Ser239) [296]. Elevated concentrations of phosphorylated VASP in hypertrophic heart models may reflect the cytoskeletal remodeling observed in the transition to heart failure.

3.2.5 Electrochemical Remodeling

Myocardium mechanical activity is driven by electrochemical wave genesis and propagation. Depolarization–repolarization cycles are produced by ion fluxes through specific ion channels, pumps, and exchangers (Vols. 3, Chap. 3. Main Sets of Ion Channels and Pumps and 5, Chap. 5. Cardiomyocytes).

Upon fast Na^+ influx, action potential duration is determined by the balance between inward depolarizing currents mainly due to Na^+ and Ca^{2+} and outward repolarizing currents (that makes cardiomyocyte more negatively charged at its intracellular leaflet of plasma membrane) principally due to K^+ efflux (Table 3.9). Redundant K^+ fluxes allow a *repolarization reserve*. When a given type of K^+ current is dysfunctional, other K^+ fluxes can compensate. However, action potential duration can heighten and weaken cardiac rhythm stability.

Congestive heart failure is often associated with ventricular *tachyarrhythmias* [298]. Repolarization change that prolongs action potential duration aims at improving cellular Ca^{2+} influx and maintaining contractility. Slowly inactivating Na^+ current augmentation contributes to action potential duration extension. Calcium uptake in the sarcoplasmic reticulum by SERCA decays. Expression of Na^+–Ca^{2+} exchanger is upregulated to redress defective cytosolic Ca^{2+} removal when intracellular Ca^{2+} concentration is high. However, Na^+–Ca^{2+} exchanger that enhances Ca^{2+} extrusion reduces sarcoplasmic reticulum Ca^{2+} store, thus further altering contractility. Synthesis of voltage-dependent K^+ channels is downregulated. Current $i_{K,s}$ especially diminishes, thereby generating longer action potential.

In addition, change in cardiomyocyte coupling by connexins of gap junctions impairs spatial distribution of action potential in the myocardium.

The activated sympathetic nervous system and renin–angiotensin–aldosterone axis also affect cardiac electrochemical command. β-Adrenoceptor stimulation increases $i_{K,s}$ and $i_{Ca,L}$ currents. Angiotensin-2 and aldosterone enhance i_{Na} and $i_{Ca,L}$ currents, but attenuate $i_{K,to}$ and i_{K1} currents.

---

[33] Three phosphorylation sites (Ser157, Ser239, and Thr278) exist. $Serine_{157}$ is targeted by PKA, Ser239 by PKG, and Thr278 by AMP-activated protein kinase. Phosphorylation of VASP is implicated in endothelial permeability, angiogenesis, and platelet aggregation.

Table 3.9 Cardiomyocyte depolarization–repolarization cycle and main associated ionic fluxes and currents

Phase	Type of ion flux	Ion current
Depolarization	Na^+ influx	i_{Na}
Initial repolarization	K^+ efflux	$i_{K,to}$
Plateau	Ca^{2+} influx	$i_{Ca,L}$
	K^+ efflux	$i_{K,r}$
Repolarization	K^+ efflux	$i_{K,r}$
	K^+ efflux	$i_{K,s}$
	K^+ influx	i_{K1}
Pacemaker depolarization	Cation influx	i_f
		(i_{HCN})

HCN hyperpolarization-activated, cyclic nucleotide-gated channel, $i_{Ca,L}$: L-type Ca^{2+} current [$Ca_V1.2$], $i_{K,to}$ transient-outward K^+ current [K_V4], $i_{K,r}$ rapid delayed-rectifier K^+ current [$K_V11.1$], $i_{K,s}$ slow delayed-rectifier K^+ current [$K_V7.1$], i_{K1} inward-rectifier background K^+ current [$K_{IR}2$], i_f depolarizing "funny" pacemaker current [HCN2/4]

Delayed repolarization can generate early *after depolarizations*. Impaired refractoriness is associated with atrial fibrillation that is related to $i_{Ca,L}$ reduction in atriomyocytes. Augmented cytosolic Ca^{2+} concentration activates PP3, as it favors calcium–calmodulin binding to PP3 phosphatase. The latter dephosphorylates nuclear factor of activated T-cells that translocates to the nucleus, where it represses expression of the gene that encodes $Ca_V1.2$ channel. Reduced inward Ca^{2+} current causes shorter cell polarity plateau. Moreover, downregulation of voltage-gated K^+ channels weakens $i_{K,to}$ current, whereas i_{K1} current increases because of heightened production of $K_{IR}2.1$ subunits.

3.2.6 *Structural Remodeling*

Cardiac remodeling is characterized by cardiomyocyte hypertrophy, i.e., an increase in cardiomyocyte size (length and/or width), which is primed by mechanical deformation detected partly by integrins and neurohumoral signaling (release of catecholamines, endothelin-1, angiotensin-2, cytokines, chemokines, and growth factors).

Myocardiocyte remodeling is associated with rearrangement of cell organelles. Calcium handling abnormalities are linked to alterations in sarcolemma, sarcoplasmic reticulum, and mitochondria.

3.2.6.1 Maladaptive Cardiomyocyte Growth

Myocardial hypertrophy can be either adaptive or maladaptive according to cardiomyocyte contractility, the former being associated with normal or even improved function, and the latter with impaired contractile function. Abnormal hypertrophy is a feature of many heart pathologies, such as congenital defects, hypertension, and cardiac ischemia. Essential hypertension is a predisposing risk factor for stroke, myocardial infarction, heart failure, arterial aneurysms, and chronic renal failure.

The development of cardiac maladaptive hypertrophy, especially left ventricular hypertrophy, in response to hypertension (i.e., increased hemodynamic load), and neurohormonal stress is initially a compensatory response, which causes ventricular dilation and heart failure when pressure overload is sustained over a long period.

Gene clusters associated with adaptive hypertrophy predominantly comprise genes involved in metabolism and cell growth, whereas maladaptive hypertrophy is characterized by changes in gene clusters involved in oxidative stress responses, inflammation, and apoptosis.

Compensated maladaptive hypertrophy leads to heart failure by recruiting additional genes (Gata4, RAB7, NRAS, GNA12, STAT3, STAT5B, FYN, CRKO, MYCN, PTEN, AKT1, and IL6ST [GP130]) [334]. Proteins that are differentially expressed during left ventricle hypertrophy in mice comprise LIM proteins, thioredoxin, myoglobin, fatty acid-binding protein-3, and myofibrillar proteins.

Maladaptive myocardial hypertrophy is characterized by increased cell size, enhanced protein synthesis, activated fetal genes, reorganized cytoskeleton, and increased cardiomyocyte apoptosis. It is combined with alterations in the extracellular matrix with possible fibrosis.

Signaling mediators include mitogen-activated protein kinase, signal transducer and activator of transduction STAT3, calmodulin-dependent kinase, and Ca^{2+}-dependent PP3 phosphatase. Regulator of G-protein signaling RGS5 synthesized in the mature heart is an inhibitor of G-protein–mediated signaling that inactivates $G\alpha_i$ and $G\alpha_q$, which mediate actions of most vasoconstrictors. Overexpression of RGS5 limits cardiac maladaptive hypertrophy and fibrosis, as it inhibits ERK1 and ERK2 signaling [299]. On the other hand, in RGS5(− / −) mice, maladaptive hypertrophy resulting from pressure overload is stronger.

Calcium- and integrin-binding protein CIB1, or calmyrin, localizes primarily to the sarcolemma (including T tubules) in human cardiomyocytes.[34] It anchors phosphatase $PP3_r$ regulatory B subunit[35] to control its activation in coordination with $Ca_V1.2a$ channel. Both CIB1 amount and membrane association rise in cardiac maladaptive hypertrophy, but not in adaptive hypertrophy [300]. It elicits the activity of the PP3–NFAT axis upon pressure overload, but not exercise.

---

[34] It can be myristoylated at its N-terminus to lodge at the plasma membrane. It interacts with multiple proteins that reside near the plasma membrane, such as $\alpha_V\beta_3$-integrin, focal adhesion kinase, P21-activated kinase, sphingosine kinase-1, and Rac3 GTPase.

[35] Phosphatase PP3 is a heterodimer made of a catalytic A and regulatory B subunit.

Table 3.10 Activated vascular endothelium in inflammation. (Source: [303])

Endothelial receptor	Partners
Chemotaxis	
CCL2	CCR2, CCR4 (monocytes and T lymphocytes)
CXCL8	CXCR1, CXCR2 (neutrophils)
Adhesion	
E/P-selectin	Sialyl-Lewis X, PSGL1
Integrins	
ICAM1, VCAM1, MAdCAM1	Integrins
ESAM	
JAM	

3.2.6.2 Inflammation

Cardiac structural remodeling is associated with inflammation, a basic cellular process in innate and adaptive immunity, followed by healing. Deregulated inflammation is implicated in cardiovascular diseases such as hypertension and atherosclerosis and associated risks (obesity and diabetes).

When the antioxidant defense is overwhelmed, reactive oxygen species stimulate production of inflammatory cytokines that activate kinases and matrix metallopeptidases involved in cardiomyocyte death and disruption of the extracellular matrix, respectively.

Molecules mainly produced by cells of the innate immunity (e.g., IL18 and MMP9 and those manufactured by cells of the adaptive immunity (e.g., TNFSF1 and TNFSF5) are involved in vascular inflammation [301].

When they are activated, IL1, IL6, TNFSF1 (proinflammatory cytokines and hence upstream mediators) trigger the release of hepatic acute-phase proteins (downstream mediators and markers of inflammation), such as C-reactive protein, fibrinogen, and plasminogen activator inhibitor-1 (serpin-E1) [302].

Vascular Endotheliocytes

The endothelium is a nonadhesive and selective barrier that control the vascular permeability, tone, growth, and remodeling, among other tasks. It maintains a quiescent state with antithrombotic, anti-inflammatory, and antiproliferative properties. It receives a set of activators that promote inflammation with increased permeability, supported leukocyte adhesion and transmigration, and promoted prothrombotic state.

Vascular endotheliocytes control the initiation, amplification, and resolution of the inflammatory response to injuries caused by chemical and physical stimuli. It controls expression of chemokines and cell adhesion molecules, thereby regulating the site, extent, and duration of inflammation (Table 3.10).

Table 3.11 NFκB in vascular inflammation. (Source: [303])

Activators	Effect
Oxidized low-density lipoprotein	Gene transcription
Very-low-density lipoprotein	Synthesis of cytokines TNFSF1, IL1/8
Lysophosphatidic acid	and adhesion molecules
Angiotensin-2	E-selectin, VCAM1, ICAM1

Table 3.12 Endothelial pattern-recognition receptors (PRR) and pathogen-associated molecular patterns (PAMP) and danger-associated molecular patterns. (DAMP; Source: [303]). The NLRP3 inflammasome is activated by various PAMPs and DAMPs

PPRs	PAMPs and DAMPs
CDSs	PAMPs:
CLRs	Lipopolysaccharide, peptidoglycan, flagellin,
NLRs	Microbial ds/ssRNAs
RLRs	DAMPs:
TLRs	Heat shock proteins,
	High-mobility group box protein HMGB1,
	Hyaluronan fragments,
	ATP, uric acid, cholesterol crystals

CDS cytosolic DNA sensor, *CLR* C-type lectin receptor, *NLR* nucleotide-binding oligomerization-domain (NOD)-containing protein-like receptor, *RLR* retinoic acid inducible gene-1 (RNA helicase RGI1) protein-like receptor, *TLR* Toll-like receptor, *dsRNA* double-stranded RNA, *ssRNA* single-stranded RNA

Endotheliocyte activation is commonly classified into 2 types [303].

Type-1 activation is a rapid but transient response that loosen the intercellular junctions, thereby increasing the permeability of the paracellular route, and favor exocytosis of Weibel-Palade bodies, hence releasing stored von Willebrand factor and P-selectin.

Type-2 activation a more sustained and delayed response that involves the synthesis of proinflammatory cytokines and adhesion molecules.

Multiproteic complex defined by their structure, composition, dynamics, and function behave as hubs in cell signaling. Numerous signalosomes integrate the proinflammatory signaling from the plasmalemmal receptors to proper transcription factors (e.g., NFκB; (Table 3.11). Inflammasomes (Vol. 5, Chap. 11. Tissue Development, Repair, and Remodeling) associated with pattern-recognition receptors (e.g., TLRs; Vol. 3, Chap. 11. Receptors of the Immune System) rely on nucleotide-binding oligomerization-domain (NOD)-like receptors (NLR) to mediate innate immunity (Table 3.12).

MicroRNAs

MicroRNAs are intra- and intercellular messengers that modulate inflammation in endotheliocytes. MicroRNA-21 is highly expressed in endotheliocytes. It targets transcripts of RhoB, PTen, PPARα (nuclear receptor NR1c1), Sprouty-1 and -2, PDCD4, BCL2, Tβ R2 receptor, and MEF2c factor, among others [303].

MicroRNA-10a, which is downregulated at the atheroprone region, suppresses NFκB activation [303]. The nuclear translocation of ^{P65}NFκB can also be inhibited by miR181b that target importin-α3.

The miR221–miR222 cluster suppress the endothelial expression of ETS1, a regulator of vascular inflammation [303].

MicroRNAs suppress the transcription of genes encoding for proinflammatory messengers and their plasmalemmal receptors. MicroRNA-125a/b-5p inhibits endothelin-1 expression and miR155 targets angiotensin receptor AT−1.

MicroRNAs also control endothelial inflammation via transcripts of proteins involved in leukocyte recruitment. MicroRNA-126, miR31, and miR17-3p target VCAM1, E-selectin, and ICAM1, respectively [303].

Last, but not least, microRNAs can be released from endotheliocytes via extracellular vesicles to smooth myocytes in particular. MicroRNA-126 is enriched in apoptotic bodies released from endotheliocytes convey paracrine signals to limit atherosclerosis evolution [303]. Conversely, monocytes secrete miR15a that target MyB in endotheliocytes.

Cytokines

Myocardial expression and blood level of IL1β rise in patients with coronary artery disease, acute myocardial infarction, dilated cardiomyopathy, and congestive heart failure [304]. Tumor-necrosis factor TNFSF1 and IL1β can ensure cardiomyocyte protection.

ROS–NFκB–TGFβ Axis

Moreover, reactive oxygen species excite the fibrogenic NFκB–TGFβ axis. Nuclear factor-κB is involved in inflammation and healing after myocardial infarction. It triggers expression of proinflammatory cytokines, such as TNFSF1 and interleukins. The complement system that participates in neutrophil and monocyte recruitment early after heart injury is activated in ischemic organs.

Toll-Like Receptors

Toll-like receptors of the innate immunity initiate inflammatory response from immunocytes. In leukocytes, TLR9 recognizes DNA from bacteria and injured host cells and operates via MyD88 adaptor. In the brain and heart, inflammation can be harmful. On the other hand, in neurons and cardiomyocytes, TLR9 is involved in an alternate, noninflammatory, protective signaling axis. It lowers cardiac contractility and energy metabolism, increasing the AMP/ATP ratio and subsequently activating AMPK kinase, hence elevating tolerance for hypoxia without intervention of NFκB and MAPK module [305].

Uncoordinated protein Unc93b1 interacts with TLR3, TLR7, and TLR9, which are sensors for double-stranded RNA, single-stranded RNA, and unmethylated DNA, respectively.[36] It is involved in their transfer within the cell. This membrane protein resides in the endoplasmic reticulum. It is mainly expressed in the heart. It regulates TLR9 subcellular localization and ensures a switch between the two different TLR9 responses. It enables TLR9 translocation from the endoplasmic reticulum to endosomes, thereby allowing TLR9 interaction with MyD88 and priming inflammatory signaling.

Peptidylprolyl Isomerase-A and Basigin

Peptidylprolyl isomerase-A (PPIa; or *cyclophilin*-A) is an abundant and widespread intracellular protein that exerts various effects.[37] Once it is acetylated, PPIa can be secreted in particular by stimulated macrophages, activated platelets, hypoxic cardiomyoctes, and vascular smooth myocytes [307].

When it is released into the extracellular space, PPIa binds to its nonspecific extracellular receptor basigin (Bsg; or EMMPrIn; Table 3.13), thereby initiating inflammation. Homophilic Bsg interactions assists in MMP activation. In particular, the expression of MMP14 and MMP9 is upregulated on monocytes upon Bsg–Bsg interaction [307]. On the other hand, PPIa–Bsg interaction can regulate production of MMP9, MMP14, and CSF1 agents.

In fact, both extra- and intracellular PPIa contribute to inflammation in the cardiovascular apparatus, especially in myocardial ischemia–reperfusion injury and remodeling [307].

Extracellular PPIa is a potent chemotactic factor for leukocytes (T lymphocytes, monocytes, and neutrophilic and eosinophilic granulocytes) via 2 operating modes, extracellular binding to basigin, which is expressed on activated vascular cells and

[36] In Caenorhabditis elegans, Uncoordinated protein Unc93 is one of five interacting gene products (Unc93, Sup9 to Sup11, and Sup18) involved in the regulation or coordination of muscle contraction. Uncoordinated protein Unc22, or twitchin, possesses a protein kinase domain and several immunoglobulin-like motifs found in myosin light-chain kinase, and titin. It interacts with Unc54 myosin heavy chain. Membrane-associated muscular Uncoordinated homolog protein may be a component of ion transport implicated in excitation–contraction coupling in myocytes or of muscle contraction coordination between myocytes via gap junctions [306].

[37] Peptidylprolyl isomerase-B (PPIb; or cyclophilin-B) is also released from various cell types. It exerts proinflammatory activities. It is constitutively secreted by cells (e.g., fibroblasts and chondrocytes). Peptidylprolyl isomerase-D (PPId; or cyclophilin-D) localizes to mitochondria, where it supports the formation of the permeability transition pore.

Table 3.13 Extracellular ligands of basigin. (Source: [307])

Ligand	Cellular sources	Effects
Basigin	Monocyte, macrophage, foam cell	NFκB activation
		Expression of MMP
		Adhesion
	Smooth myocyte	MMP2 production
GP6	Platelet, monocyte	Adhesion
PPIa	T cell, neutrophil, macrophage	Migration
	Monocyte, macrophage, foam cell	Adhesion, migration
		NFκB activation
		Synthesis of ROS, MMP
E-Selectin	Endotheliocyte	Neutrophil adhesion
PPIb	T cell, neutrophil	Migration
S100a9	Melanoma cell	Migration

GP6 glycoprotein receptor for collagen, *MMP* matrix metallopeptidase, *NFκB* nuclear factor κ light chain enhancer of activated B cells, *PPIa–g* peptidylprolyl isomerase [cyclophilin]-A–G

exerts a strong chemotactic effect, and PPiase activity [307]. Signaling launched by basigin involves activation of NFκB, PI3K, ERK1, and ERK2, the latter activating NFκB (Table 3.14) [308]. The PPIa effects (chemotaxis, ERK–NFκB activation, and cytokine release) require the proinflammatory PPIa–Bsg interaction.

The PPIa enzyme is the main ROS-induced factor that enhances the inflammatory activity of monocytes and macrophages in atherosclerotic plaques. It strongly induces the migration of monocytes as well as expression of MMP9, IL6, and TNFSF1 [308]. In atherosclerotic lesions, PPIa and Bsg colocalize with infiltrated monocytes and macrophages.

The PPIa enzyme provokes vascular smooth myocyte proliferation and recruitment of inflammatory leukocytes into the vascular wall. Once it is secreted by vascular smooth myocytes, extracellular PPIa promotes ROS generation and MMP2 activity involved in aneurysm formation.

Endotheliocytes need PPIa to express the VCAM1 molecule. In addition, endothelial NOS3 production is affected by intracellular PPIa due to a reduction of Krüppel-like factor KLF2 expression.

3.2.6.3 Fibrosis

After myocardial infarction, fibrous tissue that forms at the site of cardiomyocyte loss by necrosis preserves structural integrity, but impairs myocardium behavior.

Fibrogenesis begins with activation of transforming growth factor-β1 [309]. Because of TGFβ1 elaborated by macrophages, myofibroblasts invade the infarct site

Table 3.14 Effects of PPIa on cardiovascular cell types. (Source: [307]). Intracellular PPIa is involved in cell signaling, Ca^{2+} homoeostasis, and cell activation. Extracellular PPIa acts via basigin, among other possible receptors

Cell type	Extracellular PPIa effects		Intracellular
	Mediators	Effects	PPIa effects
Monocyte	NFκB	Chemotaxis,	LDL uptake (?)
Macrophage	PI3K, ERK1/2	adhesion, ROS and MMP production	
T lymphocyte	Ca^{2+}, ERK1/2	Chemotaxis	Polarization ($T_{H1} \ll T_{H2}$)
Neutrophil	Ca^{2+}, ERK1/2	Chemotaxis, MMP production	ND
Platelet	ND	ND	Activation, degranulation
Endotheliocyte	PI3K, ERK1/2	Activation, angiogenesis	Apoptosis, VCAM1, NOS3
Smooth myocyte	JaK–STAT, ERK1/2	Migration, proliferation	Migration, proliferation
		MMP2 production	ROS production
Cardiofibroblast	ND	Migration, proliferation	Migration, proliferation
			ROS production

ND not determined, *ERK* extracellular signal-regulated kinase, *JaK* Janus kinase, *MMP* matrix metallopeptidase, *NFκB* nuclear factor κ-light-chain–enhancer of activated B cells, *NOS* nitric oxide synthase, *PI* phosphoinositide, *ROS* reactive oxygen species, *STAT* signal transducer and activator of transcription, T_H helper T lymphocyte, *VCAM* vascular cell adhesion molecule

quickly after arrival of inflammatory cells. Interstitial and adventitial fibroblasts, pericytes, and circulating fibrocytes, monocytes, and bone marrow-derived progenitor cells may give rise to these myofibroblasts.

In the infarcted heart, interstitial fibroblasts develop interstitial and perivascular fibrosis remotely from the infarcted myocardium.

Renin–Angiotensin–Aldosterone Axis

Myofibroblasts that proliferate and accumulate in the infarcted myocardium produce collagen-1 and -3. Myofibroblasts express cathepsin-D, renin, angiotensin-converting enzyme, angiotensin-1, and angiotensin receptors.

Angiotensin-2 acts in myocardial repair and remodeling via auto- and paracrine regulation on resident cells, i.e., cardiomyocytes, fibroblasts, vascular endothelial and smooth muscle cells, and macrophages. Concentration in renin, angiotensin-converting enzyme, and angiotensin receptor AT_1 markedly rises in the infarcted myocardium [309].

Renin is produced by macrophages and myofibroblasts, angiotensin-converting enzyme by endotheliocytes, macrophages, and myofibroblasts. In macrophages, angiotensin-2 initiates an autocrine regulation to stimulate NADPH oxidase that forms ROS agents. In myofibroblasts, angiotensin-2 also primes an autocrine regulation to excite TGFβ1 and elicit collagen turnover.

Cardiac production of aldosterone rises after myocardial infarction as well as in congestive heart failure. Aldosterone promotes maladaptive cardiac remodeling, particularly excessive extracellular matrix turnover.

Apelin

Apelin is a secreted cardioprotector that attenuates the development of myocardial fibrosis as well as maldaptive cardiomyocyte hypertrophy [310]. It prevents at nanomolar doses TGFβ-mediated expression of the myofibroblast marker α-smooth muscle actin and collagen production into hypersecretory myofibroblasts as well as diminishes collagen production in cardiac fibroblasts by reducing activity of sphingosine kinase SphK1, an enzyme that synthesizes sphingosine 1-phosphate. In resting and TGFβ-stimulated cardiofibroblasts, apelin inhibits SphK1 via AMPK-dependent and -independent mechanisms, respectively, even in the absence of TGF-β [310].

In both fibroblasts and myofibroblasts, TGFβ stimulates SphK1 production, thereby stimulating collagen production via $S1P_2$ receptor and RoCK kinase.

High levels of SphK1 cause progressive myocardial degeneration and fibrosis, elevated RhoA and Rac1 activity, SMAD3 phosphorylation, and oxidative stress via $S1P_3$ transactivation.[38]

3.2.7 *Pulmonary Arterial Hypertension*

Pulmonary arterial hypertension is a severe progressive disease caused by a sustained increase in the pulmonary arterial resistance that provokes right ventricular hypertrophy and then failure.

Primary pulmonary hypertension that results from alveolar hypoxia caused by chronic obstructive pulmonary diseases deteriorates the heart function, without disturbance in myocardial β-adrenoceptor signaling.

3.2.7.1 Hypoxic Pulmonary Vasoconstriction

Hypoxic pulmonary vasoconstriction (HPV) is an important reflex that locally adapts perfusion to ventilation. It is also involved in hypoxia-induced pulmonary hypertension in respiratory disorders. In particular, hypoxic vasoconstriction contributes to

[38] Receptor $S1P_3$ is a myocardial S1P receptor subtype that couples with Rho GTPases and activates SMAD signaling.

the onset of pulmonary hypertension in response to hypoxia occurring in high altitude as well as in respiratory diseases, such as chronic obstructive pulmonary disease and fibrosis.

3.2.7.2 Chronic Hypoxia-Induced Pulmonary Hypertension

Chronic hypoxia-induced pulmonary hypertension (CHIPH) is characterized by inflammation. Leukocyte adhesion to the vascular endothelium results from transcriptional activation of cell adhesion molecules upon epigenetic switch. The latter corresponds to chromatin remodeling that enables access of condensed DNA to the regulatory transcriptional machinery. Chromatin remodeling is carried out by histone modifications by histone acetyltransferases, deacetylases, methyltransferases, and kinases, as well as ATP-dependent chromatin remodeling complexes that move, eject, or restructure nucleosomes.

Two catalytic components of the chromatin remodeling complex, the transcription activator Swi/SNF-related, matrix-associated, actin-dependent regulators of chromatin SMARCa2[39] and SMARCa4[40] implicated in NFκB-dependent transcription are activated in endotheliocytes bearing hypoxia in pulmonary arteries of an animal CHIPH model [311].

Once it stimulated by hypoxia, SMARCa4 potentiates cell adhesion molecule (CAM) transactivation in vascular endotheliocytes when it complexes with $P65_{NF\kappa B}$ (RelA), as it influences histone modifications surrounding the CAM promoters [311].

3.2.7.3 Time Constant

The time constant of a pulmonary compartment of the perfusion or ventilation circuit is the product of its resistance and compliance. It expresses the filling or emptying rate at which the compartment of interest react to a modification of pressure, using linear lumped parameter models with exponential behavior (without incorporating the effects of cardiac and breathing frequencies).

The time constant of the pulmonary circulation that is the product of the pulmonary vascular resistance (PVR) and pulmonary arterial compliance (C_{pa})

$$\mathrm{RC} = \mathrm{PVR} \times Ca_{pa} \tag{3.1}$$

remains similar in various forms and severities of pulmonary hypertension, except in left heart failure [312]. The time constant decreases in proximal chronic thromboembolic pulmonary hypertension (CTEPH) with respect to idiopathic pulmonary

[39] A.k.a. SNF2a, SNF2α, SNF2L4, SNF2Lb, BAF190b, ATP-dependent helicase, and protein Brahma homolog Brm.

[40] A.k.a. SNF2b, SNF2β, SNF2L4, SNF2Lb, BAF190, ATP-dependent helicase, mitotic growth and transcription activator, and protein Brahma homolog Brg1.

arterial hypertension (IPAH) possibly due to increased wave reflection, with a wider pulse pressure and hence greater right ventricular work for a given mean pulmonary arterial pressure.

3.2.7.4 Mediators

Reactive Oxygen Species

Reactive oxygen species (ROS) are mainly generated in cardiomyocytes by the mitochondrial electron transport chain, NADPH oxidase (NOx), and nitric oxide synthases. In cardiomyocytes, the 2 most expressed NOx isoforms are NOx2 and NOx4 [313]. The intracellular location of these 2 isoforms in cardiomyocytes differs. Whereas NOx2 lodges predominantly on the plasma membrane, NOx4 resides in organelles. The NOx4 isoform can be detected in the endoplasmic reticulum in vascular smooth muscle and endothelial cells as well as in the plasma membrane, especially in focal adhesions, and nucleus in various cell types. Mitochondrial NOx4 is a major source of oxidative stress.

In normal conditions, NOx2 is quiescent. It is activated by GPCR ligands (e.g., angiotensin-2 and endothelin-1), growth factors, and cytokines [313]. It produces superoxide ($O_2^{\bullet -}$), which is rapidly processed to hydrogen peroxide (H_2O_2) by superoxide dismutase.

On the other hand, NOx4 is not only constitutively active but also inducible as NOx4 expression rises during hypoxia, mitochondrial dysfunction, and in left ventricle pressure overload [313]. This isoform can be regulated by transcriptional and translational changes as well as posttranslational modifications. It synthesizes predominantly hydrogen peroxide.

In cardiomyocytes, NADPH oxidase is involved in the progression of cardiac hypertrophy. The NOx4 subtype increases the ROS production in mouse right ventricle pressure overload during the acute phase after pulmonary artery banding. It then stimulates mitochondrial NOx2 during the chronic hypertension phase (feedforward chain of Nox activation) [313]. The catalase activity augments at the transition between acute and chronic phase (i.e., 6 h after banding), but the expression of catalase and other ROS scavenging enzymes superoxide dismutase and glutathione peroxidase does not change [313].

Phospholamban

The diastolic dysfunction is observed in both right and left ventricles. Left ventricle diastolic dysfunction is not induced by increased afterload, but by altered activity of cardiac regulators such as phospholamban.[41]

[41] Cardiomyocyte relaxation is controlled by sarcoplasmic reticulum Ca^{2+} ATPase that pumps calcium ions from the cytosol into the sarcoplasmic reticulum. Activity of SERCA pump is regulated

Protein Ser/Thr phosphatases involved in heart functioning include:

1. catalytic PP1α subunit that has an increased expression in human heart failure;
2. catalytic PP2 subunits, the overexpression and mutations of PP2 structural regulatory A subunit being associated with cardiac hypertrophy; and
3. PP3 that causes maladaptive cardiac hypertrophy.

In the heart, PP2 dimerizes or trimerizes.

Pulmonary hypoxia can actually slow cardiac relaxation by reducing phospholamban phosphorylation. The activity of protein phosphatase-2 (but not PP1) is enhanced in sarcoplasmic reticulum membranes [314]. Hypoxia as well as heart failure lead to increased expression of PP2 catalytic subunit ($PP2_c$). Increased PP2 activity and subsequent reduced phospholamban phosphorylation can be elicited by an elevated level in circulating interleukin-18.[42] Interleukin-18 exerts its effect via protein kinase-B. In addition, protein phosphatase PP3 level rises in the right ventricle exposed to increased afterload, but not in the left ventricle, whereas PP2 activity augments in both ventricles.

Cardiac regulators, such as phospholamban and inhibitory subunit of troponin, undergo phosphorylation–dephosphorylation cycles. The impaired phosphorylation state of these proteins disturbs cardiac activity, especially excitation–contraction coupling.

Vasodilators and Vasoconstrictors

Endothelial dysfunction is an early event in pulmonary hypertension (Vol. 8, Chap. 3. Endothelial Dysfunction). Levels of vasodilators (e.g., nitric oxide and prostacyclin) lower and those of vasoconstrictors (e.g., endothelin-1) rise.

Phosphorylation of NOS3 is a posttranslational modification that ensures optimal production of nitric oxide. Activin receptor-like kinase ALK1, an endothelial-specific receptor and member of the TGF superfamily, localizes to caveolae with NOS3 enzyme. Increased NOS3 phosphorylation (Ser1177) and uncoupling elicits the production of superoxide anion, a mediator of pulmonary hypertension [315].

Endocannabinoids

In the systemic circulation, endocannabinoids cause vasodilation via primarily their specific cannabinoid CB_1 and CB_2 and other G-protein–coupled receptors. Endocannabinoids act via their intracellular metabolites engendered the fatty acid amide

---

by phospholamban. Reduced phosphorylation of phospholamban (Ser16) can also be generated via β-adrenoceptor signaling associated with the ACase–PKA cascade.

[42] Patients suffering from severe chronic obstructive pulmonary diseases have increased expression of IL18 in pulmonary alveolar macrophages. Interleukin-18 is also produced by airway and alveolar epitheliocytes.

hydrolase (FAAH) or monoacylglycerol lipase (MAGL). However, their vasoactive metabolites do not have an important role in the regulation of vascular tone in systemic vessels [316].

The endocannabinoid anandamide (^{N}arachidonoylethanolamine [AEA]) produced by pulmonary arterial smooth myocytes (but not pulmonary endotheliocytes) can control the pulmonary arterial tone. Sensors, transducers, and effectors of hypoxic pulmonary vasoconstriction reside in pulmonary vascular smooth myocytes. Anandamide is synthesized by different enzymes, the most important route being hydrolysis of phospholipid-derived ^{N}acylphosphatidylethanolamine (NAPE) by ^{NAPE}PLD.[43] The latter has an elevated expression in hypoxic pulmonary arterial smooth myocytes. Other anandamide-synthesizing enzymes include α/β-hydrolase encoded by the ABH4 gene, glycerophosphodiesterse GDE1, and cytosolic protein Tyr phosphatase PTPn22.

The fatty acid amide hydrolase that is strongly synthesized in the lung is the principal AEA-degrading enzyme, thereby limiting the AEA signaling from cannabinoid receptors [316]. Anandamide catabolism by FAAH gives rise to arachidonic acid and ethanolamine. On the other hand, the endocannabinoid 2-arachidonyl glycerol is mainly processed by MAGL that is weakly produced in the lung.

Arachidonic acid is the precursor of eicosanoids generated by COx-, LOx-, or CyP450-mediated processing. Anandamide-elicited pulmonary vasoconstriction results from the action of COx and LOx enzymes. Hence, FAAH hydrolase is an important mediator of hypoxic pulmonary vasoconstriction via the engendered metabolites [316].

Reactive oxygen species are also involved in the modulation of pulmonary vascular tone by hypoxia [316]. Anandamide leads to ROS formation.

---

[43] The ^{NAPE}PLD is a phospholipase-D-type enzyme that catalyzes the release of ^{N}acylethanolamine (NAE) from ^{N}acyl phosphatidyl ethanolamine (NAPE). Anandamide belongs to the category of long-chain ^{N}acylethanolamines. The endocannabinoid-synthesizing enzymes comprise ^{N}acyl phosphatidyl ethanolamine-hydrolysing phospholipase-D and diacylglycerol lipase. The former, a constitutively active member of the metallo β-lactamase category, specifically hydrolyzes NAPE among glycerophospholipids [317]. The endocannabinoid-degrading enzymes encompass fatty acid amidohydrolase and monoacylglycerol lipase. Anandamide is principally formed together with other NAEs from glycerophospholipid by 2 successive enzymatic reactions: (1) ^{N}acylation of phosphatidylethanolamine that generates ^{N}acylphosphatidylethanolamine (NAPE) by Ca^{2+}-dependent ^{N}acyltransferase; (2) release of NAE from NAPE by a phosphodiesterase of the phospholipase-D-type enzyme (^{NAPE}PLD).

Chapter 4
Cardiomyopathies

Cardiomyopathies are associated with cardiac dysfunction. Diverse types of cardiomyopathies include dilated, hypertrophic, restrictive, arrhythmogenic right ventricular, and unclassified cardiomyopathies [210].

Cardiac hypertrophy (Sect. 3.2) is defined by an increase in cardiomyocyte size. The quantity of sarcomeres rises to normalize the mural stress field resulting from pressure overload. However, adverse sustained cardiac hypertrophy causes dilated cardiomyopathy.

However, a given form of cardiomyopathy can progress to another. For example, hypertensive heart disease that launches cardiac hypertrophy can subsequently become a dilated cardiomyopathy. In addition, some diseases have features of more than one type of cardiomyopathy. In particular, sarcoidosis can cause restrictive and dilated cardiomyopathies at different periods in its evolution.

The cardiac functioning relies on a set of cells. Cardiomyocytes are embedded in a three-dimensional network of endotheliocytes, vascular smooth myocytes, and fibroblasts, in addition to transient populations of immunocytes. Cardiomyocytes are connected between them to coordinate their contraction as well as to the extracellular matrix to coordinate the overall myocardial contraction.

The heart can tolerate intrinsic (genetic) and extrinsic factors to maintain its contractility. However, its compensatory response mediated by signaling pathways is limited. The persistent activation of these pathways causes cardiac dysfunction that leads to cardiomyopathy.

The cardiac responses include induction of genes normally expressed during development, myocyte hypertrophy, fibrotic deposits that replace dead cardiomyocytes, and metabolic disturbances [332].

Cardiomyopathies are defined by ventricular dysfunction. Cardiomyopathies evolve with risk of arrhythmia and sudden cardiac death. *Noncompaction cardiomyopathy* refers to abnormal growth of the left ventricle. Cardiomyopathy frequently leads to heart failure.

In fact, cardiomyopathies represent a set of conditions that all finish with myocardial dysfunction. Histologic examination most often exhibits nonspecific lesions with cardiomyocyte hypertrophy, cellular necrosis, and fibrosis.

M. Thiriet, *Diseases of the Cardiac Pump*, Biomathematical and Biomechanical Modeling of the Circulatory and Ventilatory Systems 7, DOI 10.1007/978-3-319-12664-7_4

In cardiomyopathies and heart failure, cardiomyocte death is caused by BCL2 19-kDa protein-interacting protein 3-like protein (BNIP3L), which stimulates two autonomous death pathways according to its subcellular localization [319]. Mitochondrial BNIP3L activates BAX–BAK and caspase-dependent programmed apoptosis. Endoplasmic reticulum BNIP3L activates programmed necrosis via the mitochondrial permeability transition pore (MPTP), independently of the proapoptotic BCL2 family members BAX and BAK proteins.

The ubiquitin–proteasome system (UPS) is implicated in protein quality control. Dilated and hypertrophic cardiomyopathies are characterized by proteasomal dysfunction [320].

4.1 Genetic Background and Cardiomyopathies

Muscular dystrophies primarily affect striated myocytes. They are characterized by the progressive loss of muscle strength and integrity. Major forms of muscular dystrophies are caused by abnormalities of the dystrophin-associated glycoprotein complex that serves as a structural unit and scaffold for signaling molecules at the sarcolemma.

Mutations in a given gene can cause different types of cardiomyopathies. In addition, the clinical picture can be modified by diet, sex hormones, and polymorphisms in many genes such as those involved in angiotensin signaling [321].

Reduced levels of junctophilin-2 are observed in the myocardium taken from septal surgical resection in patients with inherited hypertrophic and dilated cardiomyopathies associated with disruption of T-tubule architecture [9]. The pathological cardiac remodeling is characterized by a decrease in the total number of junctional membrane complexes between the transverse tubules and sarcoplasmic reticulum as well as a reduction of transverse tubules, whereas longitudinal structures expand. The remaining junctional membrane complexes have an increased variable spacing between the juxtaposed membranes of the transverse tubules and sarcoplasmic reticulum. The JPH2 transcripts are targeted by microRNA-24, which is upregulated in cardiomyocytes from patients with dilated and ischemic cardiomyopathies. The subsequent Jph2 reduction causes mislocalized junctional membrane complexes.

4.1.1 Summary of Structures of Sarcomere and Costamere

4.1.1.1 Sarcomere

The sarcomere[1] forms the basic contractile unit of the striated myocyte, in particular the cardiomyocyte (Vol. 5, Chap. 5. Cardiomyocytes). This functional unit of contraction is an array of interdigitated actin and myosin filaments and their associated

[1] σαρξ: flesh; μερ ος: share, portion.

Table 4.1 Sarcomeric constituents

Component	Constituents
Myofilaments	Actin, tropomyosin, troponin-C/I/T
	Myosin
Hemisarcomere	Titin, nebulin,
	MyBPc3
Z disc	α-actinin,
	Tcap (telethonin), CRSP3 (MLP)
M line, H zone	Myomesin

CRSP3 (MLP) cardiac cysteine and glycine-rich protein-3 (or muscle LIM domain-containing protein), *MyBPc* myosin-binding protein-C, *Tcap* titin cap

proteins. Thin actin filaments are anchored at the Z line. They form transient sliding interactions with thick myosin filaments.

The structural features of the sarcomere in polarized light are the A (anisotropic) band with at its center the M (middle) line at the middle of the H zone[2] and I (isotropic) band. These features are defined by their composition, that is, the entire length of myosin filament, myosin attachment on crossconnecting cytoskeletal proteins, myosin without actin, and actin without myosin, respectively (Table 4.1). Titin connects the Z line to the M line.

Each thick filament in striated myocytes is constituted by two myosin heavy chains (MyHC) and four myosin light chains, more precisely, two pairs of essential (eMyLC or MyLC1) and regulatory (rMyLC or MyLC2) myosin light chains connected to the two heads of a given myosin filament (Vol. 5, Chap. 5. Cardiomyocytes).

The myosin heavy chain contains the actin- and ATP-binding sites. It thus represents the molecular nanomotor of muscle contraction. In the human heart, two MYH genes (MYH6 [αMyHC] and MYH7 [βMyHC]) are expressed. α-myosin heavy chain has a higher ATPase activity and shortening velocity than β-myosin heavy chain. The nanomotor is regulated by two myosin light chain (MyLC) isoforms. The MyLC1 subtype acts as a myosin heavy chain–actin tether. The MyLC2 isoform slows the rate of tension development of myosin [322]. The inhibition conferred by MyLC2 is relieved upon MyLC2 phosphorylation by myosin light chain kinase MLCK2. Expression of the atrial MyLC1 isoform in the hypertrophied human ventricle increases cross-bridge cycling and contractility [322].

In smooth myocytes and nonmuscle cells, phosphorylation of regulatory myosin light chains by MLCK switches on the actin-activated myosin ATPase, hence triggering contraction. In striated myocytes (i.e., skeletal and cardiac myocytes), the actin–myosin interaction is mainly regulated by the troponin–tropomyosin complex. Electrochemical stimulation increases the level of Ca^{2+} bound to the troponin–tropomyosin complex. Regulatory myosin light chains modulate the rate and

[2] German *hell*: bright.

magnitude of contractility, as MyLC2 phosphorylation increases moderately tension and Ca^{2+} sensitivity at submaximal levels of Ca^{2+} activation.

Cardiac wall torsion and compression during systole enable a more efficient blood ejection from the ventricle. Torsion depends partly on the helical orientation of the cardiomyocytes. In the epicardial layer of the ventricular myocardium, the myofibers spiral in a left-handed helix from the apex to the base of the heart (the thumb denoting the upward spiral; Vol. 5, Chap. 5. Cardiomyocytes). In the endocardial layer of the ventricular myocardium, the myofibers have a right-handed helical orientation.

A spatial gradient of myosin light chain phosphorylation in the myocardium from high epicardial to low endocardial levels facilitates torsion by changing tension production and stretch activation response [323]. This spatial gradient of MyLC2 phosphorylation increases tension and decreases the stretch activation response of epicardial myofibers and conversely in the endocardial layer. Cardiac hypertrophy can be linked to a gain-in-function mutation.

The cardiac isoform of myosin-binding protein-C (cMyBPc) is an accessory protein of sarcomeric thick myosin filaments and a modulator of cardiomyocyte contraction in response to cardiac stimulation. Mutations in the MYBPC gene generate cardiac disease. The three isoforms of human MyBPc (fast and slow skeletal and cardiac) are encoded by three genes (MYBPC1–MYBPC3).

The cMyBPc molecule localizes to the central region of the A band, the C zone (Fig. 4.1). It is contained in 7–9 bands of 43-nm spacing among the 11 structurally regular transverse C-zone stripes [324]. These bands comprise three layers of myosin head crowns. Layer 1 is the very dense MyBPc layer, in which the C-termini of three MyBPc molecules wrap around the myosin filament, and layers 2 and 3 are simple myosin head crowns.

It is involved in sarcomere assembly, as it promotes polymerization of thick filaments, the C-termini binding to specific sites on titin and light meromyosin [324].

The extent of cMyBPc phosphorylation that occurs in response to adrenergic stimulation is correlated with increased cardiac contractility. Serine phosphorylation in M-domain cMyBPc by various kinases reduces the affinity of the N terminus for actin and myosin [325]. cMyBPc affects Ca^{2+} sensitivity similarly at long and short sarcomere lengths, as it reduces the probability of myosin binding to actin [326]. It also lowers the maximum shortening velocity of myofibers in the low-velocity phase (without influencing the high-velocity phase), as it contributes to an internal load that slows the maximum shortening velocity at low levels of activation once a given amount of active shortening has occurred [327]. Slowing of actin filament sliding is a mechanism used by cMyBPc to modulate cardiac contractility.

The cMyBPc protein binds to myosin via its C-terminus and actin via its N-terminus, thereby modulating thin actin filament motility. It connects to actin close to the low Ca^{2+}-binding site of tropomyosin, hence interfering with tropomyosin regulatory movements on actin. It indeed increases actin filament calcium sensitivity by binding to actin and displaces tropomyosin from its inhibitory position toward its high Ca^{2+} position to activate actin filament interaction with myosin, promoting filament sliding [325]. At high Ca^{2+} levels, cMyBPc has little effect on tropomyosin position and slows down actin filament sliding.

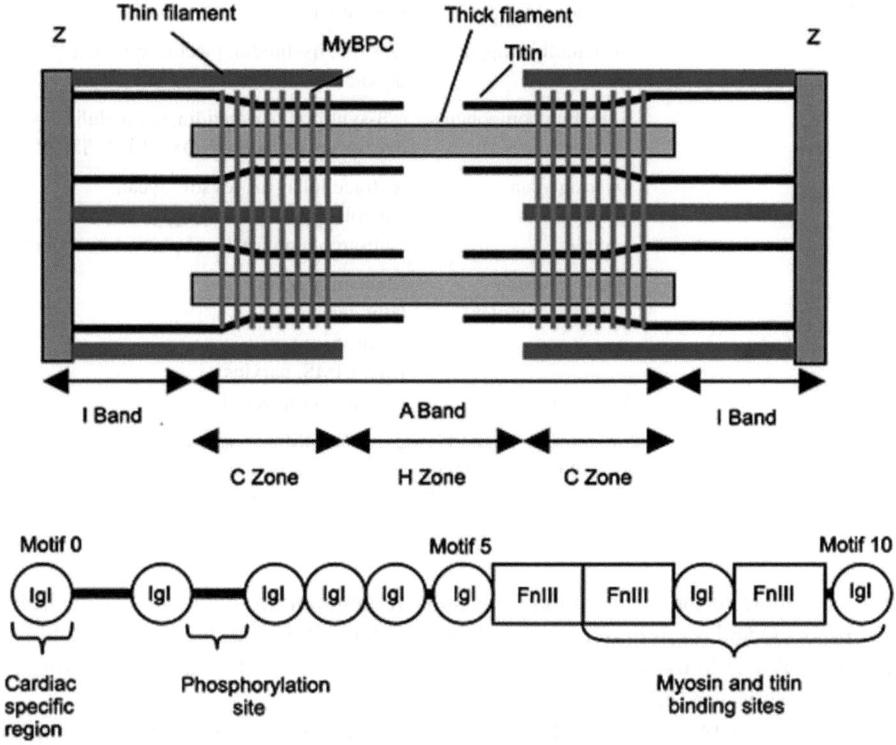

Fig. 4.1 Sarcomere striated configuration with lines, bands, and zones (Source: [324]). Position of MyBPc in the stretched sarcomere (*top*). MyBPc lodges in transverse stripes 43-nm apart in the C zone of the sarcomeric A band, where the thick myosin and thin actin filaments interact. Structure of the cardiac MyBPc isoform (*bottom*). The skeletal isoform of myosin-binding protein-C contains seven immunoglobulin-like and three fibronectin-3-like domains (C1–C10), a MyBPc-specific motif (M domain) between C1 and C2, and a Pro–Ala-rich sequence at the N terminus. The cardiac isoform has an additional N-terminal Ig-like domain (C0), 4 phosphorylation sites in the M domain (between motifs 1 and 2), and a 28-residue insert in the C5 domain. The cMyBPc protein tethers to the myosin filament in the C zone via its C-terminus (C8–C10 domains), whereas its N-terminus possesses binding sites for the myosin S2 motif and the myosin regulatory light chain. Motifs 7 to 10 link to myosin and titin

The very long elastic molecule titin (or connectin) positions the myosin filaments at the center of a sarcomere, as it links them to the Z line. Their N- and C-termini localize to the Z line and M line, respectively. The middle portion tethers to the myosin filament.

4.1.1.2 Costamere

The costamere[3] is a structural and functional component of striated myocytes. This proteic complex consists of cytoskeletal proteins and signaling kinases. It couples

[3] Latin *costa*: side, flank.

Table 4.2 Costamere constituents

Components	Constituents
Structural components	Desmin, dysbindin, plectin, syncolin, β-synemin
Signaling components	α/β-syntrophin, α-catulin, calmodulin, myospryn, caveolin, NOS1, PKA, TRPC
Dystroglycan glycoprotein complex	(extracellular) α/β-dystroglycan, (sarcolemmal) α–δ-sarcoglycan, sarcospan, (subsarcolemmal) dystrophin, α-dystrobrevin
Focal adhesion (integrin complex)	Melusin, paxilin, talin, tensin, vinculin, zyxin, ILK, LIMS, parvins
Matrix partners	Laminin, collagen-4

ILK integrin-linked kinase, *LIMS* LIM and senescent cell antigen-like domain-containing proteins (Particularly interesting new Cys–His protein (PINCH)), *NOS* nitric oxide synthase, *PKA* protein kinase-A, *TRPC* canonical transient receptor potential channel

the sarcomere to the sarcolemma. This subsarcolemmal proteic assembly encircles the myocyte and is circumferentially aligned with the Z disc, as it anchors myofibrils from the Z disc to the sarcolemma and extracellular matrix. Costameres are sites of force transmission [328].

The costamere is composed of the *dystrophin-associated glycoprotein complex* (DGC) and the *integrin complex* (Table 4.2). The dystrophin-associated glycoprotein complex is a structural unit and scaffold for signaling molecules at the sarcolemma. It is composed of dystroglycans and sarcoglycans as well as focal adhesion constituents. It maintains the structural integrity of myofibers by linking the extracellular matrix to the subsarcolemmal cytoskeleton. The costamere coordinates force transduction and intracellular signaling.

The *dystrobrevin* family comprises α- and β-dystrobrevin that are encoded by the DTNA and DTNB genes, respectively. The former is expressed predominantly in muscle and brain and the latter in nonmyocyte cells.

α-dystrobrevin is a component of the DGC that directly tethers to dystrophin. In striated myocytes, dystrobrevin and dystrophin localize to the cytoplasmic face of the sarcolemma. α-dystrobrevin also binds to intermediate filaments as well as syntrophin, a modular adaptor protein.[4]

α-syntrophin is the major isoform in skeletal and cardiac myocytes [329]. Syntrophin can coordinate the assembly of nitric oxide synthase NOS1, P38MAPKγ, transient receptor potential channels (TRPC), which modulates cation (calcium) entry, and calmodulin to the DGC.

[4] Skeletal myocytes contain several syntrophin isoforms (α, β1–β2, and γ1–γ2) encoded by the SNTA1, SNTB1, SNTB2, SNTG1, and SNTG2 genes.

α-dystrobrevin experiences alternative splicing that impacts its subcellular distribution and function in myocytes. Among the three major α-dystrobrevin isoforms, α-dystrobrevin-1 and α-dystrobrevin-2 localize to the sarcolemma. The former colocalizes with both dystrophin and utrophin and the latter only with dystrophin. α-dystrobrevin-1 to α-dystrobrevin-3 bind to the sarcoglycan complex [329]. α-dystrobrevin-1 and α-dystrobrevin-2 tether to dystrophin and syntrophin (but not α-dystrobrevin-3).

Additional α-dystrobrevin-binding partners include the constituents of the mechanical stabilizer intermediate filament syncoilin and β-synemin as well as dysbindin [329]. In striated myocytes, intermediate filaments encircle the Z disc, thereby connecting all adjacent myofibrils and linking the Z disc of the peripheral layer of cellular myofibrils to the sarcolemma. *Syncoilin* also binds to desmin, a muscle-specific intermediate filament protein. The α-dystrobrevin–syncoilin interaction provides another linkage between the DGC and cytoskeleton, which may be important for force transduction during contraction. β-*synemin* interacts with plectin, a linker protein of intermediate filaments to the Z disc. *Dysbindin* (or dystrobrevin-binding protein-1) is a ubiquitous sarcolemmal protein, the expression of which is relatively low in muscle [329]. It binds to α-dystrobrevin and myospryn.[5] *Myospryn* may function as a docking platform for structural and signaling molecules, such as α-actinin-2 and protein kinase-A [329].

α-*catulin* is a ubiquitous binding partner of α-dystrobrevin-1 that localizes to nerve bundles and blood vessels. It may regulate α1d-adrenergic receptor signaling.

α-actinin binds to: (1) actin and the CRSP3–Tcap–titin complex[6] to organize the local cell architecture and (2) to talin and vinculin to connect the sarcomeres to costameres and, hence, to integrins and the extracellular matrix constituents (e.g., collagen and laminin). The mechanosensor and mechanotransducer vinculin is also connected to intercellular junctions, not only integrin-based adhesion complexes, but also tight, adherens, and gap junctions.

The cytoplasmic domain of β-integrin binds integrin-linked kinase (ILK) and melusin. The former has structural and functional roles at the costamere, as it recruits various adaptors (e.g., LIM and senescent cell antigen-like domain-containing proteins [LIMS],[7] parvins [α-parvin, or actopaxin, and β-parvin, or affixin], and paxillin), cytoskeletal, and signaling molecules. In particular, it links to various kinases and phosphatases (e.g., protein Ser/Thr ILK-associated phosphatase [ILKAP], an inhibitor of ILK of the PPM1 category, 3-phosphoinositide-dependent [PI3K-dependent] protein kinase PDK1, which phosphorylates PKB, and PKB) [330]. Therefore, ILK is connected to the actin cytoskeleton, focal adhesion complex, and growth factor receptors.

---

[5] Also known as cardiomyopathy-associated protein CMyA5, dystrobrevin-binding protein DtnBP2, genethonin-3, and tripartite motif-containing protein TRIM76.

[6] CRSP3 refers to as cardiac cysteine and glycine-rich protein-3, also called muscle LIM domain-containing protein (MLP). Tcap stands for titin-cap protein or telethonin.

[7] Also known as Particularly interesting new Cys–His-enriched protein (PINCH).

Morerover, it phosphorylates protein kinase-B and other signaling effectors to regulate the muscular response to stretch [330].

In addition to their structural role, integrins and their associated adaptor proteins recruit to the plasma membrane signaling molecules used for the survival and growth of cardiomyocytes (e.g., focal adhesion kinase, ILK, P21-activated kinase, proline-rich tyrosine kinase PYK2, PKC, and Src) [330].

Furthermore, the focal adhesion complex can be linked to growth factor receptors, thereby enabling crosstalk between mechanotransduction and growth factor-mediated hypertrophic signals.

Costameric and sarcomeric proteins, particularly those of the Z disc, are components of the cardiac stretch sensor.

The mechanical stretch response can be impaired, whereas hypertrophic signaling triggered by humoral factors such as endothelin-1 can remain intact.

4.1.1.3 Mechanosensory Cadherin–Catenin-Mediated Adhesions

The N-cadherin complex links adjacent cardiomyocytes via adherens junctions (with α-, β-, and δ1-catenins and related partners), as do mechanosensory integrins vie focal adhesion complexes (with numerous cytoskeletal adaptors [e.g., filamin, paxillin, talin, and vinculin] and signaling effectors [e.g., focal adhesion kinase and small RhoA GTPase]), is sensitive to force transmission. Forces acting on N-cadherin–mediated adhesions elicit structural and functional changes in cardiomyocytes.

α-catenin is a major component of the mechanosensory cadherin–catenin complex that connects to the cell actin cytoskeleton. It homodimerizes or heterodimerizes with β-catenin. It localizes to regions of highest mechanical stress on cardiomyocytes [331]. Localization of α-catenin depends on N-cadherin topography, and hence on the degree of internal stress at the intercellular junction, that is, on changes in contractility. The α-catenin–cadherin complex indeed serves as a mechanosensory regulator of the cardiomyocyte cytoskeletal structure.

Sarcomeric remodeling depends on whether adhesion and force transmission is mediated by cadherins or integrins. Inhibition of myosin bound by integrins to fibronectin disrupts myofibril organization. On the other hand, inhibition of myosin bound by N-cadherins disturbs to a lower extent myofibril arrangement. These two mechanosensory cell adhesion types provide independent mechanisms for regulating sarcomeric organization. Lower myosin activity is preferential for cadherin-dependent assembly and preservation of sarcomeres [331]. Cadherin- and integrin-based junctions can sustain stresses of similar magnitudes, but subsequent cytoskeletal remodeling differs due to the specific function of proteic components and their interaction in assembling and orienting actin bundles in response to mechanical stresses.

Table 4.3 Hypertrophic cardiomyopathies, genetic causes, and cellular pathology (Source: [332]). The MYH6 and MYH7 genes encode cardiac α-myosin heavy chain-6 and cardiac β-myosin heavy chain-7, respectively. Gene mutations can be common to multiple forms of cardiomyopathies

Mutated gene	Pathology
Cardiac α-actin (ACTC1 gene)	Interstitial fibrosis
α-myosin heavy chain (MYH6 gene)	Cardiomyocyte hypertrophy
β-myosin heavy chain (MYH7 gene)	LV wall thickening
Myosin light chains	Inflammation
Myosin-binding protein-C	Increased cardiac mass
α-tropomyosin (TPM1 gene)	Myofibrillar disarrangement
Troponin-C	
Troponin-I	Enlarged nuclei
Troponin-T	↓ LV volume
MLP (muscle LIM protein)	
Telothonin	
Titin	Ventricular wall stiffness
Vinculin	

LV left ventricle, ↓ decrease

4.1.2 Gene Mutations

More than 900 mutations in genes expressed in the cardiomyocyte can cause cardiomyopathies [332]. More than 400 mutations in 13 sarcomeric proteins are linked to cardiomyopathies (Table 4.3). Mutations in genes encoding sarcomeric constituents (e.g., cardiac α-actin, myosin light chain, cardiac β-myosin heavy chain, cardiac myosin-binding protein-C, cardiac troponin-I and -T, titin, and tropomyosin) are usually inherited in an autosomal-dominant manner and are missense mutations. Mutations in both the essential and regulatory myosin light chains cause a rare form of midventricular hypertrophic cardiomyopathy.

Mutations of several sarcomeric components are associated with dilated cardiomyopathy and skeletal muscle myopathy. Major forms of muscular dystrophies are caused by abnormalities of the dystrophin-associated glycoprotein complex. Mutations in the dystrophin gene generate Duchenne and Becker muscular dystrophy. Duchenne muscular dystrophy in skeletal muscles and heart is characterized by the absence or slight production of functional dystrophin (Table 4.4). In Becker muscular dystrophy, a partly functional dystrophin is synthesized.

Mutations in the sarcoglycan genes cause several types of sarcoglycan-deficient limb-girdle muscular dystrophy (SDLGMD).

Table 4.4 Muscular dystrophy-associated cardiomyopathy, genetic causes, and cellular pathology (Source: [332]). Gene mutations can be common to multiple forms of cardiomyopathies

Mutated gene	Pathology
Dystrobrevin	LV noncompaction
Dystroglycan	Apoptosis
Dystrophin	LV hypertrophy or dilation
Sarcoglycan	Fibrosis

LV left ventricle

Mutations in several Z-disc proteins of the cardiomyocyte sarcomere, which lead to disruption and dysfunction of the contractile apparatus, cause cardiomyopathies [333].

Ischemic and nonischemic cardiomyopathies share expression of some genes compared with controls, but numerous genes are differently expressed with respect to cardiomyopathy types and controls [334]. Among extracellular stimuli (ions, hormones, and mechanical stress), pathological myocardial remodeling is mainly induced by neurohormonal factors (angiotensin-2, endothelin-1, and catecholamines) via the Gq–PLCβ axis that activates the IP_3–Ca^{2+}–PP3–NFAT and DAG–PKC pathways.

On the other hand, adaptive cardiac (exercise-induced) hypertrophy that is particularly induced by growth hormone and insulin-like growth factor is associated with the PI3K–PKB–GSK3α/β pathway. Phosphoinositide 3-kinase-α downstream from the IGF1 receptor mediates adaptive cardiac growth (normal postnatal growth as well as growth in response to chronic exercise training) and protects the heart against pressure overload, dilated cardiomyopathy, and myocardial infarction.

4.2 Causes of Cardiomyopathies

Numerous inherited and acquired cardiomyopathies has been described, in addition to unclassified cardiomyopathies. They are categorized mainly into dilated, hypertrophic, and restrictive cardiomyopathies and arrhythmogenic right ventricular cardiomyopathy.

Many causes of cardiomyopathies exist (Table 4.5). Acquired cardiomyopathies can result from stress, diabetes, some chemotherapeutic agents, pregnancy, and alcohol intake. The most common cause is ischemic cardiomyopathy.

Cardiomyopathies are generally categorized into two groups: extrinsic and intrinsic cardiomyopathies.

1. *Extrinsic cardiomyopathies* do not have myocardial origin, as they are associated with hypertension, ischemia (coronary heart disease), valvular dysfunction, metabolic and nutritional diseases, congenital heart disease, or diabetes.
2. *Intrinsic cardiomyopathies* result from myocardial weakness without identifiable external cause. Intrinsic cardiomyopathy has many causes, such as drug and alcohol toxicity, myocarditis, infections, and genetic factors.

Table 4.5 Causes of cardiomyopathies

Cause	Disorder
Cardiovascular	Ischemic heart disease, hypertension, valvular heart disease, cardiac amyloidosis, tachycardia-induced cardiomyopathy, idiopathic dilated cardiomyopathy, idiopathic restrictive cardiomyopathy
Metabolic	Starvation, vitamin deficiency, glycogen storage disease, diabetes, hypo- and hyperthyroidism, acromegaly, pheochromocytoma
Inflammatory, infectious	Secondary amyloidosis, sarcoidosis, Coxsackie-B viral infection, hepatitis C, human immunodeficiency virus infection, American trypanosomiasis (Chagas disease)
Toxic	Alcohol, cocaine, amphetamines, chemotherapy
Genetic	Familial dilated cardiomyopathy, hemochromatosis, amyloidosis, noncompacted myocardium, systolic dysfunction without dilation, arrhythmogenic right ventricular cardiomyopathy
Pregnancy	Peripartum cardiomyopathy

4.2.1 Specific Cardiomyopathies

Specific cardiomyopathies are referred to as myocardial diseases associated with specific cardiac or systemic disorders.

General system diseases, such as connective tissue disorders (e.g., systemic lupus erythematosus, polyarteritis nodosa, rheumatoid arthritis, scleroderma, and dermatomyositis). They also include infiltrations and granulomas, such as sarcoidosis and leukemia.

Hypertensive cardiomyopathy is a left ventricular hypertrophy associated with dilated or restrictive cardiomyopathy with features of cardiac failure.

Inflammatory cardiomyopathy is defined by a myocarditis and cardiac dysfunction. It has idiopathic, autoimmune, and infectious (*Trypanosoma cruzi*, human immunodeficiency virus, enterovirus, adenovirus, and cytomegalovirus) forms.

Ischemic cardiomyopathy is a dilated cardiomyopathy with impaired contractile performance not explained by the extent of coronary artery disease, that is, ischemic damage.

Metabolic cardiomyopathy includes various categories:

- *Endocrine* in the context of thyrotoxicosis, hypothyroidism, adrenal cortical insufficiency, pheochromocytoma, acromegaly, and diabetes mellitus;

- *Familial storage disease and infiltrations*, such as hemochromatosis, glycogen storage disease, Hurler's and Refsum's syndromes, Niemann–Pick, Hand–Schüller–Christian, Fabry–Anderson, and Morquio–Ullrich diseases;
- *Deficiency* and *nutritional disorders*, such as disturbances of potassium metabolism, magnesium deficiency, kwashiorkor, anemia, beri-beri, and selenium deficiency;
- *Cardiac amyloidoses* that can be primary, secondary, familial, and hereditary, including familial Mediterranean fever and senile amyloidosis.

Muscular dystrophies that comprise Duchenne, Becker-type, and myotonic dystrophies.

Neuromuscular disorders that encompass Friedreich's ataxia, Noonan's syndrome, and lentiginosis.

Peripartal cardiomyopathy that can first manifest in the peripartum period.

Sensitivity and toxic reactions to alcohol, catecholamines, anthracyclines, irradiation, among others.

Valvular cardiomyopathy is related to a ventricular dysfunction that is out of proportion to the abnormal loading conditions.

4.3 Classification of Cardiomyopathies

Intrinsic cardiomyopathies are usually classified into four main categories (Table 4.6;[210]):

1. *Dilated cardiomyopathy* with left ventricle enlargment and reduced pumping function;
2. *Hypertrophic cardiomyopathy* due to mutations in genes encoding sarcomeric proteins;
3. *Restrictive cardiomyopathy* with stiff ventricular walls; and
4. *Arrhythmogenic right ventricular cardiomyopathy* with fibrous tissue.

4.4 Mitochondria in Cardiomyopathies

Like in normal hearts, two populations of mitochondria exist in cardiomyopathies at least in hamsters [335]: (1) subsarcolemmal mitochondria beneath the sarcolemma and (2) interfibrillar mitochondria among the myofibrils.

Mitochondrial enzymes (enoylCoA hydratase, hydroxyacylCoA dehydrogenase, and citrate synthase) are depressed in cardiomyopathies. The content of coenzyme-A remains unaltered, but carnitine amount is reduced.

The two heart mitochondrial populations are distinctly affected, as they react differently in cardiomyopathy. Oxidative phosphorylation in subsarcolemmal mitochondria is normal, whereas it decays in interfibrillar mitochondria. The amount of

Table 4.6 Classification of cardiomyopathies. (Source: [210])

Type	Features	Causes
Dilated	Dilated Left or both ventricles Impaired contraction	Ischemic, immune, toxic, alcoholic, valvular, idiopathic, genetic (familial)
Hypertrophic	L(R)V hypertrophy or both	Genetic
Restrictive	Reduced diastolic filling Normal or near-normal systolic function	Amyloidosis, endomyocardial fibrosis, idiopathic
ARVC	Fibrofatty replacement of RV myocardium	Genetic, idiopathic, Naxos disease
Unclassified	Atypical	Fibroelastosis, noncompacted myocardium, systolic dysfunction with slight dilation, mitochondrial disease

ARVC arrhythmogenic right ventricular cardiomyopathy, *L(R)V* left (right) ventricle (ventricular)

mitochondrial respiratory supercomplexes mtRSC1 (^{ETC}cI–^{ETC}cIII–^{ETC}cIV) of the electron transport chain, the major form of the so-called respirasome, can decay, thereby reducing the efficiency of electron transfer and energy production.

4.5 Altered Cardiomyocyte Morphology and Function

Compensatory responses, such as changes in gene expression and cellular morphology and metabolic shifts in cardiomyocytes, can initially maintain the cardiac function in cardiomyopathy.

4.5.1 Cardiomyocyte Hypertrophy

Cardiomyocyte hypertrophy (Sect. 3.2) aimed at normalizing increased parietal stress increases the wall thickness, but decreases inner dimensions of the ventricular chambers. Whereas the initial phase of hypertrophy can be compensatory, persistent adverse hypertrophy leads to heart failure.

The addition of sarcomeres in parallel is an adaptive response to increase contractility in stressed myocytes. *Eccentric hypertrophy* (augmented myocyte length resulting from end-to-end addition of sarcomeres) is associated with ventricular dilation and advanced hypertrophic cardiomyopathy. *Concentric hypertrophy* (i.e., elevated myocyte width) distorts the cardiomyocyte alignment, hence affecting cardiac inotropy.

In cardiomyopathy, genes mainly expressed during embryogenesis are again transcribed, in addition to continuous transcription of genes normally transiently expressed. These target genes includes, in particular, those encoding natriuretic factors (natriuretic peptide precursor-A [NPPA] and -B [NPPB]), β-myosin heavy chain (MYH7), and cardiac α-actin (ACTC1).

Synthesis and secretion of natriuretic peptides by atrial and ventricular cells that favor diuresis and vasodilation and lower blood volume and pressure in response to increased wall stress is chronic rather than transient [332].

The nanomotor protein cardiac myosin is composed of two heavy chains (MyHC) and two pairs of light chains (MyLC). Cardiomyocyte contains two MyHC isoforms (α MyHC–β MyHC [encoded by the MYH6 and MYH7 genes, respectively]). The β MyHC isoform has a lower ATPase activity and hence a slower filament sliding velocity and reduced sarcomere shortening. The MyHC isoform production shifts to β MyHC, the production of α MyHC decreasing.

In the adult myocardium, phosphorylation of regulatory myosin light chains increases tension, but decreases stretch activation response of cardiomyocytes. Two major regulatory MLC isoforms are coexpressed in the early stages of murine cardiogenesis.

Mutations in the MYL2 gene that encodes the cardiac regulatory (i.e., phosphorylatable) slow myosin light chain MyLC2 required for stabilizing thick filaments cause inherited hypertrophic cardiomyopathy.

4.5.2 Cardiac Fibrosis

Interstitial and perivascular fibrosis (Sect. 3.1) associated with increased synthesis and release of TGFβ disrupts coupling between cardiomyocytes and increases wall stiffness. Focal fibrosis composed primarily of collagen-1 and -3 occurs in the early stage of cardiomyopathy [332].

4.5.3 Altered Metabolic Substrate Utilization

In the healthy heart, phosphocreatine is the main ATP source during acute need. As cardiomyopathy progresses, the phosphocreatine reserve is progressively depleted [332].

During the initial compensation, substrate utilization favors free fatty acids that yield more ATP than glucose, but require more oxygen. In dilated and end-stage hypertrophic cardiomyopathy, energy substrate utilization changes from fatty acid oxidation to glucose catabolism, as excessive cardiac growth occurs without adequate coronary angiogenesis. Synthesis and activity of fatty acid utilization enzymes (e.g., medium-chain acylCoA dehydrogenase) lower.

Table 4.7 Signaling axes in cardiomyopathy (Source: [332]). Endothelin-1 (ET1) and angiotensin-2 (ATn2) released in response to reduced contractility bind to their cognate G-protein–coupled receptors (GPCR) that raise calcium release from the sarcoplasmic reticulum. Cytosolic calcium activates calmodulin and the MEF2 (myocyte enhancer factor-2) transcription factor. Mutant sarcomeric proteins can further elevates cytosolic calcium concentration. GPCRs signal via the PKB pathway that causes fetal gene expression and cardiac hypertrophy via inhibition of glycogen synthase kinase GSK3β. Apoptosis results from cytochrome-C (CyC) release from mitochondria and activation of death receptors (e.g., TNFRSF6a) by cytokines

GPCR–PLCβ–IP_3–Ca^{2+}
GPCR–PLC–PI3K–PDK1–PKB–GSK3β–NFAT
$Ca_V1.2$–RyR–Ca^{2+}
Ca^{2+}–Cam–CamK2–HDAC$\ominus\longrightarrow$MEF2
Ca^{2+}–Cam–PP3–NFAT
Mutant sarcomeric proteins$\ominus\longrightarrow$SERCA
TNFRSF6a–PI3K–Casp8–Casp3
CyC—Casp9–Casp3

CamK calmodulin-dependent kinase, *IP_3* inositol trisphosphate, *HDAC* histone deacetylase, *NFAT* nuclear factor of activated T cells, *PI3K* phosphatidylinositol 3-kinase, *PLC* phospholipase-C, *PP3* protein phosphatase-3 (calcineurin), *RyR* ryanodine receptor (calcium channel), *SERCA* sarco(endo)plasmic reticulum calcium ATPase, $\ominus\longrightarrow$ inhibition

4.5.4 Apoptosis

Especially in dilated and hypertrophic cardiomyopathy, the apoptotic pathways and caspase are activated [332]. Hyperactivation of β-adrenergic receptors releases cytochrome-C from mitochondria. In addition, apoptosis results from activation of death receptors (e.g., TNFRSF6a) by cytokines.

4.5.5 Altered Signaling

Environmental stimuli are primarily sensed by matrix–sarcomere connectors, that is, components of the costamere and integrins, as well as mechanosensitive ion channels that signal to maintain sufficient contractility. Mechanotransduction initially leads to sarcomere addition and metabolism shift. Many pathways are associated with cardiomyopathy (Table 4.7).

4.5.5.1 Protein Kinase-B

Chronic activation of the PI3K–PKB pathway causes cardiomyopathy. In mice, 2-week duration activation of PKB1 provokes a reversible hypertrophy, whereas 6-week duration activation engenders irreversible hypertrophy with fibrosis and reduced angiogenesis [332]. The PKB effector GSK3β impedes cardiomyocyte growth.

Whereas it is most often unphosphorylated and hence constitutively active in the healthy heart, it is inactivated upon phosphorylation by PKB, thereby permitting cardiac hypertrophy.

4.5.5.2 $G\alpha_q$-Coupled Receptors

Elevated blood pressure and resulting cardiac wall stretch induce release of several auto- and paracrine factors, such as angiotensin-2, endothelin-1, and noradrenaline, which activate their specific G-protein–coupled receptors. The G-protein subunit $G\alpha_q$ favors cardiac hypertrophy via phospholipase-C.

4.5.5.3 Calcium Ions

Calcium ions are stored within the sarcoplasmic reticulum, the main intracellular reservoir. Once they are liberated in response to electrochemical stimulation of the cardiomyocyte, they bind to the sarcomeric myofilament. After contraction, sarco(endo)plasmic reticulum Ca^{2+} ATPase sequesters Ca^{2+} ions into the sarcoplasmic reticulum.

Calcium ions released from the sarcoplasmic reticulum tethers to calmodulin and activates calmodulin-dependent kinase CamK2 that phosphorylates the PP3 phosphatase (calcineurin). The latter dephosphorylates nuclear factor of activated T cells (NFAT) that then translocates into the nucleus, hence leading to cardiac hypertrophy that evolves rapidly to heart failure.

4.5.6 Nitric Oxide in Cardiomyopathies

Nitric oxide (NO) is synthesized and released by coronary and endocardial endotheliocytes as well as cardiomyocytes using the NOS3 synthase isoform. In addition, cytokine-inducible, calcium-insensitive NOS2 isoform is produced in the right ventricular myocardium of patients with dilated cardiomyopathy especially after myocarditis, whereas constitutive NOS has a low activity [336]. Cardiac production of NO by NOS2 attenuates the positive inotropic effect of β-adrenoceptor stimulation and hastens relaxation in failing human hearts [337].

The inducible NOS2 isoform has an elevated activity in cardiomyopathies. In cirrhotic cardiomyopathy in a rat model, nitric oxide acts as a negative inotropic agent [338]. Concentrations of cardiac TNFSF1, NOS2 (but not NOS3), and cGMP, as well as those of plasma IL1β, nitrite, and nitrate are significantly higher in cirrhotic rats than those in controls. Cytokine (IL1β)-induced stimulation of NOS2 operates in the pathogenesis of cirrhotic cardiomyopathy.

Nitric oxide influences myocardial contractility, as it can change myocardial relaxation (diastolic) properties, modulate the β-adrenergic inotropic response, mediate the cholinergic response, and affect the force–frequency relation [339].

In isolated ferret and cat papillary muscle preparations, isolated ejecting guinea pig hearts, and rat cardiomyocytes, upon liberation of endothelium-derived nitric oxide or administration of exogenous nitric oxide donors and cGMP analogs, as well as in normal humans, after low-dose bicoronary infusions of substance-P, a stimulator of NO release from endotheliocytes, or of sodium nitroprusside, nitric oxide enhances the myocardial relaxation and lowers the left ventricular end-diastolic stiffness probably via a reduced myofilament responsiveness to Ca^{2+} secondary to elevated cytosolic cGMP level and activated PKG, without affecting markedly systolic function [339].

Nitric oxide lowers the left ventricular filling pressure, hence diminishing myofilament responsiveness to Ca^{2+} ions, as it increases the left ventricular distensibility $(\Delta V/V)/\Delta p)$. Upon infusion of the NO donor sodium nitroprusside (≤ 4 μg/mn, hence devoid of systemic effects) during cardiac catheterization in the global coronary bed of the left ventricle of patients, the cardiac frequency decrease (from 1.30 to 1.26 Hz [78 ± 11 to $76 \pm$/mn]), as well as the left ventricular peak systolic pressure (from 21.4 to 19.4 kPa [161 ± 18 to 146 ± 18 mmHg]), and time to onset of left ventricular relaxation (interval from Q wave to LV dP/dt min; from 432 ± 36 to 419 ± 36 ms) [340]. In addition, angiograms and tip-micromanometer pressure recordings of some patients show that the left ventricular end-diastolic volume increases (from 158 ± 34 to 165 ± 40 ml), whereas the left ventricular end-diastolic pressure falls (from 2.4 to 1.6 kPa [18 ± 5 to 12 ± 3 mmHg]) [340]. When a right atrial infusion of sodium nitroprusside was performed before or after the intracoronary infusion, the decrease in left ventricular peak systolic pressure is significantly smaller than that measured during intracoronary infusion [340]. Improved left ventricular diastolic distensibility during intracoronary infusion of sodium nitroprusside is unrelated to systemic vasodilation or to pericardial constraint and may be explained by a direct myocardial effect of nitric oxide.

The basal release of nitric oxide augments the Frank–Starling response (length-dependent change in myofilament responsiveness to Ca^{2+} ions with augmented ventricular performance in response to heightened end-diastolic volume)[8] and hence preload-induced elevation of cardiac output in isolated guinea pig hearts, without relation to change in coronary flow [339].

In ischemic and nonischemic dilated cardiomyopathy, NO becomes deficient. In dilated nonischemic cardiomyopathy, the left ventricular myocardial expression of inducible (NOS2) and constitutive (NOS3) nitric oxide synthases is correlated

[8] This intrinsic factor contributes to the regulation of cardiac output, in addition to cardiac frequency, neurohumoral control, coronary flow, and cardiac load. In normal humans, it participates markedly in increase in cardiac output during postural changes and submaximal exercise. In dilated cardiomyopathy, the Frank–Starling mechanism is impaired, an acute increase in ventricular volume fails to augment stroke volume [339].

with left ventricular stroke volume and work [341]. In patients with elevated left ventricular end-diastolic pressure (> 2.13 kPa [16 mmHg]), a better correlation is observed between myocardial expression of NOS2 and left ventricular stroke volume and work and ejection fraction. In patients with elevated left ventricular end-diastolic pressure, an intracoronary infusion of substance-P (that causes NO release from the coronary endothelium) raises the left ventricular stroke volume (from 72 ± 13 to 91 ± 16 ml) and work as well as shifts the left ventricular end-diastolic pressure–volume relation to the right [341].

Altered NO availability detected from high plasma levels of methylated ^{L}arginine metabolites (monomethylarginine and symmetric and asymmetric dimethylarginines) that inhibit NOS is strongly related to left ventricular diastolic dysfunction and weakly to left ventricular systolic dysfunction [342]. ^{L}arginine can thus be used concomitantly with β-blockers in heart failure to improve myocardial NO availability and diastolic left ventricle distensibility.

4.6 Dilated Cardiomyopathies

Dilated cardiomyopathy (DCM) is the most prevalent form of cardiomyopathy. It actually represents the final common morphological outcome of various prolonged cardiac insults. Once it is symptomatic, about 50 % of DCM patients die within 5 years.

Dilated cardiomyopathy combines myocyte injury and necrosis associated with myocardial fibrosis, that is structural defects associated with an impaired mechanical function. The left ventricle or both ventricles are dilated and the contraction is impaired (ejection fraction < 40 %; left ventricular end-diastolic size > 115 % of that calculated for age and body surface area). Ventricular walls thinning is associated with an enlargement of the interior dimensions of the ventricular chamber. The left and right atria are also enlarged due to impaired ventricular relaxation.

About one-third of patients with an idiopathic DCM have a family DCM history. Familial DCM is transmitted autosomally (recessive and, most often, dominant) via maternal mitochondrial DNA or as X-linked mutation [332]. Unlike hypertrophic cardiomyopathy that is almost exclusively linked to mutations in sarcomeric proteins, familial DCM is also associated with mutations in cytoskeletal proteins and components of the mechanotransduction system (Tables 4.8 and 4.9).

Vinculin, a structural component of cardiomyocytes, is implicated in dilated cardiomyopathy. In mouse VCL-knockout cardiomyocytes, tagging magnetic resonance imaging acquired before the onset of global ventricular dysfunction shows that systolic strains transverse to the myofiber axis lower without changes along the myofibers and in tension in papillary muscles [343]. Transmission electron microscopy exhibits an enlargement of myofilament lattice spacing. Atomic force microscopy reveals a significant decrease in membrane cortical stiffness.

In addition to idiopathic and genetic causes, DCM can have a viral, immune, and/or toxic (e.g., alcoholic) origin, or be associated with a cardiovascular disease

Table 4.8 Dilated cardiomyopathy, genetic causes, and cellular pathology (Source: [332]). Gene mutations can be common to multiple forms of cardiomyopathies

Involved protein	Mutated gene	Pathology
Cardiac α-actin	ACTC1	Interstitial fibrosis
Cardiac α-actinin	ACTN1	
α-tropomyosin-1	TPM1	Ventricular wall stiffness
β-myosin heavy chain	MYH7	Ventricular wall thinning
Titin	TTN	
Troponin-C	TNNC	
Troponin-I	TNNI	
Troponin-T2	TNNT2	
δ-sarcoglycan	SGCD	Inflammation
Desmin	DES	Ventricular chamber enlargement
Vinculin	VCL	
γ-catenin	CTNNG	
Plakoglobin N-cadherin	CDH2	Cardiomyocyte apoptosis
Lamin-A/C	LMNA	
Phospholamban	PLN	
Presenilin-1/2	PSEN1	
$Na_V1.5$	SCN5A	
WWTR1	TAZ	

TAZ taffazin, *WWTR1* WW domain-containing transcription regulator-1

Table 4.9 Genetic mutations in human dilated cardiomyopathy. (Source: [321])

Locus	Involved protein	Mutation type
1q32	Cardiac troponin-T	Deletions
14q11	Cardiac β-myosin heavy chain	Missense
15q14	Cardiac actin	Missense
1p1-q21	Lamin-A/C	Missense
2q35	Desmin	Missense
Xp21	Dystrophin	Deletions
Xq28	Tafazzin	Deletions, splicing defects

in which the degree of myocardial dysfunction is not explained by the abnormal loading condition or the extent of ischemic damage.

Histology is nonspecific. Arrhythmias, thromboembolism, and sudden death are common events that can occur at any stage of dilated cardiomyopathy evolution.

The Dicer gene involved in microRNA processing is required to avoid dilated cardiomyopathy, the most common cardiomyopathy associated with heart failure [344]. Mutations in lamins-A/C are linked to dilated cardiomyopathy.

The TRPV2 carrier is a Ca^{2+} channel activated by temperatures greater than 50 °C that can also function as a mechanosensitive ion channel in sensory neurons and atriomyocytes. However, thermal and mechanical responses of $Trpv2^{-/-}$ mice are identical to those of wild-type mice. The TRPV2 channel translocates to the plasma membrane from internal cytosolic pools in response to insulin-like-growth factor IGF1 and insulin. Chemical agonists of the channel include endocannabinoids [345].

Transient receptor potential (TRP) channels are synthesized in every cardiac cell type (myocytes, fibroblasts, endotheliocytes, and vascular smooth myocytes).[9] In the myocardium, TRPV2 is responsible for the positive inotropic effect of some agonists, as it mediates Ca^{2+} release for the sarcoplasmic reticulum. The human cardiac ventricular cells of dilated cardiomyopathy contain higher TRPV2 amount at the sarcolemma [346]. In mice, overexpression of TRPV2 leads to cardiomyopathy due to Ca^{2+} overloading, phosphorylation of calmodulin-dependent protein kinase CamK2, and reactive oxygen species production. This ion channel may be activated by membrane stretch.

4.7 Restrictive Cardiomyopathies

Restrictive cardiomyopathy (RCM) is characterized by restrictive filling and reduced diastolic volume of either or both ventricles with normal or quasinormal systolic function and wall thickness. Interstitial fibrosis can be found. It represents in general less than 5 % of cardiomyopathies, but in certain populations, it can be more common. It is the least common, but most lethal form of cardiomyopathy.

Restrictive cardiomyopathy can be idiopathic or associated with other diseases, such as amyloidosis and endomyocardial disease with or without hypereosinophilia. In addition to diseases, cardiac injury caused by radiation, scarring after cardiac surgery, and infection can also cause a restrictive cardiomyopathy. Among induced initiators, RCM can have a genetic origin (Table 4.10). Mutations in troponin-I gene that participates in the regulation of calcium-mediated actin–myosin interaction, is the primary cause of familial RCM [332]. Mutations in other sarcomeric proteins, such as cardiac desmin, cardiac α-actin, and β-myosin heavy chain, as well as mutations that are not related to hypertrophic cardiomyopathy can be detected in RCM patients. Idiopathic restrictive cardiomyopathy occurs in the absence of other pathology, although it can be associated with skeletal muscle myopathies.

[9] Calcium influx through TRPC channels activates PP3–NFAT axis and hence causes cardiac hypertrophy [345]. Mutations in the human Trpm4 gene are linked to progressive familial type-1 heart block and isolated cardiac conduction block. The TRPM7 channel is involved in cardiogenesis.

Table 4.10 Restrictive cardiomyopathy, genetic causes, and cellular pathology (Source: [332]). Gene mutations can be common to multiple forms of cardiomyopathies

Mutated gene	Pathology
Cardiac α-actin (ACTC1 gene)	Fibrosis
β-myosin heavy chain (MYH7 gene)	
Troponin-I	Ventricular wall stiffness
Troponin-T	
Desmin	Apoptosis

Restrictive cardiomyopathies can be categorized as *primary* (e.g., endomyocardial fibrosis, Löffler's endocarditis, idiopathic restrictive cardiomyopathy) or *secondary* caused by infiltrative diseases (e.g., amyloidosis, sarcoidosis, radiation carditis) and storage diseases (e.g., hemochromatosis, glycogen storage disorders, Fabry's disease).

Unlike hypertrophic cardiomyopathy, altered wall deformability is observed in the absence of changes in ventricular myofibrillar arrangement [332]. Ventricular relaxation is impaired and the atria enlarge in response to the increased pressure in some patients.

4.8 Arrhythmogenic Right Ventricular Cardiomyopathy

Arrhythmogenic right ventricular cardiomyopathy (ARVC), or arrhythmogenic right ventricular dysplasia (or dystrophy [ARVD]), affects approximately 1 in 5000 individuals. It represents 20 % of deaths in individuals under 30 years old [332].

It is characterized by a fibrofatty substitution of cardiomyocytes of the right ventricle that leads to right ventricular enlargement and dysfunction with complications, such as life-threatening arrhythmias and sudden death. Patches of cell death and fatty infiltration localizes mainly to the right, but also, to a lesser extent, left ventricle. Left ventricular lesions are common in the later stage of the disease.

It is a familial disease in more than 50 % of patients, generally inherited in an autosomal dominant or recessive mode (Table 4.11). Mutations in genes that encode desmosomal constituents that mechanically couple neighboring cardiomyocytes for a coordinated contractility, account for most inherited cases [332].

Nesprin-2 may be implicated in arrythmogenic right ventricular dystrophy, both mapped to chromosome-14q23. Approximately 50 % of patients have a mutation in one of the 5 major desmosome components: desmosomal cadherins desmocollin-2 and desmoglein-2, desmoplakin, plakoglobin (or γ-catenin), and plakophilin-2 [347].

It is diagnosed usually in early adulthood by symptoms of supraventricular and ventricular arrhythmias or right-sided heart failure. It can also be discovered during family screening. However, often, sudden death is the first sign.

Table 4.11 Arrhythmogenic right ventricular cardiomyopathy (ARVC), genetic causes, and cellular pathology (Source: [332]). Arrhythmogenic right ventricular dysplasia is a genetic disease caused by mutations in desmosomal proteins characterized by a fibroadipocytic replacement of cardiomyocytes. Gene mutations can be common to multiple forms of cardiomyopathies

Mutated gene	Pathology
γ-catenin (also known as desmoplakin-3 and junctional plakoglobin [JUP gene])	Right ventricular fibrosis
Plakophilin-2 (ARVD9)	Right ventricular dilation
Desmocollin-2	Fibrofatty infiltration
Desmoglein-2	Apoptosis
Desmoplakin	Fibrofatty replacement (ARVC)
	Possible arrhythmogenic LV cardiomyopathy

LV left ventricle

Electrocardiographically, ϵ waves (slurred ST segments) are observed in V1 to V3 derivations, with possible inverted T waves in V2 and V3, in the absence of right bundle branch block. Echocardiography can detect localized right ventricular aneurysm and failure. Fatty infiltration of the right ventricle can be identified by magnetic resonance imaging.

4.9 Diabetic Cardiomyopathy

Diabetic cardiomyopathy is observed in type-1 and type-2 diabetes mellitus. It is characterized by both systolic and diastolic dysfunction. Pathogenic factors of diabetic cardiomyopathy include hyperglycemia and elevated nonesterified fatty acid levels that augment glucose flux via the hexosamine synthesis pathway and subsequent glycosylation of many proteins.

Major contributors to cardiac dysfunction in diabetes include abnormal Ca^{2+} mobilization, altered myofilament Ca^{2+} sensitivity, and slower cross-bridge cycling that lead to defective excitation–contraction coupling associated with myocardial insulin resistance [349].[10] Impaired insulin signaling alters ion carrier functioning (e.g., $Ca_V1.2$ and SERCA) and reduced activity of survival kinases (PKB), activating FoxO factors, which are initiators of cell death.

[10] Insulin binding to its receptor stimulates receptor tyrosine autophosphorylation and primes insulin receptor substrate phosphorylation to activate the phosphatidylinositol 3-kinase and mitogen-associated protein kinase pathways. The PI3K–PKB cascade leads to a set of effectors, such as glycogen synthase kinase-3, nitric oxide synthase-3, and glucose transporter GluT4. Insulin resistance is associated with defects in insulin-mediated IRS phosphorylation, PI3K activity, and glucose transport.

Hyperglycemia and hyperlipidemia increase the formation of glycated end products. Glycation of collagen, SERCA2a, and RyR2, damage cardiomyocytes. Moreover, excessive glucose and lipid levels promote ROS generation. In addition, FoxO1 stimulates NOS2, thereby causing via NO protein modification, i.e., nitration (introducing a nitro group [NO_2]) and nitrosylation (adding a nitrosyl ion [NO^-]), of transfer proteins such as VAMP2 followed by augmented membrane translocation of ScaRb3 receptor and fatty acid influx [30]. Furthermore, FoxO1 supports cell death upon increased reactive oxygen and nitrogen species generation, ROS-mediated MAPK stimulation, reduced antioxidant capability, and decreased mitochondrial function, i.e., in a caspase-dependent and caspase-independent manner.

Animal models of diabetes exhibit fetal gene expression (e.g., increased expression of atrial natriuretic peptide and α-skeletal actin; increased β-myosin heavy chain and reduced α-myosin heavy chain expression; Table 4.12).

Transient receptor potential vanilloid channel TRPV1 in vascular smooth muscle and endothelial cells, which is gated by chemical and physical stimuli (e.g., pH and heat), couples coronary blood flow with metabolism via nitric oxide and the large-conductance Ca^{2+}-sensitive K^+ channel (BK) [350]. In particular, Ca^{2+} entry via endothelial TRPV1 channel triggers NO-dependent vasodilation of coronary arteries. Myocardial blood flow regulation by TRPV1 via vasodilation and its coupling to metabolism via the NO–BK axis is disrupted in diabetic cardiomyopathy. Expression of TRPV1 in diabetic mouse hearts is markedly reduced with respect to healthy hearts [350].

The prediabetic heart is characterized by a monocyte infiltration associated with proinflammatory cytokine release, adverse cardiac remodeling, and heart dysfunction [351]. Bone morphogenetic protein BMP7 activates infiltrated monocytes into M2 macrophages that liberate anti-inflammatory cytokines, thereby reducing adverse cardiac remodeling and improving cardiac function.

4.10 Dystrophic Cardiomyopathy

Muscular dystrophy can affect the myocardium, generating a dystrophic cardiomyopathy in humans. Cardiomyopathies are then linked to muscular dystrophies caused by autosomal recessive mutations in proteins anchoring the sarcomere to the extracellular matrix.

Dystrophinopathies result from mutations in the dystrophin gene. They represent a category of diseases characterized by a cytoskeletal disarray.

Among the most severe forms of dystrophy, *Duchenne muscular dystrophy* (DMD) is associated with cardiac manifestations in the majority of adolescent DMD boys. About 20 % patients suffer from ventricular dysfunction and arrhythmias that ultimately evolve to heart failure or sudden cardiac death. Abnormal excitation–contraction coupling is linked to an impaired intracellular Ca^{2+} homeostasis.

Table 4.12 Molecular changes in diabetic cardiomyopathy (Source: [349]). The activity of PKA and PKCβ decreases and increases, respectively. Troponin-I and phospholamban phosphorylation thus decays. Mitochondrial Ca^{2+} content and ROS generation rise. Ion carrier activity changes include: (1) depressed outward transient outward K^+ and L-type Ca^{2+} currents; (2) reduced Na^+–Ca^{2+} exchange and sarcoplasmic reticulum Ca^{2+} uptake; and (3) impaired ryanodine receptor activity. β-adrenergic receptor signaling is reduced, as β1AR expression decreases. These modifications lead to the following prominent alterations in ventriculomyocyte excitation–contraction coupling: (1) prolonged action potential duration; (2) decreased cytosolic Ca^{2+} removal and sarcoplasmic reticulum Ca^{2+} uptake rate; and (3) slowed contraction and relaxation. Glycosylation of cardiomyocyte proteins (e.g., SERCA2a and RyR2) and extracellular matrix (e.g., collagen) disturbs cardiac structure and function, as it causes irreversible cross-links within and between proteins

Molecule	Activity
Ion carriers	
$Ca_V1.2$ (L-type Ca^{2+} channel) (high threshold, long lasting)	Unchanged
K_V4 (transient outward K^+ channel)	Reduced
Na^+–Ca^{2+} exchanger	Reduced (usually)
Na^+–K^+ ATPase	Reduced
Ryanodine receptor	Reduced
SERCA2a	Reduced
Regulators	
β1-adrenoceptor	Decreased expression
Angiotensin-2	Increased level
Phospholamban	Increased
Protein kinase-A	Reduced
Protein kinase-Cβ	Increased
Myofilaments	
Myosin ATPase	Reduced
MHCα	Decreased expression
MHCβ	Increased expression
Troponin-I and -T	Reduced Ca^{2+} sensitivity (PKC-mediated phosphorylation)
Actomyosin ^{Mg}ATPase	Reduced
Glucose metabolites	
Glycosylated products	Increased
ROS production	Increased

MHC myosin heavy chain, *ROS* reactive oxygen species

Although Duchenne and Becker muscular dystrophies are X-linked, up to 18 % of female carriers that do not exhibit skeletal muscle dystrophy have heart failure [332].

Changes in Ca^{2+} homeostasis and oxidative stress are associated with cardiac deterioration. Oxidative stress participates in disrupting calcium handling. Augmented sensitivity of the Ca^{2+} release channel ryanodine receptor precedes malady progression. In dystrophic cardiomyopathy, intracellular bursts of Ca^{2+} sparks and waves results from elevated Ca^{2+} leak from the sarcoplasmic reticulum and hence reduced Ca^{2+} storage [352].

Oxidative stress initiates the development of dystrophic cardiomyopathy. Oxidation of ryanodine receptors followed by phosphorylation, first by CamK2 and later by PKA, contributes to cardiac deterioration [352]. The production of NADPH oxidase NOx2 and subsequently of superoxide rises (about fivefold) [353]. Inhibition of NOx2 restores: (1) myocardial contractility; (2) amplitude of intracellular Ca^{2+} concentration transients; (2) total sarcoplasmic reticulum Ca^{2+} content, as it lowers the occurrence rate of spontaneous diastolic calcium release events. In addition, NOS1 expression rises (eightfold), but cardiac NO production decays because of NOS1 uncoupling [353]. Inhibition of NOx2 helps the NOS1 recoupling.

4.11 Hemochromatic Cardiomyopathy

Hemochromatosis results from iron overload and deposition in the sarcoplasmic reticulum of cells in many organs, including the heart. It generally follows an autosomal recessive pattern of mendelian inheritance. Hemochromatosis can provoke restrictive or dilated cardiomyopathy.

4.12 Hypertrophic Obstructive or Nonobstructive Cardiomyopathies

Hypertrophic cardiomyopathies (HCM) is a genetically inherited asymmetrical left ventricular hypertrophy (in the absence of any evident cause). They are also characterized by an increased incidence of arrhythmias and cardiac sudden death.

Obstructive and nonobstructive hypertrophic cardiomyopathy differ according to whether the distortion of the heart anatomy causes an obstruction of the blood outflow tract, especially in the left ventricle, or not. Obstruction of the left ventricular outflow tract engenders a pressure overload.

Hypertrophic obstructive cardiomyopathy (HOCM) was called idiopathic hypertrophic subaortic stenosis and asymmetric septal hypertrophy. Nonobstructive HCM is an apical hypertrophic cardiomyopathy, also termed Yamaguchi hypertrophy.

Familial hypertrophic cardiomyopathy (FHCM) ranges from the asymptomatic form to obvious cardiomyopathy. About two-thirds of cases of familial hypertrophic cardiomyopathies are attributable to mutations in genes that encode sarcomeric proteins, about half of them residing in two genes that encode cardiac myosin (e.g.,

Table 4.13 Genetic mutations in human hypertrophic cardiomyopathy (Source: [321]). A missense mutation at residue 403 in the actin-binding segment of myosin was the first mutation identified in familial hypertrophic cardiomyopathy. In mice, the heterozygous R403Q mutation in myosin heavy chain is characterized by normal systolic, but abnormal diastolic function, in the absence of hypertrophy. The mutation type include missenses (i.e., single amino acid substitutions such as in MHC mutants) and deletions (i.e., truncation of a portion of the peptidic chain such as in cMyBPc mutants)

Locus	Involved protein	Mutation type
1q32	Cardiac troponin-T	Missense, deletions, splice defects
2q31	Titin	Missense
11p11	Cardiac myosin-binding protein-C	Missense, deletions, splice defects
14q12	Cardiac β-myosin heavy chain	Missense
15q14	Cardiac actin	Missense
15q22	α-tropomyosin	Missense
19q13	Cardiac troponin-I	Missense
3p21	Essential myosin light chain	Missense
12q23-p21	Regulatory myosin light chain	Missense

MYH6 and MYH7 that encode cardiac myosin heavy chain-α and -β) and myosin-binding protein-C and most often (∼40 %) in the MYBPC3 gene. The majority of MYBPC3 mutations reduce cMyBPc expression.

The myosin tail forms the structural core of the thick filament and the myosin head has catalytic, converter, and lever arm or load-bearing functions. Myosin uses the energy released from ATP hydrolysis for contracting the sarcomere, that is, launching the movement of the thin actin–troponin–tropomyosin filament over the thick myosin filament in a ratchet-like fashion. Its intrinsic ATPase activity determines shortening velocity. The sarcomere contraction is modulated by the thick filament protein cMyBPc. The latter binds to myosin heavy chain and connects the myosin and actin filaments, thereby modifying the position of tropomyosin on actin filament [354].

Hypertrophic cardiomyopathy is inherited in an autosomal-dominant pattern in more than 50 % of patients [332]. It affects up to 0.2 % of the population (i.e., prevalence in the general population 1/500) with men affected nearly twice as much as women.

Hypertrophic obstructive cardiomyopathy is a monogenic cardiac disorder due to mutations in 13 genes (i.e., several hundred mutations) that encode sarcomeric proteins (Table 4.13). The first mutation was discovered in the MYH7 gene that encodes the myosin heavy chain cMyHCβ (cMyHC7).

Mutations in the MYH7 (β-myosin heavy chain; 30 % of cases) and MYBPC3 (cardiac myosin-binding protein-C) genes constitute 80 % of known causal mutations.

Mutations in TNNT2 (cardiac troponin-T), TNNI3, and TNNC1, as well as TPM1 (α-tropomyosin) and ACTC (cardiac actin) genes are also implicated. The cardiac troponin-I R21C mutation is linked to hypertrophic cardiomyopathy. It renders cTnnI incapable to be phosphorylated by PKA and prevents positive lusitropic effect [355]. In homozygous mice expressing the $cTnnI^{R21C}$ mutation, echocardiography shows a hypertrophy after 12 months of age. Abnormal diastolic function is characterized by longer filling times, impaired relaxation, and excitation–contraction uncoupling. Elevated cardiac frequency is associated with a reduced cardiovagal tone. Smaller (2.3-fold) Ca^{2+} transient amplitudes are linked to attenuated (2.9-fold) sarcoplamic reticulum Ca^{2+} content due to diminished (2.4-fold) calsequestrin and augmented (1.5-fold) Na^{+}–Ca^{2+} exchanger expression.

Hypertrophic cardiomyopathy is characterized morphologically by a hypertrophy of the ventricular wall and a most often asymmetrical interventricular septal thickening (> 15 mm). It leads to reduced ventricular volume and diastolic dysfunction as well as sudden cardiac death (leading cause of sudden cardiac death in young subjects).

The highly registered alignment of cardiomyocytes becomes distorted with disorientation (oblique or perpendicular alignment) of adjacent cells [321]. The myocyte disarray can be focal or widespread.

In hypertrophic obstructive cardiomyopathy, left or right ventricular hypertrophy is observed at the macroscopic scale in the absence of increased load. At the microscopic scale, it is typified by cardiomyocyte hypertrophy with myofibrillar disarray and perivascular and interstitial fibrosis.

Familial hypertrophic cardiomyopathy causes a gradual thickening of the ventricular wall. In heterozygous $MyBPc^{+/-}$ mouse hearts (thus with lowered MyBPc expression [32 %] similar to that of FHCM patients), MyBPc phosphorylation level diminishes compared with wild-type mouse hearts [356]. Attenuated MyBPc expression and phosphorylation in the sarcomere cause myofilament dysfunction and wall remodeling. Perturbations in mechanical and electrical activity in $MyBPc^{+/-}$ mice may increase the susceptibility to cardiac arrhythmia.

Arrhythmias can occur occasionally, but thromboembolic complications happen frequently. Hypertrophic cardiomyopathy is the most common cause of sudden cardiac death during adolescence.

Hypertrophic cardiomyopathy is also characterized by: (1) an impaired myocardial blood flow due to a microvasculature dysfunction and (2) an inefficient cardiac metabolism and reduced mechanical efficiency, as energy expenditure for a given work rises.

Cardiac perfusion during treadmill exercise as well as upon vasodilation (reduced coronary vasodilatory reserve) is markedly attenuated. Defective vasodilatory capacity of coronary arterioles in the absence of any stenosis of upstream large coronary arteries precludes hyperemic myocardial blood flow. Perfusion defects are generally more pronounced in the subendocardial region than in the subepicardial layer.

The microvascular dysfunction results from remodeling of intramural coronary arterioles with diminished cross-sectional area of the arteriolar lumen, hence elevation of coronary vascular resistance. The microvascular dysfunction is a predisposition for myocardial ischemia that can cause fibrosis.

Myocardial perfusion defects and abnormal energetics can be observed using nuclear magnetic resonance spectroscopy, single photon emission computed tomography (SPECT), and positron emission tomography (PET).

Myocardial oxidative metabolism per gram of myocardial tissue in hypertrophic cardiomyopathy is reduced with respect to that in healthy subjects, especially in hypertrophied segments, i.e., the interventricular septum, with respect to lateral walls [357]. Hypokinesia of hypertrophied segments revealed by impaired circumferential shortening and decreased systolic wall thickening may reduce oxygen demand. Moreover, the diffusion distance increases because the capillary density decreases. In addition, abnormal regional energy handling affects oxygen consumption. A metabolic imbalance, hence an altered oxygen consumption, occurs between the right and left ventricles.

The energetic cost of contraction is higher in the myocardium of FHC patients whatever the mutation type, that is, with missenses (single amino acid substitutions in MYH genes) or truncations (deletion of a portion of the peptidic chain upon MYBPC gene mutations). The deficit in energetics is related to a fall in phosphocreatine (PCr) concentration, whereas the ATP level can remain normal, and hence a decline of the PCr/ATP ratio with and without left ventricular hypertrophy [354].

In isolated myocardium strips from patients undergoing surgery to relieve left venticle outflow tract obstruction or heart transplantation, the cost of contraction assessed by the ratio of maximal ATPase activity to maximum tension is the highest in patients with MYH7 mutations, intermediate in patients with MYBPC3 mutations, and the lowest in patients without sarcomeric constituent-encoding gene mutations [358]. In addition, myocardial efficiency in asymptomatic prehypertrophied mutation carriers and genotype-negative relatives estimated by myocardial oxygen consumption (MVO_2) and external work is the lowest (highest tension cost) in MYH7 mutant carriers, intermediate in MYBPC3 mutant carriers, and the highest in controls [358]. Therefore, the degree of impaired efficiency of contraction depends on the affected gene.

Observed changes at the micro- and nanoscopic scales include myocyte disarray with reduced cross-bridge kinetics, diminished ATPase activity, altered calcium sensitivity, attenuated myofilament shortening, myocyte atrophy, and impaired excitation–contraction coupling with abnormal relaxation of the left ventricle. In addition, small intramyocardial vessels have thick walls.

Markers can target various involved processes: (1) inflammation (interleukin-6, tumor-necrosis factor-α, and C-reactive protein); (2) cell apoptosis and necrosis (TNFSF6 and its related molecules and troponins); (3) endothelial dysfunction (von Willebrand factor, soluble thrombomodulin, tissue factor pathway inhibitor, endothelin, and asymmetric and symmetric dimethylarginine); (4) fibrosis; (5) extracellular matrix degradation (matrix metallopeptidases [MMP1, MMP2, and MMP9] and their tissue inhibitors [TIMP1, TIMP2, and TIMP4]); (6) prothrombotic and fibrinolytic status (fibrinopeptide-A and thrombin–antithrombin-3 complex) and platelet activation (soluble TNFSF5, β-thromboglobulin, and P-selectin) [359]. Hormones (myocardial aldosterone and cardiac aldosterone synthase, as well as natriuretic peptides) and growth factors (IGF1 and its binding partner IGFBP1) can also be measured.

Obstructive sleep apnea in patients with hypertrophic obstructive cardiomyopathy heightens sympathetic activity and alters therapy efficiency [360]. Abnormal adrenergic signaling occurs in both hypertrophic obstructive cardiomyopathy[11] and obstructive sleep apnea. Moreover, hypoxia, hypercapnia, and respiratory acidosis that result from nocturnal apnea can cause pulmonary vasoconstriction that can lead to pulmonary hypertension.

The primary effect of HCM-causing mutations is a two- to threefold increase in calcium sensitivity of myofilaments that yields a hypercontractile phenotype with impaired relaxation (diastolic dysfunction). The initial hypercontractility is antagonized by pressure overload caused by the left ventricular outflow tract obstruction. The pressure overload and diastolic dysfunction engender a secondary progression to heart failure.

Myocyte contractility relies on Ca^{2+} handling and Ca^{2+} sensitivity of the contractile apparatus as well as adrenergic regulation. Troponin is the Ca^{2+}-dependent regulator of myofibrillar activity. The calcium sensitivity of myofilaments is higher when troponin-I is unphosphorylated. On the other hand, phosphorylation of troponin-I upon β-adrenergic receptor activation reduces myofilament Ca^{2+} sensitivity. In HOCM patients, the myofibrillar Ca^{2+} sensitivity is uncoupled from troponin-I phosphorylation by PKA (Ser22 and Ser23), a component of the lusitropic response to β-adrenoceptor stimulation, due to mutation-independent abnormal troponin-T [361].

Many mutations (S101R, Y141H, S165F, E169K, A405S, R436C, and G505S) in the JPH2 gene can be detected in hypertrophic cardiomyopathy [9]. These mutations can disrupt junctional membrane complexes and reduce Ca^{2+}-induced Ca^{2+} release amplitude. Hypertrophic cardiomyopathy resulting from $JPH2^{E169K}$ mutation is associated with arrhythmogenic Ca^{2+} leakage from the sarcoplasmic reticulum and increased RyR2 spark frequency. The elevated diastolic Ca^{2+} level can activate the sarcolemmal Na^{+}–Ca^{2+} exchanger and create a depolarizing current causing delayed after depolarizations that can lead to atrial fibrillation [9].

4.13 Idiopathic Dilated Cardiomyopathy

Idiopathic dilated cardiomyopathy refers to most patients with nonischemic cardiomyopathy. With progress in gene analysis, many patients should get a specific molecular determinant. Familial cardiomyopathies can follow all forms of mendelian inheritance (autosomal dominant, recessive, X-linked, and mitochondrial [matrilinear]).

[11] β-adrenoceptor inhibition is a current therapy of hypertrophic obstructive cardiomyopathy.

4.14 Infective Cardiomyopathy

In addition to acute myocarditis, various viruses are implicated in the development of cardiomyopathy, such as human immunodeficiency virus (HIV) and hepatitis C.

The protozoan *Trypanosoma cruzi* causes Chagas disease, a dilated cardiomyopathy acutely or progressively (over many years). Other parasites can cause a cardiomyopathy, such as *Toxoplasma gondii* and *Trichinella spiralis*.

4.15 Ischemic Cardiomyopathy

Ischemic cardiomyopathy (ICM) is the most common specific cause of dilated cardiomyopathy. Myocardial infarction causes localized myocyte necrosis with resultant scar formation and loss of contractile function in the ventricular segment perfused by the obstructed artery.

Adverse remodeling in the neighborhood and chamber dilation support cardiomyopathic process in adjacent nonischemic regions. In addition, areas of myocardium that are chronically underperfused and metabolically less active participate in myocardial dysfunction. Moreover, subsequent mitral valvular regurgitation caused by papillary muscle dysfunction as well as atrial and ventricular arrhythmia further worsen contractile function.

Interstitial fibrosis is often detected in ischemic cardiomyopathy. Collagen deposition indeed represents a pathogenic factor in the development of cardiac dysfunction after ischemia–reperfusion events.

Serum amyloid-P inhibits the differentiation of a circulating bone marrow-derived population of monocytes into fibroblasts that are highly proliferative and express α-smooth muscle actin, collagen-1, hematopoietic progenitor cell antigen CD34, and protein Tyr phosphatase receptor PTPRc (CD45).[12] These fibroblast precursors participate in the genesis of cardiomyopathy. Serum amyloid-P binds to Fc receptors of monocytes before transendothelial migration mediated by chemokine CCL2 (monocyte chemoattractant protein MCP1) [362].

4.16 Left Ventricular Noncompaction Cardiomyopathy

Left ventricular noncompaction cardiomyopathy (LVNCCM) is defined anatomically by deep trabeculations that create a spongiform appearance of ventricular walls, prominently in the left ventricular wall.

Clinical data comprise systolic and diastolic dysfunction, left and right ventricular hypertrophy, arrhythmias, and, in some cases, conduction block [332].

---

[12] Molecules CD34 and CD45 are markers of precursor and hematopoietic cells, respectively.

Table 4.14 Left ventricular noncompaction, genetic causes, and cellular pathology (Source: [332]). Gene mutations can be common to multiple forms of cardiomyopathies

Mutated gene	Pathology
Cardiac α-actin (ACTC1 gene)	Myofibrillar disarray
β-myosin heavy chain (MYH7 gene) Troponin	Fibrosis
α-dystrobrevin	Noncompaction of ventricular endocardium
FKBP1a	Necrotic myocytes
Taz (WWTR1)	Trabeculated ventricular wall

FKBP FK506-binding protein, *Taz* taffazin, *WWTR1* WW domain-containing transcription regulator-1

During cardiogenesis, the loose network of myofibers becomes compacted between the fifth and eighth prenatal weeks. In patients with LVNC, a normal pattern of compaction occurs in the outer band of the myocardium, but an abnormal noncompacted region is observed within the endocardial layer, mainly in the apical and midventricular regions of the heart [332].

Familial forms with autosomal-dominant or X-linked transmission account for up to 50 % of all LVNC cases (Table 4.14; [332]).

4.17 Peripartum and Postpartum Cardiomyopathy

Pregnancy is associated with cardiac hypertrophy, but peripartum and postpartum cardiomyopathy (PPCM) is a rare life-threatening dilated cardiomyopathy with left ventricular systolic dysfunction up to systolic heart failure. It happens in previously healthy women in the last month of pregnancy or puerperium (6 weeks after childbirth) and beyond, that is, within 5 months after delivery (75 % of cases in the first 2 months after delivery).

Peripartum and postpartum cardiomyopathy is defined by four criteria:

1. Development of cardiac failure in the last month of pregnancy or within 5 months after delivery;
2. Absence of an identifiable cause of cardiac failure;
3. Absence of recognizable heart disease prior to the last month of pregnancy;
4. Left ventricular systolic dysfunction with a fall in ejection fraction demonstrated by echocardiography.

Risk factors include age (>30 years), multiparity, twin pregnancy, ethnical origin, and a family history of peripartum cardiomyopathy. Recovery, usually within 6 months, occurs in 50 % of patients. However, no specific risk factor profile is available. Comorbidities comprise hypertension and inflammation detected by high levels of proinflammatory markers (e.g., TNFSF1, TNFRSF6a^{S}, C-reactive protein, interferon-γ, and interleukin-6).

This type of cardiomyopathy is associated with oxidative stress and a high loss of capillaries. Pathogenesis relies on excessive prolactin production associated with increased blood volume and circulating erythropoietin as well as decreased blood pressure, angiotensin responsiveness, and levels of water, sodium, and potassium.

In mice, the cardiomyocyte-specific deletion of STAT3 promotes production of cardiac cathepsin-D, [348].

Peripartum and postpartum cardiomyopathy is associated with abnormal levels in cardiac manganese superoxide dismutase, hence with excessive levels of reactive oxygen species, and cathepsin-D. The ^{Mn}SOD production remains insufficient; the usual upregulation of its expression is absent in the heart. Oxidative stress activates cathepsin-D. Cathepsin-D cleaves prolactin into an antiangiogenic and proapoptotic N-terminal 16-kDa fragment ($_{N}$Prl or Prl16). The latter impairs the endothelial function, causes endotheliocyte apoptosis, and reduces endotheliocyte proliferation and migration [36]. Therefore, oxidative stress in cardiomyocytes subsequently destroys the vascular endothelium. Whereas $_{N}$Prl has little effect in cardiomyocytes, in endotheliocytes, most of its adverse effects are mediated by miR146a.

Moreover, $_{N}$Prl provokes shedding of miR146a-enriched exosomes from endotheliocytes that are taken up by cardiomyocytes [36]. In cardiomyocytes, exosomal miR146a reduces metabolic activity, at least partly, by downregulating the expression of the miR146a target genes Her4, Notch1, and Irak1. Consequently, the intercellular communication is disturbed, as in cardiomyocytes the HER4 receptor cannot homo- and heterodimerize with HER2, thereby impeding the Nrg1-HER2/4 signaling, neuregulin-1 being mainly produced by endotheliocytes. Cardiomyocyte-specific deletion in signal transducer and activator of transcription STAT3 generates postpartum cardiomyopathy in mice (Sect. 1.5.1.3; [363]).

4.18 Sarcoidosis-Associated Cardiomyopathy

Sarcoidosis is characterized by the formation of noncaseating granulomas that can infiltrate the myocardium. Cardiac granulomas evolve to scar tissue. Sarcoidosis provokes a restrictive cardiomyopathy in 5 % of patients and can later progress to dilated cardiomyopathy.

4.19 Tachycardia-Induced Cardiomyopathy

Prolonged tachycardia can generate myocardial dysfunction. Persistent or permanent atrial fibrillation causes electrical and structural remodeling of the atria as well as possibly adverse ventricular remodeling and dilated cardiomyopathy.

4.20 Toxic Cardiomyopathies

Alcoholic cardiomyopathy that results from heavy drinking ($>$ 90 g/day) during an average duration of 15-year accounts for approximately 4 % of all cardiomyopathies. Diastolic dysfunction usually precedes systolic dysfunction. Alcohol abstinence can improve ejection fraction.

Cocaine and amphetamines can provoke dilated cardiomyopathy because of a direct myocyte toxicity, tachycardia-induced injury, hypertension, and myocardial infarction.

Effect of cardiotoxic chemotherapy may be antagonized by brain-type natriuretic peptide.

4.21 Valvular Cardiomyopathy

Hemodynamically significant valvular lesions, such as aortic stenosis and regurgitation as well as mitral regurgitation, produce pressure and volume overload that engenders adverse ventricular remodeling and systolic, diastolic, or combined myocardial dysfunction.

Resulting cardiomyocyte hypertrophy leads to chamber enlargement and myocardial fibrosis. Chamber dilation exacerbates existing mitral or tricuspid valvular regurgitation. Further dilation creates subendocardial ischemia and localized myocyte necrosis.

Chapter 5
Conduction and Rhythm Disorders

At the microscopic scale, multiple cell types (e.g., nodal, atrial, and ventricular myocytes, and cardiofibroblasts) and, at the nanoscopic scale, numerous signaling cascades control the generation and propagation of cardiac action potentials (i.e., temporal changes in transmembrane potential during activation of excitable cells). Cardiac excitable cells comprise successively pacemakers, conducting specialized nodal cells, atrial and ventricular myocytes.

These cells and involved signaling cascades regulate the relative regularity of the cardiac beating frequency. However, a reduced cardiac frequency variability is a marker of disease. The heart actually has a chaotic behavior.[1] Its nonperiodic

[1] Chaos, a concept defined in fluid dynamics (in particular meteorology) and patterning, is related to highly nonlinear, deterministic, phenomena (deterministic chaos). Hence, chaos is not associated with randomness or quasi-randomness, such as that involved in Brownian motion (i.e., random movement of particles suspended in a fluid resulting from their collision) and percolation. A random behavior is downright unpredictable and irreproducible. A distribution of possible aftereffects is associated with some probability. A random process behaves differently every time it is rewound. The chaos theory is linked to the behavior of dynamical systems that are highly sensitive to initial conditions, small differences in initial conditions yielding diverging outcomes. The chaotic behavior is aperiodic (irregular in time), complex, but ordered. Although a given initial conditions leads to the same final state (determinism), the final state can differ strongly for small changes in initial conditions. It is thus difficult or impossible to make a long-term prediction. Hence, it is not entirely predictable (pseudo-random), but it is determined without random by initial conditions that may not be perfectly handled. Unstable systems have finite predictability, whereas stable systems are infinitely predictable. A deterministic model incorporates initial conditions and yields an exact trajectory of the system. Because the system is chaotic, the trajectory varies when the initial conditions change even slightly. In other words, "the present determines the future, but the approximate present does not approximately determine the future." (E.N. Lorenz (1917–2008)). "If we knew exactly the laws of nature and the situation of the universe at the initial moment, we could predict exactly the situation of that same universe at a succeeding moment. But even if it were the case that the natural laws had no longer any secret for us, we could still only know the initial situation approximately. If that enabled us to predict the succeeding situation with the same approximation, that is all we require, and we should say that the phenomenon had been predicted, that it is governed by laws. But it is not always so; it may happen that small differences in the initial conditions produce very great ones in the final phenomena. A small error in the former will produce an enormous error in the latter. Prediction becomes impossible, and we have the fortuitous phenomenon"

M. Thiriet, *Diseases of the Cardiac Pump*, Biomathematical and Biomechanical Modeling of the Circulatory and Ventilatory Systems 7, DOI 10.1007/978-3-319-12664-7_5

behavior characterizes a pump able to quickly react to any changes of the body's environment. The normal heartbeat indeed exhibits complex nonlinear dynamics. As does the loss of control of generation and/or propagation of action potentials, that is, an anarchical behavior, a stable periodic cardiac dynamics yields a bad prognosis.

Cardiac conduction related to the propagation of electrochemical excitation through the heart enables the quasi-synchronous contraction of cardiomyocytes.

The QT interval is characterized by a sexual dimorphism that should be taken into account to assess arrhythmia risk and drug sensitivity and select treatment modality [365].

Action potential transmission from the natural pacemaker to ventriculomyocytes can be delayed at a point of the nodal tissue, hence causing sinoatrial, atrioventricular, and incomplete or complete right and left branch blocks. Defects in conduction disrupt synchronous activation of ventriculomyocytes and can cause life-threatening arrhythmias in many pathologies [366].

Action potentials can arise from ectopic foci, in particular after ischemia or toxic administration. The myocardium acquires a new rhythm imposed by the conduction tissue downstream from the pathological zone. Focal activity can start spontaneously or after a trigger, i.e., a preceding normal action potential, triggered rhythms being either slow (e.g., long-QT-interval-associated arrhythmias) or quick (e.g., catecholaminergic polymorphic ventricular tachycardia) [392].

Diabetes mellitus alters cardiac function. Impaired insulin signaling in cardiomyocytes disturbs ventricular repolarization [367]. In a mouse model with cardiomyocyte-restricted deletion of insulin receptors, altered insulin action reduces expression of K^+ channels involved in ventricular repolarization such as those responsible for the fast component of the repolarizing transient outward K^+ current ($i_{K,to(r)}$) through the $K_V4.2$–KChIP2 channel, thereby prolonging ventricular action potential duration (APD).[2]

Artificial pacemakers are implanted to treat severe block syndromes. The pacemaker continuously monitors the heartbeat and delivers an imperceptible electrical charge to stimulate the heartbeat if needed. Typically, a pacemaker is endowed

(J.H. Poincaré (1854–1912) [364]). Chaos is engendered by set of coupled nonlinear first-order ordinary differential equations for at least three dynamical variables:

$$\frac{dx}{dt} = \kappa_1(x - y),$$
$$\frac{dy}{dt} = \kappa_2 x - xz - y,$$
$$\frac{dz}{dt} = xy - \kappa_3 z. \tag{5.1}$$

[2] Cardiac voltage-gated potassium channel-interacting proteins (KChIP) is a small calcium-binding protein that serves as modulatory subunit for members of the pore-forming K_V4 channel category. In cardiac hypertrophy and failure, KChIP2 expression lowers. The cardiac calcium-independent transient outward potassium current relies on $K_V4.2$ and $K_V4.3$ proteins.

with two insulated wires (electrodes or leads) in the right atrium and ventricle (RA and RV leads). Cardiac resynchronization therapy (CRT) in congestive heart failure with electrical and mechanical dyssynchrony, especially in moderately to severely symptomatic patients despite medical therapy (dyspnea, edema of lower limb extremities, exercise intolerance, and depression; left ventricular ejection fraction [LVEF] ≤ 0.35), is aimed at restoring the normal coordinated pumping action of the ventricles by overcoming the delay in electrical conduction caused by a left bundle branch block (LBBB; prolonged QRS complex duration [≥ 120 ms], [368]).

5.1 Summary of Electrochemical and Mechanical Events

Cardiac electrochemical events and involved ion carriers that trigger myocardial contraction and relaxation cycles have been described in Vols. 5 (Chap. 5. Cardiomyocytes) and 6 (Chap. 3. Cardiovascular Physiology). This section summarizes and updates data.

In a broad sense, *opsins* are light-sensitive proteins that include G-protein-coupled receptors (e.g., Gq-coupled melanopsin) and ion carriers. Optogenetics uses genetically encoded light-sensitive proteins to actuate and/or sense fast biological processes. Optogenetics pacing enables remote spatiotemporal exploration as well as control of cardiac cell excitability, signal transmission, and gene expression by light-induced transcriptional effectors [370–372]. However, inserted nonnative opsins may exert yet unknown effects and implantation of modified cells may disturb electrical coupling between cell types, in addition to a different electrophysiological behavior.

5.1.1 Electromechanical Coupling

The current transmitted between cardiomyocytes during discharge and small molecular messengers cross gap junctions made up of voltage-sensitive connexins (mainly Cx40, Cx43, and Cx45) and organized in arrays in intercalated discs. Gap junctions couple cardiomyocytes axially and transversely. Gap junction conductivity depends also on intracellular Ca^{2+} level, pH, and phosphorylation.

The initial event in the cardiac cycle is the fast membrane depolarization, resulting from ion entry through gap junctions and opening of voltage-gated Na^+ channels that induces Na^+ influx (Vol. 3, Chap. 3. Main Sets of Ion Channels and Pumps).

In myocytes, the $Na_V1.5$ channel encoded by the SCN5A gene is the principal expressed voltage-gated Na^+ channel. It is responsible for the largest membrane current in excitable cells that causes the rising phase of the action potential. Most voltage-gated Na^+ channels activate and inactivate completely within the first few milliseconds of an action potential. The $Na_V1.5$ channel predominantly localizes to intercalated discs to successfully propagate action potentials.

Membrane depolarization then inactivates Na^+ channels and opens both K^+ and Ca^{2+} channels. Calcium entry into the cell triggers Ca^{2+} release from its stores (sarcoplasmic reticulum) via ryanodine-sensitive Ca^{2+} channels. Calcium ions then bind to troponin complexes and activate contractile sarcomeres.

Sarcomere relaxation follows upon Ca^{2+} removal from the cytosol back into its stores by Ca^{2+} uptake pumps of the sarcoplasmic reticulum (SERCA) and extracellular milieu by sarcolemmal Na^+–Ca^{2+} exchanger (NEX) and Ca^{2+} ATPases (PMCA). Intracellular Na^+ homeostasis is achieved by Na^+–K^+ pumps.

5.1.2 *Genesis and Propagation of Action Potentials—Nodal Tissue*

Some cardiac cells are self-excitable, that is, have their own intrinsic gating and contraction rhythm. They pertains to the sinoatrial node, or natural pacemaker that sets the rate of contraction–relaxation cycles of cardiomyocytes (Vol. 5, Chap. 5. Cardiomyocytes). The sinoatrial node generates electrochemical impulses that are sent to neighboring electrically coupled cells via adhesion sites.

Action potentials spread rapidly through atrial walls and reach the atrioventricular node located between the right atrium and ventricle, where they are delayed for about 100 ms before traveling to the ventricular walls.

Action potentials then run through the His bundle and its branches and then Purkinje fibers that conduct signals from the cardiac apex to the basis. They produce electrical currents conducted to the skin , where these signals can be recorded as an electrocardiogram trace.

5.1.2.1 Sinoatrial Node

The sinusal or sinoatrial node (SAN; nodus sinuatrialis), also called Keith and Flack node, is an extensive structure within the right atrium that acts as the natural pacemaker (impulse generator) that emits action potentials with a normal sinusal rhythm.

Cardiac automaticity arises from the integrated activity of voltage-gated ionic currents (HCN2 and HCN4 as well as $Ca_V1.3$ and $Ca_V3.1$), transporters (NCX), and sarcoplasmic reticulum Ca^{2+} release.

In addition to these classical ion carriers, some members of the TRP (transient receptor potential) superfamily of ion channels are also expressed in the myocardium (TRPC1, TRPC3–TRPC4, TRPC6, TRPM4, and TRPM7).

Structure and Location of the Sinoatrial Node

The sinoatrial node is cluster of specialized cardiomyocytes positioned on the inner layer of the right atrial wall near the entrance of the superior vena cava. It is a diffuse, but integrated aggregate of pacemaker cells that extends from the junction between the right atrium and superior vena cava (SVCRAJ) virtually to the junction between the right atrium and inferior vena cava (IVCRAJ) [373]. The right atrial pacemaker region comprises a zone of 75 $\times$ 15 mm centered about the long axis of the sulcus terminalis (or crista terminalis) posteriorly and the precaval band anteriorly.

The pacemaker is defined as unicentric when a single source of activation spreads to activate the atria. The pacemaker is considered as multicentric when several sites (≥ 2) of early activation are separated by a distance greater than or equal to 10 mm with activation time difference shorter than or equal to 5 ms [374].

Ca_V and HCN Channels

The spontaneous activity of sinoatrial myocytes determines cardiac frequency. The maximal cardiac frequency declines with aging, progressively limiting the aerobic capacity of the elderly, as the maximal cardiac frequency is a major determinant of age-dependent decrease in the maximum rate of oxygen consumption. The depressed spontaneous activity of sinoatrial myocytes results from altered ion channel activity [375]. These cells are characterized by:

- Slower firing rates
- Altered action potential waveform with changes limited to hyperpolarization of the maximum diastolic potential and slowing of the early part of the diastolic depolarization
- Reduced L- ($i_{Ca,L}$; $Ca_V1.3$ mainly)[3] and T-type ($i_{Ca,T}$; $Ca_V3.1$)[4] Ca^{2+} and hyperpolarization-activated depolarizing pacemaker funny (i_f; HCN2 and HCN4)[5] currents, which all contribute to diastolic depolarization, this reduction being associated with a hyperpolarizing shift in the voltage dependence of i_f.

These electophysiological changes are associated with cellular hypertrophy. The action potential upstroke, duration, and repolarization as well as the chronotropic response to β-adrenoceptor stimulation are preserved [375].

---

[3] The $Ca_V1.3$ channel isoform activates at more negative potentials than the $Ca_V1.2$ channel isoform that is predominant in atrial and ventricular myocytes.

[4] The voltage dependence of the $Ca_V3.1$ channel is related to a much more negative potential than that of the $Ca_V1.3$ channel, corresponding more closely to the early part of the diastolic depolarization.

[5] The activation of HCN2 is similar to that of HCN4. Both HCN1 and HCN3 activate at substantially more positive potentials.

Na^+–Ca^{2+} Exchanger

The spontaneous cardiac pacemaking relies on complex molecular mechanisms that include a coordinated interplay between intracellular Ca^{2+} cycling and plasmalemmal cation transport. Among involved ion carriers, the Na^+–Ca^{2+} exchanger NCX1 is a sarcolemmal protein involved in the maintenance of calcium homeostasis in cardiomyocytes. Selective ablation of NCX1 in nodal cells provokes a progressive reduction of cardiac frequency with irregular firing accompanied by severe bradycardia as well as various supraventricular and ventricular rhythm disorders (sinusal arrhythmia and pauses, atrioventricular block, and ventricular tachycardia, [376]). In addition, Ca^{2+} transients have a smaller magnitude and slower kinetics.

The NCX1 loss in pacemaker cells is not compensated by an modified expression of PMCA and Ca_V channels. The associated increase in the production of SERCA2 is not translated into an amplified Ca^{2+} sequestration in the sarcoplasmic reticulum. Pacemaker cells have an elevated cAMP level and a pronounced basal PKA activity very likely due to the constitutive activity of calcium-activated adenylate cyclases AC1 and AC8 [376]. Pacemaker cells lacking NCX1 have reduced, slower, and disrhythmic Ca^{2+} oscillations. In addition, the responsiveness to ligands of Gi-coupled receptors increases, but that of agonists targeting Gs-coupled receptors decreases. Hence, β-adrenergic response amplitude decays, whereas the adenosinergic and muscarinic effect intensity augment [376]. Gi- and Gs-coupled receptors inactivate and activates PKA, respectively. PKA phosphorylates phospholamban, an inhibitor of SERCA2 density, hence relieving this inhibition. The combined effect of attenuated AC activity and PKA inactivation may counterbalance increased SERCA2 level in NCX1-lacking pacemaker cells.

Transient Receptor Potential Melastatin (TRPM) Channel

TRPM7, a divalent-permeant channel-kinase, is highly expressed in the sinoatrial node, where it is required for cardiac automaticity, in addition to human atrial myocytes and fibroblasts [377]. It enables outwardly rectifying current that is inhibited by both cytoplasmic and extracellular Mg^{2+} ions.

In atrial fibrillation, the TRPM7-mediated current is elevated in human atriofibroblasts [377]. Ventricular fibroblasts are also sites of large TRPM7-conveyed currents. The TRPM7 channel concentration is the largest in myocardial cells endowed with automaticity.

Ablation of the TRPM7 channel in cultured embryonic cardiomyocytes significantly reduces spontaneous Ca^{2+} transient firing rates as well as production of Hcn4, CACNA1G ($Ca_V3.1$), and SERCA2A mRNAs. Global murine cardiac TRPM7 deletion disrupts cardiac automaticity in vivo, causing sinusal pauses and atrioventricular node block. The TRPM7 channel influences diastolic membrane depolarization and automaticity in SAN cells via the Hcn4 gene transcription [377]. It may operate via the transcription factors Tbx3 and/or MEF2.

Another TRPM family member, TRPM4, a Ca^{2+}-activated, Na^{+}-selective channal is enriched in human Purkinje cells. Mutations (gain-of-function) of the Trpm4 gene is responsible for the progressive familial heart block-I associated with an impaired TRPM4 endocytosis secondary to constitutive sumoylation.

5.1.2.2 Atrioventricular Node

The atrioventricular node conducts slowly and has a long refractory period, thereby engendering a delay between atrial and His bundle activation, controlling the ventricular impulse rate, and filtering atrial impulses especially during supraventricular tachyarrhythmias. The atrioventricular node is characterized by rate-dependent conduction and refractoriness [378]. However, the atrioventricular node with its fast and slow pathway is also the site of *atrioventricular nodal reentrant tachycardia*, the most common form of paroxysmal supraventricular tachycardia .

The atrioventricular node occupies the Koch triangle, the upper atrial and lower ventricular sides of which correspond to the tendon of Todaro and tricuspid valve insertion, respectively. The His bundle and coronary sinus ostium represent its anterosuperior apex and posteroinferior base, respectively.

The atrioventricular node contains substructures (transitional zone, posterior extension, compact node, and lower nodal bundle) that determine conduction time [378]. The connection between the transitional zone, compact node, and lower nodal bundle is related to *fast conduction pathway*. Another connection between the septal portion of the transitional zone and lower nodal bundle bypassing the compact node may also yield another fast conduction pathway. The posterior transitional zone, posterior extension, and lower bundle together provide the *slow conduction pathway*. The transitional zone that connects the atrium to the compact node and posterior extension gives the *common proximal pathway*. The lower nodal bundle is linked to both the compact node and posterior extension and furnishes the *common distal pathway* to the His bundle.

The atrioventricular node function is assessed with various protocols, each with their own drawbacks, such as the premature or extrastimulus protocol to identify the recovery and fatigue properties [378].

5.1.3 Myocardial Mechanics and Electrical Activity

Action potentials are electrochemical impulses that trigger myocardial contraction (excitation–contraction coupling [ECC]). Conversely, mechanical state of the myocardium influences its electrical activity (mechanoelectrical [MEC] or contraction–excitation [CEC] coupling, [379]).

The shape of action potentials can be modified by four basic mechanisms [379]:

1. The degree of inactivation or recovery of ionic carriers

2. The concentrations of cytosolic Ca^{2+} and external K^+ ions
3. The level of ionic exchanges, in particular through Na^+–K^+ and Na^+–Ca^{2+} exchangers
4. The stretch magnitude, which can modify the resting transmembrane potential

A diastolic elevated stretch can provoke an immediate reversible depolarization. The strain field may affect the ion carrier kinetics and change their permeability. When stretched myofibers are stimulated, the action potential from partly depolarized membrane has a reduced amplitude. In addition, the conduction velocity lowers at least in stretched rat ventricles.

5.1.4 Sarcolemmal Density of Ion Carriers

Ion channels, exchangers, and pumps yield inward excitatory and/or outward repolarizing currents. Normal cardiac electrochemical activity depends on the expression with a proper density of numerous ion carriers and their partners in specialized nanodomains of the sarcolemma and organelles of cardiomyocytes.

The plasmalemmal density of ion carriers depends on the transfer of their subunits from the endoplasmic reticulum and Golgi body to the cell surface. Exo- and endocytosis that continuously regulate the degradation or recycling of ion carriers control the density of functional ion channels in response to various stimuli.

5.1.4.1 K_V Channels

The density of K_V11.1 channels is also regulated by external potassium concentrations [380]. In LQTS2 syndrome due to a missense mutation in KCNH2 gene encoding K_V11.1 channel (responsible for the rapid delayed rectifier K^+ current [$i_{K,r}$]) associated with a prolonged ventricular repolarization, the defective channel transfer to the sarcolemma results from an elevated transport of the mutated, but functional channel to the proteasome [381].

In atriomyocytes, the surface expression of the main repolarizing channel K_V1.5 depends on the working conditions of the atrium. Hemodynamic stress stimulates the delivery of K_V1.5 channels from recycling endosomes via an Itg–FAK-mediated exocytosis [380]. During atrial pressure overload, this process is upregulated; the plasmalemmal K_V1.5 density rises and contributes to the shortening of the APD.

5.1.4.2 SK Channels

Small-conductance Ca^{2+}-activated K^+ channels (SK; $K_{Ca}2$) are widespread throughout the body. In excitable cells, they aid in processing changes in cytosolic Ca^{2+} concentration and ensuring proper membrane potentials. The SK channels are gated directly by submicromolar concentrations of calcium. In humans, cardiomyocytes express all three SK channel subtypes (SK1–SK3 or $K_{Ca}2.1$–$K_{Ca}2.3$). They can heteromultimerize. The SK2 subtype is highly sensitive to apamin, whereas both SK1 and SK3 channels show intermediate sensitivity. In atria, the SK channel contributes to repolarization.

The SK density is larger in atriomyocytes than in ventriculomyocytes [382]. The SK1 and SK2 channel subtypes are predominantly expressed in the atrium and SK3 is expressed in the atrium and the ventricle [383]. Levels of SK2 and SK3 transcripts are higher than that of the SK1 subtype in human atria [384].

The members of the SK channel family are encoded by the genes of the human KCNN family (KCNN1–KCNN4), the SK4 channel being in fact the intermediate-conductance IK channel (IK1 [$IK_{Ca}1$] or $K_{Ca}3.1$ protein). The $IK_{Ca}1$ channel is an important contributor of the repolarization maintenance, at least in a rabbit heart failure model [383].

The SK channel accelerates atrial repolarization, thereby augmenting the risk of atrial fibrillation from the pulmonary vein outlet region [385].

Both SK2 and SK3 levels are reduced in patients with chronic atrial fibrillation (CAF) with respect to patients with sinusal rhythm (SR). In isolated atriomyocytes from SR patients, a SK-selective inhibitor diminishes the inwardly rectifying K^+ current by about 15 % and prolonges APD, without affecting myocytes from CAF patients [384]. In trabeculae muscular strips from right atrial appendages from SR patients, a SK-selective inhibitor raises APD and effective refractory period and depolarizes the resting membrane potential, but only a SK-slightly selective inhibitor provokes these effects in CAF patients [384]. On the other hand, the SK channels do not contribute substantially to the ventricular action potential.

In KCNN2 (SK2) knockout mice, atrial repolarization is disturbed with elevated susceptibility to atrial fibrillation [386]. Variants in the KCNN3 transcript gene encoding the SK3 subtype ($K_{Ca}2.3$) augments the risk of atrial fibrillation [386].

The expression of SK channels in canine atria depends on the atrial region and wall remodeling. SK channel block does not cause statistically significant effects on APD in the left atrium, but affects it in pulmonary vein cardiomyocyte sleeves, especially in the presence of atrial fibrillation-related remodeling [386].

In homozygote $SK3^{T/T}$ mice, overexpression of the SK3 channel abbreviates atrial APD and facilitates occurrence of atrial fibrillation. The APD is significantly shortened at 90 % repolarization (APD_{90}) in atriomyocytes, but not at 50 % (conserved APD_{50}, [382]). In addition, overexpressed SK3 channels does not alter the resting membrane potentials.

In homozygote ($SK3^{T/T}$) and heterozygote ($SK3^{+/T}$) SK3-overexpressing mice, SK3 overexpression also disturbs conduction at and below the atrioventricular node [383]. In $SK3^{T/T}$ mice, cardiac conduction blocks and bradyarrhythmias may cause premature sudden death with abnormal atrioventricular node morphology (neurological or respiratory dysfunction can be actual cause of death). Ventricular conduction slows down by 25 % with respect to wild-type (WT) controls (0.45 ± 0.04 vs. 0.60 ± 0.09 mm/ms in WT mice). At 1 month age, the atrioventricular nodal refractory period lowers (43 ± 6 vs. 52 ± 9 ms [decrease of 18 %]); at 3 months, the atrioventricular nodal refractoriness is prolonged (61 ± 1 vs. 52 ± 6 ms [increase of 17 %]); at 5 months, the atrioventricular node enlarges and is disorganized [383].

5.1.5 Complexes Incorporating Ion Channels

The spatial dynamical organization of ion channels and adaptor proteins in molecular complexes in cardiomyocytes can confer specific roles to these complexes in the generation and/or propagation of action potentials.

Numerous PKA-sensitive ion channel complexes contain A-kinase-anchoring protein that recruits kinases, phosphatases, and phosphodiesterases to control the local phosphorylation state.

Voltage-gated sodium channels complex with the ankyrin, calmodulin, dystrophin, syntrophin, fibroblast growth factor homologous factor FHF1b,[6] and NEDD4-like protein–ubiquitin ligase. These partners are involved in the regulation of channel synthesis, localization, activity, and degradation of the Na_V α subunit [388].

Calcium–calmodulin kinase-2 (CamK2α–CamK2δ) regulates ion channel functioning, especially voltage-gated Na^+ channels primarily in cardiomyocytes. It also phosphorylates Ca_V1.2a channel β2a subunit on the T-tubule in cardiomyocytes.

In cardiomyocytes, the actin-binding protein β4-spectrin regulates the insertion of ion channels into the intercalated disc, as it connects to the adaptor protein ankyrin-G [389].

The anchoring protein β4-spectrin can have opposite effects on cardiac repolarization according to whether it interacts with K^+ channel or Na^+ channel.

β4-Spectrin links to the 2-pore K^+ channel TREK1, which is environmental and physical factors (membrane stretch, pH, and polyunsaturated fatty acids). It is activated by mechanical stress and arachidonic acid accumulation. It then participates in action potential shaping. In mice, in the absence of β4-spectrin–TREK1 binding, TREK1 channel is aberrantly located and its activity decays, hence causing delayed action potential repolarization and arrhythmia [389].

β4-Spectrin also tethers to the CamK2 kinase and the cardiac Na_V1.5 and contributes to the regulation of APD [390]. Inward Na^+ current quickly inactivate to allow repolarization. Ankyrin-G recruited by β4-spectrin brings Na_V1.5 complexed with the actin–spectrin cytoskeleton at intercalated discs. The CamK2δ kinase phosphorylates (Ser571) the Na_V1.5 channel. Deletion of the CamK2-binding site of

[6] Fibroblast growth factor (FGF) homologous factors (FHF1–FHF4 or FGF11–FGF14) interact with voltage-gated sodium channels. They bind to the kinase scaffold protein mitogen-activated protein kinase-8-interacting protein-2 (MAPK8IP2), also called islet brain IB2 [387]. FGFs operate upon binding, dimerizing, and activating members of the FGF receptor family (FGFR1–FGFR4) of receptor Tyr kinases. Both FGFs and FGFR interact with heparan sulfate glycosaminglycans for sustainable FGF–FGFR binding and dimerization. Both Na_V1.5 and Na_V1.9 bind to FHF1b.

β4-spectrin eliminates the late Na^+ current and causes a marked shortening of the action potential.

Many other scaffolding or anchoring proteins interacts with various ion channels in cardiomyocytes [380].

The membrane-associated guanylate kinase (MAGuK) protein Disc large homolog DLg1 regulates the formation of the $Na_V1.5$–$K_{IR}2.1$ complex that may be involved in depolarization (rising phase of action potential). The $K_{IR}2.1$ channel may maintain surrounding $Na_V1.5$ channels in the closed state.

Catenin-γ supports the formation of the Cx43–$Na_V1.5$ complex at the desmosome.

The cardiac $Na_V1.5$ channel can also tether to the dystrophin-associated glycoprotein complex.

5.1.6 Cardiomyocyte Tubules

The plasma membrane of cardiomyocytes invaginates to form the tubular set involved in excitation–contraction coupling. The tubular set is characterized by transverse (T) and longitudinal (or axial [A]) tubules running from a T tubule to the next. Ion carriers that respond to action potentials localize to the tubular set with different densities and isoforms from those in the remaining part of the cardiomyocyte sarcolemma. Action potentials are identical in amplitude and duration throughout the entire sarcolemma, i.e., in tubules and extratubular cell surface [391]. The myocyte tubular set rapidly conducts depolarization to trigger Ca^{2+} entry across the sarcolemma that initiates Ca^{2+} release from the sarcoplasmic reticulum, thereby priming synchronous contraction of the sarcomere set in the entire myocyte.

A structural disorganization of the tubular set alters the electrical coupling between it and the remaining part of the cardiomyocyte surface. Pathological remodeling of the tubular set can prime asynchronous calcium influx, hence dysfunctional cardiac contractility and arrhythmias [391].

5.2 Arrhythmias

Abnormal heart rhythms, notably atrial and ventricular fibrillation, cause death (10–20 % of all deaths among adults in the developed world). Arrhythmias result from interactions between a susceptible anatomical region (e.g., myocardial scar and fibrosis) and a triggering event (e.g., adrenergic surge, inflammation, acute myocardial ischemia, unusual wall stretch, and drug administration) [392]. Heart diseases with remodeling and altered expression and/or function of ion carriers frequently generate episodes of arrhythmias.

Magnetic resonance imaging with tagging can evaluate pacing protocols and locate *ectopic sites*. The temporal evolution of 3D strain fields has been computed from

MR tagging of canine hearts for three different pacing sites, the base of the left ventricle free wall, the right ventricular apex, and the right atrium [393]. Right atrium pacing shows rapid synchronous shortening. Left ventricle base pacing exhibits a slowly propagating wavefront from the pacing site. Right ventricle apex pacing induces regional variations in propagation, with rapid septal activation followed by slower than usual activation. The propagation speed corresponds to the speed of electrochemical propagation in the myocardium.

During catheterization, using either a multielectrode basket (contact mapping system) or wire mesh, twinned with a second radiofrequency emitting catheter (non-contact mapping system), magnetic resonance, and mobile X-ray imaging can be merged into a common coordinate system using tracking and registration (XMR) to collect cardiac anatomy, motion, and electrical activity [394]. In vivo observations can then be combined with numerical simulations of the heart pathological rhythms.

5.2.1 *Types of Arrhythmias*

Heterogeneities in excitability and/or conduction velocities in the heart wall can initiate arrhythmias. Furthermore, increased dispersion of recovery and/or refractoriness (Vol. 5, Chaps. 5. Cardiomyocytes and 6. Heart Wall) in nodal cells caused by tissue damage can produce a conduction block. A premature impulse from an area with a short refractory period can propagate and find a region with a longer refractory period, leading to unidirectional propagation.

5.2.1.1 Bradycardia and Tachycardia

Cardiac frequency disorders include both increase and decrease in beating rate. *Tachycardia*[7] results from various causes, in particular heart deficiency. *Bradycardia*[8] corresponds to a cardiac frequency lower than 1 Hz.

5.2.1.2 Long- or Short-QT Syndromes

Electrocardiographic QT intervals measure cardiac repolarization. Extremely long or short QT intervals[9] define long-QT syndromes (LQTS) and short-QT syndromes

---

[7] ταχος: quickness, καρδια: heart.

[8] βραδος: slowness.

[9] Interval QT measured on the electrocardiogram is the distance from the beginning of the Q wave to the end of the T-wave. Owing to its sensitivity to the cardiac frequency, this parameter is normalized by the cardiac period (RR-interval).

(SQTS). The NOS1AP gene that encodes the nitric oxide synthase NOS1 adaptor protein[10] is associated with a QT-interval variation [395].

5.2.1.3 Early and Delayed Afterdepolarizations

Alterations of the electric properties of cardiomyocytes in heart failure can favor the occurrence of atrial and ventricular arrhythmias by inducing EADs or DADs.

Arrhythmias can spontaneously arise in hearts that beat at normal or slow rates, i.e., not necessarily at fast cardiac frequency, although EADs can also appear at rapid cardiac frequencies, often in association with DADs.

In heart diseases, cardiomyocytes exhibit abnormal electrical oscillations such as EADs generated by drugs or due to genetic modifications. These coupled oscillators can synchronize and then propagate depolarizations. EADs at slow cardiac frequency actually exhibit chaotic dynamics [396].

Ventricular tachycardia or fibrillation results from spontaneous, premature ventricular complexes that typically occur irregularly (not randomly, but in the framework of dynamical chaos) and can be caused by EADs that self-organize. However, at least partial synchronization is required for propagation of EADs. Chaotic EADs globally synchronize when the tissue volume is smaller than a critical size [396].

On the other hand, when the tissue size exceeds the critical threshold, regions of partial synchronization generate premature ventricular complexes. These complexes propagate into recovered tissue without EADs.

Early Afterdepolarizations

EADs can result from a secondary activation of the $Ca_V1.2$ channel during the plateau of the action potential, as well as decreased activity from the rapid component of the delayed rectifier K^+ channels ($K_V11.1$) in damaged cardiomyocytes. EADs can result also from both $i_{Na,L}$, in addition to $i_{Ca,L}$ currents [397].

The late Na^+ current ($i_{Na,L}$) is indeed arrhythmogenic, as it induces an prolonged inward current favorable for EADs. Intracellular Na^+ cycling is coupled with Ca^{2+} homeostasis, as Na^+ modulates the transport direction of the Na^+–Ca^{2+} exchanger. Cellular Na^+ overload as well as action potential prolongation stimulates the reverse mode of Na^+–Ca^{2+} exchanger, hence Na^+ efflux in exchange of Ca^{2+} influx.[11]

In early, compensated maldaptive hypertrophy with preserved systolic contractility, $i_{Na,L}$ and APD remain unchanged. However, upon hypertension-induced heart

[10] A.k.a. C-terminal PDZ ligand of neuronal nitric oxide synthase protein (Capon).

[11] On the other hand, spontaneous sarcoplasmic reticular Ca^{2+} releases, Ca^{2+} sparks, through ryanodine receptor RyR2 elevate diastolic Ca^{2+} level that can be attenuated by the Na^+–Ca^{2+} exchanger, hence generating a depolarizing transient inward current (i_{ti}) and DADs.

failure with prolonged APD, the $i_{Na,L}$ decay time is slower, but can be normalized by inhibition of Ca^{2+}–calmodulin-dependent protein kinase CamK2 [398]. Increased CamK2 activity and subsequent $Na_V1.5$ phosphorylation (Ser571) cause afterdepolarizations.

Delayed Afterdepolarizations

Afterdepolarizations can result from aberrant calcium handling and provoke arrhythmogenesis. DADs during phase 3 of the action potential are caused by spontaneous Ca^{2+} release from the sarcoplasmic reticulum. Calcium-induced DADs are modulated, but not generated by elevated sarcoplasmic reticulum Ca^{2+} load. These afterdepolarizations require elevated calcium concentrations in the dyadic subspace and the cytosol that raise the open probability of ryanodin receptors [397].

Spontaneous Ca^{2+} release triggers a premature action potential in failing cardiomyocytes that activates Na^+–Ca^{2+} exchangers. The action potential in failing ventriculomyocytes is longer than in normal ones. The increased duration of the action potential is induced by enhanced activity of Na^+–Ca^{2+} exchangers, slowed decay in Ca^{2+} transient fluxes, and reduced activity of inwardly rectifier K^+ channels and Na^+–K^+ pumps.

Metabolic Links to Electrical Dysfunction

The sequential oxidation of nutrients, fatty acids, and glucose leads to the common substrate for the tricarboxylic acid (TCA) cycle, acetylCoA (Vol. 6, Chap. 3. Cardiovascular Physiology). The latter is used to produce reducing equivalents NADH and $FADH_2$.

Agent NADH feeds electrons to the electron transport chain. The redox reactions of the *electron transport chain* are coupled with proton translocation across the mitochondrial inner membrane. This H^+ flux establishes an electrical potential at the mitochondrial membrane and pH gradient that both determine a *proton-motive force*. The latter is used by the mitochondrial ATP synthase to produce ATP, which is exported to the cytosol via the adenine nucleotide translocase (ANT).

Although the electron transport chain efficiently utilizes O_2 in the controlled oxidation of redox carriers, superoxide anion ($O_2^{\bullet -}$) is an unavoidable by-product of oxidative phosphorylation.

Mitochondria are involved in arrhythmogenesis. Mitochondria experience waves of membrane depolarization [399]. Periodic depolarizations of mitochondrial membranes are triggered by reactive oxygen species and propagated by *ROS-induced ROS release* (RIRR). Mitochondrial membrane potential oscillations involve ROS-sensitive mitochondrial inner membrane anion channels (IMAC) as well as slow depolarization waves related to mitochondrial permeability transition pore (mtPTP) opening.

ROS-induced ROS release causes oscillations of mitochondrial inner membrane potential, altered intracellular calcium dynamics via mitochondrial Ca^{2+} uniporter, and altered mitochondrial membrane potential due to the inner membrane anion channel, mitochondrial permeability transition pore, and mitochondrial ATP-sensitive K^+ channels ($i_{K_{ATP}}$ current), which shortens action potential.

Adenosine triphosphate produced by mitochondria is mandatory for proper function of several cell determinants of electrical activity, such as ATPases SERCA and Na^+–K^+ pump, as well as $Ca_V1.2$ channel and sarcolemmal ATP-inactivated K^+ efflux channels (Vol. 6, Chap. 3. Cardiovascular Physiology). The latter increases heterogeneity of APD during metabolic stress, hence creating an arrhythmogenic substrate.

Calcium ion transfer across the sarcolemma and sarcoplasmic reticulum during the excitation–contraction process is coupled with mitochondrial energetics and ROS metabolism (Table 5.1). Reactive oxygen species are produced by the electron transport chain, carried by the inner membrane anion channel (IMAC), and scavenged by superoxide dismutase and glutathione peroxidase (among others). This ROS-induced ROS release module and mitochondrial energetics are linked to cellular electrical activity via ATP-sensitive K^+ channel, which is activated when the ADP/ATP ratio increases [397].

Myocardial ischemia and reperfusion (Sect. 8.8) provoke a metabolic stress, or ischemia–reperfusion injury, which culminates in electromechanical dysfunctions, such as ventricular tachycardia and fibrillation.

5.2.1.4 Reentry

Reentry requires a unidirectional block within a conducting path, the action potential traveling in the retrograde, but not in the direct direction, usually due to partial depolarization and suitable timing. The action potential travels into the common distal path, finds the nodal tissue excitable, runs across the block area in the reverse direction.

Reentry can occur either between the atria and ventricles, involving an accessory conduction path, or locally, within a small region of the atrium or ventricle. Reentrant circuits can involve a small (microreentry) or a large (macroreentry) cardiac region in a single chamber or in several chambers.

Since the 1980s, reentry was studied by many investigators in 2D and 3D, homogeneous, isotropic, excitable media [400–403]. An electromechanical model of a contracting excitable medium was recently proposed, the nonlinear excitation waves yielding contraction and the ensuing deformations exerting a feedback effect on the excitation properties of the medium [404]. A 3-variable FitzHugh–Nagumo-type excitation–tension finite difference model is coupled with a hyperelasticity finite element model.

Table 5.1 Modules and components of the coupling between the excitation–contraction process, mitochondrial energetics, and ROS metabolism. (Source: [397])

Components	Elements or targets
Mitochondrial energetics	
TCA–ETC–ANT	$NADH/FADH_2$, ATP
ATP	Myofilament
	Ion pumps (Na^+–K^+ ATPase, PMCA, SERCA)
	Ion channels (K_{ATP})
Phosphocreatine circuit	mtCK–cyCK–smCK
	(ATP/ADP–Cr/Cr^P)
ROS-induced ROS release	
ETC	Superoxide ($O_2^{\bullet -}$)
IMAC	Superoxide transfer
Superoxide metabolism	SOD, GPx, GR
Excitation–contraction process	
Ca^{2+} influx	$Ca_V1.2$
(entry and release)	RyR2
Mitochondrial Ca^{2+}	mtCU (Ca^{2+} influx in matrix)
	NCX (Ca^{2+} efflux from matrix)

ANT adenine nucleotide transporter, *Cr* creatine, Cr^P phosphorylated creatine, *cy/mt/smCK* cytosolic/mitochondrial/sarcomeric creatine kinase, *ETC* mitochondrial electron transport chain, *GPx* glutathione peroxidase, *GR* glutathione reductase, *mtCU* mitochondrial Ca^{2+} uniporter, *PMCA* plasma membrane Ca^{2+} ATPase, *SERCA* sarco(endo)plasmic reticulum calcium ATPase, *SOD* superoxide dismutase, *TCA* tricarboxylic acid cycle

5.2.2 Atrial Arrhythmias

The atrium contributes to about 25 % of the ventricular stroke volume. The initiation of the atrial contraction, similarly to ventricles, involves a nearly synchronous rise of intracellular Ca^{2+} concentration supported by a well-developed transverse tubule network enriched in $Ca_V1.2$ channels and Na^+–Ca^{2+} exchangers that are also concentrated along T-tubules. The former ion carrier accelerates changes in the amplitude of the systolic Ca^{2+} transient during inotropic stimuli and the latter lowers the diastolic cytosolic Ca^{2+} concentration.

Apposed $Ca_V1.2$ and RyR channels that form couplons localize close to NCX antiporters, hence favoring a positive lusitropic effect, as they lower the cytosolic Ca^{2+} level. However, the resulting inward depolarizing current can be proarrhythmogenic [405].

An excessive RyR phosphorylation by PKA or CamK causes Ca^{2+} leakage from the sarcoplasmic reticulum. Subsequent Ca^{2+} efflux through NCX carriers depolarizes the myocyte, thereby initiating and maintaining arrhythmia.

In ventriculomyocytes, the T-tubular orientation is predominantly perpendicular to the long myocyte axis at each Z line with branching between adjacent tubules. Calcium sparks occur predominantly at Z lines along the length of T-tubules.

In atriomyocytes, the T-tubular orientation is mainly axial along the long myocyte axis and less organized than in ventriculomyocytes. In addition, the density of $Ca_V1.2$–RyR couplons is smaller. Furthermore, regional differences exist between left and right atria, appendage and free wall, among other territories [405]. Narrow myocytes have a smaller density of T-tubules than wide myocytes. Hence, the spatial and temporal properties of systolic Ca^{2+} transients differ between the two myocyte types. However, T-tubules in atriomyocytes create simultaneously Ca^{2+} transient at the myocyte edge and center [405].

In addition, inositol trisphosphate receptors activated by IP_3 generated upon stimulation by endothelin-1, angiotensin-2, noradrenaline that targets α1-adrenoceptor, and acetylcholine that bins to muscarinic M_1 and M_3 receptors contribute to Ca^{2+} release from IP_3-sensitive Ca^{2+} stores. This Ca^{2+} influx can stimulate RyR-dependent Ca^{2+} release.

Atriomyocyte contraction relies on $Ca_V1.2$ and RyR channels. Atriomyocyte relaxation depends on SERCA pump, Na^+–Ca^{2+} exchanger, and, to a lesser extent, plasmalemmal PMCA pump. The difference of functioning between atrial and ventricular myocytes is related to the control of systolic Ca^{2+} transients. In the rat, atriomyocytes devoid of T-tubules are smaller and have a reduced Ca^{2+} transient amplitude, higher rate of $[Ca^{2+}]_i$ decay due to greater SERCA activity, larger Ca^{2+} buffering capacity associated with a higher SERCA/phospholamban ratio, and elevated Ca^{2+} store content than ventriculomyocytes [405].

Anchoring proteins, such as junctophilin-2, amphiphysin-2, and telothionin, connect to Z-disc and cytoskeleton constituents and partcipate in the formation of T-tubules in cardiomyocytes and trafficking of ion channels to the T-tubule membrane [405]. Altered expression of these proteins leads to a T-tubule disorganization or loss.

In atria that are the site of common cardiac arrhythmia, the T-tubular network can be remodeled with disorganization and loss in heart failure and atrial fibrillation. In atrial fibrillation, the systolic rise of Ca^{2+} level is considerably less uniform [405].

Certain regions of the atria such as pulmonary vein cardiomyocytes, in which intracellular Ca^{2+} regulation differs from that of atriomyocytes, can be involved in the genesis of atrial fibrillation. An early pulmonary vein myocardium formation under the control of the paired-like homeodomain transcription factor Pitx2c is strongly associated with future occurrence of atrial fibrillation [405].

5.2.2.1 Sinusal Bradycardia and Tachycardia

A sinus rhythm[12] with normal, evenly spaced complexes and a frequency lower than 1 Hz (60/mn) is called sinus bradycardia.

A sinus rhythm with a frequency higher than 1.67 Hz (100/mn) is termed sinus tachycardia.

Irregular sinus rhythms with the longest RR interval (exceeding the usual value by 160 ms) are called *sinus arrhythmias*.

5.2.2.2 Sinus Node Dysfunction

Sinusal node dysfunction (SND), or sick sinus syndrome (SSS), and atrioventricular node block are common causes of bradyarrhythmias that often necessitate permanent pacemaker placement. Sinus node dysfunction causes sinus bradycardia, sinus arrest, exit block, combinations of sinoatrial and atrioventricular nodal defects, and atrial tachyarrhythmias.

Sinusal node disease, a common indication for permanent cardiac pacing, results from disordered impulse generation within the sinusal node or impaired conduction of the impulse to the surrounding atrial wall, thereby engendering bradycardia. Its natural evolution is marked by recurrent syncope, heart failure, stroke, and atrial fibrillation.

Sinusal node disease is associated with diffuse atrial remodeling characterized by structural changes, conduction anomalies, and increased right atrial refractoriness. The sinusal pacemaker activity varies with a loss of the normal multicentric pattern of activation, caudal shift of the pacemaker, and abnormal conduction around lines of conduction block [374].

Whereas human atrial and ventricular arrhythmias can result from many defective ion channel genes, genetic and molecular mechanisms responsible for sinus node dysfunction are associated with ankyrin-B locus ANK2 [406]. Ankyrin-B serves in membrane organization of sinoatrial node cell channels and transporters and hence in cardiac pacing. Variants of the ANK2 gene is also responsible for congenital type-4 LQTS.

5.2.2.3 Ectopic Pacemaker

The origin of atrial contractions can localize elsewhere in the atria other than the sinusal node. If the ectopic pacemaker localizes close to the atrioventricular node, the atrial depolarization occurs in the opposite direction with respect to the normal one, with inversion of P-wave polarity. Any ectopic pacemaker located in the junction between the atria and ventricles generates a *junctional rhythm* with slow cardiac

[12] Activations originate at the sinoatrial node.

frequency (40–55/mn, i.e., 0.66–0.92 Hz), normal QRS complexes, and vanishing P waves.

5.2.2.4 Paroxysmal Atrial Tachycardia

Paroxysmal atrial tachycardia, with a frequency of 160–220/mn (2.67–3.67 Hz), results from reentrant activation. P waves are immediately followed by QRS complexes.

5.2.2.5 Atrial Flutter

Atrial flutter occurs when the atria contract two or three times for each ventricular contraction. Atrial flutter is characterized by TP interval disappearance.

5.2.2.6 Atrial Fibrillation

Atrial fibrillation (AF), the most common sustained cardiac arrhythmia, causes a rapid and irregular atrial activity. Atriomyocytes fire at a frequency ranging from 6.7 to 10 Hz (rate 400–600/mn). The atria quiver rather than truly contract.

Clinical Features

Obesity is a risk factor, especially when it coexists with hypertension, diabetes, and sleep apnea. Atrial fibrillation causes palpitations and dizziness as well as complications, such as stroke and heart failure.

Atrial fibrillations are classified into three main types:

1. Paroxysmal, intermittent, usually self-terminating
2. Persistent
3. Permanent atrial fibrillations

Main Electrophysiological Features

The ventricular frequency is also high (typically ~ 150 pulses/mn, i.e., ~ 2.5 Hz) and irregular with normal QRS complexes. The ventricular beating rate is dictated by the interactions between the atrial rate and the filtering function of the atrioventricular node.

The multiple wavelet hypothesis relies on the existence of randomly propagating waves with intermittent blockades and annihilation and regeneration of waves responsible for a self-sustained, spatiotemporally complex patterns of excitation lacking a hierarchical organization. However, wave propagation during atrial fibrillation is not random and reentry is a major mechanism of fibrillation.

The occurrence and persistence of atrial fibrillation require the combined effect of a trigger, substrate, and activation of the autonomous nervous system [68]. The *trigger* consists of focal spontaneous electrical activity (focal drivers) localized to preferential sites, mainly the pulmonary vein. The *substrate* is related to structural and functional alterations of the atrial myocardium that engender the shortening of atrial refractoriness, formation of reentry circuits, and local conduction blocks. The reentrant activity maintains the arrhythmia.

Anatomical Background

The anatomy of atria with pectinate muscles, the crista terminalis, vein orifices, and atrioventricular rings is a favorable milieu for vortices of electrochemical waves (reentrant rotors or spiral waves) [407]. The crista terminalis, pectinate muscles, and Bachman bundle have greater anisotropy ratios and conduction speeds than the remaining atrial myocardium [408]. Pulmonary veins and the ligament of Marshall act as high-frequency sources [407]. Hence, regions of the posterior left atrium can operate as AF drivers.

The Bachmann bundle and inferoposterior interatrial pathway that underlies the coronary sinus as well as the superior and inferior aspects of the fossa ovalis are routes of interatrial electrical communication that can mediate fibrillatory conduction [407].

At least some cases of paroxysmal and chronic atrial fibrillation result from uninterrupted periodic activity of reentrant sites. A single or few high-frequency reentrant sources can localize to the left atrium [407]. Sources of high-frequency periodic activity, dispersion of local activation, and stable, self-sustained rotors can exist in the atria [409].

Reentrant rotors in the left atrium are faster than those in the right atrium [410]. The acetylcholine-modulated potassium inward rectifier current (i_{KACh}) is higher in the left atrium than in the right atrium. The sensitivity of LA rotors to ACh is then greater than that of RA rotors.

In some paroxysmal AF patients, rapid impulses triggered by an ectopic pacemaker propagate promptly as succeeding wave fronts from a pulmonary vein into the left atrium to encounter heterogeneously recovered tissue [407]. When conditions of heterogeneity are appropriate, the wave fronts brake and initiate two counter-rotating vortices that may be stable. On the other hand, a single rotor persists and engenders wave fronts at an exceedingly high frequency. The AF initiation and maintenance may then depend on the formation of relatively sustained rotors in the left atrium that generate fibrillatory waves.

Histological Background

Autonomic innervation of the heart encompasses the extrinsic (ECANS) and intrinsic (ICANS) cardiac autonomic nervous system. The letter consists of a neural network with ganglionated plexi concentrated within epicardial fat pads and interconnecting ganglia and axons.[13] The ganglionated plexi intervene in the initiation and maintenance of atrial fibrillation.

The $Na_V1.8$ channel encoded by the SCN10A gene is highly expressed in thin sensory neurons of dorsal root ganglia. It also lodges in the intrinsic cardiac ganglia. It is involved in atrial fibrillation. In cardiac ganglionated plexi, blockade of $Na_V1.8$ suppresses the effects of vagus nerve stimulation on cardiac conduction [411]. This effect results most likely from inhibition of the parasympathetic ganglionic neurotransmission in the ICANS.

The electrical properties of atriomyocytes can be altered. In the myocardium, impulse propagation depends on the excitability of myocytes, discharge transmission between adjacent myocytes, and the three-dimensional arrangement of myocytes. The anisotropic electrochemical wave propagation follows myofibers with discontinuity that may be sources of reentrant rotors.

Nodal cells of the conduction paths and myocytes that are characterized by their orientation are organized in bundles,[14] in atria and ventricles (size $\mathcal{O}[100\ \mu\text{m}$–$\mathcal{O}[100\ \text{mm})$ [408]. Aligned tracts of nodal tissue such as Purkinje fibers are in fact 3D arrays of cells.

Electrochemical wave propagates preferentially along the myofiber axis. The action potential transversal spreading rate depends on the extent of lateral coupling between adjacent myocytes.

In addition, fibroblasts and myofibroblasts may influence cardiomyocyte electrophysiology via juxtacrine interactions (bidirectional coupling via gap junctions) and modulation of action potential transmission. Cardiofibroblasts are also arranged in

---

[13] ICANS neurons synthesize many types of neurotransmitters in addition to acetylcholine. A certain population of neurons produce most probably dopamine and serotonin. Some cardiac magnocellular neurons synthesize noradrenaline and adrenaline. Other neuropeptides include histamine, cocaine, and amphetamine-regulated transcript-derived protein (CART), calcitonin gene-related peptide, neuropeptide-Y, tachykinins, vasoactive intestinal polypeptide, and nitric oxide. The CART peptides are neuromodulators involved in feeding, stress, motivation, drug addiction, bone remodeling, and cardiovascular functioning.

[14] In a given tiny transmural region, axially aligned cardiomyocytes form a myofiber. Locally, adjoining myofibers are nearly parallel with a given orientation. Ventriculomyocytes are arranged in myolaminae (sheets), that is, 4-to-6-cell–thick layers. Myolaminae are separated by perimysial connective tissue with a weak intercellular coupling. A given transmural myolamina at a given site has a given lead angle with respect to the endo- or epicardial surface supposed to be locally parallel. The myocardium is thus constituted of an orthotropic material with three structural axes at any point defined by the myofiber direction and myolamina lead angle. Transmural myofiber rotation varies from approximately $-60°$ with respect to the circumferential direction at the epicardial surface to about $+90°$ in the subendocardial region (Vol. 5, Chap. 5. Cardiomyocytes).

a network that surrounds cardiomyocytes. Myocardial injury engenders myofibroblasts (absent in the normal myocardium) that have greater proliferation and collagen production rates than fibroblasts. Cardiac fibroblasts and myofibroblasts possess ion carriers, among which ion channels that carry inwardly rectifying and various voltage-dependent outward K^+ currents. Their resting transmembrane potential is less negative than that of cardiomyocytes [408]. The fibroblast (myofibroblast) network may then act as a current source for adjacent inactive cardiomyocytes and as a sink during activation. Myofibroblast density-dependent reductions in cardiomyocyte resting transmembrane potential, conduction velocity slowing, and propagation across fibroblast barriers can be observed.

Adverse atrial structural remodeling such as diffuse fibrosis causes electrophysiological disturbances due to conduction abnormalities. Atrial fibrillation is associated with cardiac and pulmonary diseases. Several cardiac disorders predispose to atrial fibrillations, such as coronary artery diseases, pericarditis, valvular diseases, congenital heart diseases, congestive heart failure, thyrotoxic heart disease, sleep apnea, and hypertension. Idiopathic atrial fibrillation can be associated with mutations in the gene encoding connexin-40.

Impulse propagation in the cardiac wall can be modeled according to the one-dimensional continuous cable theory. This framework is useful to point out the role of current source-to-sink matching responsible for conduction slowing and unidirectional block, but it cannot represent entirely electrical properties of 3D networks of myocytes. Discharge propagation in discontinuous 1D strands of myocytes coupled by discrete resistive pathways differs from the prediction of the cable theory.

Impulse transmission between cells can follow the ephaptic[15] transmission, that is, the passage of an action potential from the prejunctional cell to the postjunctional cell through their membranes across an electrochemical synapse [ephapse] in a saltatory fashion driven by the electrical potential in the cleft between two apposed cells associated with the exchange of ions between the cells. Sodium channels that are concentrated around intercalated discs support the ephaptic transmission.

Discontinuous arrangement of myocytes and the connective tissue can engender current source-to-sink mismatch, spatiotemporal distribution of refractoriness, and rate-sensitive electric instability, which contribute to the initiation and maintenance of reentry [408]. Electrochemical dysfunction and structural remodeling (fibrosis) can generate ectopic activation and local propagation delays that support reentry.

Biochemical Background

The activated renin–angiotensin–aldosterone axis, platelet-derived and connective tissue growth factor, and local inflammation secondary to endothelial dysfunction and thrombus formation favor fibrosis. The abundance of atrial fatty deposits is related to the risk and severity of atrial fibrillation.

[15] ϵφαπλ οω: spread.

The refractory period depends on the APD. Smaller inward currents ($i_{Ca,L}$) and higher outward currents (K^+ currents [i_{BKCa}]) reduce the refractory period and promote atrial fibrillation [412].

Regulation by phosphatases of $Ca_V 1.2$ can be impaired. Transcript levels of PP1 and PP2 decrease, the PP1 level is not altered, but the PP2 level increases in AF patients. The phosphatase PP1 is the major subtype of protein Ser/Thr phosphatases in the heart [413]. The inhibitor $PP1_{r14b}$ is activated by PKA to prevent PP1 target dephosphorylation, thereby enabling amplification of β-adrenoceptor signaling. The activity of both PP1 and PP2 rises in patients with chronic AF, but in a heterogeneous manner according to the subcellular compartment. Phosphorylation of myosin-binding protein-C (Ser282) by PKA lowers, that of troponin-I is preserved, and that of phospholamban (Ser16 and Thr17) augments.

Phosphorylation of the ryanodine receptor RyR2 by protein kinase-A (Ser2808) and Ca^{2+}–calmodulin-dependent protein kinase CamK2 (Ser2814) is antagonized by the protein phosphatases PP1 and PP2. Impaired local regulation of protein phosphatase PP1 leads to hyperphosphorylation (hyperactivity) of RyR2, thereby favoring atrial fibrillation [414]. PP1 dephosphorylates RyR2 that was phosphorylated by the CamK2 kinase. Spinophilin, or $PP1_{r9b}$, assists PP1 in targeting RyR2. Upon ablation of spinophilin, the interaction between PP1 and RyR2 lowers (64 %); RyR2 phosphorylation rises (43 % at Ser2814), but remains unchanged at Ser2808; and open RyR2 probability and subsequently Ca^{2+} spark frequency augments.

Transcription factors contribute to cardiac arrhythmias. They can create a proarrhythmogenic ground in the pulmonary veins and atrium. They are implicated in atrial remodeling linked to atrial fibrillation, such as TSC22, TcEb, EGR2, GTF2h2, and Fos, as well as transcription factor-related proteins, such as cardiac ankyrin repeat protein (CARP) and Four and a half LIM domain-containing protein FHL1 [415]. Mutations in genes that encode transcription factors, such as TBx5,[16] GATA4 to GATA6,[17] and NKx2-5 are responsible for familial forms of atrial fibrillation (Table 5.2; [415]).

The TBX5 gene mutation (G125R) is associated with a gain of function and increased NPPA (natriuretic peptide precursor-A), CX40 (connexin-40), and KCNJ2 ($K_{IR}2.1$) gene expression [415]. Mutations in the Gata4 gene (G16C, H28D, Y38D, S70T, P103A, and S160T), Gata5 gene (Y138F, W200G, and C210G), and Gata6 gene (Y235S) are associated with a loss of function. The NKX2-5 gene mutation (T768A) diminishes the ability of NKx2-5 to dimerize, but not of transactivating the NPPA gene.

In addition, the risk single nucleotide polymorphisms for atrial fibrillation localize to the vicinity of transcription factor PITX2, ZFHX3, and PRRX1 genes [415].

---

[16] Mutations in the TBX5 gene cause Holt–Oram syndrome with atrial and ventricular septal defects, and cardiac conduction and upper limb abnormalities.

[17] Multiple Gata gene mutations are related to congenital heart diseases, such as atrial and ventricular septal defects, pulmonary stenosis, hypoplastic right ventricle, endocardial cushion defects, persistent truncus arteriosus, and tetralogy of Fallot.

Table 5.2 Transcription factors implicated in atrial fibrillation. (Source: [415])

Transcription factors	Targets
GATA4	NPPa
GATA5	NPPa
GATA6	NPPa
NKx2-5	MinK, $Na_V1.5$, Cx40, NPPa
Pitx2	HCN4, NPPa
TBx5	$K_{IR}2.1$, $Na_V1.5$, Cx40, NPPa

Cx40 connexin-40 encode by the GJA5 gene, *HCN* hyperpolarization-activated, cyclic nucleotide-gated K^+ channel, *NPPa* natriuretic peptide precursor-A

Rapidly Discharging Ectopic Sites and Reentries

Atrial fibrillation is initiated and maintained by at least one ectopic focus with rapid activity associated with a single small reentry circuit or multiple reentries [412]. Reentrant loops in the atrial myocardium are triggered by rapid discharges of ectopic foci, commonly in sleeves of atrial tissue located within the pulmonary veins. Increased activity of Na^+–Ca^{2+} exchangers promotes afterdepolarization-related atrial ectopic firing.

Atrial fibrillation is launched by a substrate available for reentries generated by a trigger, usually spontaneous focal ectopic discharges. Focal ectopic discharges include EADs and DADs (Table 5.3). The size of reentry circuits is determined by the wavelength, that is, the product of refractory period by conduction velocity; the smaller is the wavelength, the shorter the APD, the smaller the reentry circuits and the higher their number, and the more sustained the atrial fibrillation. When the number of circuits in the atria is small, reentry is unstable and atrial fibrillation self-terminates.

Early Afterdepolarizations

EADs precede full repolarization, that is, they happen at the end of plateau or early in the final phase of repolarization. EADs occur when the action potential is excessively prolonged.

Long duration of action potentials can result from [385]: (1) increased inward $i_{Ca,L}$ current through $Ca_V1.2$ channel; (2) late $i_{Na,L}$ current through $Na_V1.5$; and/or (3) reduced K^+ currents. Recovery from inactivation of the $Ca_V1.2$ channel, enhanced inward Na^+ flux through the Na^+–Ca^{2+} exchanger, and phosphorylation by CamK2 of Ca^{2+} channels contribute to EAD occurrence.

$Na_V1.5$ channel is primarily responsible for action potential initiation and propagation in the heart. Its fast inactivation occurs within ms after membrane depolarization onset. It determines APD. Inherited mutations in the SCN5A gene that encodes $Na_V1.5$ can disrupt fast inactivation and promote persistent channel activity

Table 5.3 Mechanisms of atrial fibrillation (Source: [385]). During the systole, Ca^{2+} entry through $Ca_V1.2a$ triggers systolic Ca^{2+} release from the sarcoplasmic reticulum into the cytosol through ryanodine receptor RyR2. During diastole, normal resting cytosolic Ca^{2+} concentration is restored by Ca^{2+} export, especially reuptake by the sarcoplasmic reticulum Ca^{2+} ATPase SERCA2a and extrusion into the extracellular space via Na^+–Ca^{2+} exchange (NCX). In delayed afterdepolarization (DAD), abnormal diastolic Ca^{2+} release through RyR2s because of excess sarcoplasmic reticulum Ca^{2+} load reaching release threshold or enhanced RyR Ca^{2+} sensitivity causes a transient inward (subscript *ti*) current through NCX ($i_{Na,ti}$). In early afterdepolarizations (EADs), action potential prolongation enables $i_{Ca,L}$ current to depolarize the cell from the plateau phase

Trigger	Action potential phase	Involved ion carrier(s)
DAD	4 (rest)	Na^+–Ca^{2+} exchanger ($i_{Na,ti}$ current)
		RyR2 (abnormal activity)
Effect: postrefractory period depolarization		
EAD	2 (plateau end)	$Ca_V1.2$ ($i_{Ca,L}$ current)
		$K_V1.5$ ($i_{K,ur}$ current)
		$K_{Ca(BK)}$
		$Na_V1.5$ channel ($i_{Na,L}$ current)
	3 (final repolarization)	Na^+–Ca^{2+} exchange current (i_{NaCaX} current)
Origin: long action potential duration		
Effect: surrounding myocardium depolarization		

that can prolong APD and cause the congenital long-QT syndrome (LQT3), Brugada syndrome, and isolated conduction disease [416].

Loss-of-function mutations of genes that encode K^+ channels, either those linked to LQTS or those responsible for the atrial ultrarapid delayed rectifier K^+ current ($i_{K,ur}$), i.e., $K_V1.5$ channel, are associated with atrial fibrillation due to ectopic site activity related to EADs [385].

Atrial fibrillation relying on EADs can also be induced by gain-of-function mutations of genes that encode ion channels carrying $i_{Ca,L}$ or i_{Na} ($Na_V1.5$) current, which also cause LQTS [385].

Long APD enables the $Ca_V1.2$ channel to recover and depolarize the cell by Ca^{2+} entry. EADs cause ectopic firing by depolarizing surrounding myocardium to excitation threshold.

Delayed Afterdepolarizations

DADs occur after full repolarization. They are caused by Na^+–Ca^{2+} exchange current (i_{NaCaX} or $i_{Na,ti}$: transient inward Na^+ current) generated by a transient diastolic rise in cytosolic Ca^{2+} concentration due to abnormal RyR2 activity. Additional Ca^{2+} ions are exchanged for extracellular Na^+ ion in a 1:3 ionic ratio, creating a net inward movement of positive ions [385].

Arrhythmogenic diastolic Ca^{2+} leak through RyR2s can result from: (1) excess sarcoplasmic reticulum Ca^{2+} load reaching release threshold or (2) enhanced RyR Ca^{2+} sensitivity. Hyperphosphorylation of phospholamban, a SERCA2a inhibitor, relieves SERCA inhibition, hence increasing sarcoplasmic reticulum Ca^{2+} uptake. Downregulated expression of sarcolipin, another SERCA2a inhibitor, also offsets sarcoplasmic reticulum Ca^{2+} depletion caused by a Ca^{2+} leak. Ryanodine receptor hyperphosphorylation or RYR gene mutations associated with catecholaminergic polymorphic ventricular tachycardia enhance RyR Ca^{2+} sensitivity [385].

On the other hand, depolarization by $i_{Na,ti}$ current can be antagonized by opposing transmembrane outward repolarizing currents (e.g., i_{K1} current through K_{IR}; [385]). However, when the excitation threshold is reached, spontaneous action potentials arise.

Calcium–calmodulin-dependent protein kinase CamK2 is activated by cytosolic Ca^{2+}–calmodulin provoking autophosphorylation (Thr287) as well as angiotensin-2-related oxidation (Met281 and Met282) preventing reassociation of the catalytic subunit (inhibition) with its regulatory subunit [385]. Activated CamK2 phosphorylates multiple sarcoplasmic reticulum proteins. In particular, CamK2, in synergy with protein kinase-A also activated during atrial fibrillation, phosphorylates RyR channels. Kinase CamK2 can also contribute to atrial fibrillation by phosphorylating histone deacetylase HDAC4, thereby eliciting its nuclear export and HDAC4–14-3-3 binding and cytosolic sequestration, hence derepressing transcription of genes encoding ion carriers [385].

The Ca^{2+} sensitivity and closed state stability of RyR2 are also modulated by FK506-binding protein FKBP1b, junctophilin, triadin, calsequestrin, and junctin [385]. Altered RyR2 activity by these accessory proteins can generate a Ca^{2+} leak from the sarcoplasmic reticulum and, hence, atrial ectopic foci.

Phospholamban hyperphosphorylation and reduced sarcolipin expression disinhibit SERCA pumps and increase sarcoplasmic reticulum Ca^{2+} uptake. Phosphorylation of RyR2 depends on activity of protein phosphatases PP1 and PP2 that dephosphorylate these Ca^{2+} channels. Inhibition of PP1 by enhanced activity of sarcoplasmic reticulum compartment inhibitor-1 contributes to RyR2 and phospholamban hyperphosphorylation [385].

Reentry

Calcium-related handling participates in the control of repolarization, hence determining APD and thus the likelihood of reentry. Whereas, increased inward $i_{Ca,L}$

Table 5.4 Atrial electrical remodeling (Source: [385])

↑ $[Ca^{2+}]_c$ ⟶ $PP3_a$ ⟶ NFAT ↓ $Ca_V1.2$ expression ⟶ APD ⟶ AF
↑ $[Ca^{2+}]_c$ ⟶ $Ca_V1.2$ activity ⟶ APD ⟶ AF
↑ $[Ca^{2+}]_c$ ⟶ $calpain_a$ ⟶ ↓ PKCα ⟶ ↑ $i_{K_{ACh}}$ ⟶ HP ⟶ AF
↑ PKCϵ activity ⟶ ↑ $i_{K_{ACh}}$ ⟶ HP ⟶ AF
↓ miR26 expression ⟶ $K_{IR}2.1$ expression ⟶ ↑ i_{K1} ⟶ HP ⟶ AF

AF atrial fibrillation, *APD* action potential duration, *HP* plasmalemmal hyperpolarization, *subscript a* activated, $[\bullet]_c$ cytosolic concentration, ↑ increase, ↓ decrease

current through $Ca_V1.2$ channel raises duration of action potentials, thus probability of EADs, decreased Ca^{2+} flux reduces APD of atriomyocytes, hence strongly favoring reentry.

Adverse Atrial Functional Remodeling

Calcium ion contributes to adverse atrial electrical remodeling, i.e., change in atrial structure and/or function that supports atrial arrhythmogenesis. Atrial fibrillation causes cytosolic Ca^{2+} overload, as Ca^{2+} enters the atriomyocyte during each action potential. This event then diminishes Ca^{2+} entry to attenuate Ca^{2+} loading (Table 5.4; [385]). Rapidly, Ca^{2+} binds to the $Ca_V1.2$ channel and causes its partial inactivation, reducing APD. Moreover, augmented cytosolic Ca^{2+} concentration primes Ca^{2+}–calmodulin binding and activates PP3 phosphatase that dephosphorylates nuclear factor of activated T cells, which then translocates to the nucleus and regulates gene transcription, decreasing that of the $Ca_V1.2$ channel pore-forming α subunit. Factor NFAT also binds to a regulatory DNA region upstream from the site encoding premiR26, inhibiting its transcription and enabling that of the $K_{IR}2.1$ channel. Last, but not least, conventional Ca^{2+}-dependent PKC isoforms (PKCα and PKCβ) impede $K_{IR}3$ channel activity, whereas novel Ca^{2+}-dependent PKC isoforms (PKCδ and PKCϵ) enhance the $K_{IR}3$ channel opening. Furthermore, Ca^{2+}-dependent peptidase calpain processes PKCα, relieving its inhibition of the $K_{IR}3$ channel.

Therefore, atrial tachycardia remodeling and associated reduced APD arise from: (1) decreased inward $i_{Ca,L}$ and (2) increased outward K^+ currents, particularly inward rectifier i_{K1} (through the $K_{IR}2$ channel) and constitutively active acetylcholine-regulated $i_{K_{ACh}}$ (through the $K_{IR}3$ channel) currents (Table 5.4[385]). In addition to lowered APD, increased inward rectifier K^+ current hyperpolarizes atrial cardiomyocytes, thereby elevating Na^+ flux availability, excitability, and reentrant rotor stability during atrial fibrillation.

Adverse Atrial Structural Remodeling

Atriomyocytes exposed to long-term rapid activation lose their myofilaments due to calpain activation, accumulate glycogen, and their mitochondria modify and connexin and chromatin redistribute [385].

Atrial fibrillation directly and indirectly cause fibroblast proliferation and differentiation into myofibroblasts. Resulting fibrosis contributes to atrial fibrillation substrate. Calcium entry through nonselective transient receptor potential channels TRPM7 and/or TRPC3, the expression of which is upregulated in atrial fibrillation, enables fibroblast activation [385]. As with $K_{IR}2.1$, TRPC3 synthesis also depends on miR26.

Context

Atrial fibrillation is associated with abnormal cellular Ca^{2+} content, distribution, and handling [385]. An additional cause of atrial fibrillation may be linked to altered expression of connexin and subsequent abnormal intercellular electrochemical communication.

Increased levels of atrial extracellular signal-related kinase and angiotensin-converting enzyme are observed in patients with atrial fibrillation. Concentrations of angiotensin-2 receptors ATR1 and ATR2 decrease and increase, respectively.

ω3-polyunsaturated fatty acids (ω3PUFA)—eicosapentaenoic acid and docosahexaenoic acid—from fish oil have an antiatrial fibrillation effect. They inhibit transient outward and ultrarapid delayed rectifier K^+ channels that are important regulators of repolarization of human atrium and voltage-gated Na^+ channels [417].

Prolonged episodes of atrial fibrillation are associated with electrical and structural remodeling that favors the reoccurrence or perpetuation of atrial fibrillation. Electrical remodeling is linked to changes in synthesis and distribution of connexin-40, a major gap junction constituent in atriomyocytes [418]. Initiating firing sites of atrial fibrillation most frequently localize close to the pulmonary veins and/or superior vena cava. Arrhythmogenic framework corresponds to a reduced effective refractory period, increased spatial dispersion of refractoriness, or abnormal atrial impulse conduction.

Stroke and Thromboembolism Risk

Atrial fibrillation management requires an appropriate thromboprophylaxis, once the bleeding risk has been assessed. A common, oversimplified stroke risk assessment tool was proposed [419]: the CHADS2 (congestive heart failure or left ventricular dysfunction, hypertension, age, diabetes, stroke antecedent, gender) score. Each risk factor has equal weight (i.e., score 1 point), except age, a double-valued index above 75-year-old, and stroke, a prior stroke also yielding 2 points. Oral anticoagulant therapy is recommended in subjects with a CHADS2 score of at least 1. Other non-CHADS2 risk factors, such as history of vascular diseases and plasma level of von

Willebrand factor, an index of endothelial damage, and markers, such as proteinuria and creatinine clearance should be taken into account for decision making.

Many risk factors for bleeding are also risk factors for stroke. The HASBLED (hypertension or uncontrolled blood pressure, abnormal renal and/or hepatic function, stroke, bleeding history or predisposition, elderly [e.g., age > 65 year], drugs and/or alcohol consumption) score is a good predictor of major bleeding, but a modest predictor of cardiovascular events and death. It indicates the frequency of regular follow-up.

5.2.3 *Ventricular Arrhythmias*

Premature ventricular contractions have either a supraventricular or ventricular origin according to whether the QRS duration is smaller or greater than 100 ms, respectively.

Ventricular arrhythmia storm is an arrhythmia syndrome with its specific management and prognostic that differ from ventricular tachycardia and fibrillation unrelated to storm. A cardioverter defibrillator is implanted to prevent sudden death in patients with cardiomyopathy and life-threatening genetic cardiac disorders, but does not preclude ventricular arrhythmia storm. Radiofrequency catheter ablation can then be used.

Abnormalities of cellular electrophysiology can arise from mutations in a single ion carrier (channel, transporter, or pump). However, most ventricular tachyarrhythmias, in particular ventricular fibrillation, have a multifactorial origin, as they result from multiple mechanisms that arise from various time scales, ranging from short (s), intermediate (mn to h), and long (d to mo): abnormal automaticity, triggered activity, and reentry [420]. Various types of perturbations combine disturbed ion channel distribution and function, abnormal intracellular ion dynamics, development of dynamical heterogeneity of refractoriness, impaired cardiac innervation, and defective metabolic and signaling pathways, in addition to anatomical features[18] to produce a permissive context for initiation and maintenance of sustained arrhythmias.

Nitric oxide enables ^{S}nitrosylation of sarcoplasmic reticulum Ca^{2+} handling proteins. Reduced NOS1 activity is responsible for a deficient ^{S}nitrosylation of ryanodine receptor RyR2 [421]. This deficiency leads to attenuated RyR2 phosphorylation by Ca^{2+}–calmodulin-dependent protein kinase CamK2 (Ser2814) on the one hand and elevated ^{S}oxidation of RyR2 by reactive oxygen species produced by xanthine oxidoreductase on the other.[19] Modified RyR2 then provokes spontaneous

[18] The intrinsic transmural rotation of fibers from the epicardium to endocardium (rotational anisotropy) may generate instabilities of depolarizing waves [420].

[19] The nitrosoredox balance relies on effects of nitric oxide and reactive oxygen species that are coupled in cardiomyocytes. Some ventricular arrhythmias arise from increased spontaneous sarcoplasmic reticulum Ca^{2+} release resulting from a combination of decreased $RyR2^{SNO}$ and increased $RyR2^{SOx}$ due to increased ROS formed by xanthine oxidoreductase. Decreased $RyR2^{SNO}$ and/or increased $RyR2^{SOx}$ produces a conformational change, either alone or in combination.

Ca^{2+} release from the sarcoplasmic reticulum and subsequent augmented cytosolic Ca^{2+} concentration in cardiomyocytes, thereby causing Ca^{2+}-mediated ventricular arrhythmias [421]. In addition, NOS1 inhibition increases $Ca_V1.2$ activity ($i_{Ca,L}$ current).

Certain ventricular arrhythmias, such as catecholaminergic polymorphic ventricular tachycardia or bidirectional ventricular tachycardia, catecholaminergic idiopathic ventricular fibrillation, and arrhythmogenic right ventricular dysplasia type 2 result from either gain-of-function or loss-of-function mutations in ryanodine-sensitive Ca^{2+} channel gene. Gain-of-function RYR2 mutations are associated with enhanced propensity for spontaneous (also called store overload-induced) Ca^{2+} release that generates a Ca^{2+} leak from the sarcoplasmic reticulum) and thus DADs and subsequent ventricular arrhythmias. Loss-of-function RYR2 mutation reduces RyR2 activation by sarcoplasmic reticulum Ca^{2+} (but not cytosolic Ca^{2+}) and causes catecholaminergic idiopathic ventricular fibrillation [422].[20]

Triadin localizes to the junctional sarcoplasmic reticulum, hence maintaining the structural integrity of the $Ca_V1.2$–RyR couplon and orchestrating sarcoplasmic reticulum Ca^{2+}-release and excitation–contraction coupling. Triadin can be involved in ventricular arrhythmia [423].

5.2.3.1 Myocardial Structure

The myocardium is a heterogeneous tissue characterized by a high variability in wall thickness, fiber orientation, and electrophysiological properties (i.e., in densities and functionings of ion carriers) between and within cardiac chambers in a given subject. This heterogeneity affects the propagation of electrochemical waves and development of contractile force.

These heterogeneities may explain the breakup of rotors during fibrillation [397]. Some structures can act as barriers to electrical activation and serve as anchors for spiral waves during reentries.

5.2.3.2 Cardiac Alternans

Cardiac alternans, i.e., temporal and/or spatial oscillations , is a risk factor for cardiac arrhythmias. Beat-to-beat alternations at a constant stimulation frequency include

---

As ^{S}nitrosylation of phosphatases can inhibit their activity, NOS1 inhibition may decrease ^{S}nitrosylation of a kinase.

[20] Calcium activation of ryanodine-sensitive Ca^{2+} channels from each side of the sarcoplasmic reticulum, i.e., luminal Ca^{2+} and cytosolic Ca^{2+}, drives different gating mechanisms. The A4860G mutation diminishes luminal Ca^{2+} activation, but the sensitivity to cytosolic Ca^{2+} activation is retained. Likewise, some gain-of-function RYR2 mutations can enhance luminal Ca^{2+} activation without affecting cytosolic Ca^{2+} activation.

alternations in: (1) contraction amplitude (*mechanical alternans*); (2) APD (*electrical alternans*); and (3) Ca^{2+} transient amplitude (*calcium alternans*).

Temporal alternans is related to the steepness (slope > 1) of the APD restitution curve, i.e., the relationship between APD and preceding diastolic interval. In fact, the range of diastolic intervals over which the slope is elevated is a determinant of alternans.

APD alternans can precede a series of degenerating events leading to ventricular fibrillation. They can arise from transient calcium alternans following cellular uncoupling and SERCA failure in particular.

Afterdepolarizations that arise in a source of depolarizing current, usually a limited number of cells, can propagate to neighboring tissue that acts as a current sink, in the absence of large source–sink mismatch, i.e., a sufficient small number of surrounding, nonsusceptible cells to enable excitation of adjoining tissue. They then provoke premature ventricular complexes that serve as arrhythmia triggers. In addition to refractoriness, the source–sink mismatch, a safety factor, can explain unidirectional block and subsequent reentry [397]. Moreover, fibroblasts can affect normal excitability.

Two main processes generate Ca^{2+} alternans: fractional Ca^{2+} release from the sarcoplasmic reticulum and cytosolic Ca^{2+} sequestration. The former depends on sarcoplasmic reticulum load and the latter on the availability of adequate ATP supplies to Ca^{2+} pumps and kinases for Ca^{2+}-handling protein phosphorylation by protein kinase-A and Ca^{2+}–calmodulin-dependent protein kinase CamK2.

Between-beat alternation in active RyR availability and recovery from inactivation promotes frequency-induced pathological Ca^{2+} alternans. In particular, depressed RyR2 activity induces spatially and temporally desynchronized Ca^{2+} release from its cellular store and subcellular Ca^{2+} alternans.

Sympathetic innervation of the heart elicits positive chronotropy, inotropy, and lusitropy. In cat atriomyocytes, selective stimulation of Gs-coupled β1-adrenoceptors and Gs- and Gi-coupled β2-adrenoceptors abolish proarrhythmic Ca^{2+} alternans [424].[21] Stimulated β1-adrenoceptors suppress alternans via PKA and CamK2 kinases, whereas β2-adrenoceptors involve only PKA enzyme. Basal PKA and CamK2 activity exerts partial protection against alternans. Activated β-adrenoceptors require either mitochondrial or glycolytic ATP production [424].

[21] The β2AR–Gi–PI3K–PKB axis is coupled with NOS3, thereby producing nitric oxide. Nitric oxide: (1) activates guanylate cyclase that synthesizes cGMP and inhibits cGMP-dependent phosphodiesterase-3, thereby enhancing the cAMP–PKA signaling; and (2) inhibits the β2AR–Gs–cAMP–PKA pathway via protein nitrosylation. This signaling pathway in atriomyocytes does not influence alternans [424].

5.2.3.3 Extrasystoles

Extrasystoles, or extracardiac beats, correspond to premature ventricular contractions. This type of arrhythmia can be benign, occurring without underlying disease. Nonetheless, they can be associated with abnormalities in electrolyte blood levels, ischemia, or consumption of certain substances (smoking, alcohol, caffeine, medications, etc.).

5.2.3.4 Idioventricular Rhythms

In ventricular arrhythmias, impulses originate from the ventricular conduction system. QRS complexes are abnormal and last longer than 100 ms. *Idioventricular rhythm* is due to ventricular activation by a ventricular focus with a frequency below 40/mn (< 0.66 Hz). *Accelerated idioventricular rhythm* is defined by short bursts (< 20 s) of ventricular activity at high frequencies (40–120/mn, i.e., 0.66–2 Hz).

5.2.3.5 Ventricular Tachycardias

Ventricular tachycardias (rate $>$ 2 Hz), with wide QRS complexes, are the consequence of a slower conduction leading to reentries associated with ischemia and myocardial infarction.

Focal transfer using adenoviral vectors of a gene-encoding dominant-negative version of the KCNH2 potassium channel ($KCNH2^{G628S}$) to the infarct scar border eliminates ventricular tachyarrhythmias [425].

5.2.3.6 Ventricular Fibrillation

Ventricular fibrillation due to multiple reentry loops is the most dangerous arrhythmia. Ventricular fibrillation is characterized by irregular asynchronous undulations without QRS complexes. A hereditary factor is involved in the risk of primary ventricular fibrillation. Ventricular fibrillation may be maintained by unstable reentry with activation wavelets of changing paths and conduction blocks associated with a heterogeneous dispersion of refractoriness. The multiple-wavelet hypothesis [426] predicts activation patterns different from the observations [427, 428]. High-frequency excitation sources can produce ventricular fibrillations.

Spiral wave is a possible mechanism leading to ventricular fibrillation [429]. Spiral wave can break, generating secondary wavelets (spiral breakup; [430, 431]). Rotors are possible organizing centers of fibrillation. During ventricular fibrillation, persistent rotor activity has been observed in the anterior left ventricular wall of guinea pig hearts, the fastest activating region of the myocardium, with the shortest refractory period [432].

APD and activation rates differ between heart wall regions due to heterogeneous distribution of ion carriers, especially repolarizing K^+ channels ($i_{K,to}$, $i_{K,r}$, and $i_{K,s}$; Vols. 3, Chap. 3. Main Sets of Ion Channels and Pumps and 5, Chap. 5. Cardiomyocytes) in the normal guinea pig heart [432, 433].

5.2.4 *Genetic Risk Factors and Gene Mutations*

5.2.4.1 Genetic Risk Factors

Genome wide association studies of variation in the QT interval in population-based cohorts aim at identifying genetic risk factors, i.e., loci mapped in or near genes associated with this trait, such as NOS1AP, ATP1B1, PLN, LITAF, RNF207, and NDRG4–GINS3, in addition to gene mutations responsible for LQTS, such as the KCNH2, KCNJ2, KCNQ1, SCN5A, KCNE1, and KCNE2 genes, which encode ion channels, and the ANK2 gene, which encodes adaptor neuronal ankyrin-2 ensuring stability of Na^+–Ca^{2+} exchanger-1 in cardiomyocytes [434, 435].[22]

5.2.4.2 Gene Mutations

Mutations in the sodium channel SCN5A gene cause cardiac arrhythmia. Prolongation of the electrocardiographic QT interval, a measure of cardiac repolarization, is a risk factor for arrhythmias and sudden cardiac death.

In addition to mutations in genes that encode ion carrier proteins, mutations of genes encoding protein complexes and organelle constituents, such as sarcomere in hypertrophic cardiomyopathy, desmosome in arrhythmic right ventricular cardiomyopathy, and cytoskeleton, sarcoglycan complex, and mitochondrial structural and functional components in dilated cardiomyopathy are sources of arrhythmias (Tables 5.5 and 5.6).

[22] The KCNH2, KCNJ2, and KCNQ1 genes encode pore-forming K^+ channel subunits for different types of channels. The KCNE1 gene encodes a regulatory subunit for the channel encoded by the KCNQ1 gene. The SCN5A gene encodes the Na^+ channel pore-forming subunit. The NOS1AP gene encodes a cytosolic adaptor protein that binds to NOS1 nitric oxide synthase. The ATP1B1 gene encodes the regulatory subunit of Na^+–K^+ ATPase. Newly identified loci include the 16q21 locus near the NDRG4 and GINS3 genes, the 6q22 locus near the PLN gene, the 1p36 locus near the RNF207 gene, and the 16p13 locus near the LITAF gene. The LITAF gene encodes lipopolysaccharide-induced tumor-necrosis factor-α (TNFSF1). The PLN gene encodes a regulatory subunit of sarco(endo)plasmic reticulum calcium (SERCA) pump, i.e., inhibitory phospholamban. The RNF207 gene encodes really interesting new gene (RING) finger protein-207. The NDRG4 gene encodes a cytoplasmic protein that may be involved in the regulation of mitogenic signaling in vascular smooth myocytes.

Table 5.5 Gene mutations and arrhythmias (**Part 1**) ion carriers. (Source: [392])

Gene	Protein	Effect
Abnormal ion channel gating		
CACNA1C/B2	$Ca_V1.2$	SQTS
GJA5	Connexin-40	AF, VT
Hcn4	HCN4 K^+ channel	Pacemaker dysfunction, bradyarrhythmia
KCNA5	$K_V1.5$	AF
KCNE1/E2	MinK, MiRP1	EAD, LQTS
KCNH2	$K_V11.1$	EAD, LQTS
KCNJ2	$K_{IR}2.1$	EAD, LQTS
KCNQ1	$K_V7.1$	EAD, LQTS
SCN5A	$Na_V1.5$	VF
Defective ion channel inactivation		
CAV3	Caveolin-3	EAD, LQTS
CACNA1C	$Ca_V1.2$	EAD, LQTS
CASQ2	Calsequestrin-2	AVA, bradycardia
RYR2	RyR2 (RC)	CPVT, DAD
SCN4B	$Na_V\beta4$	EAD, LQTS
SCN5A	$Na_V1.5$	
Increased repolarizing current		
KCNH2/J2/Q1	K^+ channel	SQTS, AF, VF
PRKAG2	AMPKγ2	

AMPK AMP-activated protein kinase, *AVA* atrial and ventricular arrhythmias, *AF* atrial fibrillation, *CPVT* catecholaminergic polymorphic ventricular tachycardia, *DAD* delayed afterdepolarization, *EAD* early afterdepolarization, *LQTS* long-QT syndrome, *RC* ryanodine-sensitive calcium channel, *SQTS* short-QT syndrome, *VF* ventricular fibrillation, *VT* ventricular tachycardia

Arrhythmias result most often from mutation in genes that encode ion channels (congenital arrhythmia), such as LQTS, Brugada syndrome, and isolated cardiac conduction disorder (ICCD). Whereas LQTS results frequently from gain-of-function Na^+ channel mutations, Brugada syndrome and isolated cardiac conduction disorder are commonly caused by loss-of-function Na^+ channel mutations.

5.2.4.3 LQTS

LQTS results from an abnormal cardiac excitability characterized by prolonged repolarization and arrhythmias. LQTS causes syncope and possibly death from

Table 5.6 Gene mutations and arrhythmias (**Part 2**) Cytosolic proteins. (Source: [392])

Gene	Protein	Effect
Calcium handling		
ANK2	Ankyrin-2	AF, DAD, LQTS, VF, bradycardia
Sarcomere		
ACTC1	α-Actinin	DAD or EAD, AF, VT
MYBPC3	Myosin-binding protein-3	
MYH7	β-MHC	
MYL2/3	MLC	
TNNI3	Troponin-I	
TNNT2	Troponin-T	
TPM1	α-Tropomyosin	
TTN	Titin	
Cytoskeleton		
ACTN2	α-Actinin	Heart block, AF, VF
CLP	Cardiac LIM protein	
DES	Desmin	
DMD	Dystrophin	
Ldb3	LIM domain binding-3	
LMNA	Lamin-A	
SGCD	δ-Sarcoglycan	
VCL	Vinculin	
Desmosome		
DES	Desmin	ARVD
DSC2	Desmocollin-2	
DSG2	Desmoglein-2	
DSP	Desmoplakin	
JUP	Plakoglobin	
PKP2	Plakophilin-2	
Metabolism		
Lamp2	LAMP2	WPWS

ARVD arrythmogenic right ventricular dystrophy, *AF* atrial fibrillation, *DAD* delayed afterdepolarization, *EAD* early afterdepolarization, *LAMP2* lysosomal associated-membrane protein-2, *MHC* myosin heavy chain, *MLC* myosin light chain, *VF* ventricular fibrillation, *VT* ventricular tachycardia, *WPWS* Wolff–Parkinson–White syndrome

arrhythmia. Long-QT phenotypes are associated with mutations of genes on chromosomes 3, 4, 7, 11, and 21. Loci LQT1 to LQT3 and LQT5 to LQT6 encode for cardiac ion channels subunits (Table 5.7).

Table 5.7 Ventricular action potential and ionic currents genes. Locus LQT4 encodes adaptor ankyrin-B, an anchor to the plasmalemma for Na^+–K^+ ATPase and Na^+–Ca^{2+} exchanger. The KCNQ1 (LQT1) and KCNE1 (LQT5) genes encode for α (KvLQT1) and β (MinK) subunits of the slow delayed rectifier potassium channel ($i_{K,s}$ current), respectively. The KCNH2 (LQT2) and KCNE2 (LQT6) genes encode for α HERG) and β (MiRP1) subunits of rapid delayed rectifier potassium channel ($i_{K,r}$ current)

Phase	Ionic fluxes	Genes	LQT types
0	Rapid Na^+ influx	SCN5A	LQT3
	(i_{Na})		
1	Transient K^+ outflux		
	($i_{K,to}$)		
2	Na^+ influx		
	L-type Ca^{2+} influx	CACNA1C	LQT8
	($i_{Ca,L}$)		
	Delayed rectifiying K^+ efflux	KCNH2 (HERG)	LQT2
	($i_{K,r}$)	KCNE2 (MiRP1)	LQT6
	Delayed rectifiying K^+ efflux	KCNQ1 (KvLQT1)	LQT1
	($i_{K,s}$)	KCNE1 (minK)	LQT5
	Cl^- currents		
3	Rapid K^+ current	KCNH2, KCNE2	LQT2, LQT6
	Slow K^+ current	KCNQ1, KCNE1	LQT1, LQT5
	Inwardly rectifying K^+ influx	KCNJ2	LQT7
	(i_{K1}, $K_{IR}2.1$)		

Among 13 identified long-QT susceptibility genes (LQT1–LQT13), 5 are associated with cardiac ion channels (CACNA1C, KCNH2, KCNJ2, KCNQ1, and SCN5A) and 6 with ion channel subunits or channel-interacting proteins (Akap9, ANKB, CAV3, KCNE1, KCNE2, and SCN4B; Tables 5.8 and 5.9).

Patients with LQTS experience syncope episodes (particularly LQTS-triggered syncope secondary to torsades de pointe), in addition to vasovagally mediated (neurocardiogenic) syncope, syncope with seizures, and aborted cardiac arrest and sudden cardiac death [436].

The LQTS must be differentiated from a borderline QT prolongation observed in the normal sex- and age-dependent distribution in the *cardiac frequency-corrected QT interval* (Table 5.10).

The corrected QT interval is calculated using the Bazett's correction formula:

$$QT_c = QT/RR^{1/2}. \tag{5.2}$$

A-kinase anchoring proteins recruit signaling molecules for the formation of signaling complexes to achieve efficient spatial and temporal control of signaling. In the heart, AKAPs target several ion channels, such as $Ca_V1.2$ and ryanodine calcium (RyR) channels and slowly activating delayed rectifier potassium channels. The

Table 5.8 Long-QT syndromes (**Part 1**: LQT1–LQT6)

Type	Mutated gene	Protein effect(current)
LQT1	KCNQ1	α subunit $K_V7.1$ binds to MinK ($i_{K,s}$)
		Disruption of AKAP9 anchor binding
		Reduction of repolarizing current
		Homozygous mutations: Jervell and Lange–Nielsen syndrome
LQT2	KCNH2	α subunit $K_V11.1$ binds to MiRP1 ($i_{K,r}$, AP termination)
		Early afterdepolarization
LQT3	SCN5A	α subunit of $Na_V1.5$
		Altered channel inactivation, late Na^+ current
		Prolongation of AP duration and QT interval
		Brugada syndrome
LQT4	ANK2	Ankyrin-B anchors ion channels
		Aberrant subcellular localization of
		Na^+–K^+ ATPase, Na^+–Ca^{2+} exchanger,
		IP_3 receptor, and $PP2_{r5\alpha}$
		(PP2 regulates ankyrin-B-binding ion carriers)
		Prolonged ventricular repolarization,
		ventricular tachyarrhythmia
LQT5	KCNE1	β subunit MinK binds to $K_V7.1$ ($i_{K,s}$)
		Homozygous form: Jervell and Lange-Nielsen syndrome
LQT6	KCNE2	β subunit MiRP1 binds to $K_V11.1$ ($i_{K,r}$)

sympathetic nervous system regulates cardiac APD via β-adrenergic receptors. This control requires the formation of complexes with AKAP9 and α subunit $K_V7.1$ encoded by the KCNQ1 gene. Mutations in the KCNQ1 gene that disrupt this complex cause long-QT syndrome LQT1. Mutations in the Akap9 gene that target AKAP9 binding domain to $K_V7.1$ also affect (but not supress) interaction between $K_V7.1$ and AKAP9, hence reducing cAMP- and PKA-dependent phosphorylation of slow delayed rectifiers ($i_{K,s}$ current) [437].

α1-Syntrophin of the family of dystrophin-associated proteins that is encoded by the gene SNTA1 interacts with the cardiac $Na_V1.5$ channel encoded by the SCN5A gene, nitric oxide synthase NOS1, and NOS1 inhibitor plasma membrane Ca^{2+} ATPase subtype PMCA4b [438]. SNTA1 gene mutations selectively disrupt PMCA4b association with this complex, thereby relieving inhibition of NOS3 and leading

Table 5.9 Long-QT syndromes (**Part 2**: LQT7–LQT13)

Type	Mutated gene	Protein (current) effect
LQT7	KCNJ2	K_{IR}2.1 binds to K_{IR}2.2
		(i_{K1})
		Andersen–Tawil syndrome, ventricular arrhythmia
LQT8	CACNA1c	α subunit of Ca_V1.2 channel
		Timothy syndrome
LQT9	CAV3	Caveolae in which lodges Na_V1.5 channel
		Late Na^+ current
LQT10	SCN4B	Auxiliary subunit of Na_V1.5 channel
		Increased Na^+ current.
LQT11	Akap9	AKAP9 recruits PKA and PP1; hence defective
		phosphorylation-dependent regulation of K^+ channel
LQT12	SNTA1	Syntrophin-α1 (most abundant isoform in the heart)
		(59-kDa dystrophin-associated protein-A1 acidic component)
		Sarcolemmal binder of Na_V1.5 (SCN5A)
		Linker of Na_V1.5 to the actin cytoskeleton
		and the extracellular matrix via
		dystrophin-associated proteic complex
LQT13	KCNJ5	K_{IR}3.4, or
		G-protein-gated inwardly-rectifying K^+ channel GIRK4
		($i_{K(ACh)}$)
		Prolonged QT interval on ECG
		Polymorphic ventricular arrhythmias
		Torsade de pointes

to an increase in SCN5A nitrosylation and peak and late sodium current ($i_{Na,L}$), a dysfunction that characterizes sodium channel-mediated LQTS (LQT3).

Ventricular arrhythmias can be due to channelopathies, resulting from impaired channel function, especially targeting voltage-gated channels.

Mutations in the human ether-a-go-go-related gene (HERG) encoding K_V11.1 channel responsible for the rapid delayed rectifier K^+ current ($i_{K,r}$), cause LQTS, which can lead to arrhythmias, such as *torsade de pointes* and ventricular fibrillation [439].

Mutations in the KCNQ1 gene that encodes the α subunit for the potassium channel, which conducts the slow delayed rectifier K^+ current ($i_{K,s}$; K_V7), or in the KCNE1 gene that encodes the β subunit also lead to LQTS.

LQTS due to small, persistent inward Na^+ current during the plateau phase and membrane hyperexcitability, can be the consequence of mutations in the cardiac

Table 5.10 Long-QT syndrome diagnostic criteria in the absence of medications or disorders that affect electrocardiographic trace. (Source: [436]). A score ranging from 1.5 to 3 points means an intermediate probability of LQTS and above 3.5 points a high probability

Criterion	Points
QT_c geq480 ms	3
QT_c 460–479 ms	2
QT_c 450–459 ms (male)	1
QT_c 4 mn after exercise geq480 ms	1
Torsade de pointe	2
T-wave alternans	1
Notched T-wave in 3 leads (limb lead II or V4–V6)	1
Low cardiac frequency for aged	0.5
History of syncope with stress	2
History of syncope without stress	1
Congenital deafness	0.5
Family members with LQTS	1
Unexplained sudden cardiac death	0.5
in immediate family members ($<$30 year)	0.5

QT_c cardiac frequency-corrected QT interval, i.e., $QT/RR^{1/2}$ [*Bazett's correction formula*]

sodium channel SCN5A gene that cause slow or impair voltage-gated Na^+ channel inactivation.

Mutations in the KCNJ2 gene of the inward rectifier K^+ channel, induce ventricular arrhythmias resulting from a delay in the final stage of ventricular repolarization.

Mutations in the CACNA1C gene that encodes the cardiac Ca_V1 channel prevent channel closure.

5.2.4.4 Brugada Syndrome

The Brugada syndrome is a rare inherited arrhythmogenic disease that generates ventricular arrhythmia and ultimately sudden cardiac death. It is characterized by an ST-segment elevation in the right precordial leads unrelated to ischemic or structural heart disease.

The balance of inward (typically i_{Na} and $i_{Ca,L}$) and outward (mainly $i_{K,to}$) currents active during the early phase of the epicardial action potential determine the magnitude of the AP notch. An outward shift in the balance of currents can amplify the AP notch and predispose to loss of the AP dome, leading to the Brugada syndrome [440].

The Brugada syndrome results from numerous loss-of-function mutations of the SCN5A gene, which encodes the cardiac $Na_V1.5$ channel, reducing the magnitude of the cardiac Na^+ current (SCN5A gene mutations also engender the LQT3 syndrome).

Mutations in the SCN1B gene that encodes $Na_V\beta1$ also engender the Brugada syndrome with a loss of function of the cardiac Na^+ current. In fact, the Brugada syndrome can be caused by few mutations in the auxiliary $Na_V\beta1$ to $Na_V\beta4$ subunits [441].

A mutation in the KCNE3 gene that encodes the Isk-related, delayed rectifier, voltage-gated potassium channel accessory β subunit minimum potassium (minK) channel-related peptide MiRP2 that, in patients diagnosed with Brugada syndrome, targets the pore forming α subunit $K_V4.3$ responsible for the rapidly inactivating transient outward K^+ current ($i_{K,to(r)}$) [440]. Coexpression of the MiRP2 mutant with $K_V4.3$ subunit significantly increases the magnitude of the transient outward K^+ current with respect to wild type MiRP2–$K_V4.3$ complex.

A mutation in the Gpd1l gene that encodes the glycerol 3-phosphate dehydrogenase-1-like protein (GPD1L) also lowers the cardiac Na^+ current [440].

The Brugada syndrome (as well as SQTS) are also linked to loss-of-function mutations in the CACNA1C and CACNB2b that encode the pore-forming α1c and ancillary β2b subunits of the long-lasting $Ca_V1.2$ channel [440].

The loss of function due to SCN5A gene mutations can result from an accelerated inactivation of the sodium channel. Other loss of function are created by a reduction of the peak calcium current. On the other hand, a CACNB2B gene mutation (T11I) associated with the Brugada syndrome is linked to a loss of function caused by an accelerated inactivation of the calcium current [440].

The $Na_V\beta1b$ channel subunit transcript is expressed at higher levels than $Na_V\beta1$ transcripts in the human heart [441]. The subunits $Na_V\beta1$ and $Na_V\beta1b$ coexpressed with $Na_V1.5$ engender a negative shift on steady state of activation and inactivation with respect to $Na_V1.5$ alone. Furthermore, $Na_V\beta1b$ coexpressed with $Na_V1.5$ increases the current level.

The $Na_V\beta1b$ H162P single nucleotide polymorphism (loss-of-function mutation) in Brugada syndrome reduces peak current density (48 %) with respect to $Na_V\beta1b$ wild type coexpressed with the $Na_V1.5$ channel, in addition to a steady-state inactivation shift and slower time-dependent recovery from inactivation [441]. These electrophysiological changes lower both action potential amplitude and maximum upstroke velocity.

5.2.4.5 SQTS

The SQTS is an inherited arrhythmogenic disorder characterized by an abbreviated repolarization (i.e., abnormally short QT interval on ECG traces) and a predisposition to supraventricular and ventricular arrhythmias in the absence of detectable structural heart lesion.

Table 5.11 Gene mutations in the short-QT syndrome

Gene	Ion channel	SQTS type	Current	Function
KCNH2	$K_V11.1$	SQT1	$i_{K,r}$	Gain
KCNJ2	$K_{IR}2.1$	SQT3	i_{K1}	Gain
KCNQ1	$K_V7.1$	SQT2		
	$i_{K,s}$	Gain		
CACNA1C	$Ca_V1.2\alpha1c$	SQT4	$i_{Ca,L}$	Loss
CACNB2B	$Ca_V1.2\beta2b$	SQT5	$i_{Ca,L}$	Loss
CACNA2D1	$Ca_V1.2\alpha2/\delta1$	SQT6	$i_{Ca,L}$	Loss

Mutations in the KCNH2, KCNJ2, and KCNQ1 genes that encodes the subunits $K_V11.1$, $K_{IR}2.1$, and $K_V7.1$ of potassium ion channels, respectively. They cause short-QT syndromes SQT1, SQT3, and SQT2, respectively (Table 5.11).

In SQTS patients, mutations in the genes encoding potassium channels (i.e., SQT1–SQT3) cause gain of function. On the other hand, mutations in the CACNA1C, CACNB2B, and CACNA2D1 that encode the $\alpha1c$, $\beta2b$, and $\alpha2/\delta1$ subunit of the $Ca_V1.2$ generate loss of function [444].

Some patients remain asymptomatic. Other SQTS patients complain of palpitations and syncope, as well as paroxysmal or permanent atrial fibrillation and ventricular arrhythmias. The SQTS is associated with a risk of sudden cardiac death, most likely due to ventricular fibrillation.

Diagnostic criteria were selected to facilitate the clinical evaluation of suspected SQTS cases [442]. The *QT index* (QTI) is the ratio between the measured and predicted QT interval (QT_p) for cardiac frequency (f_C) correction due to the poor performance of the Bazett corrected QT interval (QT_c) at cardiac frequency extrema:

$$\mathrm{QTI} = \mathrm{QT}/\mathrm{QT_p} \times 100. \tag{5.3}$$

The *predicted QT interval* is calculated using the following formula [443]:

$$\mathrm{QT_p} = 656/(1 + f_C/100). \tag{5.4}$$

A QTI value lower than 88 % is considered 2 standard deviations below the mean [443]. The sex difference in the QT interval is due to QT shortening in males after puberty rather than QT prolongation in women during reproductive years.

A diagnostic scoring system was developed, taking into account ECG characteristics, clinical presentation, family history, and the genetic context (Table 5.12). Patients are deemed high (score $\geq$ 4 points), intermediate (3 points), and low probability ($\leq$2 points, [442]).

The D172N mutation of the KCNJ2 gene that encodes the strong inward rectifier $K_{IR}2.1$ channel increases outward component of the inward rectifier current i_{K1} [444]. This mutation shortens the QT interval and predisposes to reentrant arrhythmias.

The E299V mutation engendered by the A896T substitution (single nucleotide polymorphism) in the KCNJ2 variant gene (a different SQT3 type) is associated with

Table 5.12 Short-QT syndrome diagnostic criteria (Source: [442]). Using 12-lead surface ECG traces, the Bazett corrected QT interval (QT_c) and the interval extending from the J point to the peak of the T wave (J_{point}–T_{peak} interval) are measured, the latter in the precordial lead with the greatest amplitude T wave. The J-point is the point of inflection between the S wave and the ST segment; the peak of the T wave corresponds to the highest point of the waveform. Electrocardiogram must be recorded in the absence of modifiers that shorten the QT interval. Clinical history events must occur in the absence of an identifiable etiology

Criterion	Points
QT_c <370 ms	1
QT_c <350 ms	2
QT_c <330 ms	3
J_{point}–T_{peak} interval <120 ms	1
Clinical history	
Sudden cardiac arrest	2
Polymorphic VT or VF	2
Unexplained syncope	1
Atrial fibrillation	1
Family history	
First- or second-degree relative with high-probability SQTS	2
First- or second-degree relative with sudden cardiac death	1
Sudden infant death syndrome	1
Genotype	
Genotype positive	2
Mutation of undetermined significance in a culprit gene	1

VF ventricular fibrillation, *VT* ventricular tachycardia

an extremely abbreviated QT interval (200 ms) on ECG trace and paroxysmal atrial fibrillation [444]. An abnormally large outward i_{K1} current occurs at potentials above −55 mV due to a lack of inward rectification, even in the heterozygous condition, hence favoring appearance of atrial fibrillation. The homomeric assembly of E299V mutant proteins causes gain of function.

5.2.4.6 Catecholaminergic Polymorphic Ventricular Tachycardia

Calsequestrin is an abundant Ca^{2+}-storing protein in terminal cisternae of the sarcoplasmic reticulum (contacts between junctional sarcoplasmic reticulum and T tubules). Calsequestrin connects to Ca^{2+} with high capacity (40–50 mol/mol Casq) and low affinity (dissociation constant $K_d = 1$ mmol; [445]). Calsequestrin binds to ryanodine-sensitive Ca^{2+} channel RyR2 on the sarcoplasmic reticulum via triadin and junctin.

Released Ca^{2+} binds to troponin-C, thus starting the contraction. When Ca^{2+} release ceases and the cytosolic Ca^{2+} concentration decays by reuptake into stores and extrusion, relaxation starts. Calcium-induced dimerization of calsequestrin begins at concentrations of 0.5–1 mmol; at 3 mmol, Ca^{2+} elicits tetramerization [445].

Mutations in the CASQ2 gene that encodes the cardiac isoform of calsequestrin is associated with an inherited form of ventricular arrhythmia triggered by emotional or physical stress: catecholaminergic polymorphic ventricular tachycardia (CPVT) [445].

Cardiomyocytes lacking functional Casq2 protein exhibit frequent spontaneous Ca^{2+} release after exposure to catecholamines, which activates Na^+–Ca^{2+} exchanger, hence initiating an inward Naa^+ current and generating a DAD. Premature DAD-triggered action potentials, especially in the Purkinje fibers, can prime ventricular tachycardia [445]. Dysfunctional calsequestrin causes spontaneous Ca^{2+} release via: (1) loss of Ca^{2+} buffering and Ca^{2+} release refractoriness; (2) dysregulation of RyR2 by Casq2; and (3) local remodeling with reduction of Casq2-binding junctin and triadin. Loss of triadin causes structural modification of the sarcoplasmic reticulum (50 % decrease in RyR2, Casq2, and junctin), especially at terminal cisternae, thereby causing impaired excitation–contraction coupling. Junctin level attenuation alters Ca^{2+}-cycling parameters,

5.2.5 Conduction Alterations

When the PR interval is nearly constant, but shorter than normal (< 120–200 ms), either the origin of the action potential is closer to ventricles or the atrioventricular conduction uses a bypass, such as in Wolff–Parkinson–White syndrome.

5.2.5.1 Atrioventricular Blocks

First-degree atrioventricular block (Mobitz I or Wenckebach block) is signed by P waves always preceding QRS complexes with a prolonged PR interval (> 200 ms).

In *second-degree atrioventricular block* (Mobitz II or Hay block), QRS complexes sometimes do not follow the P wave. According to the anatomical site of the block, second-degree atrioventricular blocks are classified into type-1 and type-2 blocks. The second-degree type-1 atrioventricular block is related to the atrioventricular node, which is subjected to sympathetic and parasympathetic innervation. Second-degree type-2 atrioventricular block is linked to the His bundle (infranodal block). Mobitz type-1 atrioventricular block is manifested by progressive prolongation of the PR interval with dropped QRS complex. Usually, QRS complexes are narrow. Mobitz type-2 atrioventricular block is characterized by a constant PR interval before a dropped QRS complex. Commonly, QRS complexes are wide.

Table 5.13 A 2:1 atrioventricular block arises from a second-degree type-1 or -2 atrioventricular block. (Source: rezidentiat.3x.ro/eng/tulbritmeng.htm)

QRS width	Normal QRS in atrioventricular node
PR interval	>30 ms in atrioventricular nodal block
	≤160 ms in His–Purkinje nodal tissue block
Exercise	Improve conduction in atrioventricular node
	Worsen conduction in His–Purkinje nodal tissue
Carotid sinus pressure	Worsen conduction in atrioventricular node
	Improve conduction in His–Purkinje nodal tissue
Retrograde conduction	In His–Purkinje nodal tissue

According to the electrocardiographic trace (i.e., to the conduction ratio) type-1 and type-2 second-degree atrioventricular blocks can progress to a 2:1 atrioventricular block. Conversely, a 2:1 atrioventricular block can regress to type-1 (narrow QRS) or type-2 (wide QRS) block. A 2:1 atrioventricular block is a form of second-degree atrioventricular block that occurs when a P-wave over 2 is not conducted through the atrioventricular node into ventricles, i.e., is not followed by a QRS complex (Table 5.13).

Complete lack of synchronism between P waves and QRS complexes depicts *third-degree atrioventricular block* (absence of atrioventricular conduction).

First-degree and type-1 second-degree atrioventricular blocks are usually caused by delayed conduction in the atrioventricular node. Type-2 second-degree atrioventricular block usually is infranodal, especially when QRS complexes are wide.

Third-degree atrioventricular block caused by a blockage of the two branches of the His bundle prevents the propagation front from the atria from reaching ventricles. Third-degree atrioventricular blocks can occur at several anatomical levels (upstream from the bundle of His, in the bundle of His, in the upper part of both bundle branches, or even in the right bundle branch and two fascicles of the left bundle branch).

5.2.5.2 Bundle Branch Blocks

Bundle branch blocks denote conduction defects in one of the bundle branches or any fascicle of the left bundle branch. Left or right bundle branch blocks mean that ventricle activation is initiated by the opposite ventricle. Resulting QRS complexes have abnormal shapes and long duration (>120 ms).

Right bundle branch blocks are characterized by a broad S-wave in I and V6 leads, a double R-wave (RSR' complex) in the V1 lead, and inversed T-wave in the V1 and V2 leads.

Electrocardiographic manifestations of *left bundle branch blocks* are broad and tall R-wave, with M shape in I, aVL, V5, and V6 leads, inversed T-wave in left precordial leads and in I or aVL leads, a deep and large S-wave in right precordial leads.

5.2.5.3 Wolff–Parkinson–White Syndrome

This syndrome is due to action potential passage directly from the atrium to the ventricle via the bundle of Kent, hence bypassing the atrioventricular junctions. The QRS complex initially exhibits an early upstroke (δ wave), which shortens the PQ interval. Mutations affecting AMP-activated protein kinase lead to Wolff–Parkinson–White syndrome [446]. Mutations affecting MAP3K7 also cause cardiac conduction alterations [447].[23]

5.3 Experiments and Modeling of Cardiac Electrophysiology

Modeling and simulations of the generation and propagation of cardiac action potentials as well as the cardiac electrical activity recorded at the thoracic surface have been carried out in normal and pathological conditions during the past six decades [448].

5.3.1 Nodal Tissue

The myocardium is composed of cardiomyocytes that, whatever the mammalian species, are similar in size [449]. Therefore, the left ventricle of small mammals contains a fraction of the number of cardiomyocytes in the human left ventricle. The extensive subendocardial conduction tissue in the ventricle of large hearts enables a short total excitation time as well as a relative synchronization of the depolarization of the free ventricular wall and interventricular septum.

Action potential generated in the sinoatrial node propagates through atria insulated from ventricles by the atrioventricular septum of nonexcitable cells. The signal reached the atrioventricular node, where it slows down, and then enters the bundle of His that gives rise to Purkinje fibers, the fastest conducting component of the nodal tissue. Purkinje fibers form a tree-like structure that runs and ends in the subendocardial layer of the myocardium. They cyclically excite cardiomyocytes. The propagation

[23] Kinase MAP3K7 (or TGFβ-activated kinase TAK1), associated with JNK, P38MAPK, and Iκ B kinaseβ, is involved in cytokine signaling and innate immunity. Enzyme MAP3K7 itself, or an associated protein, targets kinase LKB1, which phosphorylates (activates) AMPK enzyme. Both MAP4K4 and MAP3K7 trigger AMPK activity; conversely, MAP3K7 is activated by stimulators of AMPK activity.

Table 5.14 Concentrations (mmol/kg) of main ions in the intra- and extracellular media (Source: [450]). Common ions include simple and polyatomic (e.g., ammonium NH_4^+) cations as well as simple anions, oxoanions (e.g., phosphate PO_4^{3-} and sulfate SO_4^{2-}), and anions derived from organic acids (e.g., acetate CH_3COO^- and oxalate $C_2O_4^{2-}$)

Ion	Intracellular space	Extracellular space
Simple monovalent cations		
K^+	140	4.5
Na^+	10	145
Simple divalent cations		
Ca^{2+}	0.0001	3
Simple anions		
Cl^-	4	115

of action potentials is one-dimensional in the nodal paths and three-dimensional in the myocardium.

The ventricular response results from the quality, number, and sequence of atrial impulses reaching the atrioventricular node and of the electrophysiological properties of the atrioventricular nodal junction. Adaptation of the atrioventricular conduction to changes in atrial rhythm usually occurs within one or two subsequent cycles. The ventricular rhythm during atrial fibrillation in humans is random because the excitation of atrial fibrillation is random and remains random because the atrioventricular junction operates only on few previous cardiac cycles to influence the sequence of conducted impulses [449]. The changes in frequency and a persistent randomness of the ventricular rhythm during atrial fibrillation can be explained by the direct action of drugs and possibly a vagal activity.

The nodal pacemaker and paths can be assumed to function as a nerve with its associated soma and ganglia and segments with different conduction velocity associated with sheath features.

5.3.2 Transmembrane Ion Distribution and Cardiac Ion Carriers

Typical ion concentrations on each side of the plasma membrane of a mammalian cell is given in Table 5.14).

Ion channels have two main signaling functions. They can introduce in the cell second messengers such as free Ca^{2+} ions and operate as effectors of signaling mediators.

Calcium carriers include plasmalemmal entry and cytoplasmic release channels as well as pumps and exchangers implicated in electromechanical coupling, during both cardiomyocyte contraction and relaxation.

Numerous K^+ channels also contribute to cell signaling and repolarization of nodal cells and cardiomyocytes, once they depolarized. Voltage-gated K^+ channels regulate membrane potential and excitability. Inward rectifier K^+ channel include acetylcholine-sensitive G-protein-gated inward rectifier (GIRK3.1 and GIRK3.4) and ATP-sensitive K^+ channels.

Two-pore domain K^+ channels (K_{2P}) stabilize the resting potential. They can also cause the background (or leak) K^+ current (i_{Kp}) in response to stretch and other stimuli. They are indeed regulated by mechanical stress and pH and heat, as well as lipids (fatty acids) and endocannabinoids. Both TWIK and TASK members are highly expressed in the human heart, in atria and ventricles.

Some of the actions of Ca^{2+} are achieved by Ca^{2+}-sensitive K^+ and Cl^- channels. Hyperpolarizing-activated cyclic nucleotide-gated (HCN) channels operate in pacemaker cardiac cells. Mechanosensitive channels open in response to membrane deformation and stretch.

In addition, members of the superclass of ATP-binding cassette (ABC) transporters extrude signaling components from the cell. Cystic fibrosis transmembrane conductance regulator (CFTR) conveys anions (Cl^- and HCO_3^-). The CFTR channel may be responsible for the Cl^- currents activated by protein kinase-A ($i_{Cl,PKA}$) and -C ($i_{Cl,PKC}$), and extracellular ATP ($i_{Cl,ATP}$) [451]. It also transports other molecules such as sphingosine 1-phosphate. It is involved in the release of cytokines. It partcipates to the regulation of activities of many other ion channels and transporters, such as epithelial Na^+ channels and Ca^{2+}- and volume-activated Cl^- channels.

In addition to CFTR, several chloride channels reside in the cardiomyocyte, such as voltage-gated ClC2 (responsible for the hyperpolarization-activated inwardly rectifying Cl^- current [$i_{Cl,IR}$]) and ClC3 channels (responsible for the volume-regulated outwardly rectifying Cl^- current [$i_{Cl,vol}$], including the basally activated [$i_{Cl,vol(b)}$] and swelling-activated [$i_{Cl,vol(swel}$] components), calcium-sensitive chloride channel ClCa1 (responsible for the $i_{Cl,Ca}$ current), bestrophin (a possible second component of $i_{Cl,Ca}$, and the calcium-activated chloride channel anoctamin Ano1 (or transmembrane protein TMem16a; a possible third component of $i_{Cl,Ca}$); [451]). These channels may contribute to cardiac arrhythmogenesis, myocardial hypertrophy, and heart failure, as well as cardioprotection against ischemia–reperfusion injury. A additional Cl^- current activated by extracellular acidosis ($i_{Cl,ac}$) can be observed in cardiomyocytes.

In addition, the voltage-dependent anion channel VDAC1 that lodges predominantly on the outer membrane of mitochondria also localizes to the sarcolemma [451].

Each phase of the cardiac action potential is associated with a set of ion channels (Table 5.15).

5.3.3 Hodgkin–Huxley Conductance-Based Model

In 1952, A.L. Hodgkin and A.F. Huxley proposed a mathematical formulation of ionic transfers responsible for the initiation and propagation of an action potential through the giant axon of the squid [454]. The Hodgkin–Huxley or conductance-based model is a set of differential equations that approximates the electrical characteristics of

Table 5.15 Activity of ion channels during the different phases of the cardiac action potential (Sources: [452, 453]). The $K_V1.5$ and $K_{IR}3.4$ (GIRK4) channels are expressed in atriomyocytes, but not in ventriculomyocytes

Channel	Current	Action potential phases				
		0	1	2	3	4
$Na_V1.3/1.5$	$i_{Na,f}$ (fast)	↓↓	↓	↓	↓	
	$i_{Na,l}$ (late)					
$Ca_V1.2$	$i_{Ca,L}$		↓↓	↓		
(L-type Ca^{2+} channel)						
Na^+–Ca^{2+} exchanger	i_{NCX}	↑	↑	↓↓		↓
$K_V4.2/4.3$	$i_{K,to(f)}$ (fast)		↑↑			
$K_V1.4/3.4$	$i_{K,to(s)}$ (slow)		↑↑	↑		
$K_V1.2/7.1$	$i_{K,s}$ (slow)			↑	↑↑	↑
$K_V11.1$	$i_{K,r}$ (rapid)		↑	↑	↑↑	↑
$K_V1.5/1.7/1.8/2.1$	$i_{K,ur}$ (ultrarapid)		↑	↑	↑	
$K_{IR}2.1$–2.4	i_{K1}	↑	↑		↑↑ (fin)	↑
$K_{IR}6.2$	$i_{K_{ATP}}$	↑	↑		↑↑ (fin)	↑
$K_{IR}3.4$	$i_{K_{ACh}}$		↑	↑	↑ (fin)	↑
$K_{2P}1.1/3.1/4.1$, CFTR	$i_{K,Cl}$			↑	↑	
HCN1/2/4	i_f (funny)	↓↓	↓		↓ (fin)	↓↓

↑ outward flux, ↓ inward flux, *double arrow* high-amplitude flux, *0* upstroke, *1* rapid repolarization (notch), *2* slow repolarization (plateau), *3* fast final repolarization, *4* resting potential [−80 mV]

excitable cells, such as neurons and cardiomyocytes (Vol. 6, Chap. 3. Cardiovascular Physiology).

The semipermeable plasma membrane that separates the intra- from extracellular space acts as a capacitor. Active ion transfer by ion pumps through the plasma membrane yields a transmembrane ion concentration. The resulting electrical potential is represented by a battery. The membrane at rest is polarized because the resting potential is negative with respect to the outer fluid medium. An input current $i$stim adds further charges on the capacitor or leak through ion channels. Depolarization raises the membrane potential.

In the absence of a net flux of ions, the electrochemical potential of an ion of charge z across the cell membrane can be computed using the Nernst equation:

$$E = \frac{R_g T}{zF} \ln \frac{c_e}{c_i} = 2.302 \frac{R_g T}{zF} \log \frac{c_e}{c_i}, \tag{5.5}$$

where R_g is the universal gas constant (8.31446 J/K/mol), T the absolute temperature, $\mathcal{F}$ the Faraday constant (9.64853×10^4 C/mol), and c_i and c_e the ion concentration in the intra- and extracellular space. Natural ($\ln(x)$) and common logarithm ($\log_{10}(x) \equiv \log(x)$) are related by the expression $\log_{10}(x) = \ln(x)/\ln(10)$ with $ln(10) \approx 2.302$.

The Hodgkin–Huxley model includes two voltage-gated ion channels, the conductance (resistance) of which depends on the transmembrane potential (u_{mb}; [454]). Two categories of voltage-gated ion channels can be defined according to whether their open state duration is transient (voltage-gated Na^+ channel involved in the fast depolarization) or more persistent (voltage-gated K^+ channel implicated in the relatively long repolarization).

By convention, a positive external current (i_{stim}) tends to depolarize the cell (i.e., rendering u_{mb} less negative), whereas a positive ionic current (i_ℓ) linked to the motion of ionic species ℓ tends to hyperpolarize the cell. An inward flux (into the cell) of cations causes a negative current. A potential being relative, as only potential differences can be measured directly, by convention, the intracellular potential can be defined with respect to the resting intracellular potential set to zero (the convention in the Hodgkin–Huxley model), or to the extracellular potential set to zero, then the resting intracellular potential is about -70 mV. Depolarization makes the transmembrane potential more positive (the sign of the transmembrane potential in the Hodgkin–Huxley model is negative).

A channel pore functions according to a all-or-nothing law, that is, according to whether its pore is open or closed. Each ion channel is supposed to possess one or two gates that regulate the ion flux through the channel [455]. A gate can be permissive or nonpermissive according to whether it is open or closed (another set of adjectives is used to avoid confusion between the state of a gate and that of its channel; [456]). When all of the gates of a given channel are in the permissive state, ions can pass through the channel pore; the channel is in its open state. Otherwise the channel is closed.

An open channel conveys ions with a given conductance (resistance). A transiently open channel with its two gates (activation and inactivation gates) evolves according to three states [455]: deactivated (activation gate closed and inactivation gate open); activated (both gates open); and inactivated (inactivation gate closed). On the other hand, a relatively persistently open channel with a single gate has only two states: activated and inactivated.

The standard Hodgkin–Huxley model comprises only three channel types: a sodium (index Na), potassium (index K), and unspecified leak (index L) channel. Depolarization causes a transient increase in sodium conductance and a slower, but maintained increase in potassium conductance. Repolarization close these channels, quickly (Na^+ channel) or slowly (K^+ channel).

Each component of the excitable cell is supposed to be an electrical element:

- The lipid bilayer is represented as a capacitance (C_{mb} per unit area) with a potential across the capacitor (or transmembrane voltage)
- Electrochemical gradients driving fluxes of ionic species (k) and hence current (i_k) by voltage sources, the voltage of which is determined by the ratio of the intra- and extracellular concentrations of each involved ionic species
- Voltage-gated ion channels (ℓ) by voltage- and time-dependent electrical conductances (G_ℓ per unit area)

- Unspecified leakage channel by a constant electrical conductance (G_L per unit area)

The conservation of electrical charge at the plasma membrane states that the total membrane current density is the sum of the externally applied stimulus current density (i_{stim}; capacitive current that charges the capacitor) and current densities i_ℓ that cross the ion channels.

$$i(t) = i_{stim} + \sum_1^\ell i_\ell. \tag{5.6}$$

The current through the lipid bilayer is

$$i(t) = C_{mb} \frac{du_{mb}}{dt}. \tag{5.7}$$

The current through an ion k channel ℓ is

$$i_\ell = G_\ell(u_{mb} - u_\ell). \tag{5.8}$$

When sodium, potassium, and leak channels are only considered, the total current through the membrane is given by:

$$i = C_{mb} \frac{du_{mb}}{dt} + G_K(u_{mb} - u_K) + G_{Na}(u_{mb} - u_{Na}) + G_L(u_{mb} - u_L), \tag{5.9}$$

where u_K and u_{Na} are the equilibrium potentials given by Eq. 5.5 for the potassium and sodium ions and u_L is the potential at which the leakage current due to chloride and other ions is zero.

Activation and deactivation (in the case of Na^+ channel characterized by a fast kinetics) can be supposed to evolve in time according to a statistical law mimicking the Boltzmann distribution in statistical mechanics. Using the kinetics theory, the gating function are assumed to obey the simplest first order rate equation. The channel has probabilities $P(\text{op})$ and $P(\text{cl})$ to be open or closed. As these are only two possibilities (either closed or open), $P(\text{op}) + P(\text{cl}) = 1$. The rates at which a closed and open channel opens and closes are governed by *rate constants* τ_{op} and τ_{cl}), functions of the transmembrane voltage:

$$\frac{dP(\text{op})}{dt} = \tau_{op}[1 - P(\text{op})] - \tau_{cl} P(\text{op}). \tag{5.10}$$

In a pool of ion channels, $P(\text{op}, t)$ represents the average fraction of open channels at time t. Because the transmembrane voltage is assumed constant, Eq. 5.10 is a linear differential equation with the following solution:

$$P(\text{open}; t) = P(\text{open}; 0) \exp\{-(1/\tau)t\} + \tau\tau_{op}(1 - \exp\{-(1/\tau)t\}). \tag{5.11}$$

where $\tau = 1/(\tau_{op} + \tau_{cl})$. In the resting state (i.e., at the reference transmembrane potential $u_{mb} = 0$), P(op has a resting value (initial condition). When u_{mb} changes suddenly, the rates τ_{op} and τ_{cl} instantly take up new values corresponding to the new voltage value. The solution of Eq. 5.11 that satisfies the initial condition is:

$$\begin{aligned} P(\text{open}; t) &= P_\infty - (P_\infty - P_0)\exp\{-t/\tau\}, \\ \text{where } P_0 &= P(\text{open}; 0), \\ P_\infty &= \frac{\tau_{op}}{\tau_{op} + \tau_{cl}}, \\ \text{and } \tau &= \frac{1}{\tau_{op} + \tau_{cl}}. \end{aligned} \tag{5.12}$$

In fact, the problem is not focusing on the states of the ion channel (open or closed) or, similarly, of the channel pore (permeable or not), but on the states of the channel gate (permissive or not), some channels being assumed to possess a single gate, others a couple of gates. The probability for a given gate to be in the permissive or nonpermissive state depends on the value of the membrane voltage. It equals 0 or 1 whether the gate is in the nonpermissive state or not. In fact, the probability for a given gate of a given ion channel in a permissive state corresponds to a microscopic average on a large number of ion channels of the same type and on a time scale with numerous opening and closure. In other words, a pool of a given type of ion channels is associated with the probability $P(\text{perm}(t)$ ranging from 0 to 1 that corresponds to the fraction of gates of the channel pool in the permissive state at a given instant t. The rate at which gates changes from the non-permissive state to the permissive one and vice versa is denoted by the variables α and β, respectively, in the Hodgkin–Huxley model. These rate constants depend on the transmembrane potential. When the transmembrane potential is fixed at a fixed value, then the fraction of gates in the permissive state can reach a steady state ($dP(\text{perm}/dt = 0$ as $t \rightarrow \infty$). The time course for approaching this steady-state value is described by a simple exponential with time constant τ given in Eq. 5.12.

The kinetics of the voltage-gated ion channels $\ell = 1, 2$ is taken into account by introducing the maximal conductance $G_{\ell_{max}}$ per unit area and normalized quantities (i.e., dimensionless quantities ranging from 0 to 1) that describe the probabilities of the channel gate states. More precisely, the voltage-gated K^+ and Na^+ channel gate states are controlled by the quantity n and a combined effect of the variables m and h, respectively. The quantity $n(t)$ represents the probability of the permissive K^+ channel gate state. The variables $m(t)$ and $h(t)$ represents the probability of the activation and deactivation state of the 2-gate Na^+ channel. In other words, m is the probability that both gates of the Na^+ channel are permissive (i.e., the probability that the Na^+ channel is not deactivated) and h the probability for the inactivation gate to be permissive (i.e., the probability that the Na^+ channel is not inactivated [but simply deactivated]). These model parameters are associated with empirical voltage-dependent time-independent rate functions α and β for each state of the selected ion channel gates that fit experimental data.

The Hodgkin–Huxley model then consists of a set of four ordinary differential equations:

$$\sum_{\ell=1}^{3} i_\ell = \mathrm{G}_{\mathrm{K}_{\max}} n^4(\mathrm{u}_{\mathrm{mb}} - \mathrm{u}_{\mathrm{K}}) + \mathrm{G}_{\mathrm{Na}_{\max}} m^3 h(\mathrm{u}_{\mathrm{mb}} - \mathrm{u}_{\mathrm{Na}}) + \mathrm{G}_{\mathrm{L}_{\max}}(\mathrm{u}_{\mathrm{mb}} - \mathrm{u}_{\mathrm{L}}),$$

$$\frac{\mathrm{d}n}{\mathrm{d}t} = \alpha_{\mathrm{K}}(\mathrm{u}_{\mathrm{mb}})(1-n) - \beta_{\mathrm{K}}(\mathrm{u}_{\mathrm{mb}})n,$$

$$\frac{\mathrm{d}m}{\mathrm{d}t} = \alpha_{\mathrm{Na}^{act}}(\mathrm{u}_{\mathrm{mb}})(1-m) - \beta_{\mathrm{Na}^{act}}(\mathrm{u}_{\mathrm{mb}})m,$$

$$\frac{\mathrm{d}h}{\mathrm{d}t} = \alpha_{\mathrm{Na}^{dea}}(u_{\mathrm{mb}})(1-h) - \beta_{\mathrm{Na}^{dea}}(u_{\mathrm{mb}})h. \quad (5.13)$$

5.3.4 Cardiac Structure Representation in Computational Electrophysiology

When dealing with the myocardium rather than a single cell, the first major assumption is related to the cardiac wall itself that can work as a syncytium, that is, a single cell representing the entire population of strongly connected cells via intercalated discs or a cellular pool with a complex transmural myofiber architecture[24] and distinct electrophysiological properties. The contraction of the ventricular myocardium as well as the spread of excitatory action potentials appear to be related to a syncytium. On the histological and electrophysiological points of view, discrete rod-like myofibers interconnected by intercalated discs are defined by their orientation and hence anisotropic physical properties (e.g., electrical resistivity and rheology) as well as by their electrophysiological features in a given layer of the cardiac wall. Nevertheless, inside this layer, the set of myofibers can be considered as a syncytium.

Intercalated discs are structures present only in cardiomyocytes. They ensure mechanical coupling and chemical communication between cardiomyocytes. The plasma membrane (sarcolemma) of apposed cardiomyocytes is extensively folded at intercalated discs; adjacent plasma membrane are separated by a tiny space (width ~800 nm). Intercalated discs have an egg box-like morphology, as cells interdigitate and hence form troughs and peaks. Intercalated discs contain three types of intercellular junctions:

1. Anchoring adherens junctions at the end of myofibrils, which attach thin actin myofilaments of the terminal half-sarcomere to the plasma membrane

[24] The myocardium can be subdivided into sheetlets (discontinuous sheets; thickness of four to six cells) of coupled similarly oriented cardiomyocytes [457]. The transmural distribution of cardiomyocyte orientation varies smoothly from about −60° in the epicardium to around +60° in the endocardium with respect to the ventricle horizontal plane.

2. Expanded desmosomes in nonfibrillar parts of the intercalated disc, where reside stacks of mitochondria, which maintain the connection between apposed cardiomyocytes when they contract
3. Gap junctions that facilitate fast electrochemical communication, so that the depolarisation wave passes rapidly from a given cell to another and spreads over the entire cardiac wall

The *transitional junction* is a functional subcellular domain at the intercalated disc that maintains the ordered array of terminal A band of the terminal half-sarcomere and connects myofibrils to the adherens junction [458]. It corresponds to a virtual Z disc of the terminal sarcomere; it is situated at the axial position of a sharp transition between the myofibrillar I band and intercalated disc thin filaments. It contains Z-disc proteins (e.g., α-actinin and titin N-terminus) as well as membrane-bound spectrin.[25] The transitional junction enables: (1) communication between intercalated disc and sarcomere set and (2) sarcomere adding to the cell ends during growth. In summary, the intercalated disc with its gap junctions represents a channel for ionic movement along the length of the myofiber that allows a fast spreading of the depolarization wave, at least among those of a given myocardial slice characterized by a given orientation.

The slice geometry enables the representation of a more realistic cardiac domain that includes both intracellular and interstitial properties and myofiber orientations and hence a better modeling of anisotropic conduction. The "Auckland canine heart geometry" that results from a histological analysis is associated with a mathematical representation of the fiber architecture for electromechanical coupling using finite element models. Diffusion tensor imaging allows to perform nondestructive in vivo histological analysis of the heart at a spatial resolution of several hundred microns.

5.3.5 *Application of the Hodgkin–Huxley Model to the Heart*

The work carried out by Hodgkin and Huxley triggered a set of explorations, some being devoted to the heart. In particular, experiments were performed in mammalian Purkinje fibers [460]. As the upstroke of the cardiac action potential is caused by Na^+ ion entry, the maximal rate of voltage rise is a measure of the inward sodium current. The effect of a feedback circuit that set the resting potential on some features (e.g., upstroke velocity) of the cardiac action potential supposed to result from the sodium channel permeability were thus investigated. The experimental results were fitted to get the expression of the fraction of the highest value observed for upstroke rate $h = 1/(1 + \exp\{V_h - V\}/5)$ (V: clamp potential; V_h: potential at which h is half-maximal), which is similar to the relation used by Hodgkin and Huxley, but

[25] Spectrin is a $[\alpha 2]_2[\beta 2]_3$ pentamer formed by cardiac spectrin chain isoforms (or α- and β-fodrins), α1- and β1-spectrin chains being erythrocytic isoforms.

it describes a sigmoidal curve steeper by a factor 1.4. Because the sodium channel gate kinetics in Purkinje fibers are similar to those in neurons, the sodium current equations of the Hodgkin–Huxley model were used in the Noble model.

In 1960, D. Noble developed the computational model of a cardiac excitable cell [461]. He simulated the combined sarcolemmal ion fluxes through three channels of a Purkinje fiber in the absence of full knowledge on involved ion carriers and, for those known, their accurate kinetics. However, he reproduced the shape of an action potential using a modified version of the Hodgkin–Huxley model upon model adjustment to fit experimental observations. The model was tweaked mainly by raising the voltage range of Na^+ current activation, so that it accounted both all ionic influxes.

This first mathematical model of cardiac electrophysiology incorporated two potassium currents through the inward (i_{K1}) and delayed (i_K) rectifier, the former conveying a large inward current at negative potentials and inactivating during depolarization, the latter slowly activating and inducing repolarization. The equations related to i_{K1} are empirical descriptions of the current dependence on voltage and potassium concentration. Those related to i_K are slowed-down versions of the Hodgkin–Huxley K^+ current equations, as its time-dependent activation that initiates repolarization takes a few ms in nerve, but hundreds of ms in cardiac Purkinje fibers.

The model did not take into account other important current equations. The calcium inward current implicated in the electromechanical coupling, pacemaker funny current [462], and small, but significant late sodium ($\geq$100 ms after the depolarization beginning) current with slow inactivation kinetics [463], among others, were omitted.

Since the cardiac electrophysiological model of Noble, the representation of cardiac pacemaker rhythm and action potentials has evolved. Detailed models have been integrated in ECG modeling as well as simulation platforms that couple the cardiac electrical and mechanical functions in various pathophysiological contexts (e.g., the euHeart project). These simulations incorporate phenomenological models of the multilayered myocardium with normal and altered spatiotemporal events. However, interactions between cell populations (cardiac myocytes, fibroblasts, endotheliocytes, and nodal cells, as well as microvascular endotheliocytes and neuron endings) organized in defined structures at histological and anatomical scales are not yet fully taken into account.

5.3.6 Intercellular Gap Resistance

Electrochemical signals can be transmitted between adjoining cells when the membrane resistance is sufficiently low. A model of end-to-end interaction between cardiomyocytes was developed supposing that the membrane, except at the intercalated disc, is excitable. A domain constituted by 2 cylindrical cells (diameter 15 μm; length 100 μm) is studied [464]. The two apposed end faces form the intercalated disc assumed to be inexcitable. The intercellular gap (width 800 nm) is assumed contiguous with the interstitial media and has the same conductivity. The potential is given by the Laplace equation, the solution of which relies on a combination of first-kind Bessel functions of the radial coordinate and hyperbolic functions of the axial coordinate. Signal transmission is explored using an electrical circuit representation of these two cells separated by a gap, the ionic currents traversing the plasma membrane obeying a formulation based on the Hodgkin–Huxley–Noble equations. A gap resistance equals about 4 Ω cm^2.

Each adult cardiomyocyte is enclosed by a basement membrane (thickness 80 nm), a layered assembly of collagen-4, fibronectin, and laminin, as well as glycoproteins and proteoglycans (e.g., entactin, perlecan, and heparan and chondroitin sulfate proteoglycans). The basement membrane influence the sarcomere formation and remodeling due to connection between its constituents and integrins. The blood vessel wall maturation requires the presence of laminin. Anionic glycosaminoglycans regulate the transmembrane potential [465]. In addition, the basement membrane, especially laminin, binds calcium ions, thereby participating in the coupling between Ca^{2+} entry from the extracellular space and Ca^{2+} release from its intracellular stores, especially at the sarcolemma of transverse tubules.

In neonatal cardiomyocytes cultured on an aligned collagen-1 gel, laminin deposition increases with time and cardiomyocytes have a 3D rod-like morphology similar to that of in vivo cardiomyocytes [465]. Collagen fibrils of the matrix not only stimulate myofibrillogenesis, but also regulate growth factor-mediated signaling from fibroblasts to cardiomyocytes by sequestration. After 4 days, the release of growth factors alter the cardiomyocyte morphology and alignment and provoke cardiomyocyte aggregation when collagen fiber accumulation is not prevented. Upon laminin degradation, the cardiomyocyte morphology and transmembrane potential variation changes, especially the plateau duration [465]. Consequently, the temporal evolution of the calcium influx signal is impaired, due to primarily a defective calcium entry from the extracellular space, and subsequently the contraction duration shortens.

5.3.7 *Dispersion of Refractoriness and Reentry*

The depolarization wave spreading over the myocardium is a quasi-synchronous process, as activation of a given region activates its neighboring regions. Repolarization is an asynchronous process, as the depolarization duration is determined the local electrophysiological properties and the spatial distribution of refractoriness to further stimulation.

The spatial dispersion of refractory times during the repolarization can enlarge. Hence, a depolarization wave arriving on refractory regions can lead to reentry. The *vulnerable period* roughly corresponds to the T-wave peak during half cells are refractory and half are excitable hypothermia. During this period, β-adrenoceptor stimulation and myocardial ischemia predispose to reentry and fibrillation that is defined as an irregular disorganized electrical activity of atria or ventricles. At least, two major factors intervene in fibrillation genesis in atrial or ventricular myocardium:

the total number of involved myocardiocytes and the heterogeneity of their electrical properties.

A *reentrant path* is self-sustaining when the myocardium along the path leaves the refractory period when the reentrant depolarization wave arrives. A reentry criterion can be estimated by $\ell > v_{\rm cond} t_{\rm refr}$, where ℓ is the loop length, $v_{\rm cond}$ the local conduction velocity scale, and $t_{\rm refr}$ the characteristic refractory time [466].

Computational models of ventricular conduction that incorporates the spatial dispersion of refractoriness rely on key features that disturb the refractory times. The myocardium was modeled as a network of interconnected discrete units [426]. An automata model was developed, in which rules govern the intercellular conduction of impulses.

At the macroscopic scale, the spread of excitation can be considered continuous. The cable theory uses average properties and describes the excitable cell as an intracellular space separated from the extracellular medium by a membrane through which current passes. When the propagation velocity of action potentials is modified by changing the internal resistance of cells, the cable theory predicts that the shape of the action potential upstroke does not vary. Changes in action potential upstroke associated with modification of conduction velocity are attributed to variations of plasma membrane properties. However, in healthy cardiomyocytes, experiments show that changes in action potential upstroke with conduction velocity can occur when plasma membrane properties remain constant [467]. The shape of the action potential depends on the direction of propagation with high and low propagation velocities in the axial and transverse direction, respectively. These directional differences in action potential shape cannot be explained by a continuous anisotropic model. Action potential propagation may continue in the transverse direction when a block occurs in the longitudinal direction with eventual reentrant propagation [467].

5.3.7.1 Phenomenological Models

The basic tractable phenomenological model of depolarization and repolarization consists of two variables u and v [468, 469], in which fast dynamics are coupled with slow ones (u: "fast" variable, v: "slow" variable):

$$\frac{d\mathrm{u}}{dt} = f(\mathrm{u}, \mathrm{v}), \quad \frac{d\mathrm{v}}{dt} = g(\mathrm{u}, \mathrm{v}). \tag{5.14}$$

The system behavior in the phase plane (u, v) (spiral point, stable and unstable singular points and limit cycle, and bifurcation) depends on the parameters involved in the functions f and g. Such a model has been developed for a general excitable medium, assuming spatial homogeneity. It was improved to more accurately model the electrochemical firing in the nodal pacemaker as well as impulse propagation along the nodal paths and within the myocardium. The FitzHugh–Nagumo-type 2-variable system of equations corresponds to a *monodomain model*.

The Aliev–Panfilov model is an example of a dimensionless FitzHugh–Nagumo system [470]:

$$\frac{d\mathrm{u}}{dt} = \nabla \cdot (\mathbf{D}\nabla \mathrm{u}) + k\mathrm{u}(1 - \mathrm{u})(\mathrm{u} - a) - \mathrm{uv},$$
$$\frac{d\mathrm{v}}{dt} = -\epsilon(\mathrm{u}, \mathrm{v})(k\mathrm{u}(\mathrm{u} - a - 1) + \mathrm{v}), \tag{5.15}$$

where $\epsilon(\mathrm{u}, \mathrm{v}) = \epsilon_0 + \kappa_1 \mathrm{v}/(\mathrm{u} + \kappa_2)$ is a coupling control parameter between the action potential u and the repolarization v (the coupling between the slow and fast phases), **D** is the diffusion tensor

$$\mathbf{D} : \begin{pmatrix} 1 & 0 & 0 \\ 0 & 0.25 & 0 \\ 0 & 0 & 0.25 \end{pmatrix},$$

k a repolarization control variable, and a the reaction parameter. Fitzhugh-Nagumo and Aliev-Panfilov models are two-variable models, hence not ionic models. These reduced bioelectric models are not usually well suited for cardiac electrophysiology.

5.3.8 *Ionic Current Models*

Ionic current models use transmembrane potential as variable of state, with gate variables that quickly commute between two values (permissive or not) and concentrations. The simplest ionic model is the following:

$$C\frac{d\mathrm{u}}{dt} = \mathrm{v}i_{in}(\mathrm{u}) - i_{out}(\mathrm{u}) + I_stim, \tag{5.16}$$
$$\frac{d\mathrm{v}}{dt} = G(\mathrm{u}, \mathrm{v}), \tag{5.17}$$

where the gate v switches between 0 and 1 and needs only a small amount of energy, $i_{out}(\mathrm{u})$ is the ion outflux that requires energy ($i_{out}(\mathrm{u}, q_{ATP})$; q_{ATP}: energy consumption). The two terms $i_{in}(\mathrm{u})$ et $i_{out}(\mathrm{u})$ can be decomposed in specialized ionic currents, especially when realistic APD and conduction speed are required.

A hierarchy of models of growing complexity starts from eikonal models of depolarization front that does not incorporate the coupling dynamics of chemical reaction and electrical conduction and a series of ionic models, such as the Mitchell and Schaeffer 2-variable model [471], Fenton and Karma 3-variable model [472], and Djabella and Sorine 8-variable model [473]. The latter is a reduced order version of ten Tusscher, Noble, and Panfilov 17-variable model [474] that has an improved calcium dynamics compared with Bernus and Panfilov model [475]. The most appropriate model must be selected according to the goal of numerical experiments, taking into account the well-posedness of data-assimilation problems and real-time constraints.

The Mitchell–Schaeffer model belongs to the class of ionic models with two variables: the transmembrane potential u and gate v. It also contains five parameters (τ_{in}, τ_{out}, τ_{op}, τ_{cl}, and v_{gate}). It is easy to calibrate and gives an adequate transmembrane potential. This model of electrical activity of the cardiomyocyte sarcolemma describes only an inward and outward current. Mitchell-Schaeffer two-variable model of excitable media is given by:

$$\frac{d\mathrm{u}}{dt} = \frac{1}{\tau_{in}}\mathrm{v}\mathrm{u}^2(\mathrm{u}-1) + \frac{\mathrm{u}}{\tau_{out}},$$
$$\frac{d\mathrm{v}}{dt} = \begin{cases} \frac{1}{\tau_{op}}(\mathrm{v}-1), & \mathrm{u} < v_{gate} \\ \frac{\mathrm{v}}{\tau_{cl}}, & \mathrm{u} > v_{gate} \end{cases} \tag{5.18}$$

It provides reasonable APD, but not conduction speed.

5.3.9 Bidomain Model

The bidomain model corresponds to a 3D version of the cable equations [476]. The cardiac wall structure is assumed to be constituted of two overlapping domains, every point of the myocardium residing in both the intra- and extracellular domain.

In bidomain models, both domains that correspond to intra- and extracellular media separated by the myofiber syncytium membrane have their own volume-averaged properties, especially conductivity. Cardiac myofibers have anisotropic conduction properties, the impulse propagation being faster in the axial direction than transversally. An electrical conductivity tensor $\mathbf{G}_\mathrm{e}$ is then introduced, assuming that the conductivity values are identical in all directions perpendicular to the myofiber direction [477].

The collection of cardiomyocytes, which are end-to-end or side-to-side interconnected by intercalated discs and immersed in the extracellular fluid and ground matrix, is modeled as a periodic array that leads to homogenization procedure, with homogenized conductivity tensors for both media $\mathbf{G}_{\mathrm{e_i}}$ and $\mathbf{G}_{\mathrm{e_e}}$.

The membrane current density J_{mb} is then given by:

$$J_{\mathrm{mb}} = -\nabla \cdot \boldsymbol{i}_\mathrm{i} = \nabla \cdot \mathbf{G}_{\mathrm{e_i}} \nabla \mathrm{u_i} = \nabla \cdot \boldsymbol{i}_\mathrm{e} = -\nabla \cdot \mathbf{G}_{\mathrm{e_e}} \nabla \mathrm{u_e}, \tag{5.19}$$

where $\mathrm{u_i}$ and $\mathrm{u_e}$ are the electric potentials of the intra- and extracellular spaces supposed to be passive conductors in a quasi-steady state and $\boldsymbol{i}_\mathrm{i}$ and $\boldsymbol{i}_\mathrm{e}$ the currents,

$$\nabla \cdot (\mathbf{G}_{\mathrm{e_i}} \nabla \mathrm{u_i} + \mathbf{G}_{\mathrm{e_e}} \nabla \mathrm{u_e}) = 0.$$

In its general form, the bidomain model is defined by [429]:

$$\kappa_{\mathrm{av}}(\mathrm{C_m u}_t + 1/\mathrm{R_m} f(\mathrm{u}, \mathrm{v})) = \nabla \cdot (\mathbf{G}_{\mathrm{e_i}} \nabla \mathrm{u_i}), \tag{5.20}$$

Table 5.16 Values of the bidomain model parameters. R_m corresponds to the ion transport resistance through plasmalemmal ion channels. It is currently obtained by a polynomial formulation ($\kappa_d^2(u - u_r)(u - u_t)(u - u_m)$) of the transmembrane potential using the following set of quantities: depolarization rate κ_d, rest potential u_r, threshold potential u_t, maximum potential u_m ($\mathcal{G}_{e_{(i/e)f}}$, $\mathcal{G}_{e_{(i/e)t}}$, $\mathcal{G}_{e_{(i/e)n}}$: conductivity components of the intra- and extracellular media in the main myofiber axis, normal to the main myofiber axis in the myofiber sheet and perpendicular to the myofiber sheet; Source: [478])

C_m	1 μF/cm^2
κ_{av}	2000 cm^{-1}
$\mathcal{G}_{e_{if}}$	3 mS/cm
$\mathcal{G}_{e_{it}}$	1 mS/cm
$\mathcal{G}_{e_{in}}$	0.32 mS/cm
$\mathcal{G}_{e_{ef}}$	2 mS/cm
$\mathcal{G}_{e_{et}}$	1.65 mS/cm
$\mathcal{G}_{e_{en}}$	1.35 mS/cm
κ_d	0.04
u_r	−85 mV
u_t	−65 mV
u_m	40 mV

where $u = u_i - u_e$ is the transmembrane (action) potential, v the recovery variable, κ_{av} the averaged surface area-to-volume ratio of the cardiac myofibers, C_m and R_m the plasmalemmal capacitance and resistance (Table 5.16).

5.3.10 *Intersubject Variability*

Limitations of current mathematical modeling and experimental techniques limit the knowledge of phenomena responsible for intersubject variability of cellular mechanisms and ionic fluxes involved in cardiac electrophysiology, hence prediction of the response of heart to diseases and therapies.

A given type of healthy nodal cells or cardiomyocytes of a given species trigger a given type of action potential. However, albeit qualitatively similar, quantitative intersubject differences can exist in action potential morphology and duration that explain distinct individual response to disease or drug action. Particular combinations of ionic flux properties determine the shape, amplitude, and pacing frequency dependence of action potentials.

Ionic determinants, that is, ionic current conductances and kinetics, of intersubject variability of rabbit Purkinje cell electrophysiology were investigated using model sets with randomly varied parameter ensembles [479]. From a large initial collection of 10,000 randomly varied parameter sets using Latin hypercube sampling, each parameter set initially containing 11 parameters (8 conductances and 3 time constants), the calibration reduces the number to 213 accepted parameter sets upon validation using the experimental variation intervals of 6 selected markers (resting membrane potential, plateau duration, APD at 90 % repolarization (APD_{90}), peak membrane potential, peak membrane potential of the action potential dome (because of spike and dome configuration of the rabbit Purkinje action potential); and peak velocity of the action potential upstroke) associated with 3 frequency values (0.2, 1, and 2 Hz). The model contains the following ion carriers and selected conductances and time constants: (1) the fast and late Na^+ (conductances $\mathcal{G}_{Na,f}$ and $\mathcal{G}_{Na,\ell}$; time constants $\tau_{Na,f}$ and $\tau_{Na,\ell}$); (2) Ca_V1 and Ca_V3 (conductance $\mathcal{G}_{Ca,L}$ [for L-type current]; time constant $\tau_{Ca,L}$); (3) inward rectifier K^+ (conductance $\mathcal{G}_{K1}$); (4) fast and sustained transient outward K^+ (conductances $\mathcal{G}_{K,to(f)}$ and $\mathcal{G}_{K,to(s)}$); (5) rapid and slow delayed rectifier K^+ (conductances $\mathcal{G}_{K,r}$ and $\mathcal{G}_{K,s}$); (6) funny (conductance $\mathcal{G}_f$); (8) Na^+–Ca^{2+} exchange (conductance $\mathcal{G}_{NCX}$); and (9) Na^+–K^+ pump (conductance $\mathcal{G}_{NaK}$) current; in addition to stimulus (triggering) current. The conductance $\mathcal{G}_{Na,f}$ correlates strongly and positively with markers that quantify the initial action potential upstroke; $\mathcal{G}_{K,r}$ negatively with plateau duration and APD_{90}; $\mathcal{G}_{K1}$ primarily with resting membrane potential; and $\mathcal{G}_{Na,\ell}$, $\mathcal{G}_{Ca,L}$, $\mathcal{G}_{NaK}$, and $\mathcal{G}_{K,to(s)}$, which are associated with important plateau phase currents, with most of the markers at all pacing frequencies. Nodal cell models can reproduce and predict the variability of action potential measurements under physiological conditions and after drug administration. A wide range of combinations of ionic parameter values can produce model behaviors within the limits of experimental variability [479]. The ionic current conductances and kinetics explain variability of cellular electrophysiology (shape, amplitude, duration, and rate dependence of action potential). Involved ion carrier conductances can vary considerably and still produce a proper action potential.

Chapter 6
Cardiac Valve Diseases

Cardiac valves allow a unidirectional blood flow from the atrium to the ventricle of each pump and then from the ventricle to the corresponding arterial trunk. They prevent backflow in normal conditions. They are exposed to different pressure ranges and flow features.

Four valves at the exit of each cardiac chamber encompass:

1. Two atrioventricular valves, that is, the mitral (bicuspid) valve and tricuspid valve in the left and right apposed pump, respectively
2. Two ventriculoarterial (semilunar) tricuspid valves, that is, the aortic valve and pulmonary valve in the left and right heart, respectively

The former are anchored to the ventricular wall by chordae tendineae attached to papillary muscles, which preclude valve inversion during ventricular systole (during which they take a parachute-like shape). The subvalvular apparatus comprises the papillary muscles and chordae tendineae. The latter have three cusps (or leaflets) that possess a coaptation zone that contact each other to struggle against the diastolic backflow impact.

The aortic, pulmonary, mitral, and tricuspid valves are positioned in a plane, the so-called base of the heart (early named fibrous skeleton of the heart). It contains collagenous fibers. It remains almost stationary during the cardiac cycle.

Each cardiac valve has its own anatomical and histological features. Cardiac valves are thin structures composed of connective tissue reinforced by fibers covered by endocardium. The fibrous skeleton is anchored to the myocardium. They have a similar architecture, but a specific design to ensure their function.

The two types of *valve failure* encompass:

1. *valvular stenosis*, or narrowing
2. *valvular insufficiency*, a.k.a. regurgitation, incompetence, or leaky valve

according to whether a cardiac valve does not open fully and does not close tightly, respectively.

In valvular stenosis, the impaired valve opening results from hardened (stiffer) or fused leaflets. The narrowed cardiac orifice restricts blood ejection from the upstream

M. Thiriet, *Diseases of the Cardiac Pump*, Biomathematical and Biomechanical Modeling of the Circulatory and Ventilatory Systems 7, DOI 10.1007/978-3-319-12664-7_6

cardiac chamber, thereby engendering at long-term heart failure. All four cardiac valves can be stenotic.

In valvular insufficiency, the altered valve closure is caused by inadequate valve sealing. Blood flows backward across the leaky valve. A lower blood amount is expelled from the cardiac ventricle. All four cardiac valves can generate regurgitation.

Valvular heart disease (VHD) management requires an appropriate diagnosis, evaluation of valvular dysfunction, adequate follow-up, and proper selection of intervention type (i.e., valve reconstruction or replacement) and their timing to optimize the benefit/risk ratio of the procedure and outcome.

6.1 Aortic Valve Structure

The aortic root consists of the elliptical aortic annulus (transition between the left ventricular outflow tract and aortic root), commissures of aortic valve leaflets that extend upward, sinuses of Valsalva, coronary artery ostia, and sinotubular junction. The length of the aortic valve leaflet often exceeds the distance between the annulus and the ostium of the coronary artery, in the absence of strong aortic root remodeling associated with the calcified aortic stenosis.

The aortic valve is composed of different structures: the annulus, commissures, interleaflet triangles, sinuses of Valsalva, sinotubular junction, and leaflets. The aortic valve has three leaflets nearly equally spaced around the aorta.

The aortic valve *cusps* were originally named according to their anatomical position (posterior, right, and left, or anterior and right and left posterior; vol. 6, Chap. 1. Anatomy of the Cardiovascular Apparatus). They are currently termed according to their relation to the coronary ostia (noncoronary, right, and left coronary). The noncoronary leaflet is the largest, followed by the left and right coronary leaflets, but the size differences are slight.

The *annulus* forms a crown-like fibrous structure, which crosses the ventriculoarterial junction, the ventricular myocardium being replaced by a fibroelastic wall [481]. The annulus is firmly attached distally to the media of the aortic sinuses and proximally to the muscular and membranous septa anteriorly, the fibrous triangles laterally, and the subaortic curtain posteriorly. The three upper parts of the annulus are called commissures. At the microscopic scale, the annulus is a very dense collagenous mesh. Elastic and collagenous fibrils as well as neuronal structures can be also detected. The ventricular and arterial layers are continuous with the endocardium and sinus wall.

The *commissure* corresponds to the apex of the crown-like annulus in the area where two leaflets are attached to the aortic wall in the sinotubular junction [481]. Two leaflets are hinged to the aortic wall parallel for a short distance. Three commissures exist, the first between the right- and left-coronary leaflets is positioned anteriorly, the second between the right- and noncoronary leaflets is on the right anterior, and the third one between the left- and noncoronary leaflets is usually on the right posterior aspect of the aortic root. The fibrous commissures are located above three interleaflet triangles. Microscopically, collagen fibers of the intermediate layer are orientated in a radial fashion and radiate into the media of the aortic root.

The *interleaflet triangles* are the regions of thinned fibrous wall between the insertion lines of aortic cusps to the wall up to the commissures. They are extensions of the ventricular outflow tract and reach the level of the sinotubular junction in the area of the commissures. Whereas two interleaflet triangles are fibrous walls, the triangle between the right- and left-coronary sinus in the area of the subpulmonary infundibulum is supported by muscular tissue and only fibrous at its apex [481]. Cytoskeletal and contractile proteins (e.g., vimentin, desmin, and α-smooth muscle actin) abounds. The bundle of His coming from the anterior extension of the atrioventricular node penetrates through the central fibrous body just below the inferior margin of the membranous ventricular septum at the crest of the muscular ventricular septum under the triangle between the right- and noncoronary sinuses.

The *sinuses of Valsalva* are almost symmetrical bulges confined proximally by the attachments of the valve leaflets and distally by the sinotubular junction ridge. At the base, ventricular musculature is partly incorporated. The sinus wall is predominantly made up of aortic wall. Two of the sinuses are endowed of coronary ostia. The noncoronary sinus is most often the largest. The coronary arterial orifices are rarely centrally located. The histological structure of the sinuses is composed of three layers of the aortic wall (intima, membrana elastica interna, media (with smooth myocytes, elastic fibers, collagen-2 and -3 fibers, and proteoglycans), membrana elastica externa, and adventitia with longitudinal collagen-1 fibers) [481]. The inner layer of the intima consists of endotheliocytes arranged in the streamwise direction. The wall thickness is thinner than that of the ascending aorta.

The *left coronary ostium* arises in the left posterior aortic sinus in 69 % of the normal hearts from autopsied adults, above the sinotubular junction in 22 %, and at the level of the junction in 9 % [482]. The distance of the left orifice from the zone of apposition between the left posterior and anterior aortic leaflets ranges between 13 and 61 % of the width of the aortic sinus at the sinotubular junction.

The *right coronary ostium* localizes to the anterior aortic sinus in 78 % of specimens, above the junction in 13 %, and at the level of the junction in 9 % [482]. The distance of the right orifice from the zone of apposition between the leaflets hinged from the anterior and right posterior aortic sinuses ranges between 5 and 62 % of the width of the aortic sinus at the sinotubular junction. The right coronary ostium has an accessory coronary ostium in 74 % of all the cases [482]. A third orifice in this sinus was found in 22 % of hearts.

The *sinotubular junction* is the ridge at the top of the sinus. It marks the point of transition from the aortic root to the ascending aorta. The sinotubular junction runs through the upper part of each commissure and hence the upper end of the attachment of each valve leaflet. The wall is thicker than the sinus wall. Dilation of the aortic root at this level causes aortic incompetence [481].

The three *aortic leaflets* consist of four components: the hinge (crescent or semilunar cusp insertion), the belly, the coaptation surface, and the lannula (a thin crescent-shaped portion on either side of the nodule of Arantius at the midpoint of the free edge of the coapting surface). The lannulae are attached to the wall of the aortic root (commissures). The belly is formed by a specific arrangement of collagen fibers. The thick collagenous bundles of the leaflets are hinged to the annulus.

The aortic valve leaflets are covered by an endothelium with a smooth surface on the ventricular side and numerous ridges on the arterial side. Endotheliocytes are joined by radial and axial junctions. The ventricular and aortic endothelia are separated by up to five layers of connective tissue: lamina ventricularis, radialis, spongiosa, fibrosa, and arterialis [481]. Elastic and collagen fibers are coupled in a honeycomb structure. The arterial layer contains coarse bundles of circumferential collagen fibers that form the macroscopical folds parallel to the free edge of the leaflets. The leaflets contain interstitial fibroblasts and vessels, the density of which depends on leaflet thickness and increases in the hinge area.

Aortic valve endotheliocytes operate under various shear stress patterns. Cells on the aortic side of the valve are more elongated and align along a single axis. They are softer than those on the ventricular side [483]. Nitric oxide lowers the compliance of endotheliocytes on the ventricular face of aortic valves, but does not affect that of endotheliocytes on the aortic side of the valve. Conversely, endothelin-1 raises the compliance of endotheliocytes on the aortic side of the aortic valve cells, but does not modify that of endotheliocytes on the ventricular side of the valve.

Valvular interstitial cells constitute the dominating cell population in aortic valves [484]. It represents a cell pool that can remodel with phenotypical changes according to epigenetic modifications. Activated myofibroblasts in the valve interstitium appears in aortic valve stenosis. Valvular interstitial cells can also evolve toward an osteoblastic type.

6.2 Mitral Valve Structure

The bicuspid mitral valve has a shape similar to that of a mitre of a bishop (orifice surface area 4–6 cm^2). The plane of the valve is not flat, but resembles a saddle. During the cardiac cycle, the mitral valve size and shape change. In particular, the size of the annulus and positions of the high points of the saddle vary considerably during the cardiac cycle. In normal individuals, the area of the mitral valve lowers (10–15 %) during systole. The atrial wall can contribute to mitral regurgitation when the left atrial enlarges.

The structures of the mitral valve encompass the annulus (i.e., annular attachment at the atrioventricular junction), leaflets, chordae tendinae, and posteromedial and anterolateral papillary muscles in the left ventricle.

The oval (D-shaped) flexible fibrous *mitral valve annulus* (hingeline of the valvular leaflets) enables the orifice to undergo complex shape changes during the cardiac cycle. It determines the opening area of the mitral valve. It is composed anteriorly of a fibrous component between the two fibrous trigones, the trigonum fibrosum dextrum and sinistrum, which separate the mitral valve from the aortic valve [481]. The right fibrous trigone and the membranous septum form the central fibrous body. The annulus edge opposite to the fibrous body can be altered by dilation and be the

Table 6.1 Distance from the free margin to the annulus in the central part of commissural regions and extent of the insertion zone of commissural cords along the commissural free margin (mm; Source: [485])

Geometrical parameter	Women	Men
Anterolateral commissural region		
Free margin–annulus length	5–10	5–13
Insertion length of commissural cords	3–15	6–19
Posteromedial commissural region		
Free margin–annulus length	4–11	6–12
Insertion length of commissural cords	9–22	12–26

site of calcifications. In the anterior concave part of the annulus, fibers have a parallel and circular orientation. Its lateral and posterior parts are linked to the anterior part by the left and right fibrous trigone. The lateral part of the mitral annulus consists of collagen fibers. In the annulus transition zone, where the leaflets are anchored to the myocardium, the elastin and collagen fibers radiate into the myocardium [481]. In the hinge zone, where the atrial endocardium is thickened, the number of elastin fibers increases.

The anterior leaflet is commonly semicircular and the posterior leaflet has a crescentic shape. Both leaflets merge at the posteromedial and anterolateral *commissures* (Table 6.1). The tips of the papillary muscles point toward the commissures. A commissure corresponds to long invaginations of the free margin that limit adjacent cusps, the depth of which is larger than two thirds of the maximal width of the larger of the two adjacent leaflets. An indentation that separates two valvular cusp segments have a depth ranging from one third to two thirds of the maximal width of the larger of the two adjacent leaflets. Their number and position determine the number and size of valvular segments.

The *anterior leaflet* (a.k.a. aortic and septal leaflet) is the bigger cusp (Table 6.2). It is close to the aortic root. The aortic leaflet is divided arbitrarily into three regions (A1–A3) from the cardiac base to the apex. The anterior leaflet is close to the aortic valve (left and noncoronary cusps). It is much smoother than the mural cusp, usually having no indentations along its free edge. The atrial myocardium contacts the hinge of the aortic leaflet [486].

The narrow *posterior leaflet* (a.k.a. mural leaflet) is in 91 % of cases divided into three parts (posteromedial or right, intermediate, which is wider and higher than the others, and anterolateral or left) [481]. In adults, the mural leaflet has indentations that do not extend through the leaflet to the annulus. They generally form three segments[1] (scallops due to scalloped shape) along the elongated free edge according to Carpentier's nomenclature (Fig. 6.1 and Table 6.3; [487]):

1. P1, the lateral segment adjacent to the anterolateral commissure
2. P2, the central segment that can significantly vary in size, and
3. P3, the medial segment adjacent to the posteromedial commissure

[1] The number of indentations can vary from 2 to 5 [485].

Table 6.2 Size of the anterior mitral valve cusp and of segments of the posterior mitral valve leaflet. (Source: [485])

Leaflet	Women	Men
Septal (anterior) leaflet		
Height	18–35	20–30
Width	18–42	25–48
Mural (posterior) leaflet		
P1 segment		
Height	8–14	9–20
Width	9–20	9–40
P2 segment		
Height	7–18	9–20
Width	6–26	13–38
P3 segment		
Height	5–11	6–17
Width	5–22	9–31

Each Pi segment faces Ai segment of the anterior leaflet. The posterior smaller leaflet is attached to nearly two thirds of the circumference around the left atrioventricular junction. The degree of myocardial extension toward the insertion of the hinge (or fulcrum) of the mural posterior leaflet varies between hearts and in a given heart, between regions [486].

The leaflets can be subdivided into a *basal region*, the zone of attachment to the annulus, and thin *central* translucent (clear) and thick *rough* zone from the attachment point of each leaflet at the annulus to the free edge. The chordae tendinae of the mitral valve are tethered to the ventricular face of the leaflet rough zone and free margin [481]. The rough zone is also the region of apposition (symmetrical overlap of the leaflet free edge, usually a minimum of 4–5 mm) and coaptation (region of between-leaflet contact when the mitral orifice is closed). Coaptation and correct apposition of leaflets prevent regurgitation.

Measurements of the leaflet geometry encompass:

- The circumference of the annulus
- The length of annular attachment of each leaflet
- The width of the coapted (folded) margin
- The dimensions and surface area of each leaflet (the surface area of the anterior leaflet being about 1.6 times larger than that of the posterior leaflet [490]); among others.

The two leaflets of the mitral valve differ in structure. The mitral leaflets are composed of a fibrous skeleton. The mitral valve is made up of four layers: ventricularis, fibrosa, spongiosa, and atrialis [487]. Two main layers, a *lamina spongiosa* and a *lamina fibrosa*, face the atrial and ventricular side, respectively. They are covered by the *endocardium*. The latter is continuous with that of the atrium and ventricle. The thickness of each layer varies from the cusp insertion at the annulus to the free edge.

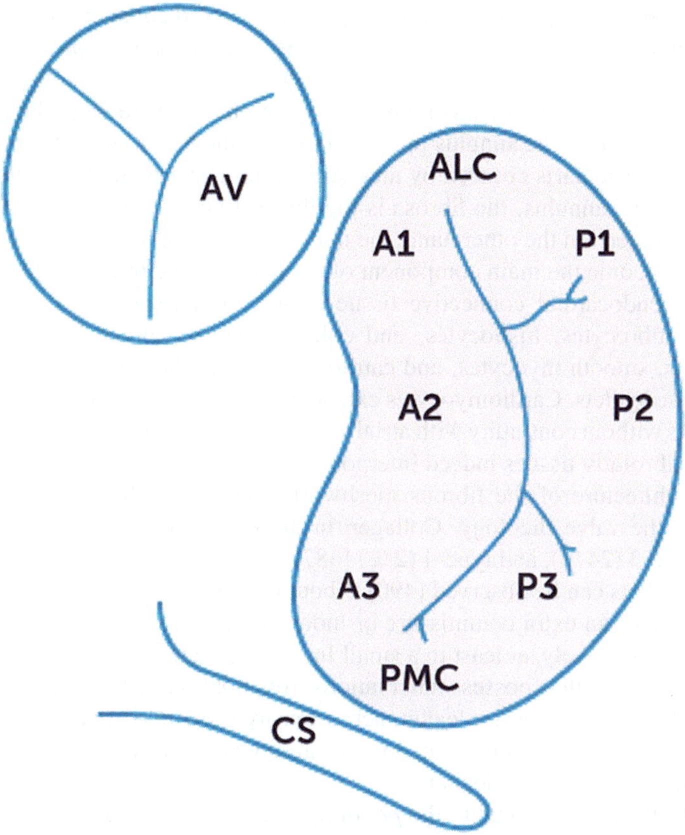

Fig. 6.1 (*Left*) Segments of the mitral valve (Carpentier's classification; Source: [488]; *A* anterior (aortic or septal) leaflet, *P* posterior (mural) leaflet, *1* lateral segment, *2* middle segment, *3* medial segment, *AV* aortic valve, *CS* coronary sinus). The line of coaptation is slightly U-shaped. The tips of the leaflets curl toward the ventricle. (*Right*) The saddle-shaped mitral valve annulus deforms during the cardiac cycle with position changes of high (*A* anterior, *P* posterior) and low points (*L* lateral, *M* medial) of the saddle

Table 6.3 Some end-systolic dimensions of the mitral valve. (Source: [489])

Between-commissure width	17.2 ± 5.8 mm
Width of the middle P2 segment	10.9 ± 2.5 mm
Distance between the mitral annulus saddle high point and the mitral orifice plane	7.3 ± 2.2 mm
Distance between the papillary muscle tip and the free margin indentation end	20.0 ± 5.8 mm
Distance between papillary muscle tips	11.6 ± 5.0 mm

The atrialis is composed of mainly aligned elastic and collagen fibers. The spongiosa, a major component of the free edge, consists principally of proteoglycans and elastic fibers. The fibrosa, a major load-bearing layer, comprises a collagenous core. The collagen fibers are aligned and provide strength and stiffness to the leaflet. The fibrosa extends from the annulus into two thirds of the leaflet; it is absent at the free edge. The ventricularis covered by an endothelium consists of elastic and collagen fibers. Near the annulus, the fibrosa is the thickest layer; close to the free edge, it becomes thinner. On the other hand, the thickness of the spongiosa and atrialis rises distally to become the main component of the leaflet at the free edge [487].

The subendocardial connective tissue of the lamina spongiosa consists of fibroblasts, fibrocytes, histiocytes, and collagen fibers (caliber 15–35 nm) [481]. Fibroblasts, smooth myocytes, and cardiomyocytes in the proximal third reside in mitral valve leaflets. Cardiomyocytes extend a short distance into the base of the mitral leaflets without continuity with atrial and ventricular walls. At the atrioventricular junction, fibrofatty tissues indeed interpose.

The architecture of the fibrous meshwork, that is, the local fiber orientation, influences the valve rheology. Collagen in the leaflet comprises primarily type-1 (74 %), type-3 (24 %), and type-4 (2 %) [487].

Extra leaflets can be observed [490]. About two thirds of heart specimens exhibit the presence of an extra commissure or indentation, hence of additional leaflets or segments, respectively, at least in a small Indian population.

Mitral valve leaflets possess innervations from both sensory and autonomic components [481]. In the anterior leaflet, nerve density is twofold greater than that in the posterior leaflet. Nerves are situated in the atrial layer and extend over the proximal and medial portions of the leaflet.

Both leaflets are attached to the *papillary muscles* by the *chordae tendinae*. The posteromedial papillary muscle gives cords to the medial half of both leaflets (i.e., posteromedial commissure, A3, P3, and posterior half of A2 and P2 segments). The anterolateral papillary muscle sends cords to the lateral half of the leaflets (i.e., anterolateral commissure, A1, P1, and anterior half of A2 and P2 segments). Their position in the left ventricle varies among patients. In some subjects, one or both papillary muscles cannot be clearly defined; they are replaced by multiple small muscle bundles connected to the ventricular wall [487].

Thin chordae have a lower average fibril bore and a greater average fibril density than thick chordae. The surface of the chordae consists of a layer of endotheliocytes and a dense layer of elastic fibers [481]. These tendinous chordae are composed of elastic and collagen fibers parallel to the axis. They transmit the tension of the contracting papillary muscles to the valve leaflets. The majority of the chordae insert at the free margin or behind the free margin at the ventricular side (rough zone) of the leaflets. The *base chordae* insert at the leaflets near their attachment at the annulus originating from ventricular myocardium. The *commissural chordae* (interleaflet or commissural cords) insert at the free margin of two adjoining leaflets. Normally, only two commissural cords exist supporting each free margin of the commissural region. These cords arise from a single stem that branches, thereby allowing adjacent leaflets to coapt and to move apart.

In fact, numerous classifications of tendinous cords exist. Three types of chordae tendinae are also described [487]:

1. *Primary cords* attached to the free edge of the rough zone of both leaflets
2. *Secondary cords* connected to the ventricular surface of the rough zone
3. *Tertiary cords* tethered to the basal zone of the mural (posterior) leaflet only

Among the secondary cords of the aortic (anterior) leaflet, two are the largest and thickest (*strut cords*). The latter originate from the tip of each papillary muscle. Hence, according to function, tendinous cords are categorized into commissural, indentation, and strut cords.

Mitral valve repair is more popular than mitral valve replacement. However, changes in cusp geometry alter valvular mechanics and outcomes, especially valve repair durability.

Annular flattening can result from myocardial infarction. On the other hand, the three-dimensional saddle shape of the mitral annulus changes during the cardiac cycle with systolic high points along the anterioposterior plane and low points at commissures. This shape enables a better leaflet curvature during systolic closure. The saddle shape influences the mechanics of the mitral leaflet. It lowers stresses on the chordae tendineae and strains on the leaflets, in particular in the posterior leaflet during systolic valve closure [492]. Maximal strain decreases from planar shape to 20 % saddle annulus.[2]

Estimation of regional stresses and strains enables to adapt the therapeutic strategy for valvular repair and replacement. In vivo deformations of the mitral valve anterior leaflet during the cardiac cycle was evaluated using sonocrystal transducers sutured to the valve [493]. Changes in annular geometry alter valvular strains in vivo.

The deformation of the insertion zone of chordae to porcine mitral valves endowed with a structured array of markers was measured during the cardiac cycle by tracking marker motions using high-speed cameras [495]. The insertion zone mechanics depends on the collagen architecture. The insertion region is stretched during the entire cardiac cycle.

Kinematics of the mitral valve annulus was explored in sheep using implanted sonocrystals [494]. Kinematic data collected can be fitted to a 3-D spline based on 5-order hermite shape functions with C^2 continuity. Axial strain varies spatially and temporally (minimum and maximum −10 and 4 %, respectively) with strain rate changing up to 100 %/s contraction and 120 %/s elongation. Most of the strain energy is related to annular axial strain. However, implantation of sonocrystals may alter the annulus deformation, in addition to differences in behavior among mammalian species.

Simulations were carried out to assess the stress field in the mitral valve anterior leaflet, using different pseudo-hyperelastic constitutive models, especially a collagen

[2] The peak major principal strain drops by about 14 % (± 13 %), 28 % (± 14 %), and 30 % (± 30 %) for saddle height–commissural diameter ratio of 10, 20, and 30 %, respectively, with respect to a flat annular configuration [491].

fiber-mapped transversely isotropic model, and a given valve microstructure [496]. However, the collagen fiber arrangement and material constant may not represent perfectly in vivo data.

6.3 Tricuspid Valve Structure

The tricuspid valve is composed of an anterior (anterosuperior, ventral, or mural), posterior (inferior or dorsal), and septal (or medial) leaflet. It builds a structural unit with the annulus, chordae tendinae, and papillary muscles. The latter vary in numbers (2–9; usually 3: anterior, posterior, and septal).

Its orifice is larger than that of the mitral valve. Its leaflets are thinner than those of the mitral valve.

The *annulus* area undergoes three-dimensional changes during the cardiac cycle. The annulus also forms a part of the triangle of Koch together with the coronary sinus and the tendon of Todaro that incorporates the nodal conduction tissue.

The three *leaflets* of the tricuspid valve differ in size. The anterior cusp is the largest. It spreads from the infundibular area downward to the inferolateral wall of the right ventricle. The septal leaflet is attached to both the membranous and muscular portions of the ventricular septum. The posterior cusp is the smallest. Sometimes, four leaflets can be identified; the posterior leaflet can be divided or an additional leaflet is positioned between the posterior and septal cusps. They are mainly composed of a fibrous skeleton and covered by an endocardium. The atrial layer of the endocardium is smooth. The lamina spongiosa is composed of loose layer of connective tissue and the lamina fibrosa of dense collagen fibers. The fibers are parallel. The innervation of human tricuspid valve leaflets is stronger than that of mitral leaflets.

Several *chordae tendinae* are attached directly to the interventricular septum. The chordae tendinae are interconnected before they attach the leaflet free margins. They are composed of a network of collagen fibers parallel to the axis.

6.4 Pulmonary Valve Structure

The pulmonary valve connects the conus arteriosus (infundibulum) of the right ventricular outflow tract to the pulmonary artery trunk. The pulmonary valve is part of the pulmonary root. Its associated structures include the sinuses (sinus trunci pulmonalis), annulus, commissures, leaflets, and sinotubular junction. The pulmonary leaflets are thinner and the noduli of Arantii smaller than those of the aortic leaflets [481].

The *annulus* of the pulmonary root is defined by the line of attachment of the leaflets to the sinus wall (crown-shaped annulus), as is the aortic annulus. It is composed of tight collagenous tissue. It is connected to the media of the pulmonary artery and the myocardium.

The *commissures* are the short parallel parts of attachment lines of the leaflets to the pulmonary wall. Three commissures exist between the left and right anterior leaflets, between the right anterior and posterior leaflet, and between the posterior and the left anterior leaflet.

The *interleaflet triangles* are the area under the commissures. They are extensions of the right ventricular outflow tract.

The three *sinuses of Valsalva* correspond to the left and right anterior and posterior leaflets. The sinus wall is much thinner than that of the pulmonary artery, especially in its middle portion.

The *sinotubular junction* of the pulmonary root is characterized by a ridge which defines the upper part of the sinuses and runs through the upper part of the commissures.

The *leaflets* of the pulmonary valve consist of the hinge, belly, coaptation surface, and lannula with the nodulus of Arantius, as those of the aortic valve. Collagen and elastin fibers serve as load-bearing components. Five layers exist between the ventricular and arterial endocardia (made up of endotheliocytes that interdigitate or overlap over a basement membrane) [481]:

- Lamina ventricularis (thickness 21–48 μm; with reticular fibers)
- Lamina radialis (thickness 58–108 μm; with radial collagen and elastin fibers)
- Lamina spongiosa (thickness 40–300 μm; with loosely arranged reticular fibers and bundles of collagen and some elastin fibers)
- Lamina fibrosa (thickness 80–170 μm; with circular collagen fibers)
- Lamina arterialis (very thin layer of reticular fibers)

The innervation arises from the ventricular endocardial plexus and localizes to the ventricular layer and lower region of each leaflet.

6.5 Left-Sided Valve Failure

Valvular diseases are mostly due to either congenital abnormalities or inflammation. Primary valve failure occurs acutely due to *leaflet perforation* or gradually from *leaflet stiffening* associated with calcifications and/or thrombus formation.

Calcific vasculopathies can be distinguished by its histological appearance as amorphous mineralized medium, without tissular organization, or as chondrobony milieu, which have architecture of cartilage or bone, when angiogenesis occurs.

6.5.1 Mitral Valve Failure

The mitral cusps are dynamically related to the left ventricular wall via the mitral annulus, chordae tendineae, and papillary muscles. Mitral diseases may be degenerative (∼ 60 %), ischemic (25 %), infective (endocarditis), or rheumatic (12 %), with or without superimposed impaired left ventricular function and calcification.

Table 6.4 Carpentier's classification of mitral valve dysfunction

Dysfunction Leaflet motion	Lesions
Type 1:	Annular dilation
Normal motion	Leaflet perforation
Central regurgitation jet	
Type 2:	Chordal rupture or elongation
Excess leaflet motion (prolapse)	Papillary muscle rupture or elongation
Eccentric regurgitation jet	
Type 3:	Commissural fusion
Restricted leaflet motion	Chordal fusion and thickening
	Valve thickening and/or calcification
Eccentric regurgitation jet	

When the mitral valve fails, partial emptying of the left atrium increases the left arterial pressure and reduces the cardiac flow rate. Pulmonary venous congestion and, ultimately, pulmonary edema occur.

A systematic analysis of the anomalies of the valvular structure and function is mandatory for selecting relevant therapeutic actions. Mitral regurgitation results from either leaflet perforation, such as that due to trauma or endocarditits, or annular dilation, usually due to left ventricular damage.

Three types are defined in the Carpentier's classification of mitral valve dysfunction (Table 6.4). The Carpentier's functional classification describes leaflet motion in relation to the mitral annular plane:

Type 1 corresponds to a normal leaflet motion.

Type 2 describes excessive leaflet motion above the annular plane into the left atrium. This cusp prolapse is usually caused by tissular degeneration.

Type 3 accounts for leaflet restriction. Two subtypes are defined: type-3a when the restriction is throughout the cardiac cycle and type-3b, when the leaflet restriction occurs only during systole.

The mitral valve *prolapse* is defined by leaflets extending above the plane of the atrioventricular junction during ventricular systole. It can result from several mechanisms, such as rupture of cords and tips of papillary muscles and abnormal wall motion secondary to myocardial ischemia. The primary prolapse is a floppy valve that domes into the left atrium and, often, with heterogeneously deformed and thickened cusps.

Two forms of prolapse are identified: (1) The leaflets protrude into the atrium, but they can coapt, ensuring valvular competence. (2) The free edge of the damaged leaflet overshoots the closure line of the opposite leaflet, thereby causing mitral regurgitation.

Table 6.5 Mitral valve stenosis grades with a cardiac frequency ranging from 1 to 1.3 Hz and in sinusal rhythm. (Source: [497])

Parameter	Mild	Moderate	Severe
Valve area (cm^2)	>1.5	1.0–1.5	<1.0
Mean pressure difference (mmHg)	<5	5–10	>10
Pulmonary arterial pressure (mmHg)	<30	30–50	>50

6.5.1.1 Mitral Valve Stenosis

Mitral valve stenosis (MVS) impedes adequate blood flow coming from the left atrium to the left ventricle. The resulting obstruction causes a pressure difference across the valve in diastole and increases left atrial and pulmonary venous pressures. Elevated left atrial pressures lead to left atrial enlargement and predispose to atrial fibrillation and arterial thromboembolism. Upstream blood stasis can engender pulmonary congestion and edema. Evolved mitral stenosis is associated with pulmonary hypertension and right-heart failure.

The main cause of mitral valve stenosis is rheumatic fever following a Streptococcus pyogenes infection, such as streptococcal pharyngitis. Myocarditis can lead to congestive heart failure with pericarditis. About half of patients with acute rheumatic fever develop inflammation of valvular endothelium.

Rheumatic valvulitis results from an immune reaction to group-A β-hemolytic streptococci that have antigens immunologically related to the human heart. The cross-reactive autoimmunity can be stimulated in some persons after streptococcal infection. $CD4^+$ T lymphocytes activated by molecular mimicry are the major effectors of cardiac autoimmune reactions.

Chronic rheumatic heart disease (CRHD) is characterized by scarring and fibrosis of valve leaflets, progressive leaflet thickening, shortening and thickening of the chordae tendineae, fusion of the commissures and chordae tendinae that reduces the size of the mitral valve opening, and calcifications of the mitral leaflets and chordae tendineae. In fact, CRHD can provoke valve stenosis or regurgitation.

Other rare causes of mitral valve stenosis include calcification of the mitral valve, valvular thrombi, tumors, radiotherapy of the chest, certain drugs, and congenital heart defects.

Mitral valve stenosis severity can be classified according to three grades (Table 6.5).

6.5.1.2 Mitral Valve Regurgitation

Mitral valve regurgitation (MVR) provokes blood leakage from the left ventricle into the left atrium during systole due to structural and/or functional abnormalities of the mitral valve, adjacent myocardium, or both.

A degenerative mitral valve is also called a floppy mitral valve, the most common cause of mitral regurgitation in the Western world. Others most common causes of mitral regurgitation are chordal rupture, rheumatic heart disease, infective endocarditis, coronary artery disease, and cardiomyopathy, in addition to myxomatous degeneration.

Mitral regurgitation is a complication of myocardial ischemia, which can be detected by echocardiography. Defective systolic contraction of the papillary muscles due to acute ischemia and regional dysfunction of the left ventricle wall leads to abnormal displacements of the papillary muscles and incomplete leaflet coaptation and causes mitral regurgitation.

Valvular regurgitation causes a ventricular overload and left ventricular hypertrophy. The left ventricular blood volume is thus the sum of the residual, regurgitant, and stroke volume. To compensate, the left ventricle dilates. In acute severe mitral regurgitation, the left atrial pressure and pulmonary venous pressure increase quickly, leading to pulmonary congestion and edema. In chronic mitral regurgitation, a gradual increase in left atrial size and compliance compensate so that left atrial and pulmonary venous pressures rise in the late course of the disease.

Progressive left ventricular dilation evolves to heart failure. Left atrial enlargement predisposes to atrial fibrillation and arterial thromboembolism. At a late stage, pulmonary hypertension and right-heart failure occur.

Quantification of mitral valve regurgitation is carried out using the *proximal isovelocity surface area* (PISA) method. When a fluid crosses a small circular orifice in a flat plate, immediately upstream from the orifice, the flow converges to the orifice and accelerates in the flow convergence zone. In this zone, the flow is assumed to be characterized by a velocity field described by a set of hemispheric isovelocity contours. Limitations of this method thus arise in the case of eccentric regurgitation that gives less accurate estimates as well as of noncircular or nonflat regurgitant orifice area, the flow convergence zone being not characterized by hemispheric isovelocity contours, hence the PISA radius cannot be used.

The PISA calculation of mitral valve regurgitation actually assumes a single central systolic jet from the left ventricle through the leaking mitral valve orifice and a proximal isovelocity surface that is completely hemispheric (raduis R; area $2\pi R^2$) [498]. The instantaneous regurgitant flow (RF) and effective regurgitant orifice area (EROA) can be calculated using the aliasing velocity (v_{alias})

$$\text{RF} = 2\pi R^2 \times v_{\text{alias}}; \tag{6.1}$$

and by continuity

$$\text{EROA} \times \hat{v}_{\text{MVR}} = \text{RF} \tag{6.2}$$

where $\hat{v}_{\text{MVR}}$ the peak velocity in mitral valve regurgitation. The orifice being typically at the apex of a conical structure, a correction term relies on the apical angle of the cone. If the base of the hemisphere is not a flat surface (180°), then the correction relies on the ratio of the angle (θ) formed by the walls adjacent to the regurgitant orifice and the flat case, that is $\theta/180$.

Table 6.6 Mitral valve regurgitation grades. (Source: [498] EROA: effective regurgitant orifice area)

Parameter	Mild	Moderate	Strong	Severe
Vena contracta width (mm)	< 3	3.0–6.9		≥ 7
Regurgitant volume (ml/beat)	< 30	30–44	45–59	≥ 60
Regurgitant fraction (%)	< 30	30–39	40–49	≥ 50
EROA (cm^2)	< 0.20	0.20–0.29	0.30–0.39	≥0.40

The regurgitant volume (V_{regur}) is computed from regurgitant flow, that is, the area of the orifice multiplied by the velocity through the orifice, knowing that regurgitant volume equals regurgitant flow multiplied by the time of regurgitation:

$$V_{regur} = \text{EROA} \times \ell_{\text{TI}} \tag{6.3}$$

where ℓ_{TI} is the time integral of the regurgitant jet velocity using continous-wave Doppler ultrasonography.

Valvular stroke volume (SV_{Vv}) at a valve annulus (the least variable anatomical area of a valvular apparatus) is calculated as the product of the sectional area (SA) and ℓ_{TI} at the annulus. The valvular orifice is assumed to be circular, which is a good approximation, except the tricuspid annulus. Hence [498],

$$\text{SV}_{\text{Vv}} = \text{SA}_{\text{Vv}} \times \ell_{\text{TI}} = 0.785\, d_{\text{Vv}}^2 \times \ell_{\text{TI}}, \tag{6.4}$$

where d_{Vv} is the diameter of the valvular annulus. Stroke volume can be computed at various sites (e.g., left ventricular outflow tract and mitral and pulmonary annulus).

The *regurgitant volume* (V_{regur}) is also the difference between the stroke volume at the annulus of the regurgitant valve ($\text{SV}_{\text{vregur}}$) and the stroke volume at the competent valve (SV_{vcomp}):

$$V_{regur} = \text{SV}_{\text{vregur}} - \text{SV}_{\text{vcomp}}. \tag{6.5}$$

The *regurgitant fraction* (RF) is the regurgitant volume divided by the forward stroke volume through the regurgitant valve:

$$\text{RF} = V_{regur}/\text{SV}_{\text{vregur}}. \tag{6.6}$$

Specific and supportive signs and quantitative parameters enables the grading of mitral valve regurgitation (Table 6.6).

6.5.2 *Aortic Valve Failure*

Acute failure of an aortic valve causes a rapidly progressive left ventricular volume overload. Increased left ventricular diastolic pressure engenders pulmonary congestion and edema. Cardiac flow rate is substantially reduced. Increased systolic wall

Table 6.7 El Khoury's classification of mitral valve dysfunction assessed by transesophageal echocardiography. (Source: [499])

Dysfunction	Lesions
Type 1: normal cusp motion	Annular dilation Leaflet perforation
Type 1a: aortic root enlargement with normal cusps	
Type 1b: aortic root and Valsalva sinus aneurysm	
Type 1c: annular dilation	
Type 1d: endocarditis and cusp perforation	
Type 2: elevated leaflet motion	Partial and complete cuspal prolapse Commissural disruption Free edge fenestrations
Flail aortic cusp	
Type 3: restricted leaflet motion	Commissural fusion Valve thickening Calcification

Table 6.8 Grading of aortic valve calcification. (Source: [500])

Grade 1	No calcification
Grade 2	Isolated small calcification spots
Grade 3	Bigger calcification spots interfering with cusp motion
Grade 4	Extensive calcifications of all cusps with restricted cusp motion

tension augments myocardial oxygen consumption. The compensatory increased sympathetic tone (positive chronotropy and inotropy) raises both the cardiac frequency and systemic vascular resistance to maintain partly the cardiac flow rate. This compensation is hampered by increased systemic vascular resistance. Consequently, diastolic filling time decreases and left ventricular flow is attenuated.

Anatomical and physiological explorations mainly based on echocardiography before conventional open-heart valve surgery are aimed at describing the leaflet anatomy and at assessing severity of valve dysfunction and hemodynamic consequences.

Commissural fusion associated with rheumatic valve disease are associated with the increased valve maximal curvatures both in direction parallel and perpendicular to the valve matrix fiber orientation [484].

As for the mitral valve, anomalies of the aortic valve structure and function can be categorized (Tables 6.7 and 6.8).

The likelihood of successful durable repair is very high in both type-1 and type-2 valvular dysfunction, in the absence of heavily calcified leaflets. Aortic repair is considered feasible when the aortic valve is moderately calcified (grade <3) with calcifications confined to the free margins. When calcifications lodge in the cusp body, the valve is considered nonrepairable [500]. Type-3 lesions are considered as nonrepairable. More than 40 % of attempted repairs in type-3 dysfunction fail over the following 4 years [499].

Table 6.9 Classification of aortic valve stenosis severity (Source: [497]; AHA/ACC: American College of Cardiology and American Heart Association; ESC: European Society of Cardiology). Discrepancies are frequently observed between the mean pressure difference and the valve area in a given patient, especially when the cardiac output lowers due to reduced left ventricular ejection fraction

Parameter	Mild	Moderate	Severe
Peak aortic jet velocity (m/s)	2.6–2.9	3.0–4.0	>4.0
Mean pressure difference (mmHg)			
ESC Guidelines	< 20	20–40	> 40
AHA/ACC Guidelines	< 30	30–50	> 50
Aortic valve area (AVA; cm^2)	> 1.5	1.0–1.5	< 1.0
Indexed AVA (cm^2/m^2)	> 0.85	0.60–0.85	< 0.6
Velocity ratio	> 0.50	0.25–0.50	< 0.25

Immediate postoperative measurements verify structural and functional lengths (e.g., sinus height and caliber, size of the annulus, sinotubular junction, and tubular ascending aorta, coaptation surface area, tip-to-annulus and cusp's belly-to-annulus lengths, and vena contracta width), coaptation quality, and jet shape.

6.5.2.1 Aortic Valve Stenosis

The severity of aortic valve stenosis (AVS) can be classified into several grades (Table 6.9). As inconsistencies arise from grading the severity with an aortic valve area smaller than 1 cm^2, four severe valvular aortic stenosis categories can be identified according to the transvalvular flow rate and pressure difference:

1. Normal transvalvular flow rate/low pressure difference (NTFR/LPD)
2. Normal transvalvular flow rate/high pressure difference (NTFR/HPD)
3. Low transvalvular flow rate/high pressure difference (LTFR/HPD)
4. Low transvalvular flow rate/low pressure difference (LTFR/LPD); low transvalvular flow rate being defined by an indexed left ventricular stroke volume lower than 35 ml/m^2 and low ressure difference by a mean transaortic pressure difference smaller than 40 mmHg [501]

To account for differences in body size in patients with aortic stenosis, the aortic valve area (AVA) is divided by the body surface area (BSA) to calculate indexed AVA (AVAI or AVA_{idx}). The threshold for severe stenosis (AVA <1 cm^2) is AVAI <0.6 cm^2/m^2. The AVAI index is a powerful predictor of adverse events in asymptomatic patients with severe aortic stenosis [502]. Furthermore, the combination of AVAI and peak aortic jet velocity provides additional prognostic information. Utilization of AVAI significantly increases the prevalence of patients with criteria for severe stenosis by including patients with a milder degree of the disease without improving the predictive accuracy for aortic valve-related events [503].

In the absence of direct operative inspection, three-dimensional, pre- and perinterventional imaging assists therapeutic planning with the detailed anatomy of the aortic valve and root as well as evaluation of vascular access strategy.

Thickening and calcification of the aortic valve reduce its opening, thereby provoking an obstruction at the left ventricular exit (aortic valve stenosis). Aortic valve stenosis is characterized by dystrophic calcification and inflammatory leukocyte infiltration [484]. The sinuses of Valsalva and sinotubular junction can also contain calcifications.

Early valve lesions in aortic valve stenosis preferentially occur on the aortic side of the cusps. The loss of cusp stretch due to valvular calcification can prevent a complete leaflet coaptation, thereby enabling a regurgitant flow from the aorta to the left ventricle during diastole.

Valvular calcifications result from the shift of valvular interstitial cells to osteoblasts. Two patterns of calcification can be observed: in the coaptation area of the aortic cusps and/or radially from the cusp attachment to the valve center [484].

Degenerative, calcified aortic stenosis is similar to atherosclerosis. Calcified aortic valve disease is defined by cusp thickening and calcification as well as lipid and lipoprotein deposition in the absence of rheumatic heart disease.

Valvular interstitial cells not only evolve toward osteoblasts, but also toward myofibroblasts in aortic valve stenosis. Calcified regions of human stenotic valves are linked to epigenetic alterations in valvular interstitial cells. In calcified valvular regions, the 5-lipoxygenase promoter is less methylated than in normal valves with subsequently increased 5-lipoxygenase transcription and leukotriene production [484].

Other features comprise activation of matrix metallopeptidases and of the renin–angiotensin axis as well as development of inflammation. Antifibrotic C-type natriuretic peptide is downregulated, whereas expression of neutral peptidases and C-type natriuretic peptide-degrading enzyme rises.

Molecules such as peptidases can be carried from the aortic to the ventricular side of the leaflet. These catalytic activities may participate in the valvular remodeling in early aortic valve sclerosis, leading to a stiffer valve.

Aortic valve stenosis is accompanied by changes in expression of endothelin-related genes. The endothelin receptor-A gene is upregulated in stenotic valves [504]. In addition, the ET_B transcript level does not change during disease progression. On the other hand, the concentration of the transcript of endothelin-converting enzyme ECE1[3] markedly decreases in stenotic valves. Transcription factor AP1 that targets the Et1 and Ece1 genes is significantly lowered. However, the number of ET1+ cells increases.

Severe aortic stenosis can be treated by endovascular procedures, a less invasive technique than surgical valve replacement. Catheter-based implantation of aortic stent grafts and balloon- or self-expandable valve prostheses use a retrograde femoral

[3] PreproET1 is processed to an inactive intermediate, big ET1, which is cleaved by endothelin-converting enzyme-1 into mature ET1 form.

or apical route. Complications of catheter-based valve insertion comprise vascular injury, impaired device positioning with paravalvular leakage, cardiac tamponade, arrhythmias, coronary artery occlusion, and prosthetic valve embolization.

6.5.2.2 Aortic Valve Regurgitation

Aortic valve regurgitation (AVR) is characterized by a diastolic reflux of blood from the aorta into the left ventricle. Acute AVR typically causes severe pulmonary edema and hypotension. Chronic severe AVR causes combined left ventricular volume and pressure overload with systolic hypertension and wide pulse pressure. The excessive afterload caused by hypertension engenders a progressive left ventricular dilation and systolic dysfunction.

Aortic valve regurgitation results from malcoaptation of the aortic leaflets due to abnormalities of the aortic leaflets, their supporting structures (aortic root and annulus), or both. Diseases that primarily affect the leaflets include bicuspid aortic valve and other congenital abnormalities, atherosclerosis, infective endocarditis, rheumatic heart disease, connective tissue or inflammatory diseases, and antiphospholipid syndrome, as well as anorectic drugs, in addition to trauma due to chest wall or deceleration injury, or a jet lesion due to dynamic or fixed subaortic stenosis [505]. Diseases that primarily affect the annulus or aortic root include idiopathic aortic root dilation, aortoannular ectasia, Marfan syndrome, Ehlers–Danlos syndrome, osteogenesis imperfecta, aortic dissection, syphilitic aortitis, and various connective tissue diseases. Ankylosing spondylitis can damage both the leaflets and aortic root. Acute AR is most commonly caused by bacterial endocarditis, aortic dissection, or blunt chest trauma.

A bicuspid aortic valve is commonly associated with dilation of the aortic root in addition to the congenital leaflet abnormality [505].

The assessment of aortic valve regurgitation is based on 2-D echocardiography, color-flow imaging, and pulsed and continuous wave Doppler ultrasonography with imaging of the regurgitant jet. Echocardiography enables the determination of AVR etiology and assessment of the AVR severity as well as its effect on left ventricular size, function, and hemodynamics. With progressing aortic valve regurgitation, both the duration and the velocity of the diastolic aortic flow reversal increase [498].

In mild aortic valve regurgitation, left ventricular size, function, and hemodynamics are normal. In acute severe form, aortic and left ventricular pressures are balanced. Elevated left atrial pressure causes pulmonary edema.

In early compensated chronic severe form, the left ventricle adapts to the volume overload by eccentric hypertrophy and begins to dilate. The left ventricular ejection fraction is often maintained in the normal range by increased preload. Eccentric hypertrophy preserves the left ventricular diastolic compliance and, hence, filling pressure remains normal or mildly increased [505]. The slope of the left ventricular pressure–volume relation (elastance) is normal. The systolic arterial hypertension is accompanied a wide pulse pressure. Left ventricular filling pressures are normal or

Table 6.10 Aortic valve regurgitation grades. (Source: [498] EROA: effective regurgitant orifice area)

Parameter	Mild	Moderate	Strong	Severe
Regurgitant volume (ml/beat)	< 30	30–44	45–59	≥ 60
Regurgitant fraction (%)	< 30	30–39	40–49	≥ 50
EROA (cm^2)	< 0.10	0.10–0.19	0.20–0.29	≥ 0.30

only slightly elevated; dyspnea is absent. Progressive left ventricular dilation and systolic hypertension increase wall stress and the volume/mass ratio.

In decompensated chronic severe form, the left ventricle is hypertrophied, dilated and stiff. Excessive afterload depresses the left ventricular function. Forward left ventricular output diminishes. Fibrosis and hypertrophy decrease the left ventricular compliance, thereby increasing filling pressure [505]. Exertional dyspnea is the most common manifestation, but angina can also occur because of a reduced coronary flow reserve with predominantly systolic coronary flow.

Echocardiographic evaluation of the anatomy of the aortic root, annulus, and leaflets enables the determination of AVR etiology and severity. In general, flow jets are composed of three distinct segments:

1. A proximal *flow convergence zone*, which is the area of flow acceleration into the orifice
2. The *vena contracta*, which is the narrowest and highest velocity region of the jet at or just downstream from the orifice and
3. The *jet* downstream from the orifice in the left ventricle

A centrally directed jet generally appears larger and wider than eccentric jets directed anteriorly toward the ventricular septum or posteriorly toward the anterior mitral leaflet.

Measurement of jet area or penetration into the left ventricle is not accurate enough to assess AVR severity. The ratio of jet width to left ventricular outflow tract width (JW/LVOTW) greater than 65 % specifies severe AVR, when the regurgitant orifice is relatively circular. When it is elliptical, as in bicuspid aortic valves, the JW/LVOTW ratio underestimates AVR severity [505].

The rate of deceleration of the diastolic regurgitant jet and the pressurehalf-time (PHT) reflects the rate of equalization of aortic and left ventricular diastolic pressures [498]. When AVR progresses, aortic diastolic pressure decreases more rapidly. The late diastolic jet velocity is lower and hence pressure half-time is shorter.

Specific and supportive signs and quantitative parameters enables the grading of aortic valve regurgitation (Table 6.10).

Aortic valve replacement is a standard surgical procedure for the treatment of aortic valve regurgitation. On the other hand, procedures repairing diseased aortic valves can be employed according to the mechanisms of valve dysfunction and aortic valve lesions identified by transesophageal echocardiography.

6.6 Right-Sided Valve Failure

Right-sided valvular disease has received less attention than left-sided valve disease, partly because of a protracted asymptomatic period. Moreover, because tricuspid regurgitation is often due to left-sided valve disease and pulmonary regurgitation is often secondary to congenital cardiac disease, the underlying disorder rather than the valve lesion tends to dominate the clinical picture [506].

6.6.1 Tricuspide Valve Failure

6.6.1.1 Tricuspide Valve Stenosis

Tricuspid valve stenosis (TVS) is an extremely rare disease, rheumatic heart disease (with commissural fusion, diastolic doming, and thickened shortened chordae tendineae) accounting for about 90 % of cases. Tricuspide valve stenosis results from [506]:

- Rheumatic heart disease
- Congenital tricuspid valve stenosis
- Right-atrial tumors
- Carcinoid heart disease
- Endomyocardial fibrosis
- Valvular vegetations
- Extracardiac tumors

The venous pressure is elevated with a prominent a wave and slow y descent. Echocardiography assesses TVS grade. PHT($t_{\hat{p}/2}$ [ms]) is defined as the time needed for the peak transvalvular pressure difference to fall to its half value. The Bernoulli equation applied to an isoheight flow of incompressible fluid states that at any arbitrary point along a streamline in a steady flow,

$$\frac{v^2}{2} + \frac{p}{\rho} = \text{cst.} \tag{6.7}$$

When this equation is improperly extrapolated to time-dependent flow and the phase lag between the pressure and velocity waveforms is neglected, at the time ($t_{\hat{p}/2}$) when the pressure (p) is halved from its peak value ($\hat{p}$), the flow velocity can be assumed to equal the peak transvalvular velocity ($\hat{v}$) divided by the square root of 2 ($\hat{v}/\sqrt{2}$). Consequently, PHT is also defined as the time for the velocity to reach 0.707 of its peak value (0.707 $\hat{v}$). The faster the transvalvular pressure difference falls, the lower the atrioventricular flow resistance through the valve, and conversely. In other words, high-diastolic transvalvular pressure difference half-time indicates a narrowed valve area. Deceleration time (DT) is proportional to pressure half-time:

$$\text{DT} = \text{PHT}/0.29 \tag{6.8}$$

Hemodynamically significant tricuspid valve stenosis is supported by various observation data [497]:

- Specific findings, such as
  - mean pressure gradient ≥ 5 mmHg
  - inflow time-velocity integral > 60 cm
  - PHT ≥ 190 ms
  - valve area ≤ 1 cm^2
- Supportive facts, such as
  - enlarged right atrium and
  - dilated inferior vena cava

6.6.1.2 Tricuspide Valve Regurgitation

Moderate tricuspide valve regurgitation (TVR) results most often from functional anomalies, that is, not related to primary tricuspid valve leaflet lesions, but secondary to a process causing right ventricular dilation, distortion of the subvalvular apparatus, tricuspid annular dilation, or a combination of these deficiencies.

Tricuspide valve regurgitation can be caused by [506]:

1. Structural damages of the tricuspide valve related to:
   - congenital defects, such as
     - Ebstein anomaly, which is characterized by apical displacement from the annulus of both the septal and posterior leaflets
     - tricuspide valve dysplasia
     - tricuspide valve hypoplasia
     - tricuspide valve cleft
     - double-orifice tricuspide valve
     - unguarded tricuspide valve orifice
   - acquired defects upon
     - endocarditis
     - trauma
     - carcinoid heart disease
     - rheumatic heart disease
     - tricuspide valve prolapse
     - iatrogenic (radiation, drugs, biopsy, device lead) origin,
2. Functional causes, that is, with morphologically normal leaflets and annular dilatation, which encompass:
   - idiopathic tricuspid annular dilatation
   - right ventricular dysplasia
   - endomyocardial fibrosis
   - primary pulmonary hypertension
   - secondary pulmonary hypertension
   - atrial septal defect
   - anomalous pulmonary venous drainage

Table 6.11 Echocardiographic and Doppler parameters in grading severity of tricuspide valve regurgitation (TVR; Source: [498]; ND: not determined; IVC: inferior vena cava; MRAML(SI)D: maximal right atrial mediolateral (superoinferior) dimension; MRAV: maximal RA volume; RA; right atrium; RV: right ventricle RVEDA: right ventricular end-diastolic area; RVMLEDD: right ventricular mediolateral end-diastolic dimension)

Parameter	Mild	Moderate	Severe
Tricuspid valve	Normal	Normal or not	Flail leaflet Poor coaptation
Cavity and vein size	RVMLEDD $\leq$ 43 mm RVEDA $\leq$35.5 cm^2 MRAMLD $\leq$46 mm MRASID $\leq$49 mm MRAV $\leq$33 ml/m^2	Normal or dilated	Usually dilated (except acute TVR) IVC dilation
Vena contracta width (mm)	ND	< 7	> 7
Hepatic vein flow	Systolic dominance	Systolic blunting	Systolic reversal

Severe tricuspide valve regurgitation has often a prolonged latent period with progressive right atrial and ventricular volume overload. Atrial arrhythmias are secondary to right atrial enlargement. Elevated right atrial pressure provokes peripheral edema. Severe forms reduce exercise capacity and cause right-cardiac failure, hepatomegaly, and anasarca, as well as jugular venous distention with a visible systolic v-wave [506].

Three-dimensional echocardiography assesses severity and cause of tricuspide valve regurgitation as well as its impact on right ventricular size and function (Table 6.11). Doppler velocimetry and tissue imaging evaluate indices of right myocardial performance, temporal change in pressure (dP/dt), myocardial acceleration during isovolumic contraction, and displacement of the tricuspid annulus. The vena contracta (the narrowest flow section surface area) indirectly reflects the effective regurgitant orifice area.

6.6.2 *Pulmonary Valve Failure*

6.6.2.1 Pulmonary Valve Stenosis

The majority of pulmonary valve stenosis (PVS) are related to congenital or genetic disorders such as autosomal dominant Noonan syndrome. The abnormal pulmonary valve can be classified as [506]:

- Acommissural with prominent systolic doming of thickened pulmonary valve cusps and an eccentric orifice
- Unicommissural
- Bicuspid with fused commissures
- Dysplastic with strongly thickened and deformed valve cusps

Table 6.12 Pulmonary valve stenosis grades. (Source: [497])

Physical variable	Mild	Moderate	Severe
Peak transvalvar pulmonary velocity (m/s)	< 3	3–4	> 4
Peak pressure difference (mmHg)	< 36	36–60	> 60

In severe pulmonary valve stenosis characterized by right atrial enlargement, right-axis deviation, and right ventricular hypertrophy, as well as dilation of the pulmonary artery, a jugular venous wave can be observed. Transthoracic echocardiography that exhibits the presence of subvalvular or supravalvular obstruction is recommended for the diagnosis and follow-up of pulmonary valve stenosis. Continuous-wave Doppler ultrasonography and catheter-connected pressure sensors measure velocity and pressure across the pulmonary valve (Table 6.12).

6.6.2.2 Pulmonary Valve Regurgitation

In adults, pulmonary valve regurgitation (PVR) most often results from prior interventions for congential heart disease such as tetralogy of Fallot repair with placement of an outflow tract patch or surgical valvotomy for isolated congenital pulmonary valve stenosis [506]. Additional causes of pulmonary valve regurgitation include rheumatic or carcinoid heart disease, trauma, endocarditis, and pulmonary artery and annular dilation. Pulmonary valve regurgitation can be the consequence of pulmonary hypertension.

Long-standing severe pulmonary valve regurgitation provokes a progressive right ventricular dilation and reduced function with possible ventricular arrhythmias evolving to ongestive heart failure and sudden death. Chest radiography can show cardiomegaly involving the right-sided chambers and pulmonary artery enlargement. Echocardiography coupled to Doppler exploration exhibits typically a wide holodiastolic jet. Cardiac magnetic resonance imaging assesses right-sided heart size and function.

6.6.3 Cardiac Valve Clinic

About 13 % of the patients older than 75year present at least a moderate valvular heart disease. The decision of intervention usually relies on several factors, such as severity of valvular dysfunction, presence and severity of symptoms, extent of myocardial impairment, comorbidities (e.g., pulmonary disease), and rapid progression, in addition to imaging features and blood marker values [507].

Table 6.13 Pulmonary valve regurgitation grading. (Source: [498])

Parameter	Mild	Moderate	Severe
Pulmonic valve	Normal	Normal or not	Abnormal
RV size	RVMLEDD ≤43 mm RVEDA ≤35.5 cm^2	Normal or dilated	Dilated
Jet length	< 10 mm Narrow origin	Intermediate	Large Wide origin
Deceleration	Slow	Variable	Steep
Pulmonic systolic flow compared to systemic flow	Slight ↑	Intermediate	Large ↑

↑ increase, *RVEDA* right ventricular end-diastolic area, *RVMLEDD* right ventricular mediolateral enddiastolic dimension

Three-dimensional echocardiography, multidetector-computed tomography, and cardiac magnetic resonance imaging help establishing etiology, quantifying its severity, following progression, and estimating repercussion on the cardiac function, as well as assessing the likelihood of successful valve repair and planning catheter-based procedures, in addition to allowing the postoperative follow-up [507]. Systematic preoperative check-up include electrocardiogram, transthoracic echocardiogram, coronary angiography, lung function testing, and carotid ultrasonography.

Because valvular heart disease progression is often clinically silent, surgical or transcatheter valve interventions must not be inappropriately delayed to obviate large irreversible myocardial damage, increased operative risk, and suboptimal long-term outcomes. Symptom onset is a key factor in the indication for intervention in valvular heart disease. Exercise testing enables the detection of latent symptoms. The occurrence of exercise-limiting symptoms during an exercise stress test may yield a strong indication of valve replacement. An old age, comorbidities, and a weak left ventricular function are not contraindications to valve intervention [507].

However, the outcome of transcatheter or surgical valve replacement is affected by prosthetic valve hemodynamics, durability, and thrombogenicity.

Chapter 7
Heart Failure

According to the New Encyclopædia Britannica (15th ed., 1985), "heart failure is the inability of either or both side of the heart to pump sufficient blood to meet the needs of the body .... Heart failure is characterized by distension of the veins serving the lungs or of those serving the rest of the body, or of both; this engorgement is expressed in the term *congestive heart failure*."

The inappropriate pumping results from myocardial weakness. The structure–function relation of the heart is impaired and bodily needs are unmatched. *Cardiac dysfunction* is defined as an alteration in the relation between preload and stroke volume. Heart failure is also frequently associated with arrhythmias and valve dysfunction.

Heart failure develops progressively after cardiac damages. It can affect the right or left pump only or, most often, the entire heart. Clinical manifestations comprise fatigue, dyspnea (breathlessness), diminished exercise capacity, and fluid retention.

Heart failure causes significant morbidity and mortality. It currently affects 0.4–2 % of the population in the Western world and ranges from 6 to 10 % patients over 65 years old, with a 4-year survival rate of 50 %. Heart failure is a frequent cause of atrial fibrillation and patients often die from sudden arrhythmia.

Diagnosis relies on either two major criteria, such as elevated jugular venous pressure, pulmonary rates, or a third heart sound, or one major criterion and two minor criteria, such as edema, dyspnea, or hepatomegaly. Acute episodes cause cardiac decompensation.

Echocardiography and magnetic resonance and nuclear medical imaging are commonly employed in the identification of ventricular systolic dysfunction.

Heart failure is not a simple contractile disorder or a single disease of the heart. It can result from maladaptive cardiac remodeling in response to pressure or volume overloading in the acute phase of cardiac injury, such as in mitral insufficiency and myocardial infarction, that leads to progressive decompensation.

Adverse cardiac remodeling is defined as an alteration in the shape, dimensions, structure, and mass of the heart in response to hemodynamic load and/or cardiac injury. Adverse cardiac remodeling evolves in a neurohormonal activation context (Chap. 3).

M. Thiriet, *Diseases of the Cardiac Pump*, Biomathematical and Biomechanical Modeling of the Circulatory and Ventilatory Systems 7, DOI 10.1007/978-3-319-12664-7_7

Whatever the cause, heart failure is associated with a chronic activation of β-adrenoceptors and elevated plasma catecholamine levels (Sect. 7.9.1). Initially, a hyperactive sympathetic nervous system is aimed at compensating contractile dysfunction. However, β-adrenoceptor overstimulation provokes cardiac hypertrophy and arrhythmias, hence progressively disorganizing the cardiac structure and deteriorating its contractile function.

Therapy of heart failure is aimed at correcting the reduced blood flow. Antineuroendocrine treatment with angiotensin-converting enzyme inhibitors, β-adrenergic blocking agents, and antialdosterone drugs can significantly reduce morbidity and mortality [208].

Therapy also targets adverse cardiac remodeling. Reverse remodeling can result from pharmacological and nonpharmacological therapies. Moderate endurance training and continuous positive airway pressure (CPAP) therapy in heart failure patients with sleep apnea induce reverse remodeling [209]. Cardiac resynchronization improves exercise capacity and quality of life in patients with ventricular dyssynchrony. Reverse remodeling by surgery, such as mitral valve replacement, aneurysmectomy, and volume reduction are associated with high perioperative mortality. Mechanical unloading of the failing ventricles by left ventricular assist systems reverses remodeling on the cellular and subcellular levels. New therapeutic procedures, such as gene or stem cell therapy, in reverse remodeling await characterization.

Heart failure with preserved left ventricular ejection fraction (HFpLVEF; with normal or quasinormal systolic function) is nearly as deadly as heart failure with reduced left ventricular ejection fraction (HFrLVEF). Metabolic risk factors of HFpLVEF include diabetes, obesity (hyperlipidemia), and hypertension.

Inhibition of the sympathetic nervous system and renin–angiotensin–aldosterone axis is effective in the treatment of HFrLVEF patients. However, medication used in HFrLVEF patients (e.g., β-blockers and angiotensin-converting enzyme inhibitors) are not efficient in HFpLVEF patients [222].

In fact, patients can be classified into three groups based on review of the left ventricular ejection fraction in the medical record [508]: (1) patients with persistently preserved (EF $\geq$ 40 %; HFpEF); (2) with recovered (EF $\geq$ 40 %; HFrEF); and (3) with low ejection fraction (EF $<$ 40 %; HFlEF). HFrEF patients have less atrial fibrillation episodes, hypertension, and diabetes, better renal function, and larger end-diastolic left ventricular size than HFpEF patients. They are younger than HFpEF and HFlEF patients. They have lower rates of coronary artery disease than HFlEF patients.

7.1 Risk Factors, Major Causes, and Natural History of Cardiac Deficiency

The likelihood of developing heart failure rises in the presence of:

- Hypertension;
- Obstructive coronary artery disease;

- Cardiomyopathies of nonischemic origin, such as idiopathic dilated cardiomyopathy;
- Cardiac rhythm and conduction disorders;
- Endo- or myocarditis; and
- Cardiac valve diseases.

Three main types of heart failure exist:

- Heart failure due to left ventricular weakness;
- Heart failure with preserved ejection fraction usually caused by a stiff left ventricle; and
- Heart failure due to valve disease.

Many etiologies can lead to the onset of heart failure. Cardiac causes of *acute decompensated heart failure*, or acute dyspnea, include left ventricular diastolic and/or systolic dysfunction, right ventricular dysfunction, and/or acute ischemia.

Chronic heart failure results from:

1. Cardiac disorders, such as acute and chronic ischemic heart disease, valvular disorders, cardiomyopathies, myocarditis, and chronic arrhythmias, among others;
2. External factors, such as hypertension, pulmonary embolism, and shunt, among others; and
3. Defects in genes that encode constituents of the sarcomere, cytoskeleton, mitochondrion, or extracellular matrix.

According to the American Heart Association and American College of Cardiology several stages of latent or patent heart failure can be defined [480]:

Stage A representing patients at risk of heart failure, without structural heart disease or symptoms of heart failure;

Stage B characterized by patients with structural heart disease, but without symptoms of heart failure

Stage C defined by patients with structural heart disease and past or present symptoms of heart failure; and

Stage D corresponding to patients with advanced (refractory) heart failure requiring specialized interventions (e.g., inotropic support).

7.2 Structure and Function of Chronic Congestive Heart Failure

Congestive heart failure, the final stage of many diseases of the heart, is a weakening of the myocardium that fails to supply adequate blood flow to the bodily organs. The ineffective blood pumping is associated with activated stress pathways by mechanical and chemical signals, myocardial dysfunction, cell death, and altered cell environment.

Chronic heart failure is linked to myocardial structural and functional abnormalities. The mass of failing hearts increases due to adverse cardiomyocyte hypertrophy as well as remodeling with fibrosis, necrosis, and cardiac dilation. Edema of the myocardial interstitium increases wall stiffness and causes collagen deposition leading to fibrosis.

Heart failures related to a preserved or reduced left ventricular ejection fraction have similar outcomes [509].

- *Systolic heart failure* is characterized by contractile failure and dilated heart chamber;
- *Diastolic heart failure* by a preserved ejection fraction, heart contractility being considered as normal, and wall thickening (cardiac hypertrophy).

Diastolic dysfunction results from myocardial remodeling, that is, cardiomyocyte hypertrophy and interstitial fibrosis. It is linked to altered relaxation (prolonged isovolumic relaxation and deceleration times and abnormal left ventricular filling) and cardiac wall stiffness (or reduced left ventricular compliance), with elevated diastolic pressure, enlarged left atrial volume and reduced emptying (A wave), whereas systolic function is preserved. It leads to heart failure with preserved left ventricular ejection fraction. Diastolic dysfunction increases the risk of mortality. Hypophosphorylation of titin that intervenes in serves in distensibility and diastolic recoil of cardiomyocytes can lead to HFpLVEF associated with hypertrophy in patients [223].

Chronic heart failure triggers general and regional neurohormonal compensatory responses, such as activation of the sympathetic nervous system, renin–angiotensin–aldosterone axis, nonosmotic release of vasopressin, and liberation of natriuretic peptides, endothelial regulators such as nitric oxide, and cytokines. These countermeasures include retention of sodium and water by the kidney, release of neurohormones, and activation of various intracellular signaling cascades in cells of the cardiovascular system.

The long-term activation of these compensatory mechanisms initially aimed at preserving hemodynamics lead progressively to cardiac dilation and decaying contractility down to decompensated stage.

Chronic heart failure is associated with impaired myocardial metabolism and signaling, altered cardiac receptors and contractile proteins, defective calcium handling, sodium retention, neurohumoral maladaptations, oxidative stress, and vascular dysfunction.

Cardiomyocytes reply to stress and abnormal loading by reprogramming gene expression, especially genes involved in metabolism and ion handling as well as genes specific of embryo- and fetogenesis.

7.3 Markers of Chronic Heart Failure

Numerous markers related to heart failure explore the role of inflammation, oxidative stress, cardiac remodeling, cardiomyocyte stretching, and renal function, among others. The most appropriate markers are selected for diagnosis, prognosis, and treatment planning. Concentrations of circulating *troponins* increase not only in myocardial infarction upon atherosclerotic plaque ulceration, fissuring, and rupture (type-1 myocardial infarction), as well as supply–demand imbalance and endothelial dysfunction (type-2 myocardial infarction), but also in acute and chronic heart failure, in particular due to cardiomyocyte apoptosis and autophagy subsequently to wall stretch, arrhythmia, infiltration such as amyloidosis, exposure to toxics (e.g., alcohol or chemotherapy drugs), myocarditis, and stress cardiomyopathy, as well as renal dysfunction [510]. Elevated concentrations of troponin-I or -T predicts an adverse outcome. *Natriuretic peptides* are currently reference markers for diagnosis and therapy monitoring of heart failure. However, natriuretic peptides are secreted in response to a nonspecific pressure overload of any cardiac chamber that can be observed not only in heart failure but also in systemic hypertension, atrial fibrillation, and valvular disease, as well as lung diseases, pulmonary embolism, and renal failure [511]. Therefore, natriuretic peptide marker specificity for diagnosis is low. However, a BNP level in the normal range excludes heart failure as the cause of complaints such as dyspnea.

In addition, the BNP release results from a neurohormomal activation cascade aimed at compensating a anatomical and/or functional defect and yields a prognosis information. Although BNP exerts vasodilatory and diuretic effects, an increase in BNP level is associated with a worse outcome in heart failure [511]. Therefore, the measurement of natriuretic peptides (type-B natriuretic peptide [BNP] and N-terminally truncated proBNP [proBNP$^{\Delta CT}$]), a marker for cardiomyocyte stretching, as well as mid-regional proatrial natriuretic peptide [proANP^{MR}]) is recommended to exclude alternative causes of dyspnea in patients with suspected heart failure and to evaluate the prognosis.

Quiescin Q6 sulfhydryl oxidases (QSOx) are thiol oxidases involved in oxidative protein folding with thiol–disulfide exchange, cell cycle control, and extracellular matrix remodeling.[1] They catalyzes the insertion of disulfide bonds into unfolded reduced proteins with concomitant reduction of oxygen to hydrogen peroxide, hence

---

[1] In humans, the QSOX1 gene pertains to chromosome 1q24. Alternative splicing generates a long (QSOx1_L with a transmembrane domain [exon 12]) and short (QSOx1_S transcript. These metalloenzymes contain at least one thioredoxin (protein disulfide isomerase) domain in the N-terminus and augmenter of liver regeneration (ALR; also called essential for respiration and vegetative growth [ERV]) in the C-terminus, a highly helical flavin-binding domain (hence the noun flavoprotein and the term FAD-linked sulfhydryl oxidase), as well as iron or copper. The flavin adenine dinucleotide (FAD) cofactor is bound noncovalently within the helical fold. The QSOx enzymes comprise an oxidoreductase and sulfhydryl oxidase moiety that act in tandem to oxidize substrates and transfer the excess electrons to oxygen. Two types of flavin-linked sulfhydryl oxidases have been identified in the secretory apparatus of higher eukaryotes: the endoplasmic reticulum-resident sulfhydryl oxidase, endoplasmic reticulum oxidoreductin ERO1, and members of the quiescin–sulfhydryl oxidase (QSOX) family of flavoproteins. In humans, QSOx1 abounds in cells with a heavy secretory load. The small dimeric sulfhydryl oxidases include mammalian growth factor augmenter of liver regeneration (GFER or ALR; a.k.a. ERV1).

altering the redox environment:[2]

$$2R^{SH} + O_2 \rightarrow R^{S--S}R + H_2O2. \quad (7.1)$$

Quiescin Q6 sulfhydryl oxidase QSOx1 identifies acute decompensated heart failure (ADHF), particularly when combined with BNP and proBNP$^{\Delta CT}$ [513].

The QSOx1 synthesis rate in the left ventricle and left atrium depends on the degree of pressure overload. Myocardial QSOx1 expression remains low in patients with acute dyspnea of noncardiac origin or with stable compensated heart failure.

In addition to C-reactive protein, pentraxin-3, and interleukin-6, tumor-necrosis factor-α receptor TNFRSF1a and growth differentiation factor GDF5 are markers of inflammation. Marker levels are linked to the pathophysiological mechanisms of heart failure, e.g. heart failure with reduced or preserved left ventricular ejection fraction [511]. With increasing age, the level of expression of inflammatory markers rises in the general population.

Markers related to extracellular matrix remodeling comprise syndecan-1 and periostin [514].

Disparities in pathophysiology and disease manifestation exist between women and men with heart failure. TNFRSF1a, GDF15, and proBNP$^{\Delta NT}$ show the strongest interaction between sex and mortality, mortality being lower in women than men, independently of differences in clinical characteristics [514]. The main etiological sex difference in heart failure is a predominance of myocardial infarctions and the presence of ischemic heart disease in men with respect to women. However, most comorbidities are preferentially observed in women with heart failure. In addition, in men, remodeling is usually characterized by left ventricular dilation and fibrosis, whereas in women, the heart more frequently remodels with a marked concentric hypertrophy and a smaller left ventricular volume [514].

7.4 Genetic Background and Chronic Heart Failure

Heart failure has a strong heritable component. Parental heart failure confers a 70 % greater disease risk than the absence of such a family history. Traditionally, inherited genetic forms of heart failure are distinguished from acquired forms. The former category encompasses familial dilated, hypertrophic, and arrhythmogenic right ventricular cardiomyopathy (Chap. 4).

[2] Quiescin Q6 sulfhydryl oxidases localize to the endoplasmic reticulum, Golgi body, secretory granules, and nuclear and plasma membranes [512]. Typically, unfolded substrates of QSOxs are obtained by exhaustive reduction of secreted proteins (e.g., RNase and lysozyme) under denaturing conditions followed by the removal of excess reductant [512]. These reduced proteins freed from the restraint of their native disulfide pairings expand, thereby allowing a rapid access of their cysteine residues to small oxidants and QSOx enzymes. Steric constraints on disulfide exchange reactions arise when substrates are large.

Mutations of numerous genes that cause inherited forms of heart failure encode proteins that function in contractility (sarcomeric, cytoskeletal, and desmosomal proteins, as well as lamin-A and -C), energy production (mitochondrial proteins), calcium handling, and transcription regulation (Table 7.1). In addition, reprogramming of gene expression represents the response of cardiomyocytes exposed to stress (heart failure dysregulome).

The contraction and relaxation at relatively high frequency (~ 1 Hz) of cardiomyocytes are mediated by highly coordinated processes that allow rapid activation and deactivation of the sarcomeric filaments. The sarcomere is the fundamental contractile unit. Its shortening results from arrangements of myosin microfilaments sliding on arrays of actin microfilaments fueled by ATP hydrolysis with a cyclic ratchet-like interaction regulated predominantly by the actin filament regulatory partners troponin and tropomyosin.

A change in myofilament isoform composition affects myocardial contractile performance in heart failure. The fast isoform of myosin heavy chain α MHC (MyHC6) accounts for approximately 10 % of the content in the healthy human cardiomyocyte myocardium (hence 90 % β MHC or MyHC7) [516]. This proportion is further reduced to nearly undetectable concentrations during heart failure, thereby lowering cardiac function. Furthermore, a shift in troponin-T isoform composition as well as in tropomyosin isoform expression also occurs in the human failing heart.

7.4.1 Troponins

Among striated myocytes, the regulation of cardiomyocyte contraction differ from that of skeletal myocytes due to distinct physiological and contractile properties. The troponin complex is a component of thin actin filaments with tropomyosin. Troponin contains three subunits:

1. The calcium-binding troponin-C,
2. The inhibitory troponin-I, and
3. The tropomyosin-binding troponin-T.

Troponin binds to tropomyosin and, in the absence of calcium, also to actin via troponin-I.

The actin–myosin interaction is activated by calcium binding to troponin-C that provokes a conformational change in troponin, thereby decreasing troponin-I affinity to actin and engendering an azimuthal shift of tropomyosin on the actin filament that partly exposes myosin-binding sites on actin. Binding of one myosin to actin filament shifts tropomyosin further on actin and fully exposes the adjacent actin-binding sites and accelerates the rate of subsequent myosin binding to actin filament. In addition, myosin binding to actin increases the calcium affinity of unsaturated troponin-C, thereby potentiating actin activation [516].

Table 7.1 Inherited forms of heart failure (Source: [515]) . Differences in contractile and regulatory function exist among isoforms of myosin, tropomyosin, and troponin-T. Mutations in a given gene causes different forms of genetic cardiomyopathies

Gene protein	Expression in heart failure
Sarcomeric constituents	
ACTC1	↑
MYH6	↓
MYH7	↑
TNNC1, TNNI3, TNNT2	↑
TPM1/2	↑
MYBPC3	Similar
Actinin-α	↑
Titin	Similar
Costameric constituents	
Desmin	↑
Dystrophin	ND
Metavinculin	ND
Muscle LIM protein	↓
Sarcoglycan-δ	Similar
Nuclear constituents	
Lamin-A/C	↓
Energetic metabolism components	
Succinate deshydrogenase	ND
Ion handling components	
Phopholamban	↑
ABCC9	ND
SCN5A	↓
Miscellaneous	
ANKRD1	↑
EYA4	ND
Ldb3	ND
Rbm20	ND
TAZ	ND
TMPO	ND

↑ increase, ↓ decrease, *ND* not described, *ANKRD1* cardiac muscle ankyrin repeat domain-containing gene-1, *EYA4* eyes absent homolog-4 [gene], *LDB3* LIM domain-binding protein-3, *RBM20* RNA-binding motif protein-20, *TAZ* tafazzin [gene], *TMPO* thymopoietin [gene]

Multiple cardiac troponin-T (cTnnT) isoforms are expressed in a developmentally regulated manner, whereas cardiac troponins cTnnC and cTnnI are expressed as single isoforms in the adult heart.

Contraction of the cardiomyocyte is carried out in many steps (Vol. 5, Chap. 5. Cardiomyocytes):

1. Increase in cytosolic Ca^{2+} concentration;
2. Ca^{2+} binding to troponin-C;
3. Troponin-C tethering to troponin-I;
4. Troponin-I dissociation from actin;
5. Tropomyosin position on actin shift;
6. Weakly bound cross-bridges transition to strongly bound cross-bridges further moving tropomyosin on actin (a rate-limiting step for contraction); and
7. Sarcomere shortening.

The rate of transition from weak to strong cross-bridge is not only influenced by myosin features but also by the Ca^{2+}-binding properties of troponin-C on the thin actin filament as well as the level of free Ca^{2+} ions [517]. The rate of contraction can rise up to fivefold with increasing Ca^{2+} concentration up to a saturating amount. The rate of contraction at submaximal concentrations of Ca^{2+} is modulated by the Ca^{2+} sensitivity of troponin-C. The myosin isozymes and ATP hydrolysis products modulate the rate of cardiomyocyte contraction independently of troponin-C. The intrinsic properties of myosin dictate the rate of contraction, with α-myosin having a faster ATPase cycle than that of β-myosin. The products of ATP hydrolysis inorganic phosphate (P_i) and ADP increase and decrease the maximal rate of myosin ATPase activity, respectively.

Relaxation of the cardiomyocyte involves several rate-limiting steps [517]:

1. Decline in cytosolic Ca^{2+} concentration;
2. Transient dissociation of Ca^{2+} from troponin-C; and
3. Cross-bridge detachment.

Cytosolic metabolites that affect the cross-bridge detachment kinetics modify the rate of relaxation independently of the fall in cytosolic Ca^{2+} level. Increase in ADP concentration diminishes cross-bridge detachment and the rate of relaxation, whereas increase in P_i concentration heightens cross-bridge detachment and the rate of relaxation. Change in myosin heavy chain expression also modulates the rate of cardiomyocyte relaxation. In addition, β-adrenoceptor stimulation accelerates the rate of relaxation, as it raises the rate of Ca^{2+} dissociation from troponin-C via phosphorylation of cTnnI by PKA. On the other hand, in TNNC1 gene mutation related to familial hypertrophic cardiomyopathy, increased Ca^{2+} sensitivity attenuates the rate of relaxation.

Slow and fast calcium-binding troponin-C are encoded by the TNNC1 and TNNC2 genes, respectively. The TnnC1 isoform corresponds to the cardiac (cTnnC) subtype. It possesses an N and C lobe. The former is the regulatory part that binds to Ca^{2+} ion and troponin-I after calcium binding, hence regulating calcium-regulated myofibrillar ATPase activity. The latter binds Ca^{2+} or Mg^{2+} ions.

Cardiac troponin-C (cTnnC) is the major regulator of myocardial contraction. The TnnC1 subtype, a member of the EF-hand family of Ca^{2+}-binding proteins, relays the Ca^{2+} signal upon acute brief Ca^{2+} influx (i.e., Ca^{2+} entry and release from the extracellular space and intracellular stores, respectively) via a conformational change to the rest of the troponin–tropomyosin complex. It possesses different functional and Ca^{2+}-binding properties compared with skeletal subtypes (skTnnC). Upon Ca^{2+} binding, the regulatory N-domain of skTnnC switches from a closed to an open conformation, thereby exposing a patch of hydrophobic residues and hence interacting with skTnnI. Unlike, other EF-hand Ca^{2+}-binding calcium sensors skTnnC and calmodulin characterized by a large conformational change upon Ca^{2+} binding as well as the EF-hand calcium buffers parvalbumin and calbindin characterized by a minor conformational changes upon Ca^{2+} tethering, cTnnC has a compact regulatory domain in the Ca^{2+}-bound structure [518]. The free energy of Ca^{2+} binding to the cTnnC-cTnnI complex is four times smaller than that to the skTnnC-skTnnI complex that enables the repetition at relatively high frequency of contraction–relaxation cycles.

Calcium binding to and dissociation from cTnnC modulate the cardiomyocyte contraction and relaxation. Calcium removal rate of cTnnC is about threefold faster than that of skTnnC. The subsequent cTnnC–cTnnI interaction determines the specific regulation of cardiomyocyte contraction. However, actin–tropomyosin and myosin influence Ca^{2+} unbinding from troponin, actin–tropomyosin increasing the rate of Ca^{2+} dissociation from 42 to 105/s and myosin slowing it to 13/s [517]. Mutations in troponin subunit genes linked to cardiomyopathies can sensitize or desensitize cTnnC to Ca^{2+} and/or affect the rates of Ca^{2+} exchange with cTnnC. Slow skeletal TnnI (sskTnnI) expressed in embryonic and early postnatal hearts and possibly reexpressed during disease increases the Ca^{2+} affinity and decreases the rate of Ca^{2+} dissociation from cTnnC in the troponin complex on the actin filament in the presence of myosin [517].

Distinct genes encode cardiac-specific (TNNI3) and fast (TNNI2) and slow (TNNI1) skeletal muscle-specific isoforms of inhibitory troponin-I. Mutations of the TNNI3 gene linked to hypertrophic or restrictive cardiomyopathies (R145G, R145W, D190H, and R192H) increase the Ca^{2+} sensitivity of the troponin complex on the actin filament (but not in isolated cTnnI) [517].

Tissue-specific subtypes of troponin-T that forms the troponin–tropomyosin complex include slow (TnnT1) and fast (TnnT3) skeletal muscle troponin-T encoded by the TNNT1 and TNNT3 genes, respectively, and cardiac troponin-T2 encoded by the TNNT2 gene.

In the human heart, four cardiac troponin-C isoforms with different calcium sensitivity are produced (cTnnT1–cTnnT4) from alternative splicing (cTnnT1: all exons present; cTnnT2: missing exon 4 [five amino acids]; cTnnT3: missing exon 5 [ten amino acids]; and cTnnT4: missing exons 4 and 5) [519]. The variable region in the N-terminus of cTnnT contributes to the determination of the Ca^{2+} sensitivity of force development in a charge-dependent manner as well as regulation of inhibitory ATPase activity. The greater the charge, the higher the Ca^{2+} sensitivity. Calcium sensitivity of force development is associated with exon 5. All cTnnT isoforms yield the same maximal actin–tropomyosin-activated myosin ATPase activity. However, cTnnT1 or cTnnT2 that both contain information from exon 5 have a reduced ability to inhibit this ATPase activity with respect to the cTnnT3 subtype [519].

The cTnnT expression differs in the normal and failing adult and fetal human heart [520]. The difference is achieved by the variable inclusion of a 15- and 30-nucleotide exon in the 5′ half of the coding region. The cTnnT1 and cTnnT2 isoforms contain both 10- and 5-residue peptides encoded by the 30- and 15-nt exons (exons 4 and 5) or the peptide encoded by the 30-nt exon alone, respectively. They are expressed in the fetal heart, cTnnT2 being produced at a very low level. The cTnnT3 isoform has the five-residue peptide (hence lacking the ten-residue peptide). It is the dominant isoform in the adult heart. The peak myofibrillar ATPase activity is correlated with cTnnT4 expression [520]. The cTnnT4 subtype lacks both peptides encoded by the 30- and 15-nt exons. It is synthesized in the fetal heart and reexpressed in the failing adult heart. Mutations in the TNNT2 gene are responsible for hypertrophic cardiomyopathy. Failed hearts from patients with human idiopathic cardiomyopathy are associated with cTnnT alterations.

7.4.2 *Tropomyosin*

In striated myocytes, tropomyosin, an actin-binding protein, interacts with the troponin complex and actin to regulate calcium-induced muscle contraction. Tropomyosin is composed of two chains coiled around each other in a parallel fashion. It lies in the groove of actin helical filament spanning seven actin monomers (7-actin–tropomyosin–troponin module). During diastole, it blocks the myosin-binding sites on actin. Each tropomyosin molecule is coupled with a paired troponin.

A change in tropomyosin flexibility (reduced stability) decays actin filament activation by calcium and myosin and governs myocardial mechanical performance, as tropomyosin drives the cooperative activation of the myofilament, that is, facilitates the extension of actin stimulation beyond the local site where operates Ca^{2+} and myosin, similarly to the cooperation induced by the binding of one oxygen molecule to one hemoglobin molecule that greatly facilitates the occupancy of the remaining three oxygen-binding sites.

In humans, isoforms (Tpm1–Tpm4 or Tpmα–Tpmδ) of tropomyosin are encoded by four genes (TPM1–TPM4). The three primary striated muscle isoforms include Tpm1 to Tpm3 (Tpmα–Tpmγ).

The isoforms Tpm1 and Tpm2 differ by sarcomeric performance, Tpm2 increasing sensitivity to calcium and decreasing the rate of relaxation and hence prolonging the relaxation time.

Alternative splicing multiplies the number of tropomyosin subtypes (>40 tropomyosin isoforms) that are not functionally redundant. Each gene indeed uses alternative splicing, alternative promoters, and distinct processing to encode multiple striated and smooth muscle and nonmuscle subtypes.

Fast contracting skeletal and cardiac muscles contain more Tpm1 homodimers, whereas slow contracting muscles express more Tpm2 homodimers. The affinities of binding the Tpm1 homodimer and Tpm1–Tpm2 heterodimer to troponin-T are equivalent, whereas that of the Tpm2 homodimer to troponin-T is lower [521]. The Tpm1 homodimer exhibits higher tropomyosin–actin–myosin (S1) ATPase activity and greater Ca^{2+}-sensitive release of troponin inhibition than the Tpm2 homodimer and Tpm1–Tpm2 heterodimer.

In murine hearts, both TPM1 and TPM2 transcript levels are elevated during embryogenesis. The TPM1 mRNA/TPM2 mRNA ratio rises from 5:1 to 60:1 during the embryonic-to-adult transition [521]. Overexpression of Tpm2 reduces the maximum rate of relaxation and heightens the time needed to complete the relaxation phase and causes a concomitant decrease in TPM1 transcript levels and a preferential formation of Tpm1–Tpm2 heterodimers, but does not alter the structure of the sarcomere and the expression of any other contractile protein genes [521].

Among Tpm1 subtypes, Tpm1α (one of ten distinct products of the TPM gene) is the predominant isoform in human hearts (> 90 % [up to 98 %] in murine hearts). The substitution of the smooth muscle-specific exon 2a by exon 2b produces Tpm1κ subtype (2–5 % of total tropomyosin expressed). Exons 2a and 2b of the TPM1 gene are spliced in a mutually exclusive manner; exon 2b is the default exon in the transcript of most cell types; exon 2a is restricted to TPM1 mRNA in smooth and striated myocytes. The Tpm1κ subtype is associated with a reduced actin activation by both Ca^{2+} and strong cross-bridges, hence altering systolic and diastolic functions and decreasing myofilament calcium sensitivity [522].

The heart may use the TPM1κ isoform to modulate sarcomeric performance during stress, as TPM1κ may increase flexibility of tropomyosin, thereby shortening the cooperative distance over which the thin filament is activated by calcium and myosin binding from a local stimulation [523]. Wrapped superhelically around actin filaments, tropomyosin moves smoothly and anisotropically, without any unfolding, chain separation, and localized kinks. Tropomyosin flexibility acts in the assembly and regulatory switching of actin filaments.

Among tropomyosin subtypes, the normal adult human heart expresses Tpm1α, Tpm1κ, and Tpm2 [522]. Overexpression of Tpm2 causes diastolic dysfunction and increased myofilament calcium sensitivity. The Tpm2 isoform-specific C-terminus does not cause morphological or pathological alterations, but lowers the rates of contraction and relaxation, elevates the time to peak pressure, diminishes intraventricular pressure, and heightens end-diastolic pressure [524]. It depresses maximal tension and myosin ATPase rate and reduces sensitivity of tropomyosin to calcium. Overexpression of Tpm3 raises systolic and diastolic function, but lowers myofilament calcium sensitivity without any morphological abnormalities.

Isoform switching of sarcomeric proteins is an adaptive mode of the heart in response to cardiac hypertrophy and heart failure in combination with modifications in the relative abundance and phosphorylation state of contractile and regulatory proteins. Substitution of nonmuscle Tpm1 C-terminal by muscle one affects tropomyosin affinities for troponin and actin.

Patients in a cardiomyopathy-caused end-stage heart failure have an increased Tpm1κ expression [522]. Incorporation of Tpm1κ in myofilaments actually engenders dilated cardiomyopathy.

Mutations in exon 2b of the TPM1 gene provoke familial hypertrophic cardiomyopathy (Glu62Gln [E62Q], Ala63Val [A63V], and Lys70Thr [K70T]) and dilated cardiomyopathy (Glu40Lys [E40K] and Glu54Lys [E54K]). Mutation Glu62Gln in exon 2b causes both dilated and familial hypertrophic cardiomyopathies.

7.4.3 Vinculin and Metavinculin

Vinculin is an actin-binding protein at specialized cytoskeletal-associated membrane nanodomains, as it localizes to focal adhesions and adherens junctions, in particular, focal adhesions in fibroblasts and fascia adherens of intercalated discs and costameres in cardiomyocytes. It is closely apposed to the plasma membrane at sites where actin-containing microfilaments terminate. It mediates the recruitment of numerous binding partners (talin, α-catenin, α-actinin, vasodilator-stimulated phosphoprotein [VaSP], ponsin [SorbS1],[3] vinexin-α and β [SorbS3], the ARP2–ARP3 complex, paxillin, raver-1[4] PIP_2, and filamentous actin) and regulates cellular responses to tension, as it links integrin receptors to the actin cytoskeleton. Vinculin mediates cell adhesion, motility, and response to force. Metavinculin[5] has, unlike vinculin, the solubility properties of an integral membrane protein [526]. It may anchor actin filaments. Metavinculin mutations are detected in some cases of dilated cardiomyopathies. Several mutations in the insert (Δ L954, A934V, and R975W) are actually associated with cardiomyopathies with improper generation of force and an impaired metavinculin–^{F}actin interaction in humans [527].

In fact, metavinculin is a splice variant of vinculin. Both are scaffold proteins that localize to cellular adhesions. Metavinculin acts in mechanotransduction. Its expression level is correlated with the force exerted on cells.

Vinculin dimerizes at high concentrations in the absence of ligands and at physiological concentrations in the presence of actin and PIP_2. It heterodimerizes as well as forms tetramers and higher-order oligomers with metavinculin. It also oligomerizes in the presence of different binding partners. Vinculin oligomerization is implicated in vinculin activation and hence scaffolding function, especially with ^{F}actin [527].

[3] Also known as Cbl-associated protein (CAP) SH3 domain-containing protein SH3D5, and sorbin and SH3 domain-containing protein SorbS1.

[4] This nuclear and costameric PTB-binding ribonucleoprotein is required for microfilament anchoring. Raver-1 colocalizes with polypyrimidine tract-binding protein (PTB; or heterogeneous nuclear ribonucleoprotein hnRNP1, a protein involved in RNA splicing of microfilament proteins [525]. Raver-1 complexes with vinculin, metavinculin, and α-actinin. It localizes at microfilament attachment sites, such as cell–cell and cell–matrix adhesions of epitheliocytes and fibroblasts, respectively, and costameres of striated myocytes.

[5] μετα: closely related to. Metavinculin is highly expressed in aortic medial smooth myocytes.

Vinculin can bundle ^{F}actin by homodimerization, each monomer binding to one filament, unlike α-actinin that bundles actin filaments by forming bridges between filaments. The actin bundling abilities of vinculin and metavinculin differ. Metavinculin may be able to form actin filament bundles, but may also cleave actin filaments into smaller filaments that form a meshwork [527]. Mutations in human metavinculin associated with cardiomyopathies alter the bundling activity of metavinculin.

7.4.4 Perturbed Transcriptome

The transcriptome is reprogrammed and changed by a fetal program. Transcription of the gene that encodes actin-α is downregulated, whereas myosin heavy chain-β production is upregulated [515].

Expression of genes that encode proteic constituents of the cytoskeleton and extracellular matrix is deregulated. Consequently, the synthesis of collagen types 1 and 3 (encoded by the COL1A1 [Col1α1] and COL1A2 [Col1α2] and COL3A1 genes, respectively), the collagen-binding keratan sulfate proteoglycan fibromudulin (encoded by the FMOD gene), fibronectin (encoded by the FN1 gene), ubiquitous actin- and calcium-binding plastin-3 (or T-plastin encoded by the PLS3 gene), and pinin (or desmosome-associated protein encoded by the PNN gene) rises [515].

The genes that encode atrial (ANP) and brain (BNP) natriuretic peptide are the most upregulated in heart failure [515]. Transforming growth factorβ1 is also overexpressed, thereby favoring interstitial fibrosis and myocytes hypertrophy.

Transcription of genes involved in fatty acid metabolism is downregulated, whereas that of genes implicated in glucose metabolism is upregulated.

Mediators of proteolysis and stress response, such as thrombospondin-4 (encoded by the THBS4 gene), are overexpressed [515]. Antiapoptotic genes are underexpressed.

MicroRNAs that regulate gene expression posttranscriptionally contribute to the pathogenesis and evolution of cardiac diseases toward heart failure, as variations of their expression alter production of target transcripts. In particular, microRNA dysregulation is linked to cardiomyocyte hypertrophy, fibrosis, and remodeling, as well as calcium handling perturbations. As microRNAs reside in the cell as well as outside them exported and conveyed in microvesicles (exosomes, microparticles, and apoptotic bodies), they can operate as intra- and auto- (self-signaling), and juxta-, para-, and endocrine regulators (intercellular signaling between members of a given cell type as well as between diverse cell types).

Long noncoding RNAs (lncRNA), which are often expressed in a development-specific manner, constitute another category of transcripts that control gene expression. Many lncRNAs are large intergenic noncoding RNAs (lincRNAs). Long noncoding RNAs can regulate gene expression via chromatin modifications and transcription and post-transcriptional processing. In particular, HoTAiR (Hox transcript antisense RNA) from the HOXC locus prevents transcription of the HOXD locus, as it induces a repressive chromatin state by recruiting Polycomb remodeling complex

PRC2 [528]. In mice, the cardiac lncRNA Braveheart (Bvht) may mediate epigenetic regulation of commitment of embryonic stem cells to cardiovascular lineages, as it interacts with suppressor of zeste homolog SuZ12, a component of Polycomb repressive complex PRC2 [529]. It functions upstream from mesoderm posterior MESP1, a master regulator of a common multipotent cardiovascular progenitor. It is required in the maintenance of cardiac fate in neonatal ventriculomyocytes. The Braveheart level in plasma of mice with induced heart failure is reduced. Long noncoding RNAs, as do certain microRNAs and cardiac transcription factors, may stimulate cellular reprogramming.

DNA methylation at CpG dinucleotides can engender transcriptional repression. Histone modifications (acetylation, methylation, and phosphorylation) also affect gene transcription. These epigenetic processes can intervene in cardiac hypertrophy and failure. A distinct DNA methylation pattern exists between patients with dilated cardiomyopathy and healthy subjects [515].

7.4.5 *Heat Shock Protein-β1*

A heart failure risk locus is linked to the chromosomal region 1p36 within the HSPB7 gene that encodes 27-kDa heat shock protein, or heat shock protein-β1 [530]. A single single-nucleotide polymorphism[6] (Arg83Gly) on the CLCNKA gene adjacent to HSPB7 that encodes the kidney-specific, voltage-gated chloride channel $ClCK_A$ (or ClCK1) is associated with the heart failure risk connected to single-nucleotide polymorphism of the HSPB7 gene. The variant $^{Gly83}ClCK_A$ chloride channel leads to a loss of function with respect to the wild-type $^{Arg83}ClCK_A$ channel.

7.4.6 *GPCR Kinase*

G-protein–coupled receptor kinases phosphorylate liganded β-adrenergic receptors that then recruit β-arrestin, uncouple from Gs proteins, and internalize, thus switching off β-adrenoceptor signaling. Whereas GRK2 can mainly operate in acute regulation of β-adrenoceptor signaling, GRK5 activity that is less fast serves in long-term regulation. The Grk5 gene variant that is common in African Americans improves survival in individuals with chronic heart failure [531].[7] The GRK5 variant has a higher kinase activity than that of wild-type enzyme.

[6] In a single-nucleotide polymorphisms (SNP), a single base (A, T, C, or G) is replaced by a another base. Fifty-one exonic CLCNKA variants are identified [530].

[7] Glutamine is substituted by leucine in position 41 (Glu41Leu).

7.4.7 Cardiorenal Syndrome

The cardiorenal syndrome combines cardiac and renal dysfunction that amplifies progression to failure of each organ. A vicious circle develops with activation of the renin–angiotensin axis, NO–ROS imbalance, stimulation of the sympathetic nervous system, and inflammation, all these factors synergizing with positive feedback in the cardiorenal connection [532].[8] Bartter and Gitelman syndromes are autosomal recessive renal disorder. They are characterized by hypokalemia, hypomagnesemia, hypochloremia, metabolic alkalosis, and normal to low blood pressure.

The *Bartter syndrome* is also associated with hyperreninemia independently of extracellular volume, hyperaldosteronemia, and hypercalciuria. It results from loss-of-function mutations in the genes that encode NKCC2, $K_{IR}1.1$ (ROMK), $ClCK_B$, and barttin in the thick ascending limb of the loop of Henle (Table 7.2).

In the loop of Henle, sodium chloride is taken up apically by the combined activity of Na^+–K^+–$2Cl^-$ cotransporter NKCC2 and potassium channel ROMK ($K_{IR}1.1$). Chloride ions exit from the cell through basolateral member of the CLC family of voltage-gated chloride channels, $ClCK_B$, or ClCK2 channel. Barttin, encoded by the BSND (Bartter syndrome, infantile, with sensorineural deafness) gene, is a β subunit of the $ClCK_A$ (ClCK1) and $ClCK_B$ channels [533]. The former resides in both the apical and basolateral plasma membrane of the thin ascending limb of Henle's loop. It is responsible for chloride permeability. The latter (ClCK2) lodges in the basal plasma membrane of cells of the thick ascending limb of Henle's loop, distal convoluted tubule, connecting tubule, and intercalated cells of collecting ducts. Barttin–ClCK heteromers are involved in renal salt (NaCl) reabsorption. Barttin regulates permeation and gating of ClCK channels. In fact, barttin–ClCK heteromers enable ClCK current due to their insertion into the plasma membrane. Type-4 Bartter syndrome caused by the R8L barttin mutation results from aberrant intracellular localization [534].

The *Gitelman syndrome*, another inherited salt-wasting disorder, causes hypocalciuria in addition to abnormal reabsorption of Na^+ and Cl^- ions from the distal convoluted tubules through thiazide-sensitive Na^+–Cl^- symporter (NCCT cotransporter or SLC12a3).[9]

---

[8] Increased production of reactive oxygen species and lower availability of nitric oxide provoke sympathetic activation, renin release, and inflammation. Stimulation of the renin–angiotensin axis causes sympathetic activation, ROS production, and NFκB-mediated inflammation. Sympathetic activation leads to ROS production, renin release, and macrophage stimulation. Inflammation causes secretion of renin, ROS, and cytokines.

[9] Thiazides are diuretics used in hypertension and edema. They inhibit the Na^+–Cl^- cotransporter in the distal convoluted tubule of the nephron, thereby impeding reabsorption of sodium and chloride, hence decreasing extracellular fluid volume. On the other hand, it increases calcium reabsorption in the proximal convoluted tubule upon thiazide-induced hypovolemia, but not through the TRPV5 channel. They are direct vasodilators, as they preclude the RhoA–RoCK pathway in vascular smooth myocytes.

Table 7.2 Different types of Bartter syndrome.. A symporter is also called a cotransporter

Bartter syndrome Type	Gene mutations	Ion carrier
Type 1	SLC12A2	Na^+–K^+–$2Cl^-$ symporter (NKCC2)
Type 2	KCNJ1	K_{IR}1.1 (ROMK)
Type 3	CLCNKB	$ClC K_B$
Type 4	BSND	Barttin
Type 5	CASR	Calcium-sensing receptor
Gitelman syndrome	SLC12A3	Na^+–Cl^- symporter (NCCT)

BSND Bartter syndrome, infantile, with sensorineural deafness, a $ClCK_A$ subunit

7.5 Cardiac Energetics

Energy metabolism in cardiomyocytes is related to cardiac contractility. The heart has a limited fuel-storing capacity. In normal conditions, ATP is efficiently and quickly produced mainly from circulating free fatty acids and, to a lesser degree, from glucose, but fatty acids require more oxygen than glucose to produce an equivalent amount of ATP. In heart failure, the ability of mitochondria to synthesize ATP ensuring a normal cardiac function is impaired.

7.5.1 Modeling

The transition between the compensatory phase and heart failure corresponds to a reduction in the total adenine nucleotide pool (TAN) of about 30 % [535]. At given values of total adenine nucleotide and the total exchangeable phosphate pool during defective cardiac remodeling, the creatine pool attains a value that is associated with optimal ATP hydrolysis potential.

A computational model incorporates the total adenine nucleotide pool, total creatine pool (Cr_T), and total exchangeable phosphate pool (TEP) [535]:

$$\begin{aligned}
\mathrm{Cr_T} &= V_c \alpha_{W_c} \left(\left[\mathrm{Cr^P}\right]_c + [\mathrm{Cr}]_c \right); \\
\mathrm{TEP} &= V_c \alpha_{W_c} \left(2\,[\mathrm{ATP}]_c + [\mathrm{ADP}]_c + [\mathrm{P_i}]_c + \left[\mathrm{Cr^P}\right]_c \right) \\
&\quad V_m \alpha_{W_{mims}} \left(2\,[\mathrm{ATP}]_{mims} + [\mathrm{ADP}]_{mims} + [\mathrm{P_i}]_{mims} + \left[\mathrm{Cr^P}\right]_{mims} \right) \\
&\quad V_m \alpha_{W_{mm}} \left([\mathrm{ATP}]_{mm} + [\mathrm{P_i}]_{mm} + [\mathrm{GTP}]_{mm} \right) \\
\mathrm{TAN} &= V_c \alpha_{W_c} \left([\mathrm{ATP}]_c + [\mathrm{ADP}]_c + [\mathrm{AMP}]_c \right) \\
&\quad V_m \alpha_{W_{mims}} \left([\mathrm{ATP}]_{mims} + [\mathrm{ADP}]_{mims} + [\mathrm{AMP}]_{mims} \right) \\
&\quad V_m \alpha_{W_{mm}} \left([\mathrm{ATP}]_{mm} + [\mathrm{ADP}]_{mm} \right)
\end{aligned} \tag{7.2}$$

where V_c and V_m are volume of cytoplasm and mitochondria, respectively, α_w water fractions in cytoplasm, mitochondrial intermembrane space (subscript $mims$), and mitochondrial matrix (subscript mm) and P_i inorganic phosphate. Myocardial density, composition, and enzyme activities are assumed to remain unchanged with respect to the normal heart.

In failing hearts, basal ATP, phosphocreatine (Cr^P), TAN, TEP, as well as Cr_T), decrease steadily as disease progresses. The model predicts a *tipping point* of the free energy of ATP hydrolysis at maximal myocardial oxygen consumption that coincides with the transition to severe cardiac dysfunction. The tipping point is associated with an approximately 30 % TAN reduction. Both increases and decreases of the creatine pool are predicted to diminish energetic state unless accompanied by appropriate simultaneous changes in other implicated pools.

7.5.2 Regulators of Cardiac Energetics

Angiogenesis does not fit the required oxygen supply for myocardium growth. Pressure overload initially stimulates production of hypoxia-inducible factor-1α involved in angiogenesis, but subsequent P53 upregulation impedes HIF1α activity in later stages of maladaptive remodeling [536]. Creatine kinase generates ATP as it reversibly converts phosphocreatine and ADP to creatine and ATP when energy is quickly needed (Vol. 1, Chap. 4. Cell Structure and Function). Phosphocreatine thus provides ATP storage.

The regulation of coupled cardiac energy supply and metabolism becomes defective in the failing heart, due to mitochondrial dysfunction and altered substrate use. In the failing heart, creatine kinase has reduced activity, which leads to dysfunctional ATP storage, and glucose oxidation rises, as PGC1α synthesis decays.

The NOS3–NO axis intervenes in basal and adaptive mitochondrial genesis as well as in the regulation of mitochondrial turnover in cardiovascular cells. Production of proteic subunits of ^{ETC}complex-I to -III and -V of the mitochondrial electron transport chain as well as voltage-dependent anion channel and manganese superoxide dismutase depend on NOS3 [47]. NOS3 supports mitochondrial adaptation to exercise and mitochondrial fusion and represses mitochondrial fission.

7.5.3 Mitochondrial Dysfunction

Mitochondrial dysfunction is a major factor in heart failure. Mitochondria are energy-producing organelles. Myocardial activity relies on energy generated in mitochondria by oxidation of fatty acids and carbohydrates to generate acetyl-CoA that enters the tricarboxylic acid cycle. This cycle yields an electron source for the electron transport chain responsible for oxidative phosphorylation and associated ATP synthesis.

The electron transport chain is composed of four major complexes:

1. NADH–CoQ reductase (^{ETC}complex-I),
2. Succinate–CoQ reductase (^{ETC}complex-II),
3. Ubiquinol–cytochrome-C reductase (^{ETC}complex-III), and
4. Cytochrome-C oxidase (^{ETC}complex-IV).

Electron transfer between these complexes is ensured by interactions between electron carriers coenzyme-Q and cytochrome-C. Like ^{ETC}complex-I, -III, and -IV (but not ^{ETC}complex-II), ATP synthase (^{ETC}complex-V) is able to form supercomplexes by oligomerization [537]. *Respirasomes* refers to these supercomplexes, such as the supercomplex made by ^{ETC}complex-I, -III dimer, and -IV (respirasome-I/III$_2$/IV). These supercomplexes reduce electron diffusion distance.

Heart failure can be linked to mitochondrial cytopathies [538]. Clusters of respiratory electron chain complexes lack and the amount of respirasomes decreases in both subsarcolemmal and interfibrillar mitochondria. Although the amount of free ^{ETC}complex-I and -III increases in subsarcolemmal mitochondria, incorporation of these complexes within respirasome-I/III$_2$/IV is reduced.

Peroxisome-proliferator–activated receptor-γ coactivator-1α increases oxidative phosphorylation to match the energy demand of cardiac growth during heart development and in response to exercise. Factor PGC1α in fact stimulates mitochondrial genesis, as it increases the production of nuclear respiratory factors NRF1 and NRF2 (GABP) that regulate mitochondrial transcription factor-A, and estrogen-related receptor-α (NR3b1), which coordinate the expression of genes encoding mitochondrial proteins. Moreover, PGC1α also controls the transcription of genes encoding proteins involved in fatty acid oxidation via coactivation of PPARα (NR1c1) and ERRα.

Damaged cardiomyocytes liberate DNA from their degraded mitochondria, thereby triggering inflammation and heart failure. Injured mitochondria are usually carried to lysosomes to be degraded by autophagy. Nuclear DNA is methylated on cytidine^{P}–guanosine oligodeoxynucleotide (C^{P}G) motifs. Toll-like receptor TLR9 senses mitochondrial DNA with unmethylated C^{P}G motifs and triggers the synthesis of proinflammatory cytokines [539, 540]. However, accumulation of mitochondrial DNA that escapes methylation activates the TLR9 receptor, thereby triggering inflammation by cardiac cells, especially cardiomyocytes, provoking myocarditis and dilated cardiomyopathy, and exacerbating heart failure [540]. Moreover, mitochondrial DNA released from dying cells into the extracellular space primes a TLR9-dependent inflammation by some immunocytes [539]. On the other hand, lysosomal deoxyribonuclease DNase2 protects the heart, at least in mice, from inflammation by degrading the mitochondrial DNA [540].

7.6 Remodeling

Cardiac remodeling is a progressive response of the heart to insults, such as ischemia, pressure overload, inflammation, and injury, among others. Fibroblast proliferation and collagen secretion engender fibrosis.

7.6.1 Transverse Tubule Remodeling in Heart Failure

Transverse (T) tubules remodel during the transition from compensated hypertrophy to heart failure, thereby affecting Ca^{2+} handling [541]. Calcium handling dysfunction in cardiomyocytes results from a loss of synchronous Ca^{2+} release.

Transverse tubule remodeling comprises [541]:

1. A reduction in T-tubule density;
2. A disorganization of the orderly T-tubule network;
3. A loss of transverse elements, but a gain in longitudinal elements; and
4. A T-tubule dilation.

Similar T-tubule alterations are also detected in end-stage dilated or ischemic cardiomyopathy.

The T-tubule network undergoes progressive deterioration from compensated hypertrophy to early and then advanced heart failure and spreads from the left to the right ventricle [541]. T-tubule remodeling leads structural, electrical, and signal transduction alterations that contribute to the progression of cardiac failure and arrhythmias.

Impaired excitation–contraction coupling in chronic heart failure with systolic and diastolic dysfunction is characterized by a smaller Ca^{2+} release amplitude and slow kinetics because of [541]:

- A loss of JP2 synthesis;
- A T-tubule loss and/or jumble and associated disorganization between $Ca_V1.2$ and RyR2 channels;
- A reduced Ca^{2+} influx through $Ca_V1.2$ channels;
- A slow secondary calcium-induced Ca^{2+} release by uncoupled leaky RyR2 channels;
- A spatiotemporal dyssynchrony of cytosolic Ca^{2+} transients;
- A diminished sarcoplasmic reticulum Ca^{2+} content following SERCA downregulation;
- Slower removal kinetics during diastole by T-tubular Na^+–Ca^{2+} exchanger and hence impaired myocardial relaxation;
- Elevated cytosolic Na^+ concentration possibly related to the late Na^+ current and/or increased activity of the Na^+–H^+ exchanger that, in combination with an upregulation of the Na^+–Ca^{2+} exchanger, favors the reverse NCX mode to import Ca^{2+} during systole and decrease Ca^{2+} efflux during diastole; and
- Metabolic inhibition.

Structural disorganization of the T-tubule system disturbs the propagation of the action potential from the cell surface to T-tubule network. In addition, T-tubule remodeling changes the distribution and organization of ion carriers, thereby altering the shape and duration of action potentials. It also causes redistribution and loss of β-adrenergic receptors of T tubules. In heart failure, dysregulation of junctophilin-2 contributes to defective Ca^{2+}-induced Ca^{2+} release and hence excitation–contraction coupling with mislocalization of $Ca_V1.2$ and RyR2 responsible for delayed weak Ca^{2+} transients, reduced systolic Ca^{2+} release, Ca^{2+} diastolic leakage, and elevated RyR2-mediated spark frequency [9].[10]

Calcium leak through RyR2 can activate the PP3–NFAT pathway. The Ca^{2+}-sensitive phosphatase PP3 activates NFAT that then stimulates hypertrophic gene transcription [9]. Calcium leak can also trigger Ca^{2+}–calmodulin-dependent kinase CamK2 activity and calmodulin-dependent dissociation of histone deacetylase HDAC4 from myocyte enhancer factor MEF2, thereby initiating hypertrophic signaling.

In addition, sustained elevated cytosolic Ca^{2+} levels provoke μ-calpain autocatalytically activates and processes Jph2 into a diffusible fragment, thereby, further raising the Ca^{2+}-induced Ca^{2+} release deficiency.

A defective kinesin-driven cytoskeletal transport is also involved in Jph2 mislocalization in the T-tubule membrane in failing hearts [9]. Loss of microtubule dynamics, densification, and destabilization are linked to adverse T-tubular remodeling and loss of cardiac contractility.

The sympathetic control of myocardial function is exerted predominantly via β1- and β2-adrenoceptors. Response to β-adrenoceptor stimulation is depressed in heart failure, irrespective of etiology. The loss of Z grooves and T tubule openings as well as T tubules provokes a regional repositioning of functional β-adrenoceptors. The β2AR–cAMP axis relocalizes from T tubules to the crests and produces diffuse cAMP signaling that propagates throughout the entire cytosol similarly to the β1AR signal [543]. Decreased β AR responsiveness is associated with uncoupling of the β2ARs from localized PKA pools that are responsible for the compartmentalization of β2AR–cAMP signals.

Mechanical load have detrimental or beneficial effects on T-tubule structure and function. Adverse T-tubular remodeling can be reversed [543, 544].

In addition, The SK channel that contribute to repolarization is upregulated in patients with heart failure and can have an antiarrhythmic effect. In ventriculomyocytes, their activation requires Ca^{2+} release from the sarcoplasmic reticulum. In adult rat ventriculomyocytes with overexpressed SK2 channels that localize to both external

[10] Junctophilin-2 is a structural protein responsible for docking of transverse (T) tubules to the junctional sarcoplasmic reticulum. In mice, cardiac-specific overexpression of junctophilin-2 raises the junctional coupling area between T tubules and the sarcoplasmic reticulum and expression of Na^+–Ca^{2+} exchanger, other excitation–contraction coupling protein levels remaining nearly constant [542]. The resting cardiac function and Ca^{2+} handling are similar to those in control mice. However, overexpression of junctophilin-2 stabilizes the T-tubule network, protects the heart upon pressure overload and prevents the progression from hypertrophy to heart failure.

sarcolemma and T-tubules, Ca^{2+} waves triggered by global spontaneous sarcoplasmic reticulum Ca^{2+} release activate the SK channel and shortens action potential duration, thereby reducing delayed after depolarizations caused by spontaneous Ca^{2+} waves [545]. The outward SK current disappears upon complete depletion of sarcoplasmic reticulum Ca^{2+} content in response to depolarizations, despite intact Ca^{2+} influx through $Ca_V1.2a$ channels.

7.6.2 Cell Apoptosis and Necrosis

Promoters of cell apoptosis and necrosis are not balanced by cardiomyocyte survival mediators. Apoptosis can be initiated by reactive oxygen species, angiotensin-2, sympathetic stimulation, and cytokines, whereas prosurvival pathways involve JaK–STAT and PI3K–PKB signaling. Necrosis that contributes to heart failure is associated with mitochondrial damage [536].

7.6.3 Cardiac Hypertrophy

The heart responds to impairment by progressive hypertrophy to preserve stroke volume at the expense of the ejection fraction.

Factor MyC is not expressed in the adult heart under normal physiological conditions. However, it can be quickly upregulated in the adult myocardium in response to stress, leading to cardiomyocyte hypertrophy, but not myocardium hyperplasia [546].

MicroRNAs miR133 and miR208 have antihypertrophic and prohypertrophic effects, respectively.

7.6.4 Cardiac Fibrosis

Cardiac fibrosis due to pressure overload is affected by many compounds, such as angiotensin-2, endothelin-1, connective tissue growth factor, transforming growth factor-β, and periostin, among others. Connective tissue growth factor is activated by several agents such as transforming growth factor-β. It promotes fibroblast proliferation, adhesion, and migration, as well as extracellular matrix formation. Angiotensin-2 and aldosterone stimulate collagen formation and fibroblast proliferation in the cardiac wall.

7.6.5 Adrenoceptor Compartmentalization

Membrane remodeling in failing cardiomyocytes disrupts compartmentalization of β1- and β2-adrenergic receptors. In normal cardiomyocytes, β1-adrenoceptors resides at the plasmalemmal domed crest (between the Z groove in which transverse tubules open), whereas β2-adrenoceptors localize to T-tubules, thus segregating signaling [547].

Transverse (T)-tubules are plasmalemmal deep invaginations at Z lines that contain many mediators coupling excitation (i.e., plasma membrane depolarization) to contraction (i.e., calcium-mediated myofilament shortening), especially calcium-induced calcium release machinery with sarcolemmal calcium channels and sarcoplasmic reticulum calcium-release ryanodine receptory channels.

β2-Adrenoceptors generate spatially restricted cAMP messenger (focused cAMP signaling). This signal is then inactivated by phosphodiesterases located in T-tubules. On the other hand, β1-adrenoceptors produce cAMP that diffuses toward sarcomeres to enhance contractility (broad cAMP signaling).

In heart failure, β1-adrenoceptors provoke cardiomyocyte remodeling and programmed death. Plasma membrane remodeling in heart failure causes T-tubule and Z-groove loss [548] (whatever etiology, ischemia, idiopathic dilation, and hypertrophic cardiomyopathy) and displaces β2-adrenoceptors to crests [547].

Moreover, adrenoceptor downregulation in human heart failure is specific to the β1-adrenoceptor; the relative proportion of β1-adrenoceptors with respect to β2-adrenoceptors decreases from about 4:1 in normal hearts to about 3:2 in heart failure. Furthermore, the β1-adrenoceptor is desensitized and thus uncouples from the Gs subunit. In addition, internalization of β2-adrenoceptors is associated with switching from the Gs to Gi subunit.

7.6.6 Sarcomeric Constituents

In rodent models of heart failure, shift in expression ratio from α- to β-myosin heavy chains alters the contractility. α-myosin heavy chain has faster cross-bridge kinetics and generates less tension than fetal β-myosin heavy chain [536].

In human fetuses, right and left ventricles contain small relative amounts of MHCα (on average $<5\,\%$ of total MHC) [549]. The ventricular level of MHCα decreases slightly between 7 and 12 weeks of gestation. On the other hand, fetal atria possess predominantly MHCα (on average $>95\,\%$), MHCβ being detected in most explored samples.

In adult humans, right and left ventricles have small content of MHCα (1–10 %), whereas the left atrium contains mainly MHCα ($\sim 90\,\%$ of total MHC) [549]. Hence, human cardiac ventricles contains mostly β-myosin heavy chain.

Failing right and left ventricles express a significantly lower level of MHCα whereas the relative amount of MHCα in the left atria of individuals with dilated or ischemic cardiomyopathy equals approximately 50 %. In cardiac failure, the gene encoding β-myosin heavy chain is re-expressed and the synthesis α-myosin heavy chain is further reduced.

7.7 Anemia

Anemia is common in heart failure [550]. Excessive *eryptosis*, i.e., RBC death following RBC shrinkage, membrane blebbing, activation of peptidases, and breakdown of phosphatidylserine asymmetry, among other features of apoptosis causes an increased turnover of erythrocytes and anemia. Eryptosis-activated Ca^{2+}-permeable cation channels enables Ca^{2+} entry that activates Ca^{2+}-sensitive K^+ channels, igniting cell shrinkage.

Oxygen transport from lungs to tissues is influenced not only by the production rate of red blood capsules as well as their hemoglobin content, but also their fate in blood circulation, i.e., environments to which they are exposed in the vascular lumens. Eryptosis prevent spontaneous hemolysis of aged and defective red blood capsules. Oxidative stress, energy depletion, and osmotic imbalance, which are events happening during cardiac failure, increase eryptosis [550]. In addition, eryptosis occurs in sepsis, malaria, iron deficiency, hepatolenticular degeneration,[11] and hemolytic uremic syndrome (HUS) [550].[12]

During eryptosis, Ca^{2+} entry stimulates phospholipid movement in the plasma membrane [550]. Phosphatidylserine moves from the inner leaflet of the cell membrane to the outer leaflet of the phospholipid bilayer and becomes a suicide marker [551]. This displacement actually calls for RBC clearance, thereby reducing RBC lifespan. In healthy individuals, the number of circulating surface phosphatidylserine+ RBCs is negligible. Phosphatidylserine-exposing erythrocytes are rapidly cleared from the blood and processed in the spleen. Phosphatidylserine indeed binds to its receptor on macrophages that engulf phosphatidylserine+ cells. On the other hand, clearance of exposed phosphatidylserine+ RBCs and eryptosis rate rise in heart failure.

Phosphatidylserine exposure on the RBC surface enables their interactions with endotheliocytes [550]. Adherence of eryptotic red blood capsules to capillary endothelium impedes blood flow and provokes bone marrow dysfunction. Phosphatidylserine on red blood capsules can tether to internalizing α_V-integrins on

[11] Hepatolenticular degeneration, or Wilson's disease, is an autosomal recessive genetic disorder due to mutations in the ATP7B gene that encodes Cu^{2+}-transporting ATPase β-polypeptide. Copper accumulates mainly in the brain and liver. It causes acute liver failure, often with a hemolytic anemia, as well as cardiomyopathy.

[12] Hemolytic uremic syndrome is characterized by hemolytic anemia, acute kidney failure, and thrombocytopenia.

endotheliocytes. Endotheliocytes can then be committed in erythrophagocytosis [551].

In addition, externalized phosphatidylserines can be clustered and then shed away from the cell surface. Shed phosphatidylserine-rich vesicles (size 100–200 nm) can be involved in thrombosis, activation of complement, and consumption of nitric oxide, among other processes [551].

Therefore, anemia in heart failure results from elevated eryptosis. Moreover, exposed phosphatidylserine− RBCs are more susceptible to eryptosis when subjected to oxidative stress, hyperosmotic shock, and energy depletion.

7.8 Cardiac Cachexia

Cachexia means weight loss and body wasting, in particular wasting of muscular mass and total adipose tissue mass and a reduction in bone mineral density. It results from an imbalance between reduced anabolism and augmented catabolism following abnormal neurohormonal control and activated proinflammatory cytokines. This complication happens during evolution of many chronic diseases, such as cancer, acquired immune deficiency syndrome, chronic obstructive pulmonary disease, and advanced chronic heart failure.

Cardiac cachexia is characterized by wasting of skeletal muscle in the lower limbs without loss of total body weight. Patients with cachectic chronic heart failure, a risk factor of mortality, have [552]: (1) an abnormal ratio between growth hormone and anabolic mediator insulin-like growth factor IGF1 (augmented plasma GH level and normal to low IGF1 level); (2) elevated plasma concentrations of adrenaline, noradrenaline, and catabolic hormone cortisol; (3) reduced level of the anabolic steroid dehydroepiandosterone; and (4) abnormal plasma concentrations of renin (hence abundant angiotensin-2) and aldosterone; and (5) high concentrations of proinflammatory cytokines (TNFSF1, IL1, and IL6). The plasma concentrations of TNFSF1 and its soluble receptors are associated with a bad short-term prognosis. Myostatin, or growth and differentiation factor GDF8, a secreted member of the TGFβ superfamily, is a strong inhibitor of skeletal muscle growth, modest abrogator of cardiomyocyte growth, and mediator of skeletal muscle atrophy in mice with heart failure [552]. It is synthesized as an inactive precursor proteic homodimer, prepromyostatin. A first cleavage generates promyostatin that is then cleaved by furin into a $_{CT}$fragment that dimerizes to form active myostatin or binds to $_{NT}$prodomain to create the latent myostatin complex. Myostatin in this complex is inhibited by the prodomain. Myostatin is activated from the latent complex after cleavage of the prodomain by BMP1–tolloid (TlL1/TlL2) metallopeptidases. It then binds to the activin receptor-2B (AcvR2b) of the TGFBR2 category, which then heterodimerizes with either the ALK5 or ALK4 receptor of the TGFBR1 category, thereby leading to phosphorylation (activation) of SMAD2 and SMAD3 factors. Myostatin precludes PKB and TOR activity.

Myostatin is mainly expressed in skeletal muscle (at a higher extent in fast muscle fibers with predominant glycolysis than in slow muscle fibers with a principally oxidative metabolism), and to a lower extent, in the heart (chiefly in fetal and postnatal cardiomyocytes of the left ventricle and adult cardiomyocytes subjected to hypertension and in ischemic or dilated cardiomyopathy) and adipose tissue. Its expression is downregulated by regular exercise training. Myostatin is mainly secreted as promyostatin by skeletal myocytes, which is sequestered in the extracellular space by latent TGFβ-binding protein LTBP3. However, in plasma, myostatin exists as a latent myostatin complex [552]. In the heart, BMP1 releases and activates myostatin from its latent complex.

In pathological conditions, the myocardium produces and secretes myostatin into the circulation [552]. Myostatin can also induce cardiac fibrosis, in addition to cardiac cachexia.

7.9 Neurohumoral Agents

Neurohumoral responses consist of the activation of the sympathetic nervous system and renin–angiotensin axis, and increased release of antidiuretic hormone (vasopressin) and natriuretic peptides.

Involved neurotransmitters of the sympathetic nervous system and hormones (adrenaline and noradrenaline) initially support compensation aimed at lowering hypoperfusion. Ultimately sustained signaling aggravates heart failure by increasing afterload.

Neuropeptides, neurohormones, cytokines, NOS2, reactive oxygen and nitrogen species, matrix metallopeptidases, and nuclear poly^{ADP}ribose polymerase are involved in the progression to heart failure. In addition to elevated activity of the sympathetic nervous system, renin–angiotensin axis, ANP and BNP peptides, and vasopressin, heart failure is also characterized by augmented levels of endothelin-1 and cytokines, such as transforming growth factor-β, interleukins, interferon-γ, and tumor-necrosis factor-α.

7.9.1 High Sympathetic Signaling

Heart failure is associated with sympathoexcitation caused by an elevation in central sympathetic output, elevated noradrenaline concentrations in the blood and tissues, reduced sensitivity of arterial baroreflex and cardiopulmonary reflexes, and enhanced sensitivity of peripheral chemoreflex and cardiac sympathetic and skeletal muscle afferent reflexes, in addition to changes in the sensitivity of central sympathetic neurons.

Circulating and local messengers (aldosterone, angiotensin-2, atrial natriuretic peptide, cytokines, endothelin-1, nitric oxide, prostaglandins, reactive oxygen species, and vasopressin, among others) modulate sympathetic output in the central nervous system.

Intermittent physiological and chronic pathological stimuli do not activate the same pathways. Increased sympathetic nerve activity launched by arterial baroreflex and chemoreceptors relies on an abnormal hypersensitivity of the carotid body in heart failure. Whereas expression of intracellular superoxide dismutase isoform $^{(Cu,Zn)}$ SOD is hindered in heart failure, increased production of superoxide by NADPH oxidase mediates angiotensin-2-augmented carotid body sensitivity by suppressing outward voltage-gated K^+ currents in glomus cells [553]. Oxidative stress causes chemoreceptor hyperexcitability.

In the sheep, sympathetic nerve activity to the heart is increased earlier and to a greater extent in heart failure than that to kidneys. In normal condition, the resting level of cardiac sympathetic nerve activity is set at a lower level than that of renal sympathetic nerve activity. In heart failure, resting levels of sympathetic nerve activity to both heart and kidneys are close to their maxima [554]. In addition, sensitivity of control by arterial baroreflex of cardiac sympathetic nerve activity is significantly higher than that of renal sympathetic nerve activity in both the normal state and heart failure. Impaired control of cardiac sympathetic nerve activity by arterial baroreflex results from depressed parasympathetic control. Whereas in the rabbit model of heart failure the baroreflex control of renal sympathetic nerve activity, in the sheep model the sympathetic excitation is not caused by desensitization of the arterial baroreflex.

7.9.1.1 Adrenoceptor Signaling and the Kinase–Phosphase Balance

β-adrenoceptor signaling relies on kinases antagonized by phosphatases. Imbalance between kinase and phosphatase effect can be involved in heart failure as well as in atrial and ventricular arrhythmias (Table 7.3). Failing heart is characterized by β AR desensitization and uncoupling from stimulatory Gs protein and elevated PP1 level and activity.

Kinases Kinases downstream from G-protein–coupled receptors control the short-term cardiac dynamics. Involved kinases include PKA, PKC, PKG, CamK, and MLCK.

Sarcomeric proteins are substrates of kinases and phosphatases that are not only activated relatively remotely by plasmalemmal receptors, but also locally. Kinases and phosphatases are docked at sarcomeric sites, especially at the Z-disk protein network and its linkage to costameres [555].

Protein Kinase-A The elevation of cytosolic Ca^{2+} concentration is sensed by troponin-C that binds to Ca^{2+} ion and undergoes a conformational change, thereby dissociating troponin-I from actin and triggering cross-bridges between sarcomeric actin and myosin filaments. Cardiac troponin-T anchors PKA regulatory subunits PKA_r1 and PKA_r2 [556].

The PKA heterotetramer is formed of two catalytic subunits kept in an inactive conformation by two regulatory subunits that facilitate tethering of the holoenzyme

Table 7.3 β-Adrenoceptor signaling and involved kinases and phosphatases. . Gs-coupled β1-adrenoceptor is the most abundant subtype in the heart. During diastole, when the cytosolic calcium concentration is low, cardiac troponin-I prevents the interaction between actin and myosin filaments. This inhibition is relieved when cytosolic calcium rises and binds to cardiac troponin-C. Phosphorylated cTnnI raises the rate of Ca^{2+} dissociation from cTnnC (positive lusitropy); bisphosphorylation is needed to enhance myofilament Ca^{2+} sensitivity with sarcomere length change. Phosphorylation of phospholamban by PKA improves Ca^{2+} reuptake by the sarcoplasmic reticulum. Upon β-adrenoceptor stimulation, activated PKA increases the rates of relaxation and contraction; it thus has chronotropic, inotropic, and lusitropic effects. (Source: [413])

Kinase	Antagonists	Substrates	Effect
PKA	PP1/2;	Ca_V1.2a (Ser1928)	Inotropy +
	PDE3/4 via	RyR2 (Ser2808)	Inotropy +
	cAMP	Phospholamban (Ser16)	Lusitropy +
			Chronotropy +
		Troponin-I (Ser23/24)	Lusitropy +
		cMyBPc (Ser282)	Chronotropy +
CamK2	PP1	RyR2 (Ser2814)	Inotropy +
		Phospholamban (Thr17)	Lusitropy +

in the vicinity of relevant substrates, thereby ensuring a rapid response to cAMP signaling.

β1-adrenoceptor is stimulated by neuronally released and circulating catecholamines. As it is coupled to Gs subunit, it activates adenylate cyclases that produces cyclic adenosine monophosphate.

In pacemaker cells, cAMP shifts the voltage dependence of the inward pacemaker current to more depolarized potentials, thereby increasing the cardiac frequency. In ventriculomyocytes, cAMP excites protein kinase-A. Phosphodiesterases regulate the cAMP concentration within cells.

In fact, both α- and β-adrenoceptor agonists induce an inotropic response in the adult heart. They promote phosphorylation of several regulators, such as myosin-binding protein-C, troponin-I, and phospholamban. Phosphorylation of cTnnI and cMyBPc controls of sarcomeric function upon β AR stimulation. The insertion of sequences with phosphorylation motifs is unique to these cardiac variants [555].

A *phosphorylation hierarchy* exists both for kinase type and, for a given kinase type, for phosphorylation site. For example, cTnnI has an N-terminal extension that houses Ser23 and Ser24 residues. The latter is more rapidly phosphorylated by PKA, but both sites must be phosphorylated to depress sarcomere sensitivity to Ca^{2+}, that is, to enhance the inactivation rate for Ca^{2+} binding to the Ca^{2+}-binding cTnnC subunit (positive lusitropy). Cardiac MyBPc has also a unique insertion at its N-terminus that has multiple phosphorylation sites. Phosphorylation by PKA of both cTnnI and cMyBPc reduces the time required to adapt the cycle to increased frequency (positive chronotropy).

In cultured neonatal rat cardiomyocytes, α AR stimulation raises cMyBPc phosphorylation, whereas β AR activation decreases it [557]. The amount of monophosphorylated cMyBPc depends on α AR activation, but not those of di- and triphosphorylated cMyBPc. The phosphorylation of cTnnI and Pln results from both α and β AR stimulation, as in adult hearts. However, β AR activation is linked to the diphosphorylated cTnnI state, whereas α AR stimulation is associated with the tetraphosphorylated state. β AR activation does not alter calcium transients. Both cTnnI and Pln play a key role in modulating the contractile function in neonatal rat cardiomyocytes in response to catecholamines than does cMyBPc. On the other hand, cMyBPc have a structural role in stabilizing myosin filament assembly rather than influencing cross-bridge formation in developing hearts.

Cyclic AMP-dependent protein kinase PKA targets the sarcomeric constituents troponin-I and myosin-binding protein-C. Phosphorylation of cTnnI by PKA decreases myofilament Ca^{2+} sensitivity whatever the sarcomere length, but enhances the impact of sarcomere length change on Ca^{2+} sensitivity (*length-dependent activation* [558].[13] Phosphorylation of cMyBPc by PKA does not alter calcium responsiveness, but lowers the spacing between thick and thin myofilaments at all sarcomere lengths.

Protein Kinase-C G-protein coupled α1-adrenoceptors activate PKC that phosphorylates the sarcomeric proteins cTnnI, cTnnT, MyRLC, cMyBPc, and titin. The PKC isoforms expressed in the human myocardium encompass PKCα, PKCβ1, PKCβ2, PKCδ, and PKCζ.

PKC crossphosphorylates cTnnI (Ser23–Ser24) and induces the same functional effects as PKA [555]. In addition, PKC phosphorylates other cTnnI residues (Thr144, Ser43, and Ser45), thereby depressing maximal tension and desensitizing the sarcomeres to Ca^{2+} ions. PKCδ preferentially phosphorylates cTnnI at Ser23 and Ser24, whereas PKCα preferentially targets cTnnI at Ser43 and Ser45.

Endothelin causes an acute and prolonged enhancement of relaxation rate associated with a time-dependent phosphorylation of different sites on cTnnI. Its Thr144 residue is phosphorylated early and Ser23 and Ser24 are phosphorylated after prolonged administration of endothelin-1 [555].

Cardiac MyBPc is a PKC substrate. PKA and PKC may crossphosphorylate cMyBPc. MyRLC2 is another PKC substrate that may explain positive inotropic effect of acute α1-adrenoceptor stimulation.

Cellular stresses, such as hypoxia, ischemia, hypertension, altered redox environment, and exposure to cytokines, excite distinct PKC isoforms. In guinea pig

---

[13] The Frank–Starling law reflects the ability of the heart to adjust the force of its contraction to changes in ventricular filling (length-dependent myofilament activation). The increase in Ca^{2+} sensitivity upon an augmented sarcomere length can result from a decrease in interfilament spacing with an increase in length, thereby influencing cross-bridge reactivity. Phosphorylation of cTnnI (Thr143) by PKC enhances myofilament Ca^{2+} sensitivity without affecting length-dependent myofilament activation). Bisphosphorylation of cTnnI (Ser23–Ser24) improves the length-dependent increase in myofilament Ca^{2+} sensitivity [559].

hearts, ischemia activates PKCα, PKCβ2, PKCγ, and PKCζ [555]. Oxidative stress simulates PKCα, PKCβ2, and PKCζ. PKCδ and PKCϵ support hypertrophy and eventually dilated cardiomyopathy. Myocardial infarction and pressure overload raise expression of PKCζ. Both PKCα and PKCβ2 are upregulated in end-stage heart failure.

Protein Kinase-D Protein Kinase-D phosphorylates cTnnI (Ser23–Ser24) [555]. At least in rat ventricular myocytes, PKD is activated downstream from PKC, particularly PKCϵ. On the other hand, endothelin-1-induced activation of PKD is suppressed by concomitant PKA activation. In addition, PKD activation rises with increased glucose uptake and GluT4 action, independently of AMPK.

Protein Kinase-G The PKG pathway operates downstream from natriuretic peptide receptors. The NO–cGMP–PKG axis pathway may also act in signaling triggered by β3-adrenoreceptors in the control of relaxation dynamics [555].

Nitric oxide synthase NOS1 is the major endogenous source of myocardial nitric oxide that facilitates cardiac relaxation and modulates contraction [560]. The NOS1 compartmentation and its connection to binding partners in the diseased heart as well as various NOS1 splicing variants explain various mechanisms of pathogenesis.

In the healthy heart, NOS1 regulates cytosolic Ca^{2+} concentration, signaling pathways, and oxidative environment. It is upregulated during the early phase of cardiac insult. NOS1-derived NO primes ^{S}nitrosylation of various Ca^{2+}-handling proteins in the failing myocardium [560].

During disease progression, PKA-dependent phospholamban phosphorylation (Ser16) is shifted to PKG-dependent phosphorylation (Ser16). Kinase signaling to the cardiac sarcomere can involve nitric oxide that engages both cAMP- and cGMP-dependent pathways. Whereas low levels of NO increase cardiac contraction, relatively high levels of NO activate guanylate cyclase and subsequently PKG that depresses cellular mechanics [555]. In early hypertension, the NOS1–NO–cGMP–PKG axis targets troponin-I (Ser23–Ser24) and cardiac myosin binding protein-C (Ser273) [560].

Calcium–Calmodulin-Dependent Kinase CamK2 The kinase CamK2 phosphorylates phospholamban (Ser16 and Thr17), hence relieving the inhibition it exerts on SERCA2a and subsequently increasing Ca^{2+} pumping back into the sarcoplasmic reticulum. It is antagonized by the PP1 phosphatase.

Myosin Light Chain Kinase Sarcomeric myosin regulatory light chain phosphorylation modulates Ca^{2+}-dependent troponin regulation of contraction. Phosphorylation of cardiac MyRLC increases the mobility of myosin cross-bridges such that they move away from the myosin filament surface toward actin filaments as well as the number of cross-bridges entering the contractile cycle, as it potentiates actin-induced phosphate release [561]. However, RLC phosphorylation in ventriculomyocytes is not a predominant contributor of positive inotropic effects of β-adrenergic agents, but is important in normal left ventricular ejection.

Cardiac Ca^{2+}–calmodulin-dependent myosin light chain kinase (cMLCK) phosphorylates MyRLC; it is the primary kinase that maintains basal MyRLC phosphorylation required for normal cardiac performance [561]. In adult mice, hearts lacking cMLCK and subsequent MyRLC phosphorylation evolve toward heart failure.

Death-Associated Protein Kinase The DAPK3 kinase targets MyRLC (Ser15) as well as $PP1_{r12a}$. It is not involved in basal MyRLC phosphorylation, but may be activated to phosphorylate RLC upon certain stimuli [561].

Rho-Activated Kinase (RoCK) The RhoA–RoCK signaling influence MyRLC phosphorylation, as it inhibits MLCP (PP1). RoCK also phosphorylate cTnnI and cTnnT [555].

Extracellular Signal-Regulated Protein Kinase The small Ras GTPase signals via ERK1 and ERK2 to ribosomal S6 kinase ^{P90}RSK that phosphorylates TnnI (Ser23–Ser24) [555].

P38MAPK Constitutive activation of P38MAPK in transgenic mouse hearts depresses cardiomyocyte mechanics without affecting Ca^{2+} fluxes [555].

P21 (CKI1a)-Activated Kinase PAK1 Downstream from Rac and CDC42 GTPases, PAK1 targets PP2 that dephosphorylates cTnnI and cMyBPc [555].

Phosphatases Activity of phosphatases relies on their production, subcellular localization, and phosphorylation of their catalytic and regulatory subunits. Their inhibitors are often active once they are dephosphorylated or phosphorylated.

In the heart, the majority of phosphatase activity is attributed to protein phosphatases PP1, PP2, and PP3. The PP1 phosphatase is the major subtype that counterbalances effects of β AR signaling.

Protein Phosphatases PP1 and Its Inhibitors The PP1 holoenzyme consists of catalytic and auxiliary subunits. The latter dictate PP1 subcellular localization, substrate specificity, and activity. The PP1 phosphatase impedes Ca^{2+} cycling and contractility in the cardiomyocyte. It restores cardiac function to the resting state after β-adrenoceptor stimulation. It counteracts the action of PKA stimulated by the β1 AR–Gs–AC–cAMP pathway.

Several inhibitors prevents its action (Table 7.4), whereas other regulators localize $PP1_c$ close to selected substrates (Tables 7.5 and 7.6).

β-adrenoceptor stimulation leads to $PP1_{r1a}$ phosphorylation and subsequent PP1 inhibition in the myocardium, hence preventing the dephosphorylation of Na^+–K^+ ATPase, phospholamban, troponin-I, and voltage-gated calcium channels [564]. This ubiquitous inhibitor precludes PP1 activity when it is phosphorylated (deactivated; Thr35) by PKA [565]. It is also phosphorylated by PKCα (Ser67), this event attenuating its inhibitory activity toward PP1 [413]. It regulates PP1 activity at the sarcoplasmic reticulum, where it is the main phosphatase dephosphorylating phospholamban that inhibits SERCA2a when it is connected to this pump. Phospholamban is phosphorylated by PKA (Ser16) and CamK2 (Thr17; positive lusitropy).

Table 7.4 Inhibitory subunits of PP1 and partners. $PP1_{r11}$ is a potent PP1 inhibitor that selectively associates with $PP1_{c\alpha}$ and $PP1_{c\gamma 1}$, but not with $PP1_{c\beta}$. It loses its inhibitory function in vitro upon phosphorylation by protein kinases PKA and PKC and casein kinase-2 [562]. Similarly, PP2 is targeted by inhibitors I_1^{PP2} and I_2^{PP2}

Agent	Other alias
PP1 inhibitors	
$PP1_{r1a}$	I1 (inhibitor-1)
$PP1_{r1b}$	DARPP32
$PP1_{r1c}$	I1L (inhibitor-1-like)
$PP1_{r2}$	I2 (inhibitor-2)
$PP1_{r11}$	I3 (Inh3; inhibitor-3)
$PP1_{r14a}$	CPI17
$PP1_{r14b}$	PHI2

Table 7.5 Substrate specifiers and partners of PP1. (Part 1; Source: [563])

Agent	Other alias
Myosin and actin targeting	
$PP1_{r12a}$	MyPT1
$PP1_{r12b}$	MyPT2
$PP1_{r12c}$	
Plasma membrane and cytoskeleton targeting	
$PP1_{r9a}$	Neurabin-1
$PP1_{r9b}$	Neurabin-2, spinophilin
RyR	
AKAP9/11	
NKCC1	
Neurofilament-L	
Nuclear membrane targeting	
AKAP1	
Spliceosomal/RNA targeting	
$PP1_{r8}$	NIPP1 (nuclear inhibitor of PP1)
$PP1_{r10}$	PNUTS
Endoplasmic reticulum targeting	
RIPP1 (ribosomal inhibitor of PP1)	
$PP1_{r15a}$	GADD34
Proteasome targeting (?)	
$PP1_{r7}$	SDS22

Table 7.6 Substrate specifiers and partners of PP1 (Part 2; Source: [563])

Agent	Other alias
Centrosome targeting	
AKAP9	
NCK2	
Microtubule targeting	
MAPT	Tau
Mitochondrial targeting	
BCL2	
Glycogen targeting	
$PP1_{r3a}$	RGL, G_M
$PP1_{r3b}$	$PP1_{r4}$, G_L
$PP1_{r3c}$	$PP1_{r5}$, PTG
$PP1_{r3d}$	$PP1_{r6}$
Other PP1 partners	
HSP5a	GRP78
PHhActR1–PHhActR14	(Phosphatase and actin regulators)
WBP11	SNP70, SIPP1 (splicing factor that interacts with PQBP-1 and PP1)
ZFYVE9/16	SARA; endofin

$PP1_{r1a}$ inactivation by dephosphorylation (Thr35) by PP2 and PP3 relieves PP1 inhibition and restores cardiac function to basal levels. Perturbations in PP1 regulation by $PP1_{r1a}$ is implicated in heart failure [565].

Phosphorylation of MyRLC is reversed by myosin light chain phosphatase (PP1). Both substrate specifiers $PP1_{r12a}$ and $PP1_{r12b}$ (expressed predominantly in striated myocytes) increase $PP1_{c\beta}$ activity, thereby decreasing MyRLC phosphorylation and cardiac contractility [561]. $PP1_{r12b}$ is the primary regulatory subunit of cardiac myosin light chain phosphatase.

In human heart failure. the concentration and activity of PP1 increase. However, desensitized β1 ARs is not necessarily associated with a diminished phosphorylation level of PKA substrates. In failing human hearts, among these substrates, the RyR2 nanoenvironment can be impoverished in PP1, PP2, and PDE4d3 enzymes [566].

The ryanodine-sensitive Ca^{2+} channel, a tetramer tethered to four FK506-binding proteins ($[RyR2]_4$–$[FKBP1b]_4$), is the major source of calcium influx into the cytosol needed for excitation–contraction coupling. Protein kinase-A phosphorylates RyR2, hence dissociating FKBP1b and augmenting the channel open probability.

In addition to FKBP1b, the RyR2 complex comprises PKA, PP1, PP2, PDE4d3, and AKAP6 [566]. Each RyR2 subunit binds to: (1) FKBP1b and AKAP6 that anchors PKA catalytic and regulatory subunits and PDE4d3; (2) PP2 and $PP2_{r3\alpha}$; and (1) PP1 and $PP1_{r9b}$.

In failing human hearts, lower amounts of PP1 and PDE4d3 connected to RyR2 potentiate the PKA action, thereby hyperphosphorylating RyR2 and depleting FKBP1b from the RyR2 complex. The resulting elevated sensitivity to Ca^{2+} ion and overactivation leads to a defective functioning that causes abnormal Ca^{2+} leak from the sarcoplasmic reticulum. Calcium overload engenders diastolic and systolic dysfunction.

Hyperphosphorylation of RyR2 coexists with hypophosphorylation of phospholamban. The activity of the antagonists PKA and PP1 differs according to the subcellular microdomains of cardiomyocytes in the failing heart.

7.9.2 Calcium

In the failing heart, defective calcium handling is characterized by: (1) a hyperphosphorylation of ryanodine receptors by protein kinase-A that causes diastolic calcium leak; (2) an impaired calcium storage in the sarcoplasmic reticulum due to downregulation of the SERCA pump; and (3) an upregulation of inhibitor phospholamban.

In other words, calcium uptake into the sarcoplasmic reticulum is impaired (*diastolic dysfunction*) due to: (1) a decline in production of isoform SERCA2a pump; (2) reduced levels of phospholamban phosphorylation; and (3) leaky ryanodine channels RyR2 (Fig. 7.1). Transient receptor potential channels TRPC1, TRPC3, and TRPC6 are involved in cardiac hypertrophy caused by pressure overload [536].

The cardiac ryanodine receptor RyR2, a Ca^{2+} release channel on the membrane of the sarcoplasmic reticulum, supplies Ca^{2+} required for myocardial contraction, thereby determining the magnitude of the heartbeat. The balance between the cytosolic and intrasarcoplasmic reticulum Ca^{2+} levels maintains the contraction–relaxation cycle. In heart failure, this role is compromised by excessive post-translational modifications of RyR2, with inadequate deactivation [567].

7.9.3 Renin–Angiotensin Axis

The renin–angiotensin axis has neural, cardiovascular, renal, and behavioral (e.g., thirst) effects. It can alter cardiac energy metabolism in heart failure.

Angiotensin-2 attenuates response to insulin and rate of glucose oxidation. On the other hand, angiotensin-2 elevates the concentration of pyruvate dehydrogenase kinase PDHK4, thereby increasing pyruvate dehydrogenase phosphorylation and favoring diastolic dysfunction [568]. Moreover, angiotensin-2 reduces the concentration of SIRT3 deacetylase, hence increasing acetylation of pyruvate dehydrogenase (PDH) and lowering its activity. The resulting drop in PDH activity combined with insulin resistance diminishes glucose oxidation, leading to cardiac function inefficiency.

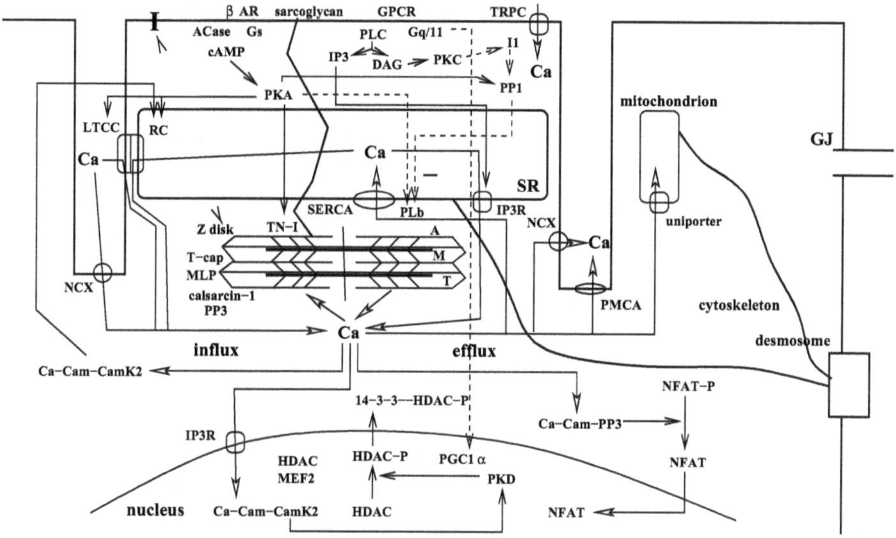

Fig. 7.1 Calcium signaling alterations in cardiomyocytes of the failing heart. (Source: [536]). Molecular components of the electrophysiological coupling, contractility, and intercellular adhesion (e.g., desmosome) are positioned correctly and anchored to each other and the cytoskeleton. Sarcolemmal integrin clusters (*I*) are connected to *Z discs* via dystrophin. Z-disc calsarcin-1 anchors *PP3* and prevents its activity. Calcium ion localizes to: (1) the cell cortex to regulate Ca^{2+} influx through RC2 (RyR2) and *TRPC* as well as β2-adrenoceptors; (2) sarcomere with actin (*A*) and myosin (*M*), regulators troponins (*TN*) and α-tropomyosin (TMy), and interlinkers (myosin-binding protein-C [MyBPc3] and titin [T]); (3) *Z disc* that couples mechanical forces with signaling via kinases and phosphatases, telethonin (titin *[T]-cap*) and muscle LIM protein (MLP) that may act as a stretch receptor; and (4) perinuclear region with IP_3R to control the nuclear pool of histone deacetylases (*HDAC*). Calcium ions interact with TN-C, changing *TN-I* conformation, releasing TMy from its inhibiting position, and allowing cross-bridges. Calcium influx occurs through $Ca_V1.2$ and RC2 (RyR2 modulated by calstabin-2), both regulated by PKA and *CamK2*. Calcium outflux happens through Na^+–Ca^{2+} exchanger (NEX), mitochondrial Ca^{2+} uniporter (*MCU*), and plasma membrane (PMCA) and SERCA2a (linked to its inhibitor phospholamban [PLb]) ATPases. Stimulated β-adrenoceptors activates *PKA* that phosphorylates $Ca_V1.2$, RC2, PLb, TN-I, MyBPc3, and titin (inotropy and lusitropy + due to improved Ca^{2+} delivery to myofilaments and then Ca^{2+} reuptake and myofilament desensitization to Ca^{2+}). In the failing heart, activity of *PKA* and SERCA2a decreases, whereas NEX functioning, RC phosphorylation and dissociation from calstabin-2, activation of Gq/11-coupled receptor signaling (exciting PKCα that represses $PP1_{r1a}$, thereby stimulating *PP1* and reducing PLb phosphorylation) increase. Activated Gq/11-coupled receptors: (1) generate IP_3, thereby causing Ca^{2+} influx and, in the nucleus, activating CamK2 and subsequently PKD, which phosphorylates *HDAC* for nuclear export; (2) reduce PGC1α action in oxidative phosphorylation. Cytosolic *Ca^{2+}–Cam–CamK2* deactivates NFAT (NFAT is activated by *Ca^{2+}–Cam–PP3*)

Angiotensin-2 and its ubiquitous AT_1 receptor modulates the sympathetic output in the central nervous system, especially in the hypothalamus and medulla, which are involved in sympathetic signaling that regulates the behavior of the cardiovascular apparatus. Angiotensin-2 level and AT_1 density in the rostral ventrolateral medulla and nucleus of the solitary tract increase in rabbits and rats with heart failure [569].

Exercise training reduces both central oxidative stress and AT_1 expression in animals with heart failure.

Angiotensin-2 stimulates ROS production via NADPH oxidase. Oxidative stress activates sympathetic neurons in the rostral ventrolateral medulla. Sustained antioxidant therapy with vitamin-C in patients with heart failure improves baroreflex.

In neurons, regulation of AT_1R transcription is redox sensitive. Several transcription factors, such as Activator protein AP1, NFκB, and CREB, as well as cRaf, activates transcription of the AT1R gene [569]. Activator protein-1 is overexpressed in the rostral ventrolateral medulla of animals with heart failure.

Receptor AT_1R stimulates the JNK–Jun–AP1 axis, hence creating a positive feedback for its own synthesis, as well as NAPDH oxidase that manufactures superoxide anion (O_2-) from which superoxide dismutase produces hydrogen peroxide (H_2O_2).

The balance between angiotensin-converting enzyme (ACE) and its homolog ACE2 may be a determinant of sympathoexcitation in heart failure. Both ACE and ACE2 are elevated in the failing heart of animals [569].

Peptidase ACE2 cleaves the octapeptide angiotensin-2 into the heptapeptide angiotensin$_{(1--7)}$ in cardiomyocytes. The latter targets Mas1 G-protein–coupled receptor. It has a positive ionotropic effect and provokes coronary vasodilation. In addition, angiotensin$_{(1--7)}$ is produced in the brain, where it regulates sympathetic activity and vasopressin release.

7.9.4 Natriuretic Peptides

B-type natriuretic peptide and proBNP$^{\Delta^{CT}}$ are used as markers in diagnosis and prognosis for patients with heart failure. Moreover, alternatively spliced intron-2-retained transcript for B-type natriuretic peptide (^{AS}BNP) is observed in failing human hearts. It lacks BNP ability to stimulate cGMP in vascular cells [570].

Although skeletal myocytes can efficiently regenerate, the regenerative capacity of the myocardium is inadequate to compensate for severe myocardium loss. Adult myocardium contains cardiac niches with a population of Isl1+, SCFR+, Sca1+ cardiac stem cells that can differentiate into cardiomyocytes, and vascular endothelial and smooth muscle cells [571]. In addition, bone marrow-derived progenitor cells could migrate to the heart and differentiate into cardiomyocytes. Endothelial progenitors that contribute to angiogenesis can also provide paracrine survival signals to cardiomyocytes. A subset of bone marrow-derived mesenchymal stem cells can differentiate into cardiomyocytes.

7.9.5 *Endothelin-1*

Endothelin-1 downregulates SERCA2 ATPase in isolated ventriculomyocytes of adult rats although it enhances bulk protein synthesis [572]. In cardiomyoyctes, ET1 binds to the ET_A receptor coupled with Gq and Gi proteins to lower cAMP production and SERCA2 gene expression by about 50 %.

7.9.6 *Relaxins*

In the cardiovascular apparatus, relaxins have beneficial effects. The prorelaxin processing enzyme proprotein convertase subtilisin/kexin PCSK1 (or simply prohormone convertase PC1) operates in the heart and vessels.

Congestive heart failure is characterized by a neurohumoral activation with elevated levels of vasoconstrictory and salt-retaining mediators (e.g., catecholamines, angiotensin-2, endothelin-1, and vasopressin) and the compensatory rise of vasodilatory and natriuretic messengers (e.g., atrial and brain natriuretic peptides and adrenomedullin). Relaxin represents another compensatory mediator [573].

Cardiomyocytes as well as interstitial cells produce relaxin in the human failing heart. The Rln1 subtype is a potent inhibitor of endothelin-1, which is the most powerful vasoconstrictor in heart failure. Moreover, in the presence of angiotensin-2, Rln1 increases endothelial ET_B receptor expression [573]. Relaxin thus promotes vasodilation via ET_B, diuresis by attenuating the renal vascular response to angiotensin-2, stimulation of atrial natriuretic peptide, collagen matrix degradation, and prevention of coronary thrombotic events by upregulating tissue plasminogen activator.

7.9.7 *Vascular Endothelium*

Subjected to various types of physical (e.g., hemodynamic stress) and chemical stimuli (change in pH, circulating hormones, cytokines, and substances released by nerves and blood cells such as platelets), endotheliocytes produce vasorelaxants (e.g., adenosine, nitric oxide, C-type natriuretic peptide, endothelium-derived hyperpolarizing factor, and prostacyclin), vasocontrictors (e.g., angiotensin-2, endothelin-1, thromboxane-A2, and isoprostanes), and regulators of vascular permeability, angiogenesis, and inflammation. Vascular endothelium maintains the balance between vasoconstriction and vasodilation, between prevention and stimulation of platelet aggregation, between blood coagulation and fibrinolysis, and between promotion and repression of smooth myocyte proliferation and migration. Disturbances of these controlled balances lead to endothelial dysfunction. Nitric oxide primes left ventricle relaxation (positive lusitropy) and raises left ventricular distensibility, whereas endothelin-1 slows down left ventricle relaxation. Coronary endothelial and left ventricular diastolic dysfunctions are thus associated.

Myocardial deformation modulates NO release from coronary endotheliocytes, thus changing myocardial NO concentration during the cardiac cycle with an end-systolic peak that can provoke the onset of left ventricle relaxation. Lusitropic effects of nitric oxide produced by nitric oxide synthase NOS3 are mediated by protein kinase-G that phosphorylates, in coordination with protein kinase-A, cardiac troponin-I, hence desensitizing troponin to Ca^{2+} ion. Moreover, phospholamban phosphorylation that augments calcium sequestration in the sarcoplasmic reticulum accounts for lusitropic effects of NO produced by NOS1 enzyme.

In inflamed coronary vessels, protein Arg methyltransferases (PRMT) and dimethylarginine dimethylaminohydrolases (DDAH) are up- and downregulated, respectively, hence causing accumulation of methylated ^{L}arginine metabolites in blood and reduced NO availability [574].[14] Reduced DDAH activity causes methylarginine accumulation, hence suppressing nitric oxide synthesis and causing vasoconstriction. The DDAH activity is also impaired by oxidative stress engendered by oxidized LDLs, inflammatory cytokines, hyperglycemia, and infectious agents.

7.9.8 Reactive Oxygen and Nitrogen Species

Reactive oxygen species are generated during repetitive ischemia–reperfusion episodes, prostaglandin synthesis, and catecholamine auto-oxidation, as well as when concentrations of inflammatory cytokines rise and when antioxidant defense is altered.

Reactive oxygen species are produced by:

1. The mitochondrial electron transport chain,
2. Xanthine oxidase,
3. NADPH oxidase activated by
4. Angiotensin-2 and other stimuli,
5. Cyclooxygenases, and
6. Monoamine oxidases, as well as
7. Uncoupled nitric oxide synthase NOS3 in an oxidative environment.

Sources of increased production of superoxide, hydrogen peroxide, and hydroxyl radical in the failing myocardium also include auto-oxidation of certain metabolites and activated neutrophils.

[14] Methyltransferase, or methylase, methylate their substrates. Dimethylargininases, or dimethylarginine dimethylaminohydrolases (DDAH1–DDAH2), degrade methylarginines, specifically asymmetric dimethylarginine (ADMA) and ^{NG}monomethyl-^{L}arginine (MMA) that prevent the production of nitric oxide synthase.

Protein QSOx1 catalyzes the oxidation of sulfhydryl groups in thiols to disulfide bonds with the reduction of oxygen to hydrogen peroxide. It is synthesized in the myocardium of the left atrium and ventricle. It is involved in protein folding, redox regulation, antioxidative protection, cell differentiation, production of extracellular matrix constituents, angiogenesis, and atherosclerosis [575].

Peroxynitrite ($ONOO^-$) causes nitrosative stress and vascular lesions, as it triggers the death of vascular endothelial and smooth muscle cells, disrupts endothelial glycocalyx, upregulates adhesion molecules on endotheliocytes, enhances neutrophils adhesion, inhibits K_V and K_{Ca} channels in coronary arterioles as well as vascular prostacyclin synthase, promotes NOS uncoupling, and modulates platelet aggregation [205].

7.9.9 Nuclear Factors of Activated T Cells

Nuclear factors of activated T cells are Ca^{2+}-sensitive transcription factors implicated in cardiac hypertrophy and failure as well as arrhythmias.

Cytosolic NFAT is dephosphorylated (activated) by Ca^{2+}-sensitive PP3 phosphatase and phosphorylated by GSK3, P38MAPK, and JNK kinases. Its dephosphorylation causes its translocation to the nucleus, where it regulates gene transcription; its rephosphorylation provokes its nuclear export.

Four NFAT isoforms (NFAT1–NFAT4) are synthesized in cardiac cells. In adult cardiomyocytes, according to their atrial or ventricular location, distinct NFAT isoforms work.

Subtype NFAT1 localizes in the nucleus of atrial and ventricular myocytes upon PP3-mediated dephosphorylation [576]. On the other hand, NFAT3 resides in the cytosol; it enters the nucleus after stimulation by angiotensin-2 and endothelin-1 in atriomyocytes, but not ventriculomyocytes. Activated Gq-coupled receptors of angiotensin-2 and endothelin-1 launch the PLC–IP_3–Ca^{2+} pathway. In rabbit ventriculomyocytes in heart failure, basal NFAT3 nuclear localization and activity rises.

7.9.10 Periostin

Periostin is an extracellular matrix molecule of the fasciclin family secreted by fibroblasts that operates in cell adhesion, growth, and migration. Periostin-mediated cell migration involves α_V-integrin, focal adhesion kinase, and protein kinase-B.

In the heart, periostin is expressed at very early stages of embryogenesis; in normal adult myocardium, it is only detected in valves. However, it is upregulated after heart injury or by environmental stimuli. Periostin is a TGFβ-responding factor that intervenes in cardiac healing after acute myocardial infarction [577]. Periostin participates in postmyocardial infarction fibrosis as well as hypertension-induced cardiac hypertrophy and fibrosis [578].

7.9.11 Nucleotides

Neurohormonal changes in congestive heart failure is associated with elevated peripheral sympathetic nerve activity and increased release of noradrenaline, neuropeptide-Y, and ATP, hence increased vascular resistance and impaired peripheral blood flow.

All known ATP-activated P2X and P2Y receptors exist in the rat left ventricle, right atrium, and sinoatrial node [579]. However, their density varies according to the cardiac cavities. The $P2X_5$ receptor is the most abundant in all three explored regions. In the rat left ventricle, $P2Y_1$, $P2Y_2$, and $P2Y_{14}$ transcript level are the highest among P2Y receptors. In the right atrium and sinoatrial node, $P2Y_2$ and $P2Y_{14}$ levels are the highest, respectively. Transcripts for $P2X_4$ to $P2X_7$ and $P2Y_1$, $P2Y_2$, $P2Y_4$, $P2Y_6$, and $P2Y_{12}$ to $P2Y_{14}$, but not $P2X_2$, $P2X_3$, and $P2Y_{11}$, are detected in human right atrium and sinoatrial node [579].

In a rat model of ischemic heart failure induced by ligation of the left coronary artery, $P2X_4$ mRNA is upregulated in the sinoatrial node [579]. In addition, the $P2X_1$ and $P2Y_2$ mRNA levels are prominently overexpressed (2.7- and 4.7-fold, respectively) in the left ventricular myocardium of rats with congestive heart failure [580]. The G-protein–coupled P2Y receptors are expressed at a higher level than the ligand-gated $P2X_1$ ion channel:

$$P2Y_6 > P2Y_1 > P2Y_2 = P2Y_4 > P2X_1.$$

In the human myocardium, $P2X_1$, $P2Y_1$, $P2Y_2$, $P2Y_6$, and $P2Y_{11}$ are detected in both right and left atria and ventricles, but the $P2Y_4$ level is weak or absent [580].

Both ATP and UDP released from cardiomyocytes through pannexin hemichannels target the $G\alpha_{12/13}$-coupled $P2Y_6$ receptor for fibrogenic gene expression [581]. The $P2Y_6$ receptor initiates a pathway that involves angiotensin-2 and then transforming growth factor-β.

7.9.12 Small GTPases

Small GTPase Rho and its effector Rho kinase isoform RoCK1 are also implicated in cardiac fibrosis (but not cardiac hypertrophy). Kinase RoCK1 increases expression of transforming growth factor-β, connective tissue growth factor, and collagen-3 [582]. Small Rho GTPase and RoCK kinase are also effectors of angiotensin-2 via $G\alpha_{12/13}$ protein and reactive oxygen species [583].

7.9.13 Kinases and Phosphatases

Maladaptive cardiac remodeling results from diverse pathways based on stress-response protein kinases and phosphatases (Sect. 3.2.6.1), that are either primed

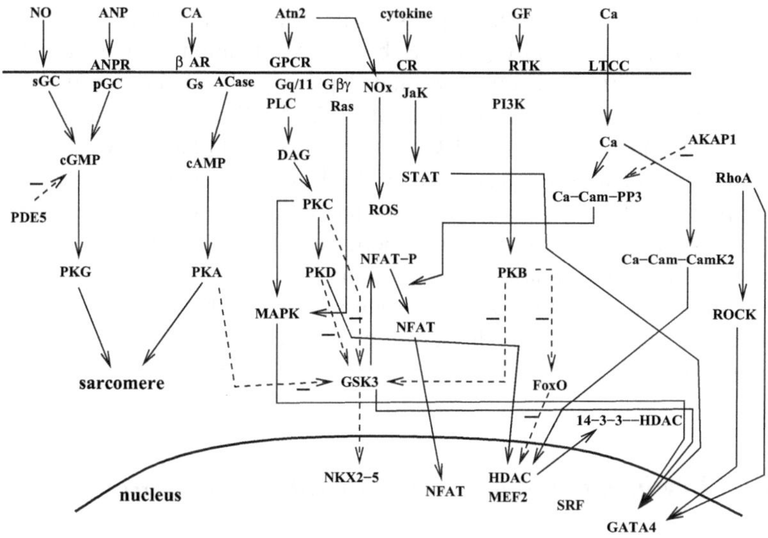

Fig. 7.2 Signaling cascades in heart failure (Source: [536]). In cardiomyocytes, stress pathways are stimulated by nitric oxide (*NO*), atrial natriuretic peptide (*ANP*), angiotensin-2 (*ATn2*), catecholamines (*CA*), cytokines, and growth factors (*GF*). *ATn2* binds to its Gq/11-coupled receptor (*GPCR*) and also stimulates NADPH oxidase (*NOx*) that produces reactive oxygen species (*ROS*). $G\beta\gamma$ subunit activates Ras GTPase and mitogen-activated protein kinase (*MAPK*). *ANP* binds to receptors coupled with guanylate cyclase. Catecholamines (*CA*) link to Gs-coupled β1-adrenoceptors (β *AR*). Nitric oxide targets soluble guanylate cyclase (*sGC*). Signaling mediators and effectors also include adenylate cyclase (*ACase*), cyclic AMP and GMP (*cAMP* and *cGMP*), phosphodiesterase-5 (*PDE5*), A-kinase anchoring protein-1 (*AKAP1*), diacylglycerol (*DAG*), Rho GTPase, RoCK kinase (*ROCK*), phosphatidylinositol 3-kinase-γ (*PI3K*); protein kinases *PKA*, *PKB*, *PKC*, *PKD*, and *PKG*, calcium–calmodulin-dependent kinase-2 (*CamK2*) and phosphatase-3 (*PP3*), Janus kinase (*JaK*), and glycogen synthase kinase-3β (*GSK3*). Targeted transcription factors comprise forkhead box-O (*FoxO*), myocyte enhancer factor-2 (*MEF2*), nuclear factor of activated T cells (*NFAT*), NK2 transcription factor related, locus-5 (*NKX2-5*), serum response factor (*SRF*), GATA-binding protein-4 (*GATA4*), and signal transducer and activator of transcription (*STAT*). In addition, histone deacetylases (*HDAC*) that promote chromatin condensation prevent transcription factor access to DNA. Growth inhibitor *FoxO* is inactivated upon phosphorylation. Phosphorylation of class-2 HDACs causes their cytosolic sequestration (binding to 14-3-3) for *MEF2* derepression and nuclear export. Class-1 *HDAC* inhibits Inpp5f gene expression. *GATA4* is activated by ERK1, ERK2, P38MAPK, and GSK3β. PKG targets troponin-I as well as phospholamban. Inhibition by GSK3β of cardiomyocyte hypertrophy is modulated by *PKA*, *PKB*, *PKC*, and *PKD* and activator phosphatase and tensin homolog (PTEN) and inositol polyphosphate 5-phosphatase-F (INPP5F) that impede the PI3K–PKB axis. On the other hand, phosphorylated GSK3α has an antihypertrophic effect

by physiological stimuli, but amplified by sustained stimulation in the failing heart, or involved in pathological stresses.

Ligand-bound plasmalemmal receptors activate signaling transmitted by diverse protein kinases and phosphatases and transcription factors, such as calcium–calmodulin-dependent kinase CamK2, protein kinases PKB and PKG, glycogen synthase kinase GSK3β, mitogen-activated protein kinases, and protein phosphatase PP3, as well as myocyte enhancer factor MEF2, nuclear factor of activated T cells, and signal transducer and activator of transcription STAT3 (Fig. 7.2) [536].

Inhibition of the cGMP-specific phosphodiesterase PDE5 prevents hypertension-resulting cardiac hypertrophy [584]. Subsequent activation of myocardial cGMP-dependent protein kinase-G represses maladaptive cardiac hypertrophy using alternative pathways in addition to precluding the PP3–NFAT axis. Calcium–calmodulin-dependent phosphatase PP3 is a PKG target.[15] Activated PKG activates regulator of G-protein signaling that is also targeted by atrial natriuretic peptide to impede cardiac hypertrophy. In addition, it can phosphorylate and block RhoA activation that operates in fibrosis. Calcium–calmodulin-dependent kinase CamK2 activates nuclear export of histone deacetylase-4 and subsequently MEF2 factor. It also phosphorylates cardiac Ca^{2+} cycling proteins, such as $Ca_V1.2$ channel, ryanodine and IP_3 receptors, and phospholamban. Its activation and deactivation can be independent of PP3 phosphatase.

7.9.14 Erythropoietin and Vascular Endothelial Growth Factor

Insufficient cardiac capillary growth causes a progressive cardiac dysfunction, i.e., the transition from an initially adaptive hypertrophy with increased capillary density to a maladaptive state with deficient oxygen supply, altered cardiac metabolism, contractility loss, inflammation, and fibrosis. Hypertension triggers the production of VEGF and angiopoietin-2 during the adaptive phase of hypertrophy. Erythropoietin elicits the differentiation and proliferation of red blood capsules from progenitors within the bone marrow in response to hypoxia and anemia. It also has an erythropoiesis-independent, proangiogenic effect. It indeed increases VEGF expression in ischemic tissues, independently of the RBC number.

In the heart, erythropoietin augments synthesis of the EPo receptor (EPoR) as well as that of VEGF via the JaK2–STAT3 axis in cardiomyocytes,[16] thereby favoring: (1) incorporation of endothelial progenitor cells into the myocardial microvasculature (independently of their mobilization)[17] and (2) proliferation of myocardial endotheliocytes by a paracrine effect, hence augmenting capillary density [585]. Vascular endothelial growth factor supports action of erythropoietin that improves the cardiac function in heart failure.

---

[15] The heart produces two PP3 isoforms, PP3α and PP3β. The latter is responsible for about 80 % of PP3 activity. Moreover, it is the dominant mediator of cardiac hypertrophy. Phosphatase PP3 is inhibited by several substances, such as protein kinase-G as well as nitric oxide, natriuretic peptides, calsarcin-1, and regulator of calcineurin-1. Phosphatase PP3 is not only a hypertrophic mediator but also a modulator of contractility via phospholamban, as it opposes PKA-mediated phospholamban phosphorylation.

[16] Erythropoietin primes VEGF transcription in cardiomyocytes predominantly in postinfarctus heart failure, but does not stimulate VEGF transcription in fibroblasts and endothelial cells during ischemia [585].

[17] VEGF may serve as a chemotactic factor for EPo-mobilized endothelial progenitor cells.

7.9.15 *Inflammatory Cytokines*

In addition to neurohormonal activation and oxidative stress, immunity (and inflammation) is implicated in the evolution of myocardial failure. In patients with heart failure, levels of inflammatory cytokines with diverse humoral and cellular immunomodulatory effects and antigens rise in the heart [586]. C-reactive protein, TNFSF1, and IL6 are inflammatory markers used in the identification of patients at high risk of developing heart failure as well as for predicting adverse events in patients with established heart failure. The immunity dysfunction results from an imbalance between T_{H1} and T_{H2} cell response [586].

Levels of inflammatory cytokines such as TNFSF1 are correlated with severity of heart failure. Moreover, circulating concentrations of TNFSF1 similar to those of patients with heart failure generate a negative inotropy. Ablation of the Tnfrsf1A gene that encodes TNFR1 receptor ameliorates heart failure, whereas removal of the Tnfrsf1B gene for TNFR2 exacerbates heart failure [586]. Interleukin-6 can be used as a predictor of the inflammatory state in the myocardium. T_{H2} cells prime IL6 production. Interleukin-6 acts via the GP130 receptor subunit. It upregulates NO synthesis production and can then provoke nitrosative stress. In fibroblasts, IL6 promotes their migration and proliferation and increases the production of collagen and IL6 receptors [586]. In cardiomyocytes, IL6 deteriorates in a dose-dependent manner contractility via peroxynitrite, promotes myocyte hypertrophy via JaK as well as apoptosis via the MAPK module, upregulates NOS2 expression via PI3K as well as xanthine oxidase and NADPH oxidase, which are the sources of free radicals, and downregulates the expression of SERCA2, myosin heavy chain, and actin [586, 587]. Xanthine oxidase also generates uric acid, another marker of severity of heart failure.

7.9.16 *Chemokines*

The membrane-bound chemokine CX_3CL1 is mainly produced in endotheliocytes in heart failure resulting from myocardial infarction or hypertension, serving as an adhesion molecule that binds to CX_3CR1 on monocytes, NK cells, and a small subset of T lymphocytes. It is also synthesized in cardiomyocytes. In inflammation, membrane-bound CX_3CL1 can be cleaved to release a soluble chemokine that attracts monocytes and T lymphocytes. Exposure to $^{S}CX_3CL1$ increases production of natriuretic peptide-A in cardiomyocytes, intercellular adhesion molecule ICAM1 in endotheliocytes, as well as matrix metallopeptidase MMP9, procollagen-1, and TGFβ in fibroblasts. After myocardial ischemia, CX_3CL1 causes cardiomyocyte damage and favors maladaptive hypertrophy as well as perivascular and myocardial fibrosis via P38MAPK and cell death [588]. In endotheliocytes, CX_3CL1 expression is induced by tumor-necrosis factor-α and hydrogen peroxide.

7.9.17 MicroRNAs in Cardiac Failure

Four types of microRNAs (miRi) arise from a given pri- and premicroRNA-i: miRi, miR$i^\star$, miRi-3p, and miRi-5p. MicroRNAs-$i^\star$ (passenger miR strand from the opposite arm of the precursor, the star strand being quasi-complementary [but not completely complementary] to the guide strand) derive from the same hairpin structure (pri- and premicroRNA), as do usual microRNAs (mature guided miR strand). When the two arms of the premicroRNA are more equal in their distribution, they are not named using a star superscript, but instead microRNA-i-3p and microRNA-i-5p, according to whether they are formed from the $3'$ or $5'$ arm of the miRNA molecule, respectively.

Search for molecular markers of nonischemic systolic heart failure (free of significant coronary artery disease) relies on changes in the human micrornome related to the pathology with a proper sensitivity in heart failure detection as well as a correlation with the left ventricular ejection fraction. MicroRNA measurements can be compared with N-terminal probrain natriuretic peptide ($_{\text{NT}}$proBNP), a heart failure marker [589].

Blood microRNAs that have altered expression levels in nonischemic heart failure with reduced ejection fraction due to genetic and inflammatory cardiomyopathy, hypertension, metabolic diseases, or toxic injury, can be used as markers [590]. However, most of the microRNAs in blood originate from circulating leukocytes, red blood capsules, and platelets. High levels of miR519*, miR622, and miR1228* are observed in CD15+ (myeloid-specific fucosyltransferase-4) granulocytes [590].

MicroRNA-622 is elevated in myocardium as well as in whole blood (not plasma) of patients with nonischemic heart failure. Moreover, its level is related to the degree of left ventricular dysfunction [590]. MicroRNA-122* belongs to the set of the most significantly upregulated microRNAs with a good discriminative power as a single marker for systolic heart failure.

In congestive heart failure, 28 microRNAs are overexpressed more than twofold with nearly complete normalization by left ventricular assist device treatment [591]. Among 444 messenger RNAs that have more than 1.3-fold abnormal expression in failing hearts, only 29 transcripts experience a change in production by as much as 25 % after left ventricular assist device implantation. MicroRNAs are more sensitive than mRNAs to label heart failure.

Seven miRNAs abound in plasma of patients with heart failure (miR18b, miR129-5p, Hs202.1 [Homo sapiens noncoding RNA], miR423-5p, miR622, miR654-3p, and miR1254), among which miR423-5p is the most strongly related to heart failure [237]. Therefore, microRNA-423-5p may be a predictor of left ventricular failure, but not right ventricular failure [238]. MicroRNA-423 level rises significantly in monocytic microparticles in cardiac failure. The circulating level of miR423-5p is inversely correlated with ejection fraction.

In the blood circulation of healthy subjects, miR423-5p specifically tethers to Argonaute-2 complexes, but not to microvesicles [237]. Elevated miR423-5p level in human failing myocardium may result from a release mechanism different from that in healthy subjects.

Concentrations of miR107, miR139, and miR142-5p decay in heart failure engendered by both nonischemic dilated and ischemic cardiomyopathies, miR125b and miR497 in ischemic cardiomyopathy only, whereas miR29b and miR142 are heightened in nonischemic dilated cardiomyopathy only [237].

Plasma concentrations of heart-specific microRNAs (miR208a–miR208b and miR499) can rise up to 140-fold in advanced heart failure with respect to normal subjects, whereas those of microRNAs (e.g., miR1-1, miR195, miR199a-1, miR199b, and miR221) increase less than 5-times in stable heart failure (levels of circulating heart-specific microRNAs and cardiac troponin-I being near the detection limit) [592]. Muscle-specific miR1-1 and miR133b that contribute to 30 % of all myocardial microRNAs increase much less during heart failure than cardiac-specific microRNAs. This augmentation disappears 3–6 months after implantation of a left ventricular assist device.

7.10 Medical Therapy of Heart Failure

The therapy of heart failure is mainly based on diuretics, inhibitors of the renin–angiotensin–aldosterone axis, and β-blockers (β-adrenoceptor antagonists) to slow progression of heart failure [593]. β-Blockers antagonize the activation by the sympathetic nervous system of β-adrenergic receptors on cardiomyocytes. Inhibitors of the renin–angiotensin–aldosterone axis are more effective than vasodilators, even though both types of drugs have similar effects on heart workload.

Decreased density in adrenal α2-adrenoceptors responsible for augmented blood levels of adrenaline and noradrenaline and increased concentrations in G-protein–coupled receptor kinase GRK2 is another target. Inhibition of adrenal GRK2 reduces the sympathetic stimulation in heart failure and improves the cardiac adrenergic responsiveness.

Dysregulated calcium-induced calcium release associated with a reduced expression of junctophilin leading to latency in activation of ryanodine channels by $Ca_V1.2$ channels as well as decaying, desynchronized calcium flux are observed in maladaptive cardiac hypertrophy, an early stage of heart failure. Treatment of such abnormalities may prevent heart failure.

Sustained hypertension causes P53 accumulation that inhibits HIF1 factor. Angiogenic factors and P53 accumulation inhibitors may restore the cardiac function.

Treatment of ischemia-induced heart failure targets G-protein–coupled receptors and nitric oxide. These two effectors converge on GPCR kinase GRK2 that acts downstream of GPCRs and provokes the death of cardiomyocytes in response to ischemia [594]. Cardioprotection ensured by ^{S}nitrosylation (inhibition) of GRK2 depends on nitric oxide synthase NOS3. The balance between NOS3 and GRK2 that mutually inhibits in the myocardium determines the outcome to ischemic injury.

Chapter 8
Coronary Artery Disease and Myocardial Infarction

Coronary artery disease (CAD), also known as atherosclerotic (AHD), coronary (CHD), and ischemic (IHD) heart disease, is the most common type of cardiac pathology. It is nearly always caused by atherosclerosis with or without luminal thrombosis and vasospasm. Chronic myocardial ischemia, or chronic ischemic heart disease (CIHD), results from coronary hypoperfusion caused by a progressive stenosis of coronary arteries (Vol. 8, Chaps. 7. Atherosclerosis—Biological Aspects, 8. Atherosclerosis—Medical Aspects, and 9. Arterial Stenosis—Mechanical and Clinical Aspects). Symptoms of cardiac ischemia comprise angina and decreased exercise tolerance. Coronary artery disease generates heart attacks, or myocardial infarctions.

Plaque rupture engenders complete remote or local arterial occlusion and death of cardiomyocytes due to oxygen deprivation. When the defective cardiac pump does not provoke death, the damaged heart replaces injured myocardium with scar tissue that maintains cardiac wall integrity, but alters the pump structure–function relation, often leading to heart failure. Promotors of concomitant coronary and myocardial growth to create a new blood supply circuit and supplant scar formation improve myocardial infarction outcome.

Rapid reperfusion by thrombolytic therapy or percutaneous coronary intervention can limit infarct size, but reperfusion contributes to tissue injury by increasing intracellular concentration of Ca^{2+} and reactive oxygen species.

Ischemic preconditioning (IPC), that is, exposure to short (5-min) episodes of hypoxia and reoxygenation, protects the heart against injury caused by a subsequent prolonged episode of ischemia and reperfusion.

8.1 Coronary Blood Flow

The coronary circulation, named for its crownlike appearance, encompasses arteries, microvessels, and veins that irrigate and drain the myocardium. When a large coronary artery is severely stenosed, the corresponding perfused myocardial region

M. Thiriet, *Diseases of the Cardiac Pump*, Biomathematical and Biomechanical Modeling of the Circulatory and Ventilatory Systems 7, DOI 10.1007/978-3-319-12664-7_8

suffers from *ischemia* (inadequate blood supply)[1] and resulting *hypoxia*[2]; a complete luminal obstruction causes oxygen deprivation and acute myocardial infarction.[3]

The coronary circulation can be subdivided into the epicardial and intramyocardial compartments, both containing arteries and veins and the latter the microvasculature. The left and right coronary arteries and their main epicardial branches run on the surface of the heart. The coronary arterial circuit generates after multiple branchings a dense mural capillary circuit that deliver oxygen and other nutrients to adjoining cardiomyocytes. Deoxygenated blood is collected by the coronary venous drainage that includes the great, middle, and small cardiac veins and anterior cardiac veins.

Heterogeneous flow distribution inside the cardiac wall depends on the architecture of the circuit as well as on perfusion pressure. Literature data demonstrate that metabolic control of flow, vasomotion, microvascular coronary α-adrenoceptor–primed vasoconstriction, oxygen delivery and consumption, glucose uptake, and glycolytic enzyme activity depend on this heterogeneous distribution.

8.1.1 Coronary Arteries

Coronary arteries supply blood to the myocardium (Vol. 6, Chap. 1. Anatomy of the Cardiovascular Apparatus). The major arteries of the coronary circulation are: the left main coronary that divides into left anterior descending and circumflex branches, and the right main coronary artery.

8.1.1.1 Formation of the Coronary Arterial Bed

Formation and maturation of the myocardium begins during embryo- and fetogenesis and continues after birth [595]. The embryonic coronary vessels initially form as a vascular plexus that covers the outer surface of the heart [596].

The primitive embryonic heart is composed of a compact layer of myocardium lined internally by the endocardium. A second trabecular layer of myocardium is then derived from and remains anchored to the outer compact muscular layer, emerging between this compact layer and endocardium [595]. It comprises a set of thin interconnected strands of muscular trabeculae that constitute fingerlike projections into the ventricular chambers. The inner trabecular layer of myocardium undergoes compaction and hence thickens prior to and after birth.

The compact myocardial layer of the embryonic heart is irrigated by a fetal coronary circulation with an intramyocardial bed. On the other hand, the trabecular layer receives oxygen by diffusion from cardiac chamber blood [595, 596]. During

[1] ισχαιμος: staunching blood flow.

[2] υπο: under, low; ox for oxygen; and -ia: suffix of medical names designating states and disorders.

[3] Latin *in-farcio*: to stuff into.

the perinatal period (compaction), the trabecular myocardial layer condenses and coalescence of the trabecular with compact myocardium traps a subset of endocardiocytes that build a vascular plexus and then new coronary vessels. This new layer of compact myocardium thus becomes richly supplied with coronary vessels. The endocardial-to-coronary endothelial lineage conversion occurs within a brief period after birth at least in mice [596]. The vasculature of the inner myocardium is thus not achieved by angiogenesis from the preexisting embryonic and fetal coronary network [596]. At least in adult mouse hearts, 60 % of the myocardium is irrigated by vessels formed from endocardiocytes during trabecular muscle compaction.

The majority of endotheliocytes of inner wall vessels originate from endocardiocytes [596]. The postnatal construction of the inner myocardial vasculature relies on hypoxia and vascular endothelial growth factor, which control the development of the fetal coronary circuit. This generation of coronary vessels in the inner wall also establishes the circulation of the muscular interventricular septum that forms prior to birth also by compaction.

The transition from fetal to postnatal circulation acutely increases the work of the left ventricle. The creation of the coronary vasculature from endocardium enables a rapid vascular and myocardial growth and hence it can match the rapidly increasing metabolic demands of the thickening myocardial wall more efficiently than the slower angiogenic expansion of the fetal coronary circulation [596].

8.1.1.2 Architecture of the Coronary Arterial Bed

In general, two coronary left and right arteries originate from coronary ostia at the root of the aorta. More precisely, the coronary ostia localize to the left and right Valsalva sinuses, respectively, just above the closed aortic valve, that is, at the level of ends of open aortic valve cusps. They then emerge between the pulmonary trunk and left and right auricle, respectively.

The left and right coronary arteries divide into primary relatively large epicardial branches. In turn, these branches give rise to parietal arteries that penetrate perpendicularly the epicardium and arterioles that progress inward to cross the myocardium and form subendocardial arterial plexus.

Small penetrating arteries (diameter 100–500 μm) traverse the cardiac wall to deliver blood from the epicardial arteries to the subendocardial microvasculature. A substantial pressure drop occurs in penetrating arteries.

Arterioles (diameter $< 100\,\mu$m) are the primary sites of vascular resistance. Coronary arterioles branch into numerous capillaries that run adjacent to cardiomyocytes. A high capillary-to-cardiomyocyte ratio, and hence a short diffusion distance, ensures adequate oxygen delivery to cardiomyocytes and removal of wastes from them.

The right coronary artery irrigates the right atrium and ventricle as well as the posterior wall of the left ventricle, thereby supplying the sinoatrial node and conus arteriosus. It can then anastomose with the left coronary artery.

The marginal branch descends to the apex along the right margin and supplies the main part of the anterior wall of the right ventricle. The posterior interventricular branch runs in the posterior interventricular sulcus toward the apex and supplies the diaphragmatic regions of both ventricles and posterior one-third of the interventricular septum as well as the atrioventricular node.

The left coronary artery supplies the left atrium and the lateral and anterior walls of the left ventricle. It divides into the anterior interventricular branch and the circumflex artery. The former descends to the cardiac apex in the anterior interventricular sulcus and yields the main supply to the anterior regions of both ventricles and anterior two-thirds of the interventricular septum. The latter anastomoses with the right coronary artery.

The sinoatrial node is often perfused by the left coronary artery rather than the right one. The atrioventricular node is occasionally irrigated by the left coronary artery.

8.1.1.3 Between-Cardiac Cavity Distribution of the Coronary Blood Flow

At rest, the right ventricular blood flow per unit mass (expressed in gram) of myocardium is typically 50–60 % of the left ventricular blood flow, at least in dogs and horses [597]. In resting swine, the right ventricular blood flow per unit mass of myocardium is about 70–90 % of the left ventricular blood flow. The lower resting flow in the right ventricle can be explained by a smaller wall thickness and oxygen consumption than those of the left ventricle.

During graded treadmill exercise in dogs, horses, and pigs, the right ventricular blood flow increases proportionally to the cardiac frequency. The right ventricular blood flow per unit mass of myocardium most often equals 75–90 % of the left ventricular blood flow during maximal exercise [597].

At rest, the right and left atrial blood flows per unit mass of myocardium are generally 20–40 % of the left ventricular blood flow in dogs, horses, and pigs [597].

During treadmill exercise, atrial flows can increase up to 15-fold reaching 50–60 % of the left ventricular flow during heavy exercise in dogs and swine, and up to 70 % in ponies [597]. Within the atria, blood flow is the lowest in the appendages, but the exercise-induced increase in blood perfusion is greater in the appendages.

8.1.1.4 Anatomical Coronary Arterial Variations

Occasionally, the left and right coronary arteries arise from a common trunk. On the other hand, three coronary arteries can originate from the aortic root. In addition, the posterior interventricular branch usually arises from the right coronary artery, but sometimes (~ 10 %) from the left coronary.

Alternative architectures of the coronary arterial circuit determine dominance of the right or left coronary artery. The artery that supplies the posterior descending artery determines the coronary dominance. The right and left dominance correspond to a main perfusion by the right coronary and circumflex artery, respectively. Codominance refers to as similar supply by both coronary arteries.

8.1.1.5 Coronary Arterial Anomalies

Some congenital coronary anomalies have no clinical repercussion. They are diagnosed fortuitously. On the other hand, major coronary anomalies have clinical impact due to resulting ischemia or volume overload (i.e., angina, dyspnea, arrhythmia, and sudden death).

Major coronary anomalies detected in children and adults by transthoracic echocardiography and coronaroangiography include [598]: (1) abnormal origins of coronary artery from the opposite sinus with interarterial course; (2) anomalous left coronary artery from the pulmonary artery; (3) single coronary ostia; and (4) coronary fistula.

8.1.1.6 Coronary Arterial Anastomoses

Intracardiac anastomoses between branches of the two coronary arteries can exist. Moreover, *extracardiac anastomoses* between coronary arteries and arteries of neighboring organs can be detected. In particular, they give rise to periadventitial arteries of the ascending aorta and other thoracic vessels, especially vasa vasorum of the ascending aorta and pulmonary artery. These extracardiac branches of the coronary arteries emerge commonly from regions near the aortic root, the base of the pulmonary arterial trunk, pulmonary veins, and ostia of the superior and inferior venae cavae, that is, at reflections sites of the pericardium that occur during development at arterial and venous ends of the heart [599]. These extracardiac anastomoses connect to pericardial, mediastinal, diaphragmatic, intercostal, and bronchial arteries, thereby irrigating up to the pulmonary hila, trachea, and esophagus. The largest communication is that with the pericardiacophrenic branches of the internal mammary arteries traveling with the phrenic nerves.

Recently, the most common anastomoses were found from bronchial and internal thoracic arteries to coronary arteries rather than from pericardiacophrenic branches of the internal thoracic arteries [600]. To a lesser extent, anastomoses exist from anterior mediastinal, intercostal, and esophageal arteries. These observations can result from interindividual anatomical variability.

8.1.1.7 Collateral Coronary Arterial Circulation

The collateral arterial circulation of the heart constitutes natural bypasses, thereby yielding an alternative source of blood supply to a ischemic myocardial region. Coronary collaterals arise from vasculo-, angio-, and arteriogenesis (Vol. 5, Chap. 10.

Vasculature Growth). They can prevent signs of myocardial ischemia during brief vascular occlusions. However, ischemic preconditioning, rather than recruitment of collateral vessels, most often heightens tolerance against myocardial ischemia [601].

The measurement of aortic and intracoronary pressure or velocity using pressure or Doppler sensor-tipped angioplasty guidewires enables the assessment of the pressure- or velocity-derived *collateral flow index* (CFI) that expresses the amount of flow through collaterals to the perfused region of interest as a fraction of the flow via the normally patent vessel. Collateral flow sufficient to prevent myocardial ischemia during coronary occlusion amounts to at least 25 % of the normal flow through the open vessel [601].

The pressure-derived CFI is determined by simultaneous measurement of mean aortic (P_{ao}), mean distal coronary occlusive (P_{occl}), and central venous (CVP) pressures [601]:

$$\mathrm{CFI} = \frac{P_{occl} - \mathrm{CVP}}{P_{ao} - \mathrm{CVP}}. \tag{8.1}$$

The velocity-derived CFI is measured by distal occlusive coronary flow velocity (v_{occl}) and coronary flow velocity in nonoccluded artery (v) taken at the same location and following occlusion-induced reactive hyperemia:

$$\mathrm{CFI} = \frac{v_{occl}}{v}. \tag{8.2}$$

In addition to arterioarterial anastomoses, generation of new arteries (angiogenesis) and enlargement of small arteries (arteriogenesis or collateral growth) contribute to the formation of collateral arterial circuits primed by ischemia. Angiogenesis is related to capillary generation and growth launched by ischemia, but as ischemia wanes, capillary growth can stop and even rarefaction can occur [603]. On the other hand, arteriogenesis continues as ischemia is resolved. Hence, a continuous arterial circuit remodeling can be maintained by mechanical forces.

The construction of coronary collaterals relies on proliferation of vascular smooth muscle and endothelial cells initiated by growth factors liberated by ischemia. Phenotypic switching of vascular smooth myocytes from a contractile to a proliferate phenotype is triggered by activation of proper signaling pathways with available energy.

Mobile and dynamical mitochondria organized along sarcomeres are thus involved in phenotypic switching under constraint resulting from ischemia-associated oxidative stress resulting from excessive production or inadequate neutralization of reactive oxygen species (ROS) [603].

The antioxidant defense eliminates ROS using mitochondrial and cytosolic free radical scavengers and resolves mismatches between outward and inward proton fluxes at the inner mitochondrial membrane by uncoupling proteins. Uncoupling proteins support proton reentry into the mitochondrial matrix, hence reducing ROS production, but also inhibiting ATP synthase.

The *Warburg effect* refers to as predominant production of energy in cancerous cells from a high rate of glycolysis followed by lactic acid fermentation in the cytosol rather than by a comparatively low rate of glycolysis followed by oxidation of pyruvate in mitochondria in normal cells.

The metabolic switch that favors aerobic glycolysis over mitochondrial oxidative phosphorylation to meet rapid cellular proliferation relies on UCP2 that counteracts P53 [603].

The magnitude of oxidative stress is correlated with the extent of mitochondrial dysfunction and thus with impaired capacity for arteriogenesis. Therefore, improved mitochondrial respiration and antioxidant defense restore at least partly coronary collateral growth arteriogenesis via AMPK signaling, as AMP/ATP ratio increases under cellular stress [603]. According to the energy status, AMPK enables or prevents phenotypic switching and hence growth and proliferation.

The AMPK kinase modulates cellular energy flux in response to an increased AMP/ATP ratio under stress. It is activated by AMP binding and phosphorylation by LKb1 and CamK kinases, in addition to AMP binding. According to the energy status, AMPK enables or prevents phenotypic switching and hence growth and proliferation.

Coronary collaterals can undergo both short- (i.e., vasomotor tone adaptation) and long-term (i.e., recruitment and growth of collateral vessels) adjustments in response to arterial occlusion.

Mature coronary collaterals are responsive to vasodilators (e.g., nitric oxide (NO) and atrial natriuretic peptide) and vasoconstrictors (e.g., angiotensin-2 and vasopressin), in addition to the platelet products serotonin and thromboxane-A_2. During effort, β-adrenergic activity and endothelium-derived NO and prostanoids exert vasodilatory influences on coronary collateral vessels [597].

When coronary collaterals can supply an adequate arterial inflow to meet the perfused myocardium needs, then the magnitude and distribution of blood flow are determined by the autoregulatory response of the microvasculature in the corresponding myocardial region.

When coronary collaterals cannot deliver sufficient arterial inflow, then the perfusion blood flow will be determined by the conductive collateral circuit with a transmural distribution of perfusion similar to that downstream from the arterial stenosis [597].

8.1.2 Coronary Veins

The venous drainage of the heart include:

1. Small veins that empty directly into the cardiac chambers such as anterior cardiac veins that collect blood from the right ventricle and open into the right atrium, the smallest cardiac veins draining into any of the four cardiac chambers; and

2. Large veins, that is, the great, middle, and small cardiac veins that empty into the coronary sinus in the coronary groove at the posterior surface of the heart, between the left atrium and ventricle.

The coronary sinus also receives the oblique vein of the left atrium, left marginal vein, and the left posterior ventricular vein. It opens into the right atrium.

Most blood from the left ventricular myocardium drains into the coronary sinus. The anterior cardiac vein receives blood from the right ventricular myocardium. These receiving veins open into the right atrium. Thebesian veins drain a small proportion of coronary blood directly into the cardiac chambers, hence creating a shunt.

8.1.3 Coronary and Cardiac Innervation

The heart is innervated by autonomic and sensory nerves of the sympathetic trunks and vagus nerves. Postganglionic parasympathetic and sympathetic fibers innervate the nodes and routes of action potential triggering and conducting (nodal) tissue and coronary vessels.

Preganglionic sympathetic nerves from the spinal cord (T1–T6) synapse in cervical and thoracic sympathetic chain ganglia. Postganglionic fibers travel in cervical and thoracic cardiac branches of the sympathetic trunk.

Preganglionic parasympathetic fibers run in cervical and thoracic cardiac branches of the vagus nerves to synapse with ganglion cells of the cardiac plexi near the heart.

Sensory fibers carrying information on blood pressure and flow and cardiac frequency travel in the vagus nerves. Pain fibers are associated with the sympathetic trunks and enter the spinal cord by dorsal roots (T1–T5).

8.1.4 Coronary Artery Physiology

The resting coronary blood flow equals about 250 ml/min (0.8 ml/min/g of myocardium [0.5–1.5 ml/min/g according to the state of alertness [597]]), that is, 5 % of cardiac output. Coronary blood flow occurs mainly during diastole. It is mainly determined by the local oxygen demand.

Exercise and pregnancy require hemodynamic adjustments, effort being the stronger physiological stimulus of increased cardiac output and arterial pressure.

Myocardial oxygen extraction is very high in basal conditions. Any additional metabolic demand thus requires an elevated myocardial blood supply. At rest, the myocardium extracts about 75 % of the oxygen delivered by coronary blood flow [604]. Only a small extraction reserve is available when myocardial oxygen consumption is augmented during exercise. Control mechanisms based on local metabolic feedback and sympathetic β-adrenoceptor–mediated feedforward arteriolar vasodilation fit coronary blood flow to myocardial oxygen consumption.

A further increase in coronary blood flow can be elicited with a pharmacological or ischemic vasodilatory stimulus. A reactive hyperemia is observed after a brief total coronary occlusion during maximal exercise in dogs and pigs. Intravenous administration of adenosine to swine provokes a 15–26 % increase in myocardial blood flow during maximum exercise and about 20 % further decrease in coronary arterial resistance [597]. The *coronary vascular resistance* is commonly given by the mean aortic pressure/mean coronary blood flow ratio.

The pharmacologically induced increase of coronary blood flow during exercise can enhance cardiac contractility (*Gregg effect*). Intracoronary administration of a selective α1-adrenergic blocker during exercise in dogs engenders a 21 % augmentation of coronary artery blood flow and maximal rate of regional myocardial segment shortening (unchanged total systolic shortening) and 26 % elevation of myocardial oxygen consumption [597]. The heightened velocity of shortening does not depend on β-adrenergic activation, as a similar response occurs after blockade of the action of adrenaline and noradrenaline on both β1- and β2-adrenergic receptors. Similarly, intracoronary administration of adenosine during exercise causes a 25–30 % increase in coronary blood flow and 27 % augmentation of the rate of systolic segment shortening and 16 % elevation of myocardial oxygen consumption without change in cardiac frequency, left ventricular systolic pressure, and myocardial end-diastolic segment length [597].

The *coronary flow reserve* (CFR; the ratio of maximum to basal coronary flow) refers to the capacity of the coronary circulation to increase blood flow using vasodilators, such as nitric oxide and prostacyclin. The *vasodilatory reserve* during exercise in atrial cavities is comparable to that in the ventricular chambers.

Hemodynamic stress and chemical messengers, such as acetylcholine, bradykinin, and histamine, are the main triggers of nitric oxide release by the endothelium. Flow-induced lectin–oligosaccharide complex formation participates in sensing shear stress applied on the wetted (luminal) endothelial surface. Flow-sensitive and lectinic substances at this surface include, at least in guinea pig hearts [605]:

1. Certain cell adhesion molecules, such as integrins and selectins;
2. Some G-protein-coupled receptors of adenosine (A_1–A_2), angiotensin-2, bradykinin (B_2), endothelin-1, noradrenaline (α1-, β1-, and β3-adrenoceptors), prolactin, thromboxane-A_2; and
3. Certain receptor protein Tyr kinases such as insulin receptor.

The amplitude of receptor-induced vascular responses upon receptor activation depends on flow and glycosylation (hyaluronate glycosaminoglycan).

Carbon dioxide, the most potent respiratory stimulant, is a vasodilator for cerebral and myocardial blood vessels. Elevated systemic p_{aCO_2} augments blood flow independently from local metabolic demands, engendering an excessive perfusion similarly to that induced by pharmacological vasodilators. In healthy individuals, voluntary breathing maneuvers modify myocardial oxygenation via CO_2. Apnea and hyperventilation change the blood gas content; capillary p_{CO_2} diminishes significantly during hyperventilation and capillary p_{O_2} declines markedly during 2-min breath hold [606]. Long breath holds prime CO_2-mediated increase in blood flow, which can compensate for O_2 desaturation. Mild hypercapnic hypoxia can further raises the myocardial blood flow, as hypercapnia and hypoxia act synergistically. p_{CO_2}-dependent vasodilation or vasoconstriction determines myocardial oxygenation.

Oxygenation-sensitive cardiovascular magnetic resonance imaging using blood oxygen level-dependent (BOLD) effect in T2*-weighted imaging sequences (^{BOLD}CMRI)[4] shows that a 2-min breath hold causes a deoxygenation in the left ventricular blood associated with an increased (8.2 %) myocardial oxygenation [606]. On the other hand, a 2-min hyperventilation that generates coronary vasoconstriction provokes a significant (7.5 %) drop in ^{BOLD}CMRI signal intensity and hence in myocardial oxygenation, deoxygenation of the myocardium resulting from an O_2 extraction outweighing coronary O_2 supply even at constant myocardial workload.

8.1.4.1 Interference Between Myocardium and Coronary Vasculature

Cardiomyocytes represent approximately 75 % of total cardiac volume, but their number accounts for less than 40 % of all cardiac cells (fibroblasts, endotheliocytes, vascular smooth myocytes, macrophages, and circulating blood cells).

The myocardium is irrigated by the coronary vasculature. Coronary perfusion ensures cardiac function. Cardiac function determines myocardial oxygen consumption and consequently coronary perfusion. The latter is regulated by extrinsic factors such as the neurohumoral regulation and intrinsic processes, such as autoregulation.

Autoregulation (Sect. 8.1.4.4) is the intrinsic mechanism by which the coronary arterial circuit maintains a quasiconstant blood flow despite changes in perfusion pressure, excluding influences of extrinsic nerves or bloodborne messengers. It is aimed at matching nutrient input supplied by coronary blood flow to myocardial metabolism.

A mechanical interaction between the contracting myocardium and the coronary vasculature that are in contact explains distribution of the coronary flow in different parietal layers (from subepicardial to subendocardial laminae) and flow variations during the cardiac cycle.

Myocardial contraction–relaxation cycles deform the mural vasculature. During systole with its two phases (isovolumic contraction and ejection), coronary artery inflow is impeded and even reversed; venous outflow is augmented. The magnitude of

---

[4] Oxygenated hemoglobin is diamagnetic (i.e., weakly stabilizes the magnetic field surrounding the molecule) and deoxygenated hemoglobin is paramagnetic (i.e., destabilizes the surrounding the magnetic field, thereby leading to a loss of magnetic field homogeneity). Hence, elevated deoxyHb levels attenuate the signal intensity in and around the myocardial capillaries [606]. On the other hand, adenosine-induced coronary vasodilation that increases the oxygen supply at constant O_2 demand elevates the signal intensity.

coronary flow depends on the local transmural pressure. The intramyocardial pressure depends on the local contraction strength as well as the magnitude of isovolumic deformation of the ventricular cavity [607]. During diastole, the collapsing effect of the myocardium on the mural coronary vasculature ceases. The contribution of the myocardium depends on initial myofiber length [607].

Conversely, the coronary vasculature and blood flow affect the myocardium and its contraction. The Gregg effect is related to the observation that an improved coronary perfusion increases myocardial oxygen consumption and contractile function. Positive inotropic factors are released by increased hemodynamic stress. During diastole, an increase in coronary perfusion pressure increases ventricular wall stiffness, but the effect is small [607]. During systole, coronary perfusion affects cardiac contractility by two mechanisms. Increased perfusion pressure increases microvascular volume, thereby opening stretch-activated ion channels that raise cytosolic Ca^{2+} concentration and hence Ca^{2+} sensitivity, enhancing myocardial contractility (Gregg effect) [607].

Thickening of the shortening myocardium reduces the space allocated to the vasculature. On the other hand, the vascular deformation, and thus vascular emptying that allows myofiber thickening during shortening, enhances myocardial contraction [607]. In other words, when the intramyocardial pump cannot displace intravascular blood, muscle contraction is reduced.

Mechanical crosstalk between the myocardium and coronary vasculature relies on short-term mutually acting mechanical factors. The contracting myocardium that shortens and thickens increases the wall stiffness and ventricular pressure. Elevated vascular filling raises vascular diameters. In addition, the vasomotor tone changes affects the rheology of the vascular wall.

Stunning refers to a transient period of ischemia with hypokinesis and a normal myocardial blood flow rate. On the other hand, sustained limitation of blood flow in myocardial regions perfused by a stenosed coronary artery can lead to a pseudohibernation state in which contractility and metabolism fall to accommodate the persistently reduced perfusion.

8.1.4.2 Myocardial Oxygen Consumption at Rest and During Exercise

The two apposed cardiac pumps continuously eject blood into the systemic and pulmonary circulations. Oxygen consumption per gram of myocardium is 20-fold higher than that of skeletal muscle, which is entirely relaxed with very low metabolic requirements at rest [597]. Cardiac oxygen consumption depends on cardiac chronotropy and inotropy and, hence, on ventricular work.

The heart has the highest oxygen consumption per unit mass among all organs. Cardiac oxygen consumption is mainly related to contraction (10–20 % for basal metabolism [597]. Arterial oxygen extraction ranges from 70 to 80 %, that is, much more than in other organs (25 %).

The high level of oxygen extraction is possible due to a high capillary density of 3000–4000/mm^2, that is, much higher than that in skeletal muscle (500–2000 capillaries/mm^2) [597].

Augmented oxygen demand during exercise, which rises up to sixfold during maximal exercise, is achieved mainly by an increase in coronary blood flow. This adaptation results from acute and chronic regulatory mechanisms that control the coronary blood flow rate. An increase in cardiac frequency decreases the end-diastolic and stroke volume and may account for 50–70 % of the elevation of myocardial oxygen consumption during exercise [597]. The exercise-induced augmentation of contractility results from β-adrenoceptor activation as well as positive inotropic effect of cardiac frequency (Bowditch–Treppe effect; Vol. 6, Chap. 3. Cardiovascular Physiology). The adrenergically mediated augmentation of contractility is estimated to be 15–25 % of the heightened myocardial oxygen during exercise [597]. Elevated ventricular work is evaluated to account for the remaining 15–25 % of the increase in oxygen consumption [597].

In the right ventricle, oxygen extraction is lower at rest than in the left ventricle and increases substantially during exercise, similarly to skeletal muscles.

In humans, unlike certain mammalian species, oxygen delivery is only slightly facilitated by an elevation of hemoglobin concentrations during exercise. Increased myocardial oxygen demands during exercise are matched chiefly by elevating coronary blood flow due to a combination of coronary vasodilation, decayed coronary vascular resistance (20–30 % from the resting level), and augmented mean arterial pressure (20–40 %) [597]. During effort, coronary blood flow rises in proportion to the cardiac frequency to reach a peak during maximal exercise from three to five times the resting level (7.5–8.5 ml/min/g myocardium for a cardiac frequency 4.3 Hz in swine [597]). However, the increased oxygen extraction indicates that the elevated myocardial blood flow does not fully compensate the increased oxygen demand during exercise.

In young healthy male humans, the coronary venous oxygen content decreases by 8 % and saturation from about 33 % at rest to approximately 24 % during heavy exercise at about 90 % of maximal cardiac frequency, whereas coronary venous oxygen partial pressure was minimally modified [597]). A rightward shift of the hemoglobin oxygen dissociation curve facilitates oxygen delivery to the myocardium during heavy exercise possibly due to a decreased blood pH linked to lactate production by working skeletal muscles.

Energy production in the normally functioning heart depends primarily on oxidative phosphorylation, with less than 5 % of ATP produced from glycolysis. Hence, increase in cardiac activity relies on instantaneous elevation of oxygen availability.

8.1.4.3 Determinants of Coronary Blood Flow

Working skeletal muscles that cyclically compress neighboring veins increase the venous return to the heart, hence stroke volume. Elevated preload and cardiac sympathetic activity increases the cardiac output.

Perfusion Time

An increase in cardiac frequency impinges on diastolic time more than systolic time, hence not only simply reducing the perfusion time but also altering the efficient perfusion period.

Coronary Perfusion Pressure and Resistance

The coronary perfusion pressure is defined as the difference between the time-dependent aortic diastolic pressure and the central venous pressure (CVP; i.e., right atrial pressure).

The coronary arterial bed can be subdivided into two compartments: (1) conductive large and mid-sized arteries (< 5 % of total coronary resistance) and (2) resistive small arteries (diameter 100–500 μm) and arterioles.

In addition, small arteries and arterioles are subjected to interaction between the intravascular distending pressure and the extravascular compressive pressure that tends to collapse parietal vessels during systole.

However, atherosclerosis in large coronary arteries or vasospasm can yield a substantial fraction of total resistance.

Systolic Compression of Parietal Coronary Vessels

However, intraparietal coronary vessels are deformable, especially during systole, during which they are more or less collapsed. Therefore, the coronary perfusion pressure is not only dictated by the pressure drop between the entry and exit of the coronary circuit compartment, but by time- and space-dependent *transmural pressure*. Coronary blood flow must be characterized by pressure–flow relations over a range of perfusion pressures. The *effective pressure drop* is the difference between inlet and outlet transmural pressures. However, the external applied pressure which varies both temporally and spatially is unknown.

The coronary pressure–flow relations depend on various factors. Increased cardiac frequency decreases ejection volume and hence coronary systolic input volume, as it reduces the total systolic time. Increased contractility raises systolic compression of intramural coronary vessels. On the other hand, increased myocardial relaxation elevates the diastolic perfusion time. Increased left ventricular diastolic filling pressure attenuates coronary blood flow.

During systole, intramyocardial blood vessels are compressed and twisted by the contracting myocardium, thereby limiting blood flow rate. The greatest throttling effect happens in the subendocardium, generating a transmural distribution of ventricular myocardial blood flow. Furthermore, systolic blood flow in compressed arteries and arterioles in the innermost ventricular wall layer is squeezed retrogradely into subepicardial arteries. Nonetheless, augmented diastolic flow can compensate systolic underperfusion, as the subendocardium has a 10 % higher arteriolar and capillary density [597].

On the other hand, intramyocardial venous blood is propelled forward toward the coronary sinus as well as retrogradely. Hence, upstream and downstream compartments of epicardial vessels are capacitors that store blood.

Anyway, the increase in external compression does not represent a very important factor in the normal coronary circulation, because the *coronary vasodilatory reserve capacity* persists even during maximal effort, the subepicardial wall layer being much less exposed. Nevertheless, when the coronary blood flow or its oxygen-carrying capacity decays upon hypoxia or anemia, then extravascular forces can significantly limit the coronary blood flow rates during exercise.

Diastolic Compensation

During diastole, when the myocardium relaxes and when the aortic valve leaflets do not impede flow at coronary artery inlets, blood flow rate rises again to reach a peak that has a higher amplitude than the systolic one. Therefore, the bulk coronary irrigation results from a balance between systolic and diastolic events.

Nervous Control of Coronary Arterial Resistance

The control of coronary vascular resistance is carried out by vasodilatory and vasoconstrictory signals transmitted remotely by the autonomic nervous and endocrine systems as well as locally by the endothelium via secreted vasoregulators and the myocardium via metabolic mediators (e.g., vasodilators adenosine, ATP, CO_2, and H^+ ion).

α-adrenergic receptors influence coronary blood flow via three mechanisms: (1) prejunctional α2-adrenoceptors and negative feedback of noradrenaline release that prevents cardiac β-adrenoceptor stimulation, both β1- and β2-adrenoceptors stimulating the rate (chronotropy) and force (inotropy) of heart contraction, thereby precluding elevation of heart rate, stroke volume, and systolic pressure; (2) α1- (predominantly) and α2-adrenoceptors on smooth myocytes of small coronary arteries and arterioles that cause vasoconstriction; and (3) α2-adrenoceptors on coronary vascular endothelium that stimulate nitric oxide release.

At rest, the cardiac sympathetic activity is minimal. During effort, α-adrenoceptor–mediated coronary vasoconstriction compete with metabolically induced coronary vasodilation. Postjunctional α1- (principally) and α2-adrenoceptors can augment subendocardial blood flow, hence enhancing the subendocardial/subepicardial blood flow ratio in an healthy coronary bed during exercise [597]. On the other hand, α-adrenoceptors generate a transmurally uniform coronary vasoconstriction, as their blockage increases blood flow uniformly in the entire wall thickness in myocardial regions perfused by a stenotic coronary artery as well as in the pressure-overload–induced hypertrophy of the left ventricular wall of dogs during exercise. Humoral adrenergic activation upon intracoronary noradrenaline injection raises blood flow uniformly in all myocardial layers.

β2-adrenergic receptors reside in coronary arterioles. As for α-adrenoceptor–mediated control, the control exerted by β-adrenergic receptors on the coronary circulation is minimal at rest. In normal human subjects or patients with angiographically normal coronary arteries, β-adrenoceptor blockade by a nonselective β-blocker decreases myocardial blood flow during bicycle exercise to a greater extent than the reduction of myocardial oxygen consumption, β-adrenoceptor activation contributing to coronary vasodilation during exercise via a feedforward mechanism [597].

Intracoronary administration of a nonselective β-adrenergic blocker in exercising dogs causes a slightly greater decrease of coronary blood flow than does a selective β1-adrenergic blocker, as β2-adrenoceptors contribute to adrenergic coronary vasodilation.

The coronary resistive vessels are richly innervated by parasympathetic fibers that engender a coronary vasodilation, as acetylcholine provokes nitric oxide release from the vascular endothelium.

Parasympathetic effects on both the myocardium and coronary arterial bed are negligible during submaximal exercise, as the myocardial vagal tone progressively withdraws during increasing levels of exercise. Vagal tone removal may contribute to β-adrenergic vasodilation at low exercise intensity. However, the global effect of vagal nerve stimulation depends on the mammalian species. For example, the acetylcholine-induced NO-mediated vasodilation, which predominates in dogs, is outweighed in swine by a direct vasoconstrictory effect of acetylcholine on coronary smooth myocytes, resulting in a net vasoconstriction.

Matabolic Control of Coronary Arterial Resistance

Myocardial oxygen and carbon dioxide partial pressures operate synergistically to raise coronary blood flow according to the cardiac frequency. Carbon dioxide dilates coronary arterioles possibly via an acidosis-induced opening of K_{ATP} channels. During exercise, coronary CO_2 partial pressure and pH remain nearly cosntant.

Adenosine that predominantly dilates arterioles is a messenger that adapts coronary resistive vessel caliber to varying myocardial metabolic needs. In normal conditions, adenosine is generated independently of the metabolic state of the cardiomyocyte mainly from extracellular AMP by 5′-ectonucleotidase and from ^{S}adenosylhomocysteine by ^{S}adenosylhomocysteine hydrolase. Cardiomyocytes use newly formed adenosine to form AMP via adenosine kinase. When the cardiac work rises, cytosolic free ADP increases, and adenosine is produced within the cell. Adenylate kinase transforms two molecules of ADP into one molecule of ATP and AMP. The latter is catabolized by AMP 5′-nucleotidase into adenosine. When the cytosolic adenosine concentration increase from the normal level of 0.8–2 μmol, adenosine is degraded by adenosine deaminase or transported out of the cell into the interstitium via nucleoside transporters where it can cause vasodilation possibly via A_1 and A_2 receptors coupled with K_{ATP} and the cAMP–PKA pathway, in particular A_{2A} on vascular smooth myocytes [597].

Coronary arteriolar smooth myocytes possess K_{ATP} channels responsible for an outward flux of K^+ ions that hyperpolarizes the plasma membrane. The subsequent decreased Ca^{2+} influx causes a vasodilation. Adenosine, prostacyclin, and β2-adrenoceptors increase K_{ATP} activity via the cAMP–PKA axis; NO activates the K_{ATP} channel via the cGMP second messenger.

Adenosine effect is small and thus does not represent a principal mechanism of coronary blood flow regulation at rest. Furthermore, adenosine is dispensable for the coronary vasodilation during exercise.

Adenosine triphosphate is a potent coronary vasodilator progressively released from red blood capsules via endothelial P2Y receptors when oxygen partial pressure decays [597].

In humans, the K_{ATP} channel contributes to coronary vasodilation during increased myocardial metabolic activity. The K_{Ca} channel that abounds in coronary smooth myocytes (prominently $BK_{V,Ca}$) also hyperpolarizes the plasma membrane, thereby closing voltage-gated calcium channels to produce vasodilation. Membrane depolarization and elevated cytosolic Ca^{2+} concentration activators of $BK_{V,Ca}$ channels, thereby providing a negative feedback that counteracts vasoconstriction. Various protein kinases phosphorylate the K_{Ca} channel, either activating (PKA and PKG) or inhibiting (PKC) the channels. The K_{ATP} and K_{Ca} can cooperate to regulate coronary resistive vessel tone during exercise.

Voltage-gated outwardly rectifying K^+ channels also lodge on the vascular smooth myocyte. Upon depolarization, they open and thus oppose vasoconstriction. In addition, they are sensitive to β-adrenoceptor stimulation, cAMP-mediated vasodilatory response, and redox signaling (ROS).

Impact of the Vasomotor Tone

The total coronary resistance is the sum of structural and vasomotor-associated active components. In the completely vasodilated arterial bed, flow in different perfused cardiac compartments is determined by the luminal area, cumulated arterial length, and the number of in-parallel arteries that supply a given territory. The total length of the arterial circuit irrigating the subendocardium is longer than that of the subepicardial supplier. Most ($\sim$ 90 %) of the coronary resistance is linked to small arteries and arterioles.

During effort, increased cardiac output and vasoconstriction in arterial beds of resting organs raises systolic blood pressure. Vasodilation in arterial beds irrigating exercising muscles buffer this increase, thereby minimizing elevation of diastolic blood pressure. When exercise is maintained, blood pressure often diminishes from its values reached during the early stage of exercise due to a redistribution of blood to the skin for heat dissipation. Cutaneous vasoconstriction at the beginning of exercise is indeed followed by vasodilation for bodily temperature regulation.

Arterial Wall Diameter

The coronary vasomotor tone determines the arterial lumen caliber. Vasoactive substances either bind to their cognate receptors (α- and β-adrenoceptors and muscarinic, adenosine, angiotensin, bradykinin, histamine, endothelin, prostanoids, and serotonin receptors; Vol. 5, Chap. 8. Smooth Myocytes) or cross the plasma membrane to activate their plasma membrane-attached and cortical specific enzymes (e.g., nitric oxide and guanylate cyclase).

Autonomic influences are generally weak. Epicardial vessels primarily have α-adrenoceptors that prime vasoconstriction, whereas intramyocardial and subendocardial vessels predominantly have β2-adrenoceptors that initiate vasodilation. Parasympathetic influences are minor and weakly vasodilatory. During exercise, adrenergic vasoconstriction restrain the increase in coronary blood flow.

Hypoxia causes coronary vasodilation directly and indirectly via adenosine and ATP-sensitive potassium channels.

In addition, the myogenic response to intraluminal pressure changes and metabolic regulation launches fast and slow reaction, respectively.

Most vasoactive messengers operate via the vascular endothelium. Atrial natriuretic peptide, vasoactive intestinal peptide, and calcitonin gene-related peptide cause endothelium-mediated vasodilation, but vasopressin has little effect on the coronary circulation, although it causes vasoconstriction in stressed subjects. Angiotensin-2 is a potent coronary vasoconstrictor. It also releases endothelin, the strongest vasoconstrictory peptide. On the other hand, angiotensin-converting enzyme inactivates bradykinin, a vasodilator.

The vascular endothelium secretes vasorelaxants (nitric oxide, endothelium-derived relaxing factor, prostacyclin, and bradykinin) as well as vasoconstrictors (endothelin-1 and thromboxane-A_2).

Vasodilation reduces coronary resistance to ensure adequate myocardial perfusion. However, vasodilatory capacity of coronary resistive arteries declines with aging, likely due to reduced availability of nitric oxide, which plays a permissive role in propagating vasomotor signals [608].

In coronary arterioles (caliber $\sim 100\ \mu$m) dissected from right atrial appendages of patients, local administration of the endothelium-dependent vasoactive agent bradykinin (100 μmol) elicits vasodilation at local and 2 distant stations (0.5 [station 1] and 1.0 mm [station 2]) from the injection site. The magnitude of vasodilation rises with the duration of stimulus (69 ± 6, 81 ± 6, 90 ± 2 % after 1, 3, and 5×100 ms, respectively). Bradykinin-induced dilation is substantial at distant sites (53 ± 7 and 46 ± 9 % at stations 1 and 2) [608]. The distant vasodilation, but not the local response, depends on gap junction. Small and intermediate conductance calcium-activated potassium channels (SK_{Ca} and IK_{Ca}) are involved in both local and distant sites.

The distant vasodilation, but not the local response, is significantly reduced in older ($\geq$64 year-old) patients [608]. Focal application of bradykinin in human coronary arterioles triggers SK_{Ca}- and IK_{Ca}-mediated hyperpolarization that spreads through gap junctions and subsequently remote vasodilation. On the other hand, in younger ($<$64 year-old) individuals, nitric oxide does not affect local response, but markedly reduces distant vasodilation.

During effort, a short-term fast adjustment relies on coronary vasodilation, hence on lowered coronary arterial resistance. Exercise triggers local, nervous, and hormonal regulatory mechanisms to match oxygen demands of the left ventricle during heavy exercise (~ sixfold increase) by augmenting the coronary blood flow (~ five times), as hemoglobin concentration and oxygen extraction elevate only modestly [597]. Various types of vasodilators are released (e.g., adenosine, atrial natriuretic peptide, carbon dioxide, lactic acid, and potassium ions).

However, the coronary microvasculature is not maximally dilated, as it retains a *vasodilatory reserve* during exercise-induced ischemia. In addition, it remains sensitive to vasoconstrictors, such as angiotensin-2, serotonin, thromboxane-A_2, and vasopressin.

Simultaneously, increased sympathetic stimulation and adrenaline secretion from the adrenal medulla cause vasoconstriction of visceral and cutaneous blood vessels as well as vasodilation of blood vessels in skeletal muscles; blood is shunted from the vasculature of the viscera and skin (high resistance) to that of skeletal muscles (low resistance).

Exercise training enables morphometric long-term adaptation of the coronary microvasculature via increased arteriolar densities and/or calibers, in addition to capillary recruitment. Moreover, regular exercise stimulates nitric oxide synthase activity.

Maintenance of α- and β-adrenergic activity despite a lower circulating catecholamine level can result from a heightened adrenoceptor responsiveness.

A significant fraction of coronary resistance is related to small arteries that are not strongly regulated by the metabolic control of the myocardium, but are sensitive to vasodilators [597].

A stenosis in an epicardial coronary artery redistributes the myocardial blood flow during exercise away from the subendocardium to the subepicardium [597].

8.1.4.4 Autoregulation and Flow Reserve

Healthy coronary arteries are characterized by autoregulation to maintain coronary blood flow and match the needs of the myocardium (Vol. 6, Chap. 3. Cardiovascular Physiology). Autoregulation is the intrinsic capacity of small arteries and arterioles to maintain a constant blood flow when the perfusion pressure changes, as they constrict in response to increased intraluminal pressure and conversely. In the physiological range, blood flow is nearly independent of the perfusion pressure. The level of the plateau of the autoregulation curve is related to the metabolic state of the heart. Blood flow adjusts to the myocardial metabolism.

Myogenic, Metabolic, and Endothelial Control

The diameter of small resistive arteries and arterioles is locally regulated by hemodynamic stress as well as by chemicals such as metabolites and gases.

Dysregulated myogenic response can cause local ischemia or vasogenic edema. Increased myogenic activity can engender an elevated peripheral resistance and systemic blood pressure.

Changes in resistive arteriole caliber can be adjusted in few seconds. Changes in microvessel diameter is related to three integrated and coordinated autoregulatory mechanisms: myogenic, metabolic, and flow-dependent endothelial control.

The endothelial control and myogenic response are activated first ($\geq$ 15 s), and then, if the myocardial contractility and metabolism rise, the metabolic regulation follows.

Myogenic Control

The myogenic mechanism associated with the vascular smooth myocyte primes vasoconstriction in response to increased stretch of the vascular smooth myocytes and conversely. It is initiated by a stretch-induced depolarization. It involves stretch-sensitive integrins and other types of receptors as well as mechanosensitive ion channels. The depolarization of the plasma membrane activates Ca_V1.2b channel, thereby allowing Ca^{2+} entry.

The vascular transmural pressure induces parietal longitudinal and circumferential stresses that is maintained quasi-invariant by the myogenic response. Once the intraluminal pressure increases, the diameter initially rises passively; subsequent smooth muscle contraction causes a diameter reduction and flow resistance elevation, thereby keeping the flow rate and wall stress nearly constant.

When the cytosolic calcium level declines, the myogenic response is maintained by calcium sensitization and actin polymerization via the Rho–PKC pathway. In addition, the PI3K–PKB pathway contributes to the regulation of myogenic tone via activation of the Ca_V1.2b channel [609].

Metabolic Control

The metabolic control is driven by metabolites such as adenosine and carbon dioxide, in addition to H^+ ions, oxygen, and K_{ATP} channel. Metabolic autoregulation relies on mutual interference between vascular cells and cardiomyocytes.

Endothelial Control

The endothelial control results from mechanotransduction, that is, sensing of mechanical forces applied to the vascular wall and its translation into chemical signals that are transmitted from endotheliocytes to adjoining smooth myocytes. Mediators include nitric oxide and endothelium-derived hyperpolarizing factor, among others.

MicroRNA-Mediated Control

An additional control mechanism of myogenic tone involves microRNAs and the miR-processing endonuclease Dicer. MicroRNAs influence vascular smooth myocyte function and intervene in the regulation of the myogenic tone via the PI3K–PKB axis and calcium influx through the Ca_V1.2b channel [609]. The loss of myogenic tone in Dicer KO vessels is associated with an increased phosphatase and tensin homolog (PTen) level, abolished stretch-sensitive PKB phosphorylation of myosin light chain, and reduced calcium influx. The PTen phosphatase, a target of several miRNAs, reduce PI3K-mediated PKB phosphorylation by dephosphorylating PIP_3 to PIP_2, thereby inhibiting phosphoinositide-dependent kinase PDK1. On the other hand, the PI3K pathway promotes $Ca_V1.2b$ transfer to the plasma membrane.

A combined activity of several miRs may operate in the regulation of the myogenic tone. However, miR26a may be a major element, as it is upregulated by mechanical stretch in airway smooth myocytes [609].

Flow Reserve

Flow reserve at any pressure is the difference between regulated and nonregulated flow. It is influenced by changes in metabolism as well as inflow magnitude, that is, according to whether, the irrigating coronary artery is healthy or stenosed.

The coronary flow reserve (CFR) is the ratio of maximal flow obtained by coronary vasodilation and the resting reference flow. This ratio is reduced in the presence of a stenosis (CFR < 2). Usually, large epicardial coronary arteries contribute slightly to total coronary resistance. When atherosclerosis narrows a large coronary artery with more than 70 % luminal cross-sectional area reduction, hence adding a proximal resistance, autoregulation can preserve basal coronary blood flow, but maximum coronary blood flow is reduced. Consequently, coronary flow reserve is attenuated. The microcirculatory resistance and the maximal dilatory capacity of the microcirculation, together with the severity of the stenosis, determine the resting and the maximal flow. Vasodilation is obtained pharmacologically.

Heterogeneity

Autoregulatory reserve is heterogeneously distributed across the left ventricular wall according to the level of exposure to extravascular compression during systole.

The subendocardium is the most vulnerable region of the cardiac wall to ischemia because of reduced *autoregulatory reserve* with respect to the subepicardium. The subendocardial coronary arteriolar dynamics exhibit a systolic retrograde flow that is much larger than in the subepicardium [602].

8.1.4.5 Hypoxemia

Ischemia primes a local vasodilator response and recruitment of capillaries within the ischemic region, thereby reducing vascular resistance and augmenting blood flow as well as minimizing the diffusion distance for nutrient transfer. In addition, coronary hypoperfusion can initiate a process within cardiomyocytes that decrease energy demands.

Myocardial ischemia that occurs during exercise in the presence of a coronary arterial stenosis does not cause maximal vasodilation of the coronary resistive vessels so that a substantial vasodilator reserve exists in the terminal arterial bed of the hypoperfused region [597]. This reserve can be recruited using small arterial (e.g., nitrovasodilators) and arteriolar (e.g., adenosine) dilators, due to a persistent vasomotor tone throughout the coronary vasculature.

Whereas adenosine does not participate in the regulation of coronary flow under physiological conditions, it contrastingly contributes to coronary vasodilation in ischemia. Intracoronary infusion of adenosine in the presence of a critical stenosis increases subepicardial flow; the resulting increase in coronary arterial flow raises the pressure drop across the stenosis, thereby further reducing the poststenotic coronary pressure and hence subendocardial blood flow.

Nitric oxide-dependent vasodilation is altered in patients with atherosclerosis, hyperlipidemia, or hypertension [597]. Endothelial dysfunction of the coronary resistive arteries can render patients more vulnerable to hypoperfusion.

In patients with coronary artery disease, inhibition of cyclooxygenase causes coronary vasoconstriction, hence repressing coronary reactive hyperemia and hypoxic coronary vasodilation [597]. Therefore, vasodilatory prostaglandins are involved in the regulation of the coronary vasomotor tone in chronic ischemia.

The K_{ATP} channel that opens and intervenes in the coronary autoregulation is an important mediator of coronary vasodilation during hypoxia and reactive hyperemia.

Cardiac nerve stimulation constricts resistive vessels in the myocardium irrigated by a stenotic coronary artery via α2-adrenoceptors. Postjunctional α2-adrenoceptor–induced vasoconstriction in ischemic myocardial regions is antagonized by simultaneous stimulation of endothelial α2-adrenoceptor–mediated NO release. α-adrenoceptor–mediated vasoconstriction limits coronary vasodilation in ischemic myocardium, especially in patients with impaired endothelial function.

Moreover, whereas α1- and α2-adrenoceptor stimulation does not affect vessels smaller than 100 μm during normal perfusion, myocardial hypoperfusion is associated with both α1- and α2-adrenoceptor–mediated vasoconstriction in these vessels [597].

Vasoconstriction potentiated by endogenous angiotensin-2 may be partly due to increased noradrenaline release from the sympathetic nerve endings [597]. In addition, hypoperfusion can augment the response of the coronary microvasculature to vasoconstrictory influence such as that of endothelin-1.

Thromboxane-A_2 is a product of prostaglandin metabolism liberated during platelet aggregation at sites of intravascular platelet activation and hence thrombus primed by ruptured atherosclerotic plaques. The level of its plasma metabolite thromboxane-B_2 and urinary (2,3)-dinorthromboxane-B_2 rise in patients with unstable angina [597]. Moreover, thromboxane-A_2 constrict small and large coronary arteries, that is, epicardial and resistive vessels when the coronary artery inflow is normal, whereas adenosine predominantly dilates arterioles. In the presence of a coronary artery stenosis that generates a subendocardial hypoperfusion and ischemic contractile dysfunction in exercising dogs, a TxA_2 receptor ligand that does not decrease the coronary flow in the normal heart can provoke vasoconstriction that further diminishes myocardial blood flow and aggravates contractile dysfunction. When a stenosis generates a slight but significant decrease in myocardial blood flow with a change from lactate consumption to production, prostacyclin does not oppose the vasoconstriction caused by agonists of thromboxane-A_2 cognate receptors, as the metabolic vasodilation is maximally activated in ischemic myocardium and thus cannot respond to a further decrease in blood flow.

Serotonin constricts epicardial arteries, acting directly on medial smooth myocytes via $5HT_{1B}$ or $5HT_{2A}$ receptors, but dilates coronary resistive vessels likely via $5HT_{2B}$ as well as $5HT_{1B}$ and $5HT_{1D}$ receptors on endotheliocytes, thereby releasing NO [597]. Thus, in epicardial arteries, endothelium-dependent vasodilation engendered by serotonin cannot compensate its direct constrictory action. In the normal heart, dilation of the resistive vessels by serotonin outweighs the effect of constriction of the penetrating arteries. However, in the presence of a coronary artery stenosis that causes arterioles to undergo metabolic vasodilation, vasoconstriction of the penetrating arteries cannot be counterbalanced by additional arteriolar vasodilation.

In summary, the residual coronary vasomotor tone in ischemic myocardium downstream from a coronary stenosis enables vasodilation engendered by adenosine, bradykinin, nitric oxide, and prostanoids, among others, via activation of K_{ATP} and K_{Ca} channels. Vasodilation is competed by vasoconstrictory influences of α-adrenoceptor, angiotensin-2, endothelin-1, thromboxane-A_2, and serotonin that can target small arteries that account for a significant fraction of total coronary resistance, but are not under metabolic control, as well as arterioles under metabolic control.

8.2 Clinical Scores

The *severity* of coronary arterial stenoses, that is, the degree of narrowing and number of arterial stenoses, their *extent*, that is, the proportion of abnormal coronary segments, and *pattern*, that is, whether plaques are discrete or diffuse, can be quantified using indexes and scores.

The angiographic analysis evaluates CAD severity and extension using the Bogaty [610], Gensini [611], Sullivan [612] scores, among others, as well as collaterals using the Rentrop score [613].

Angiographic scoring systems are strongly correlated with each other and with atherosclerotic plaque burden [614]. However, the involved arteries include according to the group of investigators either all epicardial and branch vessels, all epicardial vessels only, or some epicardial vessels (i.e., no left main coronary artery). Moreover, the severity of lesion comprises varying degrees of stenosis, a single degree of stenosis only; or one degree of stenosis only without recognizing significance criteria. In addition, the functional impact is quantified for the major epicardial and branch vessels, for selected arteries only, or a same significance is assigned to all vessels and their targets.

Apolipoprotein-B is the strongest predictor of both extent and stenosis scores [612]. However, it is more closely related to the extent score, even after correction for age and gender.

8.2.1 Bogaty Score

The severity indexes considered in [610] include: (1) the number of major epicardial vessels with a luminal narrowing $\geq 70\,\%$ (maximal number 3), a left main stem stenosis $\geq 50\,\%$ counting as 2 vessels; (2) the number of narrowings equal or exceeding 50 %, a stenosis length twice the normal lumen diameter being counted as 2 stenoses. A maximum of three stenoses per coronary arterial segment is allowed. Nonocclusive stenoses are classified as concentric, type-1 eccentric, or complex (type-2 eccentric or multiple irregularities).

The extent score of an arterial segment defined in [610] equals 0 if the segment is angiographically normal; 1 if any abnormality is confined to at most 10 % of the arterial segment length; 2 if the plaque length ranges from 10 to 50 % of the segment length; and 3 if the disease affects more than 50 % of the segment length. When a segment is occluded or suboccluded with altered antegrade flow, a score of 2 (or 3 if any lesion upstream from the occlusion involved more than half the nonoccluded portion of the segment) is arbitrarily assigned, whereas downstream segments are not taken into account. The extent score was the total score of the 15 segments of the coronary arterial tree. A dominant right coronary artery has five segments (proximal, mid, distal, posterior descending artery, and posterior left ventricular). The left main stem corresponds to a single segment. The left anterior descending artery has also five segments (proximal, mid, distal, and two diagonals). The circumflex artery (of a nondominant left coronary artery) has four segments (proximal, distal, obtuse marginal, and another marginal). The extent index was the extent score divided by the number of segments that can be properly visualized by antegrade flow. Therefore, the extent index ranges from 0 (score of 0) to 3 (score 45 divided by 15 segments).

A discrete pattern is supposed to correspond to a maximum of three diseased sites that do not involve more than half the arterial segment length, the rest of the angiogram having a normal appearance [610]. An occlusion that is considered as a discrete pattern in the absence of upstream lesion and when it involves less than half the normal segment length. When 2 segments are diseased, a discrete pattern is defined if the total length of the abnormality is smaller than 25 % of the combined length of the 2 segments. The short-length left main stem can be considered as a discrete pattern even if it is diseased over its entire length. A diffuse pattern is anything exceeding the criteria for a discrete pattern, that is, either more than three diseased sites or a lesion over more than 50 % of the segment length.

The quality of collaterals was qualified from absent-to-poor or good-to-excellent [610].

8.2.2 Gensini Score

The Gensini score considers the geometrical severity of lesions, cumulative effects of multiple obstructions, and significance of jeopardized myocardium [614]. A non-linear score is assigned to each lesion based on the reduction of luminal diameter. A multiplier is applied to each lesion score based upon its location in the coronary tree depending on the functional significance of the area supplied by that segment. The final Gensini score is the sum of the lesion scores.

8.2.3 Coronary Artery Surgery Study (CASS) Score

The Coronary Artery Surgery Study (CASS) score was developed by Ringqvist and coworkers [615]. Three simple indices encompass: (1) the number of vessels diseased; (2) the number of proximal arterial segments diseased; and (3) a left ventricular wall motion score.

Each of the 3 major epicardial vessels with at least 70 % stenosis is assigned 1 point; stenosis greater than or equal to 50 % in the left main coronary artery is considered a 2-vessel disease and assigned 2 points [614]. The final score is the sum of all points and is analogous to single-, double-, or triple-vessel disease in the coronary tree.

8.2.4 Duke–Jeopardy Score

The Duke–Jeopardy Score was developed and later validated in patients with significant coronary artery disease, but without significant left main coronary stenosis ($\leq$ 75 %) [616, 617]. The coronary tree is divided into six segments: the left anterior descending coronary artery, its diagonal branches, septal perforating branches, circumflex coronary artery, obtuse marginal branches, and the posterior descending coronary artery. Each segment distal to at least a 70 % stenosis is assigned 2 points [614]. The maximum number of points is 12.

8.2.5 Duke Coronary Artery Disease Severity Score

In the scoring system relying on the Duke coronary artery disease severity index [618], the discontinuous score ranges from 0 to 100 with higher weight given to a high number of involved vessels and numerous proximal lesions, as well as severity of left anterior descending coronary artery stenosis [614].

For the least severe CAD (i.e., one-vessel disease), the survival rate is nearly similar whatever the therapy (revascularization [PTCA or CABG] or medical therapy [618]. For intermediate CAD (i.e., two-vessel disease), revascularization increases the survival rates [618]. For less severe two-vessel CAD, PTCA yields a better prognosis than CABG; for the most severe two-vessel CAD (with a critical lesion of the proximal left anterior descending artery), CABG is a better strategy. For the most severe CAD (i.e., three-vessel disease), CABG leads to a better prognosis than PTCA and medical therapy [618].

8.2.6 Friesinger Score

The Friesinger score ranges from 0 to 5 with higher scores given to increased severity and number of luminal stenoses [619, 614].

8.2.7 Jenkins Score

The Jenkins score explores the coronary arterial tree from eight proximal segments [620]. It assigns a score to each segment based on the maximal degree of luminal stenosis. The points for each lesion are summed, thereby reflecting the extent and severity of atherosclerosis in proximal segments of the coronary tree [614].

The concentrations of several circulating lipoproteins (LDL^{Cs} and the combined effect of LDL^{Cs} and $VLDL^{TG}$) are related to the severity of coronary atherosclerosis [620].

8.2.8 Sullivan Scores

The Sullivan scores represent three distinct scoring techniques [612]. The vessel score that is related to the number of arteries with at least 70 % stenosis ranges from 0 to 3. The left main stenosis is counted as a single-vessel disease [614]. The stenosis score is identical to the Gensini score. The extent score indicates the proportion of atherosclerotic arteries identified by luminal irregularity. A weighting factor takes into account the functional impact. The final score represents the percentage of coronary intimal surface area involved by atherosclerotic plaques.

8.2.9 Agatston Score

The Agatston score is based on the amount of calcium deposits in coronary arterial walls to examine and follow up the progression of coronary artery calcifications (CAC). In the absence of calcifications, the Agatston score equals zero. An Agatston score is computed for each major artery and then summed to get the total score.

The amount of calcifications increases with age. Men develop calcifications about 10–15 years earlier than women [621]. Furthermore, in the majority of asymptomatic men over 55 years and women over 65 years of age, calcifications can be detected. Absolute Agatston scores of less than 10, 11–99, 100–400, and above 400 permit to categorize individuals into groups with minimal, moderate, increased, or extensive amounts of calcification, respectively.

8.2.10 Rentrop Score

The Rentrop score assesses the development of coronary collateralization. The collateral score is based on the opacification quality of collateral arteries after injection of a contrast agent:

- 0: no collateral vessels;
- 1: threadlike, poorly opacified collaterals with faint visualization of the distal arterial vasculature;
- 2: moderately opacified collaterals;
- 3: large, brightly filled collateral channels with immediate visualization of the entire distal artery.

Patients can then be classified according to the degree of development of collateral arteries supplying the distal arterial bed downstream from a severe occlusion: collateralization is graded as low (Rentrop score 0–1) or high (Rentrop score 2–3). The collateral score is not correlated with age, smoking habits, hypertension, or total serum, LDL, or HDL cholesterol [622].

8.3 Acute Coronary Syndrome

Acute coronary syndrome (ACS) results from atherothrombosis and myocardial ischemia. It includes *acute unstable angina*, that is, ischemic chest pain at rest or minimal exertion without cardiomyocyte necrosis, and *acute myocardial infarction* (AMI), that is, ischemic chest pain at rest with cardiomyocyte necrosis, which is subdivided into ST-segment elevation (STEMI) and non-STEMI myocardial infarction.

Rare nonatherosclerotic causes of acute coronary syndrome comprise coronary arteritis, trauma, dissection, thromboembolism, congenital anomalies, cocaine abuse, and complications of cardiac catheterization [623].

8.3.1 Atherosclerotic Plaque Rupture and Erosion

Sudden evolution of the culprit atherosclerotic plaque ranges from thrombosis with or without coronary occlusion to sudden luminal narrowing from intraplaque hemorrhage following plaque rupture and erosion.

Plaque hemorrhage can result from either a plaque rupture or fissure or from angiogenesis. The two main histopathological types of vulnerable plaques include the rupture- and erosion-prone lesions [623].

- A *plaque rupture* is a gap in the thin fibrous cap (thickness $< 65\,\mu m$) that separates the lipid-rich necrotic core from the arterial lumen. In ruptured plaques, the media is often destroyed. Plaque rupture is the main cause of coronary thrombosis.
- A *plaque fissure* is defined as a lateral tear in an eccentric plaque with a small necrotic core. The superficial tear separates a layer of the arterial intima from the underlying fibrous tissue and the hemorrhage extends into the necrotic core; this tract is usually lined by macrophages.
- A *plaque erosion* is related to thrombosis without plaque rupture. Typically, the endothelium is missing at the erosion site. The exposed intima consists predominantly of vascular smooth myocytes and proteoglycans. The plaque morphology shows pathological intimal thickening or a fibroatheroma with an intact media.
- A *calcified nodule* refers to as a disruptive nodular calcification protruding into the lumen that occurs usually in older individuals and in tortuous and heavily calcified arteries.[5]

The ruptured plaque (atheroma or thin-cap fibroatheroma) is characterized by [623]: (1) a big plaque size; (2) a large and soft lipid-rich necrotic core covered by a thin fibrous cap disrupted and infiltrated by foamy macrophages;[6] (3) an expansive remodeling (extension of the plaque to adjacent segments and moderate luminal

[5] Focal calcifications in atherosclerotic plaques are commonly obsreved (Vol. 8, Chap. 7. Atherosclerosis—Biological Aspects). Microcalcifications are described in the fibrous cap. Apoptotic cells, the extracellular matrix, and the necrotic core can calcify. Healed ruptured plaques are often heavily calcified. Spotty and dense patterns of plaque calcification are related to vulnerable and stable plaques, respectively.

[6] Macrophage apoptosis coupled with defective phagocytic clearance of apoptotic cells promotes plaque necrosis. Extravasation of red blood capsules into the necrotic core may expand it.

obstruction) with angiogenesis, plaque hemorrhage, adventitial inflammation, and a spotty pattern of calcifications.[7]

Plaque erosion is often associated with [623]: (1) thrombosis, mostly mural; (2) a scarcely calcified plaque; (3) the absence of discontinuation of the fibrous cap; and (4) a sparsely inflamed constrictive remodeling with severe luminal narrowing.

8.3.2 Atherosclerotic Plaque Hemorrhage

Plaque hemorrhage is an important factor of rapid plaque progression. Plaque hemorrhage can originate from plaque rupture. On the other hand, angiogenesis that creates fragile low-pressure neomicrovessels within the plaque derived mainly from vasa vasorum (rarely from the lumen), can be the other source of intraplaque hemorrhage. The latter can contribute to necrotic core expansion and then evolve to a vulnerable plaque. These neomicrovessels lack supporting cells and are leaky, hence allowing local extravasation of plasma proteins and red blood capsules. Angiogenesis and inflammation often coexist at the base of advanced plaques.

- *Plaque fissuring* corresponds to the formation of an opening from the arterial lumen into the intima (plaque rupture or fissure) that leads to an intraintimal thrombus containing few red blood capsules and mainly fibrin and platelets.
- *Pure plaque hemorrhage* is defined as the presence of red blood capsules within a plaque originated from small capillaries of the intima coming from the media.

The distinction between these two origins of plaque hemorrhage can have therapeutic and prognostic implications.

8.3.3 Markers of the Acute Coronary Syndrome

Markers are used in the diagnosis, categorization, and management of patients with acute coronary syndrome. Markers of plaque instability, such as myeloperoxidase, the S100 calcium-binding proteic dimer S100a8–S100a9,[8] pregnancy-associated plasma protein-A (PAPa or PAPPa),[9] and C-reactive protein, have very low diagnostic accuracy.

[7] In thin-cap fibroatheroma, proinflammatory macrophages within the cap secrete proteolytic enzymes such as matrix metallopeptidases. Mastocytes also promote degradation of the fibrous cap.

[8] Also known as myeloid-related protein-8/14, a stable heterodimer formed by MRP8 and MRP14, also termed calgranulin-A and -B, respectively.

[9] Also known as insulin-like growth factor (IGF)-dependent IGF-binding protein IGFBP4 protease and pappalysin-1.

In myocardial infarction, the cardiac troponins cTnnI and cTnnT are released from necrotic cardiomyocytes as intact proteins and degradation products. Their blood concentrations enable the estimation of cardiomyocyte damage, but with a delayed increase. A cTnn value above the 99th percentile of a normal reference population detected by sensitivity and high-sensitivity assays signs myocardial infarction [624].

Alternative markers of cardiomyocyte damage include the small soluble cardiac fatty acid-binding protein (cFABP) involved in the transfer of long-chain fatty acids into the cardiomyocyte. It is released into the blood circulation upon cardiomyocyte injury, more rapidly than bound cardiac troponins.

Copeptin is the C-terminus of provasopressin. It is secreted with vasopressin from the neurohypophysis. It quantifies the stress level and also the mortality risk in various diseases, especially myocardial infarction very early after symptom onset.

Natriuretic peptides, proadrenomedullin, and growth and differentiation factor GDF15 are powerful predictors of mortality in patients with an acute coronary syndrome.

C-reactive protein is a prognostic marker in acute coronary syndrome without significant differences between patients with and without type-2 diabetes [625].

8.4 Context

Smoking seems to promote thrombosis rather than atherosclerosis [623].

8.4.1 Genetic Background, Atherosclerosis, and Myocardial Infarction

A large proportion of the population has a genetic susceptibility for coronary artery disease, which can ultimately evolve toward myocardial infarction [173]. Identification of myocardial infarction gene loci in human subjects at high risk helps at improving prevention.

Susceptibility genes of coronary artery disease have been identified, such as the APO gene cluster and the Mef2A,[10] ALOX5AP,[11] LTA4H,[12] and Tnfsf2 genes [626].

A region in chromosome 9p21 near the CDKN2A and CDKN2B genes is associated with coronary artery disease in four Caucasian populations [627], as well as myocardial infarction [628]. At chromosomal region 9p21.3, a large antisense noncoding RNA gene (ANRIL) affects the regulation of several other genes. It is

[10] The Mef2A gene encodes myocyte enhancer factor-2A.

[11] The ALOX5AP gene on chromosome 13q12-13, also called FLAP, encodes for arachidonate 5-lipoxygenase-activating protein that promotes leukotriene-A_4 synthesis.

[12] Leukotriene-A_4 hydrolase catalyzes the rate-limiting step of LTb_4 synthesis.

expressed in cell types affected by atherosclerosis such as vascular smooth myocytes. In healthy individuals homozygous for the risk allele, RNA expression in blood cells of short ANRIL variants increases (2.2-fold) and that long ANRIL variant decreases (1.2-fold) [173]. The risk allele on chromosome 9p21.3 increases the susceptibility of coronary artery disease, stroke, peripheral arterial disease, as well as aneurysm of the aorta and cerebral vessels.

The SLC22A3-LPAL2-LPA gene cluster corresponds to a strong susceptibility locus for coronary artery disease [629]. A coronary artery disease risk locus is also found on 3q22.3 [630]. The chromosomal site 3q22.3 contains the muscular RAS (MRAS or RRAS3) gene, which is widely distributed, especially in the heart; its product (rRas3) is involved in adhesion signaling, hence cell recruitment.

Genetic polymorphism of innate immune genes that encode Toll-like receptors, nucleotide-binding oligomerization domain-like receptors (NLR), and related signal-transduction molecules, such as interleukin-1 receptor-associated kinase IRAK4, is associated with risk of infections, asthma, and atherosclerosis. Signaling from Toll-like receptors is related to the development of several diseases [631].

People with D299G (Asp299Gly) polymorphism associated with TLR4[13] have lower concentrations of circulating inflammatory cytokines (IL6) and fibrinogen, among others, and a reduced risk for atherosclerosis. The ^{D299G}TLR4 polymorphism is also correlated with a decreased risk of bronchoreactivity in patients who inhale lipopolysaccharides in house dust; asthmatic subjects have an increased severity of atopy.[14]

Variants of the ALOX5AP gene are risk factors of myocardial infarction, characterized by higher production in leukotriene-B_4 [633].[15] Leukotriene-B_4, which is synthesized from leukotriene-A_4, activates monocytes that migrate across the arterial endothelium and differentiate into macrophages, which become foam cells.

The Tnfsf4 gene is a marker of the risk of myocardial infarction in humans [634]. The myocardium infarction risk rises in coffee consumers with the CYP1A2-1F allele. Susceptibility to stroke is mapped to chromosome 5q12, in particular to the genes encoding phosphodiesterase-4D [635]. Variants in vesicle-associated membrane protein VAMP8[16] and hnRPUL1[17] are associated with early-onset myocardial infarction [636].

---

[13] The TLR4 polymorphism that replaces Asp299 with Gly (D299G) and Thr399 with Ile (T399I) modulates surface properties of TLR4 and may affect ligand binding. The D299G polymorphism compromises recruitment of myeloid differentiation primary response gene product MyD88 and Toll/IL1R resistance domain-containing adaptor inducing Ifnβ (TRIF) to TLR4 without affecting TLR4 expression, interaction between TLR4 with myeloid differentiation factor MD2, and lipopolysaccharide binding, but likely interfering with TLR4 dimerization and assembly of docking platforms for adaptor recruitment [632].

[14] Lipopolysaccharides exacerbate asthma and attenuate atopy (i.e., allergic hypersensitivity affecting body's regions that are not in contact with allergens).

[15] Leukotriene-B_4 is involved in the 5-lipoxygenase pathway.

[16] The Vamp8 gene is involved in platelet degranulation.

[17] The HNRPUL1 gene encodes heterogeneous nuclear ribonucleoprotein U-like protein-1.

In addition, the Psrc1, Mia3, and Smad3 genes, which encode the cell-growth regulators Pro/Ser-rich coiled-coil protein-1,[18] melanoma inhibitory activity protein-3,[19] and SMAD3,[20] are significantly associated with myocardial infarction risk.

Several sequence variants that affect eosinophil blood counts suggest association between blood eosinophil number and asthma or myocardial infarction [639]. Asthma (IL1RL1, WDR36, IL33, and MYB) and one myocardial infarction (SH2B3) susceptibility loci have actually been identified.

8.4.2 Inflammation

A local or systemic chronic inflammation arises when antigens are constantly replenished, as in atherosclerosis that persistently harbors antigens. The atherosclerotic lesion is infiltrated by cellular effectors of immunity, such as T and B cells as well as macrophages and dendritic and plasma cells. Inflammation involved in the progression of atherosclerosis is mainly related to macrophages and T lymphocytes (Sect. 1.5).

8.4.2.1 Lymphoid Neogenesis

To eradicate pathogens, the immune system optimizes the likelihood of encounters between antigen-specific T and B lymphocytes of the adaptive immunity with antigen-presenting cells of the innate immunity in *secondary lymphoid organs* (SLO), such as the spleen and lymph nodes.

During SLO organogenesis, CD3−, CD4+, NR1f3-2+, IL7R+, PTPRc+ lymphoid tissue inducer (LTI) cells that produce TNFSF2 and TNFSF3 interact with stromal lymphoid tissue organizer (LTO) cells and trigger a TNFRSF3-mediated cascade that increases production of adhesion molecules (e.g., ICAM1, MAdCAM1, and VCAM1) and chemokines (CCL19 and CCL21, which attract T and dendritic cells, and CXCL13, which recruits B cells) [640].

[18] This protein is regulated by P53 and may participate in P53-mediated growth suppression.

[19] Also known as transport and Golgi body organization protein TanGo1, a secreted protein widely expressed during embryogenesis and in adult cells, except in hematopoietic cells. It binds to leukocyte α_X-integrin and regulates the migration of human monocytes [637]. It belongs to a protein family that comprises the growth regulator melanoma inhibitory activity protein (MIA), also known as melanoma-derived growth regulatory protein and cartilage-derived retinoic acid-sensitive protein (CDRAP) [638], a potent tumor cell-growth inhibitor for malignant melanoma cells and some other neuroectodermal tumors, MIA2, the expression of which is inducible by IL6 and TGFβ, specifically in the liver from activated stellate cells, MIA3 (TanGo or TanGo1), and MIA-like protein (MIAL, or otoraplin [Otor], predominantly produced in the cochlea of the inner ear).

[20] This factor is a receptor-regulated SMAD (rSMAD) that modulates signals of activin and TGFβ.

The LTI–LTO interaction relies on several ligand–receptor complexes, such as $\alpha_4\beta_1$Itg–VCAM1, TNFSF1–TNFR1, and membrane-bound TNFSF2–TNFSFR3 and TNFSF3–TNFSFR3 couples. The receptors TNFR1 (TNFRSF1a), TNFR2 (TNFRSF1b), and TNFSFR3 stimulate the NFκ B pathway, involving the translocation of the P50–RelA and P52–RelB heterodimers to the nucleus, respectively. In addition, production of certain growth factors enables lymphangiogenesis.

Lymphoid neogenesis, or *tertiary lymphoid organ* (TLO) formation, refers to as the generation of B-cell follicles surrounded by T-cell regions that are anatomically and functionally similar to secondary lymphoid organs, using many of the pathways involved in secondary lymphoid organogenesis occurring before birth [640]. Tertiary lymphoid organs can be built within any nonlymphoid tissues subjected to chronic inflammation. They support adaptive immunity induction and maturation.

Lymphoid aggregates that localize all along the aorta are TLOs because they are composed of B-cell follicles surrounded by T cells, a prototypical organization of ectopic germinal centers [641]. They contain two subsets of B cells with presumably different maturation states as well as blood and lymph vessels and fibroblastic reticular cell (FRC)-like cells. They are polarized toward the media. Lymphangiogenesis and angiogenesis associated with the formation of adventitial TLOs may be due to intramural effectors; vSMCs can trigger intramural angiogenesis, as it produces VEGFa; macrophages can induce the formation of lymphatic vessels [641].

During lymphoid neogenesis in the adventitia of atherosclerotic abdominal aorta, medial smooth myocytes underlying the lesion serve as lymphoid tissue organizers. Similarly to LTO cells in SLOs, medial smooth myocytes may be activated via TNFRSF3, thereupon expressing chemokines (e.g., CCL19–CCL21, CXCL13, and CXCL16) and attracting B and T lymphocytes and dendritic cells to the adventitia [641].

Similarly to LTI cells, M1 macrophages that express high levels of TNFSF1 and membrane-bound TNFSF2 stimulate medial smooth myocytes, which then secrete chemokines (CCL19–CCL20 and CXCL16), thereby supporting the development of adventitial TLOs in arterial walls with advanced atherosclerotic plaques [641]. These macrophages liberate TNFSF1 and TNFSF2, the latter existing in membrane-bound and secreted forms. Secreted TNFSF2 binds to both TNFR1 (TNFRSF1a) and TNFR2 (TNFRSF1b) with high affinity. On the other hand, the transmembrane TNFR2–$TNFR3_2$ heterotrimer selectively binds to TNFSF3 with high affinity, signaling from which is dispensable for the action of M1 macrophages [641].

Neutralization of TNFSF1 reduces germinal center B cells and adventitial TLOs. Macrophages are major producers of TNFSF1, but B lymphocytes also manufacture TNFSF1 that takes membrane-bound and soluble forms and T lymphocytes may contributes to a complementary TNFSF1 signal [640].

8.4.2.2 Macrophage Populations

Proinflammatory M1 macrophages are antagonized by M2 macrophages that dampen inflammation and promote tissue repair. A third type of macrophage, hemoglobin-stimulated macrophage (M_{Hb}) is observed at site of hemorrhage or angiogenesis in human atherosclerotic lesions [623].

The most common macrophages in the atherosclerotic plaques are foamy M1 macrophages activated by Infγ and T_{H1}-type cytokines. M2 macrophages are activated by T_{H2}-type cytokines (i.e., IL4 and IL13). CD163+, CD206+ M_{Hb} macrophages are devoid of neutral lipids (unlike foamy macrophages) [623]. They express the scavenger receptor HbScaR (CD163; but not ScaRb3), abundantly ATP-binding cassette transporters, and mannose receptor (CD206 or CLec13d). They only slightly produced inducible nitric oxide synthase (NOS2). A decrease in intracellular iron level may protect from lipid accumulation, partly by reducing intracellular iron-driven generation of reactive oxygen species such as hydroxyl radical upon upregulation of ferroportin expression.

8.4.2.3 Myeloperoxidase-Positive Cells

The circulating inflammatory marker myeloperoxidase (MPo) is higher in patients with OCT-defined plaque erosion than rupture [623]. The density of MPo+ cells is higher in thrombi caused by eroded plaques than in ruptured plaques in fatal coronary thrombosis.

8.4.3 Oxidative Stress and PKB Kinase

Human mortality results mainly from perturbed lipid metabolism and chronic inflammation associated with atherothrombosis and myocardial infarction. High levels of reactive oxygen species are generated in the heart during ischemia, as oxygen concentration does not immediately fall to zero.

The murine model that recapitulates atherothrombosis leading to death from myocardial infarction is the mouse lacking both the high-density lipoprotein receptor scavenger receptor ScaRb1 and apolipoprotein-E ($ApoE^{-/-}$, $ScaRb1^{-/-}$ double-knockout mouse). The ScaRb1 receptor launches the main endothelial protective axis especially in hyperlipidemia, as it triggers the PKB–NOS3 pathway, which precludes NFκ B activation induced by VCAM1 on endotheliocytes [642].

The major cardiovascular PKB isoform is PKB1 that represents 50 % of the PKB activity in the heart and 70 % in endotheliocytes [642]. It is also the main isoform in other cells involved in atherothrombosis, such as vascular smooth myocytes, monocytes, and platelets.

The PKB1 subtype that is protective in the endothelium in atherosclerosis, as it supports cell survival, can have detrimental effects when it is excessively stimulated.

In particular, it is activated by oxidized low-density lipoprotein uptake. It is also involved in proinflammatory signaling in smooth myocytes and macrophages. In patients, PKB is overactivated in atherosclerotic plaques, after myocardial infarction, and in heart failure.

In $ApoE^{-/-}$ mice subjected to a Western-type diet, PKB1 deficiency that disrupts the protective NOS3 signaling increases endothelial damage and apoptosis and exacerbates atherosclerotic lesion development. Furthermore, endotheliocyte-specific deletion of the FoxO transcription factor that is degraded upon PKB activation, is atheroprotective. Knockout of three FOXO genes in endotheliocytes attenuates PKB activity. On the other hand, inhibition by rapamycin of the TOR kinase that acts both upstream and downstream of PKB, increases lifespan. In $ApoE^{-/-}$, $ScaRb1^{-/-}$ mice characterized by diet-independent changes in lipid profiles with high blood concentrations of cholesterol due to blocked cholesterol accumulation and processing in the liver, the PKB1 level is correlated with cholesterol accumulation in macrophages during atherosclerosis and cardiac dysfunction, hypertrophy, and fibrosis, as well as increased infarct region [642]. It is linked to inflammation, oxidative stress, accumulation of oxidized lipids, and an increased level of ScaRb3, a sensor of oxidative stress that binds to oxLDLs, hence promoting foam cell formation. These events create a positive feedback loop that exacerbate oxidative stress effects. On the other hand, PKB deficiency reduces ROS generation and lipid oxidation, binding, and accumulation due to low ScaRb3 density on macrophages.

8.4.4 Impaired Mitochondrial Structure and Function

Mitochondria are sensitive to ischemic insult and trigger apoptosis. Mitochondrial morphology dynamics affect the outcome of cardiac ischemia and ischemia–reperfusion injury.

8.4.4.1 Mitochondrial Morphology Dynamics

Mitochondria constantly undergo fusion or fission and present either a reticular elongated or punctate fragmented morphotype, respectively, thereby allowing elimination of dysfunctional mitochondria in the latter case.

Mitochondrial fusion involves the mitochondrial guanosine triphosphatases (GTPase) dynamin-like Optic atrophy protein OpA1 and transmembrane mitofusins Mfn1 and Mfn2. On the other hand, mitochondrial fission relies on the mitochondrial outer membrane dynamics protein fission-1 homolog (Fis1) and dynamin-related protein DRP1 (or dynamin 1-like protein Dnm1L).

The profission protein DRP1 shuttles between the cytosol and mitochondria. Dephosphorylation of DRP1 (Ser637) prevents cytosolic sequestration of DRP1, thereby launching mitochondrial fission. In addition, BCL2-binding component BBC3 (or P53 upregulated modulator of apoptosis [PUMA]), an activator of the mitochondrial apoptotic and cell death pathway, as it sequesters BCL2 and BCLxL and activates proapoptotic BAX and/or BAK, supports DRP1 activity [643].

The dynamin-related GTPase DRP1 mediates cardiac cell death during ischemic damage. On the other hand, the antiapoptotic and proproliferative Ser/Thr kinase PIM1 is an effector of PKB-mediated cardioprotection that preserves mitochondrial integrity and prevents initiation of the intrinsic mitochondrial apoptotic pathway. The kinase PIM1 phosphorylates DRP1 (Ser637), thereby precluding DRP1 localization to mitochondria and preserving reticular mitochondrial morphology in response to ischemic stress [643]. The cardioprotective PIM1 thus prevents mitochondrial fission and maintains the mitochondrial structure during cardiac ischemia, thereby promoting regeneration of the myocardium after myocardial infarction.

8.4.4.2 Oxidative Phosphorylation

The maximal oxidative phosphorylation capacity in ischemic myofibers lowers compared with nonischemic myocardium [644]. Diminished mitochondrial oxidative phosphorylation capacity and excessive production of reactive oxygen species engendered by ischemia alter the mitochondrial function in the myocardium. Chronic myocardial ischemia damages mitochondria, hence not only compromising ATP synthesis, but also producing ROS by the electron transport chain. Excessive ROS production inflicts additional damage of mitochondrial membrane constituents (lipids and proteins) and DNA, thereby opening of mitochondrial permeability transition pores that engenders mitochondrial depolarization, cytochrome-C loss, and apoptosis, ultimately leading to cardiomyopathy. Moreover, the antioxidant level is attenuated.

Blockade of ^{ETC}complex-I or -III decreases ROS production during episodes of ischemia and helps protect mitochondria against ischemic damage [644]. ^{ETC}complex-II may also manufacture ROS as in skeletal myocytes.

Alterations in the kinetic properties of cytochrome-C oxidase (^{ETC}complex-IV) can impair electron handling and generate ROS in excess. However, its capacity in the ischemic and nonischemic regions under both hyperoxic and hypoxic conditions does not vary markedly [644]. Nevertheless, the activity of cytochrome-C oxidase subunit-4 is regulated allosterically by the energetic level within the cell. In addition, the composition of cytochrome-C oxidase subunits change with hypoxia. Yet, affected cytochrome-C oxidase composition is likely a primary element in mitochondrial dysfunction in CIHD conditions.

On the other hand, ^{ETC}complex-II oxidative phosphorylation capacity diminishes in chronically ischemic left ventricular myocardium in humans [644] .

8.4.5 Proteasomal Activity

The ubiquitin–proteasome route of protein degradation is the major nonlysosomal proteolytic system that regulates protein turnover, controls the cell cycle, antigen presentation, and inflammation.

The imbalance between myocardial oxygen demand and supply during ischemia engenders a progressive depletion of the cellular ATP content (normal cardiac ATP concentration 4–6 mmol decays to 35, 16, 9, and 7 % of normal level after 15, 30, 40, and 60 min of ischemia, respectively [645]).

During protein ubiquitination catalyzed by tripartite motif-containing ubiquitin–protein ligase TRIM63,[21] double minute-2, atrogin-1, RING finger protein RNF41,[22] STIP1 homology and U-box–containing ubiquitin–protein ligase STUB1,[23] ATP activates the C-terminal glycine of ubiquitin via the generation of an adenylate intermediate by an E1 enzyme, followed by the formation of an E1–ubiquitin thiolester and release of AMP and inorganic pyrophosphate.

The 26S proteasome is constituted of the multimeric 20S core proteasome that is singly or doubly capped at its ends by a 19S regulator complex. The proteasomal core particle is composed of four stacked rings, each consisting of seven distinct α and β subunits (α1–α7 and β1–β7). The activity of the β1, β2-, and β5 proteolytic subunits are referred to as caspase-like, trypsin-like, and chymotrypsin-like, respectively [645]). The 19S regulator is made up at least of 17 subunits, including 6 ATPases associated with different cellular activities (Rpt1–Rpt6), 3 non-ATPase subunits (Rpn1, 2, and 10) that interact with the 20S proteasome, and a lid of 8 nonATPase subunits. It confers dependency on ATP and Mg^{2+} and specificity for ubiquitinated substrate [645]).

The ATP fuel is required for the 26S proteasome assembly and stability. Whereas ADP, AMP, adenosine, and inorganic pyrophosphate cannot substitute for the effect of ATP on the 26S proteasomal peptidase, the 26S proteasome has a broad specificity for nucleotides with a preference for ATP with respect to nonadenine nucleotides (ATP>CTP≫GTP>UTP) in cardiomyocytes [645]).

The 26S proteasome is split into the 19S regulator and 20S core particle upon ATP depletion. Upon readdition of ATP, the 19S regulator and 20S particle reform the 26S proteasome [645]). In addition, ATP binding and hydrolysis also regulate degradation of ubiquitinated proteins by the 26S proteasome.

Depletion of the myocardial ATP content disassembles the 26S proteasome and activates its ATP-dependent peptidase activity and, hence, abnormal degradation of ubiquitin–protein conjugates. The proteasomal peptidase activity is likely attributable to 26S proteasomes that remain intact at very low ATP concentrations [645]). Elevated proteasomal degradation of proteic substrates, such as AMPK, GRK2, RyR2, Cx43,

[21] Also known as muscle-specific RING finger protein MuRF11 and RING finger protein RNF28.

[22] Also known as neuregulin receptor degradation protein NRDP1.

[23] Also known as carboxy terminus of HSP70-interacting protein (CHIP).

and NKκ B, and possibly of myofibrillar proteins, contributes to ischemic myocardial injury [645]).

On the other hand, the proteasomal activity lowers in the postischemic heart. Ischemic preconditioning that reduces the rate of ATP depletion during subsequent sustained ischemia attenuates the increase of the myocardial proteasomal activity and preserves the 26S proteasomal function in the postischemic heart via diminished oxidative damage of the 19S regulator or interactions with protein kinases PKA and PKC [645]). However, ischemic preconditioning causes a net inhibition of the cardiac 26S proteasome during ischemia.

Proteasome inhibitors during cardiac ischemia and reperfusion can have cardioprotective effects when administered before or during ischemia. In addition, proteasome inhibitors may be beneficial, as they have antiinflammatory and immunosuppressive actions that attenuate leukocyte-mediated myocardial reperfusion injury [645]). However, proteasome inhibitors have side effects and toxicities.

8.4.6 Influence of Obesity

The coronary perivascular adipose tissue influences the coronary vasomotor tone via secreted factors, potentiating contraction of coronary vascular smooth myocytes, at least in obese swine with respect to lean swine [646]. The coronary adipose tissue (and mesenteric, but not subcutaneous adipose tissue) augments coronary contractions after exposure to KCl (20 mmol/L) as well as to prostaglandin-F2α in proportion to the amount of perivascular adipose tissue in both intact and endothelium-denuded arteries. The coronary perivascular adipose tissue also diminishes H_2O_2-mediated vasodilation in arteries of lean swine and, to a lesser extent, in arteries of obese swine. Vascular effects of the coronary perivascular adipose tissue vary according to anatomic location [646]. Augmented contractile effects of obese coronary perivascular adipose tissue are related to alterations in the perivascular adipose tissue proteome (e.g., calpastatin that elevates vSMC contraction), Rho–RoCK signaling, and K^+ and $Ca_V1.2$ channels.

8.4.7 Diabetes

Type-2 diabetes is characterized by potent proinflammatory, prooxidant, and prothrombotic stimuli, hence engendering more frequently acute vascular events than nondiabetic patients. In patient populations with a first acute coronary syndrome, diabetic patients exhibit more severe coronary atherosclerosis than nondiabetic patients [647]. Diabetic patients have a better collateral circulation observed by coronary angiography as well as a more strongly calcified plaque and smaller lipid content at the site of the minimal lumen area and more superficial calcified nodules detected by intracoronary optical coherence tomography. Hence, diabetic patients seem to experience their first event at a later stage of coronary atherosclerosis.

8.5 Obstructive Epicardial Coronary Artery Disease

Coronary artery disease results from atherosclerosis (Vol. 8, Chap. 7. Atherosclerosis—Biological Aspects). After a relatively long delay, atherosclerotic plaque forms stenosis that narrows the coronary arterial lumen and hence reduces the myocardial perfusion. Moreover, the atherosclerotic plaque can rupture; the subsequent blood clot can completely block blood flow or, most likely, can be shed by the flowing blood into emboli that prevent blood passage in downstream arterial branches. In both cases, a myocardial infarction occurs. In addition, CAD weakens the myocardium, thereby provoking heart failure and arrhythmias.

8.6 Coronary Microvascular Dysfunction

Coronary microvascular dysfunction (CMVD), or microvascular angina, resulting from structural and/or functional abnormalities can cause myocardial ischemia, thereby mimicking obstructive epicardial coronary artery disease [648]. Increased coronary microvascular resistance can impair myocardial perfusion and hence occasioning angina with ischemic electrocardiographic changes (e.g., ST-segment depression during exercise), but without arteriographic abnormalities. In other words, stress-induced ischemic ECG trace changes are associated with a normal coronary angiogram.

Abnormal coronary microvascular perfusion does not necessarily involved uniformly all coronary microvessels of a major coronary branch, but can be scattered in the myocardium. Small intramural prearteriolar coronary arteries can be the site of coronary microvascular dysfunction and microvascular ischemia.

Therefore, myocardial ischemia can engendered by three main mechanisms that can be alone or combined: atherosclerosis, vasospasm, and coronary microvascular dysfunction [649]. They can provoke transient myocardial ischemia as in patients with coronary artery disease or cardiomyopathy as well as severe acute ischemia as in stress-induced transient apical ballooning cardiomyopathy (takotsubo syndrome [TTS]).[24]

The *takotsubo syndrome* is characterized by an apical ballooning of the left ventricle associated with a basal hyperkinesia observed by echocardiography in the absence of marked coronary stenosis in coronarographic images as well as cold pressure test-induced wall motion abnormalities. Myocardial perfusion and contractility improve markedly after the administration of intravenous adenosine. The perfusion–metabolism mismatch is caused by intense microvascular constriction with subclinical CMVD persisting over time [649]. The endothelial-dependent response to acute mental stress is attenuated.

[24] Japanese, *tako*: octopus; *tsubo*: jar, pot, vase.

8.6.1 Diagnosis

A chest pain syndrome assumed to result from obstructive atherosclerosis of epicardial coronary arteries with normal angiograms is the signature of impaired function and structure of the coronary microcirculation. The coronary microcirculation has a heightened sensitivity to vasoconstrictors and a limited vasodilator capacity [649].

Coronary microcirculation can be explored during angiography by injecting a contrast medium in the coronary artery and observing intensity and speed of the opacification arrival in the myocardium. Coronary microvascular function can also be assessed by gadolinium-enhanced cardiac magnetic resonance imaging.

Positron emission tomography enables the quantification of regional myocardial blood flow at rest as well as during pharmacological, vasoactive stimuli. However, noninvasive imaging cannot differentiate between superficial epicardial and parietal defects and, hence, lack sensitivity and specificity.

Nevertheless, positron emission tomography allows the establishment in healthy volunteers of different age and gender of the normal range of the myocardial blood flow (MBF) and coronary flow reserve (CFR), that is, the ratio of MBF during near maximal coronary vasodilation assessed most often by adenosine to baseline MBF. Normal values of MBF and CFR can be used to compare results in patients with symptoms of myocardial ischemia despite normal coronary angiograms.

Intravascular ultrasonography can be combined with intracoronary Doppler velocimetry. The *index of microvascular resistance* (IMR) measured during catheterization using a pressure–temperature sensor wire is defined as the product of distal coronary pressure by the mean transit time (MTT; 1/MTT is assumed to estimate the coronary flow rate) during maximal hyperemia (after administration of adenosine) [648]. It evaluates the microvascular function. The mean transit time of saline injected into a coronary artery is obtained from thermodilution curves. Simultaneous assessment of fractional flow reserve (FFR, i.e., the ratio of distal to proximal arterial pressure during hyperemia), coronary flow reserve (CFR), which can be calculated by the ratio of the mean transit time at rest to that during hyperemia, and IMR can guide therapy. Indices FFR and IMR explore epicardial and parietal coronary arterial compartments, respectively.

Invasive coronary vasomotor testing is the gold standard for the diagnosis of the coronary microvascular disease. Diffuse coronary artery spasm in the distal epicardial coronary arteries and probably extending into the microvasculature can be provoked by intracoronary injection of acetylcholine over a short period (20 s) in sequential doses at 5-min intervals [649]. Coronary angiography is carried out 1 min after the start of each injection. Nitrates are administered at the end of tests or after distal spasm.

Microvascular spasm is revealed by the absence of epicardial artery caliber changes after provocation testing, that is, administration of acetylcholine. Intravenous or intracoronary injection of acetylcholine during angiography is aimed at assessing the coronary vasomotor. Acetylcholine causes vasodilation by releasing nitric oxide via endothelial receptors. In the presence of endothelial dysfunction, it provokes vasoconstriction via its receptors on medial smooth myocytes, once it is released by postganglionic parasympathetic nerve fibers.

Table 8.1 Classification of coronary microvascular dysfunction. (Source: [649])

Category	Main pathogenetic mechanisms
1: Absence of myocardial diseases and obstructive CAD	Endothelial dysfunction SMC dysfunction Vascular remodeling
2: Myocardial diseases	Vascular remodeling SMC dysfunction Extramural compression Luminal obstruction
3: Obstructive CAD	Endothelial dysfunction SMC dysfunction Luminal obstruction
4: Iatrogenic	Luminal obstruction Autonomic dysregulation

CAD coronary artery disease, *SMC* smooth myocyte

In addition, coronary microcirculation is less sensitive to nitroglycerin than epicardial coronary arteries.

8.6.2 Etiology

In addition to macrovascular atherosclerosis, coronary microvascular dysfunction can occur in other cardiac diseases such as cardiomyopathies. Associated with macrovascular atherosclerosis, CMVD is implicated in the failure of myocardial blood flow restoration after successful percutaneous coronary intervention. Moreover, periprocedural microemboli of plaque debris and thrombi can prime coronary microvascular dysfunction during percutaneous coronary intervention [648]. Coronary microvascular dysfunction can thus be iatrogenic.

Diabetes and metabolic syndrome, especially insulin resistance, are commonly associated with an altered coronary microcirculation function, with impaired endothelial and smooth muscle dysfunction [648]. Coronary microvascular dysfunction is also implicated in heart failure.

8.6.3 Classification

Coronary microvascular dysfunction can be classified into four main types according to the clinical setting (Table 8.1): (1) CMVD in the absence of myocardial disease and obstructive CAD; (2) CMVD in myocardial diseases; (3) CMD in obstructive CAD; and (4) iatrogenic CMVD. Several mechanisms can contribute to CMVD and can coexist.

Coronary microvascular dysfunction in myocardial diseases is observed in [649]:

1. *Hypertrophic cardiomyopathy* (HCM) with reduced luminal size, medial hypertrophy, intimal hyperplasia, and myocardial fibrosis;
2. *Dilated cardiomyopathy* (DCM) with structural vascular defect, reduced capillary density, and endothelial dysfunction;
3. *Viral myocarditis* with intense coronary vasoconstriction resulting from myocarditis-induced coronary endothelial dysfunction and/or infection of endothelial and/or smooth muscle cells;
4. *Aortic stenosis* with reduced time of diastolic coronary filling, increased LV diastolic filling pressure and intramyocardial pressure, reduced capillary density, and low coronary perfusion pressure in comparison with intracavitary pressure, and increased intramyocardial systolic pressure and delayed myocardial relaxation, as well as severely reduced subendocardial perfusion and sustained metabolic vasodilation with exhaustion of the autoregulatory capacity of the coronary microcirculation; and
5. *Infiltrative diseases*

Coronary microvascular dysfunction can coexist with obstructive coronary atherosclerosis. A preserved microvascular function and adequate collateral development limit myocardial ischemia in the presence of stable epicardial coronary artery disease. Acute coronary syndromes can be generated by plaque erosion or fissure associated with thrombus formation in epicardial arteries and paradoxical vasoconstriction in the microcirculation.

Microvascular obstruction (MVO) is defined by a state of epicardial coronary artery recanalization without myocardial reperfusion. Microvascular obstruction is caused by a variable combination of four mechanisms [649]: (1) distal atherothrombotic embolization; (2) ischemic injury; (3) reperfusion injury; and (4) individual acquired and/or genetic susceptibility of coronary microcirculation to injury, in particular after percutaneous coronary intervention.

Iatrogenic coronary microvascular dysfunction can result from percutaneous and surgical interventions, that is, distal embolization and inflammatory response to the surgical trauma.

8.6.4 Risk Factors in Type-1 CMVD

Risk factors are associated with endothelial dysfunction and abnormal nonendothelial microvascular dilation [649]. Chronic inflammation is implicated. High levels of C-reactive protein is correlated with increased frequency of ischemic episodes. Diabetes and hence chronic hyperglycemia is linked to reduced endothelial-dependent and -independent coronary vasodilator capacity. Other risk factors encompass: (1) disorders of nitric oxide metabolism; (2) dysregulation of inflammatory cytokines, estrogens, and/or adrenergic receptors; and (3) alterations in the production of local vasoactive substances, such as angiotensin-2 and endothelin.

8.7 Myocardial Ischemia and Infarction

The heart requires a constant O_2 supply to generate ATP, 95 % of which is formed by oxidative phosphorylation. Strong stenosis of coronary arteries by atherosclerotic plaques (Vol. 8, Chap. 9. Arterial Stenosis—Mechanical and Clinical Aspects) reduces perfusion and provokes myocardial ischemia, especially during conditions of increased O_2 demand, such as physical exercise and emotional stress. Atherosclerotic plaque rupture causes complete local, or more often, downstream (by emboli) arterial occlusion.

Acute coronary syndrome includes unstable angina and acute myocardial infarction with or without ST-segment elevation. It arises from the formation of platelet-rich thrombi upon rupture of the atherosclerotic plaque or erosion of endothelium.

Myocardial infarction (or heart attack) is linked to the onset of progressive death of cardiac cells, specialized such as cardiomyocytes or not (fibroblasts), within about 20 min due to O_2 deprivation.

However, prolonged myocardial ischemia is caused not only by obstruction of epicardial coronary arteries, but also reduced perfusion by diseased coronary microvessels. Microcirculatory dysfunction is associated with heart failure, diabetes, and hypertension.

Thrombi in the coronary microvasculature and microvascular spasm associated with endothelial dysfunction disturb blood flow, hence reduced oxygen and nutrient supply and catabolite removal. Cardiac blood supply by coronary arteries is interrupted due to occlusion, most often after rupture of a vulnerable atherosclerotic plaque. Boli of ruptured plaque debris and thrombus fragments block downstream smaller arteries and arterioles.

Myocardial infarction diagnosis relies on clinical symptoms,[25] electrocardiographic abnormalities,[26] increased levels of chemical markers of myocardial necrosis, and imaging data.[27] Very small regions of myocardial injury or necrosis can be detected by biochemical marker levels and/or imaging.

[25] Various combinations of chest, upper extremity, mandibular, or epigastric pain, dyspnea, and fatigue are associated with acute myocardial infarction. The discomfort usually lasts more than 20 min. Often, pain is diffuse, position-independent, nonaffected by movement. It may be accompanied by diaphoresis, nausea, or syncope. However, these symptoms are not specific to myocardial ischemia. Myocardial infarction can occur with atypical symptoms such as palpitations or even can be asymptomatic [650].

[26] The earliest manifestations of myocardial ischemia are T-wave and ST-segment changes.

[27] Usual imaging techniques to detect wall thickening and motion abnormalities in acute and chronic infarction include echocardiography, radionuclide ventriculography, myocardial perfusion scintigraphy using single photon emission computed tomography, and magnetic resonance imaging.

Table 8.2 Principal interactions between cardiac cell populations after acute ischemic injury. (Source: [35])

Source	Target transformation	Messenger	Effect
Endothelium	EMT		Fibroblast formation
Epicardium	Myocardium	Protective factors	Inhibition of cell death
	ECDC		Fibroblast formation
	Endothelium	Angiogenic factors	Angiogenesis
Myocyte	Endothelium	VEGF	Angiogenesis
Fibroblast	Endothelium	Angiogenic factors	Angiogenesis
Inflammatory leukocytes	Myocardium	Inflammatory cytokines	Fibrosis
Endothelial progenitors	Endothelium	Angiogenic factors	Angiogenesis
Cardiac progenitors	Myocardium		Regeneration

EMT endothelial–mesenchymal transition, *ECDC* epicardial-derived cells, *VEGF* vascular endothelial growth factor

8.7.1 Cardiac Cell Interactions In Myocardial Infarction

Acute cardiac injury causes myocyte death and activates fibroblasts as well as epicardiocytes that yield another source of cardiac fibroblasts (Table 8.2).

After myocardial infarction, a monocyte–macrophage infiltrate occurs in the injured region followed by a rapid proliferation of cardiac fibroblasts and endotheliocytes and deposition of matrix proteins.

Epicardiocytes surrounding the infarcted myocardium are activated. The epicardium expands and epicardial-derived cells undergo epithelial–mesenchymal transition to adopt a fibroblast phenotype and lodge in the subepicardial space [35]. Epicardiocytes secrete angiogenic and cytoprotective cytokines.

Bone marrow-derived endothelial progenitors support angiogenesis. Cardiofibroblasts and epicardiocytes also produce angiogenic factors [35]. Hypoxic cardiomyocytes secrete VEGF that activates endotheliocytes. An endotheliocyte subset undergoes an endothelial–mesenchymal transition, thereby generating cardiofibroblasts.

Bone marrow-derived cells are recruited to the heart. Inflammatory leukocytes release cytokines that promote cell death and fibrosis. Activated cardiac progenitor cells contribute to cardiac regeneration. Afterward, the cellularity in the infarcted region decays, collagen crosslink and scar contract, and a mature scar appears.

8.7.1.1 Cardiomyocyte–Leukocyte Communication

Many signaling pathways operate in cardiomyocytes and leukocytes in myocardial infarction (Table 8.3). Damaged cardiomyocytes in the infarcted region initiate

Table 8.3 Signaling pathways associated with cardiomyocyte–leukocyte communication in the ischemic heart. (Source: [69])

Signaling axes in cardiomyocytes	Mediator	Effect	Signaling axes in leukocytes
TLR4	ICAM1, VCAM1, CXC chemokines	Macrophage homing Cytokine production	
HMGB1–TLR4	Inhibition of dendritic cell recruitment		
	HMGB1	Monocyte recruitment	CXCL12–CXCR4
GP130–JaK2–STAT3	IL6, MBPa, C3	Macrophage recruitment	
TLR4–MyD88–CamK2δ–NFκ B	TNFSF1, CFb, CCL2	Monocyte recruitment	
PP3–ATF4–TORC2	GDF15	Macrophage and neutrophil recruitment and activation	β_2Itg–CDC42/Rap1
PKC	MIF	Monocyte recruitment	CXCR2
cGMP–NOS3	CXCL12	Neutrophil recruitment	CXCR4
	Predominance of $Gr1^{low}$ over $Gr1^{high}$ tissue-degrading monocytes		

ATF activating transcription factor, *CamK* calmodulin-dependent kinase, *CDC42* cell division cycle protein-42 [small monomeric GTPase], *CFb* complement factor-B, *cGMP* cyclic guanosine monophosphate, *GDF* growth differentiation factor, *GP130* glycoprotein-130 [IL6R], *HMGB* high-mobility group box protein [danger signal], *ICAM* intercellular adhesion molecule, *IL* interleukin, *Itg* integrin, *JaK* Janus kinase, *MBP* mannose-binding protein, *MIF* macrophage migration-inhibitory factor, *MyD88* myeloid differentiation primary response gene product-88, *NFκB* nuclear factor κ light chain enhancer of activated B cells, *NOS* nitric oxide synthase, *PKC* protein kinase-C, *PP3* protein phosphatase-3, *Rap1* Ras-related protein-1 [small monomeric GTPase], *STAT* signal transducer and activator of transduction, *TLR* Toll-like receptor, *TNF* tumor-necrosis factor, *TORC* target of rapamycin complex, *VCAM* vascular cell adhesion molecule

the reparative response by releasing specific damage-associated molecular pattern molecules (DAMP) recognized by Toll-like receptors on leukocytes, parenchymal cells, including healthy neighboring cardiomyocytes.

High-mobility group box protein HMGB1 is a ubiquitous danger signal released by necrotic and severely stressed cells in particular during hypoxia. Once they are secreted, DAMPs are chemoattractants in the peri-infarct zone, where they engage TLRs such TLR4 on healthy cardiomyocytes. Signaling from TLRs relies on the adaptor myeloid differentiation primary response gene product MyD88 implicated in the activation of the NFκ B transcription factor as well as reactive oxygen species production and subsequent activation of calmodulin-dependent protein kinase CamK2δ. Consequently, the expression of CXC-type chemokines and adhesion molecules (e.g., ICAM1 and VCAM1) increases, thereby facilitating monocyte homing to the infarcted zone. Once they are attracted to the infarct and peri-infarct regions in high numbers, leukocytes actively produce and release massive amounts of inflammatory mediators, such as cytokines (TNFSF1, IL1-IL2, IL4-IL-6, IL10, IL17, Ifnγ) and colony-stimulating factor CSF2 [69]. In ischemic hearts, upon TLR4 activation, a maladaptive remodeling ensues.

Danger signals assist immunocyte recruitment to sites of injury also via alternative mechanisms without receptor activation. In particular, HMGB1 can bind directly to the endothelial chemokine CXCL12 that the HMGB1–CXCL12 complexe stimulates CXCR4 and enables the migration of CXCR4+ leukocytes. However, HMGB1 favors or impedes immunocyte attraction according to the microenvironment, the cell type, and the duration and amplitude of extracellular HMGB1 level elevation [69]. Administration of exogenous HMGB1 to mice with coronary artery ligation precludes dendritic cell recruitment to the peri-infarct region.

Cardiomyocytes also recruit immunocytes to the infarct zone via secreted IL6 that binds to its receptor made up of glycoprotein-130 on neighboring cardiomyocytes. The engaged GP130–JaK–STAT3 cascade engenders a massive release of cytokines, chemokines, and factors associated with the complement system [69].

Major chemokines released by surviving cardiomyocytes in response to ischemia include CXCL12, CXCL16, and macrophage-inhibitory factor (MIF). The latter that activates cardiomyocyte and or leukocyte CXCR2 and primes an early macrophage infiltration is produced and secreted in response to oxidative stress via protein kinase-C. Engagement of myocyte CXCR2 limits the infarcted zone size, but MIF-mediated recruitment of CXCR2+ monocytes causes a maladaptive reaction [69].

The CXCL12–CXCR4 axis is initiated via a cGMP–NOS3 cascade following ischemia–reperfusion injury. Immune-modulating macrophages, but not proinflammatory macrophages, generate an efficient repair.

Among the inhibitors of monocyte homing that repress leukocyte infiltration to the infarct area and ensure a proper infarct healing, the cardiokine GDF15 has an increased production early after infarction due to nitrosative stress resulting from the activation of IL1β and Ifnγ receptors and NOS2 as well as that of the catalytic A subunit isoform $PP3_{c\beta1}$ [69]. The letter operates vie the TORC2 complex and subsequent activation of the transcription factor ATF4. The GDF15 agent prevents chemokine-triggered activation of β_2-integrins on macrophages and neutrophils via stimulation of the small GTPase CDC42 and inhibition of the small GTPase Rap1, thereby limiting macrophage and neutrophil recruitment and accelerating resolution of postinfarction inflammation [69]. Although PP3 promotes maladaptive cardiac hypertrophy, its splicing variant, $PP3_{c\beta1}$ reduces infarct expansion and myocardial remodeling [651]. It provokes VEGF synthesis in cardiomyocytes and hence stimulates angiogenesis using the PKB–TOR pathway.

8.7.1.2 Monocytes and Macrophages

In atherosclerosis, a slowly evolving lesion, bone marrow- and spleen-derived macrophages are major players only at the early stage. Resident lesional macrophages become dominant at a later phase [74].

After myocardial infarction, neutrophils and monocytes are recruited to the ischemic myocardium. Once they reside in the myocardium, monocytes differentiate to inflammatory macrophages. Both monocytes and macrophages secrete inflammatory cytokines, cathepsins, and matrix metallopeptidases.

The monocyte–macrophage response is biphasic. Initially, monocytes and macrophages accumulate in the infarct border zone. After delay that depends on whether the microvasculature remains permeable and reperfusion occurs, monocytes and macrophages invade the infarcted region.

Inflammatory, CD14+, CD16− monocytes dominate in the early stage, whereas CD14+, CD16+ monocytes lodge n the infarcted region later [73].

Cardioprotection in the postinfarcted myocardium is mediated by Ly6C^{high} phagocytic and proteolytic macrophages in the early phase and Ly6C^{low} protective macrophages in the late phase [74]. M2 macrophages help rebuild tissue and regulate angiogenesis via VEGF and myofibroblasts via TGFβ [73].

The lack of patrolling Ly6C^{low} monocytes and infarcted zone macrophages may be associated with an altered endocardium, thereby exposing the thrombogenic infarcted myocardium to blood and launching mural thrombus formation [73].

In the nonischemic myocardium, the macrophage population changes [73]. Capillaries ouside the infarcted zone enables the recruitment of monocytes. In the noninfarcted myocardium, monocytes and macrophages accumulates more slowly. Crosstalk between macrophages and other resident cell types influences postinfarcted myocardial remodeling.

8.7.2 *Prognosis*

Myocardial infarction produces two types of damage: initial ischemic and reperfusion injury. Myocardial infarction causes contractile dysfunction, cardiomyocyte death, and maladaptive remodeling. Complications can occur in the acute phase or develop after a certain delay over a given period. Myocardial infarction can cause cardiogenic shock, arrhythmias, pericarditis, myocardial rupture, and congestive heart failure. Scar can generate arrhythmias and ventricular aneurysms.

Left ventricular systolic function and volumes are independent predictors of long-term survival after myocardial infarction [652]. Myocardial infarct characteristics, such as infarct scar, microvascular obstruction, and focal myocardial hemorrhage, can be assessed by gadolinium-enhanced magnetic resonance imaging, as they are displayed as bright area, dark (hypointense) area within hyperenhanced infarct zone, and dark zone on T2- or T2*-weighted MRI, respectively. *Myocardial salvage* is the amount of ischemic and viable tissue (which can recover). Microvascular obstruction and myocardial hemorrhage are also independent predictors of outcomes, as they are associated with the highest rate of adverse cardiac events [652]. Microvascular obstruction is correlated with the infarct size. However, because bleeding occurs subsequently to therapy, myocardial hemorrhage may be a marker of treatment efficacy.

Initially, the infarct region size is large because of edema, inflammatory leukocyte infiltrates, and hemorrhage. As healing progresses, a scar tissue reduces the size of infarction. Final infarct size is an independent predictor of mortality and heart failure [652].

8.7.3 Markers of Myocardial Infarction

Markers of myocardial injury with high myocardial specificity and high clinical sensitivity comprise cardiac troponin-I or -T and MB isoform of creatine kinase. High blood levels (gender-dependent values) can remain elevated for 2 weeks or more after the onset of myocyte necrosis.

In the healthy heart, misfolded, oxidized, and some ubiquitinated proteins are degraded through both ubiquitin- and nonubiquitin-mediated pathways and constituent amino acids are recycled. The proteolytic ubiquitin–proteasome axis includes four components (proteasome, ubiquitin, ubiquitination machinery, and deubiquitinases). During cerebral ischemia, proteasomal dysfunction increases insoluble ubiquitin conjugate concentrations; the longer the ischemia, the stronger the effect, this event remaining reversible with short ischemia durations. In cardiomyopathies and myocardial ischemia, the proteasome has an impaired ability to process oxidized and ubiquitinated proteins that then accumulate [653]. On the other hand, ischemic preconditioning preserves proteasome function, as it prevents degradation of PKCϵ and increases PKA-mediated activation of proteasome.

8.7.4 Coronary Thrombosis

The primary pathophysiological mechanism responsible for acute coronary syndrome is the formation of platelet-rich thrombi. Antithrombotic therapy is then mandatory. Antiplatelet aggregation drugs such as antagonists of ADP receptor $P2Y_{12}$ and anticoagulant agents are thus used, especially those that limit bleeding risk such as antagonists of thrombin receptor PAR_1 (peptidase-activated receptor-1).

8.7.5 Precursor Cells

Hematopoietic precursors migrate into the border zone of the myocardial infarction and differentiate into cardiomyocytes and endotheliocytes. Postinfarct myocardial regeneration using stem cells still yields questions concerning cell types that produce beneficial effect, patient selection, optimal administration, cell number to deliver, and mechanisms of action of transplanted cells (Vols. 5, Chap. 11. Tissue Development, Repair, and Remodeling and 6, Chap. 1. Anatomy of the Cardiovascular Apparatus).

Resident cardiac SCA1+ (stem cell antigen) progenitor cells that express markers of proliferation such as Ki67[28] and cardiomyocyte lineage (α MHC, GATA4, and TBx5), but not cardiac troponin-T, undergo differentiation upon β-catenin depletion, thus limiting cardiac wall remodeling and heart failure after myocardial infarction [654]. Conversely, activation of the Wnt–β Ctn pathway stimulates Isl1+ cardiac progenitor cells proliferation and avoids differentiation.

Overexpression of PKB by mesenchymal stem cells restores the cardiac function after myocardial injury. Secreted Frizzled-related protein-2 is a major stem cell paracrine factor that mediates myocardium repair after ischemic injury [655].[29]

8.7.6 *Fibroblasts*

Electrical coupling between cardiac myocytes and fibroblasts contributes to the electrophysiological functioning of normal and diseased hearts. Electrophysiological properties of ventricular fibroblasts and myofibroblasts rely, at least partly, on potassium currents that can modulate membrane potential and electrical signaling.

In addition to calcium-dependent potassium channels and other nonselective ion channels, fibroblasts synthesize K_{ATP} channel subunits. The K_{ATP} channels, which are sensitive to the ATP/ADP ratio, are constituted by different combinations of pore-forming $K_{IR}6.x$ ($K_{IR}6.1$–$K_{IR}6.2$) and sulfonylurea receptor (SUR1 and SUR2a–SUR2b) subunits, which have distinct electrophysiological and pharmacological properties. The $i_{K_{ATP}}$ current is associated with the differentiation of fibroblasts into myofibroblasts that produce α-smooth muscle actin. Concentrations of both $K_{IR}6.2$ and SUR2 transcripts heighten in fibroblasts of the scar region and border zone of infarcted hearts [656]. Upon activation, these channels can modulate cardiomyocyte repolarization and arrhythmia formation after myocardial infarction.

8.7.7 *Gap Junction*

The gap junction enables the propagation of ion fluxes between cardiomyocytes that govern the heart rhythm. Most gap junctions are located in longitudinal (end-to-end) intercalated discs. Intercalated discs also contain force-transmitting adherens junctions, mostly in lateral (side-to-side) intercalated discs, and desmosomes.

---

[28] Also known as antigen identified by monoclonal antibody Ki67 (MKi67). This nuclear protein intervenes in cell proliferation.

[29] Secreted frizzled-related proteins compete with frizzled receptor for Wnt ligands by direct binding of Wnt, hence preventing activation of proapoptotoic Wnt signaling. Secreted frizzled related protein-2 increases β-catenin level within hypoxic cardiomyocyte, possibly activating antiapoptotic genes.

Many connexin types are expressed in the heart, mainly connexin-43, connexin-40, and connexin-45 that are expressed in distinct combinations and relative quantities in different cardiomyocyte subsets (nodal, atrial, ventricular, epicardial, midmyocardial, and endocardial).

Myocytes of the sinoatrial and atrioventricular nodes have small, dispersed gap junctions composed of Cx45 protein. Downstream from the His bundle, nodal cells prominently express Cx40, but distal cells abundantly produce Cx43, although Cx45 is continuously synthesized from the atrioventricular node to the ends of the Purkinje fibers.

Ventriculomyocytes are interconnected by clusters of Cx43-containing gap junctions. Connexin-43 is the major myocardial gap junction protein responsible for rapid, quasi-synchronous transmission of cardiac action potentials.

Mutations of the GJA1 gene affect phosphorylation sites in Cx43 C-terminus and lead to cardiac malformations and hypoplastic left heart syndrome. Mutations of the GJA5 gene (Cx40) can be associated with atrial fibrillation. Transplantation of embryonic cardiomyocytes and skeletal myoblasts genetically engineered to express Cx43 in mouse myocardial infarcts protects against the induction of ventricular tachycardia [657].

Connexin expression and gap junction organization change after cardiac ischemia. Gap-junctional coupling between cardiomyocytes depends on: (1) amount and types of expressed connexins; (2) proportion of each assembled connexin; (3) size and distribution of gap junctions; and (4) their gating.

Acute, focal, cardiac ischemia induces rapid Cx43 dephosphorylation and subsequent lateralization (side-to-side connections) in the distribution of gap junctions [658]. In addition, Cx43 and/or pannexin hemichannels outside gap junctions are implicated in cell swelling, ATP release, and loss of membrane potential during ischemia. Moreover, Cx43 reduction and elevated Cx45/Cx43 ratio in the failing ventricle increase susceptibility to arrhythmia.

Myocardial ischemia leads to ATP release in the interstitial space, from which ATP can stimulate the P2 receptor of cardiac sensory (afferent) nerves, more precisely ischemia-sensitive, but not ischemia-insensitive cardiac afferents [659]. Myocardial ischemia activates cardiac sympathetic afferents that prime pain and reflex cardiovascular responses.

Phosphorylation of connexin-43 by different protein kinases, such as PKC and MAPK, is implicated in connexin expression, transfer, assembly, and degradation [660]. Furthermore, Cx43 phosphorylation influences gap junctional communication.

High-density lipoprotein ensures cardioprotection, i.e., promotes cell survival and myocardial function, especially that conferred by ischemic preconditioning, i.e., by brief episodes of ischemia prior to prolonged vascular occlusion. Concentration of high-density lipoprotein–cholesterol correlates inversely with the risk of cardiovascular diseases. However, HDL cardioprotective action does not depend on its cholesterol transport activity.[30] High-density lipoprotein also acts on vascular endothelial and smooth muscle cells, thereby exerting antiapoptotic, anti-inflammatory, antioxidative, and vasodilatory effects independently of cholesterol transport via PKC as well as P38MAPK, ERK1, ERK2, and JNK on the one hand and PI3K and PKB on the other (Table 8.5).

High-density lipoprotein actually operates during ischemia via sphingosine 1-phosphate, a HDL constituent, and connexin-43 phosphorylation (Ser368)[31] by protein kinase-C, more precisely PKCδ prior to ischemia[32] and PKCϵ during ischemia–reperfusion to ensure cardioprotection [660].[33]

The sphingolipid regulator determines the balance between the ceramide-induced apoptosis and S1P-primed cardioprotection [661]. Sphingosine 1-phosphate produced by sphingosine kinase-1 and -2 and bound to plasma HDLs targets $G\alpha_i$-coupled receptors that activate PKCϵ via a PI3K–PKB– NOS–NO cascade. The HDL/S1P–PKC axis operates independently of the S1P–$S1P_3$–PI3K–PKB–NO pathway.

Protein kinase-C may target the sphingosine kinase-1 and -2, mitochondrial permeability transition pore, mitochondrial ATP-sensitive K^+ channel, proteins of the apoptosis regulation (B-cell lymphoma (leukemia) protein BCL2, BCL2 antagonist of cell death [BAD], and BCL2-associated X protein [BAX]), in addition to connexin-43 [661].

Connexin-43 and PKCϵ also reside in mitochondria. Mitochondrial reactive oxygen species activate PKCϵ that then targets P38MAPK, ERK1, ERK2, and Cx43, which also participate in ischemic preconditioning [661]. Hemichannels based on connexin-43 in the inner mitochondrial membrane may contribute to mitochondrial K^+ influx necessary for ischemic preconditioning.

Connexin-40 expression without other connexin isoforms builds high-conductance gap junction in vitro, but in cultured mouse neonatal atriomyocytes, reduction of Cx40 increases propagation velocity. Content of Cx40 augments in Purkinje cells and adjoining endocardial cardiomyocytes.

Remodeling alters not only the distribution of gap junctions, but also the amount and type of produced connexins [658]. Connexin-43 interaction with zonula occludens-1 increases, but ZO1 limits the recruitment of connexons to gap junctions.

---

[30] Reverse cholesterol transport relies on the transfer of excess cholesterol from peripheral tissues to the liver.

[31] Phosphorylation of Cx43 Ser368 reduces electrochemical conductance of gap junctions, impeding gap junctional intercellular communication, as well as transfer of cell death agents between cardiomyocytes, hence limiting the infarct size [660].

[32] Protein kinase-δ activation during reperfusion launches cytochrome-C–caspase-3 signaling, hence apoptosis [661].

[33] Current therapy of ischemia–reperfusion is based on PKCϵ activators and PKCδ inhibitors during reperfusion to minimize myocardial injury [661].

8.7.8 *Inflammation in Myocardial Infarction*

During progression of atherosclerosis, myeloid cells destabilize atheromatous plaques and cause their rupture, thus triggering myocardial infarction and stroke. A vicious cycle happens: myocardial infarction provokes acute inflammation that aggravates causal atherosclerosis. Whereas atherosclerosis is a chronic inflammatory disease, myocardial infarction triggers an acute inflammatory response. Wound healing during the first 2 postinfarct weeks is characterized by a rapid turnover of cells and structural components. Pre-existing collagen is digested and new matrix is formed.

8.7.8.1 Monocyte Infiltration

In the short-term (days–weeks), poor healing can lead to infarct expansion and ventricular dilatation, and even, in some cases, to infarct rupture and death [662]. In the long-term (months–years), adverse remodeling can induce heart failure. On the other hand, proper healing preserves ventricular geometry and prevents heart failure.

Among leukocytes involved in unspecific (w.r.t. antigen recognition), regulated, innate immunity,[34] neutrophils accumulate in the infarcted myocardium in the first hours after onset of ischemia with a peak after 1 day using CXCL8 and CXCL1 chemokines as well as L- and P-selectin and ICAM1 adhesion molecules [662]. Monocytes[35] dominate healing of injured myocardium within the first 2 postinfarct weeks using CCL2 chemokine and its CCR2 receptor as well as VCAM1 [662]. Neutrophils, monocytes, and derived macrophages release peptidases and reactive oxygen species, hence exacerbating injury.

Efficient healing after myocardial infarction depends on a proper mobilization of monocytes to the ischemic myocardium. In fact, myocardial infarction induces the liberation of hematopoietic stem and progenitor cells from bone marrow niches via sympathetic nervous system signaling provoked by pain and heart damage [663]. These progenitors then seed the spleen, yielding a sustained boost in monocyte production.

Undifferentiated monocytes reside in large amounts in the subcapsular red pulp, the spleen also containing iron-recycling red-pulp macrophages, marginal zone

---

[34] During injury, monocytes and neutrophils accumulate quickly and eliminate dead or dying cells.

[35] Monocytes are generated in the bone marrow from macrophage and dendritic cell progenitors. However, a common dendritic cell precursor gives rise to predendritic cells and classical and plasmacytoid dendritic cells monocyte intermediates. Once monocytes are matured, they enter the blood circulation using the chemokine CCR2 receptor. They then circulate freely and patrol blood vessels for several days. They can reside in lymph nodes oand spleen. Upon tissue infiltration during inflammation, they differentiate irreversibly to macrophages or inflammatory dendritic cells. In humans, monocytes can be subdivided into 2 subsets according to expression of CD14 (LPSR; dominant subset [$\sim$85 % monocyte pool]) and CD16 (Fcγ R3). Inflammatory CD16+ monocytes produce TNFSF1 and a high CCR2 level and can release myeloperoxidase [662].

macrophages, and dendritic cells. These monocytes quickly enter the blood circulation upon proper signaling primed by ischemic myocardium, relocate to the infarct region, and contribute to inflammation regulation. Migration of splenic monocytes endowed with AT_1 receptor involves angiotensin-2, the level of which rises after myocardial infarction.

Monocytes and macrophages have destructive and reparative function, as they: (1) secrete inflammatory mediators, such as Ifnγ, TNFSF1, IL1, IL6, CCL3, and ROS; (2) liberate peptidases MMP2, MMP9, and MMP13, uPA, and cathepsins that digest the collagen mesh; (3) phagocytose dead myocytes and neutrophils and other debris; (4) promote angiogenesis via FGF and VEGF secretion; (5) transport reparative enzymes and prosurvival factors such as transglutaminases; and (6) stimulate collagen synthesis and deposition by myofibroblasts, as they release FGF and TGFβ [662].

Destructive and reparative monocyte subsets that intervene in the early (d1–d4) and late phase, respectively, express different chemokine receptors, hence responding differentially to chemokines released from the cardiac wound. Chemokine CCL2 released during the early healing phase recruits inflammatory CD16+, CCR2+ monocytes (peak at 2.6 days). These cells secrete TNFSF1, IL1β, myeloperoxidase, cathepsins, MMPs, and urokinase-type plasminogen activator. Chemokine CX_3CL1, the concentration of which rises during the following phase (peak at 4.8 days), recruits reparative CD14+, CCR2− monocytes that release IL10, TGFβ, and VEGF, hence promoting creation of microvessels and deposition of collagen [662].

Monocytes and macrophages remove necrotic cells, trigger angiogenesis, and initiate collagen synthesis by myofibroblasts. If the infarct region recruits insufficient numbers of monocytes, wound healing is delayed because debris is not cleared and granulation tissue and collagen matrix are not formed. However, sustained presence of inflammatory monocytes and/or exaggerated number preclude resolution of inflammation and reparative function of CD14+ monocytes, myofibroblasts, and endothelial cells, and cause adverse remodeling.

8.7.9 Angiogenesis in the Infarcted Heart

Cardiac myocytes and fibroblasts interact with endotheliocytes to regulate angiogenesis (Table 8.4).

Both Wnt1 and Wnt3 manufactured by cardiofibroblasts can elicit VEGF expression during cardiogenesis and after ischemic injury. After cardiac injury, fibroblasts also produce matrix metallopeptidases that support endotheliocyte migration and vascular sprouting. The Wnt proteins induce several MMP types. Cardiofibroblasts also synthesize tissue inhibitor of metalloproteinases that exerts both pro- and antiangiogenic effects and operates via the Wnt–β Ctnn signaling [35].

Table 8.4 Endothelial interactions that regulate neovascularization in the infarcted heart. (Source: [35])

Cellular source	Messengers	Effect
Cardiomyocyte	VEGF	Angiogenesis
	Wnt, sFRP1	
Cardiofibroblast	Wnt, sFRP1	Angiogenesis
	TIMP	
Macrophage	VEGFR1	Inhibition of excessive angiogenesis via a Wnt–NFAT pathway
Epicardiocyte	Angiogenic cytokines	Angiogenesis
ECDC		Adoption of endotheliocyte fate
Bone marrow	Dkk1/2	EPC mobilization
	Wnt1	EPC contribution to angiogenesis

ECDC epicardial-derived cell, *EPC* endothelial progenitor cell, *TIMP* tissue inhibitor of metallopeptidase

8.7.10 Chemical Mediators in Myocardial Infarction

During and after myocardial infarction, proinflammatory and autoreactive immunocytes are formed and/or activated. These cells can continuously damage the heart. Exosomes produced by macrophages, T lymphocytes, and dendritic cells favor differentiation of antiinflammatory and prohealing immunocytes, such as M2 macrophages and regulatory T cells and suppress the formation of proinflammatory immunocytes, thereby reducing and even halting progression toward excessive inflammation and adverse remodeling [43]. In addition, microvesicles are released from activated platelets and apoptotic endotheliocytes [43].

8.7.10.1 Wnt Morphogens

In the infarcted heart, the Wnt signaling targets both myocytes and fibroblasts. The amount of Wnt1, Wnt2, Wnt4, Wnt7a, Wnt10b, and Wnt11 rises markedly in the heart the first 2 weeks of ischemic myocardial injury (Tables 1.8 and 8.5; [35]). The expression of frizzled-1, frizzled-2, frizzled-5, and frizzled-10 is augmented after myocardial infarction. Epicardial and mesenchymal cells, fibroblasts in the subepicardial region, myofibroblasts, and subsets of endothelial and smooth muscle cells in the infarcted and periinfarcted regions are Wnt responsive.

Table 8.5 Wnt-dependent interactions between cardiac myocytes and fibroblasts after myocardial infarction. (Source: [35])

Mediator	Source	Effect
Wnt1	Fibroblast	Promotion of fibrosis and myocyte death (auto- and paracrine signaling)
Wnt3a		Inhibition of cardiac progenitor self-renewal
		Enhanced hypoxia–reoxygenation-induced apoptosis of rat cardiomyoblasts
Wnt5a		Activation of proinflammatory cytokine synthesis by macrophages
		Transdifferentiation of EPCs
sFRP1		Cardioprotection, antifibrotic action
sFRP2	Fibroblast	Promotion of fibrosis and myocyte death
sFRP4		Cardioprotection, antifibrotic action
Dkk1		Mobilization of bone marrow vascular progenitors (inhibition of Wnt signaling in endosteal cells)
Dkk2		Promotion of proangiogenic ability of EPCs

EPC endothelial progenitor cell

Table 8.6 MicroRNAs in serum, plasma, whole blood, blood mononuclear cells, and platelets in coronary atherosclerosis. (Source: [237])

MicroRNAs	Source
miR19, miR21, miR146, miR155, miR223	Plasma microparticles
miR17, mir21, miR27b, miR92a, miR126, miR130a, miR145, miR155, miR210	Plasma, serum
miR19a, miR29a, miR30e-5p, miR140-3p, miR145, miR150, miR155, miR181d, miR182, miR222, miR342, miR378, miR584	Whole blood
miR340, miR624	Platelet
miR135, miR147 miR146a/b miR134, miR198, miR370	Flowing mononuclear cells (stable angina pectoris) (unstable angina pectoris)
miR221, miR222	Flowing endothelial progenitors

8.7.10.2 MicroRNAs in Coronary Artery Disease and Myocardial Infarction

Vascular endothelial and smooth muscle cells as well as macrophages that are all involved in atherosclerotic plaque formation can release microRNAs in the blood circulation (Table 8.6). They can modify their miR secretion rate. These cells can also take up microRNAs from the blood circulation.

MicroRNAs in Peripheral Arterial Atherosclerosis

Concentrations of miR21, miR27b, miR130a, and miR210 in patients with peripheral artery disease (arteriosclerosis obliterans) increase and may then serve as possible markers [237]. Both miR27b and miR130a are correlated with the disease severity. In the absence of overlap between microRNA types, levels of which are augmented in coronary and peripheral atherosclerosis, these types of microRNAs can be considered specific for the site of origin of atheroma.

MicroRNAs in Coronary Arterial Atherosclerosis

Concentrations of cardiomyocyte-enriched miR133a and miR208a are elevated in patients with stable coronary artery disease. On the other hand, concentrations of miR126 and the miR17–miR92 cluster (miR17, miR20a, and miR92a) strongly produced in endotheliocytes as well as miR145 and miR155, which are highly expressed in smooth myocytes and inflammatory cells, respectively, are significantly reduced [237, 238]. Nonetheless, miR92 level rises after completion of an exercise-based rehabilitation program carried out after surgical coronary revascularization [237]. In addition, the reduction in plasma miR155 level in patients with coronary artery disease results from a drop in miR155 concentration in plasma microparticles [237].

Concentrations of miR19, miR21, miR146, miR155, and miR223 heighten in patients with acute coronary syndrome with respect to patients with stable coronary artery disease [237].

MicroRNA-130A has a proangiogenic effect, as it targets GAX, an antiangiogenic homeobox gene product. Microparticles released by macrophages of atheromatous plaques support intraplaque angiogenesis, therefore contributing to plaque instability [239]. MicroRNA-320 associated with monocytic microparticles is involved in ischemia–reperfusion injury.

MicroRNAs in Myocardial Infarction

The time course of miR release after myocardial infarction was explored (Table 8.7). In myocardial infarction, the plasma concentration of miR1, which is highly expressed in cardiac and skeletal myocytes, reaches a maximum at 6 h and returns to baseline 3 days after infarction [238].

Circulating levels of cardiomyocyte-derived miR208a and miR499 as well as those of miR208b, expressed in the heart and skeletal muscle, miR133a, and miR133b also markedly rise. Concentrations of miR1, miR133a, and miR133b peak at 2.5 h after the onset of symptoms in patients with myocardial infarction, whereas cardiac troponin-I and miR499 have slower time courses, as they peak at 6 and 12 h, respectively [237].

However, other highly expressed microRNAs in the heart (miR24, miR26a, miR30c, and miR126) are not affected. On the other hand, concentrations of miR1, miR133a, miR208a, and miR499 are significantly lower in the infarcted myocardium.

Table 8.7 In myocardial infarction, necrosis of cardiomyocytes liberates microRNAs into the blood circulation (Source: [237]). In patients with acute coronary syndromes, levels of miR133a and miR499 increase in the coronary venous sinus plasma with respect to aorta plasma samples; differences in coronary venous sinus and aorta plasma miR208 levels are not observed

MicroRNAs	Source
miR1, miR21, miR122, miR126, miR133a/b, miR208a/b, miR328, miR375, miR423-5p, miR499, miR499-5p	Plasma
miR30c, miR133, miR145, miR328, miR663b, miR1291	Whole blood

Plasma levels of leukocyte-expressed microRNAs (miR146, miR155, and miR223) are similar in myocardial infarction or viral myocarditis [238].

Overexpression of miR125b decreases myocardial infarct size after ischemia–reperfusion injury by about 60 % and limits the reduction of ejection fraction and fractional shortening, at it represses P53, BAK1, and TRAF6 expression in the myocardium and hence apoptosis and prevents NFκ B activation [664].[36]

8.7.10.3 Transcriptional Factors

Sex comb on midleg homolog SCMH1 is a constituent of the Polycomb repressive complex PRC1 that can regulate the sequential changes in chromatin modifications, as it modifies histones. It suppresses gene transcription, thereby ensuring neuroprotection against ischemia. It associates with promoters of two genes that encode K^+ channels, hence creating tolerance to subsequent ischemia [665].

Myocardium regeneration and scar attenuation[37] can result from reprogramming cardiac-resident scar-forming fibroblasts[38] into cardiomyocytes using retroviruses to deliver genes coding for transcription factors that regulate the heart development (GATA4, MEF2c, and TBx5 [666] as well as HAND2 [667]). Additional delivery of proangiogenic and fibroblast-activating thymosin-β4 further improves cardiac function [666]. However, reprogrammed cells constituted only a fraction of cardiomyocytes in the infarct border zone, which is ill-defined and forms only a part of the injured area. Therefore, grafted or reprogrammed cells should produce growth

[36] The transcription factor NFκ B regulates the expression of miRs (e.g., miR21, miR146, and miR155). In turn, these miRs lower NFκ B binding. Lipopolysaccharide suppresses the expression of miR125b and miR125b precludes TNFSF1 expression.

[37] Fibrosis made by activated cardiac fibroblasts prevents proper cardiac regeneration and contributes to susceptibility to arrhythmias.

[38] Fibroblasts represent a major fraction of cardiac cells, as the human heart is composed of approximately 40 % cardiomyocytes and 60 % cardiac fibroblasts. Fibroblasts can be reprogrammed into pluripotent stem cells, myocytes, and neurons by combinations of lineage-enriched transcription factors (Vol. 5, Chap. 11. Tissue Development, Repair, and Remodeling).

factors, cytokines, and other messengers that improve the survival and performance of preexisting cells and local blood flow [668].

8.7.10.4 Bradykinin and Prostanoids

During myocardial ischemia, numerous mediators are simultaneously released, such as thromboxane-A2 and bradykinin, which activate cardiac spinal afferents [669]. These messengers cooperate to stimulate ischemia-sensitive cardiac afferent endings to potentiate responses.

Microsomal prostaglandin-E2 synthase-1 (mPGES1) synthesized in leukocytes as well as cardiac myocytes and fibroblasts participates in the recovery of cardiac function after myocardial infarction [670]. In particular, mPGES1 in leukocytes is associated with less left ventricular dilation, cardiomyocyte hypertrophy (without influencing infarct size and pulmonary edema), leukocyte density in the infarct region, and COx1 concentration, hence prostaglandin-E2 level. Enzyme mPGES1 in leukocytes prevents COx1 expression and prostaglandin-E2 synthesis as well as inflammation and adverse remodeling in the infarct zone.

8.7.10.5 Growth Hormone and Insulin-Like Growth Factor-1

Growth hormone-releasing hormone (GHRH) activates reparative mechanisms in the injured heart after myocardial infarction using growth hormone (GH) and insulin-like growth factor IGF1 [671]. The GHRH receptor (GHRHR) resides on cardiomyocytes; GHRH transcript can be detected in the heart. The GH–IGF1 axis is cardioprotective, promoting cardiac precursor cell proliferation and impeding apoptosis. However, GH and IGF1 can have side effects (fluid retention, hypertension, arrhythmias, and diabetes).

Both cardiac myocytes and stem cells synthesize the receptor of growth hormone-releasing hormone (GHRHR). A synthetic GHRHR agonist, analog of human GHRH, precludes maladaptive ventricular remodeling (reduction of fibrosis and apoptosis) and improves cardiac performance recovery and myocardial regeneration by recruiting SCFR+ precursor cells after myocardial infarction [672]. It also promotes myocardial angiogenesis, increasing the capillary density. Cardiac SCFR+ cells originate from circulating cells or cardiac stem cells. The synthetic GHRHR agonist augments cardiac stem cell proliferation. Regulation of cell differentiation, proliferation, and apoptosis relies on the GATA4 transcription factor. Moreover, the synthetic GHRHR agonist supports action of growth factors, such as VEGFa, IGF1, and FGF2 as well as upregulates expression of CXCL12 chemokine.

8.7.10.6 Midkine

The heparin-binding growth factor midkine (midgestation and kidney protein), or neurite growth-promoting factor NEGF2, structurally related to pleiotrophin (or neurite growth-promoting factor NEGF1), supports cell proliferation and migration as well as fibrinolysis. It improves long-term survival after myocardial infarction, mainly by enhancing angiogenesis in the peri-infarct zone [673]. Its effects are mediated by the PI3K–PKB pathway as well as ERK and P38MAPK kinases, as well as expression of syndecans.

8.7.10.7 AMPK

AMP-activated protein kinase regulates energy-generating and -consuming pathways. This master regulator of energy metabolism is activated during ischemia. It is also stimulated by endocrine (e.g., adiponectin[39] and leptin) as well as auto- and paracrine factors. Macrophage migration inhibitory factor (MIF or MMIF)[40] is released in the ischemic heart by monocytes, macrophages, and vascular smooth and cardiac myocytes from storage pools. It then activates AMPK via MIF receptor CD74[41] linked to epican (or CD44), thereby linking inflammation and metabolism in the heart during hypoxia [674]. Agent MIF controls the release of other proinflammatory cytokines. In the heart, AMPK stimulates 6-phosphofructo 2-kinase activity and GluT4 glucose transporter expression to limit myocardial injury.

In addition to regulating energetic metabolism in cardiac cells, AMPK prevents death of these cells (cardiomyocytes, fibroblasts, and vascular smooth muscle and endothelial cells). Excessive TNFSF1 stimulation provokes apoptosis via the TNFRSF1 receptor and mainly by the extrinsic pathway that sequentially involves the formation of a death-inducing signaling complex (DISC), activation of caspase-8, release of mitochondrial cytochrome-C, and activation of the caspase cascade (apoptosome-bound caspase-9 cleaves [activates] caspase-3; Vol. 2, Chap. 4. Cell Survival and Death). In cardiomyocytes, TNFSF1 launches apoptosis via increased expression of NOS2, production of ceramides, and activation of the MAPK module. It also triggers necrosis via the RIPK1–RIPK3 complex in some cell types. Cellular stress-responsive AMPK is a potent cardiac protector against TNFSF1-triggered cardiomyocyte apoptosis via proapoptotic BAD[42] phosphorylation and subsequent

[39] Adiponectin suppresses hypoxia–reoxygenation-triggered apoptosis of cardiac myocytes and fibroblasts via AMPK.

[40] Also known as glycosylation-inhibiting factor (GIF) and phenylpyruvate tautomerase [638].

[41] Also known as HLA class-2 histocompatibility antigen-γ chain and HLADR antigen-associated invariant chain [638].

[42] B-cell lymphoma (leukemia) protein BCL2 antagonist of cell death (BAD) initiates apoptosis pathway.

suppression of its interaction with antiapoptotic protein BCLxL[43] and mitochondrial apoptotic signaling, as it represses cytochrome-C release and caspase-3 activation [675]. Kinase AMPK is activated by the Ca^{2+}-triggered Cam2K–CamK1 axis.

8.7.10.8 Mitochondrial Aldehyde Dehydrogenase

Mitochondrial aldehyde dehydrogenase AlDH2 is markedly activated in hearts that are the most resistant to ischemia-induced damage in rodent models [676]. Like adenosine and protein kinase-Cϵ that mimic ischemic preconditioning, AlDh2 confers cardioprotection.

8.7.10.9 Protein Disulfide Isomerase

Protein disulfide isomerase (PDI), an enzyme in the endoplasmic reticulum that catalyzes disulfide bond formation between cysteine residues in proteins for proper folding and isomerization as well as a chaperone that assists wrongly folded proteins to reach a correct folding and inhibits aggregation, is upregulated in endotheliocytes of myocardial capillaries exposed to chronic hypoxia. It protects against myocardial infarction, because [677]: (1) it limits cell apoptosis; (2) it favors adhesion to collagen-1; and (3) it stimulates angiogenesis.

8.7.10.10 Acetylglucosamine Transferase

The posttranslational modification O-linked attachment of β^{N}acetylglucosamine ($^{O}Glc^{NAc}$; oglcnacylation) by cardiac $^{O}Glc^{NAc}$ transferase (OGT) integrates glucose metabolism with intracellular protein localization and activity reduces cardiac dysfunction in infarcted hearts [678].

8.7.10.11 Peptidases

The blood concentration of kallikrein-3 serine peptidase (KlK3)[44] can increase during acute myocardial infarction. This elevation is related to higher occurrence of cardiac events than to diminution [679].

---

[43] B-cell lymphoma extra-large protein is a prosurvival protein that prevents the release of mitochondrial elements such as cytochrome-C. BAD phosphorylation facilitates its association with 14-3-3 proteins and prevents its translocation to mitochondria, thereby hampering its interaction with BCLxL and allowing the latter to promote cell survival.

[44] Also known as prostate-specific antigen (PSA) and member hK3 of the human kallikrein family of serine peptidases.

After myocardial infarction, matrix metallopeptidases, such as collagenases (MMP1, MMP8, and MMP13), gelatinases (MMP2 and MMP9), stromelysins (MMP3, MMP10, and MMP11), and membrane-type MMPs, as well as tissue inhibitors of MMPs operate in cardiac repair and remodeling. Oxidative stress intervenes in MMP regulation [253].[45]

8.7.11 Ion Carriers

Myocardial infarction can generate arrhythmias. After myocardial infarction, amplitude of K_V current components ($i_{K,to}$, $i_{K,slow1}$, and $i_{K,slow2}$) decay, whereas NFAT3 activity rises [680]. Activated NFAT3 causes myocardial downregulation of $K_V1.5$, $K_V2.1$, $K_V4.2$, and $K_V4.3$ that are involved in three distinct K_V-mediated current components. Persistent β-adrenoceptor–mediated signaling primes the PP3–NFAT pathway to reduce K_V current.

The sodium–potassium pump not only mediates transmembrane Na^+ and K^+ gradients, but also participates in Ca^{2+} handling in cardiomyocytes, in synergy with the SERCA pump, $Ca_V1.2$ channel, ryanodine receptor, and Na^+–Ca^{2+} exchanger. Its activity is controlled indirectly by oxygen that determines the redox state as well as nitric oxide. In addition to ATP depletion, hypoxia exerts a dose-dependent inhibition of Na^+–K^+ ATPase catalytic activity. During hypoxia, oxidative stress in the heart following *uncoupling in the mitochondrial electron transport chain*, attenuated NO availability,[46] and accumulation of oxidized glutathione (GSSG) linked to an increase in H_2O_2 and $ONOO^-$ decrease ^{S}nitrosylation and increase ^{S}glutathionylation (inactivation) of regulatory thiol groups of the catalytic α subunit of the Na^+–K^+ pump [681].[47]

---

[45] Neutrophils are the primary ROS source during reperfusion, although both endotheliocytes and cardiomyocytes generate ROS.

[46] Oxygen affinity of NOS1 and NOS2 is lower than that of NOx2 and NOx4 and mitochondrial cytochromes. Nitric oxide interacts with superoxide anion four orders of magnitude faster than with superoxide dismutase.

[47] Posttranslational protein modifications serve as a signaling mechanism, reversible (most often) or not, that trigger a spatiotemporally restricted signaling. ^{S}Nitrosylation is the covalent incorporation of a nitric oxide moiety from a NO donor into thiol groups (SH) of specific cysteine residues in proteins to form ^{S}nitrosothiol (protein^{SNO}). ^{S}Nitrosylation is reversed by denitrosylation, ^{S}Glutathionylation (protein^{SSG}) is another posttranslational modification of protein cysteine residues by thiol–disulfide exchange with oxidized glutathione (GSSG), i.e., a dimer linked by a disulfide bond. Alternatively, glutathione reduced form (G^{SH}) can bind to the protein thiol groups, once they have undergone ^{S}nitrosylation or oxidation to sulfenate (RSO^-). Glutathione is a low-molecular–mass tripeptide (CysH–Gly–Glu), which yields an abundant thiol. Protein ^{S}glutathionylation is involved in oxidative and nitrosative stresses. Subunit α of Na^+–K^+ ATPase bears oxidative thiol modifications, such as ^{S}nitrosylation and ^{S}glutathionylation (Cys46), and becomes inactive. ^{S}Nitrosylation is an intermediate step of the regulatory ^{S}glutathionylation. Remodeling of ATPase function is coupled with the reversible transition of SH groups between ^{S}nitrosylated and

In the cardiomyocyte, calcium handlers, especially ion carriers, can be ^{S}glutathionylated. Nitrosothiols are formed in reaction with N_2O_3, an adduct of NO and O_2 gas. In the heart, hypoxia does not support ^{S}nitrosylation because NO is converted to $ONOO^-$ instead of the N_2O_3 agent. The resulting $ONOO^-$- and H_2O_2-induced oxidation of thiols to thiyl radicals and generation of sulfenic anions (SO^-) as well as GSSG accumulation promotes thiol ^{S}glutathionylation [681].

Hypoxia dominates over acidosis and hypoglycemia in the regulation of Na^+–K^+ ATPase activity. Hypoxia via ^{S}glutathionylation, in addition to gradual ATP deprivation, suppresses Na^+–K^+ ATPase activity.

8.7.12 Sympathoexcitatory Reflex

Myocardial ischemia activates cardiac spinal afferents that mediate sympathoexcitatory reflex. A large quantity of thromboxane-A2 that is released by activated platelets contributes to myocardial ischemia-mediated sympathoexcitatory reflex by binding to prostanoid TP receptors [682].

After myocardial infarction, activation of the cerebral renin–angiotensin axis contributes to sympathetic hyperactivity and progressive adverse remodeling and dysfunction of the left ventricle. Production of angiotensin-converting enzyme and angiotensin AT_1 receptor increases in hypothalamic nuclei, such as the paraventricular and supraoptic nuclei. Both angiotensin-2 and -3 have similar affinities for AT_1 receptors. Aminopeptidase-A and -N are involved in processing cerebral angiotensin-2 and angiotensin-3, respectively. Cerebral aminopeptidase-A that generates angiotensin-3 from angiotensin-2 and angiotensin-3 operate in the sympathetic hyperactivity [683]. Cerebral, but not peripheral angiotensin-3 increases arterial blood pressure.

8.8 Ischemia–Reperfusion Injury

Acute myocardial hypoxia and ischemia provoke a myocardial dysfunction and cellular death according to duration. Revascularization after temporary ischemia, that is, reoxygenation of hypoxic cardiomyocytes, causes myocardial ischemia–reperfusion injury.

Reperfusion is a double-edged sword. Reperfusion reestablishes blood supply, hence preventing further ischemia-induced cell damage. A prompt reperfusion is a major determinant of infarct size and prognosis. However, reperfusion injury can expand the infarct area and thus aggravate adverse cardiac wall remodeling and

---

^{S}glutathionylated forms. Deglutathionylation by glutaredoxin (GRx) in the presence of NADPH restores ATPase activity [681].

Table 8.8 Myocardial ischemia and reperfusion and associated metabolic stress (Source: [397]). Concentrations of calcium, potassium, and sodium are tightly linked to fion fluxes through Na^+–Ca^{2+} exchanger and Na^+–K^+ ATPase, which depends on the metabolic context. During ischemia, elevated intracellular Na^+ concentration can increase Ca^{2+} level via Na^+–Ca^{2+} exchanger working in the reverse mode. An elevated Ca^{2+} level inactivates $Ca_V1.2$ channel. During reperfusion, washout of the acidotic, hyperkalemic, extracellular fluid reduces extracellular H^+ and K^+ concentrations. The resulting H^+ gradient raises Na^+ influx through Na^+–H^+ exchanger, exacerbating intracellular Na^+ and Ca^{2+} overloads

Ischemia-induced events	Effect
ATP depletion	Inhibition of Na^+–K^+ ATPase Increased K^+ efflux through K_{ATP} (extracellular K^+ accumulation)
Na^+ overload	Na^+–Ca^{2+} exchanger acting in reverse mode (Na^+ efflux, Ca^{2+} influx)
Ca^{2+} overload	Abnormal sarcoplasmic reticulum Ca^{2+} handling
Anaerobic metabolism	
Metabolic acidosis	Increased H^+ efflux through Na^+–H^+ exchanger (extracellular H^+ accumulation)

dysfunction. The ischemia–reperfusion injury can eventually contribute more to myocardial damage than ischemia itself. Restoration of coronary flow adds myocardial injury mediated by an overproduction of reactive oxygen and nitrogen species (respiratory burst) as well as by rapid transcriptional activation of proinflammatory genes responsible for a rapid amplification of the initial inflammation, in addition to cytosolic Ca^{2+} overload, thereby favoring contractile dysfunction and cell death in the postischemic myocardium. Moreover, a sudden restoration of arterial flow after cardiac ischemia can also engender ventricular fibrillation and death.

Conditioning aims at attenuating tissue injury triggered by sudden and strong ischemia and/or abrupt reperfusion by intermittent and mild ischemia and/or reperfusion.

8.8.1 Metabolic Stress in Ischemia–Reperfusion Injury

Ischemia–reperfusion injury refers to metabolic stress that disturbs concentrations of calcium, hydrogen, potassium, sodium, and phosphometabolites following myocardial ischemia and reperfusion. Perturbations in ion carrier functioning alter cardiac dynamics.

In fact, microvascular dysfunction in ischemia–reperfusion injury leads to inadequate oxygen supply, reduction in cellular energy stores, accumulation of noxious metabolites, and reperfusion injury mediated by reactive oxygen species (Table 8.8).

Reperfusion of the ischemic myocardium can result from action of thrombolytic agents such as tissue plasminogen activator. Free radicals provoke reperfusion injury in the ischemic myocardium. Therefore, despite restoration of epicardial flow, reperfusion actually triggers cardiac cell damage that impairs myocardial perfusion and cardiac function.

Reperfusion injury results from several mechanisms:

1. Production of reactive oxygen species by the mitochondrial electron transport chain, NADPH oxidases, cyclooxygenases, NOS, and xanthine oxidase;
2. Alteration in intracellular calcium handling and cytosolic calcium overload;
3. Coronary endotheliocyte dysfunction;
4. Neutrophil diapedesis and activation;
5. Transient impairment of left ventricular systolic contractility (*myocardial stunning*;
6. Acute diastolic dysfunction;
7. Reenergization-induced myocyte hypercontracture;
8. Arrhythmia; and
9. Cardiomyocyte death.

Chronic intermittent hypobaric hypoxia protects against ischemia–reperfusion injury, as it reduces density and activity of β-adrenergic receptors, at least in the right ventricular papillary muscle of rats [684].

Mechanisms of ischemic tissue injury are similar whatever the organ, as they involve calpain-mediated necrotic and cytochrome-C–caspase-mediated apoptotic pathways as well as MAPK modules and PKC and survival PKB pathways.

8.8.2 *Mitochondria in Ischemia–Reperfusion Events*

Regulated necrosis results from the coexistence of two distinct pathways:

1. Cyclophilin-D (PPId)-mediated mitochondrial permeability transition; and
2. Receptor-interacting protein kinase RIPK1–RIPK3-mediated necroptosis [685].[48]

During ischemia, mitochondria become uncoupled. In the infarcted myocardium, after reopening of an occluded coronary artery, the reperfusion-linked oxygen burst engenders dysfunctional mitochondria with subsequent DNA and cell fragmentation.

In isolated rabbit hearts, mitochondrial uncoupling causes alternans of action potential duration and intracellular Ca^{2+} transients (both in amplitude and duration) that may be markers of increased propensity to ventricular fibrillation during ischemia–reperfusion injuries [686]. Calcium transients are prolonged, action potential duration increases, and conduction velocity decreases.

[48] Regulated necrosis can also be induced by ferroptosis, pyroptosis, polyadpribose-polymerase PARP1-mediated necrosis, heat stroke-associated cell death, direct lysosomal membrane permeabilization, among other processes. Necroptosis is caused by death receptor signaling upon formation of the RIPK1–RIPK3–MLKL-containing necroptosome.

Reversible phosphorylation of mitochondrial proteins regulates respiratory function, energy metabolism, and mitochondrion-mediated cell death. Various kinases (PKA, PKB, PKC, and PKG, MAPKs, GSK3β, Src, hexokinase, and creatine kinase-2) reside in or are in close association with mitochondria. Inhibitory phosphorylation of glycogen synthase kinase-3β activity in the mitochondria augments the threshold for permeability transition pore opening. PhosphoGSK3β interacts with adenine nucleotide translocase upon reperfusion and suppresses permeability transition pore opening to ensure myocardial protection [687].

Cardiac-specific ANT1 overexpression enhances mitochondrial function and structure. Phosphorylated adenine nucleotide translocase ANT1 also confers cardiomyocyte protection [688].[49]

Uncoupling of mitochondria upon mitochondrial stress during ischemia–reperfusion injury leads to alternans of action potential duration and intracytosolic calcium transient amplitude and duration that can create ventricular arrhythmias [686].

8.8.2.1 Mitophagy

Heart ischemia and reperfusion lead to irreversible loss in cardiomyocytes by both apoptosis and necrosis. The mitochondrial member BNIP3 of BH3-only subset of the proapoptotic B-cell lymphoma protein-2 (BCL2) set expressed in the adult myocardium, contributes to autophagy during myocardial ischemia–reperfusion [689].[50] Autophagic vacuoles are observed in cardiomyocytes of ischemic and cardiomyopathic hearts. In the heart, autophagy rises in response to nutrient starvation and ischemia–reperfusion. Mitophagy involves ubiquitin ligase Park2 and protein Ser/Thr kinase PInK1. Autophagy primed by BNIP3 that ensures protection involves Park2 recruitment [273].

Hypoxia combined with acidosis can activate BNIP3 in mitochondria. Cardiac fatty acid-binding protein (^{C}FABP), primarily expressed in the heart (but also at low

[49] Phosphorylation of ANT is maintained by both pre- and postconditioning, but decays during ischemia–reperfusion.

[50] Autophagy enables the clearance of cytoplasmic proteins and intracellular organelles (Vol. 2, Chap. 4. Cell Survival and Death). Three different types of autophagy, macro- and microautophagy and chaperone-mediated autophagy, differ according to the mechanism of substrates delivery to lysosomes. *Microautophagy* refers to a direct engulfment of cytosolic materials into the lysosome. In *chaperone-mediated autophagy*, proteins are translocated by heat shock protein HSP70 to the lysosome. *Macroautophagy* involves the sequestration of cytoplasmic material and organelles by an isolation membrane to form a double-membrane autophagosome that then fuses to a lysosome to form the autolysosome, in which the sequestered materials are degraded by lysosomal enzymes. Products from the autolysosome are released to the cytoplasm and recycled for macromolecular synthesis and ATP generation. Autophagy is regulated by a group of autophagy-related gene (AtG) products. Detection of microtubule-associated protein-1 light chain LC3 together with autophagosome formation characterize autophagy. Excessive autophagy causes autophagic cell death.

concentrations in other tissues), is released by the myocardium about 20 min after the beginning of heart damage. Its detection in blood samples hence allows a rapid diagnosis.

8.8.2.2 Electron Transport Chain

Consumption of ATP by mitochondrial ATPase during ischemia and generation of reactive oxygen species from ^{ETC}complex-III alter cardiomyocyte functioning.

On the other hand, a transient and reversible inhibition of the mitochondrial electron transfer chain minimizes ischemia–reperfusion injury. Blockage of electron transport at ^{ETC}complex-I by nitrosylation preserves cell respiration during reperfusion [690].

8.8.2.3 Mitochondrial NOS Uncoupling

Mitochondrial production of nitric oxide is reduced during ischemia in the absence of O_2 required by nitric oxide synthase to generate NO and due to low intracellular pH during ischemia that inhibits the enzyme. During reperfusion, mitochondrial form (mtNOS) produces a burst of NO, thereby forming peroxynitrite ($ONOO^-$) [690]. Peroxynitrite derived from NO synthesized by cardiac mtNOS (^{C}mtNOS) causes oxidative modification of mitochondrial constituents and releases mitochondrial proapoptotic cytochrome-C.

Whereas physiological levels of NO inhibit mitochondrial permeability transition pore opening, supraphysiological NO concentrations sensitize this pore. Opening of the pore causes abrupt mitochondrial depolarization, ATP depletion, and cell apoptosis or necrosis.

8.8.2.4 Mitochondrial Calcium

Mitochondria respond to hypoxia–reoxygenation primarily by increasing the intramitochondrial Ca^{2+} levels via a shift in the balance between ionized and nonionized mitochondrial calcium in favor of the ionized form [690]. Elevated intramitochondrial ionized Ca^{2+} stimulates mtNOS and increases the mitochondrial production of O_2^- and $ONOO^-$, leading to oxidative and nitrative modifications of mitochondrial lipids and proteins.

Activated Ca^{2+}–calmodulin-dependent protein kinase CamK2 increases mitochondrial Ca^{2+} entry through the inner membrane mitochondrial Ca^{2+} uniporter (MCU), hence promoting opening of the mitochondrial permeability transition pore, mitochondrial disruption, and apoptosis in response to ischemia–reperfusion injury [691].

Calcium uptake by mitochondria attenuates cytosolic Ca^{2+} overload and activation of Ca^{2+}-activated peptidases. Furthermore, mitochondrial permeability transition pores of the inner membrane devoted to exchange of matrix ATP with ADP in the intermembrane space and a voltage-dependent anion channel of the outer membrane open when cells experience stresses, leading to ATP depletion, ROS generation, and mitochondrial Ca^{2+} overload. Pore opening upon reperfusion dampens the mitochondrial membrane potential due to H^+ leak through the adenine nucleotide translocator, which also compromises ATP generation by uncoupling substrate oxidation from ATP production. The opening threshold of the mitochondrial permeability transition pore is elevated by ischemic pre- and postconditioning.

8.8.3 Inflammation and Apoptosis

Interaction between flowing leukocytes and vascular endothelium enables inflammation with increased microvascular permeability to macromolecules in ischemia–reperfusion events.

Jun N-terminal kinase controls inflammation, cell proliferation, differentiation, and apoptosis. It promotes dephosphorylation of the BCL2 (BCL: B-cell lymphoma [leukemia]) family protein BAD (BCL2 antagonist of cell death) that then heterodimerizes with BCL2 and BCLxL, inactivating them and thus allowing BAX- and BAK-triggered apoptosis. Conversely, BAD phosphorylation leaves BCL2 free to inhibit BAX-initiated apoptosis.

8.8.3.1 Toll-Like Receptors

Limitation in the innate immune system restricts myocardial ischemia–reperfusion injury. Toll-like receptors contribute to ischemic cardiomyopathy and ischemia–reperfusion injury. In addition, low plasma levels of soluble mannose-binding lectin MBL2 is associated with a reduced mortality [692].[51]

In the heart, Toll-like receptors TLR2 and TLR4 recognize not only invading pathogens, but also endogenous ligands to modulate cardiomyocyte survival. In animal models of ischemia–reperfusion injury as well as hypoxic cardiomyocytes in vitro, lipopolysaccharides that target TLR4 attenuate cardiomyocyte apoptosis [693]. In addition, deficiency in TLR2, TLR4, or adaptor myeloid differentiation primary-response gene product MyD88 reduces myocardial inflammation and adverse ventricular remodeling after ischemia.

[51] Mannose-binding lectin, or mannose-binding protein-C is a soluble factor encoded by the MBL2 gene and secreted by the liver. This trimer pertains to the collectin family of molecules that share a collagen-like and a lectin domain. This pattern recognition molecule recognizes terminal mannose groups of bacteria.

8.8.3.2 Tumor-Necrosis Factor

In the heart, myocytes, mastocytes, resident macrophages, as well as vascular smooth myocytes synthesize tumor-necrosis factor-α. Once it is released after ischemia–reperfusion injury, TNFSF1 causes inflammation and apoptosis. It elicits leukocyte infiltration into the myocardium.

This factor causes microvascular dysfunction [694]. After ischemia, but before reperfusion, administration of antiTNFSF1 antibodies diminishes TNFSF1 expression in the postischemic mouse myocardium and contributes to partial restoration of NO-associated coronary vasodilation.

Tumor-necrosis factor-α enhances generation of superoxide [694]. Expression of TNFSF1 rises fourfold after ischemia–reperfusion (30 min/90 min). The formation of oxygen-derived free radicals depends on the generation of superoxide anion by endotheliocytes that contain xanthine oxidase among other ROS-producing enzymes and leukocytes that possess NADPH oxidase. Anti-TNFSF1 lowers superoxide generation, as they preclude activity of NADPH and xanthine oxidase.

8.8.3.3 Leukotrienes

Leukotrienes produced by arachidonate 5-lipoxygenase (ALOx5, or 5-lipoxygenase [5LOx]) participate in myocardial reperfusion injury after ischemia via leukocyte transmigration, vascular inflammation, microvascular constriction and augmented permeability, and subsequent matrix damage [695]. Moreover, perivascular edema affects cardiac function.

Activated granulocytes, which are sources of leukotrienes, adhere to endotheliocytes for subsequent extravasation and initiate inflammation, as they secrete proinflammatory mediators, platelet-activating factors, and reactive oxygen species, thereby increasing vascular permeability and provoking tissue damage.

Cysteinyl leukotrienes (LTc_4, LTd_4, and LTe_4)[52] are inflammatory lipid mediators synthesized from arachidonic acid in various cell types, such as basophils, eosinophils, neutrophils, monocytes, macrophages, mastocytes, and endotheliocytes in a transcellular manner with granulocytes.[53] They reduce myocardial contractility and blood flow rate [695].

Following neutrophil infiltration and concomitant synthesis of inflammatory mediators, the expression of G-protein–coupled $CysLT_2$ receptor in the heart is upregulated 2 days after reperfusion [695]. Overexpressed $CysLT_2$ increases vascular permeability as well as production of proinflammatory transcription factors and adhesion molecules (EGR1, VCAM1, ICAM1), which enable leukocyte diapedesis. In addition, LTb_4 mediates neutrophil chemotaxis to ischemic tissue.

[52] Leukotriene-C_4 is converted extracellularly to LTd_4 by removal of glutamic acid by γ-glutamyl transpeptidase and then LTe_4 by extraction of glycine by dipeptidase.

[53] Endotheliocytes convert granulocyte-derived LTa_4 into LTc_4 via glutathione transferase.

The localization pattern of group-2 secretory $_{S}$PLA2 (Vols. 4, Chap. 2. Signaling Lipids and 8—Chap. 7. Atherosclerosis—Biological Aspects) in infarcted myocardium and its temporal course in plasma is related to those of C-reactive protein.

Two proinflammatory, secretory phospholipase-A2 types with distinct localizations intervene in myocardial infarction. Subtype $_{S}$PLA2-10 resides in neutrophils, but not cardiomyocytes [696]. Once it is released from infiltrating neutrophils in the infarct myocardium, it damages cardiomyocytes, possibly via leukotriene-B4 and reactive oxygen species.

On the other hand, $_{S}$PLA2-5 has its highest levels in the heart. Its production rises in regions of cardiac infarction [696]. It processes arachidonic acid to produce LTb_4 and TxA2, possibly in cooperation with $_{c}$PLA2-4a.

8.8.3.4 Migration Inhibitory Factor

The pleiotropic cytokine macrophage migration inhibitory factor (MIF)[54] has a proatherogenic activity via the chemokine receptor CXCR2 as well as a proinflammatory effect during atherosclerosis via the alternative MIF-binding receptor CD74 complexed with CXCR2 [697].

After prolonged ischemia, MIF augments TLR4 signaling, promotes nuclear translocation of the NFκ B subunit P65, and subsequently increases apoptosis. When extracellular MIF interacts with CXCR2+ monocytes, inflammatory cell recruitment rises, thereby impairing cardiac function.

The heparan sulfate proteoglycan epican (CD44) constitues the CD44–CD74–CXCR2 complex involved in MIF signaling in both cardiac damage and protection (CD74: ligand-binding component; CD44: signal-transducing component). In addition, MIF interacts with another chemokine receptor, CXCR4. The interaction of MIF with both CXCR2 and CXCR4 rapidly activates integrins and mediates integrin-dependent arrest and chemotaxis of monocytes and T cells [697].

However, unlike to chronically evolving atherosclerosis, MIF can ensure cardioprotection in the acute phase of myocardial infarction as well as in ischemia–reperfusion injury, as it is involved in the coordination of receptor-primed cardioprotective AMPK signaling, inhibition of proapoptotic cascades, and reduction of oxidative stress in the postischemic heart (Table 8.9; [697]). The cardioprotection of MIF are regulated by ^{S}nitros(yl)ation.

The MIF cytokine operates as an auto- and paracrine cardiac regulator. This activator of myocardial AMPK regulates the energy-generating and -consuming pathways,

---

[54] Macrophage migration inhibitory factor is one of the first soluble immune mediators secreted from T cells in delayed hypersensitivity reactions. It is stored in and secreted from the pituitary gland upon endotoxemia. It regulates innate immunity by counteracting glucocorticoids. This inflammatory cytokine endowed with a chemokine-like function is quasiubiquitous, but only secreted from a selected number of endocrine and parenchymal cells, in addition to immunocytes.

Table 8.9 Cardioprotective and detrimental effects of MIF in myocardial ischemia–reperfusion injury (Source: [697]). It is produced in rat cardiomyocytes in response to hypoxia and hydrogen peroxide (H_2O_2). It is also synthesized in fibroblasts. Extracellular MIF interacts with the CD44–CD74–CXCR2 complex on cardiomyocytes. Upon ^{S}nitros(yl)ation, intracellular MIF functions as a potent oxidoreductase

Damage
TLR4 signaling and ↑ inflammation
$P65_{NF\kappa B}$ translocation and ↑ apoptosis
Promotion of inflammatory leukocyte recruitment and subsequent exacerbation of myocardial damage
CXCR2 supports MIF-dependent monocyte infiltration and subsequent impairment of myocardial function
Cytoprotection
Stimulation of cardioprotective AMPK in the ischemic heart, ↑ GluT4 translocation and glucose uptake during reperfusion
Inhibition of the JNK pathway during myocardial ischemia–reperfusion, ↓ BAD signaling and apoptosis
Attenuation of oxidative stress in the postischemic heart ^{S}Nitros(yl)ation of MIF is cytoprotective (↓ ROS production by MIF^{SNO})

BAD BCL2 antagonist of cell death, *GluT* glucose transporter, *ROS* reactive oxygen species, *TLR* Toll-like receptor, ↑ increase, ↓ decrease

thereby protecting the heart against ischemic injury and apoptosis [697]. The AMPK pathway controls glucose and lipid uptake, storage, and use. Moreover, AMPK phosphorylates (activates) NOS3, hence augmenting the nitric oxide production.

The MIF–CD74–CXCR4 interaction causes MIF endocytosis. During myocardial ischemia–reperfusion injury, MIF hampers JNK-mediated apoptosis via CD74 and CXCR4 receptors, more precisely, JNK-dependent, proapoptotic BAD dephosphorylation. In addition, MIF precludes JNK activation via Jun N-terminal activation domain-binding JAB1 protein.

On the other hand, the interaction of MIF^{SNO}, which has an elevated oxidoreductase activity, with JAB1 lowers, thereby impeding apoptosis [697].[55] The JAB1 protein controls the autocrine MIF-mediated PKB signaling, as it can impede MIF secretion. ^{S}Nitros(yl)ation of MIF (Cys81) reduces MIF–JAB1 interaction after myocardial ischemia–reperfusion injury [697].

The cardioprotective MIF effect is, at least partly, also mediated by CXCR2, although MIF signaling via CXCR2+ leukocytes exacerbates myocardial ischemia–reperfusion injury [697].

[55] The JAB1 protein is a component of the CoP9 signalosome involved in cell cycle control and ubiquitin–proteasome-dependent protein degradation. Monomeric JAB1 is also an activator of the JNK signaling and AP1 transcriptional activity.

In addition, MIF antagonizes activity of prohypertrophic and profibrotic agents in response to hemodynamic stress, as it maintains the redox homeostasis and attenuates the effect of stress-induced activation of hypertrophic cascades [697].

The MIF cytokine has a oxidoreductase activity on protein thiols that contributes to cellular redox regulation in the postischemic heart. ^{S}Nitros(yl)ation is a post-translational protein modification that increases the cardioprotective MIF potential in vivo. MIF^{SNO} catalyzes the formation of 2-hydroxyethyldisulfide [697].

8.8.4 Mediators in Myocardial Ischemia–Reperfusion Injury

8.8.4.1 Calcium Overload

Calcium handling in cardiomyocytes is managed by a set of ion carriers, such as $Ca_V1.2$, sarcoplasmic reticulum Ca^{2+} release channel, or ryanodine receptor, sarco(endo)plasmic reticulum Ca^{2+} ATPase (SERCA), and Na^+–Ca^{2+} exchanger, as well as their regulators such as phospholamban and processing enzymes Ca^{2+}–calmodulin-dependent protein kinase CamK2 and PP3 phosphatase. Controlled Ca^{2+} handling requires a large energy expenditure.

Energy depletion during ischemia provokes a sustained augmentation of cytosolic Ca^{2+} concentration mainly upon Ca^{2+} influx through sarcolemmal $Ca_V1.2$ during ischemia and Na^+–Ca^{2+} exchanger immediately after reperfusion following an elevated cytosolic Na^+ level that may result from inhibited Na^+–K^+ ATPase or activated Na^+–H^+ exchanger and Na^+–HCO_3^- cotransporter [698].

During early reperfusion, activation of Ca^{2+}-dependent PP3 and Ca^{2+}–calmodulin kinase-2 may potentiate cytosolic Ca^{2+} overload. The predominant cardiac isoform CamK2δ targets RyR, $Ca_V1.2$, and phospholamban. Its activity increases after ischemia–reperfusion [699].

In humans, Ca^{2+} reuptake in the sarcoplasmic reticulum of postischemic atriomyocyotes decreases after ischemia–reperfusion [699]. A reduced density functional SERCA2a contributes to cytosolic Ca^{2+} overload and contractile dysfunction after ischemia–reperfusion.

In its dephosphorylated state, phospholamban inhibits SERCA2a activity. Its phosphorylation by PKA or CamK2 relieves its inhibition on SERCA2a. In PLB− mice, absence of PLB exacerbates postischemic myocardial injury [699].

The redox modification following ischemia–reperfusion of $Ca_V1.2$, RyR, and SERCA can alter functioning of these ion channels and pumps [699].

Electrogenic Na^+–Ca^{2+} exchanger is influenced by Na^+ and Ca^{2+} gradients across the plasma membrane. It operates in the forward mode to extrude Ca^{2+} from the cell and in the reverse mode to import Ca^{2+} into the cell. It increases the cytosolic Ca^{2+} level during ischemia and early reperfusion. Inhibition of NCX reverse mode at the time of reperfusion reduces contractile dysfunction and myocardial injury [699].

Myocardial Na^+–H^+ exchanger contributes to the maintenance of intracellular pH and cytosolic ion homeostasis. During myocardial ischemia, intracellular acidosis develops quickly. This exchanger then extrudes H^+ in the extracellular space and imports Na^+ ion. However, Na^+ overload stimulates Ca^{2+} influx through NCX, thus causing Ca^{2+} overload. During myocardial reperfusion, blood flow lowers the extracellular Na^+ concentration, stimulating NHE to extrude more intracellular protons. Inhibitors of NHE limit ischemia–reperfusion injury in animal models [699].

8.8.4.2 ROS Generation During Reperfusion

Reactive oxygen species are produced in large quantities in the first few minutes of postischemic reperfusion [61]. Superoxide anion and other ROS oxidize myofibers already damaged by ischemia, thus eliciting apoptosis. Moreover, superoxide anion reacts with nitric oxide to form peroxynitrite that not only reduces NO availability, but also participates in myocardial lesions and vascular dysfunction. Deficiency of $NO^\bullet$ can also cause vasoconstriction and lead to the formation of microthrombi into the lumen of small vessels. Reactive oxygen species also facilitate adhesion of leukocytes to the endothelium and hence acute postischemic inflammation.

8.8.4.3 Gaseous Mediators

The three endogenous gases, carbon monoxide, nitric oxide, and hydrogen sulfide, exert preconditioning in ischemic tissues.

Leukocyte rolling and adhesion are impeded by exogenous H_2S in mice via P38MAPK [700]. Nevertheless, exogenous H_2S activity mediated by NOS3 against leukocyte adhesion is less strong than that on leukocyte rolling.

Ischemia–reperfusion injury is characterized by impaired availability of endothelium-derived nitric oxide. Administration of an antagonist of endothelin receptor reduces the infarct size and elevates postischemic endothelium-dependent vasodilation. Modest supplementation in dietary nitrite and nitrate intake protects against myocardial ischemia–reperfusion injury [701]. Ischemic preconditioning thus relies at least partly on nitric oxide that protects during reoxygenation.

In addition to ERK and the PI3K–PKB pathway, the adenosine A_{2B} receptor and PKG are cardioprotective after reperfusion. However, NO signaling downstream from PKB and ERK does not depend on PKG [702].

On the other hand, an excessive concentration of nitric oxide is involved in the onset of cell damage in hypoxia and ischemia–reperfusion events. These events are characterized by increased levels of oxidative species associated with mitochondrial dysfunctioning.

The peroxynitrite-induced tyrosine nitration of prostacyclin synthase is responsible for the hypoxia–reoxygenation-induced coronary vasospasm. In addition, peroxynitrite-mediated myocardial protein nitration heightens myocardial stunning [205].

Nitric oxide operates in cells by covalently modifying cysteine residues of effector proteins (^{S}nitrosylation).[56] ^{S}Nitrosylated proteins (protein^{SNO}) are in equilibrium with ^{S}nitrosoglutathione (G^{SNO}), which acts as a second messenger to transduce NO activity. ^{S}Nitrosothiols (G^{SNO} and protein^{SNO}) are degraded by ^{S}nitrosoglutathione reductase (G^{SNO}R). ^{S}Nitrosothiols protect against myocardial injury. ^{S}Nitrosylated hypoxia inducible factor HIF1α^{SNO} binds to the Vegf gene, promoting angiogenesis and increasing myocardial capillary density [703].

Nitric oxide impedes oxygen consumption in mitochondria by competing with cytochrome-C oxidase (^{ETC}complex-IV), thereby slowing oxidative phosphorylation carried out by the electron transport chain in the mitochondrial inner membrane at low O_2 concentrations [704]. It ^{S}nitrosylates proteins such as ^{ETC}complex-I, thereby reversibly regulating their activity. Some mitochondrial proteins endowed with ^{S}nitrosothiols can protect against ischemia–reperfusion injury, as they can ^{S}nitrosylate mitochondrial thiol proteins by transferring a nitrosonium group (NO^+) to protein thiolates and act as a vasodilator via NO release [704].

8.8.4.4 Reperfusion Injury Salvage Kinase Pathway

Several hormones and growth factors protect cardiomyocytes against acute ischemia–reperfusion injury by activating various sarcolemmal receptors that subsequently prime signaling pathways such as the *reperfusion injury salvage kinase pathway* (RISK). Peptides, such as autacoid kinins, hormones natriuretic peptides and adrenomedullin, as well as adenosine, erythropoietin, and adipocytokines also limit myocardial infarct size when they act during myocardial reperfusion.

The RISK pathway comprises a group of prosurvival kinases, such as PI3K, PKB, MAP2K1, MAP2K2, ERK1, and ERK2, that confer cardioprotection when they are activated during myocardial reperfusion [705]. The RISK pathway is also committed during ischemic pre- and postconditioning. Ischemic preconditioning results from brief episodes of myocardial ischemia; ischemic postconditioning from a short series of repetitive cycles of brief reperfusion and reocclusion of coronary arteries immediately at the onset of reperfusion.

---

[56] In the heart, ^{S}nitrosylation of G-protein–coupled receptor kinase GRK2, β-arrestin-2, $Ca_V1.2a$ channel, and ryanodine receptor Ca^{2+} release channel RyR2 improves β-adrenergic receptor-mediated contractility and intracellular calcium handling for cardiomyocyte excitation–contraction coupling. In the vasculature, ^{S}nitrosylation of caspases, dynamin, and ^{N}ethylmaleimide sensitive factor mitigate inflammation and apoptosis; ^{S}nitrosylation of hemoglobin (Hb^{SNO}) regulates oxygen delivery [703].

The RISK pathway can operate via mitochondrial ATP-dependent K^+ channel (mtK_{ATP}; $K_{IR}6.1/2$ without sulfonylurea receptor), PKCϵ,[57] and phospholipase-A[58] to reduce mitochondrial ROS production, thereby preventing opening of the mitochondrial permeability transition pore, as well as by activating antiapoptotic mechanisms.

8.8.4.5 MicroRNAs

A given microRNA is considered linked to acute ischemia–reperfusion injury (upon 2-h reperfusion) when its expression significantly changes with respect to controls and neither preceding preconditioning (Sect. 8.9) nor subsequent postconditioning (Sect. 8.10) influence the ischemia–reperfusion-induced change in the selected microRNA production, that is, remains unaffected by cardioprotective mechanisms [706]. Eight types of microRNAs (miR92a, miR125a-5p, miR322⋆ [sister strand of miR322], miR331, miR378⋆, miR494, miR652, and miR877) are associated with ischemia–reperfusion injury.

microRNAs can be acutely produced to ensure cardioprotection in ischemic pre- and postconditioning (without determined cardioprotective signaling). They are thus called protectomicroRNAs (protectomiR) [706]. Transfection of selected protectomiRs (miR125b⋆, miR139-5p, and Let7b) and inhibitor of microRNA-487b into cardiomyocytes protect against ischemia–reperfusion injury.

In isolated hearts from male Wistar rats subjected to [706]: (1) time-matched nonischemic perfusion (controls); (2) ischemia–reperfusion (i.e., 30-min coronary occlusion followed by 120-min reperfusion); (3) preconditioning (i.e., 3 episodes of 5-mn coronary occlusion) followed by ischemia-reperfusion; or (4) ischemia–reperfusion with postconditioning (i.e., 6 phases of 10-s global ischemia–reperfusion at the onset of reperfusion),

- Five microRNAs are significantly affected by both pre- and postconditioning (miR125b⋆, miR139-3p, miR188, miR320, and miR532-3p; both pre- and postconditioning noticeably upregulates miR188 and miR532-3p [but not

[57] Both PKCϵ and PKCδ are activated during ischemia–reperfusion injury and ischemic preconditioning.

[58] Phospholipase-A2β is the primary mediator of arachidonic acid release from cellular phospholipids during ischemia. Arachidonic acid can be oxidized due to cyclooxygenases, lipoxygenases, and cytochrome-P450 monooxygenases. Prostaglandins, leukotrienes, and eicosatrienoic acids contribute to heart adaptation in acute conditions, thus yielding beneficial effects on myocardial inflammation. However, they have detrimental effects and lead to a maladaptive response when they are chronically activated. In addition, activation of both lipoxygenases and mitochondrial PLA2 favor the generation of NO-derived nitroalkene derivatives ($R–NO_2$; R: alkyl) such as nitrolinoleate. Nitrolinoleate produced during ischemic preconditioning causes mild mitochondrial uncoupling, hence resistance to ischemia–reperfusion injury.

ischemia–reperfusion injury], whereas pre- and postconditioning appreciably reverse expression changes of miR125b*, miR139-3p, and miR320 engendered by ischemia–reperfusion injury);

- Four microRNAs are appreciably influenced by preconditioning (miR139-5p, miR192, and miR212 that are upregulated by preconditioning, and miR487b, the upregulation of which by ischemia–reperfusion injury is inhibited by preconditioning); and
- Nine microRNAs are specifically regulated by postconditioning (Let7b, Let7e, Let7i, miR1, miR181a, miR208, miR328, miR335, and miR503, among which Let7b, miR181a, miR208, miR328, miR335, and miR503 are up- or downregulated by ischemia–reperfusion injury and postconditioning notably precludes production change, the synthesis of the others [miR1, Let7e, and Let7i] being not influenced by ischemia–reperfusion injury).

MicroRNA-146a suppresses production of IL1R-associated kinase IRAK1 and TNFR-associated factor TRAF6 in the myocardium, thereby impeding NFκ B activation that happens during myocardial ischemia–reperfusion injury and protecting the myocardium [707]. MicroRNA-146 is an inhibitor of innate immunity and inflammation mediated by Toll-like receptors, particularly TLR4, whereas TLR2 ligands attenuate myocardial ischemia–reperfusion injury and NFκ B activity. In addition, microRNA-320 is also involved in cardiac ischemia–reperfusion injury, as it targets the HSP20 protein.

8.8.4.6 Cardioprotective Apelin

Apelin has vasodilatory and positive inotropic effects. During ischemia–reperfusion injury, $apelin_{13}$ diminishes oxidation of the calcium carriers sarco(endo)plasmic reticulum Ca^{2+} ATPase and RyR channel [708]. It operates via G-protein–coupled APJ receptor, the kinases PI3K (the upstream kinase of the reperfusion injury survival kinase pathway) and PKC, in particular intramitochondrial PKCϵ, as well as the mitochondrial ATP-sensitive potassium channel. It stimulates glucose uptake via the PI3K–PKB axis and reduces oxidative stress. In addition, PI3K activates PKC during the cardioprotective signal pathway. The mitochondrial PKCϵ and K_{ATP} channel pertain to a activation–reactivation circuit (memory-associated protection). $Apelin_{13}$ may decrease the concentration of reactive oxygen species during reperfusion, which cause structural and functional alterations due to abnormalities in Ca^{2+} handling by the sarcoplasmic reticulum and sarcolemma.

8.8.4.7 Cardioprotective Adenosine

In the extracellular milieu, adenosine is produced from ATP hydrolysis by the sequential action of ecto-nucleoside triphosphate diphosphohydrolase-1 (CD39) and ecto-5′-nucleotidase (CD73). Adenosine activates plasmalemmal receptors, thereby: (1) priming ATP synthesis and promoting cardiac cell survival; (2) eliciting vasodilation; and (3) obviating vascular leakage and leukocyte extravasation.

On endotheliocytes, connexin-40 supports expression of ecto-5′-nucleotidase, but lowers that of adhesion molecules hence regulating leukocyte recruitment [739].

Adenosine is a recognized protector against ischemia–reperfusion injury. Adenosine accumulates in the extracellular medium in response to prolonged ischemia and cell damage.

8.8.4.8 Cardioprotective Growth Factors

Growth factors that ensure cardioprotection using the RISK pathway at the onset of myocardial reperfusion include fibroblast growth factor, insulin, insulin-like growth factor, transforming growth factor-β1, vascular endothelial growth factor, cardiotrophin-1, and urocortin (Table 8.10; [705]). Moreover, several growth factors, such as FGF2, IGF1, and bone morphogenetic protein BMP2, promote cardiomyocyte differentiation in a paracrine manner from cardiac stem cells.

VEGF receptors are detected not only on vascular endotheliocytes, but also on cardiomyocytes. Cardiomyocytes that possess both VEGF receptors VEGFR1 and VEGFR2 produce VEGF in response to hypoxia. Activated VEGF receptors initiate the MAP2K1/2–ERK1/2 axis in cardiomyocytes and the PI3K–PKB–NOS3 cascade in endotheliocytes. Both FGF1 and FGF2 are secreted by cardiomyocytes in response to myocardial ischemia. They bind to FGFR1 to target signaling mediators PKCα, PKCδ, PKCϵ, and PKCζ.

Insulin binds to its specific protein Tyr kinase receptor. Subsequently, insulin receptor substrates are recruited and phosphorylated to prime the PI3K–PKB–^{P70}RSK (S6K) and PI3K–PKB–NOS3 pathways. Insulin is also able to trigger the JaK–STAT cascade. Moreover, insulin receptor substrates interact with growth factor receptor-bound protein GRB2 that can activate preassociated guanine nucleotide-exchange factor SOS to excite small Ras GTPase, which, in turn, stimulates Raf–MAP2K1/2–ERK1/2 signaling.

Insulin-like growth factors IGF1 and IGF2 target their respective protein Tyr kinase receptors IGF1R and IGF2R. Once autophosphorylated, these receptors phosphorylate insulin receptor substrates IRS1 and IRS2 associated with the PI3K–PKB and MAP2K1/2–ERK1/2 modules. Factor IGF1 reduces cardiomyocyte death and neutrophil accumulation.

Transforming growth factor-β1 is activated within the extracellular matrix and then binds to its receptor Tβ R2 that subsequently recruits and dimerizes with Tβ R1 to stimulate SMADs, PI3K, and PKB, as well as ERK1, ERK2, P38MAPK, and JNK kinases. Cardioprotection mediated by TGFβ1 is also associated with NOS3 activity.

Cardiotrophin-1 promotes cardiomyocyte survival, thereby limiting myocardial injury. Its production is enhanced by hypoxia. Cardiotrophin-1 is a member of the

Table 8.10 Cardioprotective signal transduction cascades triggered by hormones and growth factors and in the framework of the reperfusion injury salvage kinase pathway. (Source: [705]). Cardioprotection is elicited by inhibition of the opening of the mitochondrial permeability transition pore (mtPTP) owing to phosphorylation (inhibition) of GSK3β as well as that of the opening of the mitochondrial K_{ATP} channels by the NOS3–NO–PKG–PKCϵ cascade to avoid production of mitochondrial reactive oxygen species. In addition, calcium outflux due to augmented SERCA uptake into the sarcoplasmic reticulum attenuates the calcium-induced mPTP opening

Agent	Mediators within cardiomyocyte
CT1	JaK–STAT
	PI3K–PKB–p70S6K–BAD
	PI3K–PKB–NOS3–NO–PKG–PKCϵ–K_{ATP}
	PI3K–PKB–GSK3β–mPTP
	PI3K–PKB–NOS3–NO–PKG–SERCA
	Hsp70/90
	p38 MAPK
FGF1/2	Ras–Raf–MAP2K1/2–ERK1/2
	PI3K–PKB–p70S6K–BAD
	PI3K–PKB–NOS3–NO–PKG–PKCϵ–K_{ATP}
IGF1	Ras–Raf–MAP2K1/2–ERK1/2
	PI3K–PKB–p70S6K–BAD
	PI3K–PKB–NOS3–NO–PKG–PKCϵ–K_{ATP}
Insulin	Ras–Raf–MAP2K1/2–ERK1/2
	PI3K–PKB–p70S6K–BAD
	PI3K–PKB–NOS3–NO–PKG–PKCϵ–$_{ATP}$
	JaK–STAT
TGFβ1	Ras–Raf–MAP2K1/2-ERK1/2
	PI3K–PKB–NOS3–NO–PKG–PKCϵ–$_{ATP}$
	MAPKs p38 and JNK
VEGF	Ras–Raf–MAP2K1/2–ERK1/2
	PI3K–PKB
Urocortin	Ras–Raf–MAP2K1/2-ERK1/2
	PI3K–PKB

BAD BCL2 antagonist of cell death, *CT* cardiotrophin, *ERK* extracellular signal-regulated protein kinase, *FGF* fibroblast growth factor, *GSK* glycogen synthase kinase, *IGF* insulin-like growth factor, *JaK* Janus kinase, *JNK* Jun N-terminal kinase, *MAP2K* mitogen-activated protein kinase kinase, *PI3K* phosphatidylinositol 3-kinase, *PKB* protein kinase-B, *STAT* signal transducer and activator of transduction, *TGF* transforming growth factor, *VEGF* vascular endothelial growth factor

interleukin-6 family, hence a ligand for cytokine GP130-containing receptor.[59] The acute cardioprotection elicited by cardiotrophin-1 is carried out by several distinct signal transduction axes. Cardiotrophin-1 can upregulate heat shock proteins HSP70 and HSP90. Furthermore, cardiotrophin-1 binding to its receptor activates Janus kinases that recruit transcription factors STAT and prime the Ras–Raf–MAP2K1/2–ERK1/2 and PI3K–PKB–BAD antiapoptotic cascades. Cardiotrophin-1 stimulus can also leads to P38MAPK-mediated phosphorylation.

Urocortin as well as its analogs urocortin-2 (stresscopin) and urocortin-3 (stresscopin-related peptide) exert endogenous cardioprotection. Urocortin targets G-protein–coupled receptors CRHR1 predominantly in the brain as well as CRHR2 on vascular endotheliocytes and CRHR2α in human cardiomyocytes. Urocortin binding to CRFR2α allows plasmalemmal recruitment of PI3Kγ. The latter then recruits and activates PKB. It can also launch the MAP2K1/2–ERK1/2 cascade.

8.8.4.9 Angiotensin–Aldosterone Axis and Natriuretic Peptides

Angiotensin-2, aldosterone, and endothelin-1 promote myocardium remodeling. Aldosterone is also produced in hypertension-subjected and failing ventricles [709].

Antagonists of mineralocorticoid receptor (NR3c2), the aldosterone receptor and transcriptional regulator, which causes sodium and water retention in the kidney, reduce myocardial infarct size after ischemia–reperfusion injury via adenosine and A_{2B} receptor and prosurvival kinases (PI3K, PKB, PKC, ERK1, and ERK2), which prevent the opening of mitochondrial permeability transition pores at reperfusion. [710].

Atrial natriuretic peptide precludes renin release in kidneys as well as aldosterone synthesis in adrenal glands. It represses aldosterone synthase in cardiomyocytes [709]. This antagonist of the renin–angiotensin–aldosterone axis also prevents endothelin-1 synthesis in endotheliocytes. In addition, it hinders sympathetic nerve activity. Moreover, ANP reduces the activation of NFκ B and inhibits the secretion of inflammatory mediators, such as TNFSF1 and IL1β by macrophages. It hampers superoxide, lysozyme, and matrix metallopeptidase-9 release in activated neutrophils. It limits neutrophil adhesion to endotheliocytes. Therefore, ANP has cardioprotective effects. It impedes ventricular remodeling and improves the ventricle function after acute myocardial infarction.

[59] Cytokine receptor constituted by a GP130 subunit and a leukemia-inhibitory factor receptor subunit (LIFR, or CD118) heterodimerizes upon cardiotrophin-1 binding.

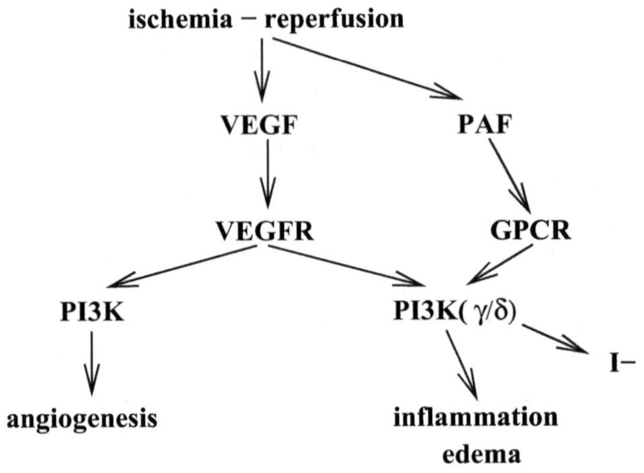

Fig. 8.1 Myocardial ischemia–reperfusion injury (Source: [713]). Two important mediators, vascular endothelial growth factor (*VEGF*) and platelet-activating factor (*PAF*), are produced by ischemic myocardium. Both act directly on endotheliocytes to increase vascular permeability. The former via different PI3K isoforms has a proangiogenic (positive) effect and edemagenic (negative) activity in myocardial infarction. The latter induces edema, promotes leukocyte adhesion to the hypoxic endothelium, activates neutrophils and platelets, and has a negative inotropic effect ($I-$) via PI3Kγ

8.8.4.10 Lipidic Mediators

Cyclooxygenase COx2 that catalyzes the conversion of arachidonic acid into prostaglandin-H2 is implicated in ischemia. Prostaglandin-E2 receptor EP_1, effector of COx2, impairs Na^+–Ca^{2+} exchanges and leads to neurotoxicity [711].

Although platelet-activating factor is a potent phospholipid mediator of inflammation, it also has a cardioprotective effect in postischemic injury via nitric oxide. The activated PAF receptor initiates nitric oxide-induced S-nitrosylation of Ca^{2+}-handling proteins (e.g., $Ca_V1.2$ channels) to attenuate Ca^{2+} overload during ischemia–reperfusion events in the heart [712].

8.8.4.11 PI3K and PKB

Phosphoinositide 3-kinases are activated either by receptor protein Tyr kinases or by G-protein–coupled receptors. Phosphoinositide 3-kinases can have a cell survival role during tissue ischemia, possibly PI3Kα and PI3Kβ isoforms. However, PI3Kγ and PI3Kδ isoforms contribute to reperfusion damaging inflammation in response to mediators, such as vascular endothelial growth factor, platelet-activating factor, cytokines, eicosanoids, histamine, thrombin, and complement factors ([713]; Fig. 8.1).

Leukemia inhibitory factor increases the association of protein kinase-B with mitochondrial hexokinase-2 to protect cardiomyocytes [714]. Both PI3K and PKB that can be activated by leukemia inhibitory factor phosphorylates hexokinase-2.

β-Arrestin–mediated AT_1 stimulation increases cardiac contractility and promotes cardioprotection via MAPK and PKB pathways in ischemia–reperfusion injury or mechanical stretch [715]. β-Arrestin–mediated AT_1 mediators are thus much more efficient then AT_1 blockers.

8.8.4.12 MAPK Module

Hypoxia–reoxygenation events rapidly and sequentially activate cell surface-associated members of the SRC family (Src and Fyn), small Ras GTPase, and mitogen-activated protein kinase kinase kinase cRaf to target extracellular signal-regulated protein kinases ERK1 and ERK2 and S6 kinase [716]. Cardioprotective agents include Janus kinases and their effectors STATs, in addition to PKCϵ, PKG, and sphingosine kinase, among others.

Deterioration of plasmalemmal permeability due to damaged membranes after even a short (5–10 min) period of ischemia causes leakage and release of various molecules, such as creatine kinase and lactate dehydrogenase. These substances have either beneficial (e.g., adenosine, bradykinin, catecholamines, and natriuretic peptides) or detrimental (e.g., tumor-necrosis factor-α, interleukin-1β and interleukin-6, and platelet-activating factor) effects on the heart.

These compounds launch an adaptive signaling that leads to resistance to a subsequent period of sustained ischemia. Chemicals, such as adenosine, adipocytokine, erythropoietin, insulin, and natriuretic peptides, reduce infarct size by activating the PI3K–PKB and ERK1/2 survival pathways.

In addition, a glycolytic enzyme, lactate dehydrogenase muscle subunit (mLDH, or LDH_m) protects cardiomyocytes against oxidative stress by increasing ERK1/2 phosphorylation during ischemia–reperfusion injury [717]. Activated ERK1 and ERK2 stimulate glycolysis and ATP generation in ischemic diseases. Erythropoietin attenuates cardiomyocyte apoptosis during myocardial ischemia–reperfusion via P38MAPK and heme oxygenase HO1 [718].

8.8.4.13 Small GTPases

Ischemia–reperfusion injury is associated with ROS production. Small Rac1 GTPase plays a dominant role in ROS generation after ischemia–reperfusion injury [719]. It also activates NFκ B and stimulates expression of several inflammatory genes (Tnfsf1 and Nos2). Moreover, activation of kRas modifies mitochondrial metabolism during hypoxia to promote growth, but also increasing ROS generation.

8.8.4.14 Calpains

The most abundant calpain isoforms, calpain-1 and -2, are ubiquitous. They predominantly localize to the cytosol, where they are inactive, bound to their endogenous inhibitor calpastatin. In response to an increase in Ca^{2+} concentration, calpains translocate to the plasma membrane, where connection to phospholipids or phosphorylation by PKA, ERK, or CamK2 may reduce the Ca^{2+}-determined activation threshold ($\sim$500 nmol) or localization near Ca^{2+} channels may trigger activation [698]. In addition to cytosolic species, mitochondrial calpains may have a proapoptotic role after Ca^{2+} overload in cardiomyocytes, as in other cell types.

The expression of calpain-2 increases 3 days after myocardial infarction, mainly in the interventricular septum, whereas calpain-1 expression peaks after 2 weeks, principally in the left ventricular free wall [272].

During ischemia, calpain remains inactive because of acidosis and increased ionic strength. At the onset of reperfusion, an uncontrolled and inappropriate activation of Ca^{2+}-dependent calpain happens [698]. During the reperfusion phase, Ca^{2+} concentration increases in sarcomere because of a reverse mode activity of sarcolemmal Na^+–Ca^{2+} exchanger, as the inactivity of Na^+–K^+ pump increases cytosolic Na^+ concentration [272]. In addition, a burst of oxygen free radicals occurs at the beginning of reperfusion; oxidative stress can modify calpain activity.

Calcium-activated calpain hydrolyzes sarcolemmal and cytoskeletal proteins. Calpain-mediated protein proteolysis contributes to [698]: (1) a dysfunction of ion carriers upon cleavage of ankyrin that binds to Na^+–K^+ ATPase, hence detaching it from the plasma membrane, in addition to eventual processing of Na^+–K^+ ATPase, Na^+–Ca^{2+} exchanger, RyR, and SERCA2a; (1) a dysfunction of the contractile cytoskeleton upon cleavage of sarcomeric titin, desmin, troponin-T and -I, as well as cytoskeleton-linked paxillin, vinculin, and talin; (3) a dysfunction of regulatory enzymes, such as PKC, CamK2, and PP3, and possible initiation of pathological cardiac signaling; (4) a sarcolemmal weakening and possible rupture upon cleavage of α-fodrin, a constituent of the membrane cytoskeleton, and subsarcolemmal dystrophin; and (5) apoptosis and necrosis. Proapoptotic calpain activates poly^{ADP}ribose polymerase and caspase-3.

Proteolysis of cell structural components for clearance can permit membrane repair and cell survival, hence cardioprotection, especially processing by calpain-4 [273]. On the other hand, in cardiomyocytes, calpain-1 and caplain-2 degrade myofibrillar proteins, such as myosin, troponin, tropomyosin, and titin, thereby provoking cell death and cardiac dysfunction. Dystrophin[60] that connects the cytoskeleton to the extracellular matrix via the plasma membrane is also altered by calpain. Destruction of dystrophin impairs sarcolemma integrity.

Ischemic preconditioning by brief episodes of myocardial ischemia limits cardiomyocyte death induced by subsequent prolonged myocardial ischemia–reperfusion. Ischemic postconditioning by prolongation of acidosis during the

[60] Dystrophin-related protein complexes consist of dystrophin, sarcoglycans, and dystroglycans.

beginning (first minute) of cardiac reperfusion protects cardiomyocytes, partly by limiting calpain activity. Protein kinase-G-dependent inhibition of Na^+–H^+ exchanger together with reduced lactate washout cause a delay in pH correction [698].

The ischemia–reperfusion event increases protein nitrosylation and carbonylation. ^{S}Nitrosylation can inhibit calpain activity in a pH-dependent manner. Ischemic preconditioning prevents calpain activation due to phosphorylation by PKA and prolonged acidosis during myocardial reperfusion. Calpain inhibition enables cardioprotection during pre- and postconditioning.

Calpain protective activity relies on cytoskeletal remodeling after membrane disruption and release of membrane microvesicles fusing with damaged membrane, this membrane patch sealing the membrane at sites of disruption. Extracellular fetuin-A facilitates plasma membrane repair by stabilizing extracellular protective calpain-2 [272].

8.8.4.15 Kallikrein

Tissue kallikrein protects against ischemia–reperfusion injury by promoting cardiomyocyte survival, as it reduces caspase-3 activation in rat myocardium after reperfusion [720]. Kallikrein binds to kinin (bradykinin) B_2 receptor and elicits phosphorylation of Src kinase, protein kinase-B (activation), glycogen synthase kinase-3β (inactivation), and BAD protein. The latter then forms a complex with 14-3-3 protein. The kallikrein–kinin pathway thus protects against cardiomyocyte apoptosis using both PKB–GSK3–caspase-3 and PKB–BAD–14-3-3 signaling cascades.

8.8.4.16 MuRF Ubiquitin Ligase

Muscle-specific ring-finger proteins (MuRF) are sarcomeric ubiquitin ligases. Subtype MuRF3 processes its partners for suitable turnover, thereby contributing to the maintenance of ventricular integrity after myocardial infarction, avoiding cardiac rupture [721].

8.8.4.17 TRPM2 Channel

Transient receptor potential melastatin channel TRPM2, a redox-sensitive, Ca^{2+}-permeable, nonselective cation channel, is expressed in the heart and vasculature. In particular, TRPM2 channels lodge in the sarcolemma and transverse tubules of adult left ventriculomyocytes. The cardiac TRPM2 channel is activated by H_2O_2, thereby causing a Ca^{2+} influx.

Neutrophils, among other immunocytes (e.g., monocytes, lymphocytes, and dendritic cells, which also produce TRPM2), are recruited to the heart during reperfusion. The neutrophil TRPM2 channel exacerbates reperfusion injury and in particular increases infarct size [722].

The TRPM2 channel (as well as TRPM6 and TRPM7) contains a catalytic domain with ^{ADP}ribose (ADPR) pyrophosphatase activity (chanzyme). Channel activation is triggered by ADPR as well as cyclic ADPR and nicotinic acid adenine dinucleotide phosphate [722].[61] It is also activated by oxidation by H_2O_2 as well as TNFSF1 involved in postischemic myocardial injury.

In the absence of the TRPM2 channel, expression of the Na^+–Ca^{2+} exchanger NCX1 rises and that of the $\alpha 1$ subunit of Na^+–K^+ ATPase decays, and action potential duration augments [723]. Compared with wild-type myocytes, after 2 h of hypoxia followed by 30 min of reoxygenation, ROS levels heighten in TRPM2 knockout left ventriculomyocytes and concentrations of radical scavenging enzymes such as superoxide dismutases and their regulators such as transcription factors forkhead box and hypoxia-inducible factor lower [723]. The TRPM2 channel protects hearts from ischemia–reperfusion injury, as it reduces ROS generation and fosters their scavenging.

8.9 Ischemic Preconditioning

Brief and intermittent episodes of ischemia and reperfusion[62] protect the heart[63] from subsequent, prolonged, severe ischemia and ischemia–reperfusion injuries with reduced infarct size and arrhythmias and improve ventricular function.

The *early preconditioning* immediately follows ischemia that ensures a first window of protection with a duration lasting several hours followed by a period without protection lasting about 12–24 h. Molecular events during this early phase rely on posttranslational modifications of pre-existing proteins. Then, the delayed, *late preconditioning* starts about 24 h after ischemia and yields a second window of protection that lasts several days (24–72 h). The late preconditioning involves gene transcription and protein synthesis.

Preconditioning can also be obtained by ischemia on distant organs or by pharmacological preconditioning. *Remote ischemic preconditioning* (rIPC) refers to as brief cycles of ischemia and reperfusion in the arm or leg by inflating and deflating a blood pressure cuff that can protect the heart against injury following a prolonged episode of coronary artery occlusion and reperfusion. Signals are transmitted by the nervous system and secreted factors.

Cardioprotection by ischemic preconditioning is triggered by adenosine, bradykinin, opioids, and platelet activating factor, produced in response to brief

[61] Stimulation of the TRPM2 channel can then result from activated polyADPR polymerase, polyADPR glycohydrolase, or cyclic^{ADP}ribose (cADPr) hydrolase-1.

[62] A first postconditioning protocol proposes 5-min ischemia and 5-min reperfusion. A second protocol consists of four cycles of 1-mn ischemia followed by 1-mn reperfusion; a third one six cycles of 10-s ischemia and reperfusion.

[63] As well as other organs, such as the kidney, small intestine, liver, and limbs.

Table 8.11 Examples of mediators of ischemic tolerance, which enhance cell survival and prevent apoptosis. (Source: [724])

Cell targets	Cell survival	Cell apoptosis
Plasmalemmal receptors	TNFSF1, adenosine, Peroxisome proliferator, $P2Y_2$	Death receptors, Nox, TNFSF6
Signaling effectors	Raf, MAPKs, PI3K–PKB, JaK2, STAT5, PKCϵ, Ceramide, NOS3, MLK3, Growth factors, Chaperones	Caspases, ASK1, GSK3β
Transcription factors	HIF, NFκ B, CREB, Activator protein-1, JNK, SRF, MEF2	FOXO
Mitochondrial molecules	BCL2, BCL2-like proteins-1/2, Thioredoxin, Uncoupling protein-2	Cytochrome-C, ROS, BCL2-associated protein X, BCL-associated death promoter
Nuclear compounds	DNA repair enzymes, Survival genes, Inhibitor of apoptosis proteins	P53, Proapoptotic genes, $Poly^{ADP}$ribose polymerase

BCL B-cell leukemia/lymphoma

ischemia–reperfusion cycles [61]. Their receptors launch signaling that ultimately prevent the opening of mitochondrial permeability transition pore during the reperfusion phase following ischemia. The mPTP opening disrupts mitochondrial structure and function and causes cell death.

8.9.1 Molecular Events in Ischemic Tolerance

Several compounds are involved in ischemic tolerance (Table 8.11).[64]

[64] The brain also resists ischemic injury. Ischemic tolerance augments with a nondeleterious preconditioning, such as hyperthermia, prolonged hypoperfusion, oxidative stress, before the ischemic event. During this adaptation, prosurvival genes can be activated to encode proteins that enhance brain resistance to ischemia. Inflammation and cell apoptosis are reduced. Survival-promoting mechanisms involve the MAPK module, nitric oxide, nuclear factor-κ B, hypoxia-inducible factor, neurotrophins, etc. Trophic factors (TGFβ, brain-derived neurotrophic factor, and glial-cell-derived neurotrophic factor) and VEGF might also contribute to the glial and vascular tolerance to ischemia. The primary adaptive response corresponds to a state of low metabolism and perfusion, reduced ion fluxes for stabilization of membrane functions, prevention of anoxic depolarization, efficient clearance of extracellular glutamate, and important glycogen storage [724].

8.9.1.1 G-Protein–Coupled Receptor Signaling and Nitric Oxide

During cardioprotective preconditioning ischemia, adenosine, bradykinin, and opioids released by ischemic myocardium bind to Gi-coupled receptors to prime signaling cascades. In particular, these cascades lead to nitric oxide production and opening of mitochondrial potassium channels. This effect indirectly activates protein kinase-C. Estrogens also contribute to preconditioning ischemia-mediated cardioprotection via PKC, especially PKCϵ [725].

In cultured newborn rat cardiomyocytes, adenosine receptors A_1 and A_3 are involved in cardioprotection by preconditioning via protein kinase-B [726]. Phosphorylation by PKB depends on Gi/o subunits, but not phosphatidylinositol 3-kinase. Released adenosine binds to its Gi-coupled receptor to activate PKC directly via phospholipases. Enzyme PKC triggers a signaling pathway that hinders opening of the mitochondrial permeability transition pore when the heart is reperfused.

Initiated pathways involve phosphatidylinositol 3-kinase that phosphorylates (activates) protein kinase-B. The latter phosphorylates (activates) nitric oxide synthase and (inactivates) glycogen synthase kinase.

Nitric oxide in turn activates protein kinase-G that stimulates mitochondrial PKCϵ and mitochondrial ATP-sensitive K^+ channels [727]. Mediators NOS3, PKG, activated bradykinin B_2 receptor, and caveolin-3 colocalize to form a signalosome. This signalosome activates mitochondrial ATP-sensitive K^+ channels via activated PKCϵ and PKG, whereas protein phosphatase PP2 prevents the stimulation of mitochondrial ATP-sensitive K^+ channels by B_2-mediated signalosome.

The infarct limitation of ischemic preconditioning relies on the NOS3–NO axis, in addition to durations of regional coronary ischemia and reperfusion ischemic periods, regardless of sex [728]. Deletion of NOS3 abolishes the cardioprotective effect of classical ischemic preconditioning, the survival rate being higher in female NOS3$^{(-/-)}$ mice.

Nitrites (concentration 0.3–1.0 μmol in plasma and 1–20 μmol in tissues) form a storage pool of nitric oxide that can be mobilized to trigger vasodilation during hypoxia. Nitrite anion is reduced to nitric oxide as oxygen concentration decays by xanthine-oxidoreductase, deoxyhemoglobin, and deoxymyoglobin, among others. Members of the heme globin family (cyto-, hemo-, myo-, and neuroglobin) can act as nitrite reductases that regulate hypoxia-induced NO generation [729].[65]

Nitric oxide inhibits cellular respiration by acting on the mitochondrial electron transfer chain, thereby limiting reactive oxygen species production and ensuring cytoprotection to myocardial ischemia–reperfusion injury. On the other hand, reactions between myoglobin and hydrogen peroxide that generate ferryl myoglobin and globin radical lead to oxidative damage. Nitric oxide reduces ferryl myoglobin to ferric myoglobin, thereby inhibiting the peroxidative activity of ferryl myoglobin.

[65] Nitric oxide production by nitric oxide synthase requires oxygen.

8.9.1.2 Hypoxia-Inducible Factor

Hypoxia-inducible factor HIF1 mediates adaptation to reduced O_2 availability (Vol. 4, Chap. 10. Other Major Types of Signaling Mediators). It regulates both O_2 delivery via vascular growth and O_2 use via the balance between oxidative and glycolytic metabolism.

This heterodimeric transcription factor consists of O_2-regulated HIF1α and HIF1β subunits, the latter being also able to dimerize with the aryl hydrocarbon receptor. Continuous hypoxia increases HIF1α stability. On the other hand, cycles of hypoxia and reoxygenation elevate both HIF1α synthesis and stability.

Both HIF1α and HIF1β in TIE2+ bone marrow and vascular endotheliocytes are required to ensure acute-phase cardioprotection during ischemic preconditioning, that is, without using its transcription regulatory role that can only be exerted in the late phase [730]. The HIF heterodimer functions as a transcriptional activator in the acute phase of ischemic preconditioning.[66]

Adenosine formed from extracellular ATP by ectonucleoside trisphosphate diphosphohydrolase-1 produced in cardiac endotheliocytes and AMP by ecto-5′-nucleotidase synthesized in vascular endotheliocytes are implicated in cardiac protection of ischemic preconditioning via HIF1 factor. The latter regulates expression of the adenosine A_{2B} receptor required for cardioprotection ensured by ischemic preconditioning.

Remote ischemic preconditioning induced by three clamping–unclamping cycles (clamping the femoral artery during 5 min followed by 5-min reperfusion) elevates plasma level of interleukin-10 and decreases myocardial infarct size in mice. The transcriptional activator hypoxia-inducible factor HIF1 binds to the Il10 gene when cardiomyocytes are subjected to hypoxia–reoxygenation cycles [731]. The HIF1–IL10 pathway is necessary and sufficient for remote ischemic preconditioning. Cardio- and neuroprotection mediated by IL10 is associated with increased cardiac PKB phosphorylation by PI3K.

ATP-sensitive, inward rectifier $K_{IR}6.2$ channel that is responsible for plasmalemmal $i_{K_{ATP}}$ current, but not mitochondrial $i_{K_{ATP}}$ current, is involved in the cardiac response to ischemia–reperfusion injury and cardioprotection ensured by ischemic preconditioning [732].

8.9.2 *Mitochondrion*

Mitochondria are involved in the generation of reactive oxygen species, ionic homeostasis, calcium storage, regulation of cellular metabolism, modulation of membrane

[66] Mice homozygous for a knockout Hif1a allele ($Hif1a^{-}/-$) die at midgestation with major cardiac malformations, whereas heterozygous mice ($Hif1a^{+}/-$) develop normally, but the acute protective effects of ischemic preconditioning is lacking.

potential, cell survival, and myocardial energetics. Mitochondria occupy about 30 % of the cardiomyocyte volume and produce more than 90 % of ATP fuel. Cardiac metabolism remains altered for an extended period of time after myocardial infarction.

Cardioprotection by preconditioning relies on opening of mitochondrial ATP-sensitive potassium channels (mtK_{ATP}; Table 8.12). Transient low concentrations of ROS and/or RNS, and hence redox signaling, can assist protective preconditioning [61]. The ROS source during ischemic preconditioning is the mitochondrion. Potassium ion influx in mitochondria supports ROS production. MtK_{ATP}-dependent ROS production requires the functioning of the electron transport chain and monoamine oxidase. Reactive nitrogen oxide species also protect the myocardium against reperfusion. Peroxynitrite can activate PKC at least in vitro, as it reacts with its thiol groups.

In cardiomyocytes, the inner mitochondrial membrane and the intermembrane space contains the cardiac 43-kDa connexin (Cx43) encoded by the GJA1 gene (gap junction protein-α1) and the protein kinase-C PKCϵ subtype.

Sarcolemmal and mitochondrial Cx43 are involved in the formation of gap junctions and production of reactive oxygen species for redox signaling, respectively. Connexin-43 phosphorylation (Ser or Tyr) regulates gap junction permeability, that is, electrical coupling between cardiomyocytes. Cx43 hemichannels in cellular membranes are involved in signal transduction, in particular cascades involving the $P2Y_1$ receptor and activator of G-protein signaling AGS8 [733].

Connexin-43 contributes to activation of a major cytoprotective signaling, that is, the PI3K–PKB-g-GSK3β pathway in cardiomyocytes. This cascade is triggered by adenosine receptor agonists, bradykinin, erythropoietin, and insulin. It is used during pre- and postconditioning. It launches ROS production by mtK_{ATP} opening.

A δ-opioid receptor agonist via endothelin-1 that bind to cognate GPCRs and insulin-like growth factor IGF1 that connects to its specific RTK provokes the phosphorylation of PKB and GSK3β, the two former via Gβ protein subunit and Cx43 [733]. In the absence of Cx43, ERK and P38MAPK are phosphorylated by ET1 [733].[67]

Preservation of the mitochondrial function ensures cardioprotection. In the mitochondrion, ischemic preconditioning suppresses H^+ leak through adenine nucleotide translocase that reduces mitochondrial membrane potential and uncouples substrate oxidation from ATP production after ischemia–reperfusion events. Moreover, ischemic preconditioning improves functional coupling between adenine nucleotide translocase and mitochondrial creatine kinase [735]. Protein kinase-Cϵ overexpression affords ANT functional capacity that favors myocardial protection.

[67] The Ras–ERK and PI3K–TOR signaling pathways are the main mechanisms for controlling cell survival, differentiation, proliferation, metabolism, and motility in response to extracellular cues [734]. These axes interfere to regulate each other (by crossinhibition, crossactivation, and pathway convergence) and coregulate cellular functions.

Table 8.12 Pathway activated prior myocardial ischemia and during reperfusion by ischemic preconditioning and postconditioning (Source: [61]). Prior to ischemia, signaling most often relies on protein kinase-C (PKC). During reperfusion, signaling cascades converge on mitochondrial permeability transition pore. G-protein–coupled receptor (GPCR) activators, such as acetylcholine, bradykinin, and opioids, trigger preconditioning-like protection via a mitochondrial ATP-sensitive potassium channels (mtK_{ATP}) and low reactive oxygen species (ROS) production. Two mitochondrial PKCϵ1 and PKCϵ2 pools exist. The former prime opening of mtK_{ATP}, mildly increasing matrix H_2O_2 that may promote further PKCϵ1 activation. The latter may inhibit mitochondrial permeability transition pore (mPTP) formation. Reactive oxygen species (ROS) released in response to mtK_{ATP} opening can limit injury. The mitochondrial protective signaling depends on PKG, PKCϵ, mtK_{ATP}, connexin-43, and ROS. PKC acts directly or not on several proteins of the mitochondrial envelope (mPTP, mtK_{ATP}, BCL2, BAX, and BAD). Adenosine acts via its cognate GPCR without requiring ROS to activate PKC. Ubiquitous glycogen synthase kinase GSK3β is inhibited by phosphatidylinositol 3-kinase (PI3K). During postconditioning tests, many signaling pathways also target the PKC effector and a low pH prevents mPTP opening

Prior to myocardial ischemia
GPCR–PI3K–PKB–NOS–NO–PKC
GPCR–PI3K–PKB–NOS–NO–GC–PKG–mtK_{ATP}–ROS–PKC
During myocardial reperfusion
Oxidative stress (massive ROS production)
Signaling during ischemic preconditioning
ROS signaling (low ROS production)
GPCR–PI3K–PKB–NOS–NO–PKC
GPCR–$PI3K_{c1\gamma}$–PKB–GSK3β–Cx43–mtK_{ATP}–ROS
GPCR–ERK1/2–BAD/BAX–mPTP
GPCR–PI3K–PKB–NOS–NO–GC–PKG–mtK_{ATP}–ROS–PKC
Signaling during ischemic postconditioning maneuvers
GPCR–PI3K–PKB–NOS–NO–GC–PKG–mtK_{ATP}–ROS–PKC
$\uparrow H^+ \ominus \longrightarrow$ mPTP
$\downarrow$ neutrophil accumulation and ROS production

$\uparrow$ increase, $\downarrow$ decrease, $\ominus \longrightarrow$ inhibition

Inhibition of the malate–aspartate transporter that links glycolysis to the electron transport chain ensures cardioprotection, as it reduces ROS production during late ischemia and early reperfusion [736].

Mitochondrial uncoupling proteins (UCP1–UCP3) are members of the superfamily of anion carriers. They localize to the mitochondrial inner membrane and regulate proton leak. Isoform UCP1 confers resistance to the hypoxia–reoxygenation event in cardiomyocytes. Subtype UCP2 also augments cardiomyocyte tolerance to oxidative stress. Isotype UCP3 ensures cardioprotection of ischemic preconditioning via modulation of the mitochondrial electrochemical gradient. Mild to moderate mitochondrial uncoupling by the small, reversible H^+ leak mediated by UCPs protects against ischemia–reperfusion injury via lowered ROS generation. Isoform UCP3 reduces the infarct zone area, incidence of arrhythmias, and ROS production as well as ameliorates recovery of the ventricular function [737]. It regulates myocardial energetics and ROS generation during ischemia–reperfusion injury.

8.9.3 Gap Junctions

Gap junctions between cardiomyocytes as well as between endotheliocytes enable the electrochemical coupling and intercellular transfer of small signaling molecules, hence supporting the function of the myocardium and coronary vessels. These between-cell channels are composed of 12 connexins (Cx). Connexins constitute a family of 21 transmembrane proteins. Endothelial connexins mainly comprise Cx40 and Cx37. Endothelial expression of gap junction proteins is influenced by oxidative stress, prothrombotic molecules, and proinflammatory cytokines. Changes in Cx43 expression, localization, and channel properties in cardiomyocytes contribute to infarction and reperfusion injury.

Gap junctions ensure a protective signaling during the trigger phase of ischemic preconditioning. Ischemic preconditioning supports suppression of chemical communications via connexin-43 phosphorylation by PKCϵ, but delays reduction of electrical communications through gap junctions after ischemia [738].

Endotheliocytes orchestrate neutrophil recruitment, as they control rolling, adhesion, crawling, and transmigration using chemokines and cell adhesion molecules. Endothelial Cx40 affects leukocyte recruitment in atherosclerosis. Neutrophils are the first inflammatory leukocytes to arrive within 24 h in the infarct zone, shortly after followed by the enrolment of monocytes. The cardioprotective effect of endothelial Cx40, but not Cx37, is associated with a decrease in neutrophil infiltration via activated 5′-nucleotidase (CD73) [739].

8.9.4 Cell Recruitment

Bone marrow-derived precursor cells give birth to smooth myocytes, inflammatory cells, and mastocytes into the injured heart. Fibrotic ischemia–reperfusion cardiomyopathy arises from daily, brief coronary occlusion. Fabrication and remodeling of connective tissue in the heart participate in cardiac repair. The production of CCL2 chemokine increases. Bone marrow-derived fibroblast precursors and monocytes are then attracted [740]. Proliferative fibroblasts express collagen-1, α-smooth muscle actin, CD34 sialomucin, and PTPRc phosphatase.

8.10 Ischemic Postconditioning

Ischemic postconditioning exerts cardioprotection, as it attenuates ischemia–reperfusion injury. Partial reperfusion reduces the infarct size. Controlled reperfusion with gradual recovery of blood flow also diminishes brain injury after infarction.

Ischemic postconditioning ensures cardioprotection by stuttering reperfusion after ischemia, that is, intermittent interruption of blood flow in the very early phase of a reperfusion. A short series of repetitive cycles of brief reperfusion and reocclusion (usually of the coronary artery) is applied immediately (i.e., less than 1–3 mn) after the onset of reperfusion. An example of postconditioning protocol is 3 cycles of 30-s reperfusion followed by 30-s occlusion at the onset of reperfusion [61].

Gradual reoxygenation after ischemia creates less cell death than abrupt reoxygenation [741]. Ischemic postconditioning with partial and gradual reperfusion (reperfusion by fits and starts or spluttering reperfusion) aims at preventing ischemia–reperfusion injury in myocardial and cerebral infarction.

As does ischemic preconditioning induced by various sublethal triggers (focal or global ischemia, inflammation, metabolic perturbations, oxidative stress, and hypo- and hyperthermia), postconditioning refers to a broad range of sublethal insults occurring immediately (*rapid postconditioning*) or up to 2 days (*delayed postconditioning*) after ischemia. Unlike preconditioning, postconditioning does not protect against stunning.

Postconditioning can reduce cell apoptosis and necrosis as well as endothelial dysfunction and activation and hence inflammation [61].

8.10.1 Mediators of Ischemic Postconditioning

Ischemic postconditioning can result exposure to noradrenaline or bradykinin [741]. Bradykinin B_2 receptors contribute to myocardial postconditioning protection. Opiods can also be involved.

The washout of adenosine is delayed. Adenosine A_1 receptor can participate in protecting the heart at reperfusion. Yet, adenosine receptor subtypes have different degrees of sensitivity during normoxia, ischemia, and reperfusion. Furthermore, adenosine is necessary, but not sufficient.

Rapid postconditioning reduces cytochrome-C release from the mitochondria to the cytosol, thereby limiting cell death. It also represses inflammation after infarction. Rapid postconditioning actually impedes myeloperoxidase activity, attenuates the synthesis of cytokines, such as IL1β and TNFSF1, as well as adhesion molecules such as ICAM1 molecule [741].

Delayed postconditioning enhances glucose uptake and/or metabolism, attenuates vascular leakage, hence edema formation [741].

Ischemic postconditioning improves the cardiac function after myocardial infarction and reduces reperfusion-induced cell damage, as it represses the aryl-hydrocarbon receptor (AHR) signaling via the expression of its effectors HSP90, AHR, aryl-hydrocarbon nuclear translocator (ANRT; or HIF1β), and β-tubulin, ultimately reducing infarct size [742]. The AHR receptor resides in the cytosol linked to an HSP90 dimer. A reduction of HSP90 level abrogates AHR function. Once it is liganded, AHR translocates to the nucleus, where it dimerizes with the ARNT protein. Ischemic postconditioning attenuates the AHR–HSP90 signaling and ARNT production.

Cardioprotection by ischemic postconditioning depends on redox signaling. The massive ROS production during myocardial reperfusion injury is reduced by ischemic postconditioning. Reactive oxygen species are protective in preischemic phase, but deleterious in the postischemic phase at high concentrations. However, at low concentrations, ROS ensure cardioprotection in ischemic postconditioning [61]. Many G-protein–coupled receptor activators (e.g., acetylcholine, adenosine, bradykinin, and opioids) trigger cardioprotection via the mtK_{ATP}–ROS axis (Table 8.12). Therefore, cardioprotection during reperfusion relies on ROS according to their concentration and/or compartmentation.

In addition to nitric oxide, superoxide, and peroxynitrite, signaling kinases PKB, PKC, MAPK, and K_{ATP} channel are also involved in ischemic postconditioning.

Reperfusion after infarction leads to ROS eruption that alters the PKB pathway, increases the activity of PKCδ, but decreases that of PKCϵ,[68] and activates JNK and ERK kinases. Postconditioning precludes cleavage of PKCδ, a cell death trigger, reducing its activity, and enhances phosphorylation (activation) of PKCϵ, a cell survival promoter [741].

The PKB pathway contributes to cell survival after infarction. Inhibition of the PI3K–PKB axis directly suppresses phosphorylation of GSK3β, hence favoring degradation of β-catenin, and indirectly the release of cytochrome-C and stimulation of caspase activity [741]. Rapid postconditioning increases PKB-mediated phosphorylation, but does not affect the activity of phosphatase and tensin homolog deleted on chromosome 10 (PTen) and phosphoinositide-dependent protein kinase PDK1 [741].

The MAPK modules (ERK, P38MAPK, and JNK pathways) intervene in ischemic injury and cell survival after infarction. Both JNKs and P38MAPKs are detrimental; ERK1 and ERK2 ensure cell protection, but also injury exacerbation, as they support inflammation and oxidative stress [741]. Rapid postconditioning reduces phosphorylated ERK levels in the penumbra.

The sarcolemmal and mitochondrial K_{ATP} channels open upon ATP depletion after ischemia, thereby ensuring cell protection.[69]

---

[68] Members of the PKC set that contribute to ischemic preconditioning depend on species: PKCα in the dog, PKCδ and PKCϵ in the rat, and PKCϵ in the rabbit [743].

[69] The mitochondrial K_{ATP} channel in cardiomyocytes is an effector of PKCϵ in preconditioning [743].

As in ischemic preconditioning, a series of interruptions of reperfusion reduces peroxynitrite and nitrotyrosine formation [205]. Redox modification of enzymes of the NO–sGC–cGMP–PKG axis and PDE causes a partial decrease in the response to endothelium-dependent and other NO-dependent vasodilators. Superoxide or peroxynitrite can inhibit sGC and impair the activity of K^+ and Ca^{2+} channels [205].

8.11 Reinfarction

Approximately 90 % of patients survive after a first myocardial infarction, but a very high risk of cardiac events exists in the following first year. More than 17 % of patients experience another myocardial infarction. In fact, new myocardial ischemia happens in 54 % of patients within the first year after myocardial infarction [663]. Reinfarction can result from evolving atherosclerosis in other segments of the coronary arterial bed after therapy of the first attack.

Mortality associated with reinfarction in remodeled hearts is at least twice that of the first attack, especially [744]: (1) when a decompensated heart failure exists (with respect to compensated hypertrophy with mild reduction in ejection fraction); (2) after a relatively long duration of ischemia (> 15 min); and (3) when the causal coronary segment is situated upstream from interventricular septum-perfusing branches.

8.11.1 Persistent Inflammation

Unstable, inflamed atherosclerotic plaques trigger ischemic injuries that in turn launch acute inflammation, thereby aggravating atherosclerosis when the early destructive phase is sustained rather than being replaced by the secong stage of reparative healing (Sect. 8.7.8). Reinfarction can then occur.

The development of larger atherosclerotic lesions with higher inflammatory cell content and protease activity is associated with a marked increase in monocyte recruitment in atherosclerotic plaques. Inflammatory monocytes infiltrate lesions and, together with their descendent macrophages, boost inflammation and deliver peptidases that digest extracellular matrix constituents and render atherosclerotic plaques unstable [663].

8.11.2 Disturbed Metabolism

Coronary revascularization of a given coronary segment increases patient survival after myocardial infarction, but also ventricular remodeling and heart failure with limited respiratory chain and tricarboxylic acid cycle capacity [745].

Postinfarct remodeled hearts are less tolerant than normal hearts to ischemia–reperfusion injury, with a marked decline in palmitate oxidation and acetylCoA energy production after ischemia–reperfusion [745].

A metabolic shift happens during development of maladaptive cardiac hypertrophy and failure, in addition to the change in intracellular signal transduction (Sects. 3.2.4 and 3.2.6.1). In healthy hearts, oxidation of fatty acids covers more than 70 % of the cardiac energy need. The remainder results mostly from glucose oxidation. Myocardial triglycerides form an energy reservoir when fatty acids are mobilized from triglyceride stores for energy production. Four enzymes of mitochondrial β-oxidation of fatty acids (fatty acid degradation; acylCoA dehydrogenase,[70] enoylCoA hydratase,[71] hydroxyacylCoA dehydrogenase,[72] and ketoacylCoA thiolase)[73] constitute an efficient processor for fatty acid oxidation. The oxidation of the fatty acid molecule yields much more ATP (~ 129 ATP from palmitate) than glucose (~ 36 ATP). However, for oxygen cost, the oxidation of glucose (3.17 ATP/O_2) is 11 % energetically more efficient than that of fatty acids (2.83 ATP/O_2) [745]. Therefore, glucose oxidation is used as a short-term adaptation in ischemia–reperfusion injury after limited oxygen supply. In ischemia–reperfusion events, increased fatty acid oxidation is detrimental, whereas elevated glucose oxidation and concomitant reduced fatty acid oxidation is beneficial [745]. In addition, fatty acid oxidation uncouples the electron from proton flux in mitochondria, increasing the formation of reactive oxygen species, and uncouples glycolysis from glucose oxidation, enhancing detrimental proton production. Increased glucose oxidation and decreased fatty acid oxidation improve functional recovery and cell survival. However, augmented fatty acid flux improves cardiac contractility during postischemic reperfusion in neonatal hearts that strongly depend on glucose oxidation for energy production [744].

During pathological heart remodeling, fetal gene reprogramming supports glycolysis for energy production. Substrate preference change from fatty acids to glucose cannot compensate the decrease in ATP synthesis of fatty acid oxidation and causes an impaired postischemic cardiac contractility.

Reversal of substrate preference in postinfarct remodeled hearts with lowered HIF1α-mediated hypoxic response as well as elevated PPARα–PGC1α activity and activation of long-chain hydroxyacylCoA dehydrogenase, which boost fatty acid mobilization and oxidation, respectively, supports postischemic recovery of cardiac contractility [745]. Sirtuin-1 and -3 deacetylate (activate) PGC1α and long-chain

---

[70] Enzyme that catalyzes the initial step of fatty acid β-oxidation in mitochondria with flavin adenine dinucleotide (FAD) as a redox cofactor, producing enoylCoA.

[71] EnoylCoA hydratase (ECH) catalyzes the second step of β-oxidation of fatty acids. This fatty acid-metabolizing enzyme manufactures hydroxyacylCoA.

[72] Oxidoreductase involved in fatty acid metabolism. It catalyzes the third step of β-oxidation. It processes two substrates, hydroxyacylCoA and NAD^+ to form three products, ketoacylCoA, NADH, and H^+.

[73] KetoacylCoA thiolase, also known as thiolase-1 and acetylCoA acetyltransferases [ACAT], synthesizes acetylCoA and acylCoA in the fourth step of β-oxidation of fatty acids.

hydroxyacylCoA dehydrogenase, respectively [745, 744]. Activity of these sirtuins is regulated not only by their concentrations, but also by their intracellular localization and NAD^+/NADH ratio. The NAD^+/NADH ratio decreases during ischemia because the limited oxygen supply prevents NADH oxidation by the mitochondrial electron transport chain [744]. In addition, activated ATP-sensitive K^+ channels may lower myocardial infarction size and enhance recovery of cardiac contractility.

Chapter 9
Interventional Medicine and Surgery of Cardiac Diseases

Preventive and curative treatment and proper medical devices are needed in chronic vascular diseases arising from exposure to various risk factors (tobacco, pollution, stressed and sedentary life, inappropriate diet, etc.) in aging populations. Current medical technology is unable to detect vascular lesions before they reach an evolved and irreversible state with sudden, unpredictable, and life-threatening complications.

Two main therapies are applied to cardiovascular diseases.

1. Surgical reconstruction aims at maintaining blood flow to irrigated tissues. Surgical procedures attempt to be minimally invasive.
2. The alternative to surgical operation is image-guided endovascular procedure. Interventional medicine reduces surgical risks, is less expensive, and can have the same efficiency as surgery.

Both techniques require unusual gestures, hence training.

Clinicians, either using catheter-based procedures for implantation of medical devices or mini-invasive surgery, thus require:

1. Appropriate visualization of the diseased region and its surroundings
2. Determination of local flow indices and flow disturbances
3. Planning tools for the treatment strategy to minimize the complication risks
4. An adequate check-up to investigate the effect of the therapy on blood flow and help in prognosis
5. Simulators to train for these new techniques

Computer sciences are involved in the development of medicosurgical simulators. Image processing coupled with virtual reality as well as numerical simulations of organ strains and stresses and blood flows lead to computer-aided medical diagnosis, treatment planning, and prognosis, as well as improved navigation during image-guided therapies. Finally, telemedicine and telesurgery require robotics and suitable communication technologies.

Any treatment procedure must limit flow disturbances, such as flow stagnation regions, local increased resistances to flow, and strong flow impacts on vascular walls with the risk of wall damage.

M. Thiriet, *Diseases of the Cardiac Pump*, Biomathematical and Biomechanical Modeling of the Circulatory and Ventilatory Systems 7, DOI 10.1007/978-3-319-12664-7_9

Simulations of blood flows have been carried out in the models of cardiac cavities and vascular segments to assess fields of applied mechanical stresses and strains. Studies were first carried out in idealized geometries. Since the end of the last century, computational domains were determined by three-dimensional reconstruction of vessels and organs.

However, medical imaging yields a model of anatomy of the explored organ of cardiovascular apparatus in a given patient at a fixed moment. Medical image acquisition is thus the first modeling stage of a multimodeling investigation leading to numerical simulations (final modeling stage) of blood circulation.

Moreover, the three-dimensional reconstruction (second modeling stage) of organs relies on image processing techniques based on a set of assumptions, hence adding another stratum of modeling. The resulting reconstruction of the cardiovascular domain of interest corresponds to a frozen state of a physiological apparatus, which, in fact, is characterized by a strongly varying configuration for adaptation to environmental conditions.

Flow is computed through reconstructed cardiac chambers and diseased vascular segments, among which are arterial grafts and cardiovascular domains with implanted medical devices. Computational investigations are also aimed at optimizing surgical procedures as well as design of medical devices.

A major objective of current research is to integrate medical imaging and computational fluid and solid dynamics. In particular, an adequate functional imaging is aimed at describing the local blood flow better than sectionwise velocity measurements by Doppler ultrasound and magnetic resonance velocimetry with limited spatial and temporal resolution.[1] Information retrieval from complex blood flow datasets may actually enable the display of most relevant flow patterns that can be used by physicians.

9.1 Medical and Surgical Robotics

Interventional computerized tools can be coupled with precise robots. *Robot-supported medical and surgical systems* (RSMCS) are aimed at placing therapeutic devices, possibly from remote locations [747, 748]. To execute the gesture with a robot, the surgical system must accurately determine the position and orientation of the target. On the other hand, the robot can use surgical manipulators with many

[1] Instantaneous blood volume flow rates can be estimated using Doppler ultrasound velocimetry, which relies on a less costly and light equipment than that of magnetic resonance velocimetry, hence a primary source of blood flow quantification. Instantaneous blood volume flow rates can be assessed from the spectral mean velocity in uniformly insonated blood vessels. As uniform insonation is rarely possible, the mean velocity is calculated from the assessed maximum velocity using a narrow beam in a small exploration volume, assuming a fully developed flow (steady Poiseuille or pulsatile Womersley velocity distribution), which is unnatural [746].

degrees of freedom to manipulate target organs on a smaller scale than the current one.

Small and tiny robots are aimed at: (1) exploring remote locations using vessels and hollow organs to target specific areas of the body and (2) at controlling therapeutic effects in the pre-, per-, and postoperation periods.

Planification of robotic surgery intervention is based on a patient-specific model of the rigid and deformable organs involved in the surgical act, which is fused to the mechanical model of the robotic system.

Augmented reality allows the real-time overlay of preoperative data with intraoperative situations, after calibration and correspondence determination [749]. Moreover, the surgeon's motions can be reduced and smoothed to increase precision and avoid hand tremor. Computerized analysis of heart wall motion is necessary during beating heart surgery [750], which is a less invasive method used for grafting in particular. The description of the motions of the heart and coronary arteries provides feeding data for robotic systems [751]. The fast movement of selected points of the left anterior descending coronary artery can range from 0.22 to 0.81 mm.

9.2 Electrical Stimulation

Electrical stimulation of cardiac tissue is used for cardiac pacing and defibrillation to provide suitable blood flow for nutrient delivery to the tissues. This type of treatment relies on a surgical or medical procedure.

Electronic devices such as implantable pacemakers and defibrillators enable the management of cardiovascular disease. Bradycardia can cause exercise intolerance. The unique treatment for symptomatic bradycardia is permanent pacemaker implantation. Implantable pacemakers have an extended battery life, contain leads that minimize inflammation and scarring as well as advanced algorithms to match exercise conditions. However, the best solution relies on highly adaptive biological pacemakers based on somatic reprogramming by reexpression of embryonic transcription regulators [752]. The human pleiotropic developmental T-box transcription factor TBx18 can reprogram cardiomyocytes into pacemaker cells.

9.2.1 Implanted Artificial Pacemaker

Cardiac frequency-responsive pacemakers can be combined with implantable defibrillators. They are implanted to regulate cardiac contractions using electrical impulses. Any pacemaker is inserted after excluding or correcting electrolyte abnormalities.

The battery of artificial cardiac pacemakers and implantable cardioverter defibrillators have a limited lifespan. Replacement surgery is hence carried out every

7–10 year (3–6 year for an implantable cardioverter defibrillator). Self-powered artificial pacemaker can convert small mechanical motions into electrical energy.

Current pacemakers treat brady- and tachyarrhythmias. Implantation of an artificial pacemaker is mandatory in the case of symptomatic bradycardia, such as sinoatrial node dysfunction (or sick sinus syndrome) with a very low atrial triggering rate, third-degree atrioventricular conduction block, and certain fascicular blocks.

A type-2 second-degree atrioventricular block can be treated with a pacemaker even in an asymptomatic patient, because it can be a precursor to complete atrioventricular block. Marked first-degree atrioventricular block (PR interval > 300 ms) can benefit from pacing.

Pacemakers also are used to treat advanced heart failure with major intraventricular conduction disorders, mainly left bundle-branch block (biventricular pacing for resynchronization).

Automatic calibration leads to adjustments of pacemaker response to normal changes in subject activity. Detection and dynamic analysis of nodal tissue rhythm provide accurate discrimination between simple acceleration and arrhythmias. An expert system associated with the pacemaker provides data for follow-up. Miniaturization of implantable pacemakers allows slightly invasive procedures. Safety features of pacemakers avoid the occurrence of induced currents during magnetic resonance imaging. Transient malfunctions of pacemakers during computed tomography can occur due to X-rays rather than small alternating electrical field (~ 150 V/m) and alternating magnetic field ($\sim 15\ \mu$T) [753].

Cardiac resynchronization therapy (CRT) is used in dyssynchronous heart failure (DHF). The CRT pacemaker possesses a third lead positioned in a vein on the outer surface of the left ventricle (RA, RV, and LV leads) [369]. Hence, both ventricles are electrically stimulated (paced) at the same time to support a synchronized contraction (biventricular pacing). Another type of implantable heart failure heart devices combines a CRT pacemaker with a defibrillator (CRTD).

The rhythm management relies on optimal left ventricle pacing lead placement. The risks and complications associated with the implantation of a CRT device are similar to those associated with the transvenous implantation of a conventional permanent pacemaker or implantable defibrillator [368]. These risks include bleeding, infection, pneumothorax, pericardial effusion, myocardial infarction, stroke, and death. Transvenous implantation of a left ventricular lead for CRT is accomplished via the coronary sinus and its tributaries. The specific risks comprise coronary sinus dissection and perforation, lead dislodgment, and extracardiac stimulation.

9.2.2 *Electroporation-Based Treatment*

Different methods have been proposed to stop the proliferation of vascular smooth myocytes, such as cryoplasty, or cryoballoon angioplasty,[2] intravascular brachytherapy, or internal radiotherapy,[3] vascular photodynamic therapy (VPDT),[4] and drug-eluting stents.

Irreversible electroporation (IRE) is a nonthermal, nonpharmacological cell ablation method that uses a sequence of electrical pulses (e.g., sequence of 10 dc pulses; amplitude 3800 V/cm; pulse duration 100 μs; frequency 0.1 Hz) to produce permanent damage to tissue within a few seconds [754]. These pulses irreversibly create pores in the plasma membrane and cause cell death. An optimal value of IRE enables vascular smooth myocyte ablation, preserves arterial scaffold, and allows enabling endothelial regeneration. Endothelium integrity recovers.

9.3 Surgery

Surgery now uses robotic and imaging systems [755]. Virtual reality-assisted surgery planing (VRASP) has been developed to provide flexible computational support preoperatively [756] and peroperatively, which can control large patient-specific datasets in real time.

9.3.1 *Cardiac Surgery*

Cardiovascular surgery deals with congenital heart defect repair, valve replacements, heart transplantation, pacemaker placement, normal rhythm and/or conduction restoration (arrhythmia surgery, i.e., Maze procedure for atrial fibrillation), and coronary revascularization.

The *ductus arteriosus* is a vessel that connects the pulmonary artery and aorta in the fetus to shunt immature lungs, gas exchanges occurring in the placenta between fetal and maternal bloods. Shortly after birth, when the neonatal lungs carry out blood

[2] Once the catheter reaches the stenosis, the balloon is filled with liquid nitrous oxide that immediately evaporates into a gas, causing the balloon to inflate and freeze the surrounding arterial wall (− 10 °C).

[3] A radiation source is inserted in the artery to treat in-stent restenosis.

[4] Nontoxic light-sensitive compounds are exposed selectively to light, whereupon they become toxic to target cells (chemical destruction). This technique relies on three components: a photosensitizer, a light source, and oxygen to produce reactive oxygen species. Some photosensitizers accumulate in vascular endotheliocytes.

oxygenation, the ductus arteriosus closes. However, in about 1 in 2000 nonpremature infants, the ductus arteriosus remains open (patent ductus arteriosus), causing vascular diseases.

Many other types of heart malformations or congenital heart diseases exist, such as *ventricular and atrial septal defects*, *atrioventricular canal*, *arterial stenosis*, *coarctation* of the aorta, *vessel transposition*, *cardiac valve atresia*, *hypoplastic left heart syndrome*, *tetralogy of Fallot*, among others. Most defects are effectively treated by surgical and heart catheterization procedures.

Cardiac pacemakers are implanted into damaged hearts to correct defects in conduction of the action potential. Sleep apnea is current in adults with pacemakers. Nearly two thirds of patients with implanted pacemakers have obstructive sleep hypopneas and apneas. Patients with pacemakers thus should be routinely checked, because untreated sleep apnea contributes to cardiovascular deterioration [757].

Cardiac transplantation is the choice therapy for evolutive heart failure. A strategy to treat severe heart failure targets cell growth and tissue regeneration or replacement, which are in a relatively early stage of development. As the donor pool for cardiac transplantation is insufficient, the alternative strategy relies on implantation of left ventricular assist devices (LVADs).

9.3.1.1 Ventricular Assist Pumps

Ventricular assist devices are mechanical circulatory support elements aimed at unloading the heart and providing adequate body perfusion, hence at improving the survival of patients suffering from heart failure and extending the time to transplantation.[5]

LVADs target patients with heart failure at a terminal stage waiting for transplantation, individuals with myocarditis and cardiomyopathy expecting complete cardiac recovery, and children in chronic stages of congenital heart disease who develop ventricular failure.

They include both portable extracorporeal and implantable pumps for patients with competent lung function. Extracorporeal membrane oxygenation is a short-term support when heart and lung failures are combined.

Device implantation can be associated with adjunctive therapy to improve ventricular recovery. Ventricular assist devices can be used not only as a bridge to transplant but also as a long-term (many years) aid.

A noticeable number of patients with LVADs later require right ventricular support. Patients with a LVAD must permanently take anticoagulants to prevent blood clot inside the pump. Complications occur, such a wire-mediated infection, device failure, hemolysis, infarction by cell fragments.

[5] Only 2100 transplants are performed in the USA every year, whereas 3500–4000 people are waiting.

Second-generation LVADs are smaller than first-generation pulsatile machines. They rely on a rotor that continuously conveys blood without requiring valves. Third-generation devices based on magnetically levitated rotors to reduce moving parts are under investigation.

A transcutaneous energy transfer system sends the energy across the skin via electromagnetic waves from an external coil to an electromagnetically coupled internal coil that charges an internal battery. This device thus avoids skin-piercing tubes.

The design of the device configuration and its wall surface must suit hemodynamic and hematological criteria [758]. LVADs were designed with an optimal size to be permanently implanted in the heart chamber, using minimally invasive procedures. Various cannula designs and diameters are proposed to match patient anatomy. LVADs must minimize hemolysis and thrombosis, as well as heat generation.

Artificial chambers are composed of two compartments, driving and blood chambers, separated by a flexible membrane [759]. The blood cavity has valved entry and exit orifices. Flow motion in the blood cavity is determined by membrane displacement controlled by the pressure in the second cavity. Numerical experiments have been performed in 2D models of these pumps [760, 761]. Once the existence and uniqueness of solutions of an elasticity problem in large displacement and small strain coupled with a viscous fluid are proved , numerical applications in an artificial heart ventricle used either a simplified arbitrary Eulerian–Lagrangian (ALE) method or the immersed boundary method. Diaphragm-type ventricular assist devices have also been studied experimentally (LDV) and numerically (using ADINA software) [762, 763].

Such a ventricular assist device is too big to be implanted in the circulatory system. Consequently, micropumps have been devised that can be implanted either in the left ventricle or between the left ventricle and the aorta. However, small pumps designed to be implanted in the left ventricle do not produce pulsatile flow (continuous-flow ventricular assist devices). Centrifugal pumps and pneumatic pulsatile ventricular assist devices have been proposed to be implanted in the body of both adults and children. Pump optimization and design methodologies have been based on 3D numerical simulations combined with design of experiments (DOE) [764], to develop a wholly implantable pump. Scaling has been set for an operating point using the Cordier diagram.

Implantation in any cardiac chamber or vascular lumen of mechanical devices, such as ventricular assist pumps, cardiac valves, and filters, among others, induces hemodynamic disturbances and damages flowing cells, causing in particular hemolysis. The implanted devices thus must minimize hemolysis.

Hemolysis indices have been proposed, such as a hemolysis index (HI), which measures the plasma-free hemoglobin (Hb) concentration [fHb], a normalized hemolysis index (NHI) relative to total blood [Hb], assuming that the hemolysis rate is small ([fHb]/[Hb] $\ll$ 1) and varies linearly with time ($\Delta \bullet / \Delta t \equiv d \bullet / dt$) [765]:

$$\mathrm{HI} \propto \frac{d[\mathrm{fHb}]}{dt} \frac{V_{\mathrm{pl}}}{q}; \tag{9.1}$$

$$\text{NHI} \propto \text{HI}/[\text{Hb}], \tag{9.2}$$

where V_{pl} is the plasma volume.

A prediction model was derived from the Giersiepen–Wurzinger blood damage relation using numerical simulations of blood flow field in the vasculature domain with the implanted device, introducing a pointwise function D_{RBC}, the damaged RBC fraction, which is supposed to depend on shear (C_{VM}) magnitude and linearly on exposure time:[6]

$$\mathsf{D}_{\text{RBC}} \propto C_{\text{VM}}^{\alpha} t; \quad \frac{\partial \mathsf{D}_{\text{RBC}}}{\partial t} + (v \cdot \nabla)\mathsf{D}_{\text{RBC}} = S, \tag{9.3}$$

where $S = \kappa_1 \kappa_2 C_{\text{VM}}^{\alpha}(1 - \mathsf{D}_{\text{RBC}})$.[7]

9.3.1.2 Valve Prostheses

> " Surgeons are not basically concerned with lesions. We care more about function. Therefore one may define the aim of a valve reconstruction as restoring normal valve fonction rather than normal valve anatomy."
> (A. Carpentier) [766]

Identification of mechanisms causing valve dysfunction is mandatory to successfully repair the valve. A complete valve analysis supports selection of appropriate repair techniques and obviates immediate and late repair failure.

Smooth, thin, and large leaflets are usually considered as repairable. Small, restrictive, fibrous, or thickened leaflets impede surgical repair. Heavily calcified valves are usually considered as nonrepairable, except when calcifications localize to the free margins.

Malfunctioning valves can be either repaired or removed and then replaced with prosthetic heart valves. Two main types of heart valve prostheses exist: mechanical valves and prosthetic stentless and stented tissue valves.

Mechanical valves are manufactured using three main designs:

1. *Caged ball valve* (e.g., Starr–Edwards valve)
2. *Tilting disc valve* (e.g., Medtronic Hall and Bjork–Shiley valves)
3. *Bileaflet valve* (e.g., St. Jude, CarboMedics, and ATS valves)

---

[6] The shear stress is computed using the von Mises criterion:

$$C_{\text{VM}}(\mathbf{x}) = \left(1/2((c_1(\mathbf{x}) - c_2(\mathbf{x}))^2 + (c_2(\mathbf{x}) - c_3(\mathbf{x}))^2 + (c_3(\mathbf{x}) - c_1(\mathbf{x}))^2)\right)^{1/2},$$

where c_i $(i = 1, \ldots, 3)$ are the stress vector components computed from stress tensor components obtained from the Navier–Stokes solution in the pump computational model.

[7] The source term $S = \kappa_1 \mathsf{D}_{\text{RBC}}(1 - \mathsf{D}_{\text{RBC}})/t$ with the hemolysis model $\mathsf{D}_{\text{RBC}} = \kappa_2 C_{\text{VM}}^{\alpha} t$. $\kappa_1 = 0$ if $C_{\text{VM}} < C_Y$ and $\kappa_1 = 1$ otherwise (C_Y: stress corresponding to the hemolysis threshold).

Valved grafts have also been proposed (e.g., St. Jude aortic valved grafts and ATS aortic valve graft prostheses). *Bioprosthetic (xenograft) valves* are made from porcine valves (e.g., Carpentier–Edwards and Hancock valves) or bovine pericardium (e.g., Ionescu–Shiley and Carpentier–Edwards valves).

A *collapsible trileaflet membrane valve* is designed with two cylindrical muffs fixed together, a stiff and a thin flexible, the latter having collapsible cusps [767]. Its mechanical behavior must minimize regurgitation and pressure drop across it.

Prosthetic valves must mimic the static and dynamic characteristics of natural human valves and flow mechanics must be preserved, avoiding turbulences, flow stagnation, or excessive mechanical stresses.

Several teams have studied in pulsatile flow conditions the mechanical features of the valve prostheses using rigid [768, 769] or deformable [770, 771] test sections, which model either the local valvular region or the heart cavities, possibly with models of existing vessels [772].[8] The gradual closure of the natural valves during flow deceleration is, in general, not reproduced.

Severe aortic stenosis are treated by catheter-based aortic valve insertion rather than surgical valve replacement. Less-invasive endovascular procedures, i.e., catheter-based implantation of aortic stent grafts and balloon- or self-expandable valve prostheses, use a retrograde femoral or apical route. Complications of catheter-based valve insertion comprise vascular injury, impaired device positioning with paravalvular leakage, cardiac tamponade, arrhythmias, coronary artery occlusion, and prosthetic valve embolization.

Anatomical and physiological explorations mainly based on echocardiography before conventional open-heart valve surgery aim at describing the leaflet anatomy and at assessing severity of valve dysfunction and hemodynamic consequences. The aortic root consists of the elliptical aortic annulus (transition between the left ventricular outflow tract and aortic root), commissures of aortic valve leaflets that extend upward, sinuses of Valsalva, coronary artery ostia, and sinotubular junction, which can also contain calcifications. The length of the aortic valve leaflet often exceeds the distance between the annulus and the ostium of the coronary artery, in the absence of strong aortic root remodeling associated with calcified aortic stenosis.

In the absence of direct operative inspection, three-dimensional, pre- and perinterventional imaging assists therapeutic planning with the detailed anatomy of the aortic valve and root as well as evaluation of vascular access strategy.

9.3.2 Coronary Arterial Surgery

In the presence of arterial *stenoses*, understanding of atherosclerosis evolution, diagnosis at relatively late stage of development, occurrence of complications in patients with moderate stenoses and not only severe ones, as well as the respective role of

[8] However, the heart geometry used by this research team is inverse.

normal and tangential hemodynamic stresses are aspects that stimulate investigations by biomechanicians and biomathematicians involved in the implementation of atherosclerosis models as well as computer-aided medicine tools.

Late-stage *intimal hyperplasia* develops after grafting at the distal anastomosis and can cause *restenosis*, that is, renarrowing of the arterial lumen (Vol. 8, Chap. 8. Atherosclerosis—Medical Aspects). Many biomechanical studies were aimed at optimizing the neovasculature design to prevent this complication.

9.4 Endovascular Therapy

Endovascular techniques are aimed at treating local wall damage in the heart or blood vessels. Percutaneous interventions are minimally invasive because they use natural paths. The catheter is inserted into a superficial artery, such as the femoral or subclavian artery, and then advanced under image guidance into the diseased segment of the cardiovascular system.

9.4.1 Thermal Ablation of Nodal Tissue

Antiarrhythmic drugs are only partly effective in treating atrial fibrillation. Moreover, they have side effects such as arrhythmia. Paroxysmal atrial fibrillation triggered by focal drivers usually situated in a pulmonary vein can be cured by a catheter-based ablation.

Tachyarrhythmias can be treated by radiofrequency ablation (RFA). A selected small patch (< 5 mm) of nodal cells of the conduction paths responsible for the abnormal cardiac rhythm is destroyed by a radiofrequency wave using an image-guided catheter-based procedure.

Cryoablation can also be used, providing cold to freeze and destroy cells.[9] Damage limited to the local vasculature (without repercussion on the organ perfusion) can lead to ischemic necrosis, thus contributing to treatment efficacy. At a few millimeters from the ice ball edge, the tissue temperature is warmer, thus avoiding lethal tissue damage.

Ablation is proposed to treat reentrant tachycardias due to an extrapath in or adjacent to the atrioventricular node, junctional tachycardias, atrial flutter and fibrillation, ventricular fibrillation, and Wolff–Parkinson–White syndrome.

---

[9] Small tumors can also be treated by thermal ablation, either cryoablation (freezing to at least $-19.4\,°C$ using liquid nitrogen or argon) or radiofrequency ablation.

Table 9.1 Strut and polymer thickness (μm) in different types of metallic stents. (Source: [773])

Type	Strut	Polymer	Total
Cypher select	140	14	154
Taxus element	132	16	148
Xience V	81	8	89
Resolute	91	6	97
Biomatrix	120	11	131

Table 9.2 Some material constant of a stent and its coating. (Source: [774])

	Elastic modulus (N/m^2)	Poisson ratio
Stent	0.193	0.3
Coating	0.01	0.3

9.4.2 *Drug-Eluting Stents*

Although the likelihood of restenosis is reduced with respect to balloon angioplasty, stents are still prone to occurrence of intimal hyperplasia and restenosis. Drugs (e.g., rapamycin and paclitaxel) are stored in stent polymeric coating and, upon implantation, are locally slowly released, further reducing the restenosis rate.

Drug-eluting stents have a dual purpose: mechanical (like BMSs) and therapeutic to prevent restenosis by preventing smooth myocyte proliferation. The first generation of drug-eluting stents (DES) were aimed at addressing the high restenosis rate after implantation of bare metal stents. The goal of following DES generation with thinner struts and biocompatible or biodegradable polymers was to tackle the late stent thrombosis.

Antiproliferative drug-eluting stents (DES) include drug (cytostatic compound) reservoirs (coating thickness 5–20 μm). The main components that determine the performance of a drug-eluting stent are its structure and rheology as well as constituent drug, polymer, and delivery mode (Tables 9.1 and 9.2). Strut embedment in the arterial wall ranges from simple contact to total inclusion, thereby mildly affecting the drug delivery in the arterial wall.

To achieve a prescribed drug release kinetics over a given therapeutic period using a given dose, the current design strategies focus on a multiphasic release from blends of biodegradable polymers. A biphasic strategy, a first phase of immediate drug release liberate an adequate dose fraction and a second extended release provides the dose fraction required to maintain an effective sustained action, once the drug pharmacokinetics and pharmacodynamics are known.

The release pattern of paclitaxel from three neat polymer matrices, polycaprolactone (PCL), poly(lactic-co-glycolic) acid (PLGA), and PLGA polyethylene glycol (PLGAPEG) was studied [775]. These matrices are representative of a broad spectrum of biodegradable hydrophobic and hydrophilic polymers [776]. In hydrophilic polymers, the internal bounds between the chains are weakened, hence facilitating surface erosion.

Although DESs reduce the restenosis rate as they prevent vascular smooth myocyte proliferation, they provoke late-stage thrombosis resulting from endothelial injury associated with delayed or absent healing.

The same cause producing the same effect, these implanted materials damage the endothelium, especially at the insertion site, and can provoke in-stent restenosis despite braking smooth myocyte proliferation as they do not support endothelium repair. Endothelium damage and removal occurring during implantation induce a loss of a local, natural controller that impedes unwanted cell proliferation and provides an interface preventing unneeded blood coagulation. Antiproliferative drugs attenuates the division rate not only of smooth myocyte, but also that of endotheliocytes.

9.4.2.1 Thrombosis

Drug-eluting stents preclude not only in-stent restenosis, but also reendothelialization, hence predisposing to thrombosis. Poststenting vascular injury comprises endothelial denudation and stretching of the locally stiff vascular wall with its risk of tear. Endotheliocyte removal exposes the underlying procoagulant materials and eliminates the local endothelial control of vascular homeostasis.

Unlike stent thrombosis after implantation of a bare metal stent, which generally occurs within the first 2 weeks, late (after >1 month) and very late (after >1 year) stent thrombosis can appear after implantation of drug-eluting stents.

The main factors of stent thrombosis include [777]: (1) stent parameters (diameter, length, strut thickness, and polymer type and thickness); (2) patient status (ejection fraction, diabetes mellitus, age), because diabetes, renal insufficiency, and heart failure, predispose to stent thrombosis; (3) endothelium state.

Repair of the endothelial monolayer is required to struggle against thrombogenicity of the stented vessel segment. Damaged endotheliocytes are renewed by healthy adjacent cells and/or circulating endothelial progenitor cells originating from the bone marrow. Implantation of drug-eluting stents causes delayed arterial healing with poorer endothelialization and persistent fibrin deposition with respect to that of bare metal stents. Furthermore, patients who experienced stent thrombosis have higher platelet reactivity evaluated by ADP-induced platelet aggregation in comparison with a matched group of patients who underwent stenting without developing thrombosis.

Endothelial progenitor cells participate in reendothelialization. Late outgrowth endothelial progenitor cells, but not early outgrowth endothelial progenitor cells, differentiate into endotheliocytes. The number of circulating VEGFR2+ and PromL1+ (prominin-like-1 or CD133+) or CD34+ (sialomucin) cells that have the potential to differentiate into mature endotheliocytes is significantly reduced in patients suffering late (> 30 d) stent thrombosis [778]. Moreover, endothelial progenitor cells are dysfunctional. Erythropoietin and vascular endothelial growth factor can enhance the number of circulating endothelial progenitor cells.

Recruitment of endothelial progenitor cells can be assisted by a GSK3β inhibitor, as their adhesion is enhanced via upregulated expression of the α_4-integrin [779].

9.4.2.2 Restenosis

After stenting, medial smooth myocytes are much more easily exposed to circulating growth factors and inflammatory cytokines, such as platelet-derived growth factor, thrombin, thromboxane-A2, and adenosine diphosphate released in particular from platelets. Thrombin and PDGF stimulate vSMC proliferation and migration. Adenosine diphosphate binds to the platelet $P2Y_{12}$ receptor and potentiates platelet activation. Furthermore, dilation of the vascular wall activates stretch-activated signaling that provokes further vSMC migration and proliferation.

Among the mechanisms that elicit smooth myocyte proliferation and migration, calcium influx through Ca^{++} ion channels such as store-operated channels is a major player. In addition to members of the family of transient receptor potential canonical (TRPC) channels, *stromal interaction molecule* STIM1 senses Ca^{++} store depletion and works with Orai1 (Vol. 3, Chap. 3. Main Sets of Ion Channels and Pumps). Sensor STIM1 regulate directly TRPC1, TRPC4, and TRPC5 and indirectly TRPC3 and TRPC6. After arterial injury caused by angioplasty, STIM1 is overexpressed [780]. Conversely, suppression of STIM1 expression prevents intimal hyperplasia. Inhibition of STIM1 lowers the amount of phosphorylated retinoblastoma protein that regulates the G1–S transition of the cell cycle, thereby blocking cell cycle progression.

Several types of drug-eluting stents have been tested in clinical trials to prevent, or at least to limit SMC migration and proliferation into the intima, which lead to restenosis. Two main types of drug-eluting stents exist: (1) a more or less uniform coating by drug–polymer matrix, with a possible screen (*topcoat*) to reduce the transport speed and loss in blood flow; and (2) a set of drug-polymer reservoirs (size $<100\ \mu m$) in the metallic structure.

Rapamycin (sirolimus) and *paclitaxel* (taxol) are used as anticancer chemotherapy. Antiproliferative stent coatings employ these drugs (sirolimus- [SES] and paclitaxel-eluting stents [PTES]). Rapamycin and paclitaxel target the G1 and M phases of the cell division cycle, respectively.

Rapamycin is an immunosuppressor used to prevent rejection in organ transplantation, as it supresses activation of B and T lymphocytes; it has potent antiproliferative properties, as it binds to FKBP12; the FKBP12-rapamycin complex then connects to and inhibits target of rapamycin, that is, the PI3K–TOR pathway.

Taxol is a microtubule-stabilizing agent with potent antiproliferative activity. Proliferation of vascular smooth myocytes can also be hindered by the inhibition of fibronectin matrix assembly [781].

Mitofusin-2, a member of transmembrane GTPases (MFn1–MFn2) of the outer mitochondrial membrane involved in mitochondrial fusion as well as maintenance and action of the mitochondrial network, has an antiproliferative effect, especially in vascular smooth myocytes, as it inhibits the Ras–ERK signaling and subsequently provokes cell cycle arrest at the G0–G1 transition [782].

Rapamycin also directly activates platelets, thereby contributing to thrombus formation. Therefore, drug-eluting stents not only repress proliferation and migration of vascular smooth muscle and endothelial cells, but also can elicit thrombosis.

Despite the reduction in restenosis rate by drug-eluting stents with respect to bare-metal stents, similar rates of overall mortality and myocardial infarction exist for treated patients whatever the stent type during long-term follow-up. However, careful selection of stent type according to patient and lesion features may optimize the therapy efficiency.

Drug-eluting stents delay healing and impair endothelialization observed with intravascular ultrasonography, as drugs target both smooth muscle and endothelial cells [783]. Furthermore, drugs released from the polymer favor thrombogenesis, as they impede the generation of thrombus inhibitor, i.e., the endothelium.[10] Early and late stent thromboses occur with similar frequency with both stent types, but very late stent thrombosis characterizes drug-eluting stents. Antiplatelet therapy duration is thus longer for drug-eluting stents.

Endothelium-like tissue-coated stents are contrived to impede the consequences of damaged endothelium during stent implantation. Endotheliocytes are fragile and do not adhere strongly to the stent material, which bears very large deformations, in particular during stent implantation. Genetically engineered chondrocytes, which have the main endotheliocyte functions and better adhesion properties than endotheliocytes, can form a stent coating.

To support reendothelialization, stents can also deliver growth factors and chemoattractants of endothelial progenitors using marker antibodies, but patients with cardiovascular diseases often have few and/or dysfunctional endothelial progenitors [785]. However, nanomatrix that mimics endothelial surface with cell adhesion molecules and nitric oxide sources aims at recruiting endothelial progenitors.

9.4.3 Depot Stents

Depot stent is a drug-eluting stent with micrometric drug containers with given spatiotemporal release kinetics and programmable drug release. Unlike DESs that have their entire surface coated, depot stent contains a set of tiny drug reservoirs, thereby reducing the contact surface between the polymer and arterial wall that may favor inflammation and thrombosis.

[10] Early, late, and very late, in-stent thromboses occur in the first 30 days, between 1 month and 1 year, and beyond 1 year after stent implantation, respectively. Probable and possible stent thromboses correspond to myocardial infarction or unexplained death within 30 days and beyond 30 days after stent implantation, respectively. Used in initial treatment, drug eluting stents with respect to bare-metal stents are associated with lower revascularization rates and higher risk of late stent thrombosis, but there is no difference in mortality up to 4 years in patients initially treated with either sirolimus-, paclitaxel-eluting, and bare-metal stents [784]. Yet, sirolimus-eluting stents has 15–20% lower risk for myocardial infarction than the other two. The ENDEAVOR programme that evaluates zotarolimus-eluting stent shows significantly lower cardiac death and myocardial infarction MI rates than those of bare-metal stents up to 3 years.

The polymeric matrix and drugs fill orifices (diameter/strut width ratio 0.5), either closed-end holes (depth 0.5 strut thickness) or microchannels open at both ends. In the latter case, different types of drugs can be stored in the polymer, such as an endothelial cicatrizant in the channel part close to the wall and an anticoagulant in the part adjacent to the arterial lumen. However, such reservoirs can weaken the stent structure.

Openings can evenly spread over the entire stent surface or localize either in connectors or in bars or apices of stent rings bridged by connectors. The spacing between two neighboring reservoirs equals 150 μm. The number of containers determine the drug dose that can be stored, hence fully replacing that in the stent coating (thickness 5–15 μm).

With respect to stent without orifices, blind-hole reservoirs on connectors or ring bars affect neither the radial strength nor plastic strain, but reduce the fatigue safety factor [786]. Blind-hole containers evenly spread on the entire stent or only located in ring apices alter both the radial strength and plastic strain and lower the fatigue safety factor to a greater extent than those situated on connectors or ring bars. Through-hole reservoirs evenly spread on the entire stent or on apices degrade further stent resistance to mechanical stress. Through-hole containers on connectors or ring bars that have the maximal drug storage capacity may kept sufficient mechanical integrity [786].

Although drug-eluting, metallic stents have reduced in-stent restenosis and thrombosis rates with respect to previous generations of stents, the permanent presence of a metallic foreign body within the artery impair the vasomotor function and can favor vascular inflammation (Table 9.3). Furthermore, these stents disturb vascular healing, are responsible for side-branch jailing, and preclude late lumen enlargement, noninvasive imaging, and possible surgical revascularization of stented segments.

Biodegradable (BP) and permanent (PP) polymer coating on rapamycin-eluting stents have quasi-equivalent efficacy that is greater than that of the polymer-free (PF) stent without differences in safety [791].

9.4.4 Resorbable Stents (Scaffolds)

Biodegradable arterial (vascular) stents (BAS [BVS]), or bioresorbable scaffolds (BRS), represent the present step of conceiving, designing, optimizing, and manufacturing stents that deliver drug and disappear when they become useless (Table 9.4). They hence restore physiological strain and stress fields and mechanotransduction capability as well as obviate long-term anticoagulant therapy.

Bioresorbable vascular scaffolds ware first implanted in patients with stable or unstable angina and silent ischemia. Second-generation resorbable arterial scaffolds are now deployed in coronary arteries upon acute ST-segment elevation myocardial infarctions (STEMI) [788, 789]. Percutaneous coronary revascularization is currently the first choice treatment for patients with , as it is better than thrombolysis.

Table 9.3 Comparison of bioresorbable scaffolds (BRS) with other percutaneous intervention devices. (Source: [787])

Event	BA	BMS	DES	BRS
Acute occlusion	−	+	+	+
Acute recoil	−	+	+	+
Acute thrombosis	−	−	−	−
Subacute thrombosis	±	−	−	−
Late thrombosis	+	−	−	±
Very late thrombosis	+	±	−/±	+/?
Intimal hyperplasia	−	−	±	+
Constrictive remodeling	−	+	+	+
Expansive remodeling	+	−	−	+
Restoration of vasomotion	+	−	−	+
Late luminal enlargement	+	−	−	+

BA balloon angioplasty, *BMS* bare metal stents, *DES* drug-eluting stents, + beneficial effect, − negative or no effect, ± neutral or uncertain effect, ? lack of definitive evidence

Table 9.4 Features of resorbable scaffolds, which are self-expanding or deploy upon heat or balloon inflation (Source: [787]). Metallic bioresorbable stents include drug-eluting magnesium-alloy scaffolds. Biodegradable polymers with variable surface properties are constituted of polyethylene glycol compounds attached to polylactic acid blocks in various types of combinations. Everolimus is a derivative of sirolimus. both inhibit target of rapamycin (TOR). Myolimus is a macrocyclic lactone and also a TOR inhibitor

Strut and coating material	Metallic (magnesium alloy) Polymeric (PDLLA, PLLA, PLAS, PTDPC, SA–AA)
Strut thickness	Metallic: 125–150 μm
	Polymeric: 150–230 μm
Eluted drug	Everolimus, myolimus, sirolimus, paclitaxel

PDLLA poly^{DL}lactic acid, *PLAS* polylactide and salicylate [with antithrombotic and anti-inflammatory action], *PLLA* poly^{L}lactic acid [biodegradable, thermoplastic, and aliphatic polyester], *PTDPC* polytyrosine-derived polycarbonate, *SA–AA* salicylic acid–adipic acid polymer used as stent coating

Yet, stent placement in acutely thrombotic lesions often causes late stent malapposition and stent thrombosis because of thrombus sequestration between struts and the arterial wall. Multislice coronary computed tomography (MSCT) as well as invasive coronary angiography, intravascular ultrasonography (IVUS), optical coherence tomography imaging (OCT), and coronary vasomotor testing check quality of implantation and absence of complications, in addition to the follow-up.

Bioresorbable scaffolds have a greater conformability and flexibility than conventional stents, hence lowering scaffold strut malapposition as well as alterations of the endothelium, mural stress field, and arterial geometry. However, thicker struts and crossing profile (scaffold-to-artery size ratio $\sim 25\,\%$ [$\sim 10\,\%$ for DES] [788]) impede stent crossing of sharp angles and its implantation in heavily calcified or tortuous arteries.

Metallic bioresorbable scaffolds have properties similar to those of conventional metallic stents (deliverability, good radial strength, and low recoil) in addition to retarded resorption. Magnesium-based alloys can be coated with by a composite two-layer film that controls both the corrosion rate of the magnesium alloy and rate of drug release.

Polymeric bioresorbable scaffolds are made of various types of polymers, each with their own chemical characteristics and resorption time [787]. Polylactic acid, or polylactide, is a thermoplastic aliphatic polyester. The degradation of polymeric devices can be optimized by combining crystalline and amorphous polymers.

A gradual resorption happens during the first 2 years after implantation, thereby restoring after a relatively long delay the arterial geometry and physiology, including vasomotion [790]. The duration of its degradation depends on the crystallization of the polymer and varies from 2 to 4 years. Magnesium is resorbed by corrosion in 2–12 months. Magnesium scaffolds are metabolized to chloride, oxide, sulfate, or phosphate salts. The by-product in the artery is hydroxyapatite that is eventually processed by macrophages [787].

A more extensive artery preparation is required before implantation of biodegradable scaffolds. Unlike drug-eluting stents, resorbable scaffolds are less amenable to postdilation because of the risk of strut fracture. Therefore, precise estimation of the vessel size is mandatory for selecting the proper scaffold size, a nontrivial task due to eventual vasoconstriction and thrombi [790].

Deposition of a uniform fibrous neointimal layer may be associated with a late lumen enlargement [790]. On the other hand, resorbable scaffolds have downsides. Early scaffold thrombosis can occur. Despite good radial strength and low recoil, acute recoil of polymeric biodegradable stents can be observed [790]. In addition, polymeric devices that have a limited stretchability can break when they are overdilated.

Due to thicker struts (150–230 μm [DES: 80–140 μm]) and larger wall surface coverage than drug-eluting stents, resorbable arterial scaffolds can entrap and squeeze more thrombi and debris between the scaffold and arterial wall, in addition to higher risk of side-branch occlusion.

Analysis of drug-eluting bioresorbable arterial scaffolds in ST-segment elevation myocardial infarction shows a mean strut malapposition equal to $2.8 \pm 3.90\,\%$, a mean incomplete scaffold apposition area 0.118 ± 0.162 mm^2, a mean intraluminal defect area to 0.013 ± 0.017 mm^2, and a mean percentage malapposed struts per patient to $2.80 \pm 3.90\,\%$ [788]. In acute ST-segment elevation myocardial infarction, implantation of bioresorbable arterial scaffolds causes acute scaffold recoil in 9.7 % patients, rate of scaffold strut malapposition in 1.1 %, and small and clinically silent edge dissections in 38 % [789]. Complete bioresorption takes at least 2 years after implantation.

9.4.5 Valvate Stents for Heart Valve Replacements

Aortic valves can be percutaneously replaced by valve-containing stents in the beating heart under image guidance, without strong damage of the aortic wall and obstruction of coronary ostia. First, percutaneous aortic valvuloplasty was proposed using balloon-expandable stents [792, 793]. Catheterized implantation uses a valve prosthesis that can be sutured into an expandable stent and keep its property after crimping and re-expansion. Bovine jugular venous valves are sutured inside the stent [794]. The valve-stent assembly is deployed by balloon inflation. Due to heavy periodic loading, self-expanding stents are preferred to balloon-expandable stents [795]. A pulmonary metal stent with a valve was very recently implanted using a catheter-based method in a child with a congenital heart defect.[11]

Percutaneous aortic valve implantation can be proposed in old patients with severe aortic stenosis, particularly patients judged inoperable or at high risk for surgical aortic valve replacement [796]. However, the global failure rate equals 33 % with 25 % periprocedural deaths. Moreover, vascular injury at the access site using femoral or iliac arteries can cause death. Periprosthetic regurgitation occurs in at least 10 % of the cases. An atrioventricular block can necessitate pacemaker implantation. An alternative approach, like the transapical implantation, overcomes these drawbacks, especially incorrect implantation of the prosthesis [797].

9.4.6 Drawbacks of Medical Devices

Any implant device may induce several biological and mechanical disturbances when device design is not appropriate and its rheology does not match the wall properties. However, the device material is selected to avoid sensitization, cytotoxicity, and carcinogenicity.

Hemolysis occurs when a part of the device is located in the flow core.

The intact and healthy endothelium possesses an antithrombotic property, hence supporting normal blood flow by separating platelets for the extracellular matrix. Moreover, endotheliocytes release nitric oxide and prostacyclin, two soluble inhibitors of platelet activation. They also secrete high concentrations of ectonucleotide pyrophosphatase–phosphodiesterase ENPP1 that rapidly hydrolyzes extracellular nucleotides such as ADP. However, a medical device, like any implanted foreign body, provides a matrix for thrombus formation, especially when endothelial cells are damaged or removed during the intravascular implantation. In particular, stenting destroys large endothelial segments. A thrombogenic response occurs after platelet deposition and clotting activation. After the initial platelet adhesion, auto- and

[11] About 20 % of the congenital heart defects involve the pulmonary valve. The mini-invasive cardiac catheterization carried out in the Hospital for Sick Children in Toronto took 90 min and required a single overnight stay.

paracrine mediators (ADP, adrenaline, thromboxane-A2, and thrombin) amplify and sustain the platelet response, as they recruit circulating platelets to form a growing hemostatic plug.

Vessel-wall damage also generates an intimal cell proliferation and quick restenosis. Growth factors can be released from platelets and fibroblasts. An inflammation due to interactions between leukocytes and the device can occur once leukocytes and complement have been activated. Reactive oxygen and nitrogen species (superoxide and peroxynitrite) and effectors (e.g., poly^{ADP}ribose polymerase)[12] participate in restenosis after vascular injury. In media and neointima, NOS2 is frequently overexpressed. The serum 3-nitrotyrosine-to-tyrosine ratio (protein Tyr nitration),[13] an index of oxidative stress mainly resulting from neutrophil activation, is correlated with the narrowing of the lumen [799]. On the other hand, the NOS–NO–sGC axis attenuates intimal hyperplasia.

Finally, the medical device disturbs blood flow. Locally, the added interface rugosity caused by the implanted device can generate local flow separations. Moreover, the abrupt transition between native vessel and stented segment induces a compliance mismatch. By means of wave reflection, it can remotely affect blood flow.

9.4.7 Design of Medical Devices

Computer-aided manufacturing allows one to test easily diverse device geometries, structures, and material properties, with an important requirement: the material matching of the mechanical properties of the biological tissues. It needs to model accurately the physical processes at the interface, because the interface between the implants and the biological tissues (blood and vascular wall) is the focus of desirable and undesirable interactions. Moreover, it must be adaptable to unsteady loading induced by cardiac contractions.

9.5 Stem Cell-Based Regenerative Therapy

The adult heart experiences a cardiomyocyte turnover and hence cannot be considered as a fully terminally differentiated organ. However, the heart can regenerate only during the first week after birth. After this short postnatal period, the majority of cardiomyocytes permanently exit the cell division cycle. The renewal by cell division

---

[12] During nutrient deprivation, ROS-induced DNA damage activates PARP1 that causes ATP depletion, hence AMPK activation. Cells then start autophagy (Vol. 2, Chap. 4. Cell Survival and Death) that can lead to adaptation or death. Enzyme PARP1 supports the prosurvival role of autophagy [798].

[13] Nitration of tyrosine to 3-nitrotyrosine is a protein modification occurring during oxidative stress.

in the human myocardium occurs at a very low rate (1 and 0.45 % annually at the age of 25 and 75 year, respectively) [800]. After ischemia, the renewal reserve is unable to rescue a failing myocardium.

At birth, the placental-to-pulmonary circulation switch inaugurates cardiac aerobic metabolism. In cardiomyocytes, during this transition, the mitochondrial genesis increases and its intracellular structure remodels under control of mitofusins Mfn1 and Mfn2 [801].

During embryo- and fetogenesis, relatively hypoxic cardiomyocytes rapidly proliferate and use anaerobic glycolysis as a main source of energy:

$$\text{Glucose} + \text{Mg}^{2+} + \text{NAD}^{+} + \text{ATP and then ADP} \rightarrow \ldots \rightarrow \text{pyruvate} + \text{NADH} + \text{ADP and then ATP}.$$

In adult cardiomyocytes, the oxidative phosphorylation is the energy source. The mitochondrial tricarboxylic (or citric) acid cycle coupled to oxidative phosphorylation by the electron transport chain associated with ATP synthase at the inner mitochondrial membrane via NADH produces 18-times more ATP than cytoplasmic glycolysis. AcetylCoA is provided by glucose and amino and fatty acid metabolisms:

$$\begin{aligned}\text{Glucose} \rightarrow \text{pyruvate} \rightarrow \text{acetylCoA} \rightarrow \text{citrate} + \text{ADP} + \text{GDP} + \text{NAD}^{+}\\ + \text{NADPH} \rightarrow \text{citrate} + \text{ATP} + \text{GTP} + \text{NADH} + \text{NADP} + \text{CO}_2;\end{aligned}$$

$$\begin{aligned} \text{O}_2 &\rightarrow \text{HO}_2, \\ \text{NADH} &\rightarrow \text{NAD}^{+} + \text{H}^{+}, \\ \text{ADP} + \text{P}_\text{i} &\rightarrow \text{ATP} \end{aligned}$$

(P_i: inorganic phosphate or free phosphate ion).

In mouse cardiac mitochondria from postnatal day 7 with respect to newborn mice, DNA-content, crista-density, and aerobic-respiration enzyme levels increase. Mitochondrial ROS are generated as a consequence of electron leak by the electron transport chain.

The transition to an oxygen-rich postnatal environment and hence to an aerobic metabolism engenders DNA damage that causes cardiomyocyte cell-cycle arrest and hence stops proliferation [802]. Concentrations of reactive oxygen species, oxidative DNA damage, and DNA damage-response markers significantly rise in the heart during the first postnatal week. Postnatal hypoxemia, diminished mitochondrial-dependent oxidative stress and ROS scavenging, or DNA damage-response inhibition prolong the postnatal cardiomyocyte proliferative time window. On the other hand, hyperoxemia and ROS generation shorten it.

As the heart has a reduced repair capacity, early reperfusion of occluded artery after acute myocardial infarction raises the long-term prognosis. However, it does not prevent chronic myocardial dysfunction and development of heart failure. Early after myocardial infarction, therapy is primarily cardioprotective, that is, salvaging the jeopardized myocardium and preventing adverse remodeling.

At a later stage, the goal is cardiorestorative, that is, aimed at reversing maladaptive remodeling and improving contractility using activation of resident cardiac progenitors and stimulation cardiomyocyte entry in the cell division cycle as well as increasing neovascularization and decreasing scar burden [803].

Transplantation of stem or progenitor cells aimed at improving revascularization of ischemic regions, reducing inflammation, and regenerating injured cardiac tissue can complement current therapy (Vol. 5, Chap. 11. *Tissue Development, Repair, and Remodeling*). This strategy relies on optimization of the progenitor cell type and engraftment site that engender the best possible tissue integration, thereby avoiding immune rejection of engrafted cells and excluding tumorigenic precursors.

During acute myocardial infarction, two peaks of VEGF release can be detected: early (24–48 h) and late ($\sim$7 day). They can influence stem cell transplantation efficiency.

Therapies of chronic heart failure target in particular afterload reduction and mineralocorticoid dysregulation to limit myocardial remodeling. Regenerative approaches struggle against pathogenesis and exploit natural repair mechanisms. They can be based on stem cell-based protocols aimed at restoring structure and function. Cell delivery is carried out after organ conditioning to prepare target tissue.

Regenerative therapy can use cardiac stem cells as well as extracardiac sources. Stem cell-based therapy provides not only a source of new functional cardiomyocytes, but also can promote tissue healing using paracrine messengers. Cardiopoietic phenotype can be primed using a set of cardiogenic growth factors. Nonetheless, stem cells and their derivatives must engraft in the organ, that is, escape rejection by immunity, unless transplanted cells derive from the patient's organism.

Cardiac repair can be supported by:

1. Activation of endogenous resident cardiac stem cells and other types of stem cells, orienting nonresident stem cells toward cardiac lineages
2. Introduction of exogenous progenitor cell pools (e.g., bone marrow-derived mesenchymal stem cells)

Stem cell transplantation for tissue regeneration must ensure:

- Directed differentiation of stem cells to specific cell types
- Survival of transplanted cells and formation of a syncytium (connected cells)[14]
- Functional coupling with host myocardium to impede arrhythmia
- Appropriate generation of extracellular matrix to avoid fibrosis and to facilitate proper electrochemical coupling
- Prevention of teratomas or cancers

[14] Prosurvival pretreatment of cardiomyocytes derived from human embryonic stem cells aimed at preventing cell death due to ischemia and inflammation enhances survival after transplantation into infarcted tissue, at least over a short term [804].

In addition, integration of stem cells into host tissues must be associated with angiogenesis to supply nutrients to both transplanted cells and cardiomyocytes that surround the damaged region.

9.5.1 Mesenchymal and Bone Marrow-Derived Stem Cells

Treatment of cardiovascular diseases can be optimized by combining various procedures. For example, coronary artery bypass grafting can be associated with intramyocardial transplantation of autologous Prom1+ (prominin-1 or CD133) bone marrow-derived stem cells in the hypokinetic infarct border zone [805]. This strategy improves regional perfusion and reduces scar size, but not the overall left ventricular function. In any case, effects of revascularization by stenting or grafting and combined cell injection are difficult to distinguish.

Mesenchymal stem cells have immunosuppressive properties in vitro. Although they express HLA class-1 molecules, they impede T-cell activation and proliferation in vitro [806]. Mesenchymal stem cells can be isolated from bone marrow as well as many tissues, including adipose tissue, umbilical cord blood, placenta, and amniotic fluid.

Autologous bone marrow-derived stem cell transplantation can be safely administered after acute myocardial infarction, but optimal infused stem cell type, infusion dose, and administration time from onset of acute myocardial infarction and with respect to primary treatment (drug and percutaneous coronary intervention) remain to be assessed [807].

Tissue-specific stem and progenitor cells usually have restricted developmental potential in vivo. As autologous adult stem cells lack multipotency, autologous pluripotent stem cells must be produced to form three germ lines (endo-, meso-, and ectoderm), i.e., to give rise to all body cell types. For example, placenta-derived stem cells can generate all three germ layers in vitro. Moreover, they do not form teratomas (noninvasive tumors with tissues from all three embryonic layers) when transplanted [806].

Cell therapy of ischemia based on late-outgrowth endothelial colony-forming cells (ECFC; i.e., endothelial progenitor cells derived from peripheral blood and mainly umbilical cord blood in opposition to early endothelial progenitor cells of myeloid origin) is aimed at promoting revascularization. However, the proportion of engrafted cells is usually low. Human bone marrow-derived mesenchymal stem cells increase the angiogenic activity of cord blood endothelial colony-forming cells via paracrine signals that elevate expression by ECFCs of sphingosine kinase SphK1, which mediates survival and proliferative signals (via the PI3K–PKB pathway), as well as the proangiogenic and nascent blood vessel stabilizer sphingosine 1-phosphate $S1P_1$ receptor [808].

9.5.2 *Cardiac Resident Progenitor Cells*

In adult human hearts, *cardiac resident progenitor cells* (CRPCs) are able to differentiate into cardiomyocytes. They express stem cell factor receptor (SCFR), stem cell antigen SCA1, and transcription factor Islet-1.

Many CRPC populations reside in the fetal and adult human heart [809, 810]:

- SCA1+, endoglin+, GATA4+, NKx2.5+, PTPRc−, PromL1−, CD14−, CD34− CRPCs, which lodge within the atrium, the interatrial septum, and the atrioventricular boundary and can differentiate into functional cardiomyocytes as well as endothelial and smooth muscle cells
- SCFR+, GATA4+, NKx2.5+, cTnnI+, Isl1−, PTPRc−, PECAM1− CRPCs, mainly located in the right atrium, which can differentiate into adipogenic and osteogenic cells as well as cardiomyocytes and smooth muscle and endothelial cells in some circumstances
- SCFR+, SCA1+, PECAM1+ CRPCs
- Isl1+, SCFR−, SCA1− CRPCs
- ABCb1+ CRPCs
- ABCg2+, SCA1+ CRPCs
- ABCg2+, cTnnT+, PECAM1−, SCFR−, SCA1− CRPCs, which localize preferentially to the right atrium of healthy and ischemic hearts at similar levels, but upon ischemia, their number rises in the left ventricle

They can include CRPCs originated from the bone marrow. They can be isolated from most patients and all four cardiac cavities. SCFR+ CRPCs exist in a higher density in ischemic than normal myocardium [810].

The human CRPC pool producing ATP-binding cassette transporter ABCg2[15] differ from ABCg2+, PECAM1+ cells that are endotheliocytes in capillaries and small arterioles [809, 810]. The ABCg2 transporter modulates the proliferation, differentiation, and survival of ABCg2+ progenitors [810]. A small number of ABCg2+ cells express NKx2.5, titin, α-actinin, cardiac troponin-T, and α-myosin heavy-chain (but neither SCA1, the pluripotency markers Oct3 and Oct4, nor glycosphingolipids stage-specific embryonic antigen SSEA3 and SSEA4 that act as cellular differentiation players, indicating the lack of pluripotency) [810]. They do not differentiate into fully functional cardiomyocytes. They may instead represent an already committed mesodermal progenitor state.

Multipotent Isl1+ cardiovascular progenitors that reside in both embryonic and adult hearts can give rise to the three main heart cell types: cardiomyocytes and smooth muscle, and endothelial cells.[16] However, regeneration capacity of cardiac stem cells is small.

[15] A.k.a. breast cancer resistance protein (BRCP).

[16] Signaling by Wnt promotes self-renewal of Islet-1+ cardiovascular progenitors.

Several signaling pathways induce or suppress differentiation of stem cells, such as cascades that involve mitogen-activated protein kinases, glycogen synthase kinase-3, and phosphatidylinositol 3-kinase, among others.

A set of growth factors participate in the generation of progenitors of cardiomyocytes (cardiac troponin-T cTnnT+ cells) and vascular endothelial and smooth muscle cells. These three lineages develop from a common, mesoderm-derived, VEGFR2+ cardiovascular progenitor.[17] Combinations of activin-A and bone morphogenetic protein-4, to which Wnt inhibitor Dickkopf homolog-1 and vascular endothelial growth factor VEGFa are secondarily added to promote expansion and maturation of VEGFR2+ cell population and later fibroblast growth factor FGF2, leads to differentiation of $VEGFR2^{low}$/SCFR− precursor cells toward cardiovascular colony-forming cells (with cardiac, endothelial, and vascular smooth muscle potential) [811].[18] SCFR+ cardiac progenitors activated with insulin-like growth factor-1 and hepatocyte growth factor before their injection in proximity of the site of occlusion of the left coronary artery in rats engraft within myocardium, then divide and differentiate into endothelial and smooth muscle cells and, to a lesser extent, cardiomyocytes [812].[19] Differentiation into vascular cells that reconstitutes a coronary network can also result from synthesis and secretion of CXCL12 chemokine by hypoxic coronary vessels owing to upregulation of hypoxia-inducible factor-1α.

On the other hand, transcription factor TCF19[20] can support self-renewal of stem cells in vitro, independently of exogenous factors [813]. Transcription factor TCF19 inhibits both Ras GTPase-activating protein (RasGAP) and extracellular signal-regulated ERK1 kinase.[21]

[17] VEGFR2+ cardiovascular progenitors are also called FLK1+ (fetal liver kinase-1) or KDR+ (kinase insert domain protein receptor) precursors.

[18] Three distinct populations of precursor cells exist: (1) $VEGFR2^{low}$/SCFR−, (2) $VEGFR2^{-}$/SCFR+, and (3) $VEGFR2^{high}$/SCFR+ cells (SCFR being also called CD117 and KIT [kinase in tyrosine]). $VEGFR2^{low}$/SCFR− cardiac progenitors express the highest level of cardiac transcription factors (NKx2-5, Isl1, TBx5, and TBx20). $VEGFR2^{-}$/SCFR+ undifferentiated cells that include cardiac stem cell express the highest level of Oct4, FoxA2, and Sox17. $VEGFR2^{high}$/SCFR+ hematopoietic and vascular progenitors express markers of a hematopoietic commitment such as GATA1, of vascular endothelium, such as PECAM1 and cadherin-5, and smooth muscle, such as calponin and smooth muscle actin.

[19] Growth factor IGF1 and its receptor IGF1R mainly act in cardiac progenitor division and survival, whereas HGF and its receptor HGFR (or Met) predominantly operate in cardiac progenitor migration.

[20] A.k.a. transcription factor SC1 and pluripotin. Self-renewal can be achieved by a specific combination of exogenous activations of leukemia inhibitory factor LIF–STAT3, bone morphogenetic protein BMP–SMAD, and Wnt–β Ctnn pathways in mouse embryonic stem cells and using exogenous factors, such as IGF1, FGF2, TGFβ, activin, and Wnt proteins, for human embryonic stem cells.

[21] Inactivation of ERK1 blocks differentiation, whereas inhibition of RasGAP activates Ras and PI3K, promoting self-renewal of mouse embryonic stem cells.

Many proteins participate in the intracrine regulation of cardiac stem cells, such as VEGF, dynorphin, soluble readthrough form of acetylcholinesterase,[22] octamer-binding transcription factor Oct3[23] pancreatic and duodenal homeobox gene product PDx1,[24] paired box protein Pax6, and high-mobility group protein-B1 (HMGB1),[25] among others [817].

Among multipotent cardiac resident progenitor cells, cardiosphere-derived cells (CDC) represent a major source of cardiac multipotent stem cells. They are obtained from a small population of cells isolated from human atrial or ventricular myocardium by biopsies. They spontaneously form clusters, the so-called cardiospheres. Cardiosphere-derived cells contain heterogeneous cell populations of cardiac multipotent progenitor cells and various cell subpopulations with mesenchymal cells. They can differentiate into various cardiac cell types, in particular cardiomyocytes or vascular endotheliocytes, in addition to neural crest-derived cells (neurons, glial cells, and smooth myocytes). They have also juxta- and paracrine effects that can enhance cardiomyogenesis and/or angiogenesis. Cardiosphere-derived cells express GATA4 and nestin. The majority of cardiosphere-derived cells are SCFR+, PECAM1+, CD34+, CD90+ (thymocyte differentiation antigen Thy1), and CD105+ (endoglin).

Human atrial appendages contain a AldH+ (aldehyde dehydrogenase), CD34+, PTPRc− cardiac stem cell population [818]. *Cardiac atrial appendage stem cells* (CASC) differ from SCFR+ cardiac stem cells. They are more related to cardiac

[22] Cholinesterases constitute a category of ubiquitous enzymes. Acetylcholinesterase (AChE) is mainly responsible for the inactivation of cholinergic neurotransmission. Different AChE variants derive from alternative splicing, which is differentially regulated according to the cell type, generating various products with the same catalytic domain, but with distinct C-termini, such as the hydrophobic ($AChE^H$, readthrough ($AChE^R$), and tailed ($AChE^T$) forms. The readthrough form can have distinct and sometimes inverse function with respect to the major $AChE^T$ variant [814]. It can also interact with specific partners such as RACK1, PKCβ2, or the glycolytic enzyme enolase. Overexpression of $AChE^R$ is associated with germ cell apoptosis and hematopoietic cell proliferation as well as various tumor types, oxidative stress, and behavioral impairment, among other pathological conditions [814]. Unlike more common splice variants, embryonic and stress-associated readthrough form of acetylcholinesterase is unable to promote cell adhesion and neurite outgrowth. It competes with AChE for ligand binding such as laminin-1 [815]. Butyrylcholinesterase (BChE) coexists with AChE in many tissues; but is controlled by distinct mechanisms. In plasma, BChE originated from hepatocytes represents the major cholinesterase. Both AChE and BChE constitute light molecular species and tetramers.

[23] Factor Oct3, a.k.a. Oct4 and POU domain, class-5 homeobox POU5F1, is targeted with SRY-related high mobility group-box protein Sox2, Krüppel-like factor KLF4, and MyC by direct reprogramming that generates induced pluripotent stem cells from neural stem cells and fibroblasts [816].

[24] A.k.a. glucose-sensitive factor (GSF), insulin promoter factor IUF1, insulin upstream factor IPF1, islet–duodenum homeobox IDx1, and somatostatin-transactivating factor STF1. Overexpression of Pdx1 can also convert hepatocytes to pancreatic cells [806].

[25] A.k.a. HMG1, HMG3, SBP1, and amphoterin. This chromatin protein can translocate to the nucleus, where it binds DNA and regulates gene expression. It interacts with nucleosomes, transcription factors, and histones. It can also be secreted by immunocytes (monocyte, macrophage, and dendritic cell). The extracellular form can bind the inflammatory receptor for advanced glycan end-products and interact with Toll-like receptor ligands and activate cells via TLR2 and TLR4 receptors.

colony-forming-unit fibroblasts. They have a better cardiac differentiation capacity than cardiosphere-derived cells that can be isolated from atrial or ventricular walls.

9.5.3 Cell Reprogramming

Cells differ according to the expressed part of the genome, that is, transcriptional and epigenetic programs. Gene expression is in fact determined by the presence of transcriptional regulators (the so-called noncoding RNAs, chromatin-binding proteins, DNA methyltransferases, histone-modifying enzymes, etc.). Protein- and miRNA-encoding genes that lead to differentiation are repressed in stem cells for self-renewal. Although adult stem cells specific to a given tissue are only able to form their tissue of origin, upon stimulation using genes and molecules that define embryonic stem cells, they can return to a more primitive cell state and then transform themselves into other tissue cells (*transdifferentiation*). These multipotent precursors derived from adult stem cells then can serve for tissue regenerating treatments.

Stem cells can be produced by different techniques, such as somatic cell nuclear transfer, cell or cell extract fusion, genetic alterations, and stimulation by exogenous molecules. Embryonic stem cell lines can be obtained from blastocyst after fusion of a sperm and oocyte. A part of the inner cell mass of the blastocyst is extracted, cultured, and maintained in an undifferentiated state. Somatic cell nuclear transfer and oocyte enucleation involve the removal of the genome of a somatic cell that is then transferred into the enucleated oocyte. The intracellular environment of the oocyte alters the epigenome of the somatic cell so that its gene-expression pattern becomes less committed. Reversion in gene expression that is also called reprogramming is based on the fact that cell differentiation as well as cellular changes during aging are reversible.

Induced pluripotent stem cells can be generated by dedifferentiation of somatic cells by introducing retroviral vectors that carry genes encoding a set of factors. Viral integration of transcription factors can be replaced by other approaches, such as a procedure based on identified exogenous factors that activate pluripotency, hinder activity of inhibitors of pluripotency, and regulate chromatin modifications.

A set of transcription factors (KLF4, Lin28, MyC, Nanog, Oct4, Sox2, and TCF2) can be used to confer stem cell pluripotency [819]. Resulting cells are able to form embryoid bodies (aggregates of embryonic stem cells that have begun to differentiate). However, gene-expression patterns of embryonic stem cells and induced pluripotent stem cells differ. In addition, reprogramming must avoid the predisposition of induced pluripotent stem cells to carcinogenesis.

The optimal combination of factors necessary and sufficient for myocardial reprogramming of human fibroblasts includes GATA-binding protein GATA4, Heart and neural crest derivative-expressed protein HAND2 (basic helix–loop–helix transcription factor bHLHa26), T-box protein TBx5, and myocardin, as well as two muscle-specific microRNAs, miR1 and miR133 that are both regulated by MEF2c [820]. They launch cardiac marker expression in neonatal and adult human fibroblasts.

9.6 Mechanophore-Linked Materials

Implantable intravascular devices must preserve mechanosensors and their adaptive signaling. Vascular endothelial and smooth muscle cells possess numerous mechanosensors (G-protein–coupled receptors, receptor Tyr kinases, in particular the VEGFR2–Cadh5–PECAM1 complex, ion channels, and integrins, including mechanosensors associated with glycocalyx and primary cilia). Transcription of multiple genes depends, at least partly, on mechanical stresses via mechanosensitive Krüppel-like factor KLF2 and nuclear factor erythroid 2-related factor (NRF2). Pulsatile flow also influences the activity of the transcription factors AP1 and NFκ B. Manifold signaling cascades are triggered, such as those using the MAPK module (JNK and P38MAPK), AMPK, PKB and PKC kinases, as well as Rho and Rac GTPases.

The body's tissue growth and remodeling are based, at least partly, on mechanically induced chemical activation (mechanochemical transduction). Similarly, mechanically responsive polymers sense and react to mechanical stress.

Polymers that contain mechanically sensitive and chemically reactive molecules—*mechanophores*—have properties that change in response to physical stimuli. Applied forces can trigger preprogrammed chemical reactions in mechanophores to cause desired responses.

Upon deformation, mechanochromic polymers change color, as light absorption, molecular interaction with incorporated *spiropyrans* that serve as strain sensors and color-generating mechanophores, and/or spiropyran conformation vary. For example, colorless mechanochromic polymers that are stretched turn red or purple when exposed to a certain level of mechanical stress, i.e., at the point where irreversible deformation and tearing appear [821]. Loading breaks and reforms covalent bonds in mechanophore-linked elastomeric and glassy polymers with appropriately located and designed spiropyrans. Mechanophore-linked materials with self-sensing and self-reinforcing capabilities can be used in medical devices for damage warning, degradation spread slowing, or even early alterations repairing (self-healing materials). Mechanosensitive materials can also be employed for drug delivery.

9.7 Nanotechnology-Based Therapy

Nanotechnology can be applied to lesion treatment. Biocompatible nanoparticles filled with suitable drugs can target diseased cells. Once inside the lesion, they can deliver their content. Drug nanovectors are designed with a flexible hydrophilic polymer coat to escape from immune cells, appropriate surface ligands to bind selectively to cognate receptors on target cells, incorporated metallic particles to activate drug release upon exposure to heating by radiofrequency or ultrasound waves.

Double-emulsion stable nanodroplets (water droplet inside of an oil droplet stabilized by a single-component surfactant; size < 100 nm) allow simultaneous delivery at a fixed ratio and given location of different pharmaceuticals, as both lipid- and water-soluble drugs can be loaded. Water-in-oil-in-water emulsions thus yield advantages over simple oil-in-water emulsions for cargo encapsulation [822].

Certain nanoparticles bind to asialoglycoprotein receptors on hepatocytes [823]. Certain macromolecular vectors, such as dendrimers, are aimed at delivering their content into determined intracellular compartments [824].[26] Vectors can carry manifold substances, such as anticancer drugs, imaging agents, and cell receptor ligand.

As cancer cells need more folic acid than normal cells, and display a higher number of folate receptors on their plasmalemma, folate can be added to dendrimers [825]. Tumor blood vessels are lined with a mesh of clotted plasma proteins that are not found in other tissues and thus can be targeted by selective nanoparticles, as they can contain tumor-homing peptides that bind to this mesh. Tumor-homing peptide accumulation in tumor vessels induces additional binding sites for more particles, mimicking platelets [826]. These nanoparticles must avoid fast clearance by the liver and spleen, except in the case of quick efficient delivery from nanoparticles. Angiogenesis that contributes to tumor malignancy also represents an additional target. Neovasculature of $\alpha_v\beta_3$-integrin–expressing tumors[27] can be degraded by specific nanoparticles that encapsulate the cytotoxic drug [827]. Antiangiogenic drug delivery mediated by $\alpha_v\beta_3$-integrin increases 15-fold the antimetastatic effect with respect to systemic administration of free drug.

Another aspect of nanoparticles is related to physical destruction rather than chemical degradation of targeted cells. With selective ligands for targeted cell receptors, these nanoparticles concentrate the energy delivered by ultrasound to kill malignant cells. Stable, biocompatible, noncytotoxic, functionalized (CD22-directed), single-walled carbon nanotubes that emit heat upon absorption of energy from near-infrared light can serve for thermal ablation of tumor cells [828]. Hyperthermia also enhances tumor cytotoxicity of chemotherapy and radiotherapy.

Lipid-coated perfluorocarbon nanodroplets can be used for therapy and ultrasound imaging. Subjected to ultrasound with acoustic energy at usual power levels, these nanoparticles enhance drug delivery. The displacement of perfluorocarbon nanoparticles can be forced in the direction of US propagation. Ultrasound (peak pressure on the order of MPa, frequency on the order of MHz) produces a particle velocity proportional to acoustic intensity, which also increases with rising center frequency [829]. Within a vessel (bore hundreds µm), a fluid motion is produced with a velocity of hundreds µm/s, which conveys nanoparticles. Furthermore, ultrasound generated by conventional ultrasound imaging devices can enhance interactions between the nanoparticle lipid layer and targeted cell plasma membrane via appropriate integrins,

[26] Dendrimers have a tree-like structure with many branches on which can be attached various molecules.

[27] Integrin-$\alpha_v\beta_3$ is preferentially expressed on endothelia of newly formed vessels in malignant or diseased tissues.

thus enhancing lipophilic drug delivery [830]. Ultrasound imaging can be combined with nanoparticle-based chemotherapy. Mixtures of drug-loaded polymeric micelles and perfluoropentane nanobubbles stabilized by the same biodegradable block copolymer are transported selectively into the tumor interstitium, where the nanobubbles coalesce to produce microbubbles with ultrasound contrast [831]. Drugs in polymeric micelles are released under the effect of tumor-directed ultrasound. Such nanoparticles have many functions: drug carriers, ultrasound contrast agents, and enhancers of ultrasound-mediated drug delivery.

9.8 Pharmacogenetics and Pharmacogenomics

Preventive medicine possesses two domains: (1) individual prediction of major risks such as rare monogenic disorders and (2) statistical risk of common diseases. Genetic tests yield risk factors as percentages, i.e., the probability to develop a given disease in the presence of a genetic variation with respect to the healthy population.

Two types of variations are common in the genome: (1) changes in single nucleotide (single nucleotide polymorphisms [SNP]) and (2) changes in DNA regions that include missing (deletion) and added blocks (copy number variation such as duplication).

However, most diseases are multifactorial. For example, type-2 diabetes is linked to a similar extent to gene mutations, inappropriate diet, and sedentary life. Hence, prediction based on the genetic context may be not very useful, whereas change in lifestyle is beneficial.

Genetic features that characterize individuals (hence differing among people) not only affect susceptibility to diseases, but also response to drugs.

Pharmacogenetics and *pharmacogenomics* refer to interindividual variation of drug response for a given medication. Individuals indeed react either strongly, in a common fashion (normally), or not; they are thus categorized as positive and negative responders. Pharmacogenetic tests identify mutations to predict responses to a medication. Prescription, i.e., drug type as well as dose, must then be adapted to be efficient and safe. Pharmacogenetic tests thus yield recommendations on dosage.

Any drug actually has therapeutic and adverse effects. When a blind therapy is used (trial-and-error strategy), adverse drug reactions cause more than 2×10^6 hospitalizations and 10^5 deaths per year in the USA and about 6 % of the new hospital admissions in Germany [832]. Hypersensitivity drug reactions represent approximately one third of all adverse drug reactions. In addition to disease determinants and environmental parameters, genetic factors participate in harmful drug side effects.

Pharmacogenetics aims at predicting drug efficiency with respect to the patient genetic and disease patterns, pointing out the relation between gene variants and drug response.

Pharmacogenomics also correlates gene expression mode with drug efficacy and toxicity. Its goal is optimization of drug therapy related to the patient's genotype, thereby identifying patients responsive to a given drug.

Whereas pharmacogenetics focuses on a single gene, pharmacogenomics uses a genome wide approach to study the entire spectrum of genes involved in drug response [832]. The technical meaning of pharmacogenetics can also be related to investigation of the individual genetic background to predict response to a drug and guide prescription; that of pharmacogenomics is analysis of the entire genome of distinct human populations to identify the genetic factors influencing the response to a given drug. As pharmacogenomics relates the genetic context to individual response to drugs, drugs are further characterized in the framework of personalized medicine.

Both pharmacogenetics and -genomics deal with subject-specific *pharmacokinetics* and *pharmacodynamics* associated with the genetic background. Pharmacokinetics studies the body's action on the administered drug. Pharmacokinetics comprises drug absorption, distribution, metabolism, and excretion. Pharmacodynamics explores the drug's effect in the body. Pharmacodynamics encompasses chemical and physiological effects of drugs as well as mechanisms of drug action, that is, chemical reactions and molecular interactions, especially between-drug interactions, and relation between drug concentration and effect.

In cardiovascular diseases, drugs include β-blockers and statins, among others. Single-nucleotide polymorphisms can affect genes that encode drug-metabolizing enzymes and drug transporters. Supermetabolizers convert drug rapidly and efficiently with the subsequent risk of poisoning when using current dose.

9.8.1 Drug-Metabolizing Enzymes

Hepatic enzymes involved in pharmacokinetics determine drug effect duration and risk of adverse reactions. Expression of hepatic ^{UDP}glucuronosyltransferase UGT1a6 that glucuronidates various drugs and toxins is associated with polymorphisms in the 5′-regulatory region and exon-1, the three most common nonsynonymous polymorphisms being S7A, T181A, and R184S. However, interindividual variability in glucuronidation may be associated with environmental factors. Acetyltransferases (e.g., butyrylcholinesterase) also comprise slow and fast acetylator variants.

Among drug-metabolizing enzymes, heme-containing, cytochrome-P450 (CYP) oxidases are responsible for oxidative degradation of chemicals taken in the diet, environment, and medications, hence clearance of drugs such as immunosuppressors (Tables 9.5 and 9.6). They thus contribute to the metabolism phenotype (rapid, intermediate, and slow) that determines the dose (high, intermediate, and low). Among the 57 CYP genes, three categories—CYP1 to CYP3—are major genes that encode enzymes contributing to the oxidative catabolism of various compounds [832]. The intestine and liver contain CyP3a enzymes responsible for the metabolism of more than half of the drugs. Their activity varies among members of a given population. Interindividual differences in CyP3a4 and Cyp3a5 and drug transporter ABCb1 should lead to individualization of immunosuppressive therapy (cyclosporine and tacrolimus).

Table 9.5 Examples of drug-processing enzymes and their target medications (Source: [832]). Captopril is an angiotensin-converting enzyme inhibitor for the treatment of hypertension. Erythromycin is a macrolide antibiotic. Labetalol is a dual α- and β-adrenoceptor antagonist. Losartan is an angiotensin receptor antagonist. Lovastatin is used in hypercholesterolemia. Metoprolol is a selective β1-blocker. Omeprazole is a proton pump inhibitor. Propranolol is a β-blocker. Tacrine is a histamine ^{N}methyltransferase and cholinesterase inhibitor

Enzyme	Catabolized drugs
CyP1a2	Propranolol, tacrine
CyP2c9	Losartan
CyP2c19	Omeprazole
CyP2d6	Metoprolol, propranolol
CyP3a	Erythromycin, losartan,
	lovastatin, omeprazole
Glucuronosyl transferase	Labetalol
Thiopurine methyltransferase	6-Mercaptopurine
^{S}Methyltransferase	Captopril

Table 9.6 Examples of individual variations in drug response associated with certain genes. (Source: [832])

Gene	Target protein
CYP1A2	ABCc1 transporter
CYP2A6	Serotonin transporter
CYP2B6	Thiopurine ^{S}methyltransferase
CYP2C8	Glutathione ^{S}transferase
CYP2C9	^{UDP}Glucuronosyl transferase
CYP2C18	Catechol ^{O}methyltransferase
CYP2C19	Sulfonylurea receptor
CYP2D6	Dihydropyrimidine dehydrogenase
CYP2E1	Epoxide hydrolase
CYP3A4	ABC transporters
CYP3A5	Dopamine receptor
CYP3A7	ABCc1 transporter

The frequency of variant alleles of the CYP genes varies according to the ethnicity, that adds differences between human races, in complement to diet and environmental and psychosocial factors. A given percentage of the population is heterozygous, carrying a single variant allele, thereby producing a reduced quantity of functional enzyme. On the other hand, homozygous individuals, those possessing two variant alleles (from both mother and father) lack enzyme activity, thereby needing only a strongly reduced quantity of the standard drug dose to avoid side effects. For example, Chinese patients require lower doses of heparin and warfarin than those usually recommended for white patients because of polymorphisms in the gene encoding vitamin-K epoxide reductase complex-1 [832].

The CYP superfamily member CyP2d6 has 78 variants, some without or poor catabolic activity, some with enhanced action [832]. Several inactivating polymorphisms have also been reported in the CYP2C19 gene. Its catalytic product processes H^+ ATPase inhibitors. Approximately 2–4 % of European- and 4 % of African-origin Americans have poor proton pump inhibitor metabolism [832].

9.8.2 Drug Transporters

Genetic variability in drug transporters plays a role in the resistance of malignant cells to anticancer drugs. Polymorphism in ABC genes that encode ATP-binding cassette transporters, or transfer ATPases, affects the function and expression of these proteins. For example, a reduced rate of methotrexate metabolism, which causes overdosing and nephrotoxicity, can result from heterozygous mutation (R412G; i.e., Gly412 mutant and Arg412 wild-type forms) in the ABCC2 gene, which encodes ABCc2 transporter (or multidrug resistant protein MRP2) [832]. This mutated region is associated with substrate affinity.

9.8.3 Antihypertensive Therapy

Variants of angiotensin-converting enzyme and nitric oxide synthase NOS3 as well as polymorphism in the promoter region of the gene that encodes Na^+ channel γ-subunit influence effects of antihypertensive medication [832]. In addition, single nucleotide polymorphisms in angiotensinogen (T1198C), apolipoprotein-B (G10108A), and α2a-adrenoceptor (A1817G) are significantly correlated with the change in left ventricular mass during antihypertensive treatment [832].

9.8.4 Drug Receptors and Antilipolytic Therapy

After a meal, excessive levels of lipids are conveyed to adipose tissue where they are processed and stored as triglycerides. During fasting, triglycerides are hydrolyzed to free fatty acids and glycerol. Lipolysis is regulated by hormones and cytokines. Insulin inhibits lipolysis. Glucagon and catecholamines (adrenaline and noradrenaline) stimulate lipolysis. Free fatty acids are released in the blood stream and used as energy substrates. In obesity, lipolysis is deregulated; the basal lipolysis rate increases and the stimulation of lipolysis by catecholamines and the antilipolytic action of insulin are inhibited. Overproduction of tumor-necrosis factor TNFSF1 in adipose tissue impairs hormonal control of lipolysis. In addition, elevated plasma levels of free fatty acids cause insulin resistance [833].

Adipocyte differentiation, maintenance, and function are partly controlled by peroxisome proliferator-activated receptor PPARγ, or ligand-inducible nuclear receptor NR1c3. Both the widespread PPARγ1 (especially in adipocytes, monocytes, and macrophages) and adipose tissue-restricted PPARγ2, which has an extended N-terminus, are involved. They heterodimerize with retinoic acid X receptors (RXR). Polyunsaturated fatty acids and eicosanoids activate PPARγ. The NR1c3 receptor regulates expression of genes implicated in lipid uptake synthesis, and storage with lipid droplet stabilization (i.e., expression of lipid-droplet protein perilipin), as well as glycerol and fatty acid recycling (reesterification of fatty acids and glycerol to triglycerides) and fatty acid oxidation [833].

The pharmacological decrease of lipolysis manifested in decreased plasma levels of free fatty acids and triglycerides in adipose tissue via inhibition of hormone-sensitive triglyceride lipase by nicotinic acid (a.k.a. niacin, vitamin-B3, and less commonly vitamin-PP [pellagra-preventive factor]) requires doses that are much higher than those provided by a normal diet [834]. Nicotinic acid raises the plasma level of high-density lipoprotein level and diminishes those of very low-density lipoprotein, low-density lipoprotein, triglycerides, and lipoproteins [835]. This antilipolytic effect results from inhibition of the AC–cAMP axis by Gi-coupled nicotinic acid receptors high-affinity NiAcR1 and human-specific, low-affinity NiAcR2 [835].[28] The reduction of circulating free fatty acids is also engendered by the antidiabetic drugs insulin-sensitizing thiazoledinediones that connect to NR1c3 with high affinity [833]. Both the antilipolytic human GPR109A and GPR109B genes that encode GPR109a (NiAcR1) and GPR109b (NiAcR2) as well as GPR81[29] are induced by thiazoledinediones in adipocytes, as NR1c3 activate transcription of these genes.

Immunomodulatory statins[30] are aimed at reducing LDL^{Cs}, the efficacy of which can be influenced by genetic variations. However, common genetic variants (LPA, APOE, CELSR2/PSRC1/SORT1, ABCC2, and SLCO1B1) do not alter the lipid response to the statin simvastatin [837].

---

[28] A.k.a. G-protein-coupled receptor GPR109a and GPR109b, as well as HM74a and HM74b, in addition to protein upregulated in macrophages by interferon-γ in mice (PUMaγ) [834]. Some aromatic ^{D}amino acids, such as ^{D}phenylalanine and ^{D}tryptophan and its metabolite ^{D}kynurenine, elicit a transient rise of intracellular Ca^{2+} concentration via GPR109b, but not GPR109a [836]. The GPR109B transcript abounds in human neutrophils. These ^{D}amino acids provokes chemotaxis of neutrophils.

[29] The GPR81, GPR109A, and GPR109B genes colocalize to the chromosome locus 12q24.31 [836]. Both GPR109a and GPR109b are expressed in adipose tissue as well as in the lung and spleen. Whereas GPR109a is also detected in trachea, GPR109b is identified in peripheral blood leukocytes.

[30] Immunomodulation refers to as a process that elicits alternative activation of immunity and precludes adverse stimulation. Regulatory and helper type-2 T lymphocytes favor the alternative macrophage phenotype. In addition, certain statins (but not simvastatin) can improve insulin sensitivity in obese subjects.

9.9 Medical and Surgical Simulators

Various minimally invasive vascular procedures exist. Mini-invasive therapy (MIT) is still improving to limit complications and mortality. Minimally invasive techniques are characterized by indirect visualization of the operation field on video monitors. Mini-invasive techniques use specially designed instruments that are manipulated either directly or via mechanical linkage. Direct manipulations must be operated over the patient, whereas robot-assisted surgery can be done from a remote location. In the operating room, video assistance can be based on the cheaper and lighter 3D ultrasound imaging, the images being merged with preoperative data, such as those acquired by CT and MR imaging.

Minimally invasive procedures are also used to treat cancers. Percutaneous image-guided cryoablation and radiofrequency ablation[31] are aimed at destroying small tumors rather than surgically removing them. These less invasive techniques must be carefully monitored using an imaging technique and suitable navigation tools.

Mini-invasive therapies in medicine and surgery can be practiced with much less risk after learning and training, using virtual reality-based simulators. Any real-time medical simulator must be extensible, scalable, maintainable, and flexible to be handled by multiple users. Any user must easily interact with the simulator. The first element deals with the detection of tool contact and both mechanical (tissue deformation) and possible associated biological responses. Such contact generates *haptic feedback* useful for depth sensation. The second element corresponds to action and its desired and unwanted consequences (tissue traction, cutting, bleeding, gaseous emboli, etc.). Computer-generated scenes are then aimed at simulating in real-time complex therapy procedures and tissue reactions.

Medical and surgical simulators are associated with subject-specific images, virtual reality hardware, and haptic devices. The training system for MIT is based on infographics and data banks of recorded forces from actual tasks. Computed-aided training must, indeed, reproduce visual and haptic senses experienced during a minimally invasive procedure. Reaction forces, calculated with a frequency on the order of 1 kHz, are sent to sensors to differentiate tissues and suitably navigate across the working field (depth sensing particularly).

Thoracoscopy and laparoscopy consist of performing surgery by introducing an endoscope and different instruments into the patient's body through small incisions [839, 840]. Mini-invasive therapy is beneficial because it reduces the surgical trauma and hospital duration, and thus the care cost, as well as the morbidity for well-mastered procedures. However, it brings new constraints on surgical practice. First, it significantly limits the surgeon's access to the organs of interest. Second, this technique requires specific hand–eye coordination, which must be acquired after a training period. Training methods for medical and surgical interventions use

[31] Mobile phones in operation emit a radiofrequency electromagnetic field. The wave energy is partially absorbed by head organs and affects the brain electrical activity. It also induces a decrease in regional cerebral blood flow [838].

either "endotrainers"[32] or living animals.[33] Limitations of such procedures stimulate development of computerized gesture-training systems. Much less sophisticated techniques can also be learned with simulators. For example, ultrasound imaging, which is a cheap, quick, and noninvasive technology, is commonly used for diagnosing many medical situations. Ultrasound-scan simulators are developed for training on patients, not only for sonography education, but also for evaluations of skills with a variety of normal and abnormal cases.

The development of MIT simulators raises important technical and scientific issues: (1) the geometry and mechanical behavior of the anatomical structures must be modeled; (2) the simulator must provide an advanced user interface including visual and force feedback.[34] Five main elements must then be introduced in a medical simulator: (1) geometric modeling,[35] (2) physical modeling,[36] (3) instrument interaction[37], and (4) visual[38] and haptic[39] feedback [748].

---

[32] Endotrainers use mannequins, inside which are placed plastic organs. The whole anatomy, organ interactions, and influence factors (cardiac pulsations and respiratory motions) are not taken into account. Consequently, these mechanical devices are of limited interest.

[33] Between-species anatomical differences are a limitation factor of animal testing.

[34] This interface can be decomposed into three distinct modules. The first module must model the interaction between surgical instruments and virtual organs. In particular, this task includes the detection and processing of the contacts that occur during the simulation. The second module aims at displaying the operating field on a video monitor in the most realistic manner. The third module must control a force-feedback device so the user can feel the applied forces that provide a 3D sense to the operator.

[35] Anatomical structures are now extracted from medical imaging. However, the automatic delineation of structures is still considered an unsolved problem. Many human interactions are required for 3DR.

[36] The mechanical behavior of organs is defined by nonlinearity, poroviscoelasticity, plasticity, and fatigue phenomena. The usual constitutive laws must be simplified and optimized for real-time computation before implementation in a surgical or medical simulator.

[37] The hardware interface driving the virtual instrument is essentially composed of one or several force-feedback systems having the same degrees of freedom and appearance as actual instruments used in minimally invasive therapy. In general, these systems are force-controlled, sending the instrument position to the simulation software and receiving force targets. Once the position of the virtual instrument is known, contacts between two instruments or between an instrument and an organ must be detected. When a contact is detected, a set of constraints is applied to soft-tissue models. However, modeling the physics of contacts can lead to complex algorithms and therefore purely geometric approaches are often preferred.

[38] Visual feedback is important in video therapy because it gives a 3D perception of the environment. In particular, the effects of shading, shadows, and textures are important clues that must be reproduced in a simulator.

[39] The touch experienced by an operator when manipulating an instrument gives 3D information. The coupling between visual feedback and force feedback produces the sense of immersion. Haptic feedback requires a greater bandwidth than visual feedback. For simulating the contact with a soft object, a refresh rate of 300 Hz should be sufficient, whereas for a hard object, a refresh rate greater than 1000 Hz should be used.

9.10 Modeling, Measurements, and Numerical Simulations

Modeling, measurements, and numerical simulations are carried out to enhance therapeutic outcome, both after anatomical reconstruction (e.g., grafting) and percutaneous intervention (e.g., stenting). Therapeutic procedures are targets of intensive research to enhance, in particular, prognosis and elaboration of new generations of implantable devices.

Modeling and simulations are aimed at:

- Exploring the local flow behavior
- Investigating the role of mechanical and physical influence parameters
- Optimizing the design of surgical reconstruction of vascular anatomy and implantable devices
- Reducing the fabrication cost of medical devices
- Predicting possible postoperative complications such as *intimal hyperplasia* (IH)

Among other implantable medical devices, drug-eluting stents have been conceived to struggle against smooth myocyte proliferation. Failure to achieve this task stimulates new explorations both experimentally and numerically. The final section of this chapter focuses on mathematical modeling of drug release from implanted drug-eluting stents. Once the stent is deployed, the drug is transferred from its storage container (transport stage 1), crosses the stent–wall interface (transport stage 2), and then distributes through the arterial wall (transport stage 3) by an advection–diffusion–reaction process.

9.10.1 Bypass Grafting

Physical and numerical experiments are aimed at investigating flow behavior in bypass grafting to determine the mechanical factors involved in postoperative complications such as *intimal hyperplasia* and to subsequently optimize anatomical reconstruction. Optimization of surgical reconstruction of arterial stenoses relies on minimization of blood flow disturbances [841].

Shape optimization methods have been applied to surgical planning [842]. Theoretical investigations based on perturbation analysis and linearized shape design provide results on existence, uniqueness of solution, and well-posedness of the problem [843]. Reduced basis approximation used for preprocessing can detect essential features of optimization, like sensitivity analysis [844].

The flow pattern depends strongly on anastomosis configurations (parallel side-to-side, diamond side-to-side, and end-to-side) and graft/host diameter ratio. In addition, compliance mismatch can lead to increased parietal stresses.

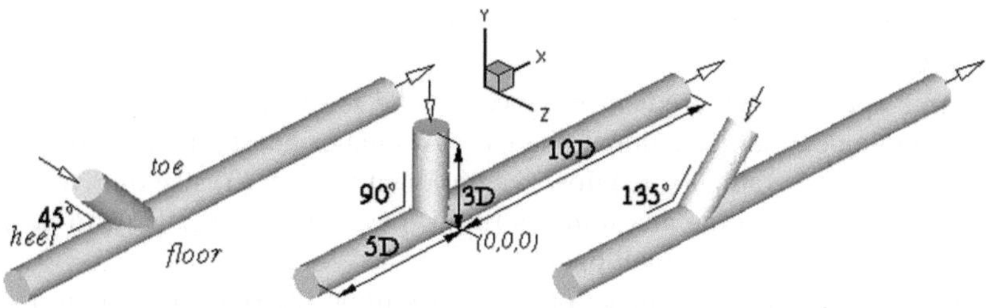

Fig. 9.1 Geometric configurations of idealized models of in-plane, end-to-side distal anastomoses used in bypass grafting. (Source: [845])

9.10.1.1 Anastomotic Geometry

Numerous anastomotic geometries have been designed to better regulate distribution of hemodynamic parameters at the distal anastomosis to enhance the patency of bypass grafts. Different geometries of the fluid domain focus on the graft–vessel anastomotic junction without dealing with any shape optimization model.

Investigations started using idealized models of anastomoses, mainly focusing on the distal anastomoses. The configuration of the distal end-to-side anastomosis is not naturally present in the arterial bed. (The merging of vertebral arteries into the basilar trunk is much more symmetrical.)

The basic configuration comprises three vascular segments of the same diameter (d) [845]. It comprises a bypass vessel (length $3\,d$) and the proximal (PSHA; length $5\,d$) and distal (DSHA, length $10\,d$) segments of the host artery (Fig. 9.1). By convention, the proximal (upstream) and distal (downstream) intersections between the graft and host artery are referred to as the *toe* and *heel*, whereas the host wall opposite to the graft outlet is the floor.

The geometrical variables of the connecting segment geometry include the anastomotic angle, toe and heel shapes, junction curvature at both the toe and heel, graft-to-host artery bore ratio, taper of the host artery, position of the suture line, out-of-plane grafts, graft wall corrugations (e.g., venous valve sinus), distance of anastomosis from the stenosis, and stenosis severity.

An MRI-based model of distal anastomosis after venous bypass surgery was used to classify bypass geometries [846]. Branching angles between centerlines of each vessel are measured. The spectrum of anastomoses can be reduced to a small subset of cases characterized by two angles: the angle between the graft and the plane of the host artery and between the graft and the proximal branch of the artery.

9.10.1.2 Hemodynamics at Distal Anastomosis

The hemodynamic factors include changes in wall shear stress (WSS) vector direction, time-averaged wall shear stress (TAWSS), spatial and temporal WSS gradients, and *oscillatory shear index* (OSI), that is, the ratio of the magnitude of time-averaged

WSS vector to the time-averaged WSS magnitude, as well as the *relative residence time* (RRT).

A large spatial WSS gradient observed at the toe of the anastomosis may assist morphological and functional changes. A smaller anastomotic angle reduces: (1) WSS peak and gradients; (2) flow separation at the toe; (3) size of reversed flow area downstream from the anastomosis [847]. In addition, large graft-to-host bore ratios (5:3) are better than small ones (1:1).

Flow features comprise impact on the artery floor, the stagnation point location fluctuating during the cardiac cycle, low shear region at the heel, and flow separation immediately downstream from the toe [847].

The effect of competitive flow, that is flow through a bypassed native artery with a low degree of stenosis, on the graft patency has been explored. The patency of bypass grafts used to treat severe stenoses is much higher than that of bypass grafts for low-grade lesions [847]. Yet, competitive flow can have a stronger role in arterial grafts than in venous grafts that had a different previous structure–function relation.

The impact of competition in the end-to-side anastomosis region between flow through the left anterior descending artery (LADA) with different stenosis degrees and that through the internal thoracic artery (ITA) bypass graft has been explored in an idealized ITA-LAD model (bypass graft diameter 4.6 mm; coronary artery diameter 4.5 mm; anastomosis angle $\sim 45°$; maximum Reynolds number ~ 1050) using the finite volume method [848]. This study suggests that grafting should be preferentially carried out, when the LAD stenosis degree is higher than 75 %, as the competitive flow in the bypass graft produces a wall shear stress distribution that may cause endothelial dysfunction and subsequent graft failure. The competitive flow in the native coronary artery is indeed related to the narrowing of the bypass graft and hence to the restriction of the graft flow. The competitive flow from saphenous vein graft in the left coronary artery is also implicated in narrowing of the arterial graft.

9.10.1.3 Measurements

The flow velocity field was measured in the symmetry plane at six stations by particle image velocimetry in a physical model of in-plane, end-to-side anastomosis (graft and host artery diameter 10 mm) with different anastomosis angle (20, 30, 45, 60, and 90°) for two values of the Reynolds number (500 and 1400) and numerical simulations were carried out with a fully developed laminar flow at the graft inlet [849]. The flow from the graft outlet impacts the floor and forms a narrow jet in the distal segment of the host artery (DSHA), with flow separation in DSHA lumen opposite to the floor. The near-wall region facing the graft outlet experiences high-velocity gradient and a strong vortex is observed in the downstream part of the proximal segment of the host artery (PSHA). The shape of this vortex depends strongly on the stenosis distance from the anastomosis.

9.10.1.4 Numerical Simulations

A computational model of steady flows in rigid vessels was used to explore flow patterns at a Reynolds number based on cross-sectional average velocity and host radius of 63, consistent with the mean value in coronary and limb arteries [845]. The vorticity brought by graft flow causes a strong vortex in the exit segment of the proximal part of the host artery when the latter is fully occluded [845]. The graft angle (45, 90, or 135°) in a fully occluded host artery affects the size and shape of the vortex at the PSHA outlet. At relatively high Reynolds number and large angle, flow separation occurs. On the floor of the anastomosis, around the stagnation point, a minimum wall shear stress is observed, whereas the nearby wall shear stress rises. The peak wall shear stress magnitude on the floor augments with the graft angle. Incorporation of curvature in the graft vessel provokes flow asymmetry, but exert a weaker influence on the flow than the graft angle [845]. An equal flow contribution from graft and host artery increases vorticity stretching and peak wall shear stress magnitude with respect to fully occluded host artery.

9.10.2 Mathematical Modeling of Drug Transfer from Stent

Once a stent is expanded into the arterial lumen, the drug is released from the polymeric coating (DESs) or perforations (depot stents). Stent implantation relies on the mechanical behavior of the stent and the reaction of the deformable, altered wall to the expanding stent.

9.10.2.1 Stent Deployment and Insertion—Wall Tunicae

Stent deployment in the arterial lumen can be investigated using the finite element method using proper stent mesh and properties as well as loading and boundary conditions. Constitutive laws for the stent and arterial wall rely on elastoplastic and hyperelastic models.

Deformation of implanted stents was investigated mostly assuming isotropic materials. However, healthy arterial walls are composed of three layers of composite materials of different rheology. Diseased arterial walls also contain these layers that are more heterogeneous due to the presence of fibrosis, lipid deposits, calcifications, and degraded elastin lamellae.

Preconditioned strips of wall tunica of segments from the midregion of the left anterior descending coronary artery subjected to cyclic quasi-static uniaxial tension in the axial and circumferential directions exhibit an anisotropic and strong nonlinear behavior in both loading directions, in addition to a small hysteresis [850]. The ratios of adventitial, medial, and intimal thickness to the total wall thickness are 0.4 ± 0.03, 0.36 ± 0.03, and 0.27 ± 0.02, respectively [850].

Table 9.7 Coefficients (MPa) of the strain energy density function of arterial wall layers (Source: [851]). The arterial wall is assumed to be made up of an isotropic, incompressible, hyperelastic material. The constitutive law coefficients are derived from measurements [850]

Wall layer	κ_{10}	κ_{20}	κ_{30}	κ_{40}	κ_{50}	κ_{60}
Intima	6.79×10^{-3}	5.4×10^{-1}	-1.11	10.65	-7.27	1.63
Media	6.52×10^{-3}	4.89×10^{-2}	9.26×10^{-3}	0.76	-0.43	8.69×10^{-2}
Adventitia	8.27×10^{-3}	1.2×10^{-2}	5.2×10^{-1}	-5.63	21.44	0

The intima is the stiffest layer, whereas the media in the longitudinal direction is the softest tunica. The average stiffness in the low-loading domain (at which the noncollagenous matrix material is mainly active) is lowest for the media and highest for the intima: 1.27 ± 0.63 and 27.90 ± 10.59 kPa, respectively). The average stiffness in the high-loading domain (dominated by the recruitment of collagenous fibers) is governed by the parameters k_1 and k_2 that are lowest for the media ($k_1 = 21.60 \pm 7.12$ kPa and $k_2 = 8.21 \pm 3.27$ kPa) and highest for the intima ($k_1 = 263.66 \pm 490.95$ kPa and $k_2 = 170.88 \pm 125.47$ kPa).

The media and intima have similar ultimate tensile stresses, which are on average three times smaller than those in the adventitia (1430 ± 604 kPa circumferentially and 1300 ± 692 kPa longitudinally). The ultimate tensile stretches are similar in all tunicae.

The structure of the arterial wall affects the drug transfer through it, that is, its effective diffusivity as well as the interstitial fluid velocity. The latter depends on the transmural pressure difference and the wall permeability. These two factors experience spatiotemporal variations due to wall deformations.

A hyperelastic material is characterized by a stress–strain relation associated with a *strain energy density function*. The latter is a measure of the energy stored in a material upon its deformation. It is generally fit to experimental data. Each constituent layer of the arterial wall can be associated with a specific strain energy density function (Table 9.7; [851]):

$$\begin{aligned} \mathcal{W} &= \kappa_{10}(I_1 - 3) + \kappa_{20}(I_1 - 3)^2 + \kappa_{30}(I_1 - 3)^3 \\ &\quad + \kappa_{40}(I_1 - 3)^4 + \kappa_{50}(I_1 - 3)^5 + \kappa_{60}(I_1 - 3)^6, \end{aligned} \tag{9.4}$$

where I_1 is the first deviatoric strain invariant.

9.10.2.2 Drug Release

Mathematical models of coated stents, which are used to prevent restenosis, have been used to study the effect of stent design on drug release and transport in the diseased wall, in particular the strut number and ratio between the coated strut area and vessel area [852].

The dose (not the drug concentration) is the quantity of interest. The delivered dose from the stent coating (drug concentration c_{coat}) depends on the strut number and strut surface density. Dose spikes are observed in front of the strut areas along the outer wall edge and dose drops in the between-strut regions for a given strut surface density. The dose is smaller at both stent ends. The higher the strut number is,[40] the lower the dose maximum and the more uniform the dose distribution in the wall for a given radial coordinate, the asymptotic stent delivering the lowest dose.

Drug release within and from a polymeric matrix depends on many factors such as the drug affinity for its surrounding medium. Paclitaxel, a natural product isolated from the bark of the Pacific yew tree, Taxus brevifolia, and a mitotic inhibitor used in cancer chemotherapy, is hydrophobic. Some of the drug blended into the polymer matrix can be retained in its storage matrix.

The existence of tiny coated struts of DESs requires highly detailed 3D models, hence a cost-prohibitive parameter identification and computations. An efficicient strategy thus relies on a phenomenological model of the drug transport inside the polymeric coating with incorporation of surface erosion and collapse of the polymeric matrix. Models of drug release from eroding surfaces currently fail to faithfully reproduce experimental data for highly degradable polymers [853].

However, drug release from a thin film of biodegradable polymer that coats a stent can be modeled using a 3D quadratic partial differential equation that enable to adapt a flux condition at the interface between the polymeric reservoir and its environment to measurements of drug release kinetics [775], hence allowing validation of numerical tests.

The experimental setup consists of a thin square film of polymeric sample (size 12×12 mm; thickness 80 μm; paclitaxel mass 4 μg). placed in a vial containing a fluid at rest. A sink is created by removing every 2 days the fluid that is analyzed and replaced by a fresh solution. The measurements were displayed by normalized release curves using the dimensionless total mass, that is the mass $M(t)$ of paclitaxel released at time t divided by the initial mass M_0.

The paclitaxel release profile can be defined by two types: sigmoidal and exponential (Fig. 9.2).

A sigmoidal pattern can be described by a simple 2-parameter quadratic ordinary differential equation model [854]:

$$\frac{dm}{dt} = \kappa_1 \, (1 - m(t)) + \kappa_2 \, (1 - m(t))^2 , \quad m(0) = 0, \tag{9.5}$$

where $0 \leq m(t) \leq 1$ ($m(t) = 1$: total release) is the ratio of the released drug mass to the medium at time t divided by the initial drug mass M_0, hence a dimensionless quantity. The two coefficients κ_1 and κ_2 are thus expressed in T^{-1} (T: time symbol). Initially ($t = 0, \; m = 0$),

$$\frac{dm}{dt}|_0 = \kappa_1 + \kappa_2 > 0. \tag{9.6}$$

The drug can be partly released from the polymeric matrix, that is, a fraction of the drug remains tethered to the polymer.

[40] Tested strut numbers are equal to 24, 48, 96, 192, 384, and ∞.

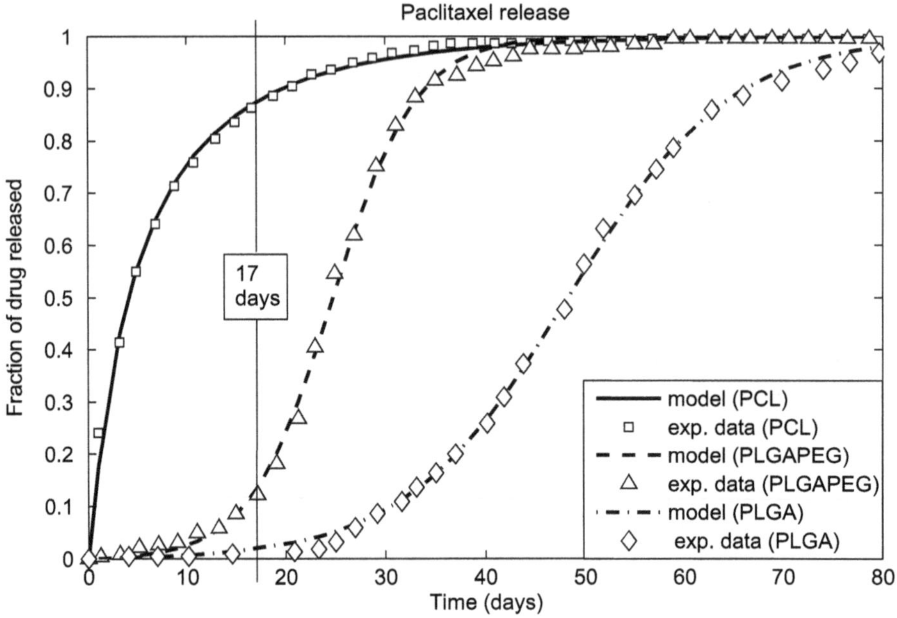

Fig. 9.2 Experimental [775] and numerical [854] (ODE-based modeling) results of normalized paclitaxel release curves for three neat polymers: polycaprolactone (PCL), poly(lactic-co-glycolic) acid (PLGA), and PLGA polyethylene glycol (PLGAPEG)

The quadratic nature enables proper capture of the microphysics and chemistry of the release. In addition, the nonlinearity is incorporated through a quadratic condition at the interface between the polymer and its surrounding medium, instead of a time-dependent or nonlinear diffusion in the polymer, thereby reducing the physical and computational complexity of mass transfer to predict the concentration of paclitaxel on the arterial wall.

When $\kappa_2 \neq 0$, the solution is given in closed form by

$$m(t) = \begin{cases} (\kappa_2 + \kappa_1) \dfrac{1 - \exp\{-\kappa_1 t\}}{\kappa_2 + \kappa_1 - \kappa_2 \exp\{-\kappa_1 t\}}, & \text{if } \kappa_1 \neq 0, \\ \dfrac{\kappa_2 t}{1 + \kappa_2 t}, & \text{if } \kappa_1 = 0. \end{cases} \tag{9.7}$$

This simple 2-parameter quadratic ODE model describes with a very good agreement the experimental data on drug release from neat PCL, PLGAPEG, and PLGA polymeric matrices (Fig. 9.2). This model can be much more easily handled and is more efficient than semiempirical models based on five to eight parameters [853].

The polymer occupies a parallelepipedic domain Ω_p surrounded by a domain Ω_m (environment) with boundaries Γ_p and Γ_m, respectively, $\Gamma_{int} = \Gamma_p \cap \Gamma_m$ being the interface between the polymer and its medium. The polymeric film has a midsurface Γ_0. It is endowed with a coordinate system $\{\xi_i\}_{i=1}^3$ with the origin at the center of the polymeric film.

The mass conservation principle states that

$$\int_{\Omega_p} c_p(\mathbf{x},t)\,d\mathbf{x} + \int_{\Omega_m} c_m(\mathbf{x},t)\,d\mathbf{x} = M_0 \tag{9.8}$$

The diffusion equation on the larger domain Ω becomes

$$\begin{cases} \lambda \dfrac{\partial c}{\partial t} = \nabla \cdot (D_m \nabla c) \text{ in } \Omega, \quad c(\mathbf{x},0) = 0 \text{ in } \Omega_m, \\ D_m \dfrac{\partial c_m}{\partial n} = F(\bar{c}_p - c) \text{ and } [c] = 0 \text{ on } \Gamma_0, \quad D_m \dfrac{\partial c}{\partial n} = 0 \text{ on } \Gamma, \end{cases} \tag{9.9}$$

where D_m is the drug diffusivity in the medium, λ the dimensionless lag coefficient, or retardation factor, which can be defined as the ratio of the velocity of the solute to the average interstitial solvent velocity and the new flux function $F(\bar{c}_p - c)$ on Γ_0 is similar to the boundary condition associated with the jump of the normal derivative along a surface in a Neumann sieve.

The flux function is given by, $\forall \boldsymbol{\xi} \in \Gamma_0$,

$$F(\bar{c}_p - c)(\boldsymbol{\xi},t) \stackrel{\text{def}}{=} 2h \frac{\lambda}{\beta} \left[A_1 \left(\bar{c}_p(\boldsymbol{\xi},t) - c(\boldsymbol{\xi},t) \right) + \frac{A_2}{c_0} \left(\bar{c}_p(\boldsymbol{\xi},t) - c(\boldsymbol{\xi},t) \right)^2 \right], \tag{9.10}$$

where β is a geometric conservation factor and the coefficients A_1 and A_2 have the same physical dimensions.

The quadratic PDE model of the 3D normalized concentrations in the polymer and medium can be studied using two versions, both completely specified by the two parameters of the ODE model [854]. (1) The drug concentration in the polymer $c_p(\mathbf{x},t)$ is a single function of the time t resulting in a nonlocal boundary condition at the interface Γ_{int} (nonlocal mass flux function at the interface). (2) A ξ_3-average of c_p is considered and the polymer domain is shrunk to the zero-thickness midsurface resulting in a local jump condition on the normal derivative across the midsurface (local flux function). An arbitrary distribution of the flux from the polymer between the top and bottom surfaces is thus avoided. Both on the mathematical and numerical implementation aspects, the second solution is preferred.

Dimensionless lengths were computed using a reference length defined by the square root of the surface area of the mid-surface. Dimensionless concentration results from the division by the initial concentration. An effective diffusion constant is introduced by dividing the diffusivity by the product λL^2.

The set of equations becomes:

$$\begin{cases} \dfrac{\partial \tilde{c}}{\partial t} = \tilde{\nabla} \cdot \left(\tilde{D}_m \tilde{\nabla} \tilde{c} \right) \text{ in } \Omega, \quad \tilde{c}(\tilde{\mathbf{x}},0) = 0 \text{ in } \Omega, \\ \tilde{D}_m \dfrac{\partial \tilde{c}}{\partial (\tilde{n}^+)} + \tilde{D}_m \dfrac{\partial \tilde{c}}{\partial (\tilde{n}^-)} = -\dfrac{2\tilde{h}}{\beta} \dfrac{\partial \tilde{c}_p}{\partial t} \quad \text{and} \quad [\tilde{c}] = 0 \text{ on } \Gamma_0, \\ \tilde{D}_m \dfrac{\partial \tilde{c}}{\partial \tilde{n}} = 0 \text{ on } \Gamma. \end{cases} \tag{9.11}$$

The results can be depicted by normalized drug release curves as a function of the thickness of the film and the diffusivity in the medium [855].

9.10.2.3 Drug Distribution in the Arterial Wall

The free drug diffusion in the coating is governed by the second Fick law of diffusion:

$$\frac{\partial c_{\text{coat}}}{\partial t} = \mathcal{D}_{\text{coat}} \nabla^2 c_{\text{coat}} \tag{9.12}$$

where c_{coat} is the concentration of the diffusing free drug (c_{coat_0}: initial concentration), t the time coordinate, and $\mathcal{D}_{\text{coat}}$ the diffusion coefficient in the stent coating.

Once it is released, the drug mural transfer depends on the wall properties for given pressure and concentration gradients as well as drug fate in the arterial wall, that is, reversible and irreversible binding, chemical processing, including catabolism and degradation, and cellular uptake.

Drugs are usually lipophilic, like elastin and collagen. The drug can then reversibly bind to these matrix constituents. Its effective diffusivity is thus lower then its diffusivity in the interstitial fluid. On the other hand, its internalization into the cell cytosol may be facilitated.

The drug total concentration (c_{tot}) is the sum of the concentrations of the free (c_f) and bound (c_b) drug form. The *binding-related partitioning coefficient* is the ratio of the free drug concentration in the wall to the drug concentration in the source (coating or incubation bath): $\kappa_b = c_f/c_{\text{input}}$. The partitioning coefficient varies according to the transmural location, but does not depend on the concentration in the incubation fluid [856].

Drug binding can be assumed to obey a first-order reversible reaction

$$\begin{aligned} [D_f] + [B] &\xrightarrow{\mathrm{k}_b} [D_b], \\ [D_b] &\xleftarrow{\mathrm{k}_d} [D_f] + [B], \end{aligned} \tag{9.13}$$

where D_f represents the free drug, B the binding partners (binding sites), D_b the bound drug, and k_b and k_d the binding and unbinding kinetic coefficient.

Drug transfer can be described by the *advection–diffusion–reaction* (ADR) equation, a type of partial differential equation (PDE), as it involves both time (t) and space ($\mathbf{x}$) variations:

$$\frac{\partial c_f}{\partial t} + \mathbf{v} \cdot \nabla c_f = \nabla \cdot \mathbf{D} \nabla c_f - \mathrm{k}_b (c_{B_0} - c_b) c_f + \mathrm{k}_d c_b + \mathcal{R}, \tag{9.14}$$

where c_{B_0} is the initial concentration of binding partners (initial free binding site concentration), $\mathbf{v}$ the interstitial fluid velocity, $\mathbf{D}$ the diffusion tensor, and $\mathcal{R}$ the reaction term. The reaction term comprises irreversible binding, catabolism, degradation, and cell uptake.

The diffusion and reaction terms are governed by the drug concentration and wall-transfer properties. Diffusion in the axial and azimuthal directions occurs in a given layer, whereas that in the radial direction cross many interfaces. Therefore, the

Table 9.8 Paclitaxel diffusivity (Source: [774]). A large variability in the literature data is observed, as the paclitaxel diffusivity in the extracellular space of the brain equals 900 μm²/s [857] and 60:40 ethanol:water mixture 427 μm²/s [855]

Medium	Diffusivity (μm²/s)
Coating	0.01
Plasma	20.3
Arterial wall	
Intima	0.22
Media	0.32
Adventitia	0.22

transmural diffusivity ($\mathcal{D}_r$) is smaller than the longitudinal ($\mathcal{D}_{ax}$) and circumferential ($\mathcal{D}_{cf}$) diffusivities:

$$\mathcal{D}_r < \mathcal{D}_{ax} \sim \mathcal{D}_{cf}.$$

The effective diffusivity of the drug that can bind in the warterial wall matrix is the ratio of the diffusivity of free drug to binding-related partitioning coefficient (κ_b): $\mathcal{D}_{eff} = \mathbf{D}/\kappa_b$ (Table 9.8).

The diffusion tensor is then expressed by a weighted effective diffusivity [774]:

$$\mathbf{D} = \mathcal{D}_{eff} \begin{bmatrix} w_r & & \\ & 1 & \\ & & 1 \end{bmatrix},$$

where the weighting factor w_r accounts for reduced radial diffusivity. The effective diffusivity takes into account drug binding ($w_r = 0.1$ for all parietal layers).

At each interface (I) within the arterial wall, the drug flux (J) depends on the *mass transfer resistance* ($\mathrm{R_{tf}}$) [858]:

$$J_I = \frac{1}{\mathrm{R}_{\mathrm{tf}_I}} \left(\frac{c_{I^+}}{\kappa_I} - c_{I^-} \right), \tag{9.15}$$

where c_{I^+} and c_{I^-} are the drug concentration at the outer and inner side of the interface, respectively, and $\kappa_I = c_{I^+}/c_{I^-}$ is the *mass transfer-related partition coefficient* on a given interface at equilibrium. This relation corresponds to a particular case of the flux function Eq. (9.10). Using this equation, the results displayed in Fig. 9.2 cannot be obtained.

The advection term is driven by the interstitial fluid velocity that depends on the interstitial fluid pressure and mural permeability ($\mathcal{P}$).

The homogeneization theory is based on a representative volumic element of the medium of interest and deals with volume-averaged quantities. The porosity, or void fraction ($\mathcal{V}$), is the ratio of pore volume to the total volume: $\mathcal{V} = V_{por}/V_{tot}$. Consequently the volume-averaged concentration of the drug and velocity of the interstitial fluid can be expressed by the following relations:

$$\breve{c} = \mathcal{V} c_f; \quad \breve{\mathbf{v}} = \mathcal{V}\mathbf{v} = -\frac{\mathbf{P}}{\mu}\nabla p, \tag{9.16}$$

where $\mathbf{P}$ is the permeability tensor and μ the fluid viscosity. The quantity $\check{\mathbf{v}}$ is also called the Darcy velocity.

The drug advection is reduced with respect to solvent advection due to additional friction represented by a *hindrance coefficient* (κ_h) that depends on the ratio between the drug size and average pore caliber.

Jump conditions (associated with a mathematical discontinuity) in drug concentration can be encountered at interfaces between stent coating and arterial wall due to the presence of a membrane as well as strong gradient of drug concentration (steep, regular variation of concentration) at mural sieves represented by elastic lamellae between constituent layers of the arterial wall with distinct porosity at both sides, in addition to thinner elastic lamellae inside the media and endothelial basement membrane.

Drug endocytosis can be represented by a reaction (sink) term in the transport equation. The volume-averaged reaction term ($\mathcal{R}$) can be approximated by a linear kinetics model using a reaction coefficient (κ_R):

$$\mathcal{R} = \kappa_R\, c. \tag{9.17}$$

The reaction coefficient is related to the drug half-life ($\kappa_R = 0.2/\mathrm{ms}$) [774].

Modeling the two-dimensional (2D) and three-dimensional (3D) diffusion from the polymeric coating on a DES on the arterial wall wetted by blood flow involves mass transfer in materials at different length scales (wall thickness 1 mm; polymeric coating thickness 10 μm).

Two-dimensional drug distribution in the arterial wall from an implanted drug-eluting stent was explored, neglecting drug convection, but assuming an anisotropic diffusivity in the wall coupled with drug binding for a given initial concentration of binding sites (c_{b_0}; Table 9.9; [858]).

The phenomenological analysis leads to various time scales, the diffusion time scale in the coating (T_Dc) and in the arterial wall in radial (T_Dwr) and azimuthal (T_Dwc) direction and the binding and unbinding time:

$$T_\mathrm{Dc} = \frac{\delta^2}{\mathcal{D}_\mathrm{coat}}, \quad T_\mathrm{Dwr} = \frac{h^2}{\mathcal{D}_\mathrm{wr}}, \quad T_\mathrm{Dwc} = \frac{d_s^2}{\mathcal{D}_\mathrm{wc}}, \quad T_\mathrm{b} = \frac{1}{\mathrm{k}_b c_{b_0} T_\mathrm{Dwr}}, \quad T_\mathrm{d} = \frac{1}{\mathrm{k}_d}, \tag{9.18}$$

where δ is the coating thickness, h the wall thickness, d_s the interstrut distance, $\mathcal{D}$ the diffusion coefficient, and k_b and k_d the binding and unbinding kinetic coefficient. The dimensionless number $G_1 = T_\mathrm{Dwr}/T_\mathrm{Dwc}$ measures the competition between intramural radial and circonferential diffusion.

The *Damköhler number* (Da) is a dimensionless parameter that relates the chemical reaction rate to the mass transfer rate, that is, either the convection time scale or diffusive rate. For the binding and dissociation reactions, they are given by:

$$\mathrm{Da}_b = T_\mathrm{Dwr}\mathrm{k}_b c_{b_0} = \frac{1}{T_\mathrm{b}}, \quad \mathrm{Da}_d = T_\mathrm{Dwr}\mathrm{k}_d \frac{c_{b_0}}{c_{\mathrm{coat}_0}} = \frac{T_\mathrm{Dwr}}{T_\mathrm{d}}\frac{c_{b_0}}{c_{\mathrm{coat}_0}}. \tag{9.19}$$

Table 9.9 Quantities involved in drug diffusion and binding in the arterial wall. (Source: [858])

Quantity	Value
Dimensions	
Strut size	140 μm
Strut coating thickness	50 μm
Interstrut spacing	1 mm
Wall thickness	200 μm
Drug diffusion	
Initial drug dose in polymer	10^{-5} l/mol/s
Isotropic drug diffusivity in coating	0.01–1 μm^2/s
Anisotropic drug diffusivity in wall radial diffusivity	0.1–10 μm^2/s
Circumferential diffusivity	0.1–10^3 μm^2/s
Drug binding kinetics	
Binding kinetic coefficient (partial reaction order 2)	10^4 l/mol/s
Detachment rate (partial reaction order 1)	10^{-2} s^{-1}

A high-Damköhler number means that the binding reaction plays a great role in the spatiotemporal drug distribution kinetics. The ratio $\mathrm{Da}_b/\mathrm{Da}_d$ specifies tethering or detachment respective role.

Conclusion

Biomechanics contribute to the development of new diagnosis methods, new measurement techniques from signal acquisition to processing, new therapeutic strategies, and new surgical or medical implantable devices.

Nanotechnology aims at improving drug delivery and medical devices. Nanomaterials[1] can be used in medicine for their ability to cross biological barriers and target specific cell populations such as cancerous cells. Nanomaterials can thus be used to develop new therapies, such as nanoparticle-based ultrasound or magnetic hyperthermia for the treatment of cancer. Nanoparticles coated with aminosilane are taken up faster by tumoral cells than by normal cells and subsequently heated and destroyed by a magnetic field. Similarly, nanoparticles can be used to concentrate the energy of ultrasound beams in cancers. Moreover, the treatment can be repeated, as nanoparticles form stable deposits within tumors.

Cybermedicine includes computer-aided procedures (CAP) and image-guided therapy (IGT). The new generation of medical tools is based upon experience in sensor fusion, computer vision, robotics, virtual reality, and image and signal processing. They include, in particular, navigation and positioning of tools prior to and during the medical and surgical procedures. Navigation systems enable the determination of the optimal patient-specific location and guide the operator to achieve the desired placement, especially in the beating heart.

Telemedicine is based on systems of electronically communicating data from one site to a distant site with data fusion by superimposing patient-specific data. Telepresence operation procedures have two major components: (1) a remote site with a 3D camera system and responsive manipulators with sensory input, and (2) an operating workstation with a 3D monitor and dexterous handles with force feedback. A remotely controlled robot may then be capable of executing the procedure at the site of the operation, where, nonetheless, specialists are ready to execute tasks. Teletaction sensors react according to the type of material with which the operator

[1] Nanomaterials usually correspond to objects with dimensions in the range of 1–100 nm. In the medical field, they include objects up to 1 μm in size.

M. Thiriet, *Diseases of the Cardiac Pump,* Biomathematical and Biomechanical Modeling of the Circulatory and Ventilatory Systems 7, DOI 10.1007/978-3-319-12664-7

is dealing, and imitation tools at the workstation correspond to actual tools on the robotic arms at the site of the operation.

Regenerative and reparative medicine is aimed at restoring the form and function of damaged tissues of the human body or replacing them. It integrates knowledge acquired from biological, biomechanical (bioengineering), and clinical research. It relies on bioreactor design that must incorporates various sources of cell signaling, that is, involved chemical, physical, and mechanical agents. Cells are conditioned in vitro and then administered to patients.

Hence, cardiac diseases provide a unique opportunity for multi- and interdisciplinary research aimed at developing computer-aided medicine and surgery.

Notation Rules: Aliases and Symbols

A given molecule usually possesses many aliases. Conversely, a given alias commonly refers to various types of molecules [638, 859–861].

Aliases that designate different types of molecules as well as those that do not have an obvious meaning should be eliminated; they are thus not used in the present text.

For example, P35 is an alias for annexin-A1, brain syntaxin-1A, ficolin-2, interleukin-12A, the cyclin-H assembly factor ménage à trois homolog-1, regulatory subunit-1 of cyclin-dependent kinase CDK5, and uroplakin-3B, among others. It is substituted by AnxA1, Stx1a, Fcn2, IL12a, MAT1, $CDK5_{r1}$, and UPk3b, respectively.

Protein P39 corresponds to the subunit D1 of the lysosomal V-type H^+ ATPase (ATP6v0d1), Jun transcription factor, a component of Activator protein AP1, and regulatory subunit-2 of cyclin-dependent kinase CDK5 ($CDK5_{r2}$).

Extracellular signal-regulated protein kinases ERK1 and ERK2, members of the mitogen-activated protein kinase (MAPK) module (last tier), are also abbreviated P44 and P42 (also P40 and P41). However, both P42 and P44 correspond to the 26S protease regulatory AAA ATPase subunit (PSMC6). Alias P42 is also utilized for cyclin-dependent kinase CDK20, cyclin-dependent kinase-like protein CDKL1, and 43-kDa NuP43 nucleoporin. Alias P44 can also refer to interferon-induced protein IFI44 (or microtubule-associated protein MTAP44) and androgen receptor cofactor P44 (a.k.a. methylosome protein MeP50 and WD repeat-containing protein WDR77).

The numbering of mitogen-activated protein kinase (MAPK) isoforms that are categorized into 3 families (ERK, JNK, and P38) is neither straighforward nor founded on unicity (ERK2 is also called MAPK1 and MAPK2 and MAPK15 refers to both ERK7 and ERK8). In the present text, stress-activated members of the P38 family (P38α–P38δ)[2] are designated as P38MAPKs to avoid confusion with other

[2] Protein P38α is also known as MAPK14, cytokine suppressive anti-inflammatory drug (CSAID)-binding protein CSBP, CSBP1, or CSBP2, and stress-activated protein kinase SAPK2a; P38β as MAPK11 and SAPK2b; P38γ as MAPK12, ERK6, and SAPK3; and P38δ as MAPK13 and SAPK4.

M. Thiriet, *Diseases of the Cardiac Pump*, Biomathematical and Biomechanical Modeling of the Circulatory and Ventilatory Systems 7, DOI 10.1007/978-3-319-12664-7

molecules, the alias of which is also P38. Alias P38 indeed stands for: (1) extracellular signal-regulated kinase ERK3 and ERK6; (2) adaptor CRK (chicken tumor virus regulator of kinase, or v-crk sarcoma virus CT10 oncogene homolog); (3) growth factor receptor-binding protein GRB2-related adaptor protein GRAP2 (a.k.a. GRID, GADS, GRB2L, GRF40, GRPL, and Mona); (4) ubiquitin ligase RING finger protein RNF19a, or dorfin; (5) 38-kDa DNA polymerase-δ-interacting protein Polδ IP2 (a.k.a. polymerase [DNA-directed] PDIP38 and PolD4); (6) activator of 90-kDa heat shock protein ATPase homolog AHSA1; and (7) aminoacyl tRNA synthase complex-interacting multifunctional protein AIMP2, or tRNA synthase complex component JTV1 [638].

Abbreviations

Aliases[3] include all written variants, that is, any abbreviation[4] such as acronyms.[5] An *acronym* corresponds to a word made from the initial letters or syllables of nouns that is pronounceable as a word. Acronyms are generally written with all letters in uppercase. Yet, some acronyms are treated as words and written in lowercase (e.g., laser [originally LASER] is an acronym for light amplification by stimulated emission of radiation; and sonar [originally SONAR] for sound navigation and ranging). A substance's name can derive from its chemical name (e.g., amphetamine: α-methylphenethylamine).

Acronyms can give rise to molecule names by adding a scientific suffix such as "-in," a common ending of molecule nouns (e.g., sirtuin, a portmanteau, that comes from the alias SIRT, which stands for silent information regulator-2 [two]). Other scientific prefixes and suffixes can be frequently detected throughout the present text. Their meaning is given in the appendix List of Currently Used Prefixes and Suffixes, particularly for readers from Asia. Many prefixes are used to specify position, configuration and behavior, quantity, direction and motion, structure, timing, frequency, and speed.

A *portmanteau* is a word that combines initials and some inner letters of at least two words (e.g., calmodulin stands for calcium modulated protein; caspase for cysteine-dependent aspartate-specific protease; chanzyme for ion channel and enzyme; chemokine for chemoattractant cytokine;[6] emilin for elastin microfibril

---

[3] Latin *alias*: at another time, at other times.

[4] In general, abbreviations exclude the initials of short function words, such as "and," "or," "of," or "to." However, they are sometimes included in acronyms to make them pronounceable (e.g., radar [originally RADAR] for radio detection and ranging). These letters are often written in lower case. In addition, both cardinal (size, molecular weight, etc.) and ordinal (isoform discovery order) numbers in names are represented by digits.

[5] ακρο-: end, tip (ακροκωλιον: extremities of body; ακροπους: extremity of the leg [πους: foot; κωλην: leg; κωλον: limb]; ακρορριvιov: tip of the nose);ονυμα: name.

[6] Cytokines are peptidic, proteic, or glycoproteic regulators that are secreted by cells of the immune system. These immunomodulating agents serve as auto- or paracrine signals.

interfacer; endorphins and endomorphins for endogenous morphines; ephrin for erythropoietin-producing hepatocyte (EPH) receptor kinase interactor; granzyme for granule enzyme; moesin for membrane-organizing extension spike protein; porin for pore-forming protein; restin for Reed–Steinberg cell-expressed intermediate filament-associated protein, an alias for cytoplasmic linker protein CLiP1 (or CLiP170); serpin for serine protease inhibitor; siglec for sialic acid-binding Ig-like lectin; transceptor for transporter-related receptor; and Prompt for promoter upstream transcript).[7]

Initialisms are abbreviations that are formed from initial letters of a single long noun or several nouns and, instead of being pronounced like an ordinary word, are read letter-by-letter (e.g., DNA stands for deoxyribonucleic acid; ASCII for American Standard Code for Information Interchange).

Some abbreviations can give rise to *alphabetisms* that are written as new words (e.g., Rho-associated, coiled-coil-containing protein kinase [RoCK] that is also called Rho kinase). In biochemistry, multiple-letter abbreviations can also be formed from a single word that can be long (e.g., Cam stands for calmodulin, which is itself a portmanteau word, Trx for thioredoxin, etc.), as well as short (e.g., Ttn for titin, etc.). In addition, single-letter symbols of amino acids are often used to define a molecule alias (e.g., tyrosine can be abbreviated as Tyr or Y, hence SYK stands for spleen tyrosine kinase).

use, in general, capital letters and can include hyphens and dots. Yet, as a given protein can represent a proto-oncogene[8] encoded by a gene that can give rise to an oncogene (tumor promoter) after gain- or loss-of-function mutations,[9] the same acronym represents three different entities.[10]

---

[7] The uppercase initial P in Prompt is used to avoid confusion with command-line interpreter prompt or prompt book to direct precise timing of actions on the theater stage.

[8] In 1911, P. Rous isolated a virus that was capable of generating tumors of connective tissue (sarcomas) in chicken. Proteins were afterward identified, the activity of which, when uncontrolled, can provoke cancer, hence the name oncogene given to genes that encode these proteins. Most of these proteins are enzymes, more precisely kinases. The first oncogene was isolated from the avian Rous virus by D. Stéhelin and called Src (from sarcoma). This investigator demonstrated that the abnormal functioning of the Src protein resulted from mutation of a normal gene, or proto-oncogene, which is involved in cell division.

[9] Loss-of-function mutations cause complete or partial loss of function of gene products that operate as tumor suppressors, whereas gain-of-function mutations generate gene products with new or abnormal function that can then act as oncogenes. Typical tumor-inducing agents are enzymes, mostly regulatory kinases and small guanosine triphosphatases, that favor proliferation of cells, which normally need to be activated to exert their activities. Once their genes are mutated, these enzymes become constitutively active. Other oncogenes include growth factors (a.k.a. mitogens) and transcription factors. Mutations can also disturb signaling axis regulation, thereby raising protein expression. Last, but not least, chromosomal translocation can also provoke the expression of a constitutively active hybrid protein.

[10] Like Latin-derived shortened expressions—as well as foreign words—that are currently written in italics, genes can be italicized. However, this usage is not required in scientific textbooks published by Springer. Italic characters are then used to highlight words within a text to target them easily. Proteins are currently romanized (ordinary print), but with a capital initial. Nevertheless, names

In addition, a given abbreviation can designate distinct molecules without necessarily erroneous consequences in a given context (e.g., PAR: poly^{ADP}ribose or protease-activated receptor and GCK: germinal center kinases or glucokinase; in the latter case, the glucokinase abbreviation should be written as GcK or, better, GK).

Molecule and Adopted Notation Rules

Numerous that designate a single molecule can result from the fact that molecules have been discovered independently several times with possibly updated functions.

Some biochemists uppercase the name of a given molecule, whereas others lowercase (e.g., cell division cycle guanosine triphosphatase of the Rho family CDC42 or Cdc42, adaptor growth factor receptor-bound protein GRB2 or Grb2, chicken tumor virus regulator of kinase CRK or Crk, guanine nucleotide-exchange factor Son of sevenless SOS or Sos, etc.).

Acronyms are then not always entirely capitalized. The printing style of should not only avoid confusion but also help one in remembering the meaning of the alias.

In the present textbook, choice of lower- and uppercase letters in molecule is dictated by the following criteria.

(1) An uppercase letter is used for initials of words that constitute molecule nouns (e.g., receptor tyrosine kinase RTK). An alias of any compound takes into account added atoms or molecules (e.g., PI: phosphoinositide and PIP: phosphoinositide phosphate) as well as their number (e.g., PIP_2: phosphatidylinositol bisphosphate, DAG: diacylglycerol, and PDE: [cyclic nucleotide] phosphodiesterases).

(2) A lowercase letter is used when a single letter denotes a subfamily or an isoform when it is preceded by a capital letter (e.g., PTPRe: protein tyrosine phosphatase receptor-like type-E). Nevertheless, an uppercase letter is used in an alias after a single or several lowercase letters to distinguish the isoform type (e.g., RhoA isoform and DNA-repair protein RecA for recombination protein-A), but OSM stands for oncostatin-M, not osmole Osm[11] to optimize molecule identification.

These criteria enable the use of differently written with the same sequence of letters for distinct molecules (e.g., CLIP for corticotropin-like intermediate peptide, CLiP: cytoplasmic CAP-Gly domain-containing linker protein, and iCliP: intramembrane-cleaving protease).

As the exception proves the rule, current , such as PKA and PLA that designate protein kinase-A and phospholipase-A, respectively, have been kept. Preceded by only two uppercase letters, a lowercase letter that should be used to specify an isoform

---

(not) of chemical species are entirely lowercased like in most (if not all) scientific articles, except to avoid confusion with a usual word (e.g., hedgehog animal vs. Hedgehog protein and raptor [bird of prey] vs. Raptor molecule).

[11] Osmole: the amount of osmotically active particles that exerts an osmotic pressure of 1 atm when dissolved in 22.4 l of solvent at 0°C.

can bring confusion with acronyms of other protein types (e.g., phospholamban alias PLb).

Nouns (e.g., hormone-like fibroblast growth factor [hFGF] and urokinase-type plasminogen activator [uPA]) or adjectives (e.g., intracellular FGF isoform [iFGF]) that categorize a subtype of a given molecule correspond to a lowercase letter to emphasize the molecule species. Hence, an uppercase letter with a commonly used hyphen (e.g., I[R]-SMAD that stands for inhibitory [receptor-regulated] SMAD; V-ATPase for vacuolar adenosine triphosphatase; MT1-MMP for membrane type-1 matrix metalloproteinase; and T[V]-SNARE for target [vesicle-associated] soluble ^{N}ethylmaleimide-sensitive factor-attachment protein receptor) is then replaced by a lowercase letter (e.g., i[r]SMAD, vATPase, mt1MMP, and t[v]SNARE), as is usual for RNA subtypes (mRNA, rRNA, snRNA, and tRNA for messenger, ribosomal, small nuclear, and transfer RNA, respectively). Similarly, membrane-bound and secreted forms of receptors and coreceptors that can derive from alternative mRNA splicing are defined by a lowercase letter (e.g., sFGFR for secreted extracellular FGFR form and sFRP for soluble Frizzled-related protein), as well as eukaryotic translation elongation (eEF) and initiation (eIF) factors.

(3) Although `l`, `r`, and `t` can stand for molecule-like, -related, and -type, respectively, when a chemical is related to another one, in general, uppercase letters are used for the sake of homogenity and to clearly distinguish between the letter L and numeral 1 (e.g., KLF: Krüppel-like factor, CTK: C-terminal Src kinase (CSK)-type kinase, and SLA: Src-like adaptor).

(4) An uppercase letter is most often used for initials of adjectives contained in the molecule name (e.g., AIP: actin-interacting protein; BAX: BCL2-associated X protein; HIF: hypoxia-inducible factor; KHC: kinesin heavy chain; LAB: linker of activated B lymphocytes; MAPK: mitogen-activated protein kinase; and SNAP: soluble N-ethylmaleimide-sensitive factor-attachment protein).

(5) Lowercase letters are used when alias letters do not correspond to initials (e.g., Fox—not fox—[forkhead box]), except for portmanteau words that are entirely written in minuscules (e.g., gadkin: γ1-`ad`aptin and `kin`esin `in`teractor).

This rule applies, whether alias letters correspond to successive noun letters (e.g., Par: partitioning defective protein and Pax: paxillin, as well as BrK: breast tumor kinase and ChK: checkpoint kinase, whereas CHK denotes C-terminal Src kinase [CSK]-homologous kinase) or not (e.g., Fz: Frizzled and HhIP: Hedgehog-interacting protein),[12] except for composite chemical species (e.g., DAG: diacylglycerol). However, some current usages have been kept for short of chemical species name (e.g., Rho for Ras homolog rather than RHo).

[12] The Hedgehog gene was originally identified in the fruit fly Drosophila melanogaster. It encodes a protein involved in the determination of segmental polarity and intercellular signaling during morphogenesis. Homologous gene and protein exist in various vertebrate species. The name of the mammal hedgehog comes from hecg and hegge (dense row of shrubs or low trees), as it resides in hedgerows, and hogg and hogge, due to its pig-like, long projecting nose (snout). The word Hedgehog hence is considered as a seamless whole.

In any case, molecule (super)family (class) as well as those of their members are written in capital letters, such as the IGSF (IGSFi: member i; immunoglobulin), KIF (KIFi; kinesin), SLC (SLCi; solute carrier), TNFSF (TNFSFi; tumor-necrosis factor), and TNFRSF (TNFRSFi; tumor-necrosis factor receptor) superfamily.

Gene names are also written with majuscules when the corresponding protein name contains at least one minuscule, otherwise only the gene name initial is written with an uppercase letter that is then followed by lowercase letters.

To highlight its function, substrate (e.g., ARF GTPases) contained in a molecule alias are partly written with lowercase letters (e.g., ArfRP, ArfGEF, ArfGAP stand for ARF-related protein, ARF guanine nucleotide-exchange factor, and ARF GTPase-activating protein, respectively).

Last, but not least, heavy and pedantic designation of protein isoforms based on roman numerals has been avoided and replaced by the usual arabic numerals (e.g., angiotensin-2 rather than angiotensin-II), except for coagulation (or clotting) factors. Moreover, the character I can mean either letter I or number 1 without obvious discrimination at first glance (e.g., GAPI that stands for Ras GTPase-activating protein GAP1, but can be used to designate a growth-associated protein inhibitor).

Unnecessary hyphenation in of substances (between an uppercase letter, which can define the molecule function, and the chemical alias, or between it and assigned isotype number) has been avoided. In any case, the Notation section serves not only to define , but also, in some instances, as disambiguation pages.

A space rather than hyphen is used in: (1) structural components at the picoscale (e.g., P loop), nanoscale (e.g., G protein [G standing for guanine nucleotide-binding]), microscale (e.g., H zone, M line, A band, I band, and Z disc of the sarcomere and T tubule of the cardiomyocyte); (2) process stages (e.g., M phase of the cell division cycle); and (3) cell types (e.g., B and T lymphocytes). When these terms are used as adjectives, a hyphen is then employed (e.g., P-loop Cys–X_5–Arg (CX_5R) motif, G-protein-coupled receptor, Z-disc ligand, M-phase enzyme, and T-cell activation).

In terms incorporating a Greek letter, similarly, a space is used in: (1) structural components (e.g., α and β chains and subunits); (2) cellular organelles (e.g., α granule); and (3) cell types (e.g., pancreatic β cell). On the other hand, terms are hyphenated when they refer to (1) structural shape (e.g., α-helix and α (β)-sheet) and (2) molecule subtype (e.g., α-actinin, β-glycan, and γ-secretase).

Symbols for Physical Variables

Unlike substance , symbols for physical quantities are most often represented by a single letter of the Latin or Greek alphabet (i: current; J: flux; L: length; m: mass; p: pressure; P: power; T: temperature; t: time; u: displacement; v: velocity; x: space; λ: wavelength; μ: dynamic viscosity; ρ: mass density; etc.). These symbols are specified using sub- and superscripts (c_p and c_v: heat capacity at constant pressure and volume, respectively; $\mathcal{D}_T$: thermal diffusivity; G_h: hydraulic conductivity; G_T: thermal conductivity; α_k: kinetic energy coefficient; α_m: momentum coefficient; etc.).

A physical quantity associated with a given point in space at a given time can be: (1) a scalar uniquely defined by its magnitude; (2) a vector characterized by a magnitude, a support, and a direction represented by an oriented line segment defined by a unit vector; and (3) a tensor specified by a magnitude and a few directions. To ensure a straightforward meaning of symbols used for scalar, vectorial, and tensorial quantities, boldface upper- ($\vec{T}$) and lowercase ($\vec{v}$) letters are used to denote a tensor and a vector, respectively, whereas both roman (plain, upright)-style upper- and lowercase letters designate a scalar.

The en dash is used rather than the hyphen to distinguish a double-barreled name from cases for which two different researchers' names as well as their derived adjectives (e.g., Newtonian) are joined up to define equations (e.g., Kedem–Katchalsky, Navier–Stokes, and Stefan–Maxwell equations); laws (e.g., Boyle–Mariotte law); chemical reactions (e.g., Michaelis–Menten enzyme kinetics); model types (e.g., Mitchell–Schaeffer model); effects (e.g., Fahraeus–Lindqvist effect); and numerical procedures (e.g., arbitrary Lagrangian–Eulerian formulation, Chorin–Temam projection scheme, and Dirichlet–Neumann domain decomposition algorithm).

List of Currently Used Prefixes and Suffixes

Prefixes (localization)

"ab-" (Latin) and "apo-" (Greek: απο): away from or off (abluminal: endothelial edge opposite to wetted surface; apolipoproteins: lipid carriers that cause egress [also ingress] from cells; aponeurosis (απονευρωσις; νευρον: sinew, tendon) muscle sheath that limits radial motion and enhances axial contraction; and apoptosis: separation ["-ptosis": fall (πτωσις): as leaves fall away from a tree], a type of programmed cell death)

"acr-" (variant "acro-" [ακρος]): top or apex

"ad-" (adfecto: to reach; adfio: to blow toward; adfluo: to flow toward): toward (ad- becomes "ac-" before c, k, or q; "af-" before f [afferent]; "ag-" before g [agglutination]; "al-" before l; "ap-" before p [approximation]; "as-" before s; and "at-" before t)

"cis-", "juxta-", and "para-" (παρα): near, beside, or alongside

"contra-": opposite side; "ipsi-" (ipse): same side; "latero-": side;

"ecto-" (εκτος), "exo-" (εξο), and "extra-": outside, outer, external, or beyond (exogenous chemicals produced by an external source, or xenobiotics ["xeno-": foreigner])

"endo-" (ενδον) and "intra-": inside (endogenous substances synthesized by the body's cells; endomembranes at organelle surfaces within the cell)

"ep-" (variant "eph-", or "epi-" [επι]): upon (epigenetics refers to the inheritance ("-genetic": ability to procreate [γεννητικος]) of variations in gene expression beyond ("epi-": on, upon, above, close to, beside, near, toward, against, among, beyond, and also) change in the DNA sequence.

"front-" and "pre-": anterior or in front of

"post-": behind

"infra-" and "sub-": under or below

"super-" and "supra-": above

"inter-": between or among

"peri-" (περι): around

"tele-" (τελε): remote

"trans-": across

Prefixes (composition)

"an-" and "aniso-" (ανισος): unequal, uneven, heterogeneous

"iso-" (ισος): equal, alike (isomer [μερος: part, portion]

"mono-" (μονος) and "uni-" (unicus): single

"oligo-" (ολιγος): few, little, small

"multi-" (multus), "pluri-" (plus, plures), and "poly-" (πολυς): many, much

"ultra-": in excess.

Prefixes (quantity)

"demi-" (dimidius) and "hemi-" (ημι): half

"sesqui-": one and a half (half more)

"di-" or "dis-" (δυο; δις) as well as "bi-" or "bis-": 2, twice

"tri" (τρεις, τρι-; tres, tria): 3

"tetra-" (τετρα), "quadri-" (variant: "quadr-" and "quadru-"): 4

"penta-" (πεντας; pentas), "quinqu-", and "quint-": 5

M. Thiriet, *Diseases of the Cardiac Pump*, Biomathematical and Biomechanical Modeling of the Circulatory and Ventilatory Systems 7, DOI 10.1007/978-3-319-12664-7

"hexa-" (ϵξ) and "sexa-": 6
"hepta-" (ϵπτα): 7
"octa-" (οκτα): 8
"nona-" (ϵννϵα): 9 (ninth part)
"deca-" (δϵκα): 10
"quadra-" (quadragenarius): 40 (elements)
"quinqua-" (quinquagenarius): 50
"sexa-" (sexagenarius [sex: 6]: 60
"septua-" (septuagenarius [septem: 7]): 70
"nona-" (nonagenarius): 90

Prefixes (motion and direction)

"af-": toward the center (single master object); e.g., nerve and vascular afferents (ferre: to carry) to brain and heart, respectively, rather than toward any slave, supplied tissue from the set of the body's organs; also affector, i.e., chemical messenger that brings a signal to the cell considered as the object of interest, this exploration focus being virtually excised from the organism with its central command system, except received signals
"ef-" (effero: to take away): from the center (efferent; effector, i.e., chemical transmitter recruited by the previous mediator of a signaling cascade at a given locus to possibly translocate to another subcellular compartment)
"antero-" (anterior): before, in front of, facing, or forward
"retro-": behind or backward
"tropo-" (τροπος): duct direction; (tropa: rotation; celestial revolution); e.g., tropomyosin (μυς, musculus: muscle; μυο-: refers to muscle [μυοτρωτος: injured at a muscle])

Prefixes (structure and size)

"macro-" (μακρος): large, long, or big
"mega-" (μϵγας): great, large
"meso-" (μϵσος): middle
"micro-" (μικρος): small
"nano-" (νανος): dwarf, tiny
"homo-" (ομο-): same (ομολογος: agreeing, corroborating; variant: "homeo" [homeostasis])
"hetero-" (ϵτϵρο-): other

Prefixes (timing)

"ana-" (ανα): culminating (anaphase of the cell division cycle), up, above (ανοδος: a way up, anode [positive electrode; οδος: way, path, road, track])
"ante-": before
"circa-": approximately, around (circadian: approximately one day)
"infra-": below, shorter (infradian: rhythm with lower frequency than that of circadian rhythm, not smaller period)
"inter-": among, between, during
"meta-" (μϵτα): after, beyond, behind, later; in the middle of (metaphase of the cell division cycle); as well as connected to, but with a change of state (metabolism) and about (metadata)
"post-": after
"pre-": earlier
"pro-" (προ): preceding, first, before (prophase of the cell division cycle)
"telo-" (τϵλος): end, completion
"ultra-": beyond, longer (ultradian: period smaller than that of 24–28-hour cycle, i.e., frequency greater than that of the circadian rhythm)

Prefixes (functioning modality)

"auto-" (αυτος): same, self
"brady-" (βραδυς): slow (decelerate)
"tachy-" (ταχος): rapid (accelerate)
"amphi-" (αμφι): both (amphiphilic substances are both hydrophilic and lipophilic; amphisomes are generated by both autophagosomes and endosomes)
"ana-" : upward (anabolism) or against (anaphylaxis)
"cata-" (κατα): downward (catabolism, cathode [negative electrode; οδος; way, path, road, track])
"anti-" (αντι): against
"pro-": favoring
"co-" (coaccedo: add itself to): together
"contra-": adverse, against, beside, next to, opposite
"de-": remove, reduce, separation after association (Latin de; e.g., deoxy-)
"dys-" (δυς): abnormal (δυσαης): ill-blowing)
"equi-" (æque): equal or alike
"hem-" or "hemat-" (αιμα: blood): related to blood

"hyper-" (υπερ): above, beyond, and large
"hypo-" (υπο): under, beneath, and low
"per-": through (e.g., percutaneous) and during (e.g., peroperative)
"pseudo-" (ψευδο): pretended, false
"re-"; again

Scientific suffixes

"-ase": enzyme (synthase, lipase, etc.)
"-ate": salt of a base
"-cyte" (κυτος): cell (erythro- [ερυθρος: red], leuko- [λευκος: light, bright, clear, white], thrombo- [θρομβος: lump, clot], adipo- [adeps: fat; adipalis, adipatus, adipeus, adipinus: fatty], fibro- [fibra: fiber, filament], myo- [μυς: muscle, mouse, mussel], myocardiocyte [κραδια: heart; cardiacus: related to heart, stomach; to have heart trouble, stomach trouble], etc.);
"-crine" (κρινω): to decide, to separate, and to secrete (e.g., endocrine regulator) (ευκρινεω: keep in order)
"-elle": small (organelle in a cell [like an organ in a body])
"-ium", "-ion", "-isk", and "-iscus": little ("-ium": tissue interface and envelope, such as endothelium and pericardium)
"-phil" (φιλια): attracted (αφιλια: want of friends)
"-phob" (φοβια): repulsed (υδροφοβια, hydrophobia [Latin]: horror of water)
"-phore" (φερω): carrier (αμφερω: to bring up)
"-yl" denotes a radical (molecules with unpaired electrons)
"-ploid" (πλοω): double, fold (diploid, twofold; διπλοω: to double; διαπλοω: unfold)
"-emia": in relation to flow (ανεμια: flatulence; ευηνεμια: fair wind), particularly blood condition
"-genesis" (γενεσις): cause, generation, life source, origin, productive force
"-iasis": for diseased condition
"-itis": inflammation
"-lemma" (λεμμα: skin): sheath
"-ole" and "-ule": small (arteriole and venule; variant "-ula" [blastula] and "-ulum")
"-plasma" (πλασμα): anything molded (plasma: creature generated from silt of earth)
"-plasia" (πλασια): formation, molding
"-podium" (ποδος: foot; podium [Latin]: small knoll, small protuberance): protrusion
"-poiesis" (ποιεω): production
"-soma" (σωμα): body
"-sclerosis" (σκλημα): hardness, induration
"-stasis" (στασις): stabilization (αποκαταστασις: restoration; ανυποστασις: migration)
"-stomosis" (στομα: mouth): equipped with an outlet
"-taxy/tactic" (ταχυ: rapid; τακτικος: to maneuver): related to motion (also prefix, i.e., ταχυκινησις: quick motion; ταχυνω: to accelerate; and ταχυπνοια: short breath; not [δια]ταξις: disposition, arrangement)
"-trophy/trophic" (τροφις: well fed): related to growth
"-oma": tumor of
"-pathy" (παθος, παθεια): disease of
"-tomy" (τομια) and "-ectomy": surgical removal (απλοτομια: simple incision; φaûrhουγγοτομια: laryngotomy)

List of Aliases and Primary Symbols

A

$\mathcal{A}$: Avogadro number
$\mathcal{A}(p)$: area–pressure relation
$\mathbf{A}$: Almansi strain tensor
A: cross-sectional area
$\mathbb{A}$: actin-binding site
A: surface area-to-volume ratio
$\mathbf{a}$: acceleration
a: major semiaxis
AA: arachidonic acid
AAA: ATPase associated with diverse cellular activities
AAA: abdominal aortic aneurysm
AAAP: aneurysm-associated antigenic protein
AAI: autonomic arousal index
AAK: adaptin-associated kinase
AAS: acute aortic syndrome
AATK: apoptosis-associated tyrosine kinase
ABC: ATP-binding cassette transporter (transfer ATPase)
AbI: Abelson kinase interactor
Abl: Abelson leukemia viral proto-oncogene product (NRTK)
ABLIM: actin-binding LIM domain-containing protein
ABP: actin-binding protein
ABR: active breakpoint cluster region (BCR)-related gene product (GEF and GAP)
AC: atrial contraction
ACAA: acetylCoA acyltransferase
ACAP: ArfGAP with coiled-coil, ankyrin repeat, PH domains
ACase: adenylate cyclase
ACi: adenylate cyclase isoform i
ACAT: acylCoA–cholesterol acyltransferase
ACC: acetyl coenzyme-A carboxylase
ACD: adrenocortical dysplasia homolog
ACE: angiotensin-converting enzyme
ACh: acetylcholine
ACK: activated CDC42-associated kinase
ACL: ATP–citrate lyase
Aco: aconitase
ACP1: acid phosphatase-1, soluble (lmwPTP)
ACS: acute coronary syndrome
ACTH: adrenocorticotropic hormone
^{F}actin: filamentous actin
(Cav–actin: caveolin-associated ^{F}actin)
^{G}actin: monomeric globular actin
AcvR: activin receptor (TGFβ receptor superfamily)
Ad: adrenaline
ADAM: a disintegrin and metallopeptidase (adamalysin)
ADAMTS: a disintegrin and metallopeptidase with thrombospondin
ADAP: adhesion and degranulation-promoting adaptor protein
ADAP: ArfGAP with dual PH domains
ADCF: adipocyte-derived constricting factor
ADF: actin-depolymerizing factor (cofilin-related destrin)
ADH: antidiuretic hormone (vasopressin)
ADHF: acute decompensated heart failure
ADMA: asymmetric dimethylarginine
ADP: adenosine diphosphate
ADRF: adipocyte-derived relaxing factor
aDuSP: atypical dual specificity phosphatase
AE: anion exchanger
AEA: N-arachidonoyl ethanolamine (anandamide)

M. Thiriet, *Diseases of the Cardiac Pump*, Biomathematical and Biomechanical Modeling of the Circulatory and Ventilatory Systems 7, DOI 10.1007/978-3-319-12664-7

AF: atrial fibrillation
AFAP: ArfGAP with phosphoinositide-binding and PH domains
aFGF: acidic fibroblast growth factor (FGF1)
AGAP: ArfGAP with GTPAse, ankyrin repeat, and PH domains
AGE: advanced glycation end product
AGF: autocrine growth factor
AGFG: ArfGAP with FG repeats
Ago: Argonaute protein
AGS: activator of G-protein signaling
AHD: atherosclerotic heart disease
AHI: apnea–hypopnea index
AHR: airway hyperresponsiveness
AHR: aryl hydrocarbon receptor
AIF: apoptosis-inducing factor
AIP: actin-interacting protein
AIRe: autoimmune regulator
AIx: augmentation index
AKAP: A-kinase (PKA)-anchoring protein
AldH: aldehyde dehydrogenase
ALE: arbitrary Eulerian Lagrangian
ALIX: apoptosis-linked gene-2-interacting protein-X
ALK: anaplastic lymphoma kinase
ALK*i*: type-*i* activin receptor-like kinase (TGFβ receptor superfamily)
ALOx5: arachidonate 5-lipoxygenase
ALOx5AP: arachidonate 5-lipoxygenase activation protein
ALP: actinin-associated LIM protein (PDLIM3)
alsin: amyotrophic lateral sclerosis protein (portmanteau)
ALX: adaptor in lymphocytes of unknown function X
AMAP: A multidomain ArfGAP protein
AMBRA: activating molecule in beclin-1-regulated autophagy protein
AMHR: anti-Müllerian hormone receptor (TGFβ receptor superfamily)
AMIS: apical membrane initiation site (lumenogenesis)
AMP: adenosine monophosphate
AMPAR: α-amino 3-hydroxy 5-methyl 4-isoxazole propionic acid receptor
AMPK: AMP-activated protein kinase
AMSH: associated molecule with SH3 domain (deubiquitinase)
AmyR: amylin receptor
Ang (AngPt): angiopoietin
AngL: angiopoietin-like molecule
Ank: ankyrin
ANP: atrial natriuretic peptide
ANPR (NP_1): atrial natriuretic peptide receptor (guanylate cyclase)
ANS: autonomic nervous system
ANT: adenine nucleotide transporter
Anx: annexin
AOC: amine oxidase copper-containing protein
AoV: aortic valve
AP: (clathrin-associated) adaptor proteic complex
AP: Activator protein (transcription factor)
AP: activating enhancer-binding protein
AP_4A: diadenosine tetraphosphate
APAF: apoptotic peptidase-activating factor
APAH: acquired arterial pulmonary hypertension
APAP: ArfGAP with PIX- and paxillin-binding domains
APC: antigen-presenting cell
APC: adenomatous polyposis coli protein (Ub ligase)
APC/C: anaphase-promoting complex (or cyclosome; Ub ligase)
APH: anterior pharynx defective phenotype homolog
aPKC: atypical protein kinase-C
APl: action potential
Apn: adiponectin
Apo: apolipoprotein
ApoER: apolipoprotein-E receptor
APPL: adaptor containing phospho-Tyr interaction, PH domain, and Leu zipper
APS: adaptor with a PH and SH2 domain
Aqp: aquaporin
AR: adrenergic receptor (adrenoceptor)
AR: androgen receptor (nuclear receptor NR3c4; transcription factor)
AR: area ratio
ARAP: ArfGAP with RhoGAP, ankyrin repeat, PH domains
ARDS: acute respiratory distress syndrome
ARE: activin-response element
ARE: androgen response element
ARE: antioxidant response element
Areg: amphiregulin (EGF superfamily member)
ARF: ADP-ribosylation factor
ArfRP: ARF-related protein
ARFTS: CKI2A-locus alternate reading frame tumor suppressor (ARF or $p14^{ARF}$)
ARH: autosomal recessive hypercholesterolemia adaptor (low-density lipoprotein receptor adaptor)
ARH: aplysia Ras-related homolog

ArhGEF: RhoGEF
ARL: ADP-ribosylation factor-like protein
ARNO: ARF nucleotide site opener
ARNT: aryl hydrocarbon nuclear receptor translocator
ARP: absolute refractory period
ARP: actin-related protein
ARPP: cAMP-regulated phosphoprotein
Arr: arrestin
ART: arrestin-related transport adaptor (α-arrestin)
ART: adpribosyltransferase
Artn: artemin
ARVCF: armadillo repeat gene deleted in velocardiofacial syndrome
ARVC: arrythmogenic right ventricular cardiomyopathy
ARVD: arrythmogenic right ventricular dystrophy
AS: Akt (PKB) substrate
ASAP: artery-specific antigenic protein
ASAP: ArfGAP with SH3, ankyrin repeat, PH domains
ASIC: acid-sensing ion channel
ASK: apoptosis signal-regulating kinase
aSMC: airway smooth muscle cell
ASO: arteriosclerosis obliterans
ASP: actin-severing protein
AT: antithrombin
ATAA: ascending thoracic aortic aneurysm
ATF: activating transcription factor
AtG: autophagy-related gene product
ATGL: adipose triacylglycerol lipase
ATMK: ataxia telangiectasia mutated kinase
ATn: angiotensin
ATng: angiotensinogen
AtOx: antioxidant protein (metallochaperone)
ATP: adenosine triphosphate
ATPase: adenosine triphosphatase
ATR ($AT_{1/2}$): angiotensin receptor
ATRK: ataxia telangiectasia and Rad3-related kinase
AVA: aortic valve area
AVAI: aortic valve area index (AVA/BSA [dimensionless])
AVB: atrioventricular node block
AVN: atrioventricular node
AVR: aortic valve regurgitation
AVS: aortic valve stenosis
AVV: atrioventricular valves
AW: analysis window

B

$\mathbf{B}$: Biot–Finger strain tensor
B: bulk modulus
$\mathcal{B}$: bilinear form
B: binding rate
b: minor semiaxis
$\mathbf{b}$: body force
$\hat{\mathbf{b}}$: unit binormal
b: birth rate
B lymphocyte (B cell): bone marrow lymphocyte
BACE: β-amyloid precursor protein-converting enzyme
BAD: BCL2 antagonist of cell death
BAF: barrier-to-autointegration factor
BAFMD: brachial artery flow-mediated dilation
BAG: BCL2-associated athanogene (chaperone regulator)
BAI: behavioural arousal index
BAI: brain-specific angiogenesis inhibitor (adhesion GPCR)
BAIAP: brain-specific angiogenesis inhibitor-1-associated protein (insulin receptor substrate)
BAK: BCL2-antagonist killer
(i)BALT: (inducible) bronchus-associated lymphoid tissue
BAMBI: BMP and activin membrane-bound inhibitor homolog
BAnk: B-cell scaffold with ankyrin repeats
Barkor: beclin-1-associated autophagy-related key regulator
BAS: biodegradable (bioresorbable) arterial scaffold (stent)
BAT: brown adipose tissue
BATF: basic leucine zipper ATF-like transcription factor (B-cell-activating transcription factor)
BAX: BCL2-associated X protein
BBB: blood–brain barrier
BBP: bilin-binding protein
BBS: Bardet–Biedl syndrome protein
BBSome: BBS coat complex (transport of membrane proteins into cilium)
BC: boundary condition
bCAM: basal cell adhesion molecule (Lutheran blood group glycoprotein)
BCAP: B-cell adaptor for phosphatidylinositol 3-kinase
BCAR: breast cancer antiestrogen resistance docking protein

BCL: B-cell lymphoma (leukemia) protein
BCLxL: B-cell lymphoma extralarge protein
BCR: B-cell receptor
BCR: breakpoint cluster region protein (GAP and GEF)
Bdk: bradykinin
BDNF: brain-derived neurotrophic factor
Be: Bejan number
Becn, beclin: BCL2-interacting protein
BEM: boundary element method
BES: biolimus-eluting stent
Best: bestrophin
bFGF: basic fibroblast growth factor (FGF2)
BFUe: burst-forming unit-erythroid
BFUmeg: burst-forming unit-megakaryocyte
BGP: bone γ-carboxyglutamate acid (Gla)-containing protein (osteocalcin)
BGT: betaine–GABA transporter
BH_4: tetrahydrobiopterin (enzyme cofactor)
BHB: blood–heart barrier
BHR: bronchial hyperresponsiveness
BID: BH3-interacting domain death agonist
BIG: brefeldin-A-inhibited GEFs for ARFs
BIK: BCL2-interacting killer
BIM: BH3-containing protein BCL2-like 11 (BCL2L11)
BK: high-conductance, Ca^{2+}-activated, voltage-gated K^+ channel
BLK: B-lymphoid tyrosine kinase
Blm: Bloom syndrome, RecQ DNA helicase-like protein
BLnk: B-cell linker protein
BLOC: biogenesis of lysosome-related organelles
BM: basement membrane
BMAL: brain and muscle ARNT-like protein (gene Bmal)
BMAT: bone marrow adipose tissue
BMF: BCL2 modifying factor
BMP: bone morphogenetic protein (TGFβ superfamily)
BMPR: bone morphogenetic protein receptor
BNIP: BCL2/adenovirus E1B 19-kDa protein-interacting protein
BNP: B-type natriuretic peptide
BMS: bare-metal stent
BMX: bone marrow Tyr kinase gene in chromosome-X product
Bo: Boltzmann constant (1.38×10^{-23} J/K)
BOC: brother of CDO
BOK: BCL2-related ovarian killer
BORG: binder of Rho GTPase
Br: Brinkman number
BRAI: breathing-related arousal index
BRAG: brefeldin-resistant ArfGEF
BrCa: breast cancer-associated (susceptibility) protein (tumor suppressor; DNA-damage repair; a.k.a. FancD1)
BrD: bromodomain-containing protein
BrK: breast tumor kinase
BRS: bioresorbable scaffold (stent)
BrSK: brain-selective kinase
BSA: body surface area
BSCB: blood–spinal cord barrier
BSP: bone sialoprotein
BSEP: bile salt export pump
BTF: basic transcription factor
BTK: Bruton Tyr kinase
BUB: budding uninhibited by benzimidazoles
BVS: biodegradable (bioresorbable) vascular scaffold (stent)

C

$\mathbf{C}$: stress tensor
C: compliance, capacitance
C: heat capacity
C: chronotropy
Cx: type-x chemokine C (γ)
C_D: drag coefficient
C_f: friction coefficient
C_L: lift coefficient
C_p: pressure coefficient
C_{VM}: van Mises stress
$\mathbf{c}$: stress vector
$\mathbf{c}_\tau$: shear stress
c_w: wall shear stress
c_X: concentration of species X
$c(p)$: wave speed
c_p: isobar heat capacity
c_v: isochor heat capacity
C1P: ceramide 1-phosphate
C-terminus: carboxy (carboxyl group COOH) terminus
C/EBP: CCAAT/enhancer-binding protein
CA: computed angiography
CAi: carbonic anhydrase isoform i
Ca: calcium
Ca_V: voltage-gated Ca^{2+} channel
Ca_V1.x: L-type high-voltage-gated Ca^{2+} channel
Ca_V2.x: P/Q/R-type Ca^{2+} channel
Ca_V3.x: T-type low-voltage-gated Ca^{2+} channel
CAAT: cationic amino acid transporter
CABG: coronary artery bypass grafting
Cables: CDK5 and Abl enzyme substrate

CACT: carnitine–acetylcarnitine transferase
CACTD: carnitine–acylcarnitine translocase deficiency
CAD: coronary artery disease
CAI: central apnea index
CAK: CDK-activating kinase (pseudokinase)
Cam: calmodulin (calcium-modulated protein)
CamK: calmodulin-dependent kinase
cAMP: cyclic adenosine monophosphate
CAN: cardiovascular autonomic neuropathy
CAP: adenylate cyclase-associated protein
CAP: carboxyalkylpyrrole protein adduct
CAP: chromosome-associated protein (BrD4)
CAPN: calpain gene
CaPON: C-terminal PDZ ligand of NOS1 (NOS1AP)
CAR: constitutive androstane receptor (NR1i3)
CaR: calcium-sensing receptor
CARD: caspase activation and recruitment domain-containing protein
CARMA: CARD and membrane-associated guanylate kinase-like (MAGuK) domain-containing protein
CARP: cell division cycle and apoptosis regulatory protein
CAS: cellular apoptosis susceptibility protein
CAS: CRK-associated substrate (or P130CAS and BCAR1)
CAs: cadherin-associated protein
CASC: cardiac atrial appendage stem cell
CASK: calcium–calmodulin-dependent serine kinase (pseudokinase)
CASL: CRK-associated substrate-related protein (CAS2)
CASP: cytohesin-associated scaffold protein
caspase: cysteine-dependent aspartate-specific peptidase (Casp)
CAT: carnitine acetyltransferase
Cav: caveolin
CBF: core-binding factor
CBL: Casitas B-lineage lymphoma adaptor and Ub ligase
CBLb: CBL-related adaptor
CBP: cap-binding protein
CBP: CREB-binding protein
CBP: C-terminal Src kinase-binding protein
CBS: cystathionine β-synthase (H_2S production)
CCD: cortical collecting duct
CCDC: coiled-coil domain-containing protein
CCE: capacitative Ca^{2+} entry channel (SOC channel)
CCHS: congenital central hypoventilation syndrome
CCICR: calcium channel-induced Ca^{2+} release
CCK4: colon carcinoma kinase-4 (PTK7)
CCL: chemokine CC-motif ligand
CCM: congenital cardiac malformation
CCN: CyR61, CTGF, and NOv (CCN1–CCN3) family
Ccn: cyclin
Ccnx–CDK*i*: type-x cyclin–type-*i* cyclin-dependent kinase dimer
CCPg: cell cycle progression protein
CCS: copper chaperone for superoxide dismutase
CCT: chaperonin containing T-complex protein
CCx: type-x chemokine CC (β)
CCR: chemokine CC motif receptor
CD: cluster determinant protein (cluster of differentiation)
CDase: ceramidase
CDC: cardiosphere-derived cell
CDC: cell division cycle protein
cDC: classical dendritic cell
CDH: CDC20 homolog
Cdh: cadherin
CDK: cyclin-dependent kinase
Cdm: caldesmon
CDO: cell adhesion molecule-related/downregulated by oncogenes
CE (CsE): cholesteryl esters
CeBF: cerebral blood flow
CEC: circulating endothelial cell
CELSR: cadherin, EGF-like, LAG-like, and seven-pass receptor
CenP: centromere protein
CEP: carboxyethylpyrrole
CeP: centrosomal protein
CEPC: circulating endothelial progenitor cell
Cer: ceramide
CerK: ceramide kinase
CerT: ceramide transfer protein
CETP: cholesterol ester transfer protein
CFD: computational fluid dynamics
CFLAR: caspase-8 and FADD-like apoptosis regulator
CFR: coronary flow reserve
CFTR: cystic fibrosis transmembrane conductance regulator
CFU: colony-forming unit
CFUb: CFU-basophil (basophil-committed stem cells)

CFUc: CFU in culture (granulocyte precursors, i.e., CFUgm)
CFUe: CFU-erythroid
CFUeo: CFU-eosinophil
CFUg: CFU-granulocyte
CFUgm: CFU-granulocyte–macrophage
CFUgemm: CFU-granulocyte–erythroid–macrophage–megakaryocyte
CFUm: CFU-macrophage
CFUmeg: CFU-megakaryocyte
CFUs: CFU-spleen (pluripotent stem cells)
CFV: cardiac frequency variability
CG: chorionic gonadotropin
CG: chromogranin
cGK: cGMP-dependent protein kinase (protein kinase-G)
cGMP: cyclic guanosine monophosphate
CGN: cis-Golgi network
CGRP: calcitonin gene-related peptide
chanzyme: ion channel and enzyme
chemokine: chemoattractant cytokine
CHD: congenital heart defect
CHD: coronary heart disease
CHI: central hypopnea index
CHIP: C-terminus heat shock cognate-70-interacting protein
CHIPH: chronic hypoxia-induced pulmonary hypertension
ChK: checkpoint kinase
CHK: CSK homologous kinase
Chn: chimerin (GAP)
CHOP: CCAAT/enhancer-binding protein homologous protein
CHREBP: carbohydrate-responsive element-binding protein
ChT: choline transporter
CI: cardiac index
CICR: calcium-induced calcium release
CIHD: chronic ischemic heart disease
Cin: chronophin
CIP: CDC42-interacting protein
CIP2a: cancerous inhibitor of protein phosphatase-2A
CIPC: CLOCK-interacting protein, circadian
CIS: cytokine-inducible SH2-containing protein
CITED: CBP/P300-interacting transactivator with glutamic (E) and aspartic acid (D)-rich C-terminus-containing protein
CK: creatine kinase
CK: casein kinase
CKI: cyclin-dependent kinase inhibitor
CLAsP: CLiP-associated protein (microtubule binder)
ClASP: clathrin-associated sorting protein
CLC: cardiotrophin-like cytokine
ClC: voltage-gated chloride channel
ClCa: calcium-activated chloride channel
CLec: C-type lectin
ClIC: chloride intracellular channel
CLINT: clathrin-interacting protein located in the trans-Golgi network
CLIP: corticotropin-like intermediate peptide
CLiP: cytoplasmic CAP-Gly domain-containing linker protein
CLK: CDC-like kinase
ClNS: Cl^- channel nucleotide-sensitive
CLOCK: circadian locomotor output cycles kaput
CLP: common lymphoid progenitor
CLS: ciliary localization signal
cmavEC: coronary macrovascular endotheliocyte
CMC: cardiomyocyte
Cmi: chylomicron
cmivEC: coronary microvascular endotheliocyte
CMLP: common myeloid–lymphoid progenitor
CMP: common myeloid progenitor
CMR^{Glc}: cerebral metabolic rate of glucose consumption
CMR^{O_2}: cerebral metabolic rate of oxygen consumption
CMVD: coronary microvascular dysfunction
CoBF: coronary blood flow
Col: collagen
CoLec: collectin
ColF: collagen fiber
CORM: carbon monoxide (CO)-releasing molecule
CNG: cyclic nucleotide-gated channel
CnK: connector enhancer of kinase suppressor of Ras
CNS: central nervous system
CNT: connecting tubule
CNTi: concentrative nucleoside transporter (SLC28ai)
CNTF: ciliary neurotrophic factor
CntnAP: contactin-associated protein
CO: cardiac output
CoA: coenzyme-A
CoBl: Cordon-bleu homolog (actin nucleator)
COLD: chronic obstructive lung disease
COOL: cloned out of library (RhoGEF6/7)

coSMAD: common (mediator) SMAD (SMAD4)
COx: cyclooxygenase (prostaglandin endoperoxide synthase)
CcOx: cytochrome-C oxidase
CcOx17: cytochrome-C oxidase copper chaperone
CoP: coat protein
CoP: constitutive photomorphogenic protein (Ub ligase)
COPD: chronic obstructive pulmonary disease
COUPTF: chicken ovalbumin upstream promoter transcription factor (NR2f1/2)
CP4H: collagen prolyl 4-hydroxylase
CPC: chromosomal passenger complex
CpG: cytidine^P–guanosine oligodeoxynucleotide (motif)
cPKC: conventional protein kinase-C
CPT: carnitine palmitoyl transferase
CPT1AD: carnitine palmitoyl transferase-1A deficiency
CPT2D: carnitine palmitoyl transferase-2 deficiency
Cpx: complexin
CR: complement component receptor
Cr: creatine
cRABP: cellular retinoic acid-binding protein
cRBP: cellular retinol-binding protein
CRAC: Ca^{2+} release-activated Ca^{2+} channel
CRACR: CRAC regulator
Crb: Crumbs homolog polarity complex
CRE: cAMP-responsive element
CREB: cAMP-responsive element-binding protein
CRF: corticotropin-releasing factor (family)
CRH: corticotropin-releasing hormone
CRHD: chronic rheumatic heart disease
CRIB: CDC42/Rac interactive-binding protein
CRIK: citron Rho-interacting, Ser/Thr kinase (STK21)
CRK: chicken tumor virus CT10 regulator of kinase
CRKL: CRK avian sarcoma virus CT10 homolog-like
CRL4: cullin-4A RING ubiquitin ligase
CRLR: calcitonin receptor-like receptor
CRP: C-reactive protein
CRPC: cardiac resident progenitor cell
CRT: cardiac resynchronization therapy
Crt: calreticulin
CRTC: CREB-regulated transcription coactivator
CRU: Ca^{2+} release unit (couplon or dyad)
Cry: cryptochrome
CS: coronary sinus
CS: citrate synthase
Cs: cholesterol
CSBP: cytokine-suppressive anti-inflammatory drug-binding protein
CSD: cortical spreading depression
CSE: cystathionine γ-lyase (H_2S production)
CSF: cerebrospinal fluid
CSF: colony-stimulating factor
CSF1: macrophage colony-stimulating factor (mCSF)
CSF2: granulocyte–macrophage colony-stimulating factors (gmCSF and sargramostim)
CSF3: granulocyte colony-stimulating factors (gCSF and filgrastim)
CSH: chorionic somatomammotropin hormone
CSHL: chorionic somatomammotropin hormone-like hormone
CSK: C-terminal Src kinase
Csk: cytoskeleton
Csq: calsequestrin
CSS: candidate sphingomyelin synthase
CT: cardiotrophin
CT: computed tomography
CTBP: C-terminal-binding protein
CTen: C-terminal tensin-like protein
CTF: C-terminal fragment
CTGF: connective tissue growth factor
CTL: cytotoxic T lymphocyte
CTLA: cytotoxic T-lymphocyte-associated protein
Ctnn: catenin
CTr: copper transporter
CtR: calcitonin receptor
CTRC: CREB-regulated transcription coactivator
CTTH: capillary transit time heterogeneity
CUD: carnitine uptake defect
Cul: cullin
CUT: cryptic unstable transcript
CVI: chronic venous insufficiency
CVLM: caudal ventrolateral medulla
CVP: central venous pressure
CVS: cardiovascular system
Cx: connexin
CXCL*i*: type-*i* CXC (C–X–C motif; α) chemokine ligand
CXCR*i*: type-*i* CXC (C–X–C motif; α) chemokine receptor
CX_3CLi: type-*i* CX_3C (δ) chemokine ligand
CX_3CRi: type-*i* CX_3C (δ) chemokine receptor

cyCK: cytosolic creatine kinase
Cyld: cylindromatosis tumor suppressor protein (deubiquitinase USPL2)
CyP: member of the cytochrome-P450 superfamily
C3G: Crk SH3-binding GEF

D

D: dromotropy
D: vessel distensibility
$\mathcal{D}$: diffusion coefficient
$\mathcal{D}_T$: thermal diffusivity
D: deformation rate tensor
d: displacement vector
D: flexural rigidity
D: demobilization function (from proliferation to quiescence)
D_{RBC}: damage rate
D_f: fractal dimension
d: death, decay, degradation rate
d: distance
d: duration
d_h: hydraulic diameter
Dab: Disabled homolog
DAD: delayed afterdepolarization
DAG: diacylglycerol
DAMP: damage-associated molecular pattern molecule
DAPC: dystrophin-associated protein complex
DAPK: death-associated protein kinase
DARC: Duffy antigen receptor for chemokine
DAT: dopamine active transporter
DAX: dosage-sensitive sex reversal, adrenal hypoplasia critical region on chromosome X (NR0b1)
DBC: deleted in breast cancer protein
DBF: dumbbell formation kinase (in Saccharomyces cerevisiae; e.g., DBF2)
DBP: albumin D-element binding protein (PAR/b–ZIP family)
DC: dendritic cell
DCA: directional coronary atherectomy
DCAF: DDB1- and Cul4-associated factor
DCC: deleted in colorectal carcinoma (netrin receptor)
DCM: dilated cardiomyopathy
DCT: distal convoluted tubule
Dctn: dynactin
DDAH: dimethylarginine dimethylaminohydrolase
DDB: damage-specific DNA-binding protein
DDEF: development and differentiation-enhancing factor (ArfGAP)
DDR: discoidin domain receptor
De: Dean number
Deb: Deborah number
DEC: differentially expressed in chondrocytes (DEC1 and DEC2 are a.k.a bHLHe40 and bHLHe41, bHLHb2 and bHLHb3, or HRT2 and HRT1)
DEC: deleted in esophageal cancer
DEG: delayed-early gene
deoxyHb: deoxyhemoglobin (deoxygenated hemoglobin)
DES: drug-eluting stent
DETC: dendritic epidermal $\gamma\delta$ T cell
DGAT: diacylglycerol acyltransferase
DH: Dbl homology
DHA: docosahexaenoic acid
DHEA: dehydroepiandosterone
DHF: dyssynchronous heart failure
DHET: dihydroxyeicosatrienoic acid
DHh: desert Hedgehog
DI: desaturation index
Dia: Diaphanous
DICOM: digital imaging and communication for medicine
DICR: depolarization-induced Ca^{2+} release
DISC: death-inducing signaling complex
Dkk: Dickkopf
DLg: Disc large homolog
DLL: Delta-like (Notch) ligand
DLx: distal-less homeobox protein
DM: double minute
DMD: Duchenne muscular dystrophy
DMM: DNA methylation modulator
DMPK: myotonic dystrophy-associated protein kinase
DMT: divalent metal transporter
DN1: double-negative-1 cell
DN2: double-negative-2 cell
DN3: double-negative-3 cell
DNA: deoxyribonucleic acid
DNAPK: DNA-dependent protein kinase
DoC2: double C2-like domain-containing protein
DOCK: dedicator of cytokinesis (GEF)
DOK: downstream of Tyr kinase docking protein
DOR: δ-opioid receptor
DPG: diphosphoglyceric acid
DPLD: diffuse parenchymal lung disease
DPTI: diastolic pressure time interval

DRAM: damage-regulated modulator of autophagy
DRF: Diaphanous-related formin (for GTPase-triggered actin rearrangement)
DRG: dorsal root ganglion
Drl: Derailed
Dsc: desmocollin
Dsg: desmoglein
Dsh: Disheveled (Wnt-signaling mediator)
DSK: dual-specificity kinase
dsRNA: double-stranded RNA
Dst: dystonin
DT: deceleration time
DUb: deubiquitinase
DuOx: dual oxidase
DUS: Doppler ultrasound
DuSP: dual-specificity phosphatase
DV: dead space volume
Dvl: Disheveled (cytoplasmic phosphoprotein; other alias Dsh)
DVT: deep-vein thrombosis
dynactin: dynein activator
DYRK: dual-specificity Tyr (Y) phosphorylation-regulated kinase

E

E: strain tensor
E: electric field
E: elastic modulus
E: elastance
$\mathcal{E}$: energy
$\{\widehat{\mathbf{e}}_i\}_{i=1}^3$: basis
e: strain vector
e: specific free energy
E-box: enhancer box sequence of DNA
E2: ubiquitin conjugase
E3: ubiquitin ligase
E_1: estrone (a single hydroxyl group in its molecule)
E_2: estradiol (2 hydroxyl groups), or 17β-estradiol
E_3: estriol (3 hydroxyl groups)
EAAT: excitatory amino acid (glutamate–aspartate) transporter
EAD: early afterdepolarization
EAR: V-ErbA-related nuclear receptor (NR2f6)
EB: end-binding protein
EBCT: electron beam CT
EBF: early B-cell factor
EC: endotheliocyte
Ec: Eckert number
ECA: external carotid artery
ECANS: extrinsic cardiac autonomic nervous system
ECDMV: endotheliocyte-derived microvesicle
ECF: extracellular fluid
ECFC: endothelial colony-forming cell
ECG: electrocardiogram
ECM: extracellular matrix
ED1L: EGF-like repeat- and discoidin-1-like domain-containing protein
EDGR: endothelial differentiation gene receptor
EDHF: endothelial-derived hyperpolarizing factor
EDIL: EGF-like repeats and discoidin-1 (I)-like domain-containing protein
EDP: epoxydocosapentaenoic acid
EDV: end-diastolic volume
EEA: early endosomal antigen
eEF: eukaryotic translation elongation factor
eEC: endocardial endotheliocyte
EEL: external elastic lamina
EET: epoxyeicosatrienoic acid
EFA6: exchange factor for ARF6 (ArfGEF)
EF-Tu: elongation factor Tu
EGF: epidermal growth factor
EGFL: EGF-like domain-containing protein
EGFR: epidermal growth factor receptor
EGR: early growth response transcription factor
EHD: C-terminal EGFR substrate-15 homology domain-containing protein
EHHADH: enoylCoA hydratase/3-hydroxyacylCoA dehydrogenase
eIF: eukaryotic translation initiation factor
EL: endothelial lipase
ELAM: endothelial–leukocyte adhesion molecules
ELCA: excimer laser coronary angioplasty
ELk: ETS-like transcription factor (ternary complex factor [TCF] subfamily)
ElMo: engulfment and cell motility adaptor
Eln: elastin
ElnF: elastin fiber
ELP: early lymphoid progenitor
EMI: early mitotic inhibitor
EMR: EGF-like module-containing, mucin-like, hormone receptor-like protein
EMRe: essential mtCU regulator
EMT: epithelial–mesenchymal transition
EMTU: epithelial–mesenchymal trophic unit

ENA–VASP: Enabled homolog and vasoactive (vasodilator)-stimulated phosphoprotein family
ENaC: epithelial Na^+ channel
EnaH: Enabled homolog
endo-siRNA: endogenous small interfering RNA
ENPP: ectonucleotide pyrophosphatase–phosphodiesterase
Ens: endosulfine
ENT: equilibrative nucleoside transporter
ENTPD: ectonucleoside triphosphate diphosphohydrolase
EPAC: exchange protein activated by cAMP
EPAS: endothelial PAS domain protein
EPC: endothelial progenitor cell
EPCR: endothelial protein-C receptor
EPDC: epicardial-derived cell
Epgn: epigen (EGF superfamily member)
EPH: erythropoietin-producing hepatocyte receptor kinase or pseudokinase (EPHa10 and EPHb6)
ephrin: EPH receptor interactor
EPo: erythropoietin
EPS: epidermal growth factor receptor pathway substrate
ER: endoplasmic reticulum
ERx: type-x estrogen receptor (NR3a1/2)
eRas: embryonic stem cell-expressed Ras (or hRas2)
ErbB: erythroblastoma viral gene product-B (HER)
ERC: elastin receptor complex
ERE: estrogen response element (DNA sequence)
Ereg: epiregulin (EGF superfamily member)
eRF: eukaryotic release factor
ERGIC: endoplasmic reticulum–Golgi intermediate compartment
ERK: extracellular signal-regulated protein kinase
ERK1/2: usually refers to ERK1 and ERK2
ERM: ezrin–radixin–moesin
ERMES: endoplasmic reticulum–mitochondrion encounter structure
EROA: effective regurgitant orifice area
ERP: effective refractory period
ERR: estrogen-related receptor (NR3b1–NR3b3)
ESCRT: endosomal sorting complex required for transport
ESL: E-selectin ligand
ESRP: epithelial splicing regulatory protein
ESV: end-systolic volume
ET: endothelin
ETC: electron transport chain
ETP: early thymocyte progenitor
ETR ($ET_{A/B}$): endothelin receptor
ETS: E-twenty six transformation-specific sequence (transcription factor; erythroblastosis virus E26 proto-oncogene product homolog)
ETV: ETS-related translocation variant
EVAR: endovascular aneurysm repair
Exo: exocyst subunit
Ext: exostosin (glycosyltransferase)

F

$\mathcal{F}$: Faraday constant
$\mathbf{F}$: transformation gradient tensor
F: function fraction of proliferating cells
F: erythrocytic rouleau fragmentation rate
$\mathbf{f}$: surface force
$\widehat{\mathbf{f}}$: fiber direction unit vector
f: binding frequency
f_C: cardiac frequency
f_R: breathing frequency
f: friction shape factor
f_v: head loss per unit length
f_X: molar fraction of gas component X
FA: fatty acid
FAAH: fatty acid amide hydrolase
FABP: fatty acid-binding protein
FABP: filamentous actin-binding protein
FACAP: ^{F}actin complex-associated protein
FACoA: fatty acylCoA
FACS: fatty acylCoA synthase
FAD: flavine adenine dinucleotide
FADD: Fas receptor-associated death domain
FAK: focal adhesion kinase
Fanc: Fanconi anemia protein
FAN: Fanconi anemia-associated nuclease
FAOD: fatty acid oxidation disorder
FAPP: phosphatidylinositol four-phosphate adaptor protein
FAST: forkhead activin signal transducer
FATP: fatty acid transport protein (SLC27a)
FB: fibroblast
Fbln (Fibl): fibulin
Fbn: fibrillin
FBS: F-box, Sec7 protein (ArfGEF)
FBx: F-box only protein (ArfGEF)
FC: fibrocyte
FCHO: FCH domain only protein
Fcα R: Fc receptor of IgA

$Fc\gamma$ R: Fc receptor of IgG
$Fc\epsilon$ R: Fc receptor of IgE
FCP: TF2F-associating C-terminal domain phosphatase
FDM: finite difference method
FEM: finite element method
FERM: four point-1, ezrin–radixin–moesin domain
FeR: FeS-related Tyr kinase
FeS: feline sarcoma kinase
FFA: free fatty acid
FFR: fractional flow reserve
FGF: fibroblast growth factor
FGFR: fibroblast growth factor receptor
FGR: viral feline Gardner–Rasheed sarcoma oncogene homolog kinase
FHCM: familial hypertrophic cardiomyopathy
FHL: four-and-a-half LIM-only protein
FHoD: formin homology domain-containing protein (FmnL)
FIH: factor inhibiting HIF1α (asparaginyl hydroxylase)
FIP: family of Rab11-interacting protein
FIP: focal adhesion kinase family-interacting protein
FIT: Fat-inducing transcript
FKBP: FK506-binding protein
FlIP: flice-inhibitory protein
FLK: fetal liver kinase
fMLP: ^{N}formyl methionyl-leucyl-phenylalanine
FN: fibronectin
Fn: fibrin
Fng: fibrinogen
Fos: Finkel–Biskis–Jinkins murine osteosarcoma virus sarcoma proto-oncogene product
Fox: forkhead box transcription factor
FPAH: familial arterial pulmonary hypertension
Fpn: ferroportin
FR: flow ratio
FRK: Fyn-related kinase
FrmD: FERM domain-containing adaptor
FRNK: FAK-related nonkinase
FRS: fibroblast growth factor receptor substrate
FSH: follicle-stimulating hormone
FSI: fluid–structure interaction
Fum: fumarase
FVM: finite volume method
FXR: farnesoid X receptor (NR1h4)
Fz: Frizzled (Wnt GPCR)

G

$\mathbf{G}$: Green–Lagrange strain tensor
G: shear modulus
G': storage modulus
G'': loss modulus
$\mathcal{G}$: Gibbs function
G: conductance
G_p: pressure gradient
$\mathcal{G}_b$: perfusion conductivity
$\mathcal{G}_e$: electrical conductivity
$\mathcal{G}_h$: hydraulic conductivity
$\mathcal{G}_T$: thermal conductivity
$\mathbf{g}$: gravity acceleration
g: physical quantity
g: gravity
g: detachment frequency
g: free enthalpy
G protein: guanine nucleotide-binding protein ($G\alpha\beta\gamma$ trimer)
$G\alpha$: α subunit (signaling mediator) of G protein
$G\alpha_{12/13}$ (G12/13): Gα subunit class 12/13
$G\alpha_i$ (Gi): inhibitory Gα subunit
$G\alpha_{i/o}$ (Gi/o): Gα subunit class
$G\alpha_{q/11}$ (Gq/11): Gα subunit class
$G\alpha_s$ (Gs): stimulatory Gα subunit
$G\alpha_t$ (Gt): transducin, Gα subunit of rhodopsin
G_{Tc}, G_{Tr}: cone, rod-transducin
Gs_{XL}: extralarge Gs protein
$G\beta\gamma$: dimeric subunit (signaling effector) of G protein
G_{gust}: gustducin, G protein α subunit (Gi/o) of taste receptor
G_{olf}: G protein α subunit (Gs) of olfactory receptor
GAB: GRB2-associated binder
GABA: γ-aminobutyric acid
$GABA_A$: GABA ionotropic receptor (Cl^- channel)
$GABA_B$: GABA metabotropic receptor (GPCR)
GABARAP: $GABA_A$ receptor-associated protein
GaBP: globular actin-binding protein
GADD: growth arrest and DNA-damage-induced protein
gadkin: γ1-adaptin and kinesin interactor
GAG: glycosaminoglycan
GAK: cyclin G-associated kinase
Gal: galanin
GAP: GTPase-activating protein

GAPDH: glyceraldehyde 3-phosphate dehydrogenase
GARP: Golgi body-associated retrograde protein complex
GAS: growth arrest-specific noncoding, single-stranded RNA
GAT: γ-aminobutyric acid transporter
GATA: DNA sequence GATA-binding transcription factor
GBF: Golgi body-associated brefeldin-A-resistant guanine nucleotide-exchange factor
GBP: guanylate-binding protein
GCAP: guanylate cyclase-activating protein
GCC: Golgi coiled-coil domain-containing protein
GCK: germinal center kinase
GCKR: GCK-related kinase
GCNF: germ cell nuclear factor (NR6a1)
GCN2: general control non-derepressible-2 (pseudokinase)
GCS: glutamylcysteine synthase
gCSF: granulocyte colony-stimulating factor (CSF3)
GCV: great cardiac vein
GD: disialoganglioside
GDP: guanosine diphosphate
GDF: growth differentiation factor
GDF: (Rab)GDI displacement (dissociation) factor
GDI: guanine nucleotide-dissociation inhibitor
GDNF: glial cell line-derived neurotrophic factor
GEF: guanine nucleotide (GDP-to-GTP)-exchange factor
GF: growth factor
GFAP: glial fibrillary acidic protein (intermediate filament)
GFL: GDNF family of ligands
GFP: geodesic front propagation
GFR: glomerular filtration rate
GFR: growth factor receptor
GFRαi: type-i GDNF family receptor-α
GGA: Golgi body-localized γ-adaptin ear-containing Arf-binding protein
Ggust: (G protein) Gα subunit gustducin
GH: growth hormone
GHR: growth hormone receptor
GHRH: growth hormone-releasing hormone
GIP: GPCR-interacting protein
GIRK: Gβγ-regulated inwardly rectifying K^+ channel
GIT: GPCR kinase-interacting protein
GKAP: G-kinase-anchoring protein
GKAP: glucokinase-associated phosphatase (DuSP12)
GKAP: guanylate kinase-associated protein
GLK: GCK-like kinase
GluK: ionotropic glutamate receptor (kainate type)
GluN: ionotropic glutamate receptor (NMDA type)
GluR: ionotropic glutamate receptor (AMPA type)
GluT: glucose transporter
GlyCAM: glycosylation-dependent cell adhesion molecule
GlyR: glycine receptor (channel)
GlyT: glycine transporter
GM: monosialoganglioside
gmCSF: granulocyte–monocyte colony-stimulating factor (CSF2)
GMP: granulocyte–monocyte progenitor
GMP: guanosine monophosphate
GnRH: gonadotropin-releasing hormone
GP: glycoprotein
Gpc: glypican
GPI: glycosylphosphatidylinositol anchor
gpiAP: GPI-anchored protein
GPCR: G-protein–coupled receptor
GPx: glutathione peroxidase
GQ: quadrisialoganglioside
GR: glucocorticoid receptor (NR3c1)
Gr: Graetz number
GRAP: GRB2-related adaptor protein (or GAds)
GRB: growth factor receptor-bound protein
GRC: growth factor-regulated, Ca^{2+}-permeable, cation channel (TRPV2)
GRE: glucocorticoid response element (DNA sequence)
GRHL: grainyhead-like transcription factor
GRK: G-protein-coupled receptor kinase
GRP: G-protein-coupled receptor phosphatase
G^{SH} (GSH): reduced form of glutathione
G^{SS} (GSSG): oxidized form of glutathione (glutathione disulfide)
GSK: glycogen synthase kinase
GSR: glutathione disulfide reductase
GsS: glutathione synthase
GST: glutathione ^{S}transferase
GT: trisialoganglioside
GTF: general transcription factor
GTP: guanosine triphosphate
GTPase: guanosine triphosphatase

GuCy: guanylate cyclase (CyG)
GWAS: genome-wide association study

H

H: height
$\mathcal{H}$: history function
H: dissipation
H: Henry parameter (solubility)
h: head loss
h: thickness
h: specific enthalpy
h_{m}: mass transfer coefficient
h_{T}: heat transfer coefficient
HA: hemagglutinin
HA: hyaluronic acid
HAD: haloacid dehalogenase
HADH: hydroxyacylCoA dehydrogenase
HAP: huntingtin-associated protein
HAT: histone acetyltransferase
HAAT: heterodimeric amino acid transporter
HAND: heart and neural crest derivatives expressed protein
Hb: hemoglobin
Hb^{SNO}: ^{S}nitrosohemoglobin
HBEGF: heparin-binding EGF-like growth factor
HCK: hematopoietic cell kinase
HCLS: hematopoietic lineage cell-specific Lyn substrate protein
HCM: hypertrophic cardiomyopathy
HCN: hyperpolarization-activated, cyclic nucleotide-gated K^+ channel
HCNP: hippocampal cholinergic neurostimulatory peptide
HCT: helical CT
HDAC: histone deacetylase complex
HDL: high-density lipoprotein
$\mathrm{HDL}^{\mathrm{Cs}}$: HDL–cholesterol
$\mathrm{HDL}^{\mathrm{CsE}}$: HDL–cholesteryl ester
HDM: human double minute (Ub ligase)
HEET: hydroxyepoxyeicosatrienoic acid
hemin: heme oxygenase-1 inducer
HERG: human ether-a-go-go related gene
HER: human epidermal growth factor receptor (HER3: pseudokinase)
HES: Hairy enhancer of split
HETE: hydroxyeicosatetraenoic acid
HETEE: HETE ethanolamide
HEV: high endothelial venule
HF: heart failure
HFlEF: heart failure with low LVEF
HFpEF: heart failure with persistently preserved LVEF
HFpLVEF: heart failure with preserved left ventricular ejection fraction
HFrEF: heart failure with recovered LVEF
HFrLVEF: heart failure with reduced left ventricular ejection fraction
HGF: hepatocyte growth factor
HGFA: hepatocyte growth factor activator (serine peptidase)
HGFR: hepatocyte growth factor receptor
HGNET: high-grade neuroendocrine tumor
HGS: HGF-regulated Tyr kinase substrate (HRS)
HhIP: Hedgehog-interacting protein
HI: hemolysis index
HIF: hypoxia-inducible factor
HIP: huntingtin-interacting protein
HIP1R: HIP1-related protein
His: histamine
Hjv: hemojuvelin
HK: hexokinase
HL: hepatic lipase
HMG: high-mobility group protein
HMGB: high-mobility group box protein
HMGCL: HMGCoA lyase
HMGCoA: 3-hydroxymethylglutarylCoA;
HMGCoAR (HMGCR): hydroxymethylglutaryl coenzyme-A reductase
HMGCS: HMGCoA synthase
HMT: histone methyltransferase
HMWK: high-molecular-weight kininogen
HNF: hepatocyte nuclear factor (NR2a1/2)
HNP: human neutrophil peptide
hnRNP: heterogeneous nuclear ribonucleoprotein
HOCM: hypertrophic obstructive cardiomyopathy
HODE: hydroxyoctadecadienoic acid
HOP: HSP70–HSP90 complex-organizing protein
HoPS: homotypic fusion and vacuole protein sorting complex
HotAIR: HOX antisense intergenic RNA (large intergenic noncoding RNA)
HOx: heme oxygenase
Hox: homeobox DNA sequence (encodes homeodomain-containing morphogens)
HPAA: hypothalamic–pituitary–adrenal axis
HpCa: hippocalcin
HPD: high transvalvular pressure difference
HPETE: hydroperoxyeicosatetraenoic acid
HPETEE: HPETE ethanolamide

HPK: hematopoietic progenitor kinase (MAP4K)
hpRNA: long hairpin RNA
HPV: hypoxic pulmonary vasoconstriction
hRas: Harvey Ras
HRE: hormone response element (DNA sequence)
HRM: hypoxia-regulated microRNA
hRNP: heterogeneous ribonucleoprotein
HRS: hepatocyte growth factor-regulated Tyr kinase substrate
HRT: Hairy and enhancer of Split-related transcription factor
HRV: heart rate variability
HS: heparan sulfate
HSC: hematopoietic stem cell
HSC: heat shock cognate
HSER: heat stable enterotoxin receptor (guanylate cyclase-2C)
HSP: heat shock protein (chaperone)
HSPG: heparan sulfate proteoglycan
Ht: hematocrit
HTR: high temperature requirement endopeptidase
HUNK: hormonally upregulated Neu-associated kinase

I

I: identity tensor
I: inotropy
i: current
IAP: inhibitor of apoptosis protein
IBABP: intestinal bile acid-binding protein
IC: isovolumetric contraction
ICA: internal carotid artery
ICAM: intercellular adhesion molecule (IgCAM member)
ICANS: intrinsic cardiac autonomic nervous system
ICBD: isobutyrylCoA dehydrogenase deficiency
ICDH: isocitrate dehydrogenase
IgCAM: immunoglobulin-like cell adhesion molecule
ICF: intracellular fluid
iCliP: intramembrane-cleaving peptidase (that clips)
ICM: ischemic cardiomyopathy
ICSA: intracranial saccular aneurysm
ID: inhibitor of DNA binding
IDL: intermediate-density lipoprotein
IDmiR: immediately downregulated microRNA
IDOL: inducible degrader of LDL receptor (Ub ligase)
IEG: immediate-early gene
IEL: internal elastic lamina
IEL: intraepithelial lymphocyte
IfIH: interferon-induced with helicase-C domain-containing protein
ifM: interfibrillar mitochondrion
Ifn: interferon
IfnAR: interferon-$\alpha/\beta/\omega$ receptor
IFT: intraflagellar transport complex
Ig: immunoglobulin
IGF: insulin-like growth factor
IGFBP: IGF-binding protein
IgHC: immunoglobulin heavy chain
IgLC: immunoglobulin light chain
iGluR: ionotropic glutamate receptor
IH: intimal hyperplasia
IHD: ischemic heart disease
IHh: indian Hedgehog
IK: intermediate-conductance Ca^{2+}-activated K^+ channel
Iκ B: inhibitor of NFκ B
IKK: Iκ B kinase
IL: interleukin
iLBP: intracellular lipid-binding protein
ILC: innate lymphoid cell
ILD: interstitial lung disease
ILK: integrin-linked (pseudo)kinase
ILKAP: integrin-linked kinase-associated Ser/Thr phosphatase-2C
IMAC: (ROS-sensitive) mitochondrial inner membrane anion channel
IMH: intramural hematoma
IMM: inner mitochondrial membrane
IMP: impedes mitogenic signal propagation
IMR: index of microvascular resistance
INADl: inactivation no after-potential D-protein
InCenP: inner centromere protein
InF: inverted formin
InsIG: insulin-induced gene product (ER anchor)
InsL: insulin-like peptide
InsR (IR): insulin receptor
InsRR: insulin receptor-related receptor
IP: inositol phosphate
IP_3: inositol (1,4,5)-trisphosphate
IP_3R: IP_3 receptor (IP_3-sensitive Ca^{2+}-release channel)
IP_4: inositol (1,3,4,5)-tetrakisphosphate
IP_5: inositol pentakisphosphate
IP_6: inositol hexakisphosphate

IPAH: idiopathic arterial pulmonary hypertension
IPC: ischemic preconditioning
IPCEF: interaction protein for cytohesin exchange factor
IPOD: (perivacuolar) insoluble protein deposit
IPP: inositol polyphosphate phosphatase
IPP: ILK–PINCH–parvin complex
iPSC: induced pluripotent stem cell
IQGAP: IQ motif-containing GTPase-activating protein (IQ: first 2 amino acids of the motif: isoleucine [I; commonly] and glutamine [Q; invariably]).
IR: isovolumetric relaxation
IRAK: IL1 receptor-associated kinase (IRAK2: pseudokinase)
IRE: irreversible electroporation
IRES: internal ribosome entry site
IRF: interferon-regulatory protein (transcription factor)
IRFF: interferon-regulatory factor family
IRP: iron regulatory protein
IRS: insulin receptor substrate
ISG: interferon-stimulated gene product
iSMAD: inhibitory SMAD (SMAD6 or SMAD7)
ISR: in-stent restenosis
ITAM: immunoreceptor tyrosine-based activation motif
Itch: Itchy homolog (Ub ligase)
Itg: integrin
ITIM: immunoreceptor tyrosine-based inhibitory motif
ITK: interleukin-2-inducible T-cell kinase
ITPK: inositol trisphosphate kinase
IVC: inferior vena cava
IVP: initial value problem
IVUS: intravascular ultrasonography

J

J: flux
J_{mb}: cell surface current density
JAM: junctional adhesion molecule
JaK: Janus (pseudo)kinase
JIP: JNK-interacting protein (MAPK8IP1 and -2)
JMy: junction-mediating and regulatory protein
JNK: Jun N-terminal kinase (MAPK8–MAPK10)
JNKBP: JNK-binding protein
JNKK: JNK kinase
JP: junctophilin
JSAP: JNK/SAPK-associated protein
jSR: junctional sarcoplasmic reticulum
Jun: avian sarcoma virus-17 proto-oncogene product (Japanese juunana: seventeen [17]; TF)
JUNQ: juxtanuclear quality-control compartment

K

$\mathbf{K}$: conductivity tensor
K: bending stiffness
K: reflection coefficient
K_d: dissociation constant (index of ligand–target affinity: ([L][T])/[C]; [L], [T], [C]: molar concentrations of the ligand, target, and created complex, respectively)
K_{M}: Michaelis constant (chemical reaction kinetics)
K_m: material compressibility
K_{R}: resistance coefficient
k: cross-section ellipticity
k_{ATP}: myosin ATPase rate
k_{B}: Boltzmann constant (1.38×10^{-23} J/K)
k_c: spring stiffness
k_i: kinetic coefficient
k_m: mass transfer coefficient
k_{P}: Planck constant
KaP: karyopherin
$\mathrm{K_{ATP}}$: ATP-sensitive K^+ channel
$\mathrm{K_{Ca}}$1.x: BK channel
$\mathrm{K_{Ca}}$2/3/4.x: SK channel
$\mathrm{K_{Ca}}$5.x: IK channel
$\mathrm{K_{IR}}$: inwardly rectifying K^+ channel
$\mathrm{K_V}$: voltage-gated K^+ channel
KAP: kinesin (KIF)-associated protein
Kap: karyopherin
KAT: lysine (K) acetyltransferase
KCC: K^+–Cl^- cotransporter
KChAP: K^+ channel-associated protein
KChIP: $\mathrm{K_V}$ channel-interacting protein
KDELR: KDEL (Lys–Asp–Glu–Leu) endoplasmic reticulum retention receptor
KDR: kinase insert domain receptor
KGDH: ketoglutarate dehydrogenase
KHC: kinesin heavy chain
KIF: kinesin family
KIR: killer cell immunoglobulin-like receptor
KIT: cellular kinase in tyrosine (SCFR)
Kk: kallikrein
KKA: kallikrein–kinin axis
KLC: kinesin light chain
KLF: Krüppel-like factor
KLR: killer cell lectin-like receptor

Kn: Knudsen number
KOR: κ-opioid receptor
kRas: Kirsten Ras
Krt: keratin
KSR: kinase suppressor of Ras (adaptor; pseudokinase)

L

L: velocity gradient tensor
L: inertance
L: length
L_e: entry length
LA: left atrium
LAB: linker of activated B lymphocyte
LAd: LCK-associated adaptor
Lam: laminin
LAMTOR: late endosomal and lysosomal adaptor, MAPK and TOR activator
LANP: long-acting natriuretic peptide
LAP: leucine-rich repeat and PDZ domain-containing protein (4-member family)
LAP: latency-associated peptide (4 isoforms LAP1–LAP4)
LAP: nuclear lamina-associated polypeptide
LAR: leukocyte common antigen-related receptor (PTPRF)
LAT: linker of activated T lymphocytes
LaTS: large tumor suppressor
LAX: linker of activated X cells (both B and T cells)
LBBB: left bundle branch block
LBR: lamin-B receptor
LCA: left coronary artery
LCAD: long-chain acylCoA dehydrogenase
LCAT: lysolecithin cholesterol acyltransferase
LCC: left coronary cusp
lcFA: long-chain fatty acid (10–16 carbon atoms)
LCHAD: long-chain 3-hydroxyacylCoA dehydrogenase
LCHADD: LCHAD deficiency
LCK: leukocyte-specific cytosolic (nonreceptor) Tyr kinase
LCP: lymphocyte cytosolic protein (adaptor SLP76)
LDH: lactate dehydrogenase
LDL: low-density lipoprotein
LDL^{Cs}: LDL–cholesterol
LDLR: low-density lipoprotein receptor
LDV: laser Doppler velocimetry
Le: entry length
LEF: lymphoid enhancer-binding transcription factor
LGalS: lectin, galactoside-binding, soluble cell adhesion molecule
LGIC: ligand-gated ion channel
LGL: lethal giant larva protein
LGNET: low-grade neuroendocrine tumor
LH: luteinizing hormone
LIF: leukemia-inhibitory factor
LIFR: leukemia-inhibitory factor receptor
LIMA: LIM domain and actin-binding protein
LIME: LCK-interacting molecule
LIMK: Lin1, Isl1, and Mec3 motif-containing kinase
LIMS: LIM and senescent cell antigen-like-containing domain protein
LiNC: linker of nucleoskeleton and cytoskeleton
lincRNA: large intergenic noncoding RNA (encoded intergenically)
LipC: hepatic lipase
LipD: lipoprotein lipase
LipE: hormone-sensitive lipase
LipG: endothelial lipase
LipH: lipase-H
liprin: LAR PTP-interacting protein
LIR: leukocyte immunoglobulin-like receptor
LIS: lissencephaly protein
LKb: liver kinase-B
LKLF: lung Krüppel-like factor
LLTC: large latent TGFβ complex
LMan: lectin, mannose-binding
LMO: LIM domain-only-7 protein
Lmod: leiomodin (actin nucleator)
LMPP: lymphoid-primed multipotent progenitor
LMR: laser myocardial revascularization
lncRNA: long noncoding RNA (encoded intragenically)
LOx: lipoxygenase
LOxLDLR: lectin-type oxidized low-density lipoprotein receptor
LP: lipoprotein
LPA: lysophosphatidic acid
LPD: low transvalvular pressure difference
lpDC: lamina propria dendritic cell
Lphn: latrophilin (adhesion GPCR)
LPL: lysophospholipid
LPLase: lipoprotein lipase
LPLase: lysophospholipase
LPP: lipid phosphate phosphatase
LPR: lipid phosphatase-related protein
LPS: lipopolysaccharide

LQTS: long-QT syndrome
LRAT: lecithin–retinol acyltransferase
LRH: liver receptor homolog (NR5a2)
LRI: lower respiratory infection
LRO: lysosome-related organelle
LRP: LDL receptor-related protein
LRRTM: leucine-rich repeat-containing transmembrane protein
LSK: Lin−, SCA1+, KIT+ cell
LST: lethal with Sec-thirteen
LSV: long saphenous vein
LT (Lkt): leukotriene
LTBP: latent TGFβ-binding protein
LTCC: L-type Ca^{2+} channel (Ca_V1)
LTFR: low transvalvular flow rate
LTI: lymphoid tissue inducer cell
LTK: leukocyte tyrosine kinase
LTO: lymphoid tissue organizer cell
LUbAC: linear ubiquitin chain assembly complex
LV: left ventricle
LVAD: left ventricular assist device
LVNCCM: left ventricular noncompaction cardiomyopathy
LX: lipoxin
LXR: liver X receptor (NR1h2/3)
LyVE: lymphatic vessel endothelial hyaluronan receptor

M

M: molar mass
M: metabolic rate
$\mathcal{M}$: moment
m: mass
M/M/DCDMV: monocyte/macrophage/dendritic cell-derived microvesicle
M/SCHAD: medium- and short-chain ^{L}3-hydroxyacylCoA dehydrogenase deficiency
Ma: Mach number
MACE: major adverse cardiovascular event
MACF: microtubule–actin crosslinking factor
mAChR: acetylcholine muscarinic receptor (metabotropic; GPCR)
MaCoA: malonylCoA
MAD: mothers against decapentaplegic homolog
MAD: mitotic arrest-deficient protein
MAdCAM: mucosal vascular addressin cell adhesion molecule
MADD: multiple acylCoA dehydrogenase deficiency
MAF: musculoaponeurotic fibrosarcoma oncogene homolog (TF)
MAGI: membrane-associated guanylate kinase-related protein with inverted domain organization
MAGL: monoacylglycerol lipase
MAGP: microfibril-associated glycoprotein
MAGuK: membrane-associated guanylate kinase
MAI: microarousal index
MAIT: mucosal-associated invariant T lymphocyte
MALT: mucosa-associated lymphoid tissue
MALT1: mucosa-associated lymphoid tissue lymphoma translocation peptidase
MAO: monoamine oxidase
MAP: microtubule-associated protein
MAP1LC3: microtubule-associated protein-1 light chain-3 (LC3)
mAP: mean arterial pressure
MAPK: mitogen-activated protein kinase
MAP2K: MAPK kinase
MAP3K: MAP2K kinase
MAP3K7IP: MAP3K7-interacting protein
MAPKAPK: MAPK-activated protein kinase
MARCKS: myristoylated alanine-rich C kinase substrate
MaRCo: macrophage receptor with collagenous structure (ScaRa2)
MARK: microtubule affinity-regulating kinase
MASTL: microtubule-associated Ser/Thr kinase-like protein
MAT: ménage à trois
MATK: megakaryocyte-associated Tyr kinase
MAVS: mitochondrial antiviral signaling protein
MAX: MyC-associated factor-X (bHLHd4–bHLHd8)
MBF: myocardial blood flow
MBP: myosin-binding protein
MBP: myeloid–B-cell progenitor
MBTPSi: membrane-bound transcription factor peptidase site i
MCAD: medium-chain acylCoA dehydrogenase
MCADD: MCAD deficiency
MCAK: mitotic centromere-associated kinesin
MCAM: melanoma cell adhesion molecule
MCC: monocarboxylate carrier
MCD: medullary collecting duct
MCD: malonylCoA decarboxylase
mcFA: medium-chain fatty acid (6–12 carbon atoms)

MCL1: BCL2-related myeloid cell leukemia sequence protein-1
MCLC: stretch-gated Mid1-related chloride channel
MCM: minichromosome maintenance protein
MCP: monocyte chemoattractant protein
mCSF: macrophage colony-stimulating factor (CSF1)
MCT: monocarboxylate–proton cotransporter
mDC: myeloid dendritic cell
MDH: malate dehydrogenase
MDM: mitochondrial distribution and morphology protein
MDR: multiple drug resistance (ABC transporter)
MEF: myocyte enhancer factor
megCSF: megakaryocyte colony-stimulating factor
MEJ: myoendothelial junction
MELK: maternal embryonic leucine zipper kinase
MEP: megakaryocyte erythroid progenitor
MEP: myeloid–erythroid progenitor
MET: mesenchymal–epithelial transition factor (proto-oncogene; HGFR)
Mfn: mitofusin
MFO: mixed-function oxidase
metHb: methemoglobin
MGIC: mechanogated ion channel
mGluR: metabotropic glutamate receptor
MGP: matrix γ-carboxyglutamate acid (Gla)-containing protein
MHC: major histocompatibility complex
MHC: myosin heavy chain
MiCU: mitochondrial Ca^{2+} uptake protein
Mid: midline
MIF: macrophage migration-inhibitory factor
MInK: MAPK-interacting protein Ser/Thr kinase
MinK: misshapen-like kinase (MAP4K6)
minK: minimal potassium channel subunit
miR: microRNA
miRNP: microribonucleoprotein
MiRP: MinK-related peptide
MIRR: multichain immune-recognition receptor
MIS: Müllerian-inhibiting substance
MIS: mini-invasive surgery
MIS: mitochondrial intermembrane space
MIST: mastocyte immunoreceptor signal transducer
MIT: mini-invasive therapy
MiV: mitral valve
MIZ: Myc-interacting zinc finger protein
MJD: Machado–Joseph disease protein domain-containing peptidase (DUb)
MKL: megakaryoblastic leukemia-1 fusion coactivator
MKnK: MAPK-interacting protein Ser/Thr kinase (MnK)
MKP: mitogen-activated protein kinase phosphatase
MLC: myosin light chain
MLCK: myosin light chain kinase
MLCP: myosin light chain phosphatase
MLK: mixed lineage kinase
MLKL: mixed lineage kinase-like pseudokinase
MLL: mixed lineage [myeloid–lymphoid] leukemia factor
MLLT: mixed lineage leukemia translocated protein
MLP: muscle LIM protein
mmCK: myofibrillar creatine kinase
MMDMV: monocyte/macrophage-derived microvesicle
MME: membrane metalloendopeptidase
MMM: maintenance of mitochondrial morphology protein
MMP: matrix metallopeptidase
MO: mouse protein
Mo: monocyte
MOMP: mitochondrial outer membrane permeabilization
MOR: μ-opioid receptor
MP: MAPK partner
MPF: mitosis (maturation)-promoting factor (CcnB–CDK1 complex)
MPG: ^{N}methylpurine (^{N}methyladenine)-DNA glycosylase
MPO: median preoptic nucleus
MPo: myeloperoxidase
MP_P: membrane protein, palmitoylated
MPP: multipotent progenitor
MR: mineralocorticoid receptor (NR3c2)
mRas: muscle Ras (or rRas3)
MRCK: myotonic dystrophy kinase-related CDC42-binding kinase
MRI: (nuclear) magnetic resonance imaging
mRNA: messenger RNA
mRNP: messenger ribonucleoprotein
MRTF: myocardin-related transcription factor
MSC: mesenchymal stem cell
MSH: melanocyte-stimulating hormone
MSIC: mechanosensitive ion channel
MSSCT: multislice spiral CT
MST: mammalian sterile-twenty-like kinase

MSt1R: macrophage-stimulating factor-1 receptor (RON)
MT: metallothionein
mtCK: mitochondrial creatine kinase
mtCU: mitochondrial calcium uniporter (Ca^{2+} uptake)
mtCUR: mitochondrial calcium uniporter regulator
mtDAMP: mitochondrial alarmin
mtDNA: mitochondrial DNA
mtTE: mitochondrial thioesterase
mtETC: mitochondrial electron transport chain
mtTerF: mitochondrial transcription termination factor
MTM: myotubularin (myotubular myopathy-associated gene product)
mtMMP: membrane-type MMP (mt*i*MMP: type-*i* mtMMP)
MTMR: myotubularin-related phosphatase
MTOC: microtubule organizing center
MTP: myeloid–T-cell progenitor
MTP: microsomal triglyceride transfer protein
MUFA: monounsaturated fatty acid
MuRF: muscle-specific RING finger (Ub ligase)
MuSK: muscle-specific kinase
MVB: multivesicular body
MVE: multivesicular endosome (MVB)
MVO: microvascular obstruction
MVO_2: myocardial oxygen consumption
MVR: mitral valve regurgitation
MVS: mitral valve stenosis
MWSS: maximal wall shear stress
MyB: myeloblastosis viral oncogene homolog (TF)
MyBPc: myosin-binding protein-C
MyC: myelocytomatosis viral oncogene homolog (TF)
MyD88: myeloid differentiation primary response gene product-88
MYH: myosin heavy chain gene
MyHC: myosin heavy chain
MYL: myosin light chain gene
MyLC: myosin light chain
MyPT: myosin phosphatase targeting subunit
MyT: myelin transcription factor

N

N: sarcomere number
$\widehat{\mathbf{n}}$: unit normal vector
n: mole number
n: PAM density with elongation x
$\mathtt{n}$: myosin head density
$\mathcal{N}_A$: Avogadro number
N-terminus: amino (amine group NH_2) terminus
NA: neuraminidase (sialidase)
NAADP: nicotinic acid adenine dinucleotide phosphate
nAChR: acetylcholine nicotinic receptor (ionotropic; LGIC)
NAD: nicotine adenine dinucleotide
NADPH: reduced form of nicotinamide adenine dinucleotide phosphate
NAd: noradrenaline
NAF: nutrient-deprivation autophagy factor
NALT: nasal-associated lymphoid tissue
NAmPT: nicotinamide phosphoribosyltransferase
Nanog: ever young (Gaelic)
NAP: NCK-associated protein (NCKAP)
NAPE: ^{N}acylphosphatidylethanolamine
NAT: nucleobase–ascorbate transporter
NAT1: noradrenaline transporter
Na_V: voltage-gated Na^+ channel
NBC: Na^+–HCO_3^- cotransporters
NCC: noncoronary cusp
NCC: Na^+–Cl^- cotransporter
Ncdn: neurochondrin
NCK: noncatalytic region of Tyr kinase adaptor
NCoA: nuclear receptor coactivator
NCoR: nuclear receptor corepressor
NCR: natural cytotoxicity-triggering receptor
ncRNA: noncoding RNA
NCS: neuronal calcium sensor
NCKX: Na^+–Ca^{2+}–K^+ exchanger
NCLX: Na^+–Ca^{2+}–Li^+ exchanger
NCX: Na^+–Ca^{2+} exchanger
NDCBE: Na^+-dependent Cl^-–HCO_3^- exchanger
NecL: nectin-like molecule
NEDD: neural precursor cell expressed, developmentally downregulated
NDFIP: NEDD4 family-interacting protein
NeK: never in mitosis gene-A (NIMA)-related kinase
NES: nuclear export signal
NESK: NIK-like embryo-specific kinase
nesprin: nuclear envelope spectrin repeat protein
NET: neuroendocrine tumor
NeuroD: neurogenic differentiation protein
NF: neurofilament protein (intermediate filament)
NF: neurofibromin (RasGAP)
NFAT: nuclear factor of activated T cells
NFe2: erythroid-derived nuclear factor-2

NFH: neurofilament, heavy polypeptide
NFκ B: nuclear factor κ light chain enhancer of activated B cells
NFL: neurofilament, light polypeptide
NFM: neurofilament, medium polypeptide
NGAL: neutrophil gelatinase-associated lipocalin
NGF: nerve growth factor
Ngn: neogenin (netrin receptor)
NHA: Na^+–H^+ antiporter
NHE: Na^+–H^+ exchanger
NHERF: NHE regulatory factor
NHI: normalized hemolysis index
NHR: nuclear hormone receptor
NIc: nucleoporin-interacting protein
NIK: NFκ B-inducing kinase
NIK: NCK-interacting kinase
NIP: neointimal proliferation
NK: natural killer cell
NKCC: Na^+–K^+–$2Cl^-$ cotransporter
NKG: NK receptor group
NKT: natural killer T cell
NKx2: NK2 transcription factor-related homeobox protein
NLR: NOD-like receptor (nucleotide-binding oligomerization domain, Leu-rich repeat-containing)
NLS: nuclear localization signal
NMDAR: ^{N}methyl ^{D}aspartate receptor
NmU: neuromedin-U
NMVOC: nonmethane volatile organic compound
NO: nitric oxide (nitrogen monoxide)
NO_x: nitrogen oxides
NOD: nucleotide-binding oligomerization domain
NonO: non-POU domain-containing octamer-binding protein
NOR: neuron-derived orphan receptor (NR4a3)
NOS: nitric oxide synthase
NOS1: neuronal NOS
NOS1AP: NOS1 adaptor protein
NOS2: inducible NOS
NOS3: endothelial NOS
NOx: NAD(P)H oxidase
Noxa: damage (Latin)
NPAS: neuronal PAS domain-containing transcription factor
NPC: nuclear-pore complex
NPc: Niemann–Pick disease type-C protein
NPc1L: Niemann–Pick protein-C1-like
nPKC: novel protein kinase-C
NPY: neuropeptide-Y
NR: nuclear receptor
NRAP: nebulin-related actinin-binding protein
nRas: neuroblastoma Ras
NRBP: nuclear receptor-binding protein
NREM: nonrapid eye movement sleep
NRF: nuclear factor erythroid-derived-2 (NF-E2)-related factor
NRF1: nuclear respiratory factor-1
Nrg: neuregulin (EGF superfamily member)
Nrgn: neuroligin
Nrp: neuropilin (VEGF-binding molecule; VEGFR coreceptor)
NRPTP: nonreceptor protein Tyr phosphatase
NRSTK: nonreceptor Ser/Thr kinase
NRTK: nonreceptor protein Tyr kinase
NRx: nucleoredoxin
Nrxn: neurexin
NSC: nonselective cation channel
NSCLC: non-small-cell lung cancer
NSF: ^{N}ethylmaleimide-sensitive factor
NSLTP: nonspecific lipid-transfer protein
NST: nucleus of the solitary tract
NT: neurotrophin
NT5E: ecto-5′-nucleotidase
NTCP: sodium–taurocholate cotransporter polypeptide
NTF: N-terminal fragment
NTFR: normal transvalvular flow rate
NTP: nucleoside triphosphate
NTPase: nucleoside triphosphate hydrolase superfamily member
NTRK: neurotrophic tyrosine receptor kinase (TRK)
NTRKR: neurotrophic protein Tyr receptor kinase-related protein ($ROR_{(RTK)}$)
NTS: nucleus tractus solitarius
Nu: Nusselt number
NuAK: nuclear AMPK-related kinase
NuP: nucleoporin (nuclear-pore complex protein)
NuRD: nucleosome remodeling and histone deacetylase
NuRR: nuclear receptor-related factor (NR4a2)
nWASP: neuronal WASP

O

^{O}Glc^{N}Ac: β^{N}acetyl ^{D}glucosamine
OAI: obstructive apnea index
OCRL: oculocerebrorenal syndrome of Lowe phosphatase
OCT: optical coherence tomography
Oct: octamer-binding transcription factor
ODE: ordinary differential equation

OEF: oxygen extraction fraction
OGA: ^{O}Glc^{N}Acase (β^{N}acetylglucosaminidase)
OHI: obstructive hypopnea index
OHS: obesity hypoventilation syndrome
OI: osteogenesis imperfecta
OMCD: outer medullary collecting duct
OMM: outer mitochondrial membrane
ONARE: obstructive nonapneic respiratory event
ORC: origin recognition complex
ORF: open reading frame
ORP: OSBP-related protein
OSA: obstructive sleep apnea
OSHAS: obstructive sleep hypopnea–apnea syndrome
OSBP: oxysterol-binding protein
OSI: oscillatory shear index
OSM: oncostatin-M
OSMR: oncostatin-M receptor
OSR (OxSR): oxidative stress-responsive kinase
OTK: off-track (pseudo)kinase
OTU: ovarian tumor superfamily peptidase (deubiquitinase)
OTUB: otubain (Ub thioesterase of the OTU superfamily)
OVLT: organum vasculosum lamina terminalis
oxyHb: oxyhemoglobin (oxygenated hemoglobin)

P

$\mathcal{P}$: permeability
P: power
P: cell division rate
$P(X)$: probability of event X
p: production rate
p: pressure
P_i: inorganic phosphate (free phosphate ion)
p_X: partial pressure of gas component X
PA: phosphatidic acid
PAAT: proton–amino acid transporter
PACS: phosphofurin acidic cluster sorting protein
PAF: platelet-activating factor
PAFAH: platelet-activating factor acetylhydrolase
PAG: phosphoprotein associated with glycosphingolipid-enriched microdomains
PAH: polycyclic aromatic hydrocarbon
PAH: pulmonary arterial hypertension
PAI: plasminogen activator inhibitor
PAK: P21 (CKI1a)-activated kinase
PALR: promoter-associated long RNA
PALS: protein associated with Lin-7
PAMP: pathogen-associated molecular pattern
PAMP: proadrenomedullin peptide
PAPC: palmitoyl arachidonoyl glycerophosphorylcholine
PAR: poly^{ADP}ribose
PAR: promoter-associated, noncoding RNA
PARi: type-i peptidase-activated receptor
Par: partitioning defective protein
PARG: poly^{ADP}ribosyl glycosidase
PARP: poly^{ADP}ribose polymerase
PASR: promoter-associated short RNA
PAT: pulse amplitude tonometry
PATJ: protein (PALS1) associated to tight junctions
PAU: penetrating atherosclerotic ulcer
Pax: paxillin
Pax$i$: paired box protein-i (transcription regulator)
PBC: pre-Bötzinger complex (ventilation frequency)
PBIP: Polo box-interacting protein
PC: phosphatidylcholine
PC: polycystin
PC: protein-C
PCD: primary ciliary dyskinesia
PCI: percutaneous coronary intervention
PCMRV: phase-contrast MR velocimetry
PCr: phosphocreatine
PCT: proximal convoluted tubule
PCTP: phosphatidylcholine-transfer protein
PD: pharmacodynamics
pDC: plasmacytoid dendritic cell
PdCD: programmed cell death protein
PdCD6IP: PdCD-6-interacting protein
PdCD1Lg: programmed cell death-1 ligand
PDE: phosphodiesterase
PDE: partial differential equation
PDGF: platelet-derived growth factor
PDGFR: platelet-derived growth factor receptor
PDHK: pyruvate dehydrogenase kinase
PDHP: pyruvate dehydrogenase phosphatase
PDI: protein disulfide isomerase
PDK: phosphoinositide-dependent kinase
PDMV: platelet-derived microvesicle
PDP: pyruvate dehydrogenase phosphatase
Pe: Péclet number
PE: phosphatidylethanolamine
PE: pulmonary embolism
PEBP: phosphatidylethanolamine-binding protein

PECAM: platelet–endothelial cell adhesion molecule
PEDF: pigment epithelium-derived factor (serpin-F1)
PEEP: positive end-expiratory pressure mechanical ventilation
PEn2: presenilin enhancer-2
PEO: proepicardial organ
Per: Period homolog
PERK: protein kinase-like endoplasmic reticulum kinase
PERP: P53 apoptosis effector related to peripheral myelin protein PMP22
PET: positron emission tomography
Pex: peroxin
PF: platelet factor
PFK: phosphofructokinase
pFRG: parafacial respiratory group
PG: prostaglandin
PGC: PPARγ coactivator
pGC: particulate guanylate cyclase
PGEA: prostaglandin ethanolamide
PGF: paracrine growth factor
PGG: prostaglandin glycerol ester
PGi2: prostacyclin
PGP: permeability glycoprotein
PGx: type-x (D, E, F, H, I) prostaglandin
PGxS: type-x prostaglandin synthase
PH: pleckstrin homology domain
PHD: prolyl hydroxylase
PhK: phosphorylase kinase
PHLPP: PH domain and Leu-rich repeat protein phosphatase
PHT: pressure half-time
PI: phosphoinositide (phosphorylated phosphatidylinositol)
PI(4)P: phosphatidylinositol 4-phosphate
PI(i)PiK: phosphatidylinositol i-phosphate i-kinase
PI(i,j)P_2: phosphatidylinositol (i,j)-bisphosphate (PIP_2)
PI(3,4,5)P_3: phosphatidylinositol (3,4,5)-trisphosphate (PIP_3)
PI3K: phosphatidylinositol 3-kinase
PI3KAP: PI3K adaptor protein
PIiK: phosphatidylinositol i-kinase
PIAS: protein inhibitor of activated STAT (SUMo ligase)
PIC: preinitiation complex
PICK: protein that interacts with C-kinase
PIDD: P53-induced protein with a death domain
PIKE: phosphoinositide 3-kinase enhancer (GTPase; ArfGAP)
PIKK: phosphatidylinositol 3-kinase-related kinase (pseudokinase)
PIM: provirus insertion of Molony murine leukemia virus gene product
PIN: peptidyl prolyl isomerase interacting with NIMA
PINCH: particularly interesting new Cys–His protein (or LIMS1)
PInK: PTen-induced kinase
PIP: phosphoinositide monophosphate
PIPiK: phosphatidylinositol phosphate i-kinase
PIP_2: phosphatidylinositol bisphosphate
PIP_3: phosphatidylinositol triphosphate
PIPP: proline-rich inositol polyphosphate 5-phosphatase
PIR: paired immunoglobulin-like receptor
piRNA: P-element-induced wimpy testis-interacting (PIWI) RNA
PIRT: phosphoinositide-interacting regulator of TRP channels
PISA: proximal isovelocity surface area
PITP: phosphatidylinositol-transfer protein
Pitx: pituitary (or paired-like) homeobox transcription factor
PIV: particle image velocimetry
PIX: P21-activated kinase (PAK)-interacting exchange factor (RhoGEF6/7)
PK: pharmacokinetics
PK: protein kinase
PKA: protein kinase-A
PKB: protein kinase-B
PKC: protein kinase-C
PKD: protein kinase-D
PKG: protein kinase-G
PKL: paxillin kinase linker
PKM: pyruvate kinase muscle isozyme
PKMYT (MYT): membrane-associated Tyr/Thr protein kinase
PKN: protein kinase novel
Pkp: plakophilin
PL: phospholipase
Pl: Planck constant ($6.62606957 \times 10^{-34}\ J \cdot s$)
PLA2: phospholipase-A2
PLC: phospholipase-C
PLD: phospholipase-D
PLd: phospholipid
PlGF: placental growth factor
PLK: Polo-like kinase
PLMAI: periodic leg movement arousal index
Pln: phospholamban
PLTP: phospholipid transfer protein

PLV: (coronary) posterolateral vein
Plxn: plexin
PM: prostamide (prostaglandin ethanolamide)
PM: particulate matter (particulates or particle pollution)
PMCA: plasma membrane Ca^{2+} ATPase
PML: promyelocytic leukemia protein
PMR: percutaneous (laser) myocardial revascularization
PMRT: protein arginine methyltransferase
Pn: plasmin
Png: plasminogen
pnM: perinuclear mitochondrion
PNS: peripheral nervous system
PoG: proteoglycan
PoM: pore membrane protein
Pon: paraoxonase
POP: persistent organic pollutant
POPx: partner of PIX
POSH: scaffold plenty of SH3 domains
POT: Protection of telomeres (single-stranded telomeric DNA-binding protein)
PP: protein phosphatase
PP3: protein phosphatase-3 (PP2b or calcineurin)
PPAR: peroxisome proliferator-activated receptor (NR1c1–3)
PPCM: peripartum and postpartum cardiomyopathies
PPG: photoplethysmography
PPHN: persistent pulmonary hypertension of the newborn
PPI: peptidylprolyl isomerase
PPIP: monopyrophosphorylated inositol phosphate
$(PP)_2IP$: bispyrophosphorylated inositol phosphate
PPK: PIP kinase
PPM: protein phosphatase (magnesium-dependent)
PPR: pathogen-recognition receptor
PPRE: PPAR response element (DNA sequence)
PPTC: protein phosphatase T-cell activation (TAPP2c)
PR: progesterone receptor (NR3c3)
Pr: Prandtl number
PRC: protein regulator of cytokinesis
PRC: Polycomb repressive complex
pre-cDC: preclassical dendritic cell
pre-miR: precursor microRNA
preBotC: pre-Bötzinger complex
preKk: prekallikrein
PREx: PIP_3-dependent Rac exchanger (RacGEF)
PRG: plasticity-related gene product
PRH: prolactin-releasing hormone
pri-miR: primary microRNA
PRL: phosphatase of regenerating liver
Prl: prolactin
PrlR: prolactin receptor
PRMT: protein arginine (R) ^{N}methyltransferase
Prompt: promoter upstream transcript
Protor: protein observed with Rictor
PROX: prospero homeobox gene
Prox: PROX gene product (transcription factor)
PrP: processing protein
PRPK: P53-related protein kinase
PRR: pattern recognition receptor
PRR: prorenin and renin receptor
PRx: peroxiredoxin
PS: presenilin
PS: protein-S
PSC: pluripotent stem cell
PSD: postsynaptic density adaptor
PsD: postsynaptic density
PSEF: pseudo-strain energy function
PSer: phosphatidylserine
PSGL: P-selectin glycoprotein ligand
PSKh: protein serine kinase-H
Psm: proteasome subunit
PSTPIP: Pro–Ser–Thr phosphatase-interacting protein
PTA: plasma thromboplastin antecedent
PTES: paclitaxel-eluting stent
Ptc: Patched receptor (Hedgehog signaling)
PTCA: percutaneous transluminal coronary angioplasty
PtcH: Patched Hedgehog receptor
PTCRA: PTC rotational burr atherectomy
PtdCho (PC): phosphatidylcholine
PtdEtn (PE): phosphatidylethanolamine
PtdSer (PS): phosphatidylserine
PtdIns (PI): phosphatidylinositol
PTen: phosphatase and tensin homolog deleted on chromosome ten (phosphatidylinositol 3-phosphatase)
PTFE: polytetrafluoroethylene
PTH: parathyroid hormone
PTHRP: parathyroid hormone-related protein
PTK: protein Tyr kinase
PTK7: pseudokinase (RTK)
(mt)PTP: (mitochondrial) permeability transition pore
PTP: protein Tyr phosphatase

PTPni: protein Tyr phosphatase nonreceptor type i
PTPR: protein Tyr phosphatase receptor
PTRF: RNA polymerase-1 and transcript release factor
PTT: pulse transit time
PUFA: polyunsaturated fatty acid
PUMA: P53-upregulated modulator of apoptosis
PuV: pulmonary valve
PVF: PDGF- and VEGF-related factor
PVNH: paraventricular nucleus of hypothalamus
PVR: pulmonary vascular resistance
PVR: pulmonary valve regurgitation
PVS: pulmonary valve stenosis
PWS: pulse wave speed
Px: pannexin
PXR: pregnane X receptor (NR1i2)
PYK: proline-rich tyrosine kinase
P2X: purinergic ligand-gated channel
P53AIP: P53-regulated apoptosis-inducing protein
$P75^{NTR}$: pan-neurotrophin receptor (TNFRSF16)

Q

$\mathtt{Q}$: material quantity
Q_e: electric current density
Q_T: thermal energy (heat)
q_T: transfer rate of thermal energy (power)
q_{met}: metabolic heat source
q: flow rate
QSOx: quiescin sulfhydryl oxidase
QTI: QT index ($QT/QT_p \times 100$; $QT_p = 656/(1 + f_C/100)$)

R

$\mathtt{R}$: resistance
$\mathcal{R}$: local reaction term
R_h: hydraulic radius
R_g: gas constant
R_R: respiratory quotient
R: recruitment function (from quiescence to proliferation)
r: cell renewal rate
$\mathtt{r}$: electrical resistivity
r: radial coordinate
RA: right atrium
RAAA: renin–angiotensin–aldosterone axis
Rab: Ras from brain
Rab11FIP: Rab11 family-interacting protein
Rac: Ras-related C3-botulinum toxin substrate
RACC: receptor-activated cation channel
RACK: receptor for activated C-kinase
RAD: recombination protein-A (RecA)-homolog DNA-repair protein
Rad: radiation sensitivity protein
Rag: Ras-related GTP-binding protein
RAI: respiratory arousal index
Ral: Ras-related protein
RAlBP: retinaldehyde-binding protein
RalGDS: Ral guanine nucleotide-dissociation stimulator
RAMP: (calcitonin receptor-like) receptor activity-modifying protein
Ran: Ras-related nuclear protein
RAP: receptor-associated protein
Rap: Ras-proximate (Ras-related) protein
Raptor: regulatory-associated protein of TOR
RAR: retinoic acid receptor (NR1b2/3)
Ras: rat sarcoma viral oncogene homolog (small GTPase)
RasA: Ras p21 protein activator
rasiRNA: repeat-associated small interfering RNA (PIWI)
RASR: rapidly adapting stretch receptor
RASSF: Ras interaction/interference protein RIN1, afadin, and Ras association domain-containing protein family member
RB: retinoblastoma protein
RBC: red blood capsule (cell, or erythrocyte [without nucleus])
RBP: retinoid-binding protein
RC: ryanodine-sensitive calcium channel (RyR)
RCA: right coronary artery
RCan: regulator of calcineurin
RCC: right coronary cusp
RCC: regulator of chromosome condensation
RCM: restrictive cardiomyopathy
RDI: respiratory disturbance index
REM: rapid eye movement
Re: Reynolds number
REDD: regulated in development and DNA-damage response gene product
Rel: reticuloendotheliosis proto-oncogene product (TF; member of NFκ B)
REM: rapid eye movement sleep
REP: Rab escort protein
ReR: renin receptor (PRR)

restin: Reed–Steinberg cell-expressed intermediate filament-associated protein (CLiP1)
RERA: respiratory-effort related arousal
ReT: rearranged during transfection (receptor Tyr kinase)
RevRE: reverse (Rev)-ErbA (NR1d1/2) response element (DNA sequence)
RF: regurgitant fraction
RFl: regurgitant flow
RFA: radiofrequency ablation
RGL: Ral guanine nucleotide-dissociation stimulator-like protein (GEF)
RGS: regulator of G-protein signaling
RHEB: Ras homolog enriched in brain
RHI: reactive hyperemia index
RHS: equation right-hand side
Rho: Ras homologous
RI: arterial resistivity (resistance) index
RIAM: $Rap1^{GTP}$-interacting adaptor molecule
RIBP: RLK- and ITK-binding protein
RICH: RhoGAP interacting with CIP4 homolog
RICK: receptor for inactive C-kinase
Rictor: rapamycin-insensitive companion of TOR
RIF: Rho in filopodium
RIN: Ras-like protein expressed in neurons (GTPase)
RIn: Ras and Rab interactor (RabGEF)
RIP: regulated intramembrane proteolysis
RIPK: receptor-interacting protein kinase
RIRR: ROS-induced ROS release
RISC: RNA-induced silencing complex
RIT: Ras-like protein expressed in many tissues
RKIP: Raf kinase inhibitor protein
RlBP: retinaldehyde-binding protein
RLC: RISC-loading complex
RLK: resting lymphocyte kinase (TXK)
RNA: ribonucleic acid
RNABP: RNA-binding protein
RNase: ribonuclease
RnBP: renin-binding protein
RNF2: RING finger protein-2 (Ub ligase)
RNP: ribonucleoprotein
RNS: reactive nitrogen species
Robo: roundabout
ROC: receptor-operated channel
RoCK: Rho-associated, coiled-coil-containing protein kinase
ROI: region of interest
ROMK: renal outer medullary potassium channel
RONS: reactive oxygen and nitrogen species
ROR: RAR-related orphan receptor (NR1f1–NR1f3)
$ROR_{(RTK)}$: receptor Tyr kinase-like orphan receptor
ROS: reactive oxygen species
Ros: ros UR2 sarcoma virus proto-oncogene product (RTK)
RPIP: Rap2-interacting protein
RPS6: ribosomal protein S6
RPTP: receptor protein Tyr phosphatase
rRas: related Ras
rRNA: ribosomal RNA
RSA: respiratory sinus arrhythmia
RSE: rapid systolic ejection
RSK: P90 ribosomal S6 kinase (^{P90}RSK)
RSKL: ribosomal protein S6 kinase-like (pseudokinase)
rSMAD: receptor-regulated SMAD (SMAD1–SMAD3, SMAD5, and SMAD9)
RSMCS: robot-supported medical and surgical system
RSpo: R-spondin
RSTK: receptor Ser/Thr kinase
RTI: respiratory tract infection
RTK: receptor protein Tyr kinase
RTN: retrotrapezoid nucleus
Rubicon: RUN domain and Cys-rich domain-containing, beclin-1-interacting protein
Runx: Runt-related transcription factor
RV: right ventricle
RVF: rapid ventricular filling
RVLM: rostral ventrolateral medulla
RVMM: rostral ventromedial medulla
RXR: retinoid X receptor (NR2b1–NR2b3)
RYK: receptor-like (or related to receptor) protein Tyr (Y) kinase (pseudokinase)
RyR: ryanodine receptor (ryanodine-sensitive Ca^{2+}-release channel)

S

$\mathbf{S}$: Cauchy–Green deformation tensor
S: hemoglobin saturation of a given gas species (%)
s: arclength
s: entropy
s: sarcomere length
s: sieving coefficient
$s_{\perp}$: normal strain
$s_{\parallel}$: shear strain
s: evolution speed
$\mathtt{s}$: solubility

SAa: serum amyloid-A
$SAC_{Cl(K)}$: stretch-activated Cl^- (K^+)-selective channel
SAc: suppressor of actin domain-containing 5-phosphatase
sAC: soluble adenylate cyclase
$SACC_{NS}$: stretch-activated cation nonselective channel
SACM1L: suppressor of actin mutation-1-like
SAH: subarachnoid hemorrhage
SAI: spontaneous arousal index
SAIC: stretch-activated ion channel
SAN: sinoatrial node
SAP: SLAM-associated protein
SAP: stress-activated protein
SAPi: synapse-associated protein i
SAPK: stress-activated protein kinase (MAPK)
SAR: secretion-associated and Ras-related protein
SARAF: SOCE-associated regulatory factor
SASR: slowly adapting stretch receptor
SBE: SMAD-binding element
SBF: SET-binding factor
Sc: Schmidt number
SCA: stem cell antigen
SCAD: short-chain acylCoA dehydrogenase
SCADD: SCAD deficiency
SCAMP: secretory carrier membrane protein
SCAP: SREBP cleavage-activating protein (SREBP escort)
SCAR: suppressor of cAMP receptor (WAVe)
ScaR: scavenger receptor
SCF: SKP1–Cul1–F-box Ub-ligase complex
SCF: stem cell factor
scFA: short-chain fatty acid (4–8 carbon atoms)
SCFR: stem cell factor receptor (KIT)
Scgb: secretoglobin
SCLC: small-cell lung cancer
scLC: squamous-cell lung cancer (NSCLC subtype)
SCN: suprachiasmatic nucleus
SCO: synthesis of cytochrome-C oxidase
SCoAS: succinylCoA synthase
SCOT: succinylCoA:3-oxoacidCoA transferase
SCP (CTDSP): small C-terminal domain (CTD)-containing phosphatase
Scp: stresscopin (urocortin-3)
Scrib: Scribble polarity protein
SDLGMD: sarcoglycan-deficient limb-girdle muscular dystrophy
SDH: succinate dehydrogenase
Sdc: syndecan
SDF: stromal cell-derived factor
SDPR: serum deprivation protein response
SE: systolic ejection
SEF: strain-energy function
SEF: similar expression to FGF genes (inhibitor of RTK signaling)
SEK: SAPK/ERK kinase
Sema: semaphorin (Sema, Ig, transmembrane, and short cytoplasmic domain)
SERCA: sarco(endo)plasmic reticulum calcium ATPase
serpin: serine peptidase inhibitor
SerT: serotonin transporter
SES: sirolimus-eluting stent
SEVR: subendocardial viability ratio
SF: steroidogenic factor (NR5a1)
SFK: SRC family kinase
SFO: subfornical organ
SFPQ: splicing factor proline and glutamine-rich
sFRP: secreted Frizzled-related protein
SftP (SP): surfactant protein
sGC: soluble guanylate cyclase
SGK: serum- and glucocorticoid-regulated kinase
SGlT: Na^+–glucose cotransporter (SLC5a)
Sgo: shugoshin (Japanese: guardian spirit)
SH: Src homology domain
Sh: Sherwood number
SH3P: Src homology-3 domain-containing adaptor protein
SHAnk: SH3 and multiple ankyrin repeat domain-containing protein
SHAX: SNF7 (VSP32) homolog associated with ALIX
SHB: Src homology-2 domain-containing adaptor
SHC: Src-homologous and collagen-like substrate
SHC: Src homology-2 domain-containing transforming protein
SHh: sonic Hedgehog
SHIP: SH-containing inositol phosphatase
SHP: SH-containing protein Tyr phosphatase (PTPn6/11)
SHP: small heterodimer partner (NR0b2)
shRNA: small (short) hairpin RNA
SIAH: Seven in absentia homolog (Ub ligase)
siglec: sialic acid-binding Ig-like lectin
SIK: salt-inducible kinase
SIn: stress-activated protein kinase-interacting protein
SIP: steroid receptor coactivator-interacting protein

siRNA: small interfering RNA
SiRP: signal-regulatory protein
SIRT: sirtuin (silent information regulator-2 [two]; histone deacetylase)
SIT: SHP2-interacting transmembrane adaptor
SK: small conductance Ca^{2+}-activated K^+ channel
SK*i*: sphingosine kinase-*i*
SKIP: sphingosine kinase-1-interacting protein
SKIP: skeletal muscle and kidney-enriched inositol phosphatase
SKP: S-phase kinase-associated protein
SLA: Src-like adaptor
SLAM: signaling lymphocytic activation molecule
SLAMF: SLAM family member
SLAP: Src-like adaptor protein
SLC: solute carrier superclass member
SLCO: solute carrier organic anion class transporter
SLK: Ste20-like kinase
Sln: sarcolipin
SLO: secondary lymphoid organ
SLPI: secretory leukocyte peptidase inhibitor
SLTC: small latent TGFβ complex
SM: sphingomyelin
SMA: smooth muscle actin
SMAD: small (son of, similar to) mothers against decapentaplegia homolog
SMAP: Small ArfGAP protein, stromal membrane-associated GTPase-activating protein
SMase: sphingomyelinase
SMC: smooth myocyte
Smo: Smoothened
SMPD: sphingomyelin phosphodiesterase
SMRT: silencing mediator of retinoic acid and thyroid hormone receptor (NCoR2)
SMS: sphingomyelin synthase
SMURF: SMAD ubiquitination regulatory factor
SNAAT: sodium-coupled neutral amino acid transporter
SNAP: soluble ^{N}ethylmaleimide-sensitive factor-attachment protein
SnAP: synaptosomal-associated protein
SNARE: SNAP receptor
SND: sinusal node dysfunction
SNF7: sucrose nonfermenting (VPS32)
SNIP: SMAD nuclear-interacting protein
snoRNA: small nucleolar RNA
snoRNP: small nucleolar ribonucleoprotein
SNP: single-nucleotide polymorphism
snRNA: small nuclear RNA
snRNP: small nuclear ribonucleoprotein
SNx: sorting nexin
SO_x: sulfur oxides
SOC: store-operated Ca^{2+} channel
SOCE: store-operated Ca^{2+} entry
SOCS: suppressor of cytokine signaling protein
SOD: superoxide dismutase
SorbS: sorbin and SH3 domain-containing adaptor
SOS: Son of sevenless (GEF)
Sost: sclerostin
SostDC: sclerostin domain-containing protein
SOX: sex-determining region Y (SRY)-box gene
Sox: SOX gene product (transcription factor)
SP1: specificity protein (transcription factor)
SPARC: secreted protein acidic and rich in cysteine
SPC: sphingosylphosphorylcholine
SPCA: secretory pathway Ca^{2+} ATPase
SPECT: single-photon emission CT
Sph: sphingosine
SphK: sphingosine kinase
SPI: spleen focus-forming virus (SFFV) proviral integration proto-oncogene product (transcription factor)
SPInt: serine peptidase inhibitor
SPN: supernormal period
SPP: sphingosine phosphate phosphatase
SpRED: Sprouty-related protein with an EVH1 domain
SPS: sleep pressure score
SPTI: systolic pressure time interval
SPURT: secretory protein in upper respiratory tract
SQTS: short-QT syndrome
SR: sarcoplasmic reticulum
SR: Arg/Ser domain-containing protein (alternative splicing)
SRA: steroid receptor RNA activator
SRC: steroid receptor coactivator
Src: `sarc`oma-associated (Schmidt–Ruppin A2 viral oncogene homolog) kinase
SREBP: sterol regulatory element-binding protein
SRF: serum response factor
SRM/SMRS: Src-related kinase lacking regulatory and myristylation sites
SRP: stresscopin-related peptide (urocortin-2)
SRPK: splicing factor RS domain-containing protein kinase
SRY: sex-determining region Y

SSAC: shear stress-activated channel
SSE: slow systolic ejection
SSEA: stage-specific embryonic antigen
Ssh: slingshot homolog phosphatase
SSI: STAT-induced STAT inhibitor
ssM: subsarcolemmal mitochondrion
SSR: sympathetic skin response
ssRNA: single-stranded RNA
SSS: sick sinus syndrome
Sst: somatostatin
SSV: short saphenous vein
St: Strouhal number
STAM: signal-transducing adaptor molecule
STAMBP: STAM-binding protein (Ub isopeptidase)
StAR: steroidogenic acute regulatory protein
StART: StAR-related lipid transfer protein
STAT: signal transducer and activator of transduction
STEAP: six transmembrane epithelial antigen of the prostate
STEMI: ST-segment elevation myocardial infarction
STICK: substrate that interacts with C-kinase
StIM: stromal interaction molecule
STK: protein Ser/Thr kinase
STK1: stem cell protein Tyr kinase receptor
STLK: Ser/Thr kinase-like (pseudo)kinase
Sto: Stokes number
StRAd: Ste20-related adaptor
STRAP: Ser/Thr kinase receptor-associated protein
StRAP: stress-responsive activator of P300
Stx: syntaxin ($SNARE^{Q}$)
SUMo: small ubiquitin-related modifier
SUn: Sad1 and Unc84 homology protein
SUR: sulfonylurea receptor
SUT: stable unannotated transcript
SV: stroke volume
SVC: superior vena cava
SVCT: sodium-dependent vitamin-C transporter
SVF: slow ventricular filling
SVP: synaptic vesicle precursor
SVR: systemic vascular resistance
SVR: surface area-to-volume ratio
SW: stroke work
SwAP70: 70-kDa switch-associated protein (RacGEF)
Swi/SNF: switch/sucrose nonfermentable complex
SYK: spleen tyrosine kinase
Synj: synaptojanin
Syp: synaptophysin
Syt: synaptotagmin
S1P: sphingosine 1-phosphate
S6K: P70 ribosomal S6 kinase (^{P70}RSK)

T

T: extrastress tensor
T: transition rate from a cell cycle phase to the next
T: temperature
$\mathcal{T}$: transport parameter
$\mathcal{T}_L$: transfer capacity of the alveolocapillary membrane for gas species
T_3: triiodothyronine
T_4: thyroxin
T_s: surface tension
T lymphocyte (T cell): thymic lymphocyte
T_C: cytotoxic T lymphocyte (CD8+ effector T cell; CTL)
T_{C1}: type-1 cytotoxic T lymphocyte
T_{C2}: type-2 cytotoxic T lymphocyte
T_{CM}: central memory T lymphocyte
T_{Conv}: conventional T lymphocyte
T_{Eff}: effector T lymphocyte
T_{EM}: effector memory T lymphocyte
T_{FH}: follicular helper T lymphocyte
T_H: helper T lymphocyte (CD4+ effector T cell)
T_{Hi}: type-i helper T lymphocyte ($i = 1/2/9/17/22$)
T_{H3}: TGFβ-secreting T_{Reg} lymphocyte
T_L: lung transfer capacity (alveolocapillary membrane)
T_{R1}: type-1, IL10-secreting, regulatory T lymphocyte
T_{Reg}: regulatory T lymphocyte
aT_{Reg}: CD45RA−, FoxP3^{hi}, activated T_{Reg} cell
iT_{Reg}: inducible T_{Reg} lymphocyte
nT_{Reg}: naturally occurring (natural) T_{Reg} lymphocyte
rT_{Reg}: CD45RA+, FoxP3^{low}, resting T_{Reg} cell
T6SS: bacterial type-6 secretion system
$\hat{\mathbf{t}}$: unit tangent vector
t: time
Tβ Ri: type-i TGFβ receptor
TAA: thoracic aortic aneurysm
TAB: TAK1-binding protein
TACE: transarterial chemoembolization
TAF: TBP-associated factor
TAI: total arousal index
TAK: TGFβ-activated kinase (MAP3K7)
TALK: TWIK-related alkaline pH-activated K^+ channel

TANK: TRAF family member-associated NFκ B activator
TASK: TWIK-related acid-sensitive K^+ channel
TASR: terminus-associated short RNA
TAP: transporter associated with antigen processing (ABC transporter)
TATN: transverse and axial tubular network
Taz: taffazin
TBC1D: Tre2 (or USP6), BUB2, CDC16 domain-containg RabGAP
TBCK: tubulin-binding cofactor kinase (pseudokinase)
TBK: TANK-binding kinase
TBP: TATA box-binding protein (subclass-4F transcription factor)
TBx: T-box transcription factor
TC: thrombocyte (platelet)
TCAC: tricarboxylic acid cycle
TCC: tricarboxylate carrier
TCF: T-cell factor
TCF: ternary complex factor
TCFA: thin-cap fibroatheroma
TcFi: type-i transcription factor
TCP: T-complex protein
TCR: T-cell receptor
TEA: transluminal extraction atherectomy
TEC: Tyr kinase expressed in hepatocellular carcinoma
TEF: thyrotroph embryonic factor (PAR/b–ZIP family)
TEK: Tyr endothelial kinase
TEM: transendothelial migration
Ten: tenascin
TF: transcription factor
Tf: transferrin
TFM: traction force microscopy
TFPD: trifunctional protein deficiency
TFPI: tissue factor pathway inhibitor
TfR: transferrin receptor
TG (TAG): triglyceride (triacylglycerol)
TGm: transglutaminase
TGF: transforming growth factor
TGFBR: TGFβ receptor gene
TGFβ RAP: TGFβ receptor-associated protein
TGN: trans-Golgi network
THET: trihydroxyeicosatrienoic acid
THIK: tandem pore-domain halothane-inhibited K^+ channel
THR: thyroid hormone receptor (NR1a1/2)
TIAM: T-lymphoma invasion and metastasis-inducing protein (RacGEF)
TICE: transintestinal cholesterol efflux
TIE: Tyr kinase with Ig and EGF homology domains (angiopoietin receptor)
TIEG: TGFβ-inducible early gene product
TIF: transcription intermediary factor (kinase and Ub. ligase)
TIGAR: TP53-inducible glycolysis and apoptosis regulator
TIM: T-cell immunoglobulin and mucin domain-containing protein
Tim: timeless homolog
TIMM: translocase of inner mitochondrial membrane
TIMP: tissue inhibitor of metallopeptidase
TIRAP: Toll–IL1R domain-containing adaptor protein
tiRNA: transcription initiation RNA
TJ: tight junction
TKR: Tyr kinase receptor
TLC: total lung capacity
TLO: tertiary lymphoid organ
TLR: Toll-like receptor
TLT: TREM-like transcript
TLX: tailless receptor (NR2e1)
TM: thrombomodulin
TMi: transmembrane segment-i of membrane protein
TMC: twisting magnetocytometry
TMePAI: transmembrane prostate androgen-induced protein
TMy: tropomyosin
Tnn (TN): troponin
Tn: thrombin
TNF: tumor-necrosis factor
TNFα IP: tumor-necrosis factor-α-induced protein
TNFR: tumor-necrosis factor receptor
TNFRSF: tumor-necrosis factor receptor superfamily member
TNFSF: tumor-necrosis factor superfamily member
TNK: Tyr kinase inhitor of NFκ B
Tns: tensin
TOR: target of rapamycin
TORC: target of rapamycin complex
TORC: transducer of regulated CREB activity (a.k.a. CRTC)
TP: thromboxane-A2 Gq/11-coupled receptor
TP53I: tumor protein P53-inducible protein
tPA: tissue plasminogen activator
TPo: thrombopoietin
TpM: tropomyosin
TPPP: tubulin polymerization-promoting protein

TPST: tyrosylprotein sulftotransferase
TR: testicular receptor (NR2c1/2)
TR: time to wave reflection
TRAAK: TWIK-related arachidonic acid-stimulated K^+ channel
TRADD: TNFR1-associated death domain adaptor
TRAF: TNFR-associated factor
TRAM: TRIF-related adaptor molecule
transceptor: transporter-related receptor
TRAP: TNFR-associated protein (HSP75)
TraPP: transport protein particle
TRAT: T-cell receptor-associated transmembrane adaptor
Trb: Tribbles homolog (pseudokinase)
TRE: TPA-response element (AP1/CREB-binding site on promoters)
TRE: trapped in endoderm
TREK: TWIK-related K^+ channel
TREM: triggering receptor expressed on myeloid cells
TRESK: TWIK-related spinal cord K^+ channel
TRF: TBP-related factor
TRF: double-stranded telomeric DNA-binding repeat-binding factor
TRH: thyrotropin-releasing hormone
TRIF: Toll–IL1R domain-containing adaptor inducing Ifnβ
TRIM: T-cell receptor-interacting molecule
TRIP: TGFβ receptor-interacting protein (eIF3S2)
TRK: tropomyosin receptor kinase (NTRK)
tRNA: transfer RNA
TRP: transient receptor potential channel
TRPA: ankyrin-like transient receptor potential channel
TRPC: canonical transient receptor potential channel
TRPM: melastatin-related transient receptor potential channel
TRPML: mucolipin-related transient receptor potential channel
TRPN: no mechanoreceptor potential C
TRPP: polycystin-related transient receptor potential channel
TRPV: vanilloid transient receptor potential channel
TrrAP: transactivation (transformation)/transcription domain-associated protein (pseudokinase)
TRT: total recording time
TrV: tricuspid valve
TRx: thioredoxin
TRxIP: thioredoxin-interacting protein
TSC: tuberous sclerosis complex
TSH: thyroid-stimulating hormone
TSLP: thymic stromal lymphopoietin
Tsp: thrombospondin
Tspan: tetraspanin
TsPO: translocator protein of the outer mitochondrial membrane
tSNARE: target SNARE
tsRNA: tRNA-derived small RNA
tssaRNA: transcription start site-associated RNA
TST: total sleep time
TTbK: Tau-tubulin kinase
TTG: tissue transglutaminase
TTK: dual-specificity Thr/Tyr kinase
Ttn: titin (pseudokinase)
TTS: takotsubo syndrome
TUT: terminal uridine transferase
TVR: tricuspide valve regurgitation
TVS: tricuspide valve stenosis
TWIK: tandem of P domains in a weak inwardly rectifying K^+ channel
TxA_2: thromboxane-A_2 (thromboxane)
TxB_2: thromboxane-B_2 (thromboxane metabolite)
TXK: Tyr kinase mutated in X-linked agammaglobulinemia
TxaS: thromboxane-A synthase
TyK: tyrosine kinase
T_3: tri-iodothyronine
T_4: thyroxine
$^+$TP: plus-end-tracking proteins

U

U: right stretch tensor
u: displacement vector
u: electrochemical command, electrical potential
u: specific internal energy
UARE: upper airway resistance episode
UAREI: UARE index
Ub: ubiquitin
UbC: ubiquitin conjugase
UbE2: E2 ubiquitin conjugase
UbE3: E3 ubiquitin ligase
UbL: ubiquitin-like protein
UCH: ubiquitin C-terminal hydrolase (DUb)
Ucn: urocortin
UCP: uncoupling protein
UDP: uridine diphosphate
^{UDP}glucose: UDP–glucose
UK: urokinase

ULK: uncoordinated-51-like kinase (pseudokinase)
Unc: uncoordinated receptor
uPA: urokinase-type plasminogen activator (urokinase)
uPAR: uPA receptor
uPARAP: uPAR-associated protein (CLec13e)
UPR: unfolded protein response
UPS: ubiquitin–proteasome system
UP4A: uridine adenosine tetraphosphate
URI: upper respiratory infection
Uro: urodilatin
US: ultrasound
USC: unipotential stem cell
USF: upstream stimulatory factor
USI: ultrasound imaging
USP: ubiquitin-specific peptidase (deubiquitinase)
UTP: uridine triphosphate
UTR: untranslated region
UVRAG: ultraviolet wave resistance-associated gene product

V

V: left stretch tensor
V: volume
$\mathcal{V}$: porosity (void fraction)
V_c: pulmonary capillary blood volume in alveolar walls
V_q: cross-sectional average velocity
V_s: specific volume
v: fluid velocity vector
v: recovery variable
v_X: volume of gas component X
V1(2)R: type-1(2) vomeronasal receptor
$V_{1A/1B/2}$: type-1A/1B/2 arginine vasopressin receptor
VAAC: volume-activated anion channel
$VAC_{Cl(K)}$: volume-activated Cl^- (K^+)-selective channel
VACamKL: vesicle-associated CamK-like (pseudokinase)
$VACC_{NS}$: volume-activated cation nonselective channel
VAChT: vesicular acetylcholine transporter
VAIC: volume-activated ion channel
VAMP: vesicle-associated membrane protein (synaptobrevin)
VanGL: Van Gogh (Strabismus)-like protein
VAP: VAMP-associated protein
VASP: vasoactive stimulatory phosphoprotein
VAI: vegetative arousal index
VAT: vesicular amine transporter
vATPasc: vesicular-type H^+ ATPase
VAV: ventriculoarterial valve
Vav: GEF named from Hebrew sixth letter
VC: vital capacity
VCAM: vascular cell adhesion molecule
VCt: vasoconstriction
VDAC: voltage-dependent anion channel (porin)
VDACL: plasmalemmal, volume- and voltage-dependent, ATP-conductive, large-conductance, anion channel
VDCC: voltage-dependent calcium channel
VDP: vesicle docking protein
VDt: vasodilation
VEGF: vascular endothelial growth factor
VEGFR: vascular endothelial growth factor receptor
VF: ventricular fibrillation
VF: ventricular filling
VGAT: vesicular GABA transporter
VGC: voltage-gated channel
VgL: Vestigial-like protein
VGluT: vesicular glutamate transporter
VHD: valvular heart disease
VHL: von Hippel–Lindau Ub ligase
VIP: vasoactive intestinal peptide
VLCAD: very-long-chain acylCoA dehydrogenase
VLCADD: VLCAD deficiency
vlcFA: very-long-chain fatty acids (17–26 carbon atoms)
VLDL: very-low-density lipoprotein
VLDLR: very-low-density lipoprotein receptor
VMAT: vesicular monoamine transporter
VN: vitronectin
VOC: volatile organic compound
VOM: vein of Marshall
VPO: vascular peroxidase
VPS: vacuolar protein sorting-associated kinase
VR: venous return
VRAC: volume-regulated anion channel
VRC: ventral respiratory column
VRK: vaccinia-related kinase
VS: vasostatin
vSMC: vascular smooth myocyte
vSNARE: vesicular SNAP receptor (SNARE)
VSOR: volume-sensitive outwardly rectifying anion channel
VSP: voltage-sensing phosphatase
VVO: vesiculovacuolar organelle
vWF: von Willebrand factor

W

W: vorticity tensor
$\mathcal{W}$: strain energy density
W: work, deformation energy
w: weight
w: computational grid velocity
WASH: WASP and SCAR homolog
WASP: Wiskott–Aldrich syndrome protein
WAT: white adipose tissue
WAVe: WASP-family verprolin homolog
WBC: white blood cell
WDR: WD repeat-containing protein
Wee: small (Scottish)
WHAMM: WASP homolog associated with actin, membranes, and microtubules
WIP: WASP-interacting protein
WIPF: WASP-interacting protein family protein
WIPI: WD repeat domain-containing phosphoinositide-interacting protein
WNK: with no K (Lys) kinase (Lys-deficient kinase)
Wnt: wingless-type
WPWS: Wolff–Parkinson–White syndrome
WNRRTK: Wnt and neurotrophin receptor-related receptor Tyr kinase ($ROR_{(RTK)}$)
WSB: WD-repeat and SOCS box-containing protein (Ub ligase)
WSS: wall shear stress
WSSTG: WSS transverse gradient
WWTR: WW domain-containing transcription regulator

X

$\mathcal{X}$: trajectory
X: reactance
X: Lagrangian position vector
x: position vector
$\{x, y, z\}$: Cartesian coordinates
XBE: X-factor-binding element
XBP: X-box-binding protein (transcription factor)
XIAP: X-linked inhibitor of apoptosis protein (Ub ligase)
XOx: xanthine oxidase

Y

Y: admittance coefficient
YAP: Yes-associated protein
YBP: Y-box-binding protein (transcription factor)
YY: yin yang (transcriptional repressor)

Z

Z: impedance
ZAP70: 70-kDa TCRζ chain-associated protein
ZBTB: zinc finger and BTB (Broad complex, Tramtrack, and bric-à-brac) domain-containing transcription factor
ZnF: zinc finger protein
ZO: zonula occludens

Miscellaneous

1D: one-dimensional
2D: two-dimensional
2-5A: 5′-triphosphorylated, (2′,5′)-phosphodiester-linked oligoadenylate
2AG: 2-arachidonyl glycerol
3D: three-dimensional
3DR: three-dimensional reconstruction
3BP2: Abl Src homology-3 domain-binding adaptor
4eBP1: inhibitory eIF4e-binding protein
5HT: 5-hydroxytryptamine (serotonin)
7TMR: 7-transmembrane receptor (GPCR)

Complementary Lists of Notations

Greek Symbols

α: volumic fraction
α: convergence/divergence angle
α: attenuation coefficient
α_k: kinetic energy coefficient
α_m: momentum coefficient
β: inclination angle
β_g: gas g solubility
$\{\beta_i\}_1^2$: myocyte parameters
β_T: coefficient of thermal expansion
Γ: domain boundary
Γ_L: local reflection coefficient
Γ_G: global reflection coefficient
γ: (specific) heat capacity ratio (adiabatic index)
γ: activation factor
γ_G: amplitude ratio (modulation rate) of G
$\dot{\gamma}$: shear rate
δ: boundary layer thickness
ϵ_T: emissivity (thermal energy radiation)
ϵ_e: electric permittivity
ϵ: strain
ε: dimensionless small quantity
ζ: singular head loss coefficient
ζ: transmural coordinate
$\{\zeta_j\}_1^3$: local coordinate
η: azimuthal spheroidal coordinate
θ: circumferential polar coordinate
θ: $(\hat{\vec{e}}_x, \hat{\vec{t}})$ angle
κ: wall curvature
κ_c: curvature ratio
κ_d: drag reflection coefficient
κ_f: frictional sieving coefficient
κ_h: hindrance coefficient
κ_o: osmotic reflection coefficient
κ_r: reflection coefficient
κ_s: size ratio
$\{\kappa_k\}_{k=1}^9$: tube law coefficients
κ_e: correction factor
Λ: head loss coefficient
λ_L: Lamé coefficient
λ: stretch ratio
λ: wavelength
λ_A: area ratio
λ_a: acceleration ratio
λ_L: length ratio
λ_{Ld}: length-to-diameter ratio
λ_p: molecule radius-to-pore radius ratio
λ_q: flow rate ratio
λ_t: time ratio
λ_v: velocity ratio
μ: dynamic viscosity
μ_L: Lamé coefficient
ν: kinematic viscosity
ν_P: Poisson ratio
Π: osmotic pressure
ρ: mass density
τ: time constant
τ: space curve torsion
Φ: potential
$\phi(t)$: creep function
φ: phase
$\vec{\chi}$: Lagrangian label
χ_i: molar fraction of species i
χ_i: wetted perimeter
$\psi(t)$: relaxation function
Ψ: porosity
Ω: computational domain
ω: angular frequency

Dual Notations

Bφ: basophil
Eφ: eosinophil

M. Thiriet, *Diseases of the Cardiac Pump*, Biomathematical and Biomechanical Modeling of the Circulatory and Ventilatory Systems 7, DOI 10.1007/978-3-319-12664-7

Lφ: lymphocyte
Mφ: macrophage
aaMφ: alternatively activated macrophage
caMφ: classically activated macrophage
Nφ: neutrophil
Σc: sympathetic
pΣc: parasympathetic

Latin Subscripts

$_A$: alveolar, atrial
$_{\overline{A}}$: mixed alveolar
$_{ACM}$: alveolocapillary membrane
$_{Ao}$: aortic
$_a$: arterial
$_{\overline{a}}$: mixed arterial
$_{ac}$: acid
$_{ao}$: airway opening (mouth or nose)
$_{app}$: apparent
$_{atm}$: atmospheric
$_{aw}$: airway
$_{ax}$: axial
$_b$: bound form of a molecule
$_b$: blood
$_{\mathbf{c}}$: contractile
$_c$: center
$_c$: point-contact
$_{cap}$: capillary
$_{cf}$: circumferential
$_{cl}$: closed
$_{coat}$: stent polymeric coating
$_{co}$: core (flow)
$_{cond}$: conduction (velocity)
$_{CW}$: chest wall
$_{cy}$: cytosolic
$_D$: Darcy (filtration)
$_D$: dead space (airway)
$_{\mathtt{dias}}$: diastolic
$_{down}$: downstream, distal
$_{dyn}$: dynamic
$_{eff}$: effective
$_{\mathtt{ed}}$: end diastolic
$_{\mathtt{es}}$: end systolic
$_E$: expiration, Eulerian
$_e$: external
$_{ECF}$: extracellular fluid
$_{\mathtt{e}}$: extremum
$_{\mathtt{syst}}$: systolic
$_{\mathtt{se}}$: systolic ejection
$_f$: free form of a molecule
$_f$: fluid
$_{fast}$: fast (inward current)
$_g$: grid
$_H$: heart
$_h$: heat
$_{he}$: hyperemic
$_{hea}$: healthy state
$_I$: inspiration
$_i$: internal
$_{ib}$: intrabronchial
$_{in}$: (ionic) influx
$_{inc}$: incremental
$_{int}$: interstitial
$_{ion}$: sum of transmembrane ionic currents
$_L$: Lagrangian
$_L$: lung
$_l$: limit
$_\ell$: line-contact
$_M$: macroscopic
$_m$: mass (e.g., q_m mass flow rate)
$_m$: mean
$_m$: muscle, mouth
$_{max}$: maximum
$_{mb}$: membrane
$_{\mathtt{md}}$: mesodiastolic
$_{mea}$: measured
$_{met}$: metabolic
$_{min}$: minimum
$_{\mathtt{ms}}$: mesosystolic
$_{\mathtt{mt}}$: mitochondrial
$_{musc}$: muscular
$_{op}$: open
$_{out}$: (ionic) outflux
$_P$: pulmonary
$_{\mathbf{p}}$: parallel
$_p$: particle
$_{por}$: pore
$_{\mathtt{pd}}$: protodiastolic
$_{\mathtt{ps}}$: protosystolic
$_{pa}$: pulmonary arterial
$_{pl}$: plasma, pleural
$_{pv}$: pulmonary veinous
$_q$: quasi-ovalization
$_r$: radial
$_{ref}$: reference
$_{refr}$: refractory (time)
$_{regur}$: regurgitant
$_{rel}$: relative
$_{rest}$: value at rest
$_S$: systemic
$_{\mathtt{s}}$: solute
$_{\mathbf{s}}$: serial
$_{sa}$: systemic arterial
$_{si}$: sink
$_{slow}$: slow (inward current)
$_{so}$: source

$_{\rm sten}$: stenosis
$_{\rm stim}$: external stimulus
$_{\rm sv}$: systemic veinous
$_{\tt syst}$: systolic
$_{\rm T}$: tidal (breathing)
$_{\rm t}$: turbulence
$_{\rm t}$: stream division
$_{t}$: time derivative of order 1
$_{tt}$: time derivative of order 2
$_{\tt td}$: telediastolic
$_{\rm tis}$: tissue
$_{\rm tors}$: torsional
$_{\rm tot}$: total
$_{\rm tf}$: related to mass transfer
$_{\tt ts}$: telesystolic
$_{\rm ung}$: ungated
$_{\rm up}$: upstream, proximal
$_{\rm V}$: ventricular
$_{\rm V_V}$: valvular
$_{\rm v}$: systemic venous blood
$_{\bar{\rm v}}$: pulmonary (mixed) venous blood
$_{\rm vregur}$: regurgitant valve
$_{\rm vcomp}$: competent valve
$_{\rm w}$: wall
$_{\rm W}$: water (solvent)

Greek Subscripts

$_{\Gamma}$: boundary
$_{\theta}$: azimuthal
$_{\mu}$: microscopic

Miscellaneous Subscripts

$_{+}$: positive command
$_{-}$: negative command
$_{*}$: at interface
$_{0}$: reference state ($\cdot_0$: unstressed or low shear rate)
$_{\infty}$: high shear rate

Latin Superscripts

$^{\rm A}$: belonging to astrocyte
$^{\rm a}$: active state
$^{\rm e}$: elastic
$^{\rm f}$: fluid
$^{\rm h}$: hypertensive
$^{\rm n}$: normotensive
$^{\rm p}$: passive state
$^{\tt P}$: power
$^{\rm s}$: solid
$^{\rm SMC}$: belonging to smooth myocyte
$^{\rm T}$: transpose
$^{\rm v}$: viscoelastic

Miscellaneous Superscripts

*: scale
*: complex variable ($z^* = \Im m\, z + \imath \Re e\, z$)
$\cdot'$: first component of complex elastic and shear moduli
$\cdot''$: second component of complex elastic and shear moduli
$^{\natural}$: static, stationary, steady variable

Mathematical Notations

$\vec{T}$: boldface capital letter means tensor
$\vec{v}$: boldface minuscule letter means vector
S, s: upper- or lowercase, lightface (italic typeface) letter means scalar
$\Delta\bullet$: difference
$\delta\bullet$: increment
$d\bullet/dt$: time gradient
∂_t: first-order time partial derivative
∂_{tt}: second-order time partial derivative
∂_i: first-order space partial derivative with respect to spatial coordinate x_i
∇: gradient operator
$\vec{\nabla} u$: displacement gradient tensor
$\vec{\nabla} v$: velocity gradient tensor
$\nabla\cdot$: divergence operator
∇^2: Laplace operator
$|\ |_+$: positive part
$|\ |_-$: negative part
$\dot{\bullet}$: time derivative
$\ddot{\bullet}$: second-order time derivative
$\bar{\bullet}$: time mean
$\breve{\bullet}$: space averaged
$\check{\bullet}$: conduit generation averaged
$\langle\bullet\rangle$: ensemble averaged
$\tilde{\bullet}$: dimensionless
$\bullet^+$: normalized ($\in [0, 1]$)
$\hat{\bullet}$: peak value
$\bullet_\sim$: modulation amplitude
det(•): determinant
cof(•): cofactor
tr(•): trace

Cranial Nerves

I: olfactory nerve (sensory)
II: optic nerve (sensory)
III: oculomotor nerve (mainly motor)
IV: trochlear nerve (mainly motor)
V: trigeminal nerve (sensory and motor)
VI: abducens nerve (mainly motor)

VII: facial nerve (sensory and motor)
VIII: vestibulocochlear (auditory-vestibular) nerve (mainly sensory)
IX: glossopharyngeal nerve (sensory and motor)
X: vagus nerve (sensory and motor)
XI: cranial accessory nerve (mainly motor)
XII: hypoglossal nerve (mainly motor)

Chemical Notations

[X]: concentration of X species
X (x): upper- and lowercase letters correspond to gene and corresponding protein or conversely (i.e., Fes, FES, and fes designate protein, a proto-oncogene product that acts as a kinase, and corresponding gene and oncogene product, respectively)
$^{\bullet}$: radical (unpaired electron[s])
$^{\Delta NT}$: truncated form without the N-terminal domain
$^{\Delta CT}$: truncated form without the C-terminal domain
^{C}X: cardiac-specific isoform of X molecule
$^{D(L)}$X: D (L)-stereoisomer of amino acids and carbohydrates (chirality prefixes for dextro- [dexter: right] and levorotation [lævus: left]), i.e., dextro(levo)rotatory enantiomer
^{G}X: globular form of X molecule
$^{F(G)}$actin: polymeric, filamentous (monomeric, globular) actin
$_{C}$X, $_{CT}$X: carboxy (carboxyl group COOH [C])-terminal cleaved part of X molecule
$_{c}$X: cytosolic molecule
$_{L,Ac}$X: lysosomal, acidic X molecule (e.g., sphingomyelinase)
$_{m}$X: membrane-bound molecule
$_{N}$X, $_{NT}$X: amino (amine group NH_2 [N])-terminal cleaved part of X molecule
$_{S}$X: secreted form of X molecule
$_{S,Ac}$X: secreted, acidic molecule X (e.g., sphingomyelinase)
$_{t}$X: truncated isoform
X_i: type-i isoform of the receptor of ligand X (i: integer)
XRi: receptor isoform i of ligand X (i: integer)
X+: molecule X expressed (X-positive)
X^+: cation; also intermediate product X of oxidation (loss of electron) from a reductant (or reducer) by an oxidant (electron acceptor that removes electrons from a reductant)
X−: molecule X absent (X-negative)
X^-: anion; also intermediate product X of reduction (gain of electron) from an oxidant (or oxidizer) by a reductant (electron donor that transfers electrons to an oxidant)
X^A: activator form of molecule X
X^a: active form of molecule X
X^{ECD}: soluble fragment correponding to the ectodomain of molecule X after extracellular proteolytic cleavage and shedding (possible extracellular messenger or sequestrator)
$X^{(ER)}$: endoplasmic reticulum type of molecule X
small GTPase$^{GTP(GDP)}$: active (inactive) form of small (monomeric), regulatory guanosine triphosphatase
$X^{GTP(GDP)}$: GTP (GDP)-loaded molecule X
X^{high}: molecule X produced at high levels
X^{ICD}: soluble fragment correponding to intracellular domain of molecule X after intracellular proteolytic cleavage (possible messenger and/or transcription factor; e.g., Notch^{ICD}: intracellular Notch fragment)
X^{low}: molecule X produced at low levels
X^{MT}: mitochondrial type of molecule X
X^{PM}: plasmalemmal type of molecule X
X^R: repressor form of molecule X
X^S: soluble form
X_{alt}: alternative splice variant
X_{FL}: full-length protein X
$X_{h(l,m)MW}$: high (low, mid)-molecular-weight isotype
$X_{L(S)}$: long (short) isoform (splice variants)
X_c: catalytic subunit
X_i: number of molecule or atom (i: integer, often 2 or 3)
$(X_1–X_2)_i$: oligomer made of i complexes constituted of molecules X_1 and X_2 (e.g., histones)
a, c, nX: atypical, conventional, novel molecule X (e.g., PKC)
al, ac, nX: alkaline, acidic, neutral molecule X (e.g., sphingomyelinase)
asX: alternatively spliced molecule X (e.g., asTF)
cX: cellular, cytosolic, constitutive (e.g., cNOS), or cyclic (e.g., cAMP and cGMP) X molecule
caX: cardiomyocyte isoform (e.g., caMLCK)
dX: deoxyX

eX: endothelial isoform (e.g., eNOS and eMLCK)
hX: human form (ortholog); heart type (e.g., hFABP); hormone-like isoform (FGF)
iX: inhibitory mediator (e.g., iSMAD) or intracellular (e.g., iFGF) or inducible (e.g., iNOS) isoform
kX: renal type (kidney) X molecule
ksX: kidney-specific isoform of X molecule
lX: lysosomal X molecule
lpX: lipoprotein-associated X molecule (e.g., lpPLA2)
mX: mammalian species or membrane-associated X molecule (e.g., mTGFβ)
mtX: mitochondrial type of X molecule
nX: neutral X; neuronal type (e.g., nWASP)
oxX: oxidized X molecule (e.g., oxLDL)
plX: plasmalemmal type of X molecule
rX: receptor-associated mediator or receptor-like enzyme; also regulatory type of molecular species (e.g., rSMAD)
skX: skeletal myocyte isoform (e.g., skMLCK)
smcX: smooth muscle cell isoform (e.g., smcMLCK)
tX: target type of X (e.g., tSNARE); tissue type (e.g., tPA)
tmX: transmembrane type of X
vX: vesicle-associated (e.g., vSNARE) or vacuolar (e.g., vATPase) type of X
GPX: glycoprotein (X: molecule abbreviation or assigned numeral)
Xx: (x: single letter) splice variants
X1: human form (ortholog)
Xi: isoform type i (paralog or splice variant; i: integer)
Xi/j: (i,j: integers) refers to either both isoforms (i.e., Xi and Xj, such as ERK1/2) or heterodimer (i.e., Xi–Xj, such as ARP2/3)
X1/X2: molecular homologs or commonly used (e.g., contactin-1/F3)
PI(i)P, PI(i,j)P_2, PI(i,j,k)P_3: i,j,k (integers): position(s) of phosphorylated OH groups of the inositol ring of phosphatidylinositol mono-, bis-, and trisphosphates

Post-translational Modification

X_A: acetylated molecule X
acX: acetylated molecule X (e.g., acLDL)
X^M: methylated molecule X
X_M: myristoylated molecule X
X_P: palmitoylated molecule X
X^P: phosphorylated molecule X
pAA: phosphorylated amino acid (pSer, pThr, and pTyr)
X^{SNO}: ^{S}nitrosylated molecule X
X^{SSG}: ^{S}glutathionylated molecule X
X^U: ubiquitinated protein X

Amino Acids

Ala (A): alanine
Arg (R): arginine
Asn (N): asparagine
Asp (D): aspartic acid
Asp^{COO^-}: aspartate
CysH (C): cysteine
Cys: cystine
Gln (Q): glutamine
Glu (E): glutamic acid
Glu^{COO^-}: glutamate
Gly (G): glycine
His (H): histidine
Iso, Ile (I): isoleucine
Leu (L): leucine
Lys (K): lysine
Met (M): methionine
Orn: ornithine (not encoded by DNA, but use in the urea cycle)
Phe (F): phenylalanine
Pro (P): proline
Ser (S): serine
Thr (T): threonine
Trp (W): tryptophan
Tyr (Y): tyrosine
Val (V): valine

Ions

Asp^-: aspartate (carboxylate anion of aspartic acid)
ADP^{3-}: ADP anion
ATP^{4-}: ATP anion
Ca^{2+}: calcium cation
Cl^-: chloride anion
Co^{2+}: cobalt cation
Cu^+: copper monovalent cation
Cu^{2+}: copper divalent cation
Fe^{2+}: ferrous iron cation
Fe^{3+}: ferric iron cation
Glu^-: glutamate (carboxylate anion of glutamic acid)
H^+: hydrogen cation (proton)

H_3O^+: hydronium (oxonium or hydroxonium) cation
HCO_3^-: bicarbonate anion
HPO_4^{2-}: hydrogen phosphate divalent anion (inorganic phosphate species)
$H_2PO_4^-$: dihydrogen phosphate monovalent ion (inorganic phosphate species)
HS^-: hydrosulfide (hydrogen sulfide) anion (sulfanide)
K^+: potassium cation
Mg^{2+}: magnesium cation
${}^{Mg}ATP^{2-}$: ATP anion
Mn^{2+}: manganese cation
Na^+: sodium cation
Ni^{2+}: nickel cation (common oxidation state)
OH^-: hydroxide anion
PO_4^{3-}: phosphate anion (inorganic phosphate species)
S^{2-}: sulfide anion
S_n^{2-}: polysulfide anion
SO_3^{2-}: sulfite anion
SO_4^{2-}: sulfate anion
$S_2O_3^{2-}$: thiosulfate
Zn^{2+}: zinc cation (common oxidation state)

Atmospheric Pollutants

CH_4: methane
HNO_2: nitrous acid
HNO_3: nitric acid
H_2SO_4: sulfuric acid
NO_x: nitrogen oxides
NO_2: nitrogen dioxide
O_3: ozone
PM_{10}: inhalable coarse particulate matter ($2.5 < \text{size} < 10\,\mu m$)
$PM_{2.5}$: fine particulate matter ($0.1 < \text{size} \leq 2.5\,\mu m$)
$PM_{0.1}$: ultrafine particulate matter (aerodynamic diameter $\leq 0.1\,\mu m$)
SO_x: sulfur oxides
SO_2: sulfur dioxide

Inhaled and Signaling Gas

CO: carbon monoxide (or carbonic oxide; signaling gas and pollutant [air level ~ 0.1 ppm])
CO_2: carbon dioxide (cell waste)
H_2S: hydrogen sulfide (signaling gas) [air level ~ 0.0001 ppm])
He: helium (inert monatomic gas)
N_2: nitrogen (inert diatomic gas)
NH_3: ammonia (trihydrogen nitride; trace quantities in air)
NO: nitric oxide (or nitrogen monoxide; signaling gas and pollutant [air level ~ 0.1 ppm])
O_2: oxygen (cell energy producer)

Nitric Oxide Derivatives

HNO: protonated nitroxyl anion
$NO^\bullet$: free radical form
NO^+: nitrosyl or nitrosonium cation
NO^-: nitroxyl or hyponitrite anion (inodilator)
NO_2^-: nitrite anion
NO_3^-: nitrate anion

Reactive Oxygen and Nitrogen Species

$CO_3^{\bullet -}$: carbonate radical
H_2O_2: hydrogen peroxide
HOCl: hypochlorous acid
$HS^\bullet$: sulfanyl or hydrosulfide radical
N_2O_3: dinitrogen trioxide
$NO_2^\bullet$: nitrogen dioxide
${}^1O_2$: singlet oxygen
O_2^-: superoxide ($O_2^{\bullet -}$)
$O{=}C(O^\bullet)O^-$: carbonate radical
$OH^\bullet$: hydroxyl radical (hydroxide ion neutral form)
$ONOO^-$: peroxynitrite
$RO^\bullet$: alkoxyl
$RO_2^\bullet$: peroxyl

Moieties (R denotes an organic group)

R: alkyl group (with only carbon and hydrogen atoms linked exclusively by single bonds)
R–CH_3: methyl group (with 3 forms: methanide anion [CH_3^-], methylium cation [CH_3^+], and methyl radical [$CH_3^\bullet$])
R–CHO: aldehyde group
R–CN: nitrile group
R–CO: acyl group
R–CO–R: carbonyl group
R–COO^-: carboxylate group
R–COOH: carboxyl group
R–NC: isonitrile group
R–NCO: isocyanate group
R–NH_2: amine group
R–NO: nitroso group
R–NO_2: nitro group
R–O: alkoxy group
R=O: oxo group
R–OCN: cyanate group

R–OH: hydroxyl group
R–ONO: nitrosooxy group
R–ONO_2: nitrate group
R–OO–R: peroxy group
R–OOH: hydroperoxy group
R–S–R: sulfide group
R–SH: thiol (or sulfhydryl) moiety
R–SN: sulfenyl-amide moiety
R–SNO: nitrosothiol (or thionitrite) moiety
R–SO: sulfinyl
R–SO–R: sulfoxide group
R–SO_2: sulfonyl group
R–SO_2H: sulfinic acid (sulfinyl moiety)
R–SO_2N: sulfonyl-amide moiety
R–SO_3H: sulfonic acid (sulfonyl moiety)
R–SOH: sulfenic acid (sulfenyl moiety)
R–SON: sulfinyl-amide moiety
R–SSH: hydropersulfide moiety
R–SS–R: disulfide group
R–$S(S)_nS$–R: polysulfide

Lung Function Testing

ERV: expiratory reserve volume
f_C: cardiac frequency
f_R: breathing frequency
FEF_f: forced expiratory flow at a fraction (f[%]) of forced expiration (FEF_{25}, FEF_{50}, and FEF_{75})
FEV_1: volume expired at the end of the first second of forced expiration
FEV_τ: forced expiratory volume at time τ (fraction of a second over which maximally fast exhaled volume is measured) starting from full inspiration
FRC: functional residual capacity (lung volume at end of rest expiration)
IC: inspiratory capacity (IRV + V_T)
IRV: inspiratory reserve volume
MBC: maximum breathing capacity (per mn of effort)
MVV: maximal voluntary ventilation (volume of air breathed in a specified period during repetitive maximal exercise)
PEF: peak expiratory flow
R_R: respiratory quotient ($\dot{V}_{CO_2}/\dot{V}_{O_2}$)
RV: residual volume
TLC: total lung capacity
V_A: alveolar gas volume
V_D: dead space volume
V_L: lung volume
V_T: tidal volume
$\dot{V}$: total ventilation (air volume exhaled par mn)
$\dot{V}_A$: alveolar ventilation ($f_R(V_T - V_D)$)
$\dot{V}_{O_2}$: oxygen consumption
$\dot{V}_{CO_2}$: carbon dioxide production
VC: vital capacity (air volume quietly expelled from full inspiration)

Time Units

s: second
mn: minute
h: hour
d: day
wk: week
mo: month
yr: year

SI-Based and Non-SI Units of Quantity

mmHg: millimeter of mercury (133.322 Pa [$\sim$0.1333 kPa])
mmol, nmol, μ mol: milli-, nano-, micromoles
(amount of a chemical species, one mole containing about $6.02214078 \times 10^{23}$ molecules)
mosm: milliosmole
(osm: number of moles of a osmotically active chemical compound)
kDa: kiloDalton
(Da: atomic or molecular mass unit)
ppm: parts per million
l: liter

Temperature and Pressure Conditions

ATPS: ambient temperature and pressure, saturated with water at body temperature, i.e., at 37°C, $p_{H_2O} = 6.27$ kPa (47 mmHg)
BTPS: body temperature and ambient pressure, saturated with water
STPD: standard temperature (0°C) and pressure (101 kPa [760 mmHg]), dry air

$$V_{BTPS} = V_{ATPS} \times \frac{273+37}{273+T} \times \frac{p - p_{H_2O}}{p - 47}$$

$$V_{STPD} = V_{ATPS} \times \frac{273}{273+T} \times \frac{p - p_{H_2O}}{760}$$

References

Chap. 1. Pathogenesis of Cardiac Diseases—Biological Context

1. Mathiyalagan P, Keating ST, Du XJ, El-Osta A (2014) Chromatin modifications remodel cardiac gene expression. Cardiovasc Res 103:7–16
2. Demos-Davies KM, Ferguson BS, Cavasin MA, Mahaffey JH, Williams SM, Spiltoir JI, Schuetze KB, Horn TR, Chen B, Ferrara C, Scellini B, Piroddi N, Tesi C, Poggesi C, Jeong MY, McKinsey TA (2014) HDAC6 contributes to pathological responses of heart and skeletal muscle to chronic angiotensin-II signaling. Am J Physiol Heart Circ Physiol 307:H252–H258
3. Evans PC, Kwak BR (2013) Biomechanical factors in cardiovascular disease. Cardiovasc Res 99:229–231
4. Frueh J, Maimari N, Homma T, Bovens SM, Pedrigi RM, Towhidi L, Krams R (2013) Systems biology of the functional and dysfunctional endothelium. Cardiovasc Res 99:334–341
5. Zhang C, Zeng L, Emanueli C, Xu Q (2013) Blood flow and stem cells in vascular disease. Cardiovasc Res 99:251–259
6. Oakley RH, Ren R, Cruz-Topete D, Bird GS, Myers PH, Boyle MC, Schneider MD, Willis MS, Cidlowski JA (2013) Essential role of stress hormone signaling in cardiomyocytes for the prevention of heart disease. Proc Natl Acad Sci U S A 110:17035–17040
7. Zacchigna S, Giacca M (2014) Extra- and intracellular factors regulating cardiomyocyte proliferation in postnatal life. Cardiovasc Res 102:312–320
8. Hong TT, Smyth JW, Gao D, Chu KY, Vogan JM, Fong TS, Jensen BC, Colecraft HM, Shaw RM (2010) BIN1 localizes the L-type calcium channel to cardiac T-tubules. PLoS Biol 2010 8:e1000312
9. Beavers DL, Landstrom AP, Chiang DY, Wehrens XHT (2014) Emerging roles of junctophilin-2 in the heart and implications for cardiac diseases. Cardiovasc Res 103:198–205
10. Gorospe G, Zhu R, Millrod MA, Zambidis ET, Tung L, Vidal R (2014) Automated grouping of action potentials of human embryonic stem cell-derived cardiomyocytes. IEEE Trans Biomed Eng 61:2389–2396
11. Nemir M, Metrich M, Plaisance I, Lepore M, Cruchet S, Berthonneche C, Sarre A, Radtke F, Pedrazzini T (2014) The Notch pathway controls fibrotic and regenerative repair in the adult heart. Eur Heart J 35:2174–2185
12. Wackerhage H, Del Re DP, Judson RN, Sudol M, Sadoshima J (2014) The Hippo signal transduction network in skeletal and cardiac muscle. Sci Signal 7:re4
13. Zouein FA, Kurdi M, Booz GW (2013) HSPA12B and repairing the heart: beauty in simplicity. Cardiovasc Res 99:587–589
14. Li J, Zhang Y, Li C, Xie J, Liu Y, Zhu W, Zhang X, Jiang S, Liu L, Ding Z (2013) HSPA12B attenuates cardiac dysfunction and remodelling after myocardial infarction through an eNOS-dependent mechanism. Cardiovasc Res 99:674–684

M. Thiriet, *Diseases of the Cardiac Pump*, Biomathematical and Biomechanical Modeling of the Circulatory and Ventilatory Systems 7, DOI 10.1007/978-3-319-12664-7

15. Tarone G, Brancaccio M (2014) Keep your heart in shape: molecular chaperone networks for treating heart disease. Cardiovasc Res 102:346–361
16. Ferretti R, Sbroggiò M, Di Savino A, Fusella F, Bertero A, Michowski W, Tarone G, Brancaccio M (2011) Morgana and melusin: two fairies chaperoning signal transduction. Cell Cycle 10:3678–3683
17. Jian Z, Han H, Zhang T, Puglisi J, Izu LT, Shaw JA, Onofiok E, Erickson JR, Chen YJ, Horvath B, Shimkunas R, Xiao W, Li Y, Pan T, Chan J, Banyasz T, Tardiff JC, Chiamvimonvat N, Bers DM, Lam KS, Chen-Izu Y (2014) Mechanochemotransduction during cardiomyocyte contraction is mediated by localized nitric oxide signaling. Sci Signal 7:ra27
18. Collins HE, Zhu-Mauldin X, Marchase RB, Chatham JC (2013) STIM1/Orai1-mediated SOCE: current perspectives and potential roles in cardiac function and pathology. Am J Physiol Heart Circ Physiol 305:H446–H458
19. Yarotskyy V, Dirksen RT (2012) Temperature and RyR1 regulate the activation rate of store-operated Ca^{2+} entry current in myotubes. Biophys J 103:202–211
20. Grigoriev I, Gouveia SM, van der Vaart B, Demmers J, Smyth JT, Honnappa S, Splinter D, Steinmetz MO, Putney JW, Hoogenraad CC, Akhmanova A (2008) STIM1 is a MT-plus-end-tracking protein involved in remodeling of the ER. Curr Biol 18:177–182
21. Zeiger W, Ito D, Swetlik C, Oh-hora M, Villereal ML, Thinakaran G (2011) Stanniocalcin 2 is a negative modulator of store-operated calcium entry. Mol Cell Biol 31:3710–3722
22. Prins D, Groenendyk J, Touret N, Michalak M (2011) Modulation of STIM1 and capacitative Ca^{2+} entry by the endoplasmic reticulum luminal oxidoreductase ERp57. EMBO Rep 12:1182–1188
23. Srikanth S, Jung HJ, Kim KD, Souda P, Whitelegge J, Gwack Y (2010) A novel EF-hand protein, CRACR2A, is a cytosolic Ca^{2+} sensor that stabilizes CRAC channels in T cells. Nat Cell Biol 12:436–446
24. Fujii Y, Shiota M, Ohkawa Y, Baba A, Wanibuchi H, Kinashi T, Kurosaki T, Baba Y (2012) Surf4 modulates STIM1-dependent calcium entry. Biochem Biophys Res Commun 422:615–620
25. Palty R, Raveh A, Kaminsky I, Meller R, Reuveny E (2012) SARAF inactivates the store operated calcium entry machinery to prevent excess calcium refilling. Cell 149:425–438
26. Jha A, Ahuja M, Maléth J, Moreno CM, Yuan JP, Kim MS, Muallem S (2013) The STIM1 CTID domain determines access of SARAF to SOAR to regulate Orai1 channel function. J Cell Biol 202:71–79
27. Darbellay B, Arnaudeau S, Bader CR, Konig S, Bernheim L (2011) STIM1L is a new actin-binding splice variant involved in fast repetitive Ca^{2+} release. J Cell Biol 194:335–346
28. Ogawa A, Firth AL, Smith KA, Maliakal MV, Yuan JX (2012) PDGF enhances store-operated Ca^{2+} entry by upregulating STIM1/Orai1 via activation of Akt/mTOR in human pulmonary arterial smooth muscle cells. Am J Physiol Cell Physiol 302:C405-C411
29. Xiang SY, Ouyang K, Yung BS, Miyamoto S, Smrcka AV, Chen J, Brown JH, (2013) PLCϵ, PKD1, and SSH1L transduce RhoA signaling to protect mitochondria from oxidative stress in the heart. Sci Signal 6:ra108
30. Puthanveetil P, Wan A, Rodrigues B (2013) FoxO1 is crucial for sustaining cardiomyocyte metabolism and cell survival. Cardiovasc Res 97:393–403
31. Liu S, Zhang J, Xiang YK (2011) FRET-based direct detection of dynamic protein kinase A activity on the sarcoplasmic reticulum in cardiomyocytes. Biochem Biophys Res Commun 404:581–586
32. Haj Slimane Z, Bedioune I, Lechêne P, Varin A, Lefebvre F, Mateo P, Domergue-Dupont V, Dewenter M, Richter W, Conti M, El-Armouche A, Zhang J, Fischmeister R, Vandecasteele G (2014) Control of cytoplasmic and nuclear protein kinase A by phosphodiesterases and phosphatases in cardiac myocytes. Cardiovasc Res 102:97–106
33. Hirsch E, Nagai R, Thum T (2014) Heterocellular signalling and crosstalk in the heart in ischaemia and heart failure. Cardiovasc Res 102:191–193

34. Carbe CJ, Cheng L, Addya S, Gold JI, Gao E, Koch WJ, Riobo NA (2014) Gi proteins mediate activation of the canonical hedgehog pathway in the myocardium. Am J Physiol Heart Circ Physiol 307:H66–H72
35. Deb A (2014) Cell–cell interaction in the heart via Wnt/β-catenin pathway after cardiac injury. Cardiovasc Res 102:214–223
36. Haghikia A, Ricke-Hoch M, Stapel B, Gorst I, Hilfiker-Kleiner D (2014) STAT3, a key regulator of cell-to-cell communication in the heart. Cardiovasc Res 102:281–289
37. Mohamed TM, Zi M, Prehar S, Maqsood A, Abou-Leisa R, Nguyen L, Pfeifer GP, Cartwright EJ, Neyses L, Oceandy D (2014) The tumour suppressor Ras-association domain family protein 1A (RASSF1A) regulates TNF-α signalling in cardiomyocytes. Cardiovasc Res 103:47–59 '
38. Roberge S, Roussel J, Andersson DC, Meli AC, Vidal B, Blandel F, Lanner JT, Le Guennec JY, Katz A, Westerblad H, Lacampagne A, Fauconnier J (2014) TNF-α-mediated caspase-8 activation induces ROS production and TRPM2 activation in adult ventricular myocytes. Cardiovasc Res 103:90–99
39. Emde A, Hornstein E (2014) miRNAs at the interface of cellular stress and disease. EMBO J 33:1428–1437
40. Viereck J, Bang C, Foinquinos A, Thum T (2014) Regulatory RNAs and paracrine networks in the heart. Cardiovasc Res 102:290–301
41. Tian FJ, An LN, Wang GK, Zhu JQ, Li Q, Zhang YY, Zeng A, Zou J, Zhu RF, Han XS, Shen N, Yang HT, Zhao XX, Huang S, Qin YW, Jing Q (2014) Elevated microRNA-155 promotes foam cell formation by targeting HBP1 in atherogenesis. Cardiovasc Res 103:100–110
42. Hartmann P, Schober A2, Weber C (2014) MicroRNA-155 and macrophages: a fatty liaison. Cardiovasc Res 103:5–6
43. Sluijter JP, Verhage V, Deddens JC, van den Akker F, Doevendans PA (2014) Microvesicles and exosomes for intracardiac communication. Cardiovasc Res 102:302–311
44. Hollander JM, Thapa D, Shepherd DL (2014) Physiological and structural differences in spatially distinct subpopulations of cardiac mitochondria: influence of cardiac pathologies. Am J Physiol Heart Circ Physiol 307:H1–H14
45. Chen L, Gong Q, Stice JP, Knowlton AA (2009) Mitochondrial OPA1, apoptosis, and heart failure. Cardiovasc Res 84:91–99
46. Norton M, Ng ACH, Baird S, Dumoulin A, Shutt T, Mah N, Andrade-Navarro MA, McBride HM, Screaton RA (2014) ROMO1 is an essential redox-dependent regulator of mitochondrial dynamics. Sci Signal 7:ra10
47. Miller MW, Knaub LA, Olivera-Fragoso LF, Keller AC, Balasubramaniam V, Watson PA, Reusch JE (2013) Nitric oxide regulates vascular adaptive mitochondrial dynamics. Am J Physiol Heart Circ Physiol 304:H1624–H1633
48. Wenceslau CF, McCarthy CG, Szasz T, Spitler K, Goulopoulou S, Webb RC, Working Group on DAMPs in Cardiovascular Disease (2014) Mitochondrial damage-associated molecular patterns and vascular function. Eur Heart J 35:1172–1177
49. Lopaschuk GD, Ussher JR, Folmes CD, Jaswal JS, Stanley WC (2010) Myocardial fatty acid metabolism in health and disease. Physiol Rev 90:207–258
50. Lapuente-Brun E, Moreno-Loshuertos R, Acín-Pérez R, Latorre-Pellicer A, Colás C, Balsa E, Perales-Clemente E, Quirós PM, Calvo E, Rodríguez-Hernández MA, Navas P, Cruz R, Carracedo A, López-Otín C, Pérez-Martos A, Fernández-Silva P, Fernández-Vizarra E, Enríquez JA (2013) Supercomplex assembly determines electron flux in the mitochondrial electron transport chain. Science 340:1567–1570
51. Garcia-Dorado D, Ruiz-Meana M, Rodríguez-Sinovas A (2014) Connexin 43 phosphorylation in subsarcolemmal mitochondria: a general cardioprotective signal targeted by fibroblast growth factor-2?. Cardiovasc Res 103:1–2
52. Srisakuldee W, Makazan Z, Nickel BE, Zhang F, Thliveris JA, Pasumarthi KB, Kardami E (2014) The FGF-2-triggered protection of cardiac subsarcolemmal mitochondria from calcium overload is mitochondrial connexin 43-dependent. Cardiovasc Res 103:72–80

53. Foskett JK, Madesh M (2014) Regulation of the mitochondrial Ca^{2+} uniporter by MICU1 and MICU2. Biochem Biophys Res Commun 449:377–383
54. Rizzuto R, De Stefani D, Raffaello A, Mammucari C (2012) Mitochondria as sensors and regulators of calcium signalling. Nat Rev Mol Cell Biol 13:566–578
55. Williams GS, Boyman L, Chikando AC, Khairallah RJ, Lederer WJ (2013) Mitochondrial calcium uptake. Proc Natl Acad Sci U S A 110:10479–10486
56. Carafoli E (2014) Historical introduction. Biochem Biophys Res Commun 449:365–366
57. De Stefani D, Rizzuto R (2014) Molecular control of mitochondrial calcium uptake. Biochem Biophys Res Commun 449:373–376
58. Chaudhuri D, Clapham DE (2014) Outstanding questions regarding the permeation, selectivity, and regulation of the mitochondrial calcium uniporter. Biochem Biophys Res Commun 449:367–369
59. Takeuchi A, Kim B, Matsuoka S (2013) The mitochondrial Na^{+}–Ca^{2+} exchanger, NCLX, regulates automaticity of HL-1 cardiomyocytes. Sci Rep 3:2766
60. Alavian KN, Beutner G, Lazrove E, Sacchetti S, Park HA, Licznerski P, Li H, Nabili P, Hockensmith K, Graham M, Porter GA, Jonas EA (2014) An uncoupling channel within the c-subunit ring of the F_1F_O ATP synthase is the mitochondrial permeability transition pore. Proc Natl Acad Sci U S A 111:10580–10585
61. Penna C, Mancardi D, Rastaldo R, Pagliaro P (2009) Cardioprotection: a radical view: free radicals in pre and postconditioning. Biochim Biophys Acta 1787:781–793. (Bioenergetics)
62. Carpi A, Menabò R, Kaludercic N, Pelicci P, Di Lisa F, Giorgio M (2009) The cardioprotective effects elicited by p66(Shc) ablation demonstrate the crucial role of mitochondrial ROS formation in ischemia/reperfusion injury. Biochim Biophys Acta 1787:774–780. (Bioenergetics)
63. Rudat C, Norden J, Taketo MM, Kispert A (2013) Epicardial function of canonical Wnt-, Hedgehog-, Fgfr1/2-, and Pdgfra-signalling. Cardiovasc Res 100:411–421
64. Zhang M, Shah AM (2014) ROS signalling between endothelial cells and cardiac cells. Cardiovasc Res 102:249–257
65. Brutsaert DL (2003) Cardiac endothelial-myocardial signaling: its role in cardiac growth, contractile performance, and rhythmicity. Physiol Rev 83:59–115
66. Hendrickx J, Doggen K, Weinberg EO, Van Tongelen P, Fransen P, De Keulenaer GW (2004) Molecular diversity of cardiac endothelial cells in vitro and in vivo. Physiol Genomics 19:198–206
67. Parodi EM, Kuhn B (2014) Signalling between microvascular endothelium and cardiomyocytes through neuregulin. Cardiovasc Res 102:194–204
68. Hatem SN, Sanders P (2014) Epicardial adipose tissue and atrial fibrillation. Cardiovasc Res 102:205–213
69. Ghigo A, Franco I, Morello F, Hirsch E (2014) Myocyte signalling in leucocyte recruitment to the heart. Cardiovasc Res 102:270–280
70. Van Linthout S, Miteva K, Tschöpe C (2014) Crosstalk between fibroblasts and inflammatory cells. Cardiovasc Res 102:258–269
71. Fairweather DL, Frisancho-Kiss S, Coronado MJ, Frisancho JA, Cihakova D, Rose NR (2009) Two populations of alternatively activated M2 macrophages (TLR4+ vs. Tim-3+) regulate inflammation and fibrosis in the heart during coxsackievirus-induced myocarditis. J Immun 182:137.20. (Meeting Abstract Supplement)
72. Kane LP (2010) T cell Ig and mucin domain proteins and immunity. J Immun 184:2743–2749
73. Frantz S, Nahrendorf M (2014) Cardiac macrophages and their role in ischaemic heart disease. Cardiovasc Res 102:240–248
74. Fujiu K, Wang J, Nagai R (2014) Cardioprotective function of cardiac macrophage. Cardiovasc Res 102:232–239
75. Martinez FO, Gordon S (2014) The M1 and M2 paradigm of macrophage activation: time for reassessment. F1000Prime Rep 6:13

76. Inoue T, Plieth D, Venkov CD, Xu C, Neilson EG (2005) Antibodies against macrophages that overlap in specificity with fibroblasts. Kidney International 67:2488–2493
77. Ramkisoensing AA, de Vries AA, Atsma DE, Schalij MJ, Pijnappels DA (2014) Interaction between myofibroblasts and stem cells in the fibrotic heart: balancing between deterioration and regeneration. Cardiovasc Res 102:224–231
78. Chng SC, Ho L, Tian J, Reversade B (2013) ELABELA: a hormone essential for heart development signals via the apelin receptor. Dev Cell 27:672–680
79. Pauli A, Norris ML, Valen E, Chew GL, Gagnon JA, Zimmerman S, Mitchell A, Ma J, Dubrulle J, Reyon D, Tsai SQ, Joung JK, Saghatelian A, Schier AF (2014) Toddler: an embryonic signal that promotes cell movement via Apelin receptors. Science 343:1248636
80. Magny EG, Pueyo JI, Pearl FM, Cespedes MA, Niven JE, Bishop SA, Couso JP (2013) Conserved regulation of cardiac calcium uptake by peptides encoded in small open reading frames. Science 341:1116–1120
81. Traaseth NJ, Ha KN, Verardi R, Shi L, Buffy JJ, Masterson LR, Veglia G (2008) Structural and dynamic basis of phospholamban and sarcolipin inhibition of Ca^{2+}-ATPase. Biochemistry 473–13
82. Smith WS, Broadbridge R, East JM, Lee AG (2002) Sarcolipin uncouples hydrolysis of ATP from accumulation of Ca^{2+} by the Ca^{2+}-ATPase of skeletal-muscle sarcoplasmic reticulum. Biochem J. Jan 15;361(Pt 2):277-86.
83. Hirose T, Mishima Y, Tomari Y (2014) Elements and machinery of non-coding RNAs: Toward their taxonomy. EMBO Rep 15:489–507
84. Cohen SM (2014) Everything old is new again: (linc)RNAs make proteins! EMBO J 33:937–1085
85. Bazzini AA, Johnstone TG, Christiano R, Mackowiak SD, Obermayer B, Fleming ES, Vejnar CE, Lee MT, Rajewsky N, Walther TC, Giraldez AJ (2014) Identification of small ORFs in vertebrates using ribosome footprinting and evolutionary conservation. EMBO J 33:981–993
86. Nelson BR, Anderson DM, Olson EN (2014) Small open reading frames pack a big punch in cardiac calcium regulation. Circ Res 114:18–20
87. Shukla P, Ghatta S, Dubey N, Lemley CO, Johnson ML, Modgil A, Vonnahme K, Caton JS, Reynolds LP, Sun C, O'Rourke ST (2014) Maternal nutrient restriction during pregnancy impairs an endothelium-derived hyperpolarizing factor-like pathway in sheep fetal coronary arteries. Am J Physiol Heart Circ Physiol 307:H134–H142
88. Lesoon LA, Mahesh VB (1992) Stimulatory and inhibitory effects of progesterone on FSH secretion by the anterior pituitary. J Steroid Biochem Mol Biol 42:479–491
89. Giguère V (2002) To ERR in the estrogen pathway. Trends Endocrinol Metab 13:220–225
90. Schreiber SN, Emter R, Hock MB, Knutti D, Cardenas J, Podvinec M, Oakeley EJ, Kralli A (2004) The estrogen-related receptor α (ERRα) functions in PPARγ coactivator 1α (PGC-1α)-induced mitochondrial biogenesis. Proc Natl Acad Sci U S A 101:6472–6477
91. Ochnik AM, Yee D (2012) Estrogen-related receptor α: an orphan finds a family. Breast Cancer Res 14:309
92. Gadkar-Sable S, Shah C, Rosario G, Sachdeva G, Puri C (2005) Progesterone receptors: various forms and functions in reproductive tissues. Front Biosci 10:2118–2130
93. Moussatche P, Lyons TJ (2012) Non-genomic progesterone signalling and its non-canonical receptor. Biochem Soc Trans 40:200–204
94. Hales DB (2009–2014) Hormones of pregnancy. Southern Illinois University School of Medicine (www.siumed.edu/d̃bhaleslab)
95. Bergeron C (2000) Morphological changes and protein secretion induced by progesterone in the endometrium during the luteal phase in preparation for nidation. Hum Reprod 15:119–128
96. Koistinen H, Seppälä M (2011) PAEP (progestagen-associated endometrial protein) Atlas of Genetics and Cytogenetics in Oncology and Haematology 15:576–581 (atlasgeneticsoncology.org)

97. Rutanen EM, Koistinen R, Seppälä M, Julkunen M, Suikkari AM, Huhtala ML (1987) Progesterone-associated proteins PP12 and PP14 in the human endometrium. J Steroid Biochem 27:25–31
98. Cocchiara R, Lampiasi N, Albeggiani G, Azzolina A, Bongiovanni A, Gianaroli L, Di Blasi F, Geraci D (2006) A factor secreted by human embryo stimulates cytokine release by uterine mast cell. Mol Hum Reprod 2:781–791
99. Susini L, Besse S, Duflaut D, Lespagnol A, Beekman C, Fiucci G, Atkinson AR, Busso D, Poussin P, Marine JC, Martinou JC, Cavarelli J, Moras D, Amson R, Telerman A (2008) TCTP protects from apoptotic cell death by antagonizing bax function. Cell Death Differ 15:1211–1220
100. Kashiwakura JC, Ando T, Matsumoto K, Kimura M, Kitaura J, Matho MH, Zajonc DM, Ozeki T, Ra C, MacDonald SM, Siraganian RP, Broide DH, Kawakami Y, Kawakami T (2012) Histamine-releasing factor has a proinflammatory role in mouse models of asthma and allergy. J Clin Investig 122:218–228
101. MacLeod JN, Lee AK, Liebhaber SA, Cooke NE (1992) Developmental control and alternative splicing of the placentally expressed transcripts from the human growth hormone gene cluster. J Biol Chem 267:14219–14226
102. Lowndes K, Amano A, Yamamoto SY, Bryant-Greenwood GD (2006) The human relaxin receptor (LGR7): expression in the fetal membranes and placenta. Placenta 27:610–618
103. Tregear GW, Bathgate RA, Layfield S, Ferraro T, Gundlach A, Ma S, Lin F, Hanson NF, Summers RJ, Rosengren J, Craik DJ, Wade JD (2005) The chemistry and biology of human relaxin-3. Ann N Y Acad Sci 1041:40–46
104. Liu LX, Arany Z (2014) Maternal cardiac metabolism in pregnancy. Cardiovasc Res 101:545–553
105. Pierlot CM, Lee JM, Amini R, Sacks MS, Wells SM (2014) Pregnancy-induced remodeling of collagen architecture and content in the mitral valve. Ann Biomed Eng 42:
106. Nada MA, Chace DH, Sprecher H, Roe CR (1995) Investigation of β-oxidation intermediates in normal and MCAD-deficient human fibroblasts using tandem mass spectrometry. Biochem Mol Med 54:59–66
107. Ricke-Hoch M, Bultmann I, Stapel B, Condorelli G, Rinas U, Sliwa K, Scherr M, Hilfiker-Kleiner D (2014) Opposing roles of Akt and STAT3 in the protection of the maternal heart from peripartum stress. Cardiovasc Res 101:587–596
108. Lok SI, Kirkels JH, Klöpping C, Doevendans PA, de Jonge N (2011) Peripartum cardiomyopathy: The need for a national database. Neth Heart J 19:126–133
109. Chung E, Yeung F, Leinwand LA (2013) Calcineurin activity is required for cardiac remodelling in pregnancy. Cardiovasc Res 100:402–410
110. Gonzalez AM, Osorio JC, Manlhiot C, Gruber D, Homma S, Mital S (2007) Hypertrophy signaling during peripartum cardiac remodeling. Am J Physiol Heart Circ Physiol 293: H3008–H3013

Chap. 2. Context of Cardiac Diseases

111. Faergeman O (2013) Genes and cardiovascular risk. Eur Heart J 34:949–950
112. Navasiolava NM, Dignat-George F, Sabatier F, Larina IM, Demiot C, Fortrat JO, Gauquelin-Koch G, Kozlovskaya IB, Custaud MA (2010) Enforced physical inactivity increases endothelial microparticle levels in healthy volunteers. Am J Physiol Heart Circ Physiol 299:H248–H256
113. Hoshi T, Wissuwa B, Tian Y, Tajima N, Xu R, Bauer M, Heinemann SH, Hou S (2013) Omega-3 fatty acids lower blood pressure by directly activating large-conductance Ca^{2+}-dependent K^{+} channels. Proc Natl Acad Sci U S A 110:4816–4821

114. Hoshi T, Tian Y, Xu R, Heinemann SH, Hou S (2013) Mechanism of the modulation of BK potassium channel complexes with different auxiliary subunit compositions by the omega-3 fatty acid DHA. Proc Natl Acad Sci U S A 110:4822–4827
115. Heiss C, Keen CL, Kelm M (2010) Flavanols and cardiovascular disease prevention. Eur Heart J 31:2583–2592
116. Heiss C, Kelm M (2010) Chocolate consumption, blood pressure, and cardiovascular risk. Eur Heart J 31:1554–1556
117. Corder R, Mullen W, Khan NQ, Marks SC, Wood EG, Carrier MJ, Crozier A (2006) Oenology: red wine procyanidins and vascular health. Nature 444:566
118. Hammerstone JF, Lazarus SA, Schmitz HH (2000) Procyanidin content and variation in some commonly consumed foods. Journal of Nutrition 130:2086S–2092S
119. Buijsse B, Weikert C, Drogan D, Bergmann M, Boeing H (2010) Chocolate consumption in relation to blood pressure and risk of cardiovascular disease in German adults. Eur Heart J 31:1616–1623
120. Benavides GA, Squadrito GL, Mills RW, Patel HD, Isbell TS, Patel RP, Darley-Usmar VM, Doeller JE, Kraus DW (2007) Hydrogen sulfide mediates the vasoactivity of garlic. Proc Natl Acad Sci U S A 104:17977–17982
121. Petyaev IM, Bashmakov YK (2012) Could cheese be the missing piece in the French paradox puzzle? Med Hypotheses 79:746–749
122. Trostchansky A, Rubbo H (2008) Nitrated fatty acids: mechanisms of formation, chemical characterization, and biological properties. Free Radic Biol Med 44:1887–1896
123. Charles RL, Rudyk O, Prysyazhna O, Kamynina A, Yang J, Morisseau C, Hammock BD, Freeman BA, Eaton P (2014) Protection from hypertension in mice by the Mediterranean diet is mediated by nitro fatty acid inhibition of soluble epoxide hydrolase. Proc Natl Acad Sci U S A 111:8167–8172
124. Artim DE, Bazely F, Daugherty SL, Sculptoreanu A, Koronowski KB, Schopfer FJ, Woodcock SR, Freeman BA, de Groat WC (2011) Nitro-oleic acid targets transient receptor potential (TRP) channels in capsaicin sensitive afferent nerves of rat urinary bladder. Exp Neurol 232:90–99
125. Cotter DG, Schugar RC, Crawford PA (2013) Ketone body metabolism and cardiovascular disease. Am J Physiol Heart Circ Physiol 304:H1060–H1076
126. Ketteler M, Wolf M, Hahn K, Ritz E (2013) Phosphate: a novel cardiovascular risk factor. Eur Heart J 34:1099–1101
127. Heine GH, Nangaku M, Fliser D (2013) Calcium and phosphate impact cardiovascular risk. Eur Heart J 34:1112–1121
128. Gattineni J, Bates C, Twombley K, Dwarakanath V, Robinson ML, Goetz R, Mohammadi M, Baum M (2009) FGF23 decreases renal NaPi-2a and NaPi-2c expression and induces hypophosphatemia in vivo predominantly via FGF receptor 1. Am J Physiol Ren Physiol 297:F282–F291
129. Wierup N, Björkqvist M, Weström B, Pierzynowski S, Sundler F, Sjölund K (2007) Ghrelin and motilin are cosecreted from a prominent endocrine cell population in the small intestine. J Clin Endocrinol Metab 92:3573–3581
130. Zizzari P, Hassouna R, Grouselle D, Epelbaum J, Tolle V (2011) Physiological roles of preproghrelin-derived peptides in GH secretion and feeding. Peptides 32:2274–2282
131. Hassouna R, Zizzari P, Tolle VJ (2010) The ghrelin/obestatin balance in the physiological and pathological control of growth hormone secretion, body composition and food intake. J Neuroendocrinol 22:793–804
132. Rogler G, Rosano G (2014) The heart and the gut. Eur Heart J 35:426–430
133. Yang IA, Barton SJ, Rorke S, Cakebread JA, Keith TP, Clough JB, Holgate ST, Holloway JW (2004) Toll-like receptor 4 polymorphism and severity of atopy in asthmatics. Genes Immun 5:41–45
134. Kiechl S, Lorenz E, Reindl M, Wiedermann CJ, Oberhollenzer F, Bonora E, Willeit J, Schwartz DA (2002) Toll-like receptor 4 polymorphisms and atherogenesis. N Engl J Med 347:185–192

135. Yasuno K, Bakircioglu M, Low SK, Bilgüvar K, Gaál E, Ruigrok YM, Niemelä M, Hata A, Bijlenga P, Kasuya H, Jääskeläinen JE, Krex D, Auburger G, Simon M, Krischek B, Ozturk AK, Mane S, Rinkel GJ, Steinmetz H, Hernesniemi J, Schaller K, Zembutsu H, Inoue I, Palotie A, Cambien F, Nakamura Y, Lifton RP, Günel M (2011) Common variant near the endothelin receptor type A (EDNRA) gene is associated with intracranial aneurysm risk. Proc Natl Acad Sci U S A 108:19707–19712
136. Sugamura K, Keaney JF (2012) Nicotine: linking smoking to abdominal aneurysms. Nat Med 18:856–858
137. Yu E, Mercer J, Bennett M (2012) Mitochondria in vascular disease. Cardiovasc Res 95:173–182
138. Peluffo G, Calcerrada P, Piacenza L, Pizzano N, Radi R (2009) Superoxide-mediated inactivation of nitric oxide and peroxynitrite formation by tobacco smoke in vascular endothelium: studies in cultured cells and smokers. Am J Physiol Heart Circ Physiol 296:H1781–H1792
139. Vazquez-Padron RI, Mateu D, Rodriguez-Menocal L, Wei Y, Webster KA, Pham SM (2010) Novel role of Egr-1 in nicotine-related neointimal formation. Cardiovasc Res 88:296–303
140. Fukuda N (2010) Cigarette smoking induces vascular proliferative disease through the activation of Egr-1. Cardiovasc Res 88:207–208
141. Thorin E, Thorin-Trescases N (2009) Vascular endothelial ageing, heartbeat after heartbeat. Cardiovasc Res 84(1):24-32
142. Patel KP, Li YF, Hirooka Y (2001) Role of nitric oxide in central sympathetic outflow. Exp Biol Med 226:814–824
143. Broeders MA, Doevendans PA, Bekkers BC, Bronsaer R, van Gorsel E, Heemskerk JW, Egbrink MG, van Breda E, Reneman RS, van Der Zee R (2000) Nebivolol: A third-generation β-blocker that augments vascular nitric oxide release: endothelial β_2-adrenergic receptor-mediated nitric oxide production. Circulation 102:677–684
144. Lundby A, Andersen MN, Steffensen AB, Horn H, Kelstrup CD, Francavilla C, Jensen LJ, Schmitt N, Thomsen MB, Olsen JV (2013) In vivo phosphoproteomics analysis reveals the cardiac targets of β-adrenergic receptor signaling. Sci Signal 6:rs11
145. Wirth A, Benyó Z, Lukasova M, Leutgeb B, Wettschureck N, Gorbey S, Örsy P, Horváth B, Maser-Gluth C, Greiner E, Lemmer B, Schütz G, Gutkind JS, Offermanns S (2008) G12-G13–LARG–mediated signaling in vascular smooth muscle is required for salt-induced hypertension. Nat Med 14:64–68
146. Pandolfi A, Di Pietro N (2010) High glucose, nitric oxide, and adenosine: a vicious circle in chronic hyperglycaemia? Cardiovasc Res 86:9–11
147. Farías M, Puebla C, Westermeier F, Jo MJ, Pastor-Anglada M, Casanello P, Sobrevia L (2010) Nitric oxide reduces SLC29A1 promoter activity and adenosine transport involving transcription factor complex hCHOP-C/EBPα in human umbilical vein endothelial cells from gestational diabetes. Cardiovasc Res 86:45–54
148. Xu A, Vanhoutte PM (2012) Adiponectin and adipocyte fatty acid binding protein in the pathogenesis of cardiovascular disease. Am J Physiol Heart Circ Physiol 302:H1231–H1240
149. Ferrante AW (2013) Improving metabolism by throwing out all the JNK. Science 339:147–148
150. Han MS, Jung DY, Morel C, Lakhani SA, Kim JK, Flavell RA, Davis RJ (2013) JNK expression by macrophages promotes obesity-induced insulin resistance and inflammation. Science 339:218–222
151. Mani A, Radhakrishnan J, Wang H, Mani A, Mani MA, Nelson-Williams C, Carew KS, Mane S, Najmabadi H, Wu D, Lifton RP (2007) LRP6 mutation in a family with early coronary disease and metabolic risk factors. Science 315:1278–1282
152. Hao HX, Cardon CM, Swiatek W, Cooksey RC, Smith TL, Wilde J, Boudina S, Abel ED, McClain DA, Rutter J (2007) PAS kinase is required for normal cellular energy balance. Proc Natl Acad Sci U S A 104:15466–15471
153. Office of Communications and Public Liaison (2014) Brain basics: understanding sleep. National Institute of Neurological Disorders and Stroke, National Institutes of Health, Bethesda, MD. www.ninds.nih.gov/disorders/brain_basics

154. Markwald RR, Melanson EL, Smith MR, Higgins J, Perreault L, Eckel RH, Wright KP (2013) Impact of insufficient sleep on total daily energy expenditure, food intake, and weight gain. Proc Natl Acad Sci U S A 110:5695–5700
155. Möller-Levet CS, Archer SN, Bucca G, Laing EE, Slak A, Kabiljo R, Lo JC, Santhi N, von Schantz M, Smith CP, Dijk DJ (2013) Effects of insufficient sleep on circadian rhythmicity and expression amplitude of the human blood transcriptome. Proc Natl Acad Sci U S A 110:E1132–E1141
156. Chouchou F, Pichot V, Pépin JL, Tamisier R, Celle S, Maudoux D, Garcin A, Lévy P, Barthélémy JC, Roche F, PROOF Study Group (2013) Sympathetic overactivity due to sleep fragmentation is associated with elevated diurnal systolic blood pressure in healthy elderly subjects: the PROOF-SYNAPSE study. Eur Heart J 34:2122–2131
157. Argod J, Pépin JL, Smith RP, Lévy P (2000) Comparison of esophageal pressure with pulse transit time as a measure of respiratory effort for scoring obstructive nonapneic respiratory events. Am J Respir Crit Care Med 162:87–93
158. Chen WF, Maguire S, Sowcik M, Luo W, Koh K, Sehgal A (2014) A neuron–glia interaction involving GABA transaminase contributes to sleep loss in sleepless mutants. Mol Psychiatr (dx.doi.org/10.1038/mp.2014.11)
159. Stanke-Labesque F, Pépin JL, Gautier-Veyret E, Lévy P, Bäck M (2014) Leukotrienes as a molecular link between obstructive sleep apnoea and atherosclerosis. Cardiovasc Res 101:187–193
160. Kohler M, Stradling JR (2010) Mechanisms of vascular damage in obstructive sleep apnea. Nat Rev Cardiol 7:677–685
161. Nanduri J, Makarenko V, Reddy VD, Yuan G, Pawar A, Wang N, Khan SA, Zhang X, Kinsman B, Peng YJ, Kumar GK, Fox AP, Godley LA, Semenza GL, Prabhakar NR (2102) Epigenetic regulation of hypoxic sensing disrupts cardiorespiratory homeostasis. Proc Natl Acad Sci U S A 109:2515–2520
162. Pakkanen TA, Kerminen VM, Korhonen CH, Hillamo RE, Aarnio P, Koskentalo T, Maenhaut W (2001) Urban and rural ultrafine ($PM_{0.1}$) particles in the Helsinki area. Atmos Environ 35:4593–4607
163. World Health Organization (WHO), Public Health and Environment (PHE). Database 2011: outdoor air pollution in cities. www.who.int/phe/health_topics/outdoorair/databases/cities/en
164. Araujo JA, Barajas B, Kleinman M, Wang X, Bennett BJ, Gong KW, Navab M, Harkema J, Sioutas C, Lusis AJ, Nel AE (2008) Ambient particulate pollutants in the ultrafine range promote early atherosclerosis and systemic oxidative stress. Circ Res 102:589–596
165. Ying Z, Yue P, Xu X, Zhong M, Sun Q, Mikolaj M, Wang A, Brook RD, Chen LC, Rajagopalan S (2009) Air pollution and cardiac remodeling: a role for RhoA/Rho-kinase. Am J Physiol Heart Circ Physiol 296:H1540–H1550
166. Mannucci PM (2013) Airborne pollution and cardiovascular disease: burden and causes of an epidemic. Eur Heart J 34:1251–1253
167. Tonne C, Wilkinson P (2013) Long-term exposure to air pollution is associated with survival following acute coronary syndrome. Eur Heart J 34:1306–1311
168. Kälsch H, Hennig F, Moebus S, Möhlenkamp S, Dragano N, Jakobs H, Memmesheimer M, Erbel R, Jöckel KH, Hoffmann B, Heinz Nixdorf Recall Study Investigative Group (Roggenbuck U, Slomiany U, Beck EM, Offner A, Münkel S, Schrader S, Peter R, Hirche H, Meinertz T, Bode C, Defeyter PJ, Güntert B, Halli T, Gutzwiller F, Heinen H, Hess O, Klein B, Löwel H, Reiser M, Schmidt G, Schwaiger M, Steinmüller C, Theorell T, Willich SN) (2014) Are air pollution and traffic noise independently associated with atherosclerosis: the Heinz Nixdorf Recall Study. Eur Heart J 35:853–860
169. Cairns BJ, Baigent C (2014) Air pollution and traffic noise: do they cause atherosclerosis? Eur Heart J 35:826–828
170. Pope CA, Burnett RT, Krewski D, Jerrett M, Shi Y, Calle EE, Thun MJ (2009) Cardiovascular mortality and exposure to airborne fine particulate matter and cigarette smoke: shape of the exposure–response relationship. Circulation 120:941–948

171. Rodó X, Curcoll R, Robinson M, Ballester J, Burns JC, Cayan DR, Lipkin WI, Williams BL, Couto-Rodriguez M, Nakamura Y, Uehara R, Tanimoto H, Morguí JA (2014) Tropospheric winds from northeastern China carry the etiologic agent of Kawasaki disease from its source to Japan. Proc Natl Acad Sci U S A 111:7952–7957
172. Bennett JW, Klich M (2003) Mycotoxins. Clin Microbiol Rev 16:497–516
173. Schunkert H, Erdmann J, Samani NJ (2010) Genetics of myocardial infarction: a progress report. Eur Heart J 31:918–925
174. Shin SY, Fauman EB, Petersen AK, Krumsiek J, Santos R, Huang J, Arnold M, Erte I, Forgetta V, Yang TP, Walter K, Menni C, Chen L, Vasquez L, Valdes AM, Hyde CL, Wang V, Ziemek D, Roberts P, Xi L, Grundberg E, Multiple Tissue Human Expression Resource (MuTHER) Consortium, Waldenberger M, Richards JB, Mohney RP, Milburn MV, John SL, Trimmer J, Theis FJ, Overington JP, Suhre K, Brosnan MJ, Gieger C, Kastenmüller G, Spector TD, Soranzo N (2014) An atlas of genetic influences on human blood metabolites. Nat Genet 46:543–550
175. Coucke PJ, Willaert A, Wessels MW, Callewaert B, Zoppi N, De Backer J, Fox JE, Mancini GMS, Kambouris M, Gardella R, Facchetti F, Willems PJ, Forsyth R, Dietz HC, Barlati S, Colombi M, Loeys B, De Paepe A (2006) Mutations in the facilitative glucose transporter GLUT10 alter angiogenesis and cause arterial tortuosity syndrome. Nat Genet 38:452–457
176. Durier S, Fassot C, Laurent S, Boutouyrie P, Couetil J-P, Fine E, Lacolley P, Dzau VJ, Pratt RE (2003) Physiological genomics of human arteries: quantitative relationship between gene expression and arterial stiffness. Circulation 108:1845–1851
177. Cowley AW (2006) The genetic dissection of essential hypertension. Nat Rev Genet 7:829–840
178. Ware JS, Petretto E, Cook SA (2013) Integrative genomics in cardiovascular medicine. Cardiovasc Res 97:623–630
179. Schroeder C, Jordan J (2012) Norepinephrine transporter function and human cardiovascular disease. Am J Physiol Heart Circ Physiol 303:H1273–H1282
180. Emilsson V, Thorleifsson G, Zhang B, Leonardson AS, Zink F, Zhu J, Carlson S, Helgason A, Walters GB, Gunnarsdottir S, Mouy M, Steinthorsdottir V, Eiriksdottir GH, Bjornsdottir G, Reynisdottir I, Gudbjartsson D, Helgadottir A, Jonasdottir A, Jonasdottir A, Styrkarsdottir U, Gretarsdottir S, Magnusson KP, Stefansson H, Fossdal R, Kristjansson K, Gislason HG, Stefansson T, Leifsson BG, Thorsteinsdottir U, Lamb JR, Gulcher JR, Reitman ML, Kong A, Schadt EE, Stefansson K (2008) Genetics of gene expression and its effect on disease. Nature 452:423–428
181. Rader DJ, Daugherty A (2008) Translating molecular discoveries into new therapies for atherosclerosis. Nature 451:904–913
182. Vega Genome Browser release 51 (2012) vega.sanger.ac.uk/Homo_sapiens
183. Barriot R, Breckpot J, Thienpont B, Brohée S, Van Vooren S, Coessens B, Tranchevent LC, Van Loo P, Gewillig M, Devriendt K, Moreau Y (2010) Collaboratively charting the gene-to-phenotype network of human congenital heart defects. Genome Med 2:16
184. Ching YH, Ghosh TK, Cross SJ, Packham EA, Honeyman L, Loughna S, Robinson TE, Dearlove AM, Ribas G, Bonser AJ, Thomas NR, Scotter AJ, Caves LS, Tyrrell GP, Newbury-Ecob RA, Munnich A, Bonnet D, Brook JD (2005) Mutation in myosin heavy chain 6 causes atrial septal defect. Nat Genet 37:423–428
185. Zhu Y, Gramolini AO, Walsh MA, Zhou YQ, Slorach C, Friedberg MK, Takeuchi JK, Sun H, Henkelman RM, Backx PH, Redington AN, MacLennan DH, Bruneau BG (2008) Tbx5-dependent pathway regulating diastolic function in congenital heart disease. Proc Natl Acad Sci U S A 105:5519–5524
186. Garg V, Muth AN, Ransom JF, Schluterman MK, Barnes R, King IN, Grossfeld PD, Srivastava D (2005) Mutations in NOTCH1 cause aortic valve disease. Nature 437:270–274
187. Rodriguez-Viciana P, Tetsu O, Tidyman WE, Estep AL, Conger BA, Santa Cruz M, McCormick F, Rauen KA (2006) Germline mutations in genes within the MAPK pathway cause cardio-facio-cutaneous syndrome. Science 311:1287–1290

188. d'Udekem Y, Iyengar AJ, Cochrane AD, Grigg LE, Ramsay JM, Wheaton GR, Penny DJ, Brizard CP (2007) The Fontan procedure: contemporary techniques have improved long-term outcomes. Circulation 116:I157–I164
189. Sheu TWH, Tsai SF, Hwang WS, Chang TM (1999) A finite element study of the blood flow in total cavopulmonary connection. Comput Fluids 28:19–39
190. Tsai SF, Sheu TWH, Chang TM (2001) Lung effect on the hemodynamics in pulmonary artery. Int J Numer Methods Fluids 36:249–263
191. Dubini G, de Leval MR, Pietrabissa R, Montevecchi FM, Fumero R (1996) A numerical fluid mechanical study of repaired congenital heart defects. Application to the total cavopulmonary connection. J Biomech 29:111–121
192. Troianowski G, Taylor CA, Feinstein JA, Vignon-Clementel IE (2011) Three-dimensional simulations in Glenn patients: clinically based boundary conditions, hemodynamic results and sensitivity to input data. J Biomech Eng 133:111006
193. Baretta A, Corsini C, Yang W, Vignon-Clementel IE, Marsden AL, Feinstein JA, Hsia TY, Dubini G, Migliavacca F, Pennati G, Modeling of Congenital Hearts Alliance (MOCHA) Investigators (2011) Virtual surgeries in patients with congenital heart disease: a multi-scale modelling test case. Philos Trans R Soc Lond Math Phys Eng Sci 369:4316–4330
194. Wu J, Chen P, Li Y, Ardell C, Der T, Shohet R, Chen M, Wright GL (2013) HIF-1α in heart: protective mechanisms. Am J Physiol Heart Circ Physiol 305:H821–H828
195. Yang F, Zhang H, Mei Y, Wu M (2014) Reciprocal regulation of HIF-1α and lincRNA-p21 modulates the Warburg effect. Mol Cell 53:88–100
196. Bartels K, Grenz A, Eltzschig HK (2013) Hypoxia and inflammation are two sides of the same coin. Proc Natl Acad Sci U S A 110:18351–18352
197. Scholz CC, Cavadas MA, Tambuwala MM, Hams E, Rodríguez J, Kriegsheim Av, Cotter P, Bruning U, Fallon PG, Cheong A, Cummins EP, Taylor CT (2013) Regulation of IL-1β-induced NF-κ B by hydroxylases links key hypoxic and inflammatory signaling pathways. Proc Natl Acad Sci U S A 110:18490–18495
198. de Bernard M, Rizzuto R (2014) Toll-like receptors hit calcium. EMBO Rep 15:455–617
199. Shintani Y, Drexler HC, Kioka H, Terracciano CM, Coppen SR, Imamura H, Akao M, Nakai J, Wheeler AP, Higo S, Nakayama H, Takashima S, Yashiro K, Suzuki K (2014) Toll-like receptor 9 protects non-immune cells from stress by modulating mitochondrial ATP synthesis through the inhibition of SERCA2. EMBO Rep 15:438–445
200. Lee HS, Hwang CY, Shin SY, Kwon KS, Cho KH (2014) MLK3 is part of a feedback mechanism that regulates different cellular responses to reactive oxygen species. Sci Signal 7:ra52
201. Maher TJ, Ren Y, Li Q, Braunlin E, Garry MG, Sorrentino BP, Martin CM (2014) ATP-binding cassette transporter Abcg2 lineage contributes to the cardiac vasculature after oxidative stress. Am J Physiol Heart Circ Physiol 306:H1610–H1618
202. Billia F, Hauck L, Grothe D, Konecny F, Rao V, Kim RH, Mak TW (2013) Parkinson-susceptibility gene DJ-1/PARK7 protects the murine heart from oxidative damage in vivo. Proc Natl Acad Sci U S A 110:6085–6090
203. Sumandea MP, Steinberg SF (2011) Redox signaling and cardiac sarcomeres. J Biol Chem 286:9921–9927
204. Sastri M, Haushalter KJ, Panneerselvam M, Chang P, Fridolfsson H, Finley JC, Ng D, Schilling JM, Miyanohara A, Day ME, Hakozaki H, Petrosyan S, Koller A, King CC, Darshi M, Blumenthal DK, Ali SS, Roth DM, Patel HH, Taylor SS (2013) A kinase interacting protein (AKIP1) is a key regulator of cardiac stress. Proc Natl Acad Sci U S A 110:E387–E396
205. Pacher P, Beckman JS, Liaudet L (2007) Nitric oxide and peroxynitrite in health and disease. Physiol Rev 87:315–424
206. Danson EJ, Paterson DJ (2006) Reactive oxygen species and autonomic regulation of cardiac excitability. J Cardiovasc Electrophysiol 17:S104–S112
207. Iida N (1999) Nitric oxide mediates sympathetic vasoconstriction at supraspinal, spinal, and synaptic levels. Am J Physiol Heart Circ Physiol 276:H918–H925

Chap. 3. Adverse Cardiac Remodeling

208. Cohn JN, Ferrari R, Sharpe N (2000) Cardiac remodeling—concepts and clinical implications: a consensus paper from an international forum on cardiac remodeling. J Am Coll Cardiol 35:569–582
209. Pieske B (2004) Reverse remodeling in heart failure—fact or fiction? Eur Heart J Suppl 6:D66–D78
210. Richardson P, McKenna W, Bristow M, Maisch B, Mautner B, O'Connell J, Olsen E, Thiene G, Goodwin J, Gyarfas I, Martin I, Nordet P (1996) Report of the 1995 World Health Organization/International Society and Federation of Cardiology Task Force on the definition and classification of cardiomyopathies. Circulation 93:841–842
211. Bradshaw AD (2009) The role of SPARC in extracellular matrix assembly. J Cell Commun Signal 3:239–246
212. Rentz TJ, Poobalarahi F, Bornstein P, Sage EH, Bradshaw AD (2007) SPARC regulates processing of procollagen I and collagen fibrillogenesis in dermal fibroblasts. J Biol Chem 282:22062–22071
213. Harris BS, Zhang Y, Card L, Rivera LB, Brekken RA, Bradshaw AD (2011) SPARC regulates collagen interaction with cardiac fibroblast cell surfaces. Am J Physiol Heart Circ Physiol 301:H841–H847
214. Satoh M, Nakamura M, Akatsu T, Shimoda Y, Segawa I, Hiramori K (2005) Myocardial osteopontin expression is associated with collagen fibrillogenesis in human dilated cardiomyopathy. Eur J Heart Fail 7:755–762
215. Traianedes K, Martin TJ, Findlay DM (1996) Regulation of osteopontin expression by type I collagen in preosteoblastic UMR201 cells. Connect Tissue Res 34:63–74
216. Kaartinen MT, Pirhonen A, Linnala-Kankkunen A, Mäenpää PH (1999) Cross-linking of osteopontin by tissue transglutaminase increases its collagen binding properties. J Biol Chem 274:1729–1735
217. Driesen RB, Nagaraju CK, Abi-Char J, Coenen T, Lijnen PJ, Fagard RH, Sipido KR, Petrov V (2014) Reversible and irreversible differentiation of cardiac fibroblasts. Cardiovasc Res 101:411–422
218. Rahmutula D, Marcus GM, Wilson EE, Ding CH, Xiao Y, Paquet AC, Barbeau R, Barczak AJ, Erle DJ, Olgin JE (2013) Molecular basis of selective atrial fibrosis due to overexpression of transforming growth factor-β1. Cardiovasc Res 99:769–779
219. Fabritz L, Kirchhof P (2013) Selective atrial profibrotic signalling in mice and man. Cardiovasc Res 99:592–594
220. Zhao T, Zhao W, Chen Y, Li VS, Meng W, Sun Y (2013) Platelet-derived growth factor-D promotes fibrogenesis of cardiac fibroblasts. Am J Physiol Heart Circ Physiol 304:H1719–H1726
221. Zhao T, Zhao W, Meng W, Liu C, Chen Y, Sun Y (2014) Vascular endothelial growth factor-C: Its unrevealed role in fibrogenesis. Am J Physiol Heart Circ Physiol 306:H789–H796
222. Westermann D, Heymans S (2014) Fibrosis or hypertrophy: let TIMPs decide. Cardiovasc Res 103:196–197
223. Fan D, Takawale A, Basu R, Patel V, Lee J, Kandalam V, Wang X, Oudit GY, Kassiri Z (2014) Differential role of TIMP2 and TIMP3 in cardiac hypertrophy, fibrosis, and diastolic dysfunction. Cardiovasc Res 103:268–280
224. Sirish P, Li N, Liu JY, Lee KS, Hwang SH, Qiu H, Zhao C, Ma SM, López JE, Hammock BD, Chiamvimonvat N (2013) Unique mechanistic insights into the beneficial effects of soluble epoxide hydrolase inhibitors in the prevention of cardiac fibrosis. Proc Natl Acad Sci U S A 110:5618–5623
225. van Bilsen M, van Nieuwenhoven FA, van der Vusse GJ (2009) Metabolic remodelling of the failing heart: beneficial or detrimental? Cardiovasc Res 81:420–428
226. Ingwall JS (2009) Energy metabolism in heart failure and remodelling. Cardiovasc Res 81:412–419

227. Pellieux C, Montessuit C, Papageorgiou I, Lerch R (2009) Angiotensin II downregulates the fatty acid oxidation pathway in adult rat cardiomyocytes via release of tumour necrosis factor-α. Cardiovasc Res 82:341–350
228. Tsutsui H, Kinugawa S, Matsushima S (2009) Mitochondrial oxidative stress and dysfunction in myocardial remodelling. Cardiovasc Res 81:449–456
229. Ferguson BS, Harrison BC, Jeong MY, Reid BG, Wempe MF, Wagner FF, Holson EB, McKinsey TA (2013) Signal-dependent repression of DUSP5 by class I HDACs controls nuclear ERK activity and cardiomyocyte hypertrophy. Proc Natl Acad Sci U S A 110:9806–9811
230. Pandya K, Kim HS, Smithies O (2006) Fibrosis, not cell size, delineates beta-myosin heavy chain reexpression during cardiac hypertrophy and normal aging in vivo. Proc Natl Acad Sci U S A 103:16864–16869
231. Fu J, Yoon HG, Qin J, Wong J (2007) Regulation of P-TEFb elongation complex activity by CDK9 acetylation. Molecular and Cellular Biology 27:4641–4651
232. Catalucci D, Condorelli G (2013) HEXIM1: a new player in myocardial hypertrophy? Cardiovasc Res 99:1–3
233. Montano MM, Desjardins CL, Doughman YQ, Hsieh YH, Hu Y, Bensinger HM, Wang C, Stelzer JE, Dick TE, Hoit BD, Chandler MP, Yu X, Watanabe M (2013) Inducible re-expression of HEXIM1 causes physiological cardiac hypertrophy in the adult mouse. Cardiovasc Res 99:74–82
234. Balligand JL (2013) Reducing damage through Nrf-2. Cardiovasc Res 100:1–3
235. Kannan S, Muthusamy VR, Whitehead KJ, Wang L, Gomes AV, Litwin SE, Kensler TW, Abel ED, Hoidal JR, Rajasekaran NS (2013) Nrf2 deficiency prevents reductive stress-induced hypertrophic cardiomyopathy. Cardiovasc Res 100:63–73
236. Sano M, Minamino T, Toko H, Miyauchi H, Orimo M, Qin Y, Akazawa H, Tateno K, Kayama Y, Harada M, Shimizu I, Asahara T, Hamada H, Tomita S, Molkentin JD, Zou Y, Komuro I (2007) p53-induced inhibition of Hif-1 causes cardiac dysfunction during pressure overload. Nature 446:444–448
237. Tijsen AJ, Pinto YM, Creemers EE (2012) Circulating microRNAs as diagnostic biomarkers for cardiovascular diseases. Am J Physiol Heart Circ Physiol 303:H1085–H1095
238. Zampetaki A, Willeit P, Drozdov I, Kiechl S, Mayr M (2012) Profiling of circulating microRNAs: from single biomarkers to re-wired networks. Cardiovasc Res 93:555–562
239. Diehl P, Fricke A, Sander L, Stamm J, Bassler N, Htun N, Ziemann M, Helbing T, El-Osta A, Jowett JB, Peter K (2012) Microparticles: major transport vehicles for distinct microRNAs in circulation. Cardiovasc Res 93:633–644
240. van Rooij E, Sutherland LB, Liu N, Williams AH, McAnally J, Gerard RD, Richardson JA, Olson EN (2006) A signature pattern of stress-responsive microRNAs that can evoke cardiac hypertrophy and heart failure. Proc Natl Acad Sci U S A 103:18255–18260
241. Lin Z, Murtaza I, Wang K, Jiao J, Gao J, Li PF (2009) miR-23a functions downstream of NFATc3 to regulate cardiac hypertrophy. Proc Natl Acad Sci U S A 106:12103–12108
242. Bernardo BC, Gao XM, Winbanks CE, Boey EJ, Tham YK, Kiriazis H, Gregorevic P, Obad S, Kauppinen S, Du XJ, Lin RC, McMullen JR (2012) Therapeutic inhibition of the miR-34 family attenuates pathological cardiac remodeling and improves heart function. Proc Natl Acad Sci U S A 109:17615–17620
243. Huang GN, Thatcher JE, McAnally J, Kong Y, Qi X, Tan W, DiMaio JM, Amatruda JF, Gerard RD, Hill JA, Bassel-Duby R, Olson EN (2012) C/EBP transcription factors mediate epicardial activation during heart development and injury. Science 338:1599–1603
244. Harzheim D, Movassagh M, Foo RS, Ritter O, Tashfeen A, Conway SJ, Bootman MD, Roderick HL (2009) Increased InsP3Rs in the junctional sarcoplasmic reticulum augment Ca^{2+} transients and arrhythmias associated with cardiac hypertrophy. Proc Natl Acad Sci U S A 106:11406–11411
245. Seo K, Rainer PP, Shalkey Hahn V, Lee DI, Jo SH, Andersen A, Liu T, Xu X, Willette RN, Lepore JJ, Marino JP Jr, Birnbaumer L, Schnackenberg CG, Kass DA (2014) Combined TRPC3 and TRPC6 blockade by selective small-molecule or genetic deletion inhibits pathological cardiac hypertrophy. Proc Natl Acad Sci U S A 111:1551–1556

246. Minami T (2014) Calcineurin-NFAT activation and DSCR-1 auto-inhibitory loop: how is homoeostasis regulated? J Biochem 155:217–226
247. Keune WJ, Jones DR, Divecha N (2013) PtdIns5P and Pin1 in oxidative stress signaling. Adv Biol Regul 53:179–189
248. Yang B, Rizzo V (2007) TNFα potentiates protein-tyrosine nitration through activation of NADPH oxidase and eNOS localized in membrane rafts and caveolae of bovine aortic endothelial cells. Am J Physiol Heart Circ Physiol 292:H954–H962
249. Mehta PK, Griendling KK (2006) Angiotensin II cell signaling: physiological and pathological effects in the cardiovascular system. Am J Physiol Cell Physiol 292:C82–C97
250. Griendling KK, Sorescu D, Lassegue B, Ushio-Fukai M (2000) Modulation of protein kinase activity and gene expression by reactive oxygen species and their role in vascular physiology and pathophysiology. Arterioscler Thromb Vasc Biol 20:2175–2183
251. Griendling KK, Sorescu D, Ushio-Fukai M (2000) NAD(P)H oxidase: Role in cardiovascular biology and disease. Circ Res 86:494–501
252. Keune WJ, Jones DR, Bultsma Y, Sommer L, Zhou XZ, Lu KP, Divecha N (2012) Regulation of phosphatidylinositol-5-phosphate signaling by pin1 determines sensitivity to oxidative stress. Sci Signal 5:ra86
253. Hori M, Nishida K (2009) Oxidative stress and left ventricular remodelling after myocardial infarction. Cardiovasc Res 81:457–464
254. Camici GG, Schiavoni M, Francia P, Bachschmid M, Martin-Padura I, Hersberger M, Tanner FC, Pelicci P, Volpe M, Anversa P, Luscher TF, Cosentino F (2007) Genetic deletion of p66(Shc) adaptor protein prevents hyperglycemia-induced endothelial dysfunction and oxidative stress. Proc Natl Acad Sci U S A 104:5217–5222
255. McMahon KA, Zajicek H, Li WP, Peyton MJ, Minna JD, Hernandez VJ, Luby-Phelps K, Anderson RG (2009) SRBC/cavin-3 is a caveolin adapter protein that regulates caveolae function. EMBO J 28:1001–1015
256. Hernandez VJ, Weng J, Ly P, Pompey S, Dong H, Mishra L, Schwarz M, Anderson RG, Michaely P (2013) Cavin-3 dictates the balance between ERK and Akt signaling. eLIFE 2:e00905
257. Ogata T, Naito D, Nakanishi N, Hayashi YK, Taniguchi T, Miyagawa K, Hamaoka T, Maruyama N, Matoba S, Ikeda K, Yamada H, Oh H, Ueyama T (2014) MURC/Cavin-4 facilitates recruitment of ERK to caveolae and concentric cardiac hypertrophy induced by α1-adrenergic receptors. Proc Natl Acad Sci U S A 111:3811–3816
258. Kazakov A, Müller P, Jagoda P, Semenov A, Böhm M, Laufs U (2012) Endothelial nitric oxide synthase of the bone marrow regulates myocardial hypertrophy, fibrosis, and angiogenesis. Cardiovasc Res 93:397–405
259. Kobayashi N, Takeshima H, Fukushima H, Koguchi W, Mamada Y, Hirata H, Machida Y, Shinoda M, Suzuki N, Yokotsuka F, Tabei K, Matsuoka H (2009) Cardioprotective effects of pitavastatin on cardiac performance and remodeling in failing rat hearts. Am J Hypertens 22:176–182
260. Duncker DJ, van Deel ED (2012) Endothelial nitric oxide synthase and cardiac remodelling: Location, location, location? Cardiovasc Res 93:383–385
261. Gao S, Long CL, Wang RH, Wang H (2009) K_{ATP} activation prevents progression of cardiac hypertrophy to failure induced by pressure overload via protecting endothelial function. Cardiovasc Res 83:444–456
262. Lorenz K, Schmitt JP, Schmitteckert EM, Lohse MJ (2009) A new type of ERK1/2 autophosphorylation causes cardiac hypertrophy. Nat Med 15:75–83
263. Gul R, Park JH, Kim SY, Jang KY, Chae JK, Ko JK, Kim UH (2009) Inhibition of ADP-ribosyl cyclase attenuates angiotensin II-induced cardiac hypertrophy. Cardiovasc Res 81:582–591
264. Ainscough JF, Drinkhill MJ, Sedo A, Turner NA, Brooke DA, Balmforth AJ, Ball SG (2009) Angiotensin II type-1 receptor activation in the adult heart causes blood pressure-independent hypertrophy and cardiac dysfunction. Cardiovasc Res 81:592–600

265. Woodcock EA, Matkovich SJ (2005) Ins(1,4,5)P_3 receptors and inositol phosphates in the heart—evolutionary artefacts or active signal transducers? Pharmacol Ther 107:240–251
266. Abdellatif M, Packer SE, Michael LH, Zhang D, Charng MJ, Schneider MD (1998) A Ras-dependent pathway regulates RNA polymerase II phosphorylation in cardiac myocytes: Implications for cardiac hypertrophy. Mol Cell Biol 18:6729–6736
267. Chang L, Zhang J, Tseng YH, Xie CQ, Ilany J, Brüning JC, Sun Z, Zhu X, Cui T, Youker KA, Yang Q, Day SM, Kahn CR, Chen YE (2007) Rad GTPase deficiency leads to cardiac hypertrophy. Circulation 116:2976–2983
268. Bannister RA, Colecraft HM, Beam KG (2008) Rem inhibits skeletal muscle EC coupling by reducing the number of functional L-type Ca^{2+} channels. Biophys J 94:2631–2638
269. Mahalakshmi RN, Nagashima K, Ng MY, Inagaki N, Hunziker W, Béguin P (2007) Nuclear transport of Kir/Gem requires specific signals and importin α5 and is regulated by calmodulin and predicted serine phosphorylations. Traffic 8:1150–1163
270. Kuwahara K, Teg Pipes GC, McAnally J, Richardson JA, Hill JA, Bassel-Duby R, Olson EN (2007) Modulation of adverse cardiac remodeling by STARS, a mediator of MEF2 signaling and SRF activity. J Clin Investig 117:1324–1334
271. Ai D, Pang W, Li N, Xu M, Jones PD, Yang J, Zhang Y, Chiamvimonvat N, Shyy JY, Hammock BD, Zhu Y (2009) Soluble epoxide hydrolase plays an essential role in angiotensin II-induced cardiac hypertrophy. Proc Natl Acad Sci U S A 106:564–569
272. Letavernier E, Zafrani L, Perez J, Letavernier B, Haymann JP, Baud L (2012) The role of calpains in myocardial remodelling and heart failure. Cardiovasc Res 96:38–45
273. Yamaguchi O, Taneike M, Otsu K (2012) Cooperation between proteolytic systems in cardiomyocyte recycling. Cardiovasc Res 96:46–52
274. Matsuda T, Zhai P, Maejima Y, Hong C, Gao S, Tian B, Goto K, Takagi H, Tamamori-Adachi M, Kitajima S, Sadoshima J (2008) Distinct roles of GSK-3α and GSK-3β phosphorylation in the heart under pressure overload. Proc Natl Acad Sci U S A 105:20900–20905
275. Muraski JA, Fischer KM, Wu W, Cottage CT, Quijada P, Mason M, Din S, Gude N, Alvarez R, Rota M, Kajstura J, Wang Z, Schaefer E, Chen X, MacDonnel S, Magnuson N, Houser SR, Anversa P, Sussman MA (2008) Pim-1 kinase antagonizes aspects of myocardial hypertrophy and compensation to pathological pressure overload. Proc Natl Acad Sci U S A 105:13889–13894
276. Rigor DL, Bodyak N, Bae S, Choi JH, Zhang L, Ter-Ovanesyan D, He Z, McMullen JR, Shioi T, Izumo S, King GL, Kang PM (2009) Phosphoinositide 3-kinase Akt signaling pathway interacts with protein kinase Cβ2 in the regulation of physiologic developmental hypertrophy and heart function. Am J Physiol Heart Circ Physiol 296:H566–H572
277. Lukowski R, Rybalkin SD, Loga F, Leiss V, Beavo JA, Hofmann F (2010) Cardiac hypertrophy is not amplified by deletion of cGMP-dependent protein kinase I in cardiomyocytes. Proc Natl Acad Sci U S A 107:5646–5651 and 107:E99
278. Kass DA, Takimoto E (2010) Regulation and role of myocyte cyclic GMP-dependent protein kinase-1. Proc Natl Acad Sci U S A 107:E98
279. Kramann N, Hasenfuβ G, Seidler T (2014) B-RAF and its novel negative regulator reticulocalbin 1 (RCN1) modulates cardiomyocyte hypertrophy. Cardiovasc Res 102:88–96
280. Honoré B (2009) The rapidly expanding CREC protein family: members, localization, function, and role in disease. Bioessays 31:262–277
281. Hauck L, Harms C, An J, Rohne J, Gertz K, Dietz R, Endres M, von Harsdorf R (2008) Protein kinase CK2 links extracellular growth factor signaling with the control of p27(Kip1) stability in the heart. Nat Med 14:315–324
282. Fassett JT, Xu X, Hu X, Zhu G, French J, Chen Y, Bache RJ (2009) Adenosine regulation of microtubule dynamics in cardiac hypertrophy. Am J Physiol Heart Circ Physiol 297:H523–H532
283. Kreis TE (1987) Microtubules containing detyrosinated tubulin are less dynamic. EMBO J 6:2597–2606
284. Rachmin I, Tshori S, Smith Y, Oppenheim A, Marchetto S, Kay G, Foo RS, Dagan N, Golomb E, Gilon D, Borg JP, Razin E (2014) Erbin is a negative modulator of cardiac hypertrophy. Proc Natl Acad Sci U S A 111:5902–5907

285. Akazawa H, Komuro I (2014) Dickkopf-3: a stubborn protector of cardiac hypertrophy. Cardiovasc Res 102:6–8
286. Zhang Y, Liu Y, Zhu XH, Zhang XD, Jiang DS, Bian ZY, Zhang XF, Chen K, Wei X, Gao L, Zhu LH, Yang Q, Fan GC, Lau WB, Ma X, Li H (2014) Dickkopf-3 attenuates pressure overload-induced cardiac remodelling. Cardiovasc Res 102:35–45
287. Zhou Y, Bourcy K, Kang YJ (2009) Copper-induced regression of cardiomyocyte hypertrophy is associated with enhanced vascular endothelial growth factor receptor-1 signalling pathway. Cardiovasc Res 84:54–63
288. Shi JJ, Wei L (2012) Regulation of JAK/STAT signalling by SOCS in the myocardium. Cardiovasc Res 96:345–347
289. González A, López B, Ravassa S, Beaumont J, Arias T, Hermida N, Zudaire A, Díez J (2009) Biochemical markers of myocardial remodelling in hypertensive heart disease. Cardiovasc Res 81:509–518
290. Weinberg EO, Shimpo M, Hurwitz S, Tominaga S, Rouleau JL, Lee RT (2003) Identification of serum soluble ST2 receptor as a novel heart failure biomarker. Circulation 107:721–726
291. Cittadini A, Monti MG, Iaccarino G, Castiello MC, Baldi A, Bossone E, Longobardi S, Marra AM, Petrillo V, Saldamarco L, During MJ, Saccà L, Condorelli G (2012) SOCS1 gene transfer accelerates the transition to heart failure through the inhibition of the gp130/JAK/STAT pathway. Cardiovasc Res 96:381–390
292. Villeneuve C, Caudrillier A, Ordener C, Pizzinat N, Parini A, Mialet-Perez J (2009) Dose-dependent activation of distinct hypertrophic pathways by serotonin in cardiac cells. Am J Physiol Heart Circ Physiol 297:H821–H828
293. Walther S, Pluteanu F, Renz S, Nikonova Y, Maxwell JT, Yang LZ, Schmidt K, Edwards JN, Wakula P, Groschner K, Maier LS, Spiess J, Blatter LA, Pieske B, Kockskämper J (2014) Urocortin 2 stimulates nitric oxide production in ventricular myocytes via Akt- and PKA-mediated phosphorylation of eNOS at serine 1177. Am J Physiol Heart Circ Physiol 307:H689–H700
294. Liu YH, D'Ambrosio M, Liao TD, Peng H, Rhaleb NE, Sharma U, André S, Gabius HJ, Carretero OA (2009) N-acetyl-seryl-aspartyl-lysyl-proline prevents cardiac remodeling and dysfunction induced by galectin-3, a mammalian adhesion/growth-regulatory lectin. Am J Physiol Heart Circ Physiol 296:H404–H412
295. Chen YX, Wang XQ, Fang CF, Wang JF, Tang LJ (2008) Value of BNP and tumour marker CA125 in patients with heart failure. Acta Cardiol 63:501–506
296. Sartoretto JL, Jin BY, Bauer M, Gertler FB, Liao R, Michel T (2009) Regulation of VASP phosphorylation in cardiac myocytes: differential regulation by cyclic nucleotides and modulation of protein expression in diabetic and hypertrophic heart. Am J Physiol Heart Circ Physiol 297:H1697–H1710
297. Pillai VB, Sundaresan NR, Kim GH, Samant S, Moreno-Vinasco L, Garcia JG, Gupta M (2013) Nampt secreted from cardiomyocytes promotes development of cardiac hypertrophy and adverse ventricular remodeling. Am J Physiol Heart Circ Physiol 304:H415–H426
298. Michael G, Xiao L, Qi XY, Dobrev D, Nattel S (2009) Remodelling of cardiac repolarization: how homeostatic responses can lead to arrhythmogenesis. Cardiovasc Res 81:491–499
299. Li H, He C, Feng J, Zhang Y, Tang Q, Bian Z, Bai X, Zhou H, Jiang H, Heximer SP, Qin M, Huang H, Liu PP, Huang C (2010) Regulator of G protein signaling 5 protects against cardiac hypertrophy and fibrosis during biomechanical stress of pressure overload. Proc Natl Acad Sci U S A 107:13818–13823
300. Heineke J, Auger-Messier M, Correll RN, Xu J, Benard MJ, Yuan W, Drexler H, Parise LV, Molkentin JD (2010) CIB1 is a regulator of pathological cardiac hypertrophy. Nat Med 16:872–879
301. Kaptoge S, Seshasai SR, Gao P, Freitag DF, Butterworth AS, Borglykke A, Di Angelantonio E, Gudnason V, Rumley A, Lowe GD, Jørgensen T, Danesh J (2014) Inflammatory cytokines and risk of coronary heart disease: new prospective study and updated meta-analysis. Eur Heart J 35:578–589
302. Ridker PM (2014) Targeting inflammatory pathways for the treatment of cardiovascular disease. Eur Heart J 35:540–543

303. Xiao L, Liu Y, Wang N (2014) New paradigms in inflammatory signaling in vascular endothelial cells. Am J Physiol Heart Circ Physiol 306:H317-H325
304. Frantz S, Bauersachs J, Ertl G (2009) Post-infarct remodelling: contribution of wound healing and inflammation. Cardiovasc Res 81:474–481
305. Shintani Y, Kapoor A, Kaneko M, Smolenski RT, D'Acquisto F, Coppen SR, Harada-Shoji N, Lee HJ, Thiemermann C, Takeshima S, Yashiro K, Suzuki K (2013) TLR9 mediates cellular protection by modulating energy metabolism in cardiomyocytes and neurons. Proc Natl Acad Sci U S A 110:5109–5114
306. Levin JZ, Horvitz HR (1992) The Caenorhabditis elegans unc-93 gene encodes a putative transmembrane protein that regulates muscle contraction. J Cell Biol 117:143–155
307. Seizer P, Gawaz M, May AE (2014) Cyclophilin A and EMMPRIN (CD147) in cardiovascular diseases. Cardiovasc Res 102:17–23
308. Yuan W, Ge H, He B (2010) Pro-inflammatory activities induced by CyPA-EMMPRIN interaction in monocytes. Atherosclerosis 213:415–421
309. Sun Y (2009) Myocardial repair/remodelling following infarction: roles of local factors. Cardiovasc Res 81:482–490
310. Pchejetski D, Foussal C, Alfarano C, Lairez O, Calise D, Guilbeau-Frugier C, Schaak S, Seguelas MH, Wanecq E, Valet P, Parini A, Kunduzova O (2012) Apelin prevents cardiac fibroblast activation and collagen production through inhibition of sphingosine kinase 1. Eur Heart J 33:2360–2369
311. Chen D, Fang F, Yang Y, Chen J, Xu G, Xu Y, Gao Y (2013) Brahma-related gene 1 (Brg1) epigenetically regulates CAM activation during hypoxic pulmonary hypertension. Cardiovasc Res 100:363–373
312. Mackenzie Ross RV, Toshner MR, Soon E, Naeije R, Pepke-Zaba J (2013) Decreased time constant of the pulmonary circulation in chronic thromboembolic pulmonary hypertension. Am J Physiol Heart Circ Physiol 305:H259–H264
313. Frazziano G, Al Ghouleh I, Baust J, Shiva S, Champion HC, Pagano PJ (2014) Nox-derived ROS are acutely activated in pressure overload pulmonary hypertension: indications for a seminal role for mitochondrial Nox4. Am J Physiol Heart Circ Physiol 306:H197–H205
314. Larsen KO, Lygren B, Sjaastad I, Krobert KA, Arnkvaern K, Florholmen G, Ruud Larsen AK, Levy FO, Taskén K, Skjønsberg OH, Christensen G (2008) Diastolic dysfunction in alveolar hypoxia: a role for interleukin-18-mediated increase in protein phosphatase 2A. Cardiovasc Res 80:47–54
315. Jerkic M, Kabir MG, Davies A, Yu LX, McIntyre BA, Husain NW, Enomoto M, Sotov V, Husain M, Henkelman M, Belik J, Letarte M (2011) Pulmonary hypertension in adult Alk1 heterozygous mice due to oxidative stress. Cardiovasc Res 92:375–384
316. Wenzel D, Matthey M, Bindila L, Lerner R, Lutz B, Zimmer A, Fleischmann BK (2013) Endocannabinoid anandamide mediates hypoxic pulmonary vasoconstriction. Proc Natl Acad Sci U S A 110:18710–18715
317. Okamoto Y, Wang J, Morishita J, Ueda N (2007) Biosynthetic pathways of the endocannabinoid anandamide. Chem Biodivers 4:1842–1857

Chap. 4. Cardiomyopathies

318. Bruneau BG (2008) The developmental genetics of congenital heart disease. Nature 451:943–948
319. Chen Y, Lewis W, Diwan A, Cheng EH, Matkovich SJ, Dorn GW (2010) Dual autonomous mitochondrial cell death pathways are activated by Nix/BNip3L and induce cardiomyopathy. Proc Natl Acad Sci U S A 107:9035–9042
320. Day SM (2013) The ubiquitin proteasome system in human cardiomyopathies and heart failure. Am J Physiol Heart Circ Physiol 304:H1283–H1293

321. Seidman JG, Seidman C (2001) The genetic basis for cardiomyopathy: from mutation identification to mechanistic paradigms. Cell 104:557–567
322. Morano I (1999) Tuning the human heart molecular motors by myosin light chains. J Mol Med (Berlin, Germany) 77:544–555
323. Davis JS, Hassanzadeh S, Winitsky S, Lin H, Satorius C, Vemuri R, Aletras AH, Wen H, Epstein ND (2001) The overall pattern of cardiac contraction depends on a spatial gradient of myosin regulatory light chain phosphorylation. Cell 107:631–641
324. Oakley CE, Hambly BD, Curmi PM, Brown LJ (2004) Myosin binding protein C: structural abnormalities in familial hypertrophic cardiomyopathy. Cell Res 14:95–110
325. Mun JY, Previs MJ, Yu HY, Gulick J, Tobacman LS, Beck Previs S, Robbins J, Warshaw DM, Craig R (2014) Myosin-binding protein C displaces tropomyosin to activate cardiac thin filaments and governs their speed by an independent mechanism. Proc Natl Acad Sci U S A 111:2170–2175
326. Hofmann PA, Hartzell HC, Moss RL (1991) Alterations in Ca^{2+} sensitive tension due to partial extraction of C-protein from rat skinned cardiac myocytes and rabbit skeletal muscle fibers. J Gen Physiol 97:1141–1163
327. Hofmann PA, Greaser ML, Moss RL (1991) C-protein limits shortening velocity of rabbit skeletal muscle fibres at low levels of Ca^{2+} activation. J Physiol 439:701–715
328. Danowski BA, Imanaka-Yoshida K, Sanger JM, Sanger JW (1992) Costameres are sites of force transmission to the substratum in adult rat cardiomyocytes. J Cell Biol 118:1411–1420
329. Nakamori M, Takahashi MP (2011) The role of α-dystrobrevin in striated muscle. Int J Mol Sci 12:1660–1671
330. Srivastava D, Yu S (2006) Stretching to meet needs: integrin-linked kinase and the cardiac pump. Genes Dev 20:2327–2331
331. Chopra A, Patel A, Shieh AC, Janmey PA, Kresh JY (2012) α-Catenin localization and sarcomere self-organization on N-cadherin adhesive patterns are myocyte contractility driven. PLoS One 7:e47592
332. Harvey PA, Leinwand LA (2011) Cellular mechanisms of cardiomyopathy. J Cell Biol 194:355–365
333. Frank D, Kuhn C, Katus HA, Frey N (2006) The sarcomeric Z-disc: a nodal point in signalling and disease. J Mol Med 84:446–468
334. Samuel JL, Schaub MC, Zaugg M, Mamas M, Dunn WB, Swynghedauw B (2008) Genomics in cardiac metabolism. Cardiovasc Res 79:218–227
335. Hoppel CL, Tandler B, Parland W, Turkaly JS, Albers LD (1982) Hamster cardiomyopathy. A defect in oxidative phosphorylation in the cardiac interfibrillar mitochondria. J Biol Chem 257:1540–1548
336. de Belder AJ, Radomski M, Why H, Richardson PJ, Bucknall CA, Salas E, Martin JF (1993) Nitric oxide synthase activities in human myocardium. Lancet 341:84–85
337. Drexler H, Kästner S, Strobel A, Studer R, Brodde OE, Hasenfuss G (1998) Expression, activity and functional significance of inducible nitric oxide synthase in the failing human heart. J Am Coll Cardiol 32:955–963
338. Liu H, Ma Z, Lee SS (2000) Contribution of nitric oxide to the pathogenesis of cirrhotic cardiomyopathy in bile duct-ligated rats. Gastroenterology 118:937–944
339. Prendergast BD, Sagach VF, Shah AM (1997) Basal release of nitric oxide augments the Frank–Starling response in the isolated heart. Circulation 96:1320–1329
340. Paulus WJ, Vantrimpont PJ, Shah AM (1994) Acute effects of nitric oxide on left ventricular relaxation and diastolic distensibility in humans. Assessment by bicoronary sodium nitroprusside infusion. Circulation 89:2070–2078
341. Heymes C, Vanderheyden M, Bronzwaer JG, Shah AM, Paulus WJ (1999) Endomyocardial nitric oxide synthase and left ventricular preload reserve in dilated cardiomyopathy. Circulation 99:3009-3016
342. Wilson Tang WH, Tong W, Shrestha K, Wang Z, Levison BS, Delfraino B, Hu B, Troughton RW, Klein AL, Hazen SL (2008) Differential effects of arginine methylation on diastolic

dysfunction and disease progression in patients with chronic systolic heart failure. Eur Heart J 29:2506
343. Tangney JR, Chuang JS, Janssen MS, Krishnamurthy A, Liao P, Hoshijima M, Wu X, Meininger GA, Muthuchamy M, Zemljic-Harpf A, Ross RS, Frank LR, McCulloch AD, Omens JH (2013) Novel role for vinculin in ventricular myocyte mechanics and dysfunction. Biophys J 104:1623–1633
344. Chen JF, Murchison EP, Tang R, Callis TE, Tatsuguchi M, Deng Z, Rojas M, Hammond SM, Schneider MD, Selzman CH, Meissner G, Patterson C, Hannon GJ, Wang DZ (2008) Targeted deletion of Dicer in the heart leads to dilated cardiomyopathy and heart failure. Proc Natl Acad Sci U S A 105:2111–2116
345. Vennekens R (2013) A TRiP to heart failure. Cardiovasc Res 99:590–591
346. Iwata Y, Ohtake H, Suzuki O, Matsuda J, Komamura K, Wakabayashi S (2013) Blockade of sarcolemmal TRPV2 accumulation inhibits progression of dilated cardiomyopathy. Cardiovasc Res 99:760–768
347. Awad MM, Calkins H, Judge DP (2008) Mechanisms of disease: molecular genetics of arrhythmogenic right ventricular dysplasia/cardiomyopathy. Nat Clin Pract Cardiovasc Med 5:258–267
348. Hilfiker-Kleiner D, Kaminski K, Podewski E, Bonda T, Schaefer A, Sliwa K, Forster O, Quint A, Landmesser U, Doerries C, Luchtefeld M, Poli V, Schneider MD, Balligand JL, Desjardins F, Ansari A, Struman I, Nguyen NQ, Zschemisch NH, Klein G, Heusch G, Schulz R, Hilfiker A, Drexler H (2007) A cathepsin D-cleaved 16 kDa form of prolactin mediates postpartum cardiomyopathy. Cell 128:589–600
349. Lebeche D, Davidoff AJ, Hajjar RJ (2008) Interplay between impaired calcium regulation and insulin signaling abnormalities in diabetic cardiomyopathy. Nat Clin Pract Cardiovasc Med 5:715–724
350. Guarini G, Ohanyan VA, Kmetz JG, DelloStritto DJ, Thoppil RJ, Thodeti CK, Meszaros JG, Damron DS, Bratz IN (2012) Disruption of TRPV1-mediated coupling of coronary blood flow to cardiac metabolism in diabetic mice: role of nitric oxide and BK channels. Am J Physiol Heart Circ Physiol 303:H216–H223
351. Urbina P, Singla DK (2014) BMP-7 attenuates adverse cardiac remodeling mediated through M2 macrophages in prediabetic cardiomyopathy. Am J Physiol Heart Circ Physiol 307:H762–H772
352. Kyrychenko S, Poláková E, Kang C, Pocsai K, Ullrich ND, Niggli E, Shirokova N (2013) Hierarchical accumulation of RyR post-translational modifications drives disease progression in dystrophic cardiomyopathy. Cardiovasc Res 97:666–675
353. Gonzalez DR, Treuer AV, Lamirault G, Mayo V, Cao Y, Dulce RA, Hare JM (2014) NADPH oxidase-2 inhibition restores contractility and intracellular calcium handling and reduces arrhythmogenicity in dystrophic cardiomyopathy. Am J Physiol Heart Circ Physiol 307:H710–H721
354. Ingwall JS (2014) The energetic cost of contraction is higher in the myocardium of patients with hypertrophic cardiomyopathy. Cardiovasc Res 103:192–193
355. Dweck D, Sanchez-Gonzalez MA, Chang AN, Dulce RA, Badger CD, Koutnik AP, Ruiz EL, Griffin B, Liang J, Kabbaj M, Fincham FD, Hare JM, Overton JM, Pinto JR (2014) Long-term ablation of PKA-mediated cardiac troponin I phosphorylation leads to excitation-contraction uncoupling and diastolic dysfunction in a knock-in mouse model of hypertrophic cardiomyopathy. J Biol Chem 289:23097–23111
356. Cheng Y, Wan X, McElfresh TA, Chen X, Gresham KS, Rosenbaum DS, Chandler MP, Stelzer JE (2013) Impaired contractile function due to decreased cardiac myosin binding protein C content in the sarcomere. Am J Physiol Heart Circ Physiol 305:H52–H65
357. Timmer SA, Knaapen P (2013) Coronary microvascular function, myocardial metabolism, and energetics in hypertrophic cardiomyopathy: Insights from positron emission tomography. Eur Heart J Cardiovasc Imaging 14:95–10

358. Witjas-Paalberends ER, Güçlü A, Germans T, Knaapen P, Harms HJ, Vermeer AM, Christiaans I, Wilde AA, Dos Remedios C, Lammertsma AA, van Rossum AC, Stienen GJ, van Slegtenhorst M, Schinkel AF, Michels M, Ho CY, Poggesi C, van der Velden J (2014) Gene-specific increase in the energetic cost of contraction in hypertrophic cardiomyopathy caused by thick filament mutations. Cardiovasc Res 103:248–257
359. Cambronero F, Marín F, Roldán V, Hernández-Romero D, Valdés M, Lip GYH (2009) Biomarkers of pathophysiology in hypertrophic cardiomyopathy: implications for clinical management and prognosis. Eur Heart J 30:139–151
360. Sengupta PP, Sorajja D, Eleid MF, Somers VK, Ommen SR, Parish JM, Khandheria B, Tajik AJ (2008) Hypertrophic obstructive cardiomyopathy and sleep-disordered breathing: an unfavorable combination. Nat Clin Pract Cardiovasc Med 1038:14–15
361. Bayliss CR, Jacques AM, Leung MC, Ward DG, Redwood CS, Gallon CE, Copeland O, McKenna WJ, Dos Remedios C, Marston SB, Messer AE (2013) Myofibrillar Ca^{2+} sensitivity is uncoupled from troponin I phosphorylation in hypertrophic obstructive cardiomyopathy due to abnormal troponin T. Cardiovasc Res 97:500–508
362. Haudek SB, Trial J, Xia Y, Gupta D, Pilling D, Entman ML (2008) Fc receptor engagement mediates differentiation of cardiac fibroblast precursor cells. Proc Natl Acad Sci U S A 105:10179–10184
363. Hilfiker-Kleiner D, Kaminski K, Podewski E, Bonda T, Schaefer A, Sliwa K, Forster O, Quint A, Landmesser U, Doerries C, Luchtefeld M, Poli V, Schneider MD, Balligand JL, Desjardins F, Ansari A, Struman I, Nguyen NQ, Zschemisch NH, Klein G, Heusch G, Schulz R, Hilfiker A, Drexler H (2007) A cathepsin D-cleaved 16 kDa form of prolactin mediates postpartum cardiomyopathy. Cell 128:589–600

Chap. 5. Conduction and Rhythm Disorders

364. Poincaré H (1903) Science and method. Thomas Nelson and Sons, London (Maitland F translator)
365. Salama G, Bett GCL (2014) Sex differences in the mechanisms underlying long QT syndrome. Am J Physiol Heart Circ Physiol 307:H640–H648
366. Veeraraghavan R, Gourdie RG, Poelzing S (2014) Mechanisms of cardiac conduction: a history of revisions. Am J Physiol Heart Circ Physiol 306:H619–H627
367. Lopez-Izquierdo A, Pereira RO, Wende AR, Punske BB, Abel ED, Tristani-Firouzi M (2014) The absence of insulin signaling in the heart induces changes in potassium channel expression and ventricular repolarization. Am J Physiol Heart Circ Physiol 306:H747–H754
368. Strickberger SA, Conti J, Daoud EG, Havranek E, Mehra MR, Piña IL, Young J, Council on Clinical Cardiology Subcommittee on Electrocardiography and Arrhythmias and the Quality of Care and Outcomes Research Interdisciplinary Working Group, Heart Rhythm Society (2005) Patient selection for cardiac resynchronization therapy. Circulation 111:2146–2150
369. Shea JB, Sweeney MO (2003) Cardiac resynchronization therapy. A patient's guide. Circulation 108:e64–e66
370. Entcheva E (2014) Fiat lux in understanding cardiac pacing, resynchronization and signalling by way of optogenetics. Cardiovasc Res 102:342–343
371. Beiert T, Bruegmann T, Sasse P (2014) Optogenetic activation of Gq signalling modulates pacemaker activity of cardiomyocytes. Cardiovasc Res 102:507–516
372. Nussinovitch U, Shinnawi R, Gepstein L (2014) Modulation of cardiac tissue electrophysiological properties with light-sensitive proteins. Cardiovasc Res 102:176–187
373. Boineau JP, Canavan TE, Schuessler RB, Cain ME, Corr PB, Coxet JL (1988) Demonstration of a widely distributed atrial pacemaker complex in the human heart. Circulation 77:1221–1237

374. Sanders P, Morton JB, Kistler PM, Spence SJ, Davidson NC, Hussin A, Vohra JK, Sparks PB, Kalman JM (2004) Electrophysiological and electroanatomic characterization of the atria in sinus node disease: evidence of diffuse atrial remodeling. Circulation 109:1514–1522
375. Larson ED, St Clair JR, Sumner WA, Bannister RA, Proenza C (2013) Depressed pacemaker activity of sinoatrial node myocytes contributes to the age-dependent decline in maximum heart rate. Proc Natl Acad Sci U S A 110:18011–18016
376. Herrmann S, Lipp P, Wiesen K, Stieber J, Nguyen H, Kaiser E, Ludwig A (2013) The cardiac sodium-calcium exchanger NCX1 is a key player in the initiation and maintenance of a stable heart rhythm. Cardiovasc Res 99:780–788
377. Sah R, Mesirca P, Van den Boogert M, Rosen J, Mably J, Mangoni ME, Clapham DE (2013) Ion channel–kinase TRPM7 is required for maintaining cardiac automaticity. Proc Natl Acad Sci U S A 110:E3037–E3046
378. Billette J, Tadros R (2014) Integrated rate-dependent and dual pathway AV nodal functions: principles and assessment framework. Am J Physiol Heart Circ Physiol 306:H173–H183
379. Lab MJ (1982) Contraction-excitation feedback in myocardium. Physiological basis and clinical relevance. Circ Res 50:757–766
380. Balse E, Hatem SN (2014) A glimpse at cardiac ion channel macromolecular complexes. Cardiovasc Res 102:344–345
381. Mehta A, Sequiera GL, Ramachandra CJ, Sudibyo Y, Chung Y, Sheng J, Wong KY, Tan TH, Wong P, Liew R, Shim W (2014) Re-trafficking of hERG reverses long QT syndrome 2 phenotype in human iPS-derived cardiomyocytes. Cardiovasc Res 102:497–506
382. Zhang XD, Timofeyev V, Li N, Myers RE, Zhang DM, Singapuri A, Lau VC, Bond CT, Adelman J, Lieu DK, Chiamvimonvat N (2014) Critical roles of a small conductance Ca^{2+}-activated K^+ channel (SK3) in the repolarization process of atrial myocytes. Cardiovasc Res 101:317–325
383. Mahida S, Mills RW, Tucker NR, Simonson B, Macri V, Lemoine MD, Das S, Milan DJ, Ellinor PT (2014) Overexpression of KCNN3 results in sudden cardiac death. Cardiovasc Res 101:326–334
384. Skibsbye L, Poulet C, Diness JG, Bentzen BH, Yuan L, Kappert U, Matschke K, Wettwer E, Ravens U, Grunnet M, Christ T, Jespersen T (2014) Small-conductance calcium-activated potassium (SK) channels contribute to action potential repolarization in human atria. Cardiovasc Res 103:156–167
385. Nattel S, Dobrev D (2012) The multidimensional role of calcium in atrial fibrillation pathophysiology: mechanistic insights and therapeutic opportunities. Eur Heart J 33:1870–1877
386. Nattel S, Qi XY (2014) Calcium-dependent potassium channels in the heart: clarity and confusion. Cardiovasc Res 101:185–186
387. Olsen SK, Garbi M, Zampieri N, Eliseenkova AV, Ornitz DM, Goldfarb M, Mohammadi M (2003) Fibroblast growth factor (FGF) homologous factors share structural but not functional homology with FGFs. J Biol Chem 278:34226–34236
388. Sampson KJ, Kass RS (2010) Location, location, regulation: a novel role for β-spectrin in the heart. J Clin Investig 120:3434–3437
389. Hund TJ, Snyder JS, Wu X, Glynn P, Koval OM, Onal B, Leymaster ND, Unudurthi SD, Curran J, Camardo C, Wright PJ, Binkley PF, Anderson ME, Mohler PJ (2014) β_{IV}-Spectrin regulates TREK-1 membrane targeting in the heart. Cardiovasc Res 102:166–175
390. Hund TJ, Koval OM, Li J, Wright PJ, Qian L, Snyder JS, Gudmundsson H, Kline CF, Davidson NP, Cardona N, Rasband MN, Anderson ME, Mohler PJ (2010) A β_{IV}-spectrin/CaMKII signaling complex is essential for membrane excitability in mice. J Clin Investig 120:3508–3519
391. Sacconi L, Ferrantini C, Lotti J, Coppini R, Yan P, Loew LM, Tesi C, Cerbai E, Poggesi C, Pavone FS (2012) Action potential propagation in transverse–axial tubular system is impaired in heart failure. Proc Natl Acad Sci U S A 109:5815–5819
392. Knollmann BC, Roden DM (2008) A genetic framework for improving arrhythmia therapy. Nature 451:929–936

393. Wyman BT, Hunter WC, Prinzen FW, McVeigh ER (1999) Mapping propagation of mechanical activation in the paced heart with MRI tagging. Am J Physiol Heart Circ Physiol 276:881–891
394. Sermesant M, Rhode KS, Anjorin A, Hegde S, Sanchez-Ortiz G, Rueckert D, Lambiase P, Bucknall C, Hill D, Razavi R (2004) Simulation of the elctromechanical activity of the heart using XMR interventional imaging. In : Barillot C, Haynor DR, Hellier P (eds) Medical image computing and computer-assisted intervention—MICCAI 2004. Springer, Berlin
395. Arking DE, Pfeufer A, Post W, Kao WHL, Newton-Cheh C, Ikeda M, West K, Kashuk C, Akyo M, Perz S, Jalilzadeh S, Illig T, Gieger C, Guo CY, Larson MG, Wichmann HE, Marbán E, O'Donnell CJ, Hirschhorn JN, Kääb S, Spooner PM, Meitinger T, Chakravarti A (2006) A common genetic variant in the NOS1 regulator NOS1AP modulates cardiac repolarization. Nat Genet 38:644–651
396. Sato D, Xie LH, Sovari AA, Tran DX, Morita N, Xie F, Karagueuzian H, Garfinkel A, Weiss JN, Qu Z (2009) Synchronization of chaotic early afterdepolarizations in the genesis of cardiac arrhythmias. Proc Natl Acad Sci U S A 106:2983–2988
397. Roberts BN, Yang PC, Behrens SB, Moreno JD, Clancy CE (2012) Computational approaches to understand cardiac electrophysiology and arrhythmias. Am J Physiol Heart Circ Physiol 303:H766–H783
398. Toischer K, Hartmann N, Wagner S, Fischer TH, Herting J, Danner BC, Sag CM, Hund TJ, Mohler PJ, Belardinelli L, Hasenfuss G, Maier LS, Sossalla S (2013) Role of late sodium current as a potential arrhythmogenic mechanism in the progression of pressure-induced heart disease. J Mol Cell Cardiol 61:111–122
399. Yang L, Korge P, Weiss JN, Qu Z (2010) Mitochondrial oscillations and waves in cardiac myocytes: insights from computational models. Biophys J 98:1428–1438
400. Winfree AT, Strogatz SH (1984) Organizing centres for three-dimensional chemical waves. Nature 311:611–615
401. Panfilov AV, Pertsov AM (1984) Vortex ring in a 3-dimensional active medium described by reaction-diffusion equations (Russian). Dokl Akad Nauk SSSR 274:1500–1503
402. Panfilov AV, Keener JP (1993) Generation of reentry in anisotropic myocardium. J Cardiovasc Electrophysiol 4:412–421
403. Panfilov AV, Kerkhof PL (2004) Quantifying ventricular fibrillation: in silico research and clinical implications. IEEE Trans Biomed Eng 51:195–196. (Erratum. 558)
404. Nash MP, Panfilov AV (2004) Electromechanical model of excitable tissue to study reentrant cardiac arrhythmias. Prog Biophys Mol Biol 85:501–522
405. Trafford AW, Clarke JD, Richards MA, Eisner DA, Dibb KM (2013) Calcium signalling microdomains and the t-tubular system in atrial mycoytes: potential roles in cardiac disease and arrhythmias. Cardiovasc Res 98:192–203
406. Le Scouarnec S, Bhasin N, Vieyres C, Hund TJ, Cunha SR, Koval O, Marionneau C, Chen B, Wu Y, Demolombe S, Song LS, Le Marec H, Probst V, Schott JJ, Anderson ME, Mohler PJ (2008) Dysfunction in ankyrin-B-dependent ion channel and transporter targeting causes human sinus node disease. Proc Natl Acad Sci U S A 105:15617–15622
407. Jalife J, Berenfeld O, Mansour M (2002) Mother rotors and fibrillatory conduction: a mechanism of atrial fibrillation. Cardiovasc Res 54:204–216
408. Smaill BH, Zhao J, Trew ML (2013) Three-dimensional impulse propagation in myocardium: arrhythmogenic mechanisms at the tissue level. Circ Res 112:834–848
409. Mandapati R, Skanes A, Chen J, Berenfeld O, Jalife J (2000) Stable microreentrant sources as a mechanism of atrial fibrillation in the isolated sheep heart. Circulation 101:194–199
410. Berenfeld O (2010) Ionic and substrate mechanisms of atrial fibrillation: rotors and the excitation frequency approach. Arch cardiol Méx 80:301–314
411. Qi B, Wei Y, Chen S, Zhou G, Li H, Xu J, Ding Y, Lu X, Zhao L, Zhang F, Chen G, Zhao J, Liu S (2014) $Na_V1.8$ channels in ganglionated plexi modulate atrial fibrillation inducibility. Cardiovasc Res 102:480–486
412. Nattel S (2002) New ideas about atrial fibrillation 50 years on. Nature 415:219–226

413. Wittköpper K, Dobrev D, Eschenhagen T, El-Armouche A (2011) Phosphatase-1 inhibitor-1 in physiological and pathological β-adrenoceptor signalling. Cardiovasc Res 91:392–401
414. Chiang DY, Li N, Wang Q, Alsina KM, Quick AP, Reynolds JO, Wang G, Skapura D, Voigt N, Dobrev D, Wehrens XH (2014) Impaired local regulation of ryanodine receptor type 2 by protein phosphatase 1 promotes atrial fibrillation. Cardiovasc Res 103:178–187
415. Mahida S (2014) Transcription factors and atrial fibrillation. Cardiovasc Res 101:194–202
416. Fredj S, Sampson KJ, Liu H, Kass RS (2006) Molecular basis of ranolazine block of LQT-3 mutant sodium channels: evidence for site of action. Br J Pharmacol 148:16–24
417. Li GR, Sun HY, Zhang XH, Cheng LC, Chiu SW, Tse HF, Lau CP (2009) Omega-3 polyunsaturated fatty acids inhibit transient outward and ultra-rapid delayed rectifier K^+ currents and Na^+ current in human atrial myocytes. Cardiovasc Res 81:286–293
418. Chaldoupi SM, Loh P, Hauer RN, de Bakker JM, van Rijen HV (2009) The role of connexin40 in atrial fibrillation. Cardiovasc Res 84:15–23
419. Lip GYH (2013) Stroke and bleeding risk assessment in atrial fibrillation: when, how, and why? Eur Heart J 34:1041–1049
420. Cherry EM, Fenton FH, Gilmour RF (2012) Mechanisms of ventricular arrhythmias: a dynamical systems-based perspective. Am J Physiol Heart Circ Physiol 302:H2451–H2463
421. Cutler MJ, Plummer BN, Wan X, Sun QA, Hess D, Liu H, Deschenes I, Rosenbaum DS, Stamler JS, Laurita KR (2012) Aberrant S-nitrosylation mediates calcium-triggered ventricular arrhythmia in the intact heart. Proc Natl Acad Sci U S A 109:18186–18191
422. Jiang D, Chen W, Wang R, Zhang L, Chen SRW (2007) Loss of luminal Ca^{2+} activation in the cardiac ryanodine receptor is associated with ventricular fibrillation and sudden death. Proc Natl Acad Sci U S A 104:18309–18314
423. Chopra N, Knollmann BC (2013) Triadin regulates cardiac muscle couplon structure and microdomain Ca^{2+} signalling: a path towards ventricular arrhythmias. Cardiovasc Res 98:187–191
424. Florea SM, Blatter LA (2102) Regulation of cardiac alternans by β-adrenergic signaling pathways. Am J Physiol Heart Circ Physiol 303:H1047–H1056
425. Sasano T, McDonald AD, Kikuchi K, Donahue JK (2006) Molecular ablation of ventricular tachycardia after myocardial infarction. Nat Med 12:1256–1258
426. Moe GK, Rheinboldt WC, Abildskov JA (1964) A computer model of atrial fibrillation. Am Heart J 67:200–220
427. Gray RA, Pertsov AM, Jalife J (1998) Spatial and temporal organization during cardiac fibrillation. Nature 392:75–78
428. Witkowski FX, Leon LJ, Penkoske PA, Giles WR, Spano ML, Ditto WL, Winfree AT (1998) Spatiotemporal evolution of ventricular fibrillation. Nature 392:78–82
429. Bourgault Y, Éthier M, LeBlanc VG (2003) Simulation of electrophysiological waves with an unstructured finite element method. Math Model Numer Anal 37:649–661
430. Winfree AT (1989) Electrical instability in cardiac muscle: phase singularities and rotors. J Theor Biol 138:353–405
431. Panfilov AV (1998) Spiral breakup as a model of ventricular fibrillation. Chaos 8:57–64
432. Samie FH, Berenfeld O, Anumonwo J, Mironov SF, Udassi S, Beaumont J, Taffet S, Pertsov AM, Jalife J (2001) Rectification of the background potassium current: a determinant of rotor dynamics in ventricular fibrillation. Circ Res 89:1216–1223
433. Choi BR, Liu T, Salama G (2001) The distribution of refractory periods influences the dynamics of ventricular fibrillation. Circ Res 88:e49–e58
434. Newton-Cheh C, Eijgelsheim M, Rice KM, de Bakker PI, Yin X, Estrada K, Bis JC, Marciante K, Rivadeneira F, Noseworthy PA, Sotoodehnia N, Smith NL, Rotter JI, Kors JA, Witteman JC, Hofman A, Heckbert SR, O'Donnell CJ, Uitterlinden AG, Psaty BM, Lumley T, Larson MG, Stricker BHC (2009) Common variants at ten loci influence QT interval duration in the QTGEN Study. Nat Genet 41:399–406
435. Pfeufer A, Sanna S, Arking DE, Müller M, Gateva V, Fuchsberger C, Ehret GB, Orrú M, Pattaro C, Köttgen A, Perz S, Usala G, Barbalic M, Li M, Pütz B, Scuteri A, Prineas RJ,

Sinner MF, Gieger C, Najjar SS, Kao WH, Mühleisen TW, Dei M, Happle C, Möhlenkamp S, Crisponi L, Erbel R, Jöckel KH, Naitza S, Steinbeck G, Marroni F, Hicks AA, Lakatta E, Müller-Myhsok B, Pramstaller PP, Wichmann HE, Schlessinger D, Boerwinkle E, Meitinger T, Uda M, Coresh J, Kääb S, Abecasis GR, Chakravarti A (2009) Common variants at ten loci modulate the QT interval duration in the QTSCD Study. Nat Genet 41:407–414
436. Schwartz PJ, Ackerman MJ (2013) The long QT syndrome: a transatlantic clinical approach to diagnosis and therapy. Eur Heart J 34:3109–3116
437. Chen L, Marquardt ML, Tester DJ, Sampson KJ, Ackerman MJ, Kass RS (2007) Mutation of an A-kinase-anchoring protein causes long-QT syndrome. Proc Natl Acad Sci U S A 104:20990–20995
438. Ueda K, Valdivia C, Medeiros-Domingo A, Tester DJ, Vatta M, Farrugia G, Ackerman MJ, Makielski JC (2008) Syntrophin mutation associated with long QT syndrome through activation of the nNOS–SCN5A macromolecular complex. Proc Natl Acad Sci U S A 105:9355–9360
439. Sanguinetti MC, Tristani-Firouzi M (2006) hERG potassium channels and cardiac arrhythmia. Nature 440:463–469
440. Cordeiro JM, Marieb M, Pfeiffer R, Calloe K, Burashnikov E, Antzelevitch C (2009) Accelerated inactivation of the L-type calcium current due to a mutation in CACNB2b underlies Brugada syndrome. J Mol Cell Cardiol 46:695–703
441. Yuan L , Koivumäki JT, Liang B, Lorentzen LG, Tang C, Andersen MN, Svendsen JH, Tfelt-Hansen J, Maleckar M, Schmitt N, Olesen MS, Jespersen T (2014) Investigations of the $Na_V\beta1b$ sodium channel subunit in human ventricle; functional characterization of the H162P Brugada syndrome mutant. Am J Physiol Heart Circ Physiol 306:H1204–H1212
442. Gollob M, Redpath C, Roberts J (2011) The short QT syndrome: proposed diagnostic criteria. J Am Coll Cardiol 57:802–812
443. Rautaharju PM, Zhou SH, Wong S, Calhoun HP, Berenson GS, Prineas R, Davignon A (1992) Sex differences in the evolution of the electrocardiographic QT interval with age. Can J Cardiol 8:690–695
444. Deo M, Ruan Y, Pandit SV, Shah K, Berenfeld O, Blaufox A, Cerrone M, Noujaim SF, Denegri M, Jalife J, Priori SG (2013) KCNJ2 mutation in short QT syndrome 3 results in atrial fibrillation and ventricular proarrhythmia. Proc Natl Acad Sci U S A 110:4291–4296
445. Faggioni M, Knollmann BC (2012) Calsequestrin 2 and arrhythmias. Am J Physiol Heart Circ Physiol 302:H1250–H1260
446. Sidhu J, Roberts R (2003) Genetic basis and pathogenesis of familial WPW syndrome. Indian Pacing Electrophysiol J 3:197–201
447. Xie M, Zhan D, Dyck JRB, Li Y, Zhang H, Morishima M, Mann DL, Taffet GE, Baldini A, Khoury DS, Schneider MD (2006) A pivotal role for endogenous TGF-beta-activated kinase-1 in the LKB1/AMP-activated protein kinase energy-sensor pathway. Proc Natl Acad Sci U S A 103:17378–17383
448. Henriquez CS (2014) A brief history of tissue models for cardiac electrophysiology. IEEE Trans Biomed Eng 61:1457–1465
449. Meijler FL (1983) Atrial fibrillation: a new look at an old arrhythmia. J Am Coll Cardiol 2:391–393
450. Despopoulos A, Silbernagl S (2009) Color atlas of physiology, 6th ed. Thieme, Stuttgart
451. Duan DD (2013) Phenomics of cardiac chloride channels. Compr Physiol 3:667–692
452. Jost N (2009) Transmembrane ionic currents underlying cardiac action potential in mammalian hearts. In: Nánási PP (ed) Advances in cardiomyocyte research. Transworld Research Network, Kerala
453. Tripathi ON (2011) Cardiac ion channels and heart rate and rhythm. In: Tripathi ON, Ravens U, Sanguinetti MC (eds) Heart rate and rhythm. Molecular basis, pharmacological modulation and clinical implications. Part I: normal cardiac rhythm and pacemaker activity. Springer, Heidelberg

454. Hodgkin AL, Huxley AF (1952) A quantitative description of membrane current and its application to conduction and excitation in nerve. J Physiol 117:500–544
455. Wells RB (2010) The Hodgkin–Huxley Model. Biological signal processing. www.mrc.uidaho.edu/r̃wells/techdocs
456. Nelson ME (2004) Electrophysiological models. In: Koslow S, Subramaniam S (eds) Databasing the brain: from data to knowledge. Wiley, New York
457. Quinn TA, Kohl P (2013) Combining wet and dry research: experience with model development for cardiac mechano-electric structure-function studies. Cardiovasc Res 97:601–611
458. Bennett PM, Maggs AM, Baines AJ, Pinder JC (2006) The transitional junction: a new functional subcellular domain at the intercalated disc. Mole Biol Cell 17:2091–2100
459. LeGrice IJ, Smaill BH, Chai LZ, Edgar SG, Gavin JB, Hunter PJ (1995) Laminar structure of the heart: ventricular myocyte arrangement and connective tissue architecture in the dog. Am J Physiol Heart Circ Physiol 269:H571–H582
460. Weidmann S (1955) The effect of the cardiac membrane potential on the rapid availability of the sodium-carrying system. J Physiol 127:213–224
461. Noble D (1960) Cardiac action and pacemaker potentials based on the Hodgkin-Huxley equations. Nature 188:495–497
462. Brown HF, DiFrancesco D, Noble SJ (1979) How does adrenaline accelerate the heart? Nature 280:235–236
463. Kiyosue T, Arita M (1989) Late sodium current and its contribution to action potential configuration in guinea pig ventricular myocytes. Circ Res 64:389–397
464. Heppner DB, Plonsey R (1970) Simulation of electrical interaction of cardiac cells. Biophys J 10:1057–1075
465. Yang H, Borg TK, Wang Z, Ma Z, Gao BZ (2014) Role of the basement membrane in regulation of cardiac electrical properties. Ann Biomed Eng 42:1148–1157
466. Smith JM, Cohen RJ (1984) Simple finite-element model accounts for wide range of cardiac dysrhythmias. Proc Natl Acad Sci U S A 81:233–237
467. Spach MS, Miller WT, Geselowitz DB, Barr RC, Kootsey JM, Johnson EA (1981) The discontinuous nature of propagation in normal canine cardiac muscle. Evidence for recurrent discontinuities of intracellular resistance that affect the membrane currents. Circ Res 48:39–54
468. FitzHugh R (1961) Impulses and physiological states in theoretical models of nerve membrane. Biophys J 1:445–466
469. Nagumo J, Arimoto S, Yoshizawa S (1962) An active pulse transmission line simulating nerve axons. Proc IRE 50:2061–2070
470. Aliev RR, Panfilov AV (1996) A simple two-variable model of cardiac excitation. Chaos Solitons Fractals 7:293–301
471. Mitchell CC, Schaeffer DG (2003) A two-current model for the dynamics of cardiac membrane. Bull Math Biol 65:767–793
472. Fenton F, Karma A (1998) Vortex dynamics in three-dimensional continuous myocardium with fiber rotation: filament instability and fibrillation. Chaos 8:20-47
473. Djabella K, Sorine M (2006) A reduced differential model for cardiac action potentials. SIAM Conference on the Life Sciences, Raleigh, USA
474. ten Tusscher KH, Noble D, Noble PJ, Panfilov AV (2004) A model for human ventricular tissue. Am J Physiol Heart Circ Physiol 286:H1573–H1589
475. Bernus O, Wilders R, Zemlin CW, Verschelde H, Panfilov AV (2002) A computationally efficient electrophysiological model of human ventricular cells. Am J Physiol Heart Circ Physiol 282:H2296–H2308
476. Geselowitz DB, Miller WT (1983) A bidomain model for anisotropic cardiac muscle. Ann Biomed Eng 11:191–206
477. Colli-Franzone P, Guerri L, Tentoni S (1990) Mathematical modeling of the excitation process in myocardial tissue: influence of fiber rotation on the wavefront propagation and potential field. Math Biosci 101:155–235

478. Sundnes J, Lines GT, Cai X, Nielsen BF, Mardal KA, Tveito A (2006) Computing the electrical activity in the heart. Springer, Berlin
479. Britton OJ, Bueno-Orovio A, Van Ammel K, Lu HR, Towart R, Gallacher DJ, Rodriguez B (2013) Experimentally calibrated population of models predicts and explains intersubject variability in cardiac cellular electrophysiology. Proc Natl Acad Sci U S A 110:E2098–E2105

Chap. 6. Cardiac Valve Diseases

480. Hunt SA, Abraham WT, Chin MH, Feldman AM, Francis GS, Ganiats TG, Jessup M, Konstam MA, Mancini DM, Michl K, Oates JA, Rahko PS, Silver MA, Warner Stevenson L, Yancy CW (2005) ACC/AHA 2005 Guideline update for the diagnosis and management of chronic heart failure in the adult: a report of the American College of Cardiology and American Heart Association Task Force on Practice Guidelines (Writing Committee to Update the 2001 Guidelines for the Evaluation and Management of Heart Failure) in collaboration with the American College of Chest Physicians and the International Society for Heart and Lung Transplantation and endorsed by the Heart Rhythm Society. Circulation 112:e154–e235.
481. Misfeld M, Sievers HH (2007) Heart valve macro- and microstructure. Philos Trans R Soc Lond B Biol Sci 362:1421–1436
482. Muriago M, Sheppard MN, Ho SY, Anderson RH (1997) Location of the coronary arterial orifices in the normal heart. Clin Anat 10:297–302
483. Miragoli M, Yacoub MH, El-Hamamsy I, Sanchez-Alonso JL, Moshkov A, Mongkoldhumrongkul N, Padala M, Paramagurunathan S, Sarathchandra P, Korchev YE, Gorelik J, Chester AH (2014) Side-specific mechanical properties of valve endothelial cells. Am J Physiol Heart Circ Physiol 307:H15–H24
484. Bäck M, Gasser TC, Michel JB, Caligiuri G (2013) Biomechanical factors in the biology of aortic wall and aortic valve diseases. Cardiovasc Res 99:232–241
485. Ranganathan N, Lam JHC, Wigle ED, Silver MD (1970) Morphology of the human mitral valve. II. The valve leaflets. Circulation 41:459–467
486. Ho SY (2002) Anatomy of the mitral valve. Heart 4:iv5–iv10
487. McCarthy KP, Ring L, Rana BS (2010) Anatomy of the mitral valve: understanding the mitral valve complex in mitral regurgitation. Eur J Echocardiogr 11:i3–i9
488. Binder T, Goliasch G, Hübler O, Erbschwendner S, Wagner MS, Bergler-Klein J, Osis S, Fassbach M, Vorraber I, Ksica I, Vadehra A (2014) Anatomy and function of the mitral valve. Medical University of Vienna. 123sonography.com/node/20857
489. Nonaka M, Marui A, Shimamoto T, Fukuoka M, Masuyama S, Ikeda T, Komeda M (2006) Saline injection test (SIT) may provide inaccurate evaluation of mitral-left ventricular (LV) dimension during mitral valve plasty (MVP). Circulation 114:II–565
490. Deopujari R, Sinha U, Athavale SA (2013) Anatomy of left atrioventricular valve (mitral valve) leaflets in adult Indian cadavers. Int J Morphol 31:1276–1281
491. Jimenez JH, Liou SW, Padala M, He Z, Sacks M, Gorman RC, Gorman JH 3rd, Yoganathan AP (2007) A saddle-shaped annulus reduces systolic strain on the central region of the mitral valve anterior leaflet. J Thorac Cardiovasc Surg 134:1562–1568
492. Padala M, Hutchison RA, Croft LR, Jimenez JH, Gorman RC, Gorman JH, Sacks MS, Yoganathan AP (2009) Saddle shape of the mitral annulus reduces systolic strains on the P2 segment of the posterior mitral leaflet. Ann Thorac Surg 88:1499–1504
493. Amini R, Eckert CE, Koomalsingh K, McGarvey J, Minakawa M, Gorman JH, Gorman RC, Sacks MS (2012) On the in vivo deformation of the mitral valve anterior leaflet: effects of annular geometry and referential configuration. Ann Biomed Eng 40:1455–1467
494. Eckert CE, Zubiate B, Vergnat M, Gorman JH, Gorman RC, Sacks MS (2009) In vivo dynamic deformation of the mitral valve annulus. Ann Biomed Eng 37:1757–1771
495. Padala M, Sacks MS, Liou SW, Balachandran K, He Z, Yoganathan AP (2010) Mechanics of the mitral valve strut chordae insertion region. J Biomech Eng Trans ASME 132:081004

496. Lee CH, Amini R, Gorman RC, Gorman JH, Sacks MS (2014) An inverse modeling approach for stress estimation in mitral valve anterior leaflet valvuloplasty for in-vivo valvular biomaterial assessment. J Biomech 47:2055–2063
497. Baumgartner H, Hung J, Bermejo J, Chambers JB, Evangelista A, Griffin BP, Iung B, Otto CM, Pellikka PA, and Quiñones M (2009) Echocardiographic assessment of valve stenosis: EAE/ASE recommendations for clinical practice. J Am Soc Echocardiogr 22:1–23
498. Zoghbi WA, Enriquez-Sarano M, Foster E, Grayburn PA, Kraft CD, Levine RA, Nihoyannopoulos P (2003)
American Society of Echocardiography: recommendations for evaluation of the severity of native valvular regurgitation with two-dimensional and Doppler echocardiography. A report from the American Society of Echocardiography's Nomenclature and Standards Committee and The Task Force on Valvular Regurgitation, developed in conjunction with the American College of Cardiology Echocardiography Committee, the Cardiac Imaging Committee, Council on Clinical Cardiology, the American Heart Association, and the European Society of Cardiology Working Group on Echocardiography. Eur J Echocardiogr 4:237–261
499. Vanoverschelde JLJ (2012) Functional anatomy of aortic regurgitation. www.escardio.org
500. le Polain de Waroux JB, Pouleur AC, Goffinet C, Vancraeynest D, Van Dyck M, Robert A, Gerber BL, Pasquet A, El Khoury G, Vanoverschelde JL (2007) Functional anatomy of aortic regurgitation: accuracy, prediction of surgical repairability, and outcome implications of transesophageal echocardiography. Circulation 116:I264–I269
501. Lancellotti P (2012) Grading aortic stenosis severity when the flow modifies the gradientvalve area correlation. Cardiovasc Diagn Ther 2:6–9
502. Saito T, Muro T, Takeda H, Hyodo E, Ehara S, Nakamura Y, Hanatani A, Shimada K, Yoshiyama M (2012) Prognostic value of aortic valve area index in asymptomatic patients with severe aortic stenosis. Am J Cardiol 110:93–97
503. Jander N, Gohlke-Bärwolf C, Bahlmann E, Gerdts E, Boman K, Chambers JB, Egstrup K, Nienaber CA, Pedersen TR, Ray S, Rossebø AB, Willenheimer R, Kienzle RP, Wachtell K, Neumann FJ, Minners J (2014) Indexing aortic valve area by body surface area increases the prevalence of severe aortic stenosis. Heart 100:28–33
504. Peltonen T, Taskinen P, Näpänkangas J, Leskinen H, Ohtonen P, Soini Y, Juvonen T, Satta J, Vuolteenaho O, Ruskoaho H (2009) Increase in tissue endothelin-1 and ETA receptor levels in human aortic valve stenosis. Eur Heart J 30:242–249
505. Bekeredjian R, Grayburn PA (2005) Contemporary reviews in cardiovascular medicine—valvular heart disease—aortic regurgitation. Circulation 112:125–134
506. Bruce CJ, Connolly HM (2009) Valvular heart disease: changing concepts in disease management. Right-sided valve disease deserves a little more respect. Circulation 119:2726–2734
507. Lancellotti P, Rosenhek R, Pibarot P, Iung B, Otto CM, Tornos P, Donal E, Prendergast B, Magne J, La Canna G, Piérard LA, Maurer G (2013) ESC Working Group on Valvular Heart Disease Position Paper–heart valve clinics: organization, structure, and experiences. Eur Heart J 34:1597–1606

Chap. 7. Heart Failure

508. Punnoose LR, Givertz MM, Lewis EF, Pratibhu P, Stevenson LW, Desai AS (2011) Heart failure with recovered ejection fraction: a distinct clinical entity. J Card Fail 17:527–532
509. Bui AL, Horwich TB, Fonarow GC (2011) Epidemiology and risk profile of heart failure. Nat Rev Cardiol 8:30–41
510. Januzzi JL, Filippatos G, Nieminen M, Gheorghiade M (2012) Troponin elevation in patients with heart failure: on behalf of the third Universal Definition of Myocardial Infarction Global Task Force: Heart Failure Section. Eur Heart J 33:2265–2271
511. Lopez-Sendon J (2013) To BNP or not to BNP. Eur Heart J 34:2498–2500

512. Heckler EJ, Rancy PC, Kodali VK, Thorpe C (2008) Generating disulfides with the Quiescin-sulfhydryl oxidases. Biochim Biophys Acta (Mol Cell Res) 1783:567–577
513. Mebazaa A, Vanpoucke G, Thomas G, Verleysen K, Cohen-Solal A, Vanderheyden M, Bartunek J, Mueller C, Launay JM, Van Landuyt N, D'Hondt F, Verschuere E, Vanhaute C, Tuytten R, Vanneste L, De Cremer K, Wuyts J, Davies H, Moerman P, Logeart D, Collet C, Lortat-Jacob B, Tavares M, Laroy W, Januzzi JL, Samuel JL, Kas K (2012) Unbiased plasma proteomics for novel diagnostic biomarkers in cardiovascular disease: identification of quiescin Q6 as a candidate biomarker of acutely decompensated heart failure. Eur Heart J 33:2317–2324
514. Meyer S, van der Meer P, van Deursen VM, Jaarsma T, van Veldhuisen DJ, van der Wal MH, Hillege HL, Voors AA (2013) Neurohormonal and clinical sex differences in heart failure. Eur Heart J 34:2538–2547
515. Pattini L, Sassi R, Cerutti S (2014) Dissecting heart failure through the multiscale approach of systems medicine. IEEE Trans Biomed Eng 61:1593–1603
516. VanBuren P, Palmer BM (2010) Cooperative activation of the cardiac myofilament. The pivotal role of tropomyosin. Circulation 121:351–353
517. Davis JP, Tikunova SB (2008) Ca^{2+} exchange with troponin C and cardiac muscle dynamics. Cardiovasc Res 77:619–626
518. Sia SK, Li MX, Spyracopoulos L, Gagné SM, Liu W, Putkey JA, Sykes BD (1997) Structure of cardiac muscle troponin C unexpectedly reveals a closed regulatory domain. J Biol Chem 272:18216–18221
519. Gomes AV, Guzman G, Zhao J, Potter JD (2002) Cardiac troponin T isoforms affect the Ca^{2+} sensitivity and inhibition of force development. Insights into the role of troponin T isoforms in the heart. J Biol Chem 277:35341–35349
520. Anderson PA, Greig A, Mark TM, Malouf NN, Oakeley AE, Ungerleider RM, Allen PD, Kay BK (1995) Molecular basis of human cardiac troponin T isoforms expressed in the developing, adult, and failing heart. Circ Res 76:681–686
521. Muthuchamy M, Grupp IL, Grupp G, O'Toole BA, Kier AB, Boivin GP, Neumann J, Wieczorek DF (1995) Molecular and physiological effects of overexpressing striated muscle β-tropomyosin in the adult murine heart. J Biol Chem 270:30593–30603
522. Rajan S, Jagatheesan G, Karam CN, Alves ML, Bodi I, Schwartz A, Bulcao CF, D'Souza KM, Akhter SA, Boivin GP, Dube DK, Petrashevskaya N, Herr AB, Hullin R, Liggett SB, Wolska BM, Solaro RJ, Wieczorek DF (2010) Molecular and functional characterization of a novel cardiac-specific human tropomyosin isoform. Circulation 121:410–418
523. Li XE, Holmes KC, Lehman W, Jung H, Fischer S (2010) The shape and flexibility of tropomyosin coiled coils: implications for actin filament assembly and regulation. J Mol Biol 395:327–39
524. Jagatheesan G, Rajan S, Petrashevskaya N, Schwartz A, Boivin G, Vahebi S, DeTombe P, Solaro RJ, Labitzke E, Hilliard G, Wieczorek DF (2003) Functional importance of the carboxyl-terminal region of striated muscle tropomyosin. J Biol Chem 278:23204–23211
525. Hüttelmaier S, Illenberger S, Grosheva I, Rüdiger M, Singer RH, Jockusch BM (2001) Raver1, a dual compartment protein, is a ligand for PTB/hnRNPI and microfilament attachment proteins. J Cell Biol 155:775–786
526. Siliciano JD, Craig SW (1982) Meta-vinculin—a vinculin-related protein with solubility properties of a membrane protein. Nature 300:533–535
527. Thompson PM, Tolbert CE, Campbell SL (2013) Vinculin and metavinculin: Oligomerization and interactions with F-actin. FEBS Lett 587:1220–1229
528. Mercer TR, Dinger ME, Mattick JS (2009) Long non-coding RNAs: insights into functions. Nat Rev Genet 10:155–159
529. Klattenhoff CA, Scheuermann JC, Surface LE, Bradley RK, Fields PA, Steinhauser ML, Ding H, Butty VL, Torrey L, Haas S, Abo R, Tabebordbar M, Lee RT, Burge CB, Boyer LA (2013) Braveheart, a long noncoding RNA required for cardiovascular lineage commitment. Cell 152:570–583

530. Cappola TP, Matkovich SJ, Wang W, van Booven D, Li M, Wang X, Qu L, Sweitzer NK, Fang JC, Reilly MP, Hakonarson H, Nerbonne JM, Dorn GW (2011) Loss-of-function DNA sequence variant in the CLCNKA chloride channel implicates the cardio-renal axis in interindividual heart failure risk variation. Proc Natl Acad Sci U S A 108:2456–2461
531. Liggett SB, Cresci S, Kelly RJ, Syed FM, Matkovich SJ, Hahn HS, Diwan A, Martini JS, Sparks L, Parekh RR, Spertus JA, Koch WJ, Kardia SLR, Dorn GW (2008) A GRK5 polymorphism that inhibits β-adrenergic receptor signaling is protective in heart failure. Nat Med 14:510–517
532. Bongartz LG, Cramer MJ, Doevendans PA, Joles JA, Braam B (2005) The severe cardiorenal syndrome: 'Guyton revisited'. Eur Heart J 26:11–17
533. Estévez R, Boettger T, Stein V, Birkenhäger R, Otto E, Hildebrandt F, Jentsch TJ (2001) Barttin is a Cl^- channel β-subunit crucial for renal Cl^- reabsorption and inner ear K^+ secretion. Nature 414:558–561
534. Nomura N, Tajima M, Sugawara N, Morimoto T, Kondo Y, Ohno M, Uchida K, Mutig K, Bachmann S, Soleimani M, Ohta E, Ohta A, Sohara E, Okado T, Rai T, Jentsch TJ, Sasaki S, Uchida S (2011) Generation and analyses of R8L barttin knockin mouse. Am J Physiol Ren Physiol 301:F297–F307
535. Wu F, Zhang J, Beard DA (2009) Experimentally observed phenomena on cardiac energetics in heart failure emerge from simulations of cardiac metabolism. Proc Natl Acad Sci U S A 106:7143–7148
536. Mudd JO, Kass DA (2008) Tackling heart failure in the twenty-first century. Nature 451:919–928
537. Schägger H (2002) Respiratory chain supercomplexes of mitochondria and bacteria. Biochim Biophys Acta 1555:154–159
538. Rosca MG, Vazquez EJ, Kerner J, Parland W, Chandler MP, Stanley W, Sabbah HN, Hoppel CL (2008) Cardiac mitochondria in heart failure: decrease in respirasomes and oxidative phosphorylation. Cardiovasc Res 80:30–39
539. Konstantinidis K, Kitsis RN (2012) Cardiovascular biology: Escaped DNA inflames the heart. Nature 485:179–180
540. Oka T, Hikoso S, Yamaguchi O, Taneike M, Takeda T, Tamai T, Oyabu J, Murakawa T, Nakayama H, Nishida K, Akira S, Yamamoto A, Komuro I, Otsu K (2012) Mitochondrial DNA that escapes from autophagy causes inflammation and heart failure. Nature 485:251–255
541. Guo A, Zhang C, Wei S, Chen B, Song LS (2013) Emerging mechanisms of T-tubule remodelling in heart failure. Cardiovasc Res 98:204–215
542. Guo A, Zhang X, Iyer VR, Chen B, Zhang C, Kutschke WJ, Weiss RM, Franzini-Armstrong C, Song LS (2014) Overexpression of junctophilin-2 does not enhance baseline function but attenuates heart failure development after cardiac stress. Proc Natl Acad Sci U S A 111:12240–12245
543. Gorelik J, Wright PT, Lyon AR, Harding SE (2013) Spatial control of the Ŏ3b2AR system in heart failure: the transverse tubule and beyond. Cardiovasc Res 98:216–224
544. Ibrahim M, Terracciano CM (2013) Reversibility of T-tubule remodelling in heart failure: mechanical load as a dynamic regulator of the T-tubules. Cardiovasc Res 98:225–232
545. Terentyev D, Rochira JA, Terentyeva R, Roder K, Koren G, Li W (2014) Sarcoplasmic reticulum Ca^{2+} release is both necessary and sufficient for SK channel activation in ventricular myocytes. Am J Physiol Heart Circ Physiol 306:H738–H746
546. Zhong W, Mao S, Tobis S, Angelis E, Jordan MC, Roos KP, Fishbein MC, de Alboran IM, MacLellan WR (2006) Hypertrophic growth in cardiac myocytes is mediated by Myc through a cyclin D2-dependent pathway. EMBO J 25:3869–3879
547. Dorn GW (2010) Refugee receptors switch sides. Science 327:1586–1587
548. Lyon AR, MacLeod KT, Zhang Y, Garcia E, G Kikonda Kanda, Lab MJ, Korchev YE, Harding SE, Gorelik J (2009) Loss of T-tubules and other changes to surface topography in ventricular myocytes from failing human and rat heart. Proc Natl Acad Sci U S A 106:6854–6859

549. Reiser PJ, Portman MA, Ning XH, Schomisch Moravec C (2001) Human cardiac myosin heavy chain isoforms in fetal and failing adult atria and ventricles. Am J Physiol Heart Circ Physiol 280:H1814–H1820
550. Mahmud H, Ruifrok WP, Westenbrink BD, Cannon MV, Vreeswijk-Baudoin I, van Gilst WH, Silljé HH, de Boer RA (2013) Suicidal erythrocyte death, eryptosis, as a novel mechanism in heart failure-associated anaemia. Cardiovasc Res 98:37–46
551. Gaillard CA, Schiffelers RM (2013) Red blood cell: barometer of cardiovascular health? Cardiovasc Res 98:3–4XS
552. Breitbart A, Auger-Messier M, Molkentin JD, Heinekel J (2011) Myostatin from the heart: local and systemic actions in cardiac failure and muscle wasting. Am J Physiol Heart Circ Physiol 300:H1973–H1982
553. Ding Y, Li YL, Zimmerman MC, Davisson RL, Schultz HD (2009) Role of CuZn superoxide dismutase on carotid body function in heart failure rabbits. Cardiovasc Res 81:678–685
554. Ramchandra R, Hood SG, Denton DA, Woods RL, McKinley MJ, McAllen RM, May CN (2009) Basis for the preferential activation of cardiac sympathetic nerve activity in heart failure. Proc Natl Acad Sci U S A 106:924–928
555. Solaro RJ (2008) Multiplex kinase signaling modifies cardiac function at the level of sarcomeric proteins. J Biol Chem 283: 26829–26833
556. Sumandea CA, Garcia-Cazarin ML, Bozio CH, Sievert GA, Balke CW, Sumandea MP (2011) Cardiac troponin T, a sarcomeric AKAP, tethers protein kinase A at the myofilaments. J Biol Chem 286:530–541
557. Decker RS, Rines AK, Nakamura S, Naik TJ, Wassertsrom JA, Ardehali H (2010) Phosphorylation of contractile proteins in response to α- and β-adrenergic stimulation in neonatal cardiomyocytes. Trans Res 155:27–34
558. Konhilas JP, Irving TC, Wolska BM, Jweied EE, Martin AF, Solaro RJ, de Tombe PP (2003) Troponin I in the murine myocardium: influence on length-dependent activation and interfilament spacing. J Physiol 547:951–961
559. Wijnker PJ, Sequeira V, Foster DB, Li Y, Dos Remedios CG, Murphy AM, Stienen GJ, van der Velden J (2014) Length-dependent activation is modulated by cardiac troponin I bisphosphorylation at Ser23 and Ser24 but not by Thr143 phosphorylation. Am J Physiol Heart Circ Physiol 306:H1171–H1181
560. Zhang YH, Jin CZ, Jang JH, Wang Y (2014) Molecular mechanisms of neuronal nitric oxide synthase in cardiac function and pathophysiology. J Physiol. jp.physoc.org
561. Kamm KE, Stull JT (2011) Signaling to myosin regulatory light chain in sarcomeres. J Biol Chem 286:9941–9947
562. Huang HS, Lee EY (2008) Protein phosphatase-1 inhibitor-3 is an in vivo target of caspase-3 and participates in the apoptotic response. J Biol Chem 283:18135–46
563. Cohen PTW (2002) Protein phosphatase 1—targeted in many directions. J Cell Sci 115:241–256
564. Oliver CJ, Shenolikar S (1998) Physiologic importance of protein phosphatase inhibitors. Front Biosci 3:D961–D972
565. Nicolaou P, Hajjar RJ, Kranias EG (2009) Role of protein phosphatase-1 inhibitor-1 in cardiac physiology and pathophysiology. J Mol Cell Cardiol 47:365–371
566. Marx SO, Reiken S, Hisamatsu Y, Jayaraman T, Burkhoff D, Rosemblit N, Marks AR (2000) PKA phosphorylation dissociates FKBP12.6 from the calcium release channel (ryanodine receptor): defective regulation in failing hearts. Cell 101:365–376
567. Belevych AE, Radwański PB, Carnes CA, Györke S (2013) 'Ryanopathy': causes and manifestations of RyR2 dysfunction in heart failure. Cardiovasc Res 98:240–247
568. Mori J, Alrob OA, Wagg CS, Harris RA, Lopaschuk GD, Oudit GY (2013) ANG II causes insulin resistance and induces cardiac metabolic switch and inefficiency: a critical role of PDK4. Am J Physiol Heart Circ Physiol 304:H1103–H1113

569. Zucker IH, Schultz HD, Patel KP, Wang W, Gao L (2009) Regulation of central angiotensin type 1 receptors and sympathetic outflow in heart failure. Am J Physiol Heart Circ Physiol 297:H1557–H1566
570. Pan S, Chen HH, Dickey DM, Boerrigter G, Lee C, Kleppe LS, Hall JL, Lerman A, Redfield MM, Potter LR, Burnett JC, Simari RD (2009) Biodesign of a renal-protective peptide based on alternative splicing of B-type natriuretic peptide. Proc Natl Acad Sci U S A 106:11282–11287
571. Segers VFM, Lee RT (2008) Stem-cell therapy for cardiac disease. Nature 451:937–942
572. Hilal-Dandan R, He H, Martin JL, Brunton LL, Dillmann WH (2009) Endothelin downregulates SERCA2 gene and protein expression in adult rat ventricular myocytes: Regulation by pertussis toxin-sensitive Gi protein and cAMP. Am J Physiol Heart Circ Physiol 296:H728–H734
573. Dschietzig T, Richter C, Bartsch C, Laule M, Armbruster FP, Baumann G, Stangl K (2001) The pregnancy hormone relaxin is a player in human heart failure. FASEB J 15:2187–2195
574. Bronzwaer JGF, Paulus WJ (2008) Nitric oxide: the missing lusitrope in failing myocardium. Eur Heart J 29:2453–2455
575. Doehner W (2012) Diagnostic biomarkers in cardiovascular disease: the proteomics approach. Eur Heart J 33:2249–2251
576. Rinne A, Kapur N, Molkentin JD, Pogwizd SM, Bers DM, Banach K, Blatter LA (2010) Isoform- and tissue-specific regulation of the Ca^{2+}-sensitive transcription factor NFAT in cardiac myocytes and heart failure. Am J Physiol Heart Circ Physiol 298:H2001–H2009
577. Shimazaki M, Nakamura K, Kii I, Kashima T, Amizuka N, Li M, Saito M, Fukuda K, Nishiyama T, Kitajima S, Saga Y, Fukayama M, Sata M, Kudo A (2008) Periostin is essential for cardiac healing after acute myocardial infarction. J Exp Med 205:295–303
578. Ruiz-Ortega M, Rodríguez-Vita J, Sanchez-Lopez E, Carvajal G, Egido J (2007) TGF-β signaling in vascular fibrosis. Cardiovasc Research 74:196–206
579. Musa H, Tellez JO, Chandler NJ, Greener ID, Maczewski M, Mackiewicz U, Beresewicz A, Molenaar P, Boyett MR, Dobrzynski H (2009) P2 purinergic receptor mRNA in rat and human sinoatrial node and other heart regions. Naunyn Schmiedeberg's Arch Pharmacol 379:541–549
580. Hou M, Malmsjö M, Möller S, Pantev E, Bergdahl A, Zhao XH, Sun XY, Hedner T, Edvinsson L, Erlinge D. (1999) Increase in cardiac $P2X_1$- and $P2Y_2$-receptor mRNA levels in congestive heart failure. Life Sci 65:1195–1206
581. Nishida M, Sato Y, Uemura A, Narita Y, Tozaki-Saitoh H, Nakaya M, Ide T, Suzuki K, Inoue K, Nagao T, Kurose H (2008) P2Y6 receptor-Gα12/13 signalling in cardiomyocytes triggers pressure overload-induced cardiac fibrosis. EMBO J 27:3104–3115
582. Rikitake Y, Oyama N, Wang CY, Noma K, Satoh M, Kim HH, Liao JK (2005) Decreased perivascular fibrosis but not cardiac hypertrophy in ROCK1+/− haploinsufficient mice. Circulation 112:2959–2965
583. Nishida M, Tanabe S, Maruyama Y, Mangmool S, Urayama K, Nagamatsu Y, Takagahara S, Turner JH, Kozasa T, Kobayashi H, Sato Y, Kawanishi T, Inoue R, Nagao T, Kurose H (2005) Gα12/13- and reactive oxygen species-dependent activation of c-Jun NH2-terminal kinase and p38 mitogen-activated protein kinase by angiotensin receptor stimulation in rat neonatal cardiomyocytes. J Biol Chem 280:18434–18441
584. Hsu S, Nagayama T, Koitabashi N, Zhang M, Zhou L, Bedja D, Gabrielson KL, Molkentin JD, Kass DA, Takimoto E (2009) Phosphodiesterase 5 inhibition blocks pressure overload-induced cardiac hypertrophy independent of the calcineurin pathway. Cardiovasc Res 81:301–309
585. Westenbrink BD, Ruifrok WP, Voors AA, Tilton RG, van Veldhuisen DJ, Schoemaker RG, van Gilst WH, de Boer RA (2010) Vascular endothelial growth factor is crucial for erythropoietin-induced improvement of cardiac function in heart failure. Cardiovasc Res 87:30–39
586. Tamariz L, Hare JM (2010) Inflammatory cytokines in heart failure: roles in aetiology and utility as biomarkers. Eur Heart J 31:768–770
587. Yan RT, Cushman M, Redheuil A, Tracy RP, Arnett DK, Rosen BD, McClelland RL, Bluemke DA, Lima JA (2010) Relationship of interleukin-6 with regional and global left-ventricular function in asymptomatic individuals without clinical cardiovascular disease: insights from the multi-ethnic study of atherosclerosis. Eur Heart J 31:875–882

588. Xuan W, Liao Y, Chen B, Huang Q, Xu D, Liu Y, Bin J, Kitakaze M (2011) Detrimental effect of fractalkine on myocardial ischaemia and heart failure. Cardiovasc Res 92:385–393
589. Mayr M, Zampetaki A, Kiechl S (2013) MicroRNA biomarkers for failing hearts? Eur Heart J 34:2782–2783
590. Vogel B, Keller A, Frese KS, Leidinger P, Sedaghat-Hamedani F, Kayvanpour E, Kloos W, Backe C, Thanaraj A, Brefort T, Beier M, Hardt S, Meese E, Katus HA, Meder B (2013) Multivariate miRNA signatures as biomarkers for non-ischaemic systolic heart failure. Eur Heart J 34:2812–2823
591. Matkovich SJ, Van Booven DJ, Youker KA, Torre-Amione G, Diwan A, Eschenbacher WH, Dorn LE, Watson MA, Margulies KB, Dorn GW (2009) Reciprocal regulation of myocardial microRNAs and messenger RNA in human cardiomyopathy and reversal of the microRNA signature by biomechanical support. Circulation 119:1263–1271
592. Akat KM, Moore-McGriff D, Morozov P, Brown M, Gogakos T, Correa Da Rosa J, Mihailovic A, Sauer M, Ji R, Ramarathnam A, Totary-Jain H, Williams Z, Tuschl T, Schulze PC (2014) Comparative RNA-sequencing analysis of myocardial and circulating small RNAs in human heart failure and their utility as biomarkers. Proc Natl Acad Sci U S A 111:11151–11156
593. Kaye DM, Krum H (2007) Drug discovery for heart failure: a new era or the end of the pipeline? Nat Rev Drug Discov 6:127–139
594. Huang ZM, Gao E, Fonseca FV, Hayashi H, Shang X, Hoffman NE, Chuprun JK, Tian X, Tilley DG, Madesh M, Lefer DJ, Stamler JS, Koch WJ (2013) Convergence of G protein-coupled receptor and S-nitrosylation signaling determines the outcome to cardiac ischemic injury. Sci Signal 6:ra95

Chap. 8. Obstructive Coronary Artery Disease and Infarction

595. Burns CG, Burns CE (2014) A crowning achievement for deciphering coronary origins. Science 345:28–29
596. Tian X, Hu T, Zhang H, He L, Huang X, Liu Q, Yu W, He L, Yang Z, Yan Y, Yang X, Zhong TP, Pu WT, Zhou B (2014) De novo formation of a distinct coronary vascular population in neonatal heart. Science 345:90–94
597. Duncker DJ, Bache RJ (2008) Regulation of coronary blood flow during exercise. Physiol Rev 88:1009–1086
598. Labombarda F, Coutance G, Pellissier A, Mery-Alexandre C, Roule V, Maragnes P, Milliez P, Saloux E (2014) Major congenital coronary artery anomalies in a paediatric and adult population: a prospective echocardiographic study. Eur Heart J Cardiovasc Imaging 15:761–768
599. Hudson CL, Moritz AR, Wearn JT (1932) The extracardiac anastomoses of the coronary arteries. J Exp Med 56:919–925
600. Loukas M, Hanna M, Chen J, Tubbs RS, Anderson RH (2011) Extracardiac coronary arterial anastomoses. Clin Anat 24:137–142
601. Seiler C (2003) The human coronary collateral circulation. Heart 89:1352–1357
602. Toyota E, Ogasawara Y, Hiramatsu O, Tachibana H, Kajiya F, Yamamori S, Chilian WM (2005) Dynamics of flow velocities in endocardial and epicardial coronary arterioles. Am J Physiol Heart Circ Physiol 288:H1598–H1603
603. Pung YF, Sam WJ, Hardwick JP, Yin L, Ohanyan V, Logan S, Di Vincenzo L, Chilian WM (2013) The role of mitochondrial bioenergetics and reactive oxygen species in coronary collateral growth. Am J Physiol Heart Circ Physiol 305:H1275–H1280
604. Tune JD, Gorman MW, Feigl EO (2004) Matching coronary blood flow to myocardial oxygen consumption. J Appl Physiol 97:404–415
605. Perez-Aguilar S, Torres-Tirado D, Martell-Gallegos G, Velarde-Salcedo J, Barba de la Rosa AP, Knabb M, Rubio R (2014) G protein-coupled receptors mediate coronary flow- and agonist-induced responses via lectin-oligosaccharide interactions. Am J Physiol Heart Circ Physiol 306:H699–H708

606. Guensch DP, Fischer K, Flewitt JA, Yu J, Lukic R, Friedrich JA, Friedrich M (2013) Breathing manoeuvre-dependent changes in myocardial oxygenation in healthy humans. Eur Heart J Cardiovasc Imaging 15:409–414
607. Westerhof N, Boer C, Lamberts RR, Sipkema P (2006) Cross-talk between cardiac muscle and coronary vasculature. Physiol Rev 86:1263–1308
608. Feher A, Broskova Z, Bagi Z (2014) Age-related impairment of conducted dilation in human coronary arterioles. Am J Physiol Heart Circ Physiol 306:H1595–H1601
609. Bhattachariya A, Dahan D, Turczynska KM, Swärd K, Hellstrand P, Albinsson S (2014) Expression of microRNAs is essential for arterial myogenic tone and pressure-induced activation of the PI3-kinase/Akt pathway. Cardiovasc Res 101:288–296
610. Bogaty P, Brecker SJ, White SE, Stevenson RN, el-Tamimi H, Balcon R, Maseri A (1993) Comparison of coronary angiographic findings in acute and chronic first presentation of ischemic heart disease. Circulation 87:1938–1946
611. Gensini GG (1983) A more meaningful scoring system for determining the severity of coronary heart disease. Am J Cardiol 51:606
612. Sullivan DR, Marwick TH, Freedman SB (1990) A new method of scoring coronary angiograms to reflect extent of coronary atherosclerosis and improve correlation with major risk factors. Am Heart J 119:1262–1267
613. Rentrop KP, Cohen M, Blanke H, Phillips RA (1985) Changes in collateral channel filling immediately after controlled coronary artery occlusion by an angioplasty balloon in human subjects. J Am Coll Cardiol 5:587–592
614. Neeland IJ, Patel RS, Eshtehardi P, Dhawan S, McDaniel MC, Rab ST, Vaccarino V, Zafari AM, Samady H, Quyyumi AA (2012) Coronary angiographic scoring systems: an evaluation of their equivalence and validity. Am Heart J 164:547–552
615. Ringqvist I, Fisher LD, Mock M, Davis KB, Wedel H, Chaitman BR, Passamani E, Russell RO, Alderman EL, Kouchoukas NT, Kaiser GC, Ryan TJ, Killip T, Fray D (1983) Prognostic value of angiographic indices of coronary artery disease from the Coronary Artery Surgery Study (CASS). J Clin Invest 71:1854–1866
616. Dash H, Johnson RA, Dinsmore RE, Harthorne JW (1977) Cardiomyopathic syndrome due to coronary artery disease: I. Relation to angiographic extent of coronary disease and to remote myocardial infarction. Br Heart J 39:733–739
617. Califf RM, Phillips HR, Hindman MC, Mark DB, Lee KL, Behar VS, Johnson RA, Pryor DB, Rosati RA, Wagner GS (1985) Prognostic value of a coronary artery jeopardy score. J Am Coll Cardiol 5:1055–1063
618. Mark DB, Nelson CL, Califf RM, Harrell FE, Lee KL, Jones RH, Fortin DF, Stack RS, Glower DD, Smith LR (1994) Continuing evolution of therapy for coronary artery disease. Initial results from the era of coronary angioplasty. Circulation 89:2015–2025
619. Friesinger GC, Page EE, Ross RS (1970) Prognostic significance of coronary arteriography. Trans Assoc Am Physicians 83:78–92
620. Jenkins PJ, Harper RW, Nestel PJ (1978) Severity of coronary atherosclerosis related to lipoprotein concentration. Br Med J 2:388–391
621. Hoffmann U, Brady TJ, Muller J (2003) Cardiology patient page. Use of new imaging techniques to screen for coronary artery disease. Circulation 108:e50–3
622. Abaci A, Oğuzhan A, Kahraman S, Eryol NK, Ünal S, Arinç H, Ergin A (1999) Effect of diabetes mellitus on formation of coronary collateral vessels. Circulation 99:2239–2242
623. Falk E, Nakano M, Bentzon JF, Finn AV, Virmani R (2013) Update on acute coronary syndromes: the pathologists' view. Eur Heart J 34:719–728
624. Mueller C (2014) Biomarkers and acute coronary syndromes: an update. Eur Heart J 35:552–556
625. Biasucci LM, Liuzzo G, Della Bona R, Leo M, Biasillo G, Angiolillo DJ, Abbate A, Rizzello V, Niccoli G, Giubilato S, Crea F (2009) Different apparent prognostic value of hsCRP in type 2 diabetic and nondiabetic patients with acute coronary syndromes. Clin Chem 55:365–368
626. Watkins H, Farrall M (2006) Genetic susceptibility to coronary artery disease: from promise to progress. Nat Rev Genet 7:163–173

627. McPherson R, Pertsemlidis A, Kavaslar N, Stewart A, Roberts R, Cox DR, Hinds DA, Pennacchio LA, Tybjaerg-Hansen A, Folsom AR, Boerwinkle E, Hobbs HH, Cohen JC (2007) A common allele on chromosome 9 associated with coronary heart disease. Science 316:1488–1491
628. Helgadottir A, Thorleifsson G, Manolescu A, Gretarsdottir S, Blondal T, Jonasdottir A, Jonasdottir A, Sigurdsson A, Baker A, Palsson A, Masson G, Gudbjartsson DF, Magnusson KP, Andersen K, Levey AI, Backman VM, Matthiasdottir S, Jonsdottir T, Palsson S, Einarsdottir H, Gunnarsdottir S, Gylfason A, Vaccarino V, Hooper WC, Reilly MP, Granger CB, Austin H, Rader DJ, Shah SH, Quyyumi AA, Gulcher JR, Thorgeirsson G, Thorsteinsdottir U, Kong A, Stefansson K (2007) A common variant on chromosome 9p21 affects the risk of myocardial nfarction. Science 316:1491–1493
629. Trégouët DA, König IR, Erdmann J, Munteanu A, Braund PS, Hall AS, Grosshennig A, Linsel-Nitschke P, Perret C, DeSuremain M, Meitinger T, Wright BJ, Preuss M, Balmforth AJ, Ball SG, Meisinger C, Germain C, Evans A, Arveiler D, Luc G, Ruidavets JB, Morrison C, van der Harst P, Schreiber S, Neureuther K, Schäfer A, Bugert P, El Mokhtari NE, Schrezenmeir J, Stark K, Rubin D, Wichmann HE, Hengstenberg C, Ouwehand W, Wellcome Trust Case Control Consortium, Cardiogenics Consortium, Ziegler A, Tiret L, Thompson JR, Cambien F, Schunkert H, Samani NJ (2009) Genome-wide haplotype association study identifies the SLC22A3-LPAL2-LPA gene cluster as a risk locus for coronary artery disease. Nat Genet 41:283–285
630. Erdmann J, Grosshennig A, Braund PS, König IR, Hengstenberg C, Hall AS, Linsel-Nitschke P, Kathiresan S, Wright B, Trégouët DA, Cambien F, Bruse P, Aherrahrou Z, Wagner AK, Stark K, Schwartz SM, Salomaa V, Elosua R, Melander O, Voight BF, O'Donnell CJ, Peltonen L, Siscovick DS, Altshuler D, Merlini PA, Peyvandi F, Bernardinelli L, Ardissino D, Schillert A, Blankenberg S, Zeller T, Wild P, Schwarz DF, Tiret L, Perret C, Schreiber S, El Mokhtari NE, Schäfer A, März W, Renner W, Bugert P, Klüter H, Schrezenmeir J, Rubin D, Ball SG, Balmforth AJ, Wichmann HE, Meitinger T, Fischer M, Meisinger C, Baumert J, Peters A, Ouwehand WH, Italian Atherosclerosis, Thrombosis, and Vascular Biology Working Group, Myocardial Infarction Genetics Consortium, Wellcome Trust Case Control Consortium, Cardiogenics Consortium, Deloukas P, Thompson JR, Ziegler A, Samani NJ, Schunkert H (2009) New susceptibility locus for coronary artery disease on chromosome 3q22.3. Nat Genet 41:280–282
631. Cook DN, Pisetsky DS, Schwartz DA (2004) Toll-like receptors in the pathogenesis of human disease. Nat Immunol 5:975–979
632. Figueroa L, Xiong Y, Song C, Piao W, Vogel SN, Medvedev AE (2012) The Asp299Gly polymorphism alters TLR4 signaling by interfering with recruitment of MyD88 and TRIF. J Immunol 188:4506–4515
633. Helgadottir A, Manolescu A, Thorleifsson G, Gretarsdottir S, Jonsdottir H, Thorsteinsdottir U, Samani NJ, Gudmundsson G, Grant SF, Thorgeirsson G, Sveinbjornsdottir S, Valdimarsson EM, Matthiasson SE, Johannsson H, Gudmundsdottir O, Gurney ME, Sainz J, Thorhallsdottir M, Andresdottir M, Frigge ML, Topol EJ, Kong A, Gudnason V, Hakonarson H, Gulcher JR, Stefansson K (2004) The gene encoding 5-lipoxygenase activating protein confers risk of myocardial infarction and stroke. Nat Genet 36:233–239
634. Wang X, Ria M, Kelmenson PM, Eriksson P, Higgins DC, Samnegård A, Petros C, Rollins J, Bennet AM, Wiman B, de Faire U, Wennberg C, Olsson PG, Ishii N, Sugamura K, Hamsten A, Forsman-Semb K, Lagercrantz J, Paigen B (2005) Positional identification of TNFSF4, encoding OX40 ligand, as a gene that influences atherosclerosis susceptibility. Nat Genet 37:365–372
635. Gretarsdottir S, Thorleifsson G, Reynisdottir ST, Manolescu A, Jonsdottir S, Jonsdottir T, Gudmundsdottir T, Bjarnadottir SM, Einarsson OB, Gudjonsdottir HM, Hawkins M, Gudmundsson G, Gudmundsdottir H, Andrason H, Gudmundsdottir AS, Sigurdardottir M, Chou TT, Nahmias J, Goss S, Sveinbjörnsdottir S, Valdimarsson EM, Jakobsson F, Agnarsson U, Gudnason V, Thorgeirsson G, Fingerle J, Gurney M, Gudbjartsson D, Frigge ML, Kong A, Stefansson K, Gulcher JR (2003) The gene encoding phosphodiesterase 4D confers risk of ischemic stroke. Nat Genet 35:131–138.

636. Shiffman D, Rowland CM, Louie JZ, Luke MM, Bare LA, Bolonick JI, Young BA, Catanese JJ, Stiggins CF, Pullinger CR, Topol EJ, Malloy MJ, Kane JP, Ellis SG, Devlin JJ (2006) Gene variants of VAMP8 and HNRPUL1 are associated with early-onset myocardial infarction. Arterioscler Thromb Vasc Biol 26:1613–1618
637. Ibelgaufts H (2010) Cytokines and Cells Online Pathfinder Encyclopaedia. www.copewithcytokines.de/cope.cgi
638. Hoffmann R, Valencia A (2004) A gene network for navigating the literature. Nat Genet 36:664. (Information Hyperlinked over Proteins (www.ihop-net.org/)
639. Gudbjartsson DF, Bjornsdottir US, Halapi E, Helgadottir A, Sulem P, Jonsdottir GM, Thorleifsson G, Helgadottir H, Steinthorsdottir V, Stefansson H, Williams C, Hui J, Beilby J, Warrington NM, James A, Palmer LJ, Koppelman GH, Heinzmann A, Krueger M, Boezen HM, Wheatley A, Altmuller J, Shin HD, Uh ST, Cheong HS, Jonsdottir B, Gislason D, Park CS, Rasmussen LM, Porsbjerg C, Hansen JW, Backer V, Werge T, Janson C, Jönsson UB, Ng MC, Chan J, So WY, Ma R, Shah SH, Granger CB, Quyyumi AA, Levey AI, Vaccarino V, Reilly MP, Rader DJ, Williams MJ, van Rij AM, Jones GT, Trabetti E, Malerba G, Pignatti PF, Boner A, Pescollderungg L, Girelli D, Olivieri O, Martinelli N, Ludviksson BR, Ludviksdottir D, Eyjolfsson GI, Arnar D, Thorgeirsson G, Deichmann K, Thompson PJ, Wjst M, Hall IP, Postma DS, Gislason T, Gulcher J, Kong A, Jonsdottir I, Thorsteinsdottir U, Stefansson K (2009) Sequence variants affecting eosinophil numbers associate with asthma and myocardial infarction. Nat Genet 41:342–347
640. Bobik A, Kyaw TS, Tipping P, Toh BH (2014) M1 macrophages, key contributors to lymphoid neogenesis in atherosclerotic aorta. Cardiovasc Res 101:339–341
641. Guedj K, Khallou-Laschet J, Clement M, Morvan M, Gaston AT, Fornasa G, Dai J, Gervais-Taurel M, Eberl G, Michel JB, Caligiuri G, Nicoletti A (2014) M1 macrophages act as LTβ R-independent lymphoid tissue inducer cells during atherosclerosis-related lymphoid neogenesis. Cardiovasc Res 101:434–443
642. Kerr BA, Ma L, West XZ, Ding L, Malinin NL, Weber ME, Tischenko M, Goc A, Somanath PR, Penn MS, Podrez EA, Byzova TV (2013) Interference with Akt signaling protects against myocardial infarction and death by limiting the consequences of oxidative stress. Sci Signal 6:ra67
643. Din S, Mason M, Völkers M, Johnson B, Cottage CT, Wang Z, Joyo AY, Quijada P, Erhardt P, Magnuson NS, Konstandin MH, Sussman MA (2013) Pim-1 preserves mitochondrial morphology by inhibiting dynamin-related protein 1 translocation. Proc Natl Acad Sci U S A 110:5969–5974
644. Stride N, Larsen S, Hey-Mogensen M, Hansen CN, Prats C, Steinbrüchel D, Køber L, Dela F (2013) Impaired mitochondrial function in chronically ischemic human heart. Am J Physiol Heart Circ Physiol 304:H1407–H1414
645. Majetschak M (2013) Regulation of the proteasome by ATP: implications for ischemic myocardial injury and donor heart preservation. Am J Physiol Heart Circ Physiol 305:H267–H278
646. Owen MK, Witzmann FA, McKenney ML, Lai X, Berwick ZC, Moberly SP, Alloosh M, Sturek M, Tune JD (2013) Perivascular adipose tissue potentiates contraction of coronary vascular smooth muscle: Influence of obesity. Circulation 128:9–18
647. Niccoli G, Giubilato S, Di Vito L, Leo A, Cosentino N, Pitocco D, Marco V, Ghirlanda G, Prati F, Crea F (2013) Severity of coronary atherosclerosis in patients with a first acute coronary event: a diabetes paradox. Eur Heart J 34:729–741
648. Papadopoulou S, Kaski JC (2014) Microcirculation—coronary—a clinical perspective. In: Lanzer CA (ed) PanVascular medicine II, Part IV, Springer, Heidelberg
649. Crea F, Camici PG, Bairey Merz CN (2014) Coronary microvascular dysfunction: an update. Eur Heart J 35:1101–1111
650. Thygesen K, Alpert JS, Jaffe AS, Simoons ML, Chaitman BR, White HD, The Writing Group on Behalf of the Joint ESC/ACCF/AHA/WHF Task Force for the Universal Definition of Myocardial Infarction, Authors/Task Force Members Chairpersons: Thygesen K, Alpert JS, White HD, Biomarker Subcommittee: Jaffe AS, Katus HA, Apple FS, Lindahl B, Morrow

DA; ECG Subcommittee: Chaitman BR, Clemmensen PM, Johanson P, Hod H, Imaging Subcommittee, Underwood R, Bax JJ, Bonow RO, Pinto F, Gibbons RJ, Classification Subcommittee: Fox KA, Atar D, Newby LK, Galvani M, Hamm CW, Intervention Subcommittee: Uretsky BF, Gabriel Steg P, Wijns W, Bassand JP, Menasché P, Ravkilde J; Trials and Registries Subcommittee: Ohman EM, Antman EM, Wallentin LC, Armstrong PW, Simoons ML, Heart Failure Subcommittee: Januzzi JL, Nieminen MS, Gheorghiade M, Filippatos G, Epidemiology Subcommittee: Luepker RV, Fortmann SP, Rosamond WD, Levy D, Wood D, Global Perspective Subcommittee: Smith SC, Hu D, Lopez-Sendon JL, Robertson RM, Weaver D, Tendera M, Bove AA, Parkhomenko AN, Vasilieva EJ, Mendis S, ESC Committee for Practice Guidelines: Bax JJ, Baumgartner H, Ceconi C, Dean V, Deaton C, Fagard R, Funck-Brentano C, Hasdai D, Hoes A, Kirchhof P, Knuuti J, Kolh P, McDonagh T, Moulin C, Popescu BA, Reiner Z, Sechtem U, Sirnes PA, Tendera M, Torbicki A, Vahanian A, Windecker S, Document Reviewers: Morais J, Aguiar C, Almahmeed W, Arnar DO, Barili F, Bloch KD, Bolger AF, Bøtker HE, Bozkurt B, Bugiardini R, Cannon C, de Lemos J, Eberli FR, Escobar E, Hlatky M, James S, Kern KB, Moliterno DJ, Mueller C, Neskovic AN, Pieske BM, Schulman SP, Storey RF, Taubert KA, Vranckx P, Wagner DR (2012) Third universal definition of myocardial infarction. Eur Heart J 33:2551–2567

651. López-Olañeta MM, Villalba M, Gómez-Salinero JM, Jiménez-Borreguero LJ, Breckenridge R, Ortiz-Sánchez P, García-Pavía P, Ibáñez B, Lara-Pezzi E (2014) Induction of the calcineurin variant CnAβ1 after myocardial infarction reduces post-infarction ventricular remodelling by promoting infarct vascularization. Cardiovasc Res 102:396–406
652. Carrick D, Berry C (2013) Prognostic importance of myocardial infarct characteristics. Eur Heart J Cardiovasc Imaging 14:313–315
653. Calise J, Powell SR (2013) The ubiquitin proteasome system and myocardial ischemia. Am J Physiol Heart Circ Physiol 304:H337–H349
654. Zelarayán LC, Noack C, Sekkali B, Kmecova J, Gehrke C, Renger A, Zafiriou MP, van der Nagel R, Dietz R, de Windt LJ, Balligand JL, Bergmann MW (2008) β-Catenin downregulation attenuates ischemic cardiac remodeling through enhanced resident precursor cell differentiation. Proc Natl Acad Sci U S A 105:19762–19767
655. Mirotsou M, Zhang Z, Deb A, Zhang L, Gnecchi M, Noiseux N, Mu H, Pachori A, Dzau V (2007) Secreted frizzled related protein 2 (Sfrp2) is the key Akt-mesenchymal stem cell-released paracrine factor mediating myocardial survival and repair. Proc Natl Acad Sci U S A 104:1643–1648
656. Benamer N, Vasquez C, Mahoney VM, Steinhardt MJ, Coetzee WA, Morley GE (2013) Fibroblast K_{ATP} currents modulate myocyte electrophysiology in infarcted hearts. Am J Physiol Heart Circ Physiol 304:H1231–H1239
657. Roell W, Lewalter T, Sasse P, Tallini YN Choi BR, Breitbach M, Doran R, Becher UM, Hwang SM, Bostani T, von Maltzahn J, Hofmann A, Reining S, Eiberger B, Gabris B, Pfeifer A, Welz A, Willecke K, Salama G, Schrickel JW, Kotlikoff MI, Fleischmann BK (2007) Engraftment of connexin 43-expressing cells prevents post-infarct arrhythmia. Nature 450:819–824
658. Severs NJ, Bruce AF, Dupont E, Rothery S (2008) Remodelling of gap junctions and connexin expression in diseased myocardium. Cardiovasc Res 80:9–19
659. Fu LW, Longhurst JC (2010) A new function for ATP: activating cardiac sympathetic afferents during myocardial ischemia. Am J Physiol Heart Circ Physiol 299:H1762–H1771
660. Morel S, Frias MA, Rosker C, James RW, Rohr S, Kwak BR (2012) The natural cardioprotective particle HDL modulates connexin43 gap junction channels. Cardiovasc Res 93:41–49
661. Veenstra RD (2012) Sphingosine-1-phosphate signals the way for Cx43-mediated cardioprotection. Cardiovasc Res 93:8–9
662. Nahrendorf M, Pittet MJ, FK Swirski (2010) Monocytes: protagonists of infarct inflammation and repair. Circulation 121:2437–2445
663. Dutta P, Courties G, Wei Y, Leuschner F, Gorbatov R, Robbins CS, Iwamoto Y, Thompson B, Carlson AL, Heidt T, Majmudar MD, Lasitschka F, Etzrodt M, Waterman P, Waring MT, Chicoine AT, van der Laan AM, Niessen HWM, Piek JJ, Rubin BB, Butany J, Stone JR,

Katus HA, Murphy SA, Morrow DA, Sabatine MS, Vinegoni C, Moskowitz MA, Pittet MJ, Libby P, Lin CP, Swirski FK, Weissleder R, Nahrendorf M (2012) Myocardial infarction accelerates atherosclerosis. Nature 487:325–329
664. Wang X, Ha T, Zou J, Ren D, Liu L, Zhang X, Kalbfleisch J, Gao X, Williams D, Li C (2014) MicroRNA-125b protects against myocardial ischaemia/reperfusion injury via targeting p53-mediated apoptotic signalling and TRAF6. Cardiovasc Res 102:385–395
665. Stapels M, Piper C, Yang T, Li M, Stowell C, Xiong ZG, Saugstad J, Simon RP, Geromanos S, Langridge J, Lan JQ, Zhou A (2010) Polycomb group proteins as epigenetic mediators of neuroprotection in ischemic tolerance. Sci Signal 3:ra15
666. Qian L, Huang Y, Spencer CI, Foley A, Vedantham V, Liu L, Conway SJ, Fu JD, Srivastava D (2012) In vivo reprogramming of murine cardiac fibroblasts into induced cardiomyocytes. Nature 485:593–598
667. Song K, Nam YJ, Luo X, Qi X, Tan W, Huang GN, Acharya A, Smith CL, Tallquist MD, Neilson EG, Hill JA, Bassel-Duby R, Olson EN (2012) Heart repair by reprogramming non-myocytes with cardiac transcription factors. Nature 485:599–604
668. Palpant NJ, Murry CE (2012) Regenerative medicine: reprogramming the injured heart. Nature 485:585–586
669. Fu LW, Longhurst JC (2010) Bradykinin and thromboxane A2 reciprocally interact to synergistically stimulate cardiac spinal afferents during myocardial ischemia. Am J Physiol Heart Circ Physiol 298:H235–H244
670. Degousee N, Simpson J, Fazel S, Scholich K, Angoulvant D, Angioni C, Schmidt H, Korotkova M, Stefanski E, Wang XH, Lindsay TF, Ofek E, Pierre S, Butany J, Jakobsson PJ, Keating A, Li RK, Nahrendorf M, Geisslinger G, Backx PH, Rubin BB (2012) Lack of microsomal prostaglandin E2 synthase-1 in bone marrow-derived myeloid cells impairs left ventricular function and increases mortality after acute myocardial infarction. Circulation 125:2904–2913
671. Kanashiro-Takeuchi RM, Tziomalos K, Takeuchi LM, Treuer AV, Lamirault G, Dulce R, Hurtado M, Song Y, Block NL, Rick F, Klukovits A, Hu Q, Varga JL, Schally AV, Hare JM (2010) Cardioprotective effects of growth hormone-releasing hormone agonist after myocardial infarction. Proc Natl Acad Sci U S A 107:2604–2609
672. Kanashiro-Takeuchi RM, Takeuchi LM, Rick FG, Dulce R, Treuer AV, Florea V, Rodrigues CO, Paulino EC, Hatzistergos KE, Selem SM, Gonzalez DR, Block NL, Schally AV, Hare JM (2012) Activation of growth hormone releasing hormone (GHRH) receptor stimulates cardiac reverse remodeling after myocardial infarction (MI). Proc Natl Acad Sci U S A 109:559–563
673. Takenaka H, Horiba M, Ishiguro H, Sumida A, Hojo M, Usui A, Akita T, Sakuma S, Ueda Y, Kodama I, Kadomatsu K (2009) Midkine prevents ventricular remodeling and improves long-term survival after myocardial infarction. Am J Physiol Heart Circ Physiol 296:H462–H469
674. Miller EJ, Li J, Leng L, McDonald C, Atsumi T, Bucala R, Young LH (2008) Macrophage migration inhibitory factor stimulates AMP-activated protein kinase in the ischaemic heart. Nature 451:578–582
675. Kewalramani G, Puthanveetil P, Wang F, Kim MS, Deppe S, Abrahani A, Luciani DS, Johnson JD, Rodrigues B (2009) AMP-activated protein kinase confers protection against TNF-α-induced cardiac cell death. Cardiovasc Res 84:42–53
676. Chen CH, Budas GR, Churchill EN, Disatnik MH, Hurley TD, Mochly-Rosen D (2008) Activation of aldehyde dehydrogenase-2 reduces ischemic damage to the heart. Science 321:1493–1495
677. Tian F, Zhou X, Wikström J, Karlsson H, Sjöland H, Gan LM, Borén J, Akyürek LM (2009) Protein disulfide isomerase increases in myocardial endothelial cells in mice exposed to chronic hypoxia: a stimulatory role in angiogenesis. Am J Physiol Heart Circ Physiol 297:H1078–H1086
678. Watson LJ, Facundo HT, Ngoh GA, Ameen M, Brainard RE, Lemma KM, Long BW, Prabhu SD, Xuan YT, Jones SP (2010) O-linked β-N-acetylglucosamine transferase is indispensable in the failing heart. Proc Natl Acad Sci U S A 107:17797–17802

679. Patanè S, Marte F (2009) Prostate-specific antigen kallikrein: from prostate cancer to cardiovascular system. Eur Heart J 30:1169–1170
680. Rossow CF, Minami E, Chase EG, Murry CE, Santana LF (2004) NFATc3-induced reductions in voltage-gated K^+ currents after myocardial infarction. Circ Res 94:1340–1350
681. Yakushev S, Band M, Tissot van Patot MC, Gassmann M, Avivi A, Bogdanova A (2012) Cross talk between S-nitrosylation and S-glutathionylation in control of the Na, K-ATPase regulation in hypoxic heart. Am J Physiol Heart Circ Physiol 303:H1332–H1343
682. Fu LW, Phan A, Longhurst JC (2008) Myocardial ischemia-mediated excitatory reflexes: a new function for thromboxane A2? Am J Physiol Heart Circ Physiol 295:H2530–H2540
683. Huang BS, Ahmad M, White RA, Marc Y, Llorens-Cortes C, Leenen FH (2013) Inhibition of brain angiotensin III attenuates sympathetic hyperactivity and cardiac dysfunction in rats post-myocardial infarction. Cardiovasc Res 97:424–431
684. Guan Y, Gao L, Ma HJ, Li Q, Zhang H, Yuan F, Zhou ZN, Zhang Y (2010) Chronic intermittent hypobaric hypoxia decreases β-adrenoceptor activity in right ventricular papillary muscle. Am J Physiol Heart Circ Physiol 298:H1267–H1272
685. Linkermann A, Bräsen JH, Darding M, Jin MK, Sanz AB, Heller JO, De Zen F, Weinlich R, Ortiz A, Walczak H, Weinberg JM, Green DR, Kunzendorf U, Krautwald S (2013) Two independent pathways of regulated necrosis mediate ischemia-reperfusion injury. Proc Natl Acad Sci U S A 110:12024–12029
686. Smith RM, Visweswaran R, Talkachova I, Wothe JK, Tolkacheva EG (2013) Uncoupling the mitochondria facilitates alternans formation in the isolated rabbit heart. Am J Physiol Heart Circ Physiol 305:H9–H18
687. Nishihara M, Miura T, Miki T, Tanno M, Yano T, Naitoh K, Ohori K, Hotta H, Terashima Y, Shimamoto K (2007) Modulation of the mitochondrial permeability transition pore complex in GSK-3β-mediated myocardial protection. J Mol Cell Cardiol 43:564–570
688. Feng J, Zhu M, Schaub MC, Gehrig P, Roschitzki B, Lucchinetti E, Zaugg M (2008) Phosphoproteome analysis of isoflurane-protected heart mitochondria: phosphorylation of adenine nucleotide translocator-1 on Tyr194 regulates mitochondrial function. Cardiovasc Res 80:20–29
689. Hamacher-Brady A, Brady NR, Logue SE, Sayen MR, Jinno M, Kirshenbaum LA, Gottlieb RA, Gustafsson AB (2007) Response to myocardial ischemia/reperfusion injury involves Bnip3 and autophagy. Cell Death Differ 14:146–157
690. Zaobornyj T, Ghafourifar P (2012) Strategic localization of heart mitochondrial NOS: a review of the evidence. Am J Physiol Heart Circ Physiol 303:H1283–H1293
691. Joiner ML, Koval OM, Li J, He BJ, Allamargot C, Gao Z, Luczak ED, Hall DD, Fink BD, Chen B, Yang J, Moore SA, Scholz TD, Strack S, Mohler PJ, Sivitz WI, Song LS, Anderson ME (2012) CaMKII determines mitochondrial stress responses in heart. Nature 491:269–273
692. Trendelenburg M, Theroux P, Stebbins A, Granger C, Armstrong P, Pfisterer M (2010) Influence of functional deficiency of complement mannose-binding lectin on outcome of patients with acute ST-elevation myocardial infarction undergoing primary percutaneous coronary intervention. Eur Heart J 31:1181–1187
693. Chao W (2009) Toll-like receptor signaling: a critical modulator of cell survival and ischemic injury in the heart. Am J Physiol Heart Circ Physiol 296:H1–H12
694. Zhang C, Xu X, Potter BJ, Wang W, Kuo L, Michael L, Bagby GJ, Chilian WM (2006) TNF-α contributes to endothelial dysfunction in ischemia/reperfusion injury. Arter Thromb Vasc Biol 26:475–480
695. Poeckel D, Funk CD (2010) The 5-lipoxygenase/leukotriene pathway in preclinical models of cardiovascular disease. Cardiovasc Res 86:243–253
696. Murakami M, Taketomi Y, Sato H, Yamamoto K (2011) Secreted phospholipase-A2 revisited. J Biochem 150:233–255
697. Rassaf T, Weber C, Bernhagen J (2014) Macrophage migration inhibitory factor in myocardial ischaemia/reperfusion injury. Cardiovasc Res 102:321–328
698. Inserte J, Hernando V, Garcia-Dorado D (2012) Contribution of calpains to myocardial ischaemia/reperfusion injury. Cardiovasc Res 96:23–31

699. Talukder MAH, Zweier JL, Periasamy M (2009) Targeting calcium transport in ischaemic heart disease. Cardiovasc Res 84:345–352
700. Yusof M, Kamada K, Kalogeris T, Gaskin FS, Korthuis RJ (2009) Hydrogen sulfide triggers late-phase preconditioning in postischemic small intestine by an NO- and p38 MAPK-dependent mechanism. Am J Physiol Heart Circ Physiol 296:H868–H876
701. Bryan NS, Calvert JW, Elrod JW, Gundewar S, Ji SY, Lefer DJ (2007) Dietary nitrite supplementation protects against myocardial ischemia-reperfusion injury. Proc Natl Acad Sci U S A 104:19144–19149
702. Cohen MV, Yang XM, Liu Y, Solenkova NV, Downey JM (2010) Cardioprotective PKG-independent NO signaling at reperfusion. Am J Physiol Heart Circ Physiol 299:H2028–H2036
703. Lima B, Lam GK, Xie L, Diesen DL, Villamizar N, Nienaber J, Messina E, Bowles D, Kontos CD, Hare JM, Stamler JS, Rockman HA (2009) Endogenous S-nitrosothiols protect against myocardial injury. Proc Natl Acad Sci U S A 106:6297–6302
704. Prime TA, Blaikie FH, Evans C, Nadtochiy SM, James AM, Dahm CC, Vitturi DA, Patel RP, Hiley CR, Abakumova I, Requejo R, Chouchani ET, Hurd TR, Garvey JF, Taylor CT, Brookes PS, Smith RA, Murphy MP (2009) A mitochondria-targeted S-nitrosothiol modulates respiration, nitrosates thiols, and protects against ischemia-reperfusion injury. Proc Natl Acad Sci U S A 106:10764–10769
705. Hausenloy DJ, Yellon DM (2009) Cardioprotective growth factors. Cardiovasc Res 83:179–194
706. Varga ZV, Zvara A, Faragó N, Kocsis GF, Pipicz M, Gáspár R, Bencsik P, Görbe A, Csonka C, Puskás LG, Thum T, Csont T, Ferdinandy P (2014) MicroRNAs associated with ischemia-reperfusion injury and cardioprotection by ischemic pre- and postconditioning: ProtectomiRs. Am J Physiol Heart Circ Physiol 307:H216–H227
707. Wang X, Ha T, Liu L, Zou J, Zhang X, Kalbfleisch J, Gao X, Williams D, Li C (2013) Increased expression of microRNA-146a decreases myocardial ischaemia/reperfusion injury. Cardiovascular Research 97:432–442
708. Wang C, Liu N, Luan R, Li Y, Wang D, Zou W, Xing Y, Tao L, Cao F, Wang H (2013) Apelin protects sarcoplasmic reticulum function and cardiac performance in ischaemia–reperfusion by attenuating oxidation of sarcoplasmic reticulum Ca^{2+}-ATPase and ryanodine receptor. Cardiovasc Res 100:114–124
709. Kasama S, Furuya M, Toyama T, Ichikawa S, Kurabayashi M (2008) Effect of atrial natriuretic peptide on left ventricular remodelling in patients with acute myocardial infarction. Eur Heart J 29:1485–1494
710. Schmidt K, Tissier R, Ghaleh B, Drogies T, Felix SB, Krieg T (2010) Cardioprotective effects of mineralocorticoid receptor antagonists at reperfusion. Eur Heart J 31:1655–1662
711. Kawano T, Anrather J, Zhou P, Park L, Wang G, Frys KA, Kunz A, Cho S, Orio M, Iadecola C (2006) Prostaglandin E2 EP1 receptors: downstream effectors of COX-2 neurotoxicity. Nature – Medicine 12, 225–229
712. Leary PJ, Rajasekaran S, Morrison RR, Tuomanen EI, Chin TK, Hofmann PA (2008) A cardioprotective role for platelet-activating factor through NOS-dependent S-nitrosylation. Am J Physiol Heart Circ Physiol 294:H2775–H2784
713. Doukas J, Wrasidlo W, Noronha G, Dneprovskaia E, Fine R, Weis S, Hood J, DeMaria A, Soll R, Cheresh D (2006) Phosphoinositide 3-kinase gamma/delta inhibition limits infarct size after myocardial ischemia/reperfusion injury. Proc Natl Acad Sci U S A 103:19866–19871
714. Miyamoto S, Murphy AN, Brown JH (2008) Akt mediates mitochondrial protection in cardiomyocytes through phosphorylation of mitochondrial hexokinase-II. Cell Death Differ 15:521–529
715. Kim KS, Abraham D, Williams B, Violin JD, Mao L, Rockman HA (2012) β-Arrestin-biased AT1R stimulation promotes cell survival during acute cardiac injury. Am J of Physiol Heart Circ Physiol 303:H1001–H1010
716. Seko Y, Tobe K, Takahashi N, Kaburagi Y, Kadowaki T, Yazaki Y (1996) Hypoxia and hypoxia/reoxygenation activate Src family tyrosine kinases and p21ras in cultured rat cardiac myocytes. Biochem Biophys Res Commun 226:530–535

717. Mizukami Y, Ono K, Du CK, Aki T, Hatano N, Okamoto Y, Ikeda Y, Ito H, Hamano K, Morimoto S (2008) Identification and physiological activity of survival factor released from cardiomyocytes during ischaemia and reperfusion. Cardiovasc Res 79:589–599
718. Burger D, Xiang F, Hammoud L, Lu X, Feng Q (2009) Role of heme oxygenase-1 in the cardioprotective effects of erythropoietin during myocardial ischemia and reperfusion. Am J Physiol Heart Circ Physiol 296:H84–H93
719. Ferro E, Goitre L, Retta SF, Trabalzini L (2012) The interplay between ROS and Ras GTPases: physiological and pathological implications. J Signal Transduct 2012:365769
720. Yin H, Chao L, Chao J (2005) Kallikrein/kinin protects against myocardial apoptosis after ischemia/reperfusion via Akt-glycogen synthase kinase-3 and Akt-Bad.14-3-3 signaling pathways. J Biol Chem 280:8022–8030
721. Fielitz J, van Rooij E, Spencer JA, Shelton JM, Latif S, van der Nagel R, Bezprozvannaya S, de Windt L, Richardson JA, Bassel-Duby R, Olson EN (2007) Loss of muscle-specific RING-finger 3 predisposes the heart to cardiac rupture after myocardial infarction. Proc Natl Acad Sci U S A 104:4377–4382
722. Hiroi T, Wajima T, Negoro T, Ishii M, Nakano Y, Kiuchi Y, Mori Y, Shimizu S (2013) Neutrophil TRPM2 channels are implicated in the exacerbation of myocardial ischaemia/reperfusion injury. Cardiovasc Res 97:271–281
723. Miller BA, Wang J, Hirschler-Laszkiewicz I, Gao E, Song J, Zhang XQ, Koch WJ, Madesh M, Mallilankaraman K, Gu T, Chen SJ, Keefer K, Conrad K, Feldman AM, Cheung JY (2013) The second member of transient receptor potential-melastatin channel family protects hearts from ischemia-reperfusion injury. Am J Physiol Heart Circ Physiol 304:H1010–H1022
724. Gidday JM (2006) Cerebral preconditioning and ischaemic tolerance. Nat Rev Neurosci 7:437–448
725. Cohen MV, Downey JM (2008) Oestrogen plays a permissive role in cardioprotection. Cardiovasc Res 79:353–354
726. Germack R, Griffin M, Dickenson JM (2004) Activation of protein kinase B by adenosine A1 and A3 receptors in newborn rat cardiomyocytes. J Mol Cell Cardiol 37:989–999
727. Quinlan CL, Costa ADT, Costa CL, Pierre SV, Dos Santos P, Garlid KD (2008) Conditioning the heart induces formation of signalosomes that interact with mitochondria to open mitoKATP channels. Am J Physiol Heart Circ Physiol 295:H953–H961
728. Talukder MA, Yang F, Shimokawa H, Zweier JL (2010) eNOS is required for acute in vivo ischemic preconditioning of the heart: effects of ischemic duration and sex. Am J Physiol Heart Circ Physiol 299:H437–H445
729. Hendgen-Cotta UB, Merx MW, Shiva S, Schmitz J, Becher S, Klare JP, Steinhoff HJ, Goedecke A, Schrader J, Gladwin MT, Kelm M, Rassaf T (2008) Nitrite reductase activity of myoglobin regulates respiration and cellular viability in myocardial ischemia-reperfusion injury. Proc Natl Acad Sci U S A 105:10256–10261
730. Sarkar K, Cai Z, Gupta R, Parajuli N, Fox-Talbot K, Darshan MS, Gonzalez FJ, Semenza GL (2012) Hypoxia-inducible factor 1 transcriptional activity in endothelial cells is required for acute phase cardioprotection induced by ischemic preconditioning. Proc Natl Acad Sci U S A 109:10504–10509
731. Cai Z, Luo W, Zhan H, Semenza GL (2013) Hypoxia-inducible factor 1 is required for remote ischemic preconditioning of the heart. Proc Natl Acad Sci U S A 110:17462–17467
732. Wojtovich AP, Urciuoli WR, Chatterjee S, Fisher AB, Nehrke K, Brookes PS (2013) Kir6.2 is not the mitochondrial KATP channel but is required for cardioprotection by ischemic preconditioning. Am J Physiol Heart Circ Physiol 304:H1439–H1445
733. Ishikawa S, Kuno A, Tanno M, Miki T, Kouzu H, Itoh T, Sato T, Sunaga D, Murase H, Miura T (2012) Role of connexin-43 in protective PI3K–Akt–GSK-3β signaling in cardiomyocytes. Am J Physiol Heart Circ Physiol 302:H2536–H2544
734. Mendoza MC, Er EE, Blenis J (2011) The Ras-ERK and PI3K-mTOR pathways: cross-talk and compensation. Trends Biochem Sci 36:320–328

735. Laclau MN, Boudina S, Thambo JB, Tariosse L, Gouverneur G, Bonoron-Adèle S, Saks VA, Garlid KD, Dos Santos P (2001) Cardioprotection by ischemic preconditioning preserves mitochondrial function and functional coupling between adenine nucleotide translocase and creatine kinase. J Mol Cell Cardiol 33:947–956
736. Støttrup NB, Løfgren B, Birkler RD, Nielsen JM, Wang L, Caldarone CA, Kristiansen SB, Contractor H, Johannsen M, Bøtker HE, Nielsen TT (2010) Inhibition of the malate-aspartate shuttle by pre-ischaemic aminooxyacetate loading of the heart induces cardioprotection. Cardiovasc Res 88:257–266
737. Ozcan C, Palmeri M, Horvath TL, Russell KS, Russell RR (2013) Role of uncoupling protein 3 in ischemia-reperfusion injury, arrhythmias, and preconditioning. Am J Physiol Heart Circ Physiol 304:H1192–H1200
738. Miura T, Miki T, Yano T (2010) Role of the gap junction in ischemic preconditioning in the heart. Am J Physiol Heart Circ Physiol 298:H1115–H1125
739. Morel S, Braunersreuther V, Chanson M, Bouis D, Rochemont V, Foglia B, Pelli G, Sutter E, Pinsky DJ, Mach F, Kwak BR (2014) Endothelial Cx40 limits myocardial ischaemia/reperfusion injury in mice. Cardiovasc Res 102:329–337
740. Haudek SB, Xia Y, Huebener P, Lee JM, Carlson S, Crawford JR, Pilling D, Gomer RH, Trial J, Frangogiannis NG, Entman ML (2006) Bone marrow-derived fibroblast precursors mediate ischemic cardiomyopathy in mice. Proc Natl Acad Sci U S A 103:18284–18289
741. Zhao H (2009) Ischemic postconditioning as a novel avenue to protect against brain injury after stroke. J Cereb Blood Flow Metab 29:873–885
742. Vilahur G, Cubedo J, Casani L, Padro T, Sabate-Tenas M, Badimon JJ, Badimon L (2013) Reperfusion-triggered stress protein response in the myocardium is blocked by post-conditioning. Systems biology pathway analysis highlights the key role of the canonical aryl-hydrocarbon receptor pathway. Eur Heart J 34:2082–2093
743. Ohnuma Y, Miura T, Miki T, Tanno M, Kuno A, Tsuchida A, Shimamoto K (2002) Opening of mitochondrial K_{ATP} channel occurs downstream of PKC-ϵ activation in the mechanism of preconditioning. Am J Physiol Heart Circ Physiol 283:H440–H447
744. Tanno M, Kuno A (2013) Reversal of metabolic shift in post-infarct-remodelled hearts: possible novel therapeutic approach. Cardiovasc Res 97:195–196
745. Lou PH, Zhang L, Lucchinetti E, Heck M, Affolter A, Gandhi M, Kienesberger PC, Hersberger M, Clanachan AS, Zaugg M (2013) Infarct-remodelled hearts with limited oxidative capacity boost fatty acid oxidation after conditioning against ischaemia/reperfusion injury. Cardiovasc Res 97:251–261

Chap. 9. Interventional Medicine and Surgery of Cardiac Diseases

746. Thiriet M (2008) Biology and mechanics of blood flows, part II: mechanics and medical aspects of blood flows, CRM series in mathematical physics, Springer, New York
747. Marescaux J, Leroy J, Gagner M, Rubino F, Mutter D, Vix M, Butner SE, Smith MK (2001) Transatlantic robot-assisted telesurgery. Nature 413:379–380
748. Krupa A, Gangloff J, Doignon C, de Mathelin M, Morel G, Leroy J, Soler L, Marescaux J (2003) Autonomous 3D positioning of surgical instruments in robotized laparoscopic surgery using visual serving. IEEE Trans Robot Autom 19:842–853
749. Coste-Manière E, Adhami L, Mourgues F, Bantiche O (2004) Optimal planning of robotically assisted heart surgery: transfer precision in the operating room. Int J Robot Res 23:539–548
750. Borst C (2000) Operating on a beating heart. Sci Am 283:58–63
751. Cattin P, Dave H, Grunenfelder J, Szekely G, Turina M, Zuend G (2004) Trajectory of coronary motion and its significance in robotic motion cancellation. Eur J Cardio-Thorac Surg 25:786–790
752. Hu YF, Dawkins JF, Cho HC, Marbán E, Cingolani E (2014) Biological pacemaker created by minimally invasive somatic reprogramming in pigs with complete heart block. Sci– Transl Med 6:245ra94

753. Yamaji S, Imai S, Saito F, Yagi H, Kushiro T, Uchiyama T (2006) Does high-power computed tomography scanning equipment affect the operation of pacemakers? Circ J 70:190–197
754. Maor E, Ivorra A, Leor J, Rubinsky B (2008) Irreversible electroporation attenuates neointimal formation after angioplasty. IEEE Trans Biomed Eng 55:2268–2274
755. Falk V, Mourgues F, Adhami L, Jacobs S, Thiele H, Nitzsche S, Mohr F, Coste-Manière E (2005) Cardionavigation: planning, simulation and augmented reality to robotic assisted endoscopic bypass grafting. Ann Thorac Surg 79:2040–2047
756. Soler L, Delingette H, Malandain G, Montagnat J, Ayache N, Koehl C, Dourthe O, Malassagne B, Smith M, Mutter D, Marescaux J (2001) Fully automatic anatomical, pathological, and functional segmentation from CT scans for hepatic surgery. Comput Aided Surg 6:131–142
757. Garrigue S, Pépin JL, Defaye P, Murgatroyd F, Poezevara Y, Clémenty J, Lévy P (2007) High prevalence of sleep apnea syndrome in patients with long-term pacing. The European multicenter polysomnographic study. Circulation 115:1703–1709
758. Burgreen GW, Antaki JF, Wu ZJ, Holmes AJ (2001) Computational fluid dynamics as a development tool for rotary blood pumps. Artif Org 25:336–340
759. Jarvik RK (1981) The total artificial heart. Sci Am 244:74–80
760. Grandmont C, Maday Y, Métier P (2002) Existence of a solution for a unsteady elasticity problem in large displacement and small perturbation. Comptes Rendus Mathématique Académie des Sciences Paris 334:521–526
761. Métier P (2003) Modélisation, analyse mathématique et applications numériques de problèmes d'interaction fluide-structure instationnaires [Modeling, mathematical analysis, and numerical applications of unsteady fluid-structure interaction problems]. PhD thesis, University Pierre & Marie Curie, Paris
762. Doyle MG, Tavoularis S, Bourgault Y (2004) Computation of blood flow in a diaphragme-type ventricular assist device. In: Chen S, McIllwain S (eds) Proceedings of the 12th Annual Conference of the CFD Society of Canada
763. Doyle MG, Tavoularis S, Bourgault Y (2005) Closed-loop simulation of a ventricular assist device coupled with a circulatory system model. In: Bathe KJ (ed) Computational fluid and solid mechanics 2005. Elsevier, Amsterdam
764. Farinas MI, Garon A (2004) Application of DOE for optimal turbomachinery design. AIAA-2004-2139 34th AIAA fluid dynamics conference and exhibit, Portland, Oregon
765. Garon A, Farinas MI (2004) Fast three-dimensional numerical hemolysis approximation. Artif Organs 28:1016–1025
766. Carpentier A (1983) Cardiac valve surgery—the "French correction". J Thorac Cardiovasc Surg 86:323–337
767. Comolet R, Miraoui M (1987) Performance de valves à feuillets onguiformes [Performance of nail-shaped membrane valves]. Journal de Biophysique et de Biomécanique 11:55–61
768. Bruss KH, Reul H, van Gilse J, Knott E (1983) Pressure drop and velocity fields at four mechanical heart valve prostheses: Bjork–Shiley Standard, Bjork–Shiley Concave-Convex, Hall–Kaster and St. Jude Medical. Life Support Sys 1:3–22
769. Woo YR, Sung HW, Williams FP, Yoganathan AP (1986) In vitro fluid dynamic characteristics of aortic bioprostheses: old versus new. Life Support Sys 4:63–85
770. van Steenhoven AA van Duppen TJ Cauwenberg JW van Renterghem RJ (1982) In vitro closing behaviour of Bjork-Shiley, St Jude and Hancock heart valve prostheses in relation to the in vivo recorded aortic valve closure. J Biomech 15:841–848
771. Chandran KB, Cabell GN, Khalighi B, Chen CJ (1984) Pulsatile flow past aortic valve bioprostheses in a model human aorta. J Biomech 17:609–619
772. Garitey V (1994) Étude expérimentale de l'écoulement intraventriculaire en aval de différent types de prothèses valvulaires [Experimental study of the intraventricular flow downstream from different types of valvular prothesis]. PhD thesis, University Aix–Marseille II
773. Gutiérrez-Chico JL, Alegría-Barrero E, Teijeiro-Mestre R, Chan PH, Tsujioka H, de Silva R, Viceconte N, Lindsay A, Patterson T, Foin N, Akasaka T, di Mario C (2012) Optical coherence tomography: from research to practice. Eur Heart J Cardiovasc Imaging 13:370–384

774. Feenstra PH, Taylor CA (2009) Drug transport in artery walls: a sequential porohyperelastic-transport approach. Comput Methods Biomech Biomed Eng 12:263–76
775. Lao LL, Venkatraman SS (2008) Adjustable paclitaxel release kinetics and its efficacy to inhibit smooth muscle cells proliferation. J Controll Release 130:9–14
776. Lao LL, Venkatraman SS, Peppas NA (2009) A novel model and experimental analysis of hydrophilic and hydrophobic agent release from biodegradable polymers. J Biomed Mater Res A 90:1054–1065
777. Zimmer S, Nickenig G (2010) Prediction and prevention by progenitors? Stent thrombosis and EPCs. Eur Heart J 31:2569–2571
778. Lev EI, Leshem-Lev D, Mager A, Vaknin-Assa H, Harel N, Zimra Y, Bental T, Greenberg G, Dvir D, Solodky A, Assali A, Battler A, Kornowski R (2010) Circulating endothelial progenitor cell levels and function in patients who experienced late coronary stent thrombosis. Eur Heart J 31:2625
779. Ma X, Hibbert B, Dhaliwal B, Seibert T, Chen YX, Zhao X, O'Brien ER (2010) Delayed re-endothelialization with rapamycin-coated stents is rescued by the addition of a glycogen synthase kinase-3beta inhibitor. Cardiovasc Res 86:338–345
780. Guo RW, Wang H, Gao P, Li MQ, Zeng CY, Yu Y, Chen JF, Song MB, Shi YK, Huang L (2009) An essential role for stromal interaction molecule 1 in neointima formation following arterial injury. Cardiovasc Res 81:660–668
781. Mercurius KO, Morla AO (1998) Inhibition of vascular smooth muscle cell growth by inhibition of fibronectin matrix assembly. Circ Res 82:548–556
782. Chen KH, Guo X, Ma D, Guo Y, Li Q, Yang D, Li P, Qiu X, Wen S, Xiao RP, Tang J (2004) Dysregulation of HSG triggers vascular proliferative disorders. Nat Cell Biol 6:872–883
783. Windecker S, Jüni P (2008) Safety of drug-eluting stents. Nat Clin Pract Cardiovasc Med 5:316–328
784. Daemen J, Simoons ML, Wijns W, Bagust A, Bos G, Bowen JM, Braunwald E, Camenzind E, Chevalier B, DiMario C, Fajadet J, Gitt A, Guagliumi G, Hillege HL, James S, Jüni P, Kastrati A, Kloth S, Kristensen SD, Krucoff M, Legrand V, Pfisterer M, Rothman M, Serruys PW, Silber S, Steg PG, Tariah I, Wallentin L, Windecker SW (2009) Report on ESC Forum on Drug Eluting Stents (European Heart House, Nice, 27–28 September 2007). Eur Heart J 30:152–161
785. Van der Heiden K Gijsen FJ Narracott A Hsiao S Halliday I Gunn J Wentzel JJ Evans PC (2013) The effects of stenting on shear stress: relevance to endothelial injury and repair. Cardiovasc Res 99:269–275
786. Hsiao HM, Chiu YH (2013) Assessment of mechanical integrity for drug-eluting renal stent with micro-sized drug reservoirs. Comput Methods Biomech Biomed Eng 16:1307–1318
787. Iqbal J, Onuma Y, Ormiston J, Abizaid A, Waksman R, Serruys P (2014) Bioresorbable scaffolds: rationale, current status, challenges, and future. Eur Heart J 35:765–776
788. Diletti R, Karanasos A, Muramatsu T, Nakatani S, van Mieghem NM, Onuma Y, Nauta ST, Ishibashi Y, Lenzen MJ, Ligthart J, Schultz C, Regar E, de Jaegere PP, Serruys PW, Zijlstra F, van Geuns RJ (2014) Everolimus-eluting bioresorbable vascular scaffolds for treatment of patients presenting with ST-segment elevation myocardial infarction: BVS STEMI first study. Eur Heart J 35:777–786
789. Kocka V, Maly M, Tousek P, Budesínsky T, Lisa L, Prodanov P, Jarkovsky J, Widimsky P (2014) Bioresorbable vascular scaffolds in acute ST-segment elevation myocardial infarction: a prospective multicentre study 'Prague 19'. Eur Heart J 35:787–794
790. Barbato E, Wijn W (2014) Bioresorbable coronary scaffolds: a novel device-based solution in search of its clinical need. Eur Heart J 35:753–757
791. Mehilli J, Byrne RA, Wieczorek A, Iijima R, Schulz S, Bruskina O, Pache J, Wessely R, Schömig A, Kastrati A for the intracoronary stenting and angiographic restenosis investigators – test efficacy of Rapamycin-eluting Stents with different polymer coating strategies [ISAR-TEST-3] (2008) Randomized trial of three rapamycin-eluting stents with different coating strategies for the reduction of coronary restenosis. Eur Heart J 29:1975–1982

792. Cribier A, Savin T, Saoudi N, Behar P, Rocha P, Mechmeche R, Berland J, Letac B (1986) Percutaneous transluminal aortic valvuloplasty using a balloon catheter. A new therapeutic option in aortic stenosis in the elderly. Archives des Maladies du Coeur et des Vaisseaux 79:1678–1686
793. Cribier A, Eltchaninoff H, Tron C, Bauer F, Agatiello C, Sebagh L, Bash A, Nusimovici D, Litzler PY, Bessou JP, Leon MB (2004) Early experience with percutaneous transcatheter implantation of heart valve prosthesis for the treatment of end-stage inoperable patients with calcific aortic stenosis. J Am Coll Cardiol 43:698–703
794. Boudjemline Y, Bonhoeffer P (2002) Steps towards percutaneous aortic valve replacement. Circulation 105:775–778
795. Ferrari M, Figulla HR, Schlosser M, Tenner I, Damm C, Guyenot V, Werner GS, Hellige G (2004) Transarterial aortic valve replacement with a self expanding stent in pigs. Heart 90:1326–1331
796. Descoutures F, Himbert D, Lepage L, Iung B, Détaint D, Tchetche D, Brochet E, Castier Y, Depoix JP, Nataf P, Vahanian A (2008) Contemporary surgical or percutaneous management of severe aortic stenosis in the elderly. Eur Heart J 29:1410–1417
797. Antunes MJ (2008) Percutaneous aortic valve implantation. The demise of classical aortic valve replacement? Eur Heart J 29:1339–1341
798. Leal A, Rodríguez MI, Muñoz-Gámez JA, de Almodóvar MR, Siles E, Rivas AL, Jäättela M, Oliver FJ (2012) ROS-induced DNA damage and PARP-1 are required for optimal induction of starvation-induced autophagy. Cell Res 22:1181–1198
799. Inoue T, Kato T, Hikichi Y, Hashimoto S, Hirase T, Morooka T, Imoto Y, Takeda Y, Sendo F, Node K (2006) Stent-induced neutrophil activation is associated with an oxidative burst in the inflammatory process, leading to neointimal thickening. Thromb Haemost 95:43–48
800. Bergmann O, Bhardwaj RD, Bernard S, Zdunek S, Barnabé-Heider F, Walsh S, Zupicich J, Alkass K, Buchholz BA, Druid H, Jovinge S, Frisén J (2009) Evidence for cardiomyocyte renewal in humans. Science 324:98–102
801. Papanicolaou KN, Kikuchi R, Ngoh GA, Coughlan KA, Dominguez I, Stanley WC, Walsh K (2012) Mitofusins 1 and 2 are essential for postnatal metabolic remodeling in heart. Circ Res 111:1012–1026
802. Puente BN, Kimura W, Muralidhar SA, Moon J, Amatruda JF, Phelps KL, Grinsfelder D, Rothermel BA, Chen R, Garcia JA, Santos CX, Thet S, Mori E, Kinter MT, Rindler PM, Zacchigna S, Mukherjee S, Chen DJ, Mahmoud AI, Giacca M, Rabinovitch PS, Aroumougame A, Shah AM, Szweda LI, Sadek HA (2014) The oxygen-rich postnatal environment induces cardiomyocyte cell-cycle arrest through dna damage response. Cell 157:565–579
803. Terzic A, Behfar A (2014) Regenerative heart failure therapy headed for optimization. Eur Heart J 35:1231–1234
804. Laflamme MA, Chen KY, Naumova AV, Muskheli V, Fugate JA, Dupras SK, Reinecke H, Xu C, Hassanipour M, Police S, O'Sullivan C, Collins L, Chen Y, Minami E, Gill EA, Ueno S, Yuan C, Gold J, Murry CE (2007) Cardiomyocytes derived from human embryonic stem cells in pro-survival factors enhance function of infarcted rat hearts. Nat Biotechnol 25:1015–1024
805. Nasseri BA, Ebell W, Dandel M, Kukucka M, Gebker R, Doltra A, Knosalla C, Choi YH, Hetzer R, Stamm C (2014) Autologous CD133+ bone marrow cells and bypass grafting for regeneration of ischaemic myocardium: the Cardio133 trial. Eur Heart J 35:1263–1274
806. ChidgeyAP Layton D Trounson A Boyd RL (2008) Tolerance strategies for stem-cell-based therapies. Nature 453:330–337
807. Martin-Rendon E, Brunskill SJ, Hyde CJ, Stanworth SJ, Mathur A, Watt SM (2008) Autologous bone marrow stem cells to treat acute myocardial infarction: a systematic review. Eur Heart J 29:1807–1818
808. Poitevin S, Cussac D, Leroyer AS, Albinet V, Sarlon-Bartoli G, Guillet B, Hubert L, Andrieu-Abadie N, Couderc B, Parini A, Dignat-George F, Sabatier F (2014) Sphingosine kinase 1 expressed by endothelial colony-forming cells has a critical role in their revascularization activity. Cardiovasc Res 103:121–130

809. Guan K, Hasenfuss G (2013) Cardiac resident progenitor cells: evidence and functional significance. Eur Heart J 34:2784–2787
810. Emmert MY, Emmert LS, Martens A, Ismail I, Schmidt-Richter I, Gawol A, Seifert B, Haverich A, Martin U, Gruh I (2013) Higher frequencies of BCRP+ cardiac resident cells in ischaemic human myocardium. Eur Heart J 34:2830–2838
811. Yang L, Soonpaa MH, Adler ED, Roepke TK, Kattman SJ, Kennedy M, Henckaerts E, Bonham K, Abbott GW, Linden RM, Field LJ, Keller GM (2008) Human cardiovascular progenitor cells develop from a KDR+ embryonic-stem-cell-derived population. Nature 453:524–528
812. Tillmanns J, Rota M, Hosoda T, Misao Y, Esposito G, Gonzalez A, Vitale S, Parolin C, Yasuzawa-Amano S, Muraski J, De Angelis A, Lecapitaine N, Siggins RW, Loredo M, Bearzi C, Bolli R, Urbanek K, Leri A, Kajstura J, Anversa P (2008) Formation of large coronary arteries by cardiac progenitor cells. Proc Natl Acad Sci U S A 105:1668–1673
813. Xu Y, Shi Y, Ding S (2008) A chemical approach to stem-cell biology and regenerative medicine. Nature 453:338–344
814. García-Ayllón MS, Millán C, Serra-Basante C, Bataller R, Sáez-Valero J (2012) Readthrough acetylcholinesterase is increased in human liver cirrhosis. PLoS One 7:e44598
815. Johnson G, Moore SW (2007) Acetylcholinesterase readthrough peptide shares sequence similarity to the 28–53 peptide sequence of the acetylcholinesterase adhesion-mediating site and competes for ligand binding in vitro. J Mol Neurosci 31:113–126
816. Kim JB, Zaehres H, Wu G, Gentile L, Ko K, Sebastiano V, Araúzo-Bravo MJ, Ruau D, Han DW, Zenke M, Schöler HR (2008) Pluripotent stem cells induced from adult neural stem cells by reprogramming with two factors. Nature 454:646–650
817. Re RN, Cook JL (2008) The physiological basis of intracrine stem cell regulation. Am J Physiol Heart Circ Physiol 295:H447–H453
818. Koninckx R, Daniëls A, Windmolders S, Mees U, Macianskiene R, Mubagwa K, Steels P, Jamaer L, Dubois J, Robic B, Hendrikx M, Rummens JL, Hensen K (2013) The cardiac atrial appendage stem cell: a new and promising candidate for myocardial repair. Cardiovasc Res 97:413–423
819. Yu J, Vodyanik MA, Smuga-Otto K, Antosiewicz-Bourget J, Frane JL, Tian S, Nie J, Jonsdottir GA, Ruotti V, Stewart R, Slukvin II, Thomson JA (2007) Induced pluripotent stem cell lines derived from human somatic cells. Science 318:1917–1920
820. Nam YJ, Song K, Luo X, Daniel E, Lambeth K, West K, Hill JA, DiMaio JM, Baker LA, Bassel-Duby R, Olson EN (2013) Reprogramming of human fibroblasts toward a cardiac fate. Proc Natl Acad Sci U S A 110:5588–5593
821. Davis DD, Hamilton A, Yang J, Cremar LD, Van Gough D, Potisek SL, Ong MT, Braun PV, Martínez TJ, White SR, Moore JS, Sottos NR (2009) Force-induced activation of covalent bonds in mechanoresponsive polymeric materials. Nature 459:68–72
822. Hanson JA, Chang CB, Graves SM, Li Z, Mason TG, Deming TJ (2008) Nanoscale double emulsions stabilized by single-component block copolypeptides. Nature 455:85–88
823. Kim KS, Lei Y, Stolz DB, Liu D (2007) Bifunctional compounds for targeted hepatic gene delivery. Gene Ther 14:704–708
824. Huang K, Voss B, Kumar D, Hamm HE, Harth E (2007) Dendritic molecular transporters provide control of delivery to intracellular compartments. Bioconjugate Chem 18:403–409
825. Patri AK, Kukowska-Latallo JF, Baker JR (2005) Targeted drug delivery with dendrimers: comparison of the release kinetics of covalently conjugated drug and non-covalent drug inclusion complex. Adv Drug Deliv Rev 57:2203–2214
826. Simberg D, Duza T, Park JH, Essler M, Pilch J, Zhang L, Derfus AM, Yang M, Hoffman RM, Bhatia S, Sailor MJ, Ruoslahti E (2007) Biomimetic amplification of nanoparticle homing to tumors. Proc Natl Acad Sci U S A 104:932–936
827. Murphy EA, Majeti BK, Barnes LA, Makale M, Weis SM, Lutu-Fuga K, Wrasidlo W, Cheresh DA (2008) Nanoparticle-mediated drug delivery to tumor vasculature suppresses metastasis. Proc Natl Acad Sci U S A 105:9343–9348

828. Chakravarty P, Marches R, Zimmerman NS, Swafford ADE, Bajaj P, Musselman IH, Pantano P, Draper RK, Vitetta ES (2008) Thermal ablation of tumor cells with antibody-functionalized single-walled carbon nanotubes. Proc Natl Acad Sci U S A 105:8697–8702
829. Dayton PA, Zhao S, Bloch SH, Schumann P, Penrose K, Matsunaga TO, Zutshi R, Doinikov A, Ferrara KW (2006) Application of ultrasound to selectively localize nanodroplets for targeted imaging and therapy. Mol Imaging 5:160–174
830. Crowder KC, Hughes MS, Marsh JN, Barbieri AM, Fuhrhop RW, Lanza GM, Wickline SA (2005) Sonic activation of molecularly-targeted nanoparticles accelerates transmembrane lipid delivery to cancer cells through contact-mediated mechanisms: implications for enhanced local drug delivery. Ultrasound Med Biol 31:1693–1700
831. Rapoport N, Gao Z, Kennedy A (2007) Multifunctional nanoparticles for combining ultrasonic tumor imaging and targeted chemotherapy. J Natl Cancer Inst 99:1095–1106
832. Shastry BS (2006) Pharmacogenetics and the concept of individualized medicine. Pharmacogn J 6:16–21
833. Jeninga EH, Bugge A, Nielsen R, Kersten S, Hamers N, Dani C, Wabitsch M, Berger R, Stunnenberg HG, Mandrup S, Kalkhoven E (2009) Peroxisome proliferator-activated receptor γ regulates expression of the anti-lipolytic G-protein-coupled receptor 81 (GPR81/Gpr81). J Biol Chem 284:26385–26393
834. Tunaru S, Kero J, Schaub A, Wufka C, Blaukat A, Pfeffer K, Offermanns S (2003) PUMA-G and HM74 are receptors for nicotinic acid and mediate its anti-lipolytic effect. Nat Med 9:352–355
835. Wise A, Foord SM, Fraser NJ, Barnes AA, Elshourbagy N, Eilert M, Ignar DM, Murdock PR, Steplewski K, Green A, Brown AJ, Dowell SJ, Szekeres PG, Hassall DG, Marshall FH, Wilson S, Pike NB (2003) Molecular identification of high and low affinity receptors for nicotinic acid. J Biol Chem 278:9869–9874
836. Irukayama-Tomobe Y, Tanaka H, Yokomizo T, Hashidate-Yoshida T, Yanagisawa M, Sakurai T (2009) Aromatic D-amino acids act as chemoattractant factors for human leukocytes through a G protein-coupled receptor, GPR109B. Proc Natl Acad Sci U S A 106:3930–3934
837. Hopewell JC, Parish S, Offer A, Link E, Clarke R, Lathrop M, Armitage J, Collins R, MRC/BHF Heart Protection Study Collaborative Group (2003) Impact of common genetic variation on response to simvastatin therapy among 18705 participants in the Heart Protection Study. Eur Heart J 34:982–992
838. Aalto S, Haarala C, Brück A, Sipilä H, Hämäläinen H, Rinne JO (2006) Mobile phone affects cerebral blood flow in humans. J Cereb Blood Flow Metab 26:885–890
839. Cotin S, Delingette H, Clément JM, Marescaux J, Ayache N (1997) Simulation active de chirurgie endoscopique [Active simulation of endoscopic surgery]. Rev Eur Technol Bioméd 19:167–172
840. Marescaux J, Clément JM, Tassetti V, Koehl C, Cotin S, Russier Y, Mutter D, Delingette H, Ayache N (1998) Virtual reality applied to hepatic surgery simulation: the next revolution. Ann Surg 228:627–634
841. Lei M, Kleinstreuer C, Archie JP (1997) Hemodynamic simulations and computer-aided designs of graft-artery junctions. J Biomech Eng 119:343–348
842. Quarteroni A, Rozza G (2003) Optimal control and shape optimization in aorto-coronaric bypass anastomoses. Math Models Methods Appl Sci 13:1801–1823
843. Agoshkov V, Quarteroni A, Rozza G (2006) Shape design in aorto-coronaric bypass anastomoses using perturbation theory. SIAM J Numer Anal 44:367–384
844. Rozza G (2005) Real-time reduced-basis techniques in arterial bypass geometries (1283–1287). In: Bathe KJ (ed) Computational fluid and solid mechanics. Elsevier, Amsterdam
845. Sherwin SJ, Doorly DJ (2003) Flow Dynamics within Model Distal Arterial Bypass Grafts. In: Tura A (ed) Vascular grafts: experiment and modelling. WIT Press, Southampton
846. Giordana S, Sherwin SJ, Peiro J, Doorly DJ, Papaharilaou Y, Caro CG, Watkins N, Cheshire N, Jackson M, Bicknall C, Zervas V (2005) Automated classification of peripheral distal by-pass geometries reconstructed from medical data. J Biomech 38:47–62

847. Ghista DN, Kabinejadian F (2013) Coronary artery bypass grafting hemodynamics and anastomosis design: a biomedical engineering review. Biomed Eng Online 12:129
848. Ding J, Liu Y, Wang F, Bai F (2012) Impact of competitive flow on hemodynamics in coronary surgery: numerical study of ITA-LAD model. Comput Math Methods Med 2012:356187
849. Grus T, Lindner J, Vik K, Tošovský J, Matěcha J, Netřebská H, Tůma J, Adamec J (2007) Particle image velocimetry measurement in the model of vascular anastomosis. Prague Med Rep 108:75–86
850. Holzapfel GA, Sommer G, Gasser CT, Regitnig P (2005) Determination of layer-specific mechanical properties of human coronary arteries with nonatherosclerotic intimal thickening and related constitutive modeling. Am J Physiol Heart Circ Physiol 289:H2048–H2058
851. Ní Ghriallais R Bruzzia M (2014) Self-expanding stent modelling and radial force accuracy. Comput Methods Biomech Biomed Eng 17:318–333
852. Delfour MC, Garon A, Longo V (2005) Modeling and design of coated stents to optimize the effect of the dose. SIAM J Appl Math 65:858–881
853. Lao LL, Venkatraman SS, Peppas NA (2008) Modeling of drug release from biodegradable polymer blends. Eur J Pharm Biopharm 70:796–803
854. Blanchet G, Delfour MC, Garon A (2011) Quadratic models to fit experimental data of paclitaxel release kinetics from biodegradable polymers. SIAM J Appl Math 71:2269–2286
855. Garon A, Delfour MC (2014) Three-dimensional quadratic model of paclitaxel release from biodegradable polymer films. SIAM J Appl Math 74:1354–1374
856. Creel CJ, Lovich MA, Edelman ER (2000) Arterial paclitaxel distribution and deposition. Circ Res 86:879–884
857. Fung LK, Ewend MG, Sills A, Sipos EP, Thompson R, Watts M, Colvin OM, Brem H, Saltzman WM (1998) Pharmacokinetics of interstitial delivery of carmustine, 4-hydroperoxycyclophosphamide, and paclitaxel from a biodegradable polymer implant in the monkey brain. Cancer Res 58:672–684
858. Zhu X, Pack DW, Braatz RD (2014) Modelling intravascular delivery from drug-eluting stents with biodurable coating: investigation of anisotropic vascular drug diffusivity and arterial drug distribution. Comput Methods Biomech Biomed Eng 17:187–198
859. BioGRID: General Repository for Interaction Datasets; database of physical and genetic interactions for model organisms (www.thebiogrid.org)
860. GeneCards human gene database. Crown Human Genome Center, Department of Molecular Genetics, the Weizmann Institute of Science (www.genecards.org)
861. Universal Protein Resource (UniProt) Consortium (European Bioinformatics Institute, Swiss Institute of Bioinformatics, and Protein Information Resource. www.uniprot.org)

Index

M. Thiriet, *Diseases of the Cardiac Pump*, Biomathematical and Biomechanical Modeling of the Circulatory and Ventilatory Systems 7, DOI 10.1007/978-3-319-12664-7

MIX
Papier aus verantwortungsvollen Quellen
Paper from responsible sources
FSC® C105338

If you have any concerns about our products,
you can contact us on
ProductSafety@springernature.com

In case Publisher is established outside the EU,
the EU authorized representative is:
Springer Nature Customer Service Center GmbH
Europaplatz 3, 69115 Heidelberg, Germany

Printed by Libri Plureos GmbH
in Hamburg, Germany